Classical Electrodynamics

Classical Electrodynamics captures Schwinger's inimitable lecturing style, in which everything flows inexorably from what has gone before. This anniversary edition offers a refreshing update while still maintaining Schwinger's voice.

The book provides the student with a thorough grounding in electrodynamics in particular, and in classical field theory in general. An essential resource for both physicists and their students, the book includes a Reader's Guide, which describes the major themes in each chapter, suggests a possible path through the book, and identifies topics for inclusion in, and exclusion from, a given course, depending on the instructor's preference.

Carefully constructed problems complement the material of the text. Classical Electrodynamics should be of great value to all physicists, from first-year graduate students to senior researchers, and to all those interested in electrodynamics, field theory, and mathematical physics.

The original text for the graduate classical electrodynamics course was left unfinished upon Julian Schwinger's death in 1994, but was completed by his former students and co-authors, who have brilliantly recreated the excitement of Schwinger's novel approach. This anniversary edition has been revised by one of those original co-authors, Kimball Milton.

Selling Points:

- Based on Nobel laureate Julian Schwinger's lectures at UCLA and MIT.
- Starts from inference of full Maxwell theory, including energy and momentum conservation, before specializing to electrostatics and magnetostatics.
- Action and variational principles take center stage throughout, as do Green's functions.
- Mathematics is integrated into the physics, so mathematical topics emerge in proper physical context.
- Updated to include two new topics: relativity in four-vector notation and variational principles for steady currents.
- Includes appendix devoted to converting between different systems of units, Gaussian, Heaviside-Lorentz, and SI.
- Many new problems have been added in this edition, illustrating and expanding on topics discussed in the text.

The late Julian Schwinger shared the 1965 Nobel Prize for Physics with Richard Feynman and Shin'ichirō Tomonaga for their work on the theory of quantum electrodynamics. Kimball A. Milton is George Lynn Cross Research Professor of Physics, Emeritus, at the University of Oklahoma, currently residing in Memphis, Tennessee.

Frontiers in Physics

Feynman and Computation
Anthony Hey

The Physics of Laser Plasma Interactions
William Kruer

Plate Tectonics: An Insider's History of the Modern Theory of the Earth
Naomi Oreskes

The Framework of Plasma Physics
Richard D. Hazeltine, Francois L. Waelbroeck

Introduction to Ultrahigh Energy Cosmic Ray Physics
Pierre Sokolsky

Statistical Plasma Physics, Volume I: Basic Principles
Setsuo Ichimaru

Statistical Plasma Physics, Volume II: Condensed Plasmas
Setsuo Ichimaru

Computational Plasma Physics: With Applications To Fusion And Astrophysics
Toshi Tajima

The Field Theoretic Renormalization Group in Critical Behavior Theory and Stochastic Dynamics
A.N. Vasil'ev

Statistical Physics of Dense Plasmas: Elementary Processes and Phase Transitions
Setsuo Ichimaru

Are There Really Neutrinos?
An Evidential History
Allan Franklin, Alysia Marino

Introduction to Ultrahigh Energy Cosmic Ray Physics
2nd Edition
Pierre Sokolsky, Gordon B. Thomson

Feynman Lectures On Computation: Anniversary Edition
Richard P. Feynman
Edited by Tony Hey

Classical Electrodynamics: Second Edition
Kimball Milton, Julian Schwinger

For more information about this series, please visit: www.crcpress.com/Frontiers-in-Physics/book-series/FRONTIERSPHYS

Classical Electrodynamics
Second Edition

Kimball Milton
and Julian Schwinger

CRC Press is an imprint of the
Taylor & Francis Group, an **informa** business

Designed cover image: Agreed with Jonathan Pennell

Second edition published 2024
by CRC Press
2385 NW Executive Center Drive, Suite 320, Boca Raton FL 33431

and by CRC Press
4 Park Square, Milton Park, Abingdon, Oxon, OX14 4RN

CRC Press is an imprint of Taylor & Francis Group, LLC

First edition published by CRC Press 1998

ISBN: 978-0-367-50207-2 (hbk)
ISBN: 978-0-367-52298-8 (pbk)
ISBN: 978-1-003-05736-9 (ebk)

DOI: 10.1201/9781003057369

Typeset in CMR10 font
by KnowledgeWorks Global Ltd.

I dedicate this book to my devoted and ever-loving wife, Margarita Baños-Milton, who has enthusiastically supported my work on book projects over the years.

Contents

Preface to Second Edition

This new edition of *Classical Electrodynamics* is a modest revision of the volume published a quarter of a century ago. The original edition, in turn, was based on lectures given by Julian Schwinger at UCLA in the 1970s. So why revisit such a mature treatise?

The answer to Harold's[1] question is that these pages still contain much that is fresh and useful to both beginning graduate students and established researchers. I and my earlier co-authors, Wu-yang Tsai and Lester DeRaad, Jr., tried to keep the flavor of the apparently spontaneous lecturing style of Schwinger.[2] The informality of the style belies the rigor of the treatment of electrodynamics and the associated mathematical development.

In this new edition, I have corrected the (very few) errors of the first, added a few chapters: one on the four-dimensional treatment of relativity (Chapter 12) and another on variational principles for steady currents (Chapter 32), and moved the chapter on dispersion relations to its rightful place (Chapter 6). Most of the figures have been redone. The most significant addition in the inclusion of a large number of new problems drawn from my years of teaching this course at the University of Oklahoma (and once at Washington University). It also includes material from *Electromagnetic Radiation* [1], derived largely from Schwinger's lectures on electromagnetic radiation and waveguides at MIT's wartime Radiation Laboratory. I have also extensively expanded the index.

I hope that this revision will enhance the original book and bring to it a new generation of readers. I am aware that I am fighting the precedent of revised editions of classic books being more ponderous and less accessible than the originals, for which reason I have endeavored to keep changes in the text as minimal as possible. I hope that this volume will be useful to those who will explore the implications of electromagnetic theory, which in its quantum form underlies most of our understanding of the world.

I would be remiss if I did not mention classic treatises that have informed our work. These include Stratton, *Electromagnetic Theory* [2]; Sommerfeld, *Electrodynamics* [3]; Landau and Lifshitz, *The Classical Theory of Fields* [4] and *Electrodynamics of Continuous Media* [5]; and Jackson, *Classical Electrodynamics* [6]. This is not meant to be exhaustive, and the readers will have their own favorites.

Finally, I must thank those who assisted me in this project: This includes my many students who participated in my graduate courses on classical electrodynamics, doing much more than detecting errors of typographical and factual matters, but greatly helped in the exposition. Many colleagues from' around the world offered helpful insights, among whom I must single out Roman Jackiw. As noted in the preface, the UCLA archives provided extensive material, and I especially thank the curator of Special Collections Charlotte Brown. I thank the staff of Taylor & Francis for proposing this revision, and for putting up with the slowness of the project. Lastly, I thank anonymous reviewers of Part I of this volume for suggesting numerous improvements, and particularly Kirk McDonald for correcting many historical details.

[1] Harold stood for the Hypothetical Reader of Limitless Dedication but was a tribute to Schwinger's elder brother Harold.

[2] In fact, Schwinger rehearsed his lectures carefully, recording them to his prodigious memory, one of the reasons he hated interruptions to his lectures.

During the revision of this book, I have been supported in part by the US National Science Foundations, grant numbers PHY-2008417 and PHY-1707511, for research related to Casimir physics.

Kimball A. Milton

September, 2023
Memphis, TN USA

Preface to First Edition

This book is an outgrowth of lectures that Julian Schwinger gave at UCLA in 1976, when, for the first time in many years, he was asked to teach Classical Electrodynamics. Of course, Schwinger had contributed greatly to the development of the subject, partly through his profound work on the theory of waveguides and microwave cavities at the MIT Radiation Laboratory during World War II. Only a small part of that material was ever published in *Discontinuities in Waveguides* [7]. After the war, Schwinger played a key role in the perfection of the synchrotron and, especially, was largely responsible for the theory of synchrotron radiation [8, 9]. His first course at Harvard in 1946, Applied Science 33, was on advanced applications of electromagnetic theory, particularly to waveguides. (We have incorporated four problems from the final exam to that course in the present volume.) Schwinger's discoveries in classical electrodynamics led directly to his solution of the difficulties in quantum electrodynamics a few years later.

As his former graduate students at Harvard and postdoctoral fellows working with him, we attended the UCLA lectures and found them so novel and exciting that we proposed turning them into a textbook. Schwinger agreed and cooperated to the extent of supplying us with his detailed notes. By 1979, a typed manuscript existed, and a publication contract was signed. At that point, Schwinger began to read the manuscript carefully, and not finding his voice there, began an extensive process of revision. That revision, of the first half of the manuscript, continued until 1984. He used the partially revised manuscript as the basis for his Classical Electrodynamics course, which he taught at UCLA again in 1983. We had gone our separate ways long before that point, so the project lay dormant for a decade.

Shortly after Schwinger's death in 1994, one of us (KAM) began teaching Classical Electrodynamics, and used the opportunity to begin completion of the manuscript using the extant materials. The present volume is the result of that effort. We offer it now, not merely as a homage to a great physicist and teacher, but as a vital approach to a fundamental, and still not closed, subject. We have retained our original organization of the manuscript, and not the few, very long chapters of Schwinger's later revision, because we feel that this makes the contents more accessible to the student and teacher. We have tried to retain Schwinger's inimitable lecturing style in this book, in which everything flows inexorably from what has gone before. However, as an aid to the reader, we have included a *Reader's Guide*, which identifies major themes in each chapter, suggests a possible path through the book, and identifies topics for inclusion or exclusion from a given course, depending upon circumstances.

We dedicate this book to the memory of Julian Schwinger. We are indebted to Clarice Schwinger for her gracious permission to pursue this project, and for her encouragement. We are also grateful to Professor Michael Strauss, who attended Schwinger's 1983 course, and made available his notes, and the corresponding book manuscript at that time. Further materials have been obtained from the Julian Schwinger Papers (Collection 371), Department of Special Collections, University Research Library, University of California, Los Angeles. Many students at the University of Oklahoma contributed by finding innumerable errors in the typescript, so we thank them as well.

Finally, we acknowledge the Alfred P. Sloan Foundation and the U.S. National Science Foundation for partial financial support during the early stages of this project, and the U.S. Department of Energy for partial support of KAM during the completion of this project.

Lester L. DeRaad, Jr. — Tustin, CA
Kimball A. Milton — Norman, OK
Wu-yang Tsai — Torrance, CA

December 1997

Reader's Guide (Updated from First Edition)

As noted in the Preface, this book originated in lectures given in 1976. It further included, in its original incarnation, two chapters, now Chapters 6 and 54, which were based on lectures Schwinger gave, around the same time, on the energy loss by electrically and magnetically charged particles when they pass through matter. This latter work was evidently based on the excitement generated by the purported discovery of magnetic charge in cosmic ray experiments [10] evidence that was subsequently interpreted in terms of a conventional but rare nuclear fragmentation event [11]. The manuscript prepared during the 1977–1979 period remains the core of the present volume.

However, as we have described, Schwinger undertook extensive revisions beginning in 1979, and continuing at least through 1983. Indeed, the UCLA archives contain an extensive revision of the book through electrostatics, corresponding to the material in the present Chapters 1–27. In the interest of preserving the more spontaneous flavor of the original lectures, we have retained most of the original material, but incorporated new content at the appropriate points. A few chapters were not in the original version and are based closely on the 1979–83 revision; they are Chapter 11, "Einsteinian Relativity," Chapter 20, "Modified Bessel Functions," Chapter 21, "Cylindrical Conductors," and Chapter 27, "Modes and Variations." Schwinger ceased work on the project in 1984, but he did present new material on radiation in his UCLA lectures in the previous year, and that new material now appears in Chapter 47, "Waveguides," and Chapter 49, "Partial-Wave Analysis of Scattering." Of course, the present authors have incorporated numerous improvements and additional material. The new edition includes two new chapters: Chapter 12, "Relativistic Formulation," which introduces covariant notation, and Chapter 32, "Steady Currents and Dissipation," which presents additional variational methods for magnetostatics. Most significant is the inclusion of a great many additional problems, of a wide range of difficulty, some of which had been prepared by us in the late 1970s, but many of which are new to this edition. The problems constitute an integral part of the text; indeed many important concepts appear only in the problems, and an instructor might feel it appropriate to base lectures on those topics. Examples include radiation by a particle undergoing hyperbolic motion, in Chapter 40, and vector spherical harmonics, introduced in Chapter 53.

In view of this long history, the reader will not be surprised to detect a certain variation in the level of the material. This seems to us entirely appropriate. Some of the topics are quite elementary, but there are many which are more advanced, and which therefore could be omitted in a first course. Indeed, it is the authors' experience that there is far more material here than can be covered in a typical two-semester graduate electrodynamics course. (It might be borne in mind that Schwinger was a very fast lecturer.) It is hoped that the more advanced material will be of utility to practicing researchers, and indeed, the authors have found the insights contained herein extremely valuable in their own research in electromagnetic theory and quantum field theory.

The balance of this Guide will sketch major subjects covered in each chapter, and make suggestions for material that is essential, and inessential, for a first course or reading.

Chapter 1. The central point of this chapter is a heuristic derivation of Maxwell's equations starting from Coulomb's law together with the imposition of Galilean invariance. This is not to be taken as a rigorous derivation, especially because the correct relativity group is that of Einstein. *Essential.*

Chapter 2. Here we show how Maxwell's equations are modified if magnetic charge is present. *Inessential.*

Chapter 3. We discuss the conservation of energy, momentum, and angular momentum, thereby establishing the consistency of Maxwell's equations. *Essential.*

Chapter 4. This chapter is devoted to the inference of the macroscopic Maxwell equations. *Essential.*

Chapter 5. Here we present simple classical models for the electric permeability and conductivity, and discuss the Clausius-Mossotti equation. *Essential.*

Chapter 6. Dispersion relations are the subject of this chapter, including Kramers-Kronig relations, but this may be *inessential* for a first reading.

Chapter 7. We give models for the magnetic properties of matter here, and introduce the vector potential. *Essential.*

Chapter 8. We treat the somewhat subtle issues of how to construct the energy and momentum for macroscopic electrodynamics. *Essential.*

Chapter 9. Mechanical action principles are the subject here. Although these should be familiar, the viewpoint is somewhat novel and crucial in the following. *Essential.*

Chapter 10. Here we present the action principle for electrodynamics. *Essential* except for Section 10.6.

Chapter 11. We discuss the modification of the particle Lagrangian required by Einstein's relativity, as well as the usual relativistic kinematics and field transformations. Probably only Section 11.1 is *essential.*

Chapter 12. The covariant formulation of electrodynamics is the subject here. Very useful for more advanced courses, but *inessential* for the rest of the book.

Chapter 13. Here the study of electrostatics begins. We show that the general action principle implies the stationary principles for electrostatics. *Essential.*

Chapter 14. We introduce Green's functions. *Essential.*

Chapter 15. We discuss Green's function for free space. *Essential.*

Chapter 16. We derive Green's function for a semi-infinite dielectric. Only Sections 16.1 and 16.4 are *essential.*

Chapter 17. Here we use the previous Green's function to compute the force between a point charge and a dielectric slab. *Inessential.*

Chapter 18. Now we introduce Bessel functions. As they are used throughout the following, this is *essential.*

Chapter 19. We give the derivation of Green's function for parallel conducting plates, along with an extensive discussion of image charges in this case. Only Section 19.1 is *essential.*

Chapter 20. Now we introduce modified Bessel functions. This material is very useful, but perhaps *inessential* for a first reading.

Chapter 21. We discuss cylindrical conductors, of rectangular, triangular, and circular cross section, and derive useful mathematical results. *Inessential.*

Chapter 22. Now we introduce spherical harmonics. *Essential.*

Chapter 23. Legendre's polynomials and spherical Bessel functions are the principal topics here. Only the beginning through Section 23.1, and perhaps Section 23.3, are *essential.*

Chapter 24. The general multipole expansion for the energy of interaction of two arbitrary bounded charge distributions is the subject. At least the beginning of this chapter is *essential.*

Chapter 25. We consider Green's functions for conducting and dielectric spheres. The first parts of Sections 25.1 and 25.4 seem *essential.*

Chapter 26. We give a general treatment of electrostatics in the presence of dielectrics and conductors. We prove Thomson's theorem, and give the general expression for capacitance. At least the latter, given in Section 26.6, is *essential.*

Chapter 27. Variational methods for estimating eigenvalues are the subjects treated here. Elegant, but *inessential.*

Chapter 28. We now introduce the variational principle for magnetostatics. *Essential.*

Chapter 29. The energy of interaction of steady currents, and inductance, are the subjects here. *Essential.*

Chapter 30. Here the topic in magnetic dipoles. *Essential.*

Chapter 31. We introduce the magnetic scalar potential. *Inessential.*

Chapter 32. Variational principles applied to situations with steady currents is the subject here. *Inessential.*

Chapter 33. Here, the subject is the "string" required for the definition of the vector potential when magnetic charge is present. *Inessential.*

Chapter 34. Now the treatment of radiation begins, with the retarded Green's function and the Liénard-Wiechert potentials. *Essential.*

Chapter 35. We discuss the asymptotic fields, and dipole radiation. *Essential.*

Chapter 36. Here we study radiation from the point of view of energy transferred from the source, and derive the Darwin-Breit Hamiltonian for the charges. *Inessential.*

Chapter 37. We discuss radiation from simple models of antennas. *Inessential.*

Chapter 38. We derive the general formula for the spectral distribution of radiation. *Essential.*

Chapter 39. Now we obtain a formula for the power spectrum, and apply it to Čerenkov radiation. *Essential.*

Chapter 40. We discuss radiation in the extreme situations of constant acceleration, and of impulsive scattering. In the problems, we give general formulas for the energy radiated by a particle undergoing arbitrary acceleration for a finite amount of time. *Inessential.*

Chapter 41. Synchrotron radiation is the subject here. *Essential.*

Chapter 42. We derive the power spectrum for the different polarization states in synchrotron radiation. *Inessential.*

Chapter 43. The high energy regime for synchrotron radiation is the subject here. Sections 43.1 and 43.5 are *essential.*

Chapter 44. We derive Snell's law, and the reflection and transmission coefficients for light incident on a plane dielectric interface. Mostly *essential.*

Chapter 45. We treat reflection by a conductor having finite conductivity here. *Inessential.*

Chapter 46. Here we perform the $2+1$-dimensional break-up of Green's function for Helmholtz' equation leading to Hankel functions. *Inessential.*

Chapter 47. Here we give a general treatment of cylindrical waveguides, harking back to Chapters 21 and 27. *Inessential.*

Chapter 48. We present simple models of scattering, including Thomson scattering. *Essential.*

Chapter 49. We give here a partial-wave approach to scattering. Although this is very useful for quantum mechanics, it is here *inessential.*

Chapter 50. The elementary theory of diffraction is the subject. At least Sections 50.1 and 50.2 are *essential.*

Chapter 51. Here we show that the diffraction of a normally incident plane wave on a straight edge is exactly solvable. *Inessential.*

Chapter 52. We derive Babinet's principle in the case of a plane with an aperture or slit to give the diffraction for the opposite polarization state. *Inessential.*

Chapter 53. The general formulation of scattering, the Born approximation, and the optical theorem are the subjects. *Essential.*

Chapter 54. We give a general treatment of energy loss by an electrically or magnetically charged particle traversing matter. *Inessential.*

Part I

Formulation of Electrodynamics

1

Maxwell's Equations

The teaching of electromagnetic theory is something like that of American History in school; you get it again and again. Well, this is the end of the line. Here is where we put it all together, and yet, not quite, since it is still classical electrodynamics and the final goal is quantum electrodynamics. This preoccupation reflects the all-pervasive nature of electromagnetism, with implications ranging from the farthest galaxies to the interiors of the fundamental particles. In particular, the properties of ordinary matter, including those properties classified as chemical and biological, depend only on electromagnetic forces, in conjunction with the microscopic laws of quantum mechanics.

1.1 Electrostatics

Our intention is to move toward the general picture as quickly as possible, starting with a review of electrostatics. We take for granted the phenomenology of electric charge, including the Coulomb law of force[1] between charges of dimensions that are small in comparison with their separation. This is expressed by the interaction energy, E, of a system of such charges in otherwise empty space, a vacuum:

$$E = \frac{1}{2} \sum_{\substack{a,b \\ a \neq b}} \frac{e_a e_b}{r_{ab}}, \tag{1.1}$$

where e_a is the charge of the ath particle while

$$r_{ab} = |\mathbf{r}_a - \mathbf{r}_b| \tag{1.2}$$

is the separation between the ath and bth particles. (Throughout this book we use the Gaussian system of units. Connection with the SI units will be given in Appendix A.) As we shall see, this starting point, the Coulomb energy (1.1), summarizes all the experimental facts of electrostatics. The energy of interaction of an individual charge with the rest of the system can be emphasized by rewriting (1.1) as

$$E = \frac{1}{2} \sum_a e_a \sum_{b \neq a} \frac{e_b}{r_{ab}} = \frac{1}{2} \sum_a e_a \phi_a, \tag{1.3}$$

where we have introduced the electrostatic potential at the location of the ath charge that is due to all the other charges,

$$\phi_a = \sum_{b \neq a} \frac{e_b}{r_{ab}}. \tag{1.4}$$

This is an action-at-a-distance point of view, in which the charge at a given point

[1] First published by Charles-Augustin de Coulomb in 1785, although Joseph Priestley stated it in 1767.

DOI: 10.1201/9781003057369-1

interacts with charges at other distant points. Another approach, which generalizes and transcends action at a distance, employs the field concept (due to Faraday[2]), a field being a local quantity, defined at every point of space. We take a first step in this direction by considering the potential as a field, which is defined everywhere, not just where the point charges are located. This generalized potential function, or simply the potential, $\phi(\mathbf{r})$, is

$$\phi(\mathbf{r}) = \sum_b \frac{e_b}{|\mathbf{r} - \mathbf{r}_b|}, \tag{1.5}$$

where we now treat every charge on an equal footing, which means that in (1.5) we sum over all charges e_b. In terms of this potential, which is different from ϕ_a, the energy E can be written as

$$E = \frac{1}{2}\sum_a e_a \phi(\mathbf{r}_a) - \sum_a E_a. \tag{1.6}$$

The last part of (1.6) is not to be understood numerically but rather as an injunction to remove those terms in the first sum that refer to a single particle. In other words, we remove "self-action," leaving the mutual interactions between particles. The field concept naturally leads to self-action.

The notion of force is derived from that of energy, as we can see by considering the work done as a result of a spatial displacement. If we displace the ath charge by an amount $\delta\mathbf{r}_a$, the energy changes by an amount

$$\delta E = (\boldsymbol{\nabla}_a E) \cdot \delta\mathbf{r}_a = -\mathbf{F}_a \cdot \delta\mathbf{r}_a, \tag{1.7}$$

where $\mathbf{F}_a$ is the force acting on the ath point charge. Comparing this with the energy expression (1.1), we find the force on the ath particle to be

$$\mathbf{F}_a = -\boldsymbol{\nabla}_a \sum_{b \neq a} \frac{e_a e_b}{r_{ab}} = -\boldsymbol{\nabla}_a e_a \sum_{b \neq a} \frac{e_b}{r_{ab}} = -\boldsymbol{\nabla}_a e_a \phi(\mathbf{r}_a). \tag{1.8}$$

In the last form, we have substituted $\phi(\mathbf{r}_a)$ for ϕ_a, so it would appear that an extra self-action contribution has been introduced. To see that this is not true, we first argue physically that the difference between $\phi(\mathbf{r}_a)$ and ϕ_a is independent of position, and so self-action does not contribute to the force. Mathematically, what is this additional, unwanted, term? It is the negative gradient of the self-energy:

$$-\boldsymbol{\nabla} e_a \frac{e_a}{|\mathbf{r} - \mathbf{r}_a|}\bigg|_{\mathbf{r}\to\mathbf{r}_a} = e_a^2 \frac{\mathbf{r} - \mathbf{r}_a}{|\mathbf{r} - \mathbf{r}_a|^3}\bigg|_{\mathbf{r}\to\mathbf{r}_a}. \tag{1.9}$$

Can we make sense of this? We could define the limit here by arbitrarily adding a displacement vector $\boldsymbol{\epsilon}$ of fixed direction to $\mathbf{r}_a$ and letting its length approach zero:

$$\mathbf{r} = \mathbf{r}_a + \boldsymbol{\epsilon}, \quad \boldsymbol{\epsilon} \to \mathbf{0}, \tag{1.10a}$$

but at the cost of picking out a particular direction. In order to remove the most blatant aspect of this directional dependence, let us also approach $\mathbf{r}_a$ from the opposite direction,

$$\mathbf{r} = \mathbf{r}_a - \boldsymbol{\epsilon}, \quad \boldsymbol{\epsilon} \to \mathbf{0}, \tag{1.10b}$$

and average over the two possibilities so that the additional term (1.9) becomes

$$\longrightarrow e_a^2 \frac{1}{2}\left(\frac{\boldsymbol{\epsilon}}{|\boldsymbol{\epsilon}|^3} - \frac{\boldsymbol{\epsilon}}{|\boldsymbol{\epsilon}|^3}\right) = \mathbf{0}. \tag{1.11}$$

[2]Michael Faraday, 1791–1867.

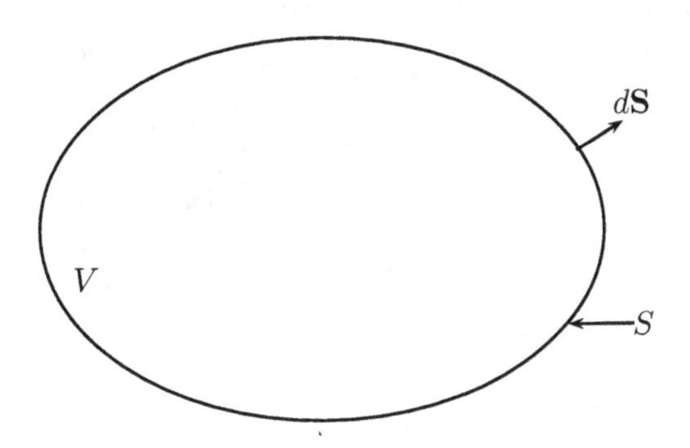

FIGURE 1.1
A surface S bounding a volume V used in computing the electric flux.

More elaborate limiting procedures, such as an average over all directions, can be used, but the simple procedure of (1.11) suffices. Therefore, we can employ $\phi(\mathbf{r})$ in (1.8), with the implicit use of the two-sided limit, (1.11), to calculate the force.

With the force given in terms of the gradient of a field (the potential), the electric field $\mathbf{E}$ can now be defined by

$$\mathbf{E}(\mathbf{r}) = -\boldsymbol{\nabla}\phi(\mathbf{r}), \tag{1.12}$$

so that the force on a point charge e_a located at $\mathbf{r}_a$ is

$$\mathbf{F}_a = e_a\mathbf{E}(\mathbf{r}_a). \tag{1.13}$$

The electric field $\mathbf{E}$ so introduced is a function calculable at $\mathbf{r}$ in terms of the point charges located at $\mathbf{r}_b$,

$$\mathbf{E}(\mathbf{r}) = \sum_b e_b \frac{\mathbf{r}-\mathbf{r}_b}{|\mathbf{r}-\mathbf{r}_b|^3}. \tag{1.14}$$

As such, it remains an action-at-a-distance description, whereas for many purposes, it would be much more convenient to be able to completely characterize the electric field by *local* properties. Such local statements will lead to differential equations, which, of course, must be supplemented by boundary conditions.

From its definition as the negative gradient of the potential, (1.12), the electric field has zero curl:

$$\boldsymbol{\nabla}\times\mathbf{E}(\mathbf{r}) = -\boldsymbol{\nabla}\times\boldsymbol{\nabla}\phi(\mathbf{r}) = \mathbf{0}. \tag{1.15}$$

Besides the curl, the other elementary differential operation that can be applied to a vector field is the divergence. To find $\boldsymbol{\nabla}\cdot\mathbf{E}$, we consider a related integral statement. The integral of the normal component of $\mathbf{E}$ over a closed surface S bounding a volume V is the electric flux (see Fig. 1.1):

$$\oint_S d\mathbf{S}\cdot\mathbf{E}(\mathbf{r}) = \sum_b e_b \oint_S d\mathbf{S}\cdot\frac{\mathbf{r}-\mathbf{r}_b}{|\mathbf{r}-\mathbf{r}_b|}\frac{1}{|\mathbf{r}-\mathbf{r}_b|^2} = \sum_b e_b \oint_S d\Omega_b. \tag{1.16}$$

Here, $d\mathbf{S}$ is an area element, directed normally to the surface, and $d\Omega_b$ is an element of solid angle, which is defined in the following manner. The element of area perpendicular to the line from the bth charge is (see Fig. 1.2):

$$d\mathbf{S}\cdot\frac{\mathbf{r}-\mathbf{r}_b}{|\mathbf{r}-\mathbf{r}_b|}, \tag{1.17}$$

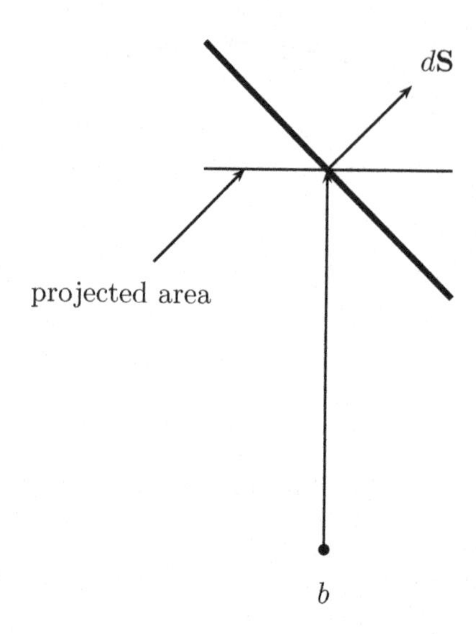

FIGURE 1.2
Geometrical definition of solid angle.

which when divided by the square of the distance from the bth charge gives the solid angle $d\Omega_b$ subtended by $d\mathbf{S}$ as seen from the bth charge. There are now two possible situations: either e_b is inside, or it is outside the closed surface S. Correspondingly, the integral over all solid angles in the two cases is

$$\oint_S d\Omega_b = \begin{cases} 4\pi \text{ if } e_b \text{ is inside } S, \\ 0 \quad \text{if } e_b \text{ is outside } S. \end{cases} \tag{1.18}$$

Hence, the electric flux through a closed surface S is proportional to the enclosed charge:

$$\oint_S d\mathbf{S} \cdot \mathbf{E}(\mathbf{r}) = \sum_{b \text{ in } V} 4\pi e_b. \tag{1.19}$$

This is the theorem of Carl Friedrich Gauss (1777–1855).[3]

With our aim of deriving local statements in mind, we generalize the idea of point charges to that of a continuous distribution of charge, as measured by $\rho(\mathbf{r})$, the volume density of charge at the point $\mathbf{r}$. Then, the total charge in a volume V is obtained by integrating the charge density over that region:

$$\sum_{b \text{ in } V} e_b = \int_V (d\mathbf{r})\, \rho(\mathbf{r}). \tag{1.20}$$

[Throughout this book we use the following notation for the element of volume:

$$(d\mathbf{r}) = dx\, dy\, dz.] \tag{1.21}$$

For point charges, the charge density is zero except at the location of the charges,

$$\rho(\mathbf{r}) = \sum_b e_b \delta(\mathbf{r} - \mathbf{r}_b), \tag{1.22}$$

[3]First stated by Joseph-Louis Lagrange around 1773.

where the three-dimensional (Dirac[4]) δ function is defined by

$$\int_V (d\mathbf{r})\,\delta(\mathbf{r}-\mathbf{r}_b) = \begin{cases} 0 \text{ if } \mathbf{r}_b \text{ is outside V,} \\ 1 \text{ if } \mathbf{r}_b \text{ is inside V.} \end{cases} \tag{1.23}$$

Then, the flux statement (1.19) becomes

$$4\pi \int_V (d\mathbf{r})\,\rho(\mathbf{r}) = \oint_S d\mathbf{S}\cdot\mathbf{E}(\mathbf{r}) = \int_V (d\mathbf{r})\,\boldsymbol{\nabla}\cdot\mathbf{E}(\mathbf{r}), \tag{1.24}$$

by use of the divergence theorem relating surface and volume integrals. (See Problem 1.3.) Since (1.24) is true for an arbitrary volume V, the integrands of the volume integrals must be equal, so we obtain the equation satisfied by the divergence of $\mathbf{E}$,

$$\boldsymbol{\nabla}\cdot\mathbf{E}(\mathbf{r}) = 4\pi\rho(\mathbf{r}). \tag{1.25}$$

These differential equations for the curl and divergence of $\mathbf{E}$, (1.15) and (1.25), respectively, completely characterize $\mathbf{E}$ when appropriate *boundary conditions* are imposed. It is evident from (1.14) that, for a localized charge distribution, the magnitude of the electric field becomes vanishingly small with increasing distance from the collection of charges:

$$|\mathbf{E}| \to 0 \quad \text{as} \quad \mathbf{r}\to\infty. \tag{1.26}$$

One can also specify how rapidly this occurs. But it is remarkable that the weak boundary condition (1.26) already implies a unique solution to the differential equations (1.15) and (1.25). To show this, we suppose that $\mathbf{E}_1$ and $\mathbf{E}_2$ are two such solutions. The difference, $\boldsymbol{\mathcal{E}} = \mathbf{E}_1 - \mathbf{E}_2$ satisfies

$$\boldsymbol{\nabla}\cdot\boldsymbol{\mathcal{E}} = 0, \quad \boldsymbol{\nabla}\times\boldsymbol{\mathcal{E}} = \mathbf{0} \quad \text{everywhere}, \tag{1.27a}$$

$$|\boldsymbol{\mathcal{E}}| \to 0 \quad \text{as} \quad \mathbf{r}\to\infty, \tag{1.27b}$$

from which we must prove that $\boldsymbol{\mathcal{E}} = \mathbf{0}$. The identity

$$\boldsymbol{\nabla}\times(\boldsymbol{\nabla}\times\boldsymbol{\mathcal{E}}) = \boldsymbol{\nabla}(\boldsymbol{\nabla}\cdot\boldsymbol{\mathcal{E}}) - \nabla^2\boldsymbol{\mathcal{E}}, \tag{1.28}$$

combined with the vanishing of $\boldsymbol{\nabla}\times\boldsymbol{\mathcal{E}}$ and $\boldsymbol{\nabla}\cdot\boldsymbol{\mathcal{E}}$, implies that

$$-\nabla^2\boldsymbol{\mathcal{E}} = \mathbf{0}. \tag{1.29}$$

Let the single function $\mathcal{E}(\mathbf{r})$ be any Cartesian component of the vector field $\boldsymbol{\mathcal{E}}$; it obeys

$$-\nabla^2\mathcal{E}(\mathbf{r}) = 0. \tag{1.30}$$

We present this as the everywhere valid statement

$$0 = -\mathcal{E}\nabla^2\mathcal{E} = -\boldsymbol{\nabla}\cdot(\mathcal{E}\boldsymbol{\nabla}\mathcal{E}) + (\boldsymbol{\nabla}\mathcal{E})^2, \tag{1.31a}$$

or

$$(\boldsymbol{\nabla}\mathcal{E})^2 - \nabla^2\frac{1}{2}\mathcal{E}^2 = 0. \tag{1.31b}$$

Now we integrate this over the interior volume $V(R)$ of a sphere of radius R centered about an arbitrary point, which we take as the origin. The integral of the second term in (1.31b)

[4]Paul Adrien Maurice Dirac (1902–1984).

is turned into an integral over the surface $S(R)$ of the sphere by means of the divergence theorem,

$$-\int_{V(R)} (d\mathbf{r})\, \boldsymbol{\nabla} \cdot (\boldsymbol{\nabla} \frac{1}{2}\mathcal{E}^2) = -\oint_{S(R)} d\mathbf{S} \cdot \boldsymbol{\nabla} \frac{1}{2}\mathcal{E}^2 = -\oint_{S(R)} dS \frac{\partial}{\partial R} \frac{1}{2}\mathcal{E}^2. \tag{1.32}$$

Using the relation between an element of area and an element of solid angle, $dS = R^2 d\Omega$, we can present this surface integral in terms of the average value of $\mathcal{E}^2$ over the surface of the sphere,

$$\langle \mathcal{E}^2 \rangle_R = \frac{1}{4\pi} \oint d\Omega\, \mathcal{E}^2. \tag{1.33}$$

And so the integral of (1.31b) is

$$\int_{V(R)} (d\mathbf{r})\, (\boldsymbol{\nabla}\mathcal{E})^2 - 4\pi R^2 \frac{d}{dR} \frac{1}{2} \langle \mathcal{E}^2 \rangle_R = 0. \tag{1.34}$$

The decisive step now is to divide by the area $4\pi R^2$, and then integrate (1.34) over R from 0 to ∞:

$$\int_0^\infty dR \frac{1}{4\pi R^2} \int_{V(R)} (d\mathbf{r})\, (\boldsymbol{\nabla}\mathcal{E})^2 + \frac{1}{2} \langle \mathcal{E}^2 \rangle_0 = 0, \tag{1.35}$$

which finally incorporates the boundary condition (1.27b), that $\mathcal{E}$ vanishes at all infinitely remote points. Everything on the left side of (1.35) is non-negative, yet it all adds up to zero. Accordingly, every individual contribution must be zero. This tells us quite explicitly that $\mathcal{E} = 0$ at the origin, which is anywhere, and, consistently, that $\boldsymbol{\nabla}\mathcal{E} = 0$ everywhere, or, that $\mathcal{E}$ is a constant, which is required to be zero by the boundary condition. This being true of any component, we conclude that the vector $\boldsymbol{\mathcal{E}} = \mathbf{0}$. This completes our proof of the "uniqueness theorem" of electrostatics,[5, 6] that the differential equations (1.15) and (1.25) have a unique solution when the boundary condition (1.26) is imposed. (See Problem 1.3.)

From the Coulomb energy, we have thus derived the equations of *electrostatics*:

$$\boldsymbol{\nabla} \cdot \mathbf{E} = 4\pi\rho, \qquad \frac{\partial}{\partial t}\rho = 0, \tag{1.36a}$$

$$\boldsymbol{\nabla} \times \mathbf{E} = \mathbf{0}, \qquad \frac{\partial}{\partial t}\mathbf{E} = \mathbf{0}, \tag{1.36b}$$

where the time independence has been made explicit. We are now going to remove the restriction to static conditions by letting the charges move in a particularly simple way. The equations of electromagnetism that emerge from this discussion will then be accepted as applicable to more general motions, as justified by various tests of internal consistency.

1.2 Inference of Maxwell's Equations

We introduce time dependence in the simplest way by assuming that all charges are in uniform motion with a common velocity $\mathbf{v}$ as produced by transforming a static arrangement of charges to a coordinate system moving with velocity $-\mathbf{v}$. (We insist that the same physics

[5] Due to George Green (1793–1841).

[6] Hermann von Helmholtz (1821–1894) emphasized the importance of the divergence and curl in establishing this theorem.

applies in the two situations—this is Galilean relativity.) At first we will take $|\mathbf{v}|$ to be very small in comparison with a critical speed c, which will be identified with the speed of light. To catch up with the moving charges, one would have to move with their velocity, $\mathbf{v}$. Accordingly, the time derivative in the co-moving coordinate system, in which the charges are at rest, is the sum of explicit time-dependent and coordinate dependent contributions,

$$\frac{d}{dt} = \frac{\partial}{\partial t} + \mathbf{v}\cdot\boldsymbol{\nabla}, \tag{1.37}$$

so, in going from the static system to the uniformly moving system, we make the replacement

$$\frac{\partial}{\partial t} \to \frac{d}{dt} = \frac{\partial}{\partial t} + \mathbf{v}\cdot\boldsymbol{\nabla}. \tag{1.38}$$

The equation for the constancy of the charge density in (1.36a) becomes, in the moving system

$$0 = \frac{\partial\rho}{\partial t} \to \frac{d\rho}{dt} = \frac{\partial\rho}{\partial t} + \mathbf{v}\cdot\boldsymbol{\nabla}\rho, \tag{1.39}$$

or, since $\mathbf{v}$ is constant,

$$\frac{\partial\rho}{\partial t} + \boldsymbol{\nabla}\cdot(\mathbf{v}\rho) = 0. \tag{1.40}$$

We recognize here a particular example of the charge flux vector or the (electric) current density $\mathbf{j}$,

$$\mathbf{j} = \rho\mathbf{v}. \tag{1.41}$$

The relation between charge density and current density,

$$\frac{\partial}{\partial t}\rho(\mathbf{r},t) + \boldsymbol{\nabla}\cdot\mathbf{j}(\mathbf{r},t) = 0, \tag{1.42}$$

is the general statement of the *conservation of charge*. Conservation demands that the rate of decrease of the charge within an arbitrary volume V must equal the rate at which the charge flows out of the bounding surface S, that is

$$-\frac{d}{dt}\int_V (d\mathbf{r})\,\rho(\mathbf{r},t) = \oint_S d\mathbf{S}\cdot\mathbf{j}(\mathbf{r},t) = \int_V (d\mathbf{r})\,\boldsymbol{\nabla}\cdot\mathbf{j}(\mathbf{r},t). \tag{1.43}$$

Since V is arbitrary, the local conservation law, (1.42), follows. We also note that the expression for the current density, (1.41), continues to be valid even when $\mathbf{v}$ is dependent upon position, $\mathbf{v} \to \mathbf{v}(\mathbf{r},t)$. (See Problem 1.6.)

We can perform a similar transformation on the equation (1.36b) for the electric field $\partial\mathbf{E}/\partial t = \mathbf{0}$; namely,

$$\mathbf{0} = \frac{d}{dt}\mathbf{E} = \frac{\partial\mathbf{E}}{\partial t} + (\mathbf{v}\cdot\boldsymbol{\nabla})\mathbf{E}. \tag{1.44}$$

Making use of a vector identity, together with (1.25) and (1.41), ($\mathbf{v}$ is constant),

$$\boldsymbol{\nabla}\times(\mathbf{v}\times\mathbf{E}) = \mathbf{v}(\boldsymbol{\nabla}\cdot\mathbf{E}) - (\mathbf{v}\cdot\boldsymbol{\nabla})\mathbf{E} \tag{1.45a}$$

$$= \mathbf{v}4\pi\rho - (\mathbf{v}\cdot\boldsymbol{\nabla})\mathbf{E}$$

$$= 4\pi\mathbf{j} - (\mathbf{v}\cdot\boldsymbol{\nabla})\mathbf{E}, \tag{1.45b}$$

we find an equation relating $\mathbf{E}$ to the current density,

$$\mathbf{0} = \frac{\partial\mathbf{E}}{\partial t} + 4\pi\mathbf{j} - \boldsymbol{\nabla}\times(\mathbf{v}\times). \tag{1.46}$$

[Notice that by taking the divergence of (1.46), we recover the local charge conservation equation (1.42), so that the conservation of charge is not an independent statement.] The quantity $\mathbf{v} \times \mathbf{E}$ represents a new phenomenon combining the effects of motion with those of electric charge. To describe this new, induced effect, we define the magnetic induction[7] **B** by

$$\mathbf{v}\times\mathbf{E} = c\mathbf{B}, \tag{1.47}$$

where c is a constant having the dimensions of velocity (which will turn out to be the speed of light). Expressed in terms of the magnetic field, (1.46) becomes an equation determining the curl of **B**,

$$\boldsymbol{\nabla}\times\mathbf{B} = \frac{1}{c}\frac{\partial}{\partial t}\mathbf{E} + \frac{4\pi}{c}\mathbf{j}. \tag{1.48}$$

Next, we naturally ask for the divergence of **B**. According to the definition, (1.47), we have

$$\boldsymbol{\nabla}\cdot\mathbf{B} = \boldsymbol{\nabla}\cdot\left(\frac{\mathbf{v}}{c}\times\mathbf{E}\right) = -\left(\frac{\mathbf{v}}{c}\times\boldsymbol{\nabla}\right)\cdot\mathbf{E} = -\frac{\mathbf{v}}{c}\cdot(\boldsymbol{\nabla}\times\mathbf{E}) = 0, \tag{1.49}$$

or

$$\boldsymbol{\nabla}\cdot\mathbf{B} = 0. \tag{1.50}$$

Moreover, in the co-moving coordinate system where the charges are at rest—static—the magnetic field should also not change in time:

$$\frac{d}{dt}\mathbf{B} = \frac{\partial}{\partial t}\mathbf{B} + (\mathbf{v}\cdot\boldsymbol{\nabla})\mathbf{B} = \mathbf{0}, \tag{1.51}$$

which becomes, when we use the identity in (1.45a) as well as (1.50),

$$\frac{\partial\mathbf{B}}{\partial t} = \boldsymbol{\nabla}\times(\mathbf{v}\times\mathbf{B}), \tag{1.52}$$

consistent with $\boldsymbol{\nabla}\cdot\mathbf{B} = 0$.

What do we do now? We need one experimental fact. Light is an electromagnetic oscillation. The evidence for this is overwhelming. As examples, we note that electric and magnetic fields are known to influence the emission, propagation, and absorption of light; and that radio and infrared waves, which differ only in wavelength from visible light, are emitted by electric charge oscillations. What must be done so that this fact is built into the equations we are inferring? The existence of electromagnetic waves means that the equations determining the electric field have solutions of the form

$$\mathbf{E} \sim \mathbf{f}(z - ct), \tag{1.53}$$

where c is the speed of the waves. Such waves, propagating in the z direction, satisfy the second-order differential equation

$$\frac{\partial^2}{\partial z^2}\mathbf{E} = \frac{1}{c^2}\frac{\partial^2}{\partial t^2}\mathbf{E}; \tag{1.54}$$

for an arbitrary direction of propagation, the corresponding wave equation is

$$\nabla^2\mathbf{E} = \frac{1}{c^2}\frac{\partial^2}{\partial t^2}\mathbf{E}. \tag{1.55}$$

More precisely, we require that this equation should hold far from the charges that produce the field. The left side of this equation can be written as [cf. (1.28)]

$$\nabla^2\mathbf{E} = -\boldsymbol{\nabla}\times(\boldsymbol{\nabla}\times\mathbf{E}), \tag{1.56}$$

[7] We will usually call **B** the magnetic *field*, but see Chapter 4.

since $\boldsymbol{\nabla}\cdot\mathbf{E} = 0$ outside the charge distribution, while, by means of (1.48) and (1.52), the right side becomes ($\mathbf{j}$ is zero outside the charge distribution)

$$\frac{1}{c^2}\frac{\partial^2}{\partial t^2}\mathbf{E} = \frac{1}{c}\frac{\partial}{\partial t}\boldsymbol{\nabla}\times\mathbf{B} = \frac{1}{c}\boldsymbol{\nabla}\times[\boldsymbol{\nabla}\times(\mathbf{v}\times\mathbf{B})]. \tag{1.57}$$

This shows that the desired differential equation will hold if

$$\mathbf{E} = -\frac{\mathbf{v}}{c}\times\mathbf{B}. \tag{1.58}$$

But this cannot be a completely correct statement since then $\mathbf{v}\to\mathbf{0}$ would require $\mathbf{E}\to\mathbf{0}$. No electrostatics! However, all that is really necessary is that the curl of this tentative identification be valid:

$$\boldsymbol{\nabla}\times\mathbf{E} = -\boldsymbol{\nabla}\times\left(\frac{\mathbf{v}}{c}\times\mathbf{B}\right), \tag{1.59}$$

or, if we use (1.52),

$$\boldsymbol{\nabla}\times\mathbf{E} = -\frac{1}{c}\frac{\partial}{\partial t}\mathbf{B}. \tag{1.60}$$

This is consistent with electrostatics since it generalizes $\boldsymbol{\nabla}\times\mathbf{E} = \mathbf{0}$ to the time-dependent situation. The fact that $\boldsymbol{\nabla}\times\mathbf{E} = \mathbf{0}$ has been used before to derive $\boldsymbol{\nabla}\cdot\mathbf{B} = 0$ is consistent here since the error is now seen to be of order $(v/c)^2$. [See (1.49).]

Collecting the above relations, you will recognize that we have arrived at Maxwell's equations,

$$\boldsymbol{\nabla}\times\mathbf{B} = \frac{1}{c}\frac{\partial}{\partial t}\mathbf{E} + \frac{4\pi}{c}\mathbf{j}, \qquad \boldsymbol{\nabla}\cdot\mathbf{E} = 4\pi\rho, \tag{1.61a}$$

$$-\boldsymbol{\nabla}\times\mathbf{E} = \frac{1}{c}\frac{\partial}{\partial t}\mathbf{B}, \qquad \boldsymbol{\nabla}\cdot\mathbf{B} = 0. \tag{1.61b}$$

These equations of *electromagnetism*, as local, differential field equations, are no longer restricted to the initial assumption of a common small velocity for all charges.

To complete the dynamical picture, we ask: What replaces (1.13) to describe the force on an electric charge, when that charge moves with some velocity $\mathbf{v}$ in given electric and magnetic fields $\mathbf{E}$ and $\mathbf{B}$? We consider two coordinate systems, one in which the particle is at rest (co-moving coordinate system) and one in which it moves at velocity $\mathbf{v}$. Suppose in the latter coordinate system, the electric and magnetic fields are given by $\mathbf{E}$ and $\mathbf{B}$, respectively. In the co-moving frame, the force on the particle is

$$\mathbf{F} = e\mathbf{E}_{\text{eff}}, \tag{1.62}$$

where $\mathbf{E}_{\text{eff}}$ is the electric field in this frame. In transforming to the co-moving frame, all the other charges—those responsible for $\mathbf{E}$ and $\mathbf{B}$—have been given an additional counter velocity $-\mathbf{v}$. We then infer from (1.58) that $(\mathbf{v}/c)\times\mathbf{B}$ has the character of an additional electric field in the co-moving frame. Hence, the suggested $\mathbf{E}_{\text{eff}}$ is

$$\mathbf{E}_{\text{eff}} = \mathbf{E} + \frac{\mathbf{v}}{c}\times\mathbf{B}, \tag{1.63}$$

leading to the force law, attributed to Hendrik Antoon Lorentz (1853–1928),

$$\mathbf{F} = e\left(\mathbf{E} + \frac{\mathbf{v}}{c}\times\mathbf{B}\right). \tag{1.64}$$

These results, Maxwell's equations, (1.61), and the Lorentz force law, (1.64), have *not* been *derived*, but *inferred* from a special circumstance. We will adopt these equations as describing the electromagnetic fields produced by, and acting on, charges possessing arbitrary velocities, although the above discussion does allow room for additional terms if v/c is no longer small. The fact that no such terms are actually required is part of the implication of the special theory of relativity (see Problem 1.6). We will prefer, instead, to show the physical consistency of the equations as they stand (see Chapter 3).

1.3 Discussion

We have arrived at the Maxwell-Lorentz electrodynamics by combining three ingredients: the laws of electrostatics; the Galileo-Newton principle of relativity (charges at rest, and charges with a common velocity viewed by a co-moving observer, are physically indistinguishable); and the existence of electromagnetic waves that travel in a vacuum at the speed c. The historical line of development was otherwise. Until the beginning of the nineteenth century, electricity and magnetism were unrelated phenomena. The discovery in 1820 by Hans Christian Oersted (1777–1851) that an electric current influences a magnet—creates a magnetic field—is formulated, for stationary currents, in the field equation

$$\boldsymbol{\nabla}\times\mathbf{B} = \frac{4\pi}{c}\mathbf{j}. \tag{1.65}$$

The symbol c that appears in this equation is the ratio of electromagnetic and electrostatic units of electricity (see Appendix A). Then, in 1831, Michael Faraday discovered that the relative motion of a wire and a magnet induces a voltage in the wire—creates an electric field. Such is the content of

$$-\boldsymbol{\nabla}\times\mathbf{E} = \frac{1}{c}\frac{\partial}{\partial t}\mathbf{B}, \tag{1.66}$$

which extends the magnetostatic relation

$$\boldsymbol{\nabla}\cdot\mathbf{B} = 0, \tag{1.67}$$

that expresses the empirical absence of single magnetic poles. Finally, in 1864, James Clerk Maxwell (1831–1879) recognized that the restriction to stationary currents in (1.65), as expressed by $\boldsymbol{\nabla}\cdot\mathbf{j} = 0$, was removed in

$$\boldsymbol{\nabla}\times\mathbf{B} = \frac{4\pi}{c}\mathbf{j} + \frac{1}{c}\frac{\partial}{\partial t}\mathbf{E}, \tag{1.68}$$

when joined to the electrostatic equation

$$\boldsymbol{\nabla}\cdot\mathbf{E} = 4\pi\rho. \tag{1.69}$$

The deduction of the existence of electromagnetic waves that travel at the speed c, in remarkable numerical agreement with the speed of light, was confirmed in 1887 by Heinrich Rudolf Hertz (1857–1894). It was the conflict between the existence of this absolute speed c and the relativity concept of Newtonian mechanics that set the stage for Einsteinian relativity. Already at the age of 16, Albert Einstein (1879–1955) had recognized this paradox: To a co-moving Newtonian observer, light waves should oscillate in space but not move; however, Maxwell's equations admit no such solutions. Einsteinian relativity is an outgrowth of Maxwellian electrodynamics, not the other way about. That is the spirit in which electrodynamics is developed as a self-contained subject in this book.[8]

1.4 Problems for Chapter 1

1. Verify the following identities explicitly:

$$\mathbf{A}\times(\mathbf{B}\times\mathbf{C})+\mathbf{B}\times(\mathbf{C}\times\mathbf{A})+\mathbf{C}\times(\mathbf{A}\times\mathbf{B})=\mathbf{0}, \tag{1.70a}$$

$$\begin{aligned}\boldsymbol{\nabla}\times(\mathbf{A}\times\mathbf{B})&=\mathbf{A}\times(\boldsymbol{\nabla}\times\mathbf{B})-\mathbf{B}\times(\boldsymbol{\nabla}\times\mathbf{A})\\&\quad-(\mathbf{A}\times\boldsymbol{\nabla})\times\mathbf{B}+(\mathbf{B}\times\boldsymbol{\nabla})\times\mathbf{A},\end{aligned} \tag{1.70b}$$

$$\boldsymbol{\nabla}\cdot(\lambda\mathbf{A}\times\mathbf{B})=\lambda(\mathbf{B}\cdot\boldsymbol{\nabla}\times\mathbf{A}-\mathbf{A}\cdot\boldsymbol{\nabla}\times\mathbf{B})+\mathbf{A}\times\mathbf{B}\cdot\boldsymbol{\nabla}\lambda. \tag{1.70c}$$

2. Use the three-dimensional Levi-Civita symbol, which has the properties of being totally antisymmetric in all three indices,

$$\epsilon_{ijk}=\epsilon_{jki}=\epsilon_{kij}=-\epsilon_{jik}=-\epsilon_{kji}=-\epsilon_{ikj}, \tag{1.71}$$

and is normalized to unity, $\epsilon_{123}=+1$, to define the cross product in terms of the Cartesian components of vectors,

$$\mathbf{A}=\mathbf{B}\times\mathbf{C} \iff A_i=\epsilon_{ijk}B_jC_k, \tag{1.72}$$

which uses the summation convention for repeated indices, for example,

$$A_iB_i=\sum_{i=1}^{3}A_iB_i=\mathbf{A}\cdot\mathbf{B}. \tag{1.73}$$

Prove the identity

$$\epsilon_{ijk}\epsilon_{klm}=\delta_{il}\delta_{jm}-\delta_{im}\delta_{jl}, \tag{1.74}$$

where δ_{ij} is the Kronecker δ symbol

$$\delta_{ij}=\begin{cases}1,\, i=j,\\ 0,\, i\neq j\end{cases} \tag{1.75}$$

Show that this is equivalent to the familiar identity

$$\mathbf{A}\times(\mathbf{B}\times\mathbf{C})=\mathbf{B}(\mathbf{A}\cdot\mathbf{C})-\mathbf{C}(\mathbf{A}\cdot\mathbf{B}), \tag{1.76}$$

provided **A**, **B**, and **C** are commuting vectors.

[8] Although nearly invariably attributed to Maxwell, he certainly did not express these equations in a very comprehensible form. Oliver Heaviside (1850–1925), through his great work on telegraphy, recast the equations into their modern vector form, including the force law, so it entirely appropriate to refer to them as Maxwell-Heaviside equations. However, independently, Josiah Willard Gibbs (1839–1903), in the backwater of New England, invented vector calculus and the modern formulation of the theory of electromagnetism, and, due to his outstanding contributions to statistical thermodynamics, was the worthy heir of Maxwell.

3. Verify, using Cartesian coordinates, the divergence theorem,

$$\int_V (d\mathbf{r})\, \boldsymbol{\nabla} \cdot \mathbf{E} = \oint_S d\mathbf{S} \cdot \mathbf{E}, \tag{1.77}$$

where V is the volume contained within the closed surface S, $d\mathbf{S}$ being the surface element in the direction of the outward normal, and Stokes' theorem [George Gabriel Stokes, 1819–1903],

$$\int_S d\mathbf{S} \cdot (\boldsymbol{\nabla} \times \mathbf{E}) = \oint_C d\mathbf{l} \cdot \mathbf{E}, \tag{1.78}$$

where C is the closed boundary of the open surface S, and $d\mathbf{l}$ is the tangentially directed line element. The sense of the line integration is given by the right-hand rule. [That is, if the contour C is traversed in the sense of the fingers of the right hand, the thumb points in the sense of the orientation of the surface.]

4. Show that the following functions represent the delta function in one dimension, by showing that

$$\delta_\epsilon(x \neq 0) \to 0, \quad \text{as} \quad \epsilon \to 0+, \tag{1.79a}$$

and

$$\int_{-\infty}^{\infty} dx\, \delta_\epsilon(x) = 1. \tag{1.79b}$$

(a)

$$\delta_\epsilon(x) = \frac{1}{\sqrt{\pi\epsilon}} e^{-x^2/\epsilon}, \tag{1.80a}$$

(b)

$$\delta_\epsilon(x) = \frac{1}{\pi} \frac{\epsilon}{x^2 + \epsilon^2}, \tag{1.80b}$$

(c)

$$\delta_\epsilon(x) = \int_{-\infty}^{\infty} \frac{dk}{2\pi} e^{ikx - |k|\epsilon}. \tag{1.80c}$$

5. This question has to do with the uniqueness theorem, which follows from (1.34).
   (a) Directly from that equation, what assumption about $|\mathcal{E}(\mathbf{r})|$, $|\mathbf{r}| \to \infty$, will produce the conclusion that $\mathcal{E} = 0$ everywhere?
   (b) How would it work out if one had integrated this equation from $R = 0$ to ∞, without dividing by R^2?
   (c) How fast would $\langle \mathcal{E}^2 \rangle_R$ have to fall off with R so that we could conclude $\mathcal{E} = 0$ everywhere by simply taking $R \to \infty$ in (1.34)?

6. For an arbitrarily moving charge, the charge and current densities are

$$\rho(\mathbf{r}, t) = e\delta(\mathbf{r} - \mathbf{R}(t)), \quad \mathbf{j}(\mathbf{r}, t) = e\frac{d\mathbf{R}}{dt}\delta(\mathbf{r} - \mathbf{R}(t)), \tag{1.81}$$

where $\mathbf{R}(t)$ is the position vector of the charged particle. Verify the statement of conservation of charge,

$$\frac{\partial}{\partial t}\rho(\mathbf{r}, t) + \boldsymbol{\nabla} \cdot \mathbf{j}(\mathbf{r}, t) = 0. \tag{1.82}$$

7. In a region where no charges are present, the potential satisfies Laplace's equation,

$$\nabla^2\phi = 0. \tag{1.83}$$

Such a function is called harmonic. Show that in a region where the potential is harmonic, the potential nowhere assumes a maximum or minimum value. Use this result to give another proof of the uniqueness theorem of electrostatics proved in Section 1.1.

8. A one-dimensional continuous distribution of charge is placed, in vacuum, along the z-axis, consisting of a uniform charge per unit length λ extended from $z = -L$ to $z = +L$. Find the potential, $\phi(\rho, z)$, in terms of cylindrical coordinates ρ and z, by direct integration, for $\rho, |z| \ll L$. Does the potential obey Laplace's equation for $\rho > 0$? What is the exact potential?

9. Show that Maxwell's equations are invariant, up to terms of order $(\delta v/c)^2$, under a Galilean transformation, where $\delta\mathbf{v}$ is constant:

$$\mathbf{r} \to \mathbf{r} + \delta\mathbf{r}, \quad \delta\mathbf{r} = \delta\mathbf{v}t; \qquad t \to t, \quad (\delta t = 0); \tag{1.84a}$$

$$\mathbf{E} \to \mathbf{E} + \frac{\delta\mathbf{v}}{c}\times\mathbf{B}; \qquad \mathbf{B} \to \mathbf{B} - \frac{\delta\mathbf{v}}{c}\times\mathbf{E}; \tag{1.84b}$$

$$\rho \to \rho, \quad (\delta\rho = 0); \qquad \mathbf{J} \to \mathbf{J} - \delta\mathbf{v}\rho. \tag{1.84c}$$

Verify that charge conservation is preserved by the above Galilean transformation.

10. In this chapter, we "derived" Maxwell's equations by exploiting approximate Galilean invariance. However, we cannot push Galilean invariance further since it is not valid in $O(v^2/c^2)$. The correct relativity is that of Einstein. Verify that Maxwell's equations are invariant under the transformations of Einstein's special relativity, as follows. Consider a Lorentz transformation corresponding to a boost in the x direction, which on the space-time coordinates is defined by

$$x_0' = \gamma(x_0 - \beta x_1), \tag{1.85a}$$

$$x_1' = \gamma(x_1 - \beta x_0), \tag{1.85b}$$

$$x_2' = x_2, \tag{1.85c}$$

$$x_3' = x_3. \tag{1.85d}$$

Here $x_0 = ct$, $x_1 = x$, $x_2 = y$, $x_3 = z$, and

$$\beta = \frac{v}{c}, \quad \gamma = (1 - \beta^2)^{-1/2}, \tag{1.86}$$

$\mathbf{v}$ being the relative velocity of the two coordinate frames. We can regard the four quantities x_μ, $\mu = 0, 1, 2, 3$, as forming a four-vector. The four-current j_μ, $j_0 = c\rho$, j_i, $i = 1, 2, 3$, constructed from the electric charge and current densities, transforms by the same law:

$$j_0' = \gamma(j_0 - \beta j_1), \tag{1.87a}$$

$$j_1' = \gamma(j_1 - \beta j_0), \tag{1.87b}$$

$$j_2' = j_2, \tag{1.87c}$$

$$j_3' = j_3. \tag{1.87d}$$

On the other hand, the electric and magnetic field vectors are components of a four-tensor, and so they have a more complicated transformation law. Consider a boost by an arbitrary velocity $\mathbf{v}$. Then the components of the electric and magnetic fields in the direction of $\mathbf{v}$ do not change, while the components in directions perpendicular to $\mathbf{v}$ are entangled:

$$\mathbf{E}'_{\parallel} = \mathbf{E}_{\parallel}, \quad \mathbf{E}'_{\perp} = \gamma\left(\mathbf{E} + \frac{\mathbf{v}}{c}\times\mathbf{B}\right)_{\perp}, \tag{1.88a}$$

$$\mathbf{B}'_{\parallel} = \mathbf{B}_{\parallel}, \quad \mathbf{B}'_{\perp} = \gamma\left(\mathbf{B} - \frac{\mathbf{v}}{c}\times\mathbf{E}\right)_{\perp}. \tag{1.88b}$$

For $\mathbf{v} = (v, 0, 0)$, verify explicitly that if Maxwell's equations hold in the unprimed frame, they hold in the primed frame as well, no matter how near v may approach c. This was essentially the path by which Lorentz and (Henri) Poincaré (1854–1912) derived the transformation equations (but not the physics) of special relativity. A more complete treatment of Einsteinian relativity will be given in Chapters 11 and 12.

11. A charge e moves in the vacuum under the influence of uniform fields $\mathbf{E}$ and $\mathbf{B}$. Assume that $\mathbf{E}\cdot\mathbf{B} = 0$ and that $\mathbf{v}\cdot\mathbf{B} = 0$. At what velocity does the charge move without acceleration? What is its speed when $|\mathbf{E}| = |\mathbf{B}|$?

12. Consider the Newton-Lorentz equations. The equations of motion for the vectors $\mathbf{r}$ and $\mathbf{v}$ are effectively referred to a coordinate frame rotating with angular velocity $\boldsymbol{\omega}$ by the substitutions

$$\frac{d}{dt} \to \frac{d}{dt} + \boldsymbol{\omega}\times. \tag{1.89}$$

Show this by using the Galilean transformations (1.84a). Then, from (1.84b) determine how the electric and magnetic fields transform.

(a) Find the choice of $\boldsymbol{\omega}$ (this is the cyclotron frequency) that, for $\mathbf{E} = \mathbf{0}$, removes the effect on $\mathbf{v}(t)$ of a *weak* homogeneous $\mathbf{B}$.

(b) What is that value of $\boldsymbol{\omega}$ that removes the effect on $\mathbf{r}(t)$ of $\mathbf{B}$ to first order?

2

Magnetic Charge I

Our discussion in Chapter 1 contains a certain implicit assumption. When it came to (1.58),

$$\mathbf{E} = -\frac{\mathbf{v}}{c}\times\mathbf{B}, \tag{2.1}$$

with its implication that static electric charges produce no electric field, we knew better than to accept this and altered it to

$$\boldsymbol{\nabla}\times\mathbf{E} = -\boldsymbol{\nabla}\times\left(\frac{\mathbf{v}}{c}\times\mathbf{B}\right), \tag{2.2}$$

thereby admitting, for $\mathbf{v} = \mathbf{0}$, a static electric field, one obeying $\boldsymbol{\nabla}\times\mathbf{E} = \mathbf{0}$. Why then did we earlier accept without question the relation (1.47),

$$\mathbf{B} = \frac{\mathbf{v}}{c}\times\mathbf{E}, \tag{2.3}$$

with its implication that all magnetic fields are due to the motion of electric charges? This is the (1820) hypothesis of André Marie Ampère (1775–1836). But is it true? An affirmative response is conventional, but the mathematical development allows a more general possibility. Again, all that was really required in the above was the curl relation

$$\boldsymbol{\nabla}\times\mathbf{B} = \boldsymbol{\nabla}\times\left(\frac{\mathbf{v}}{c}\times\mathbf{E}\right), \tag{2.4}$$

admitting the possibility, for $\mathbf{v} = \mathbf{0}$, of a static magnetic field obeying $\boldsymbol{\nabla}\times\mathbf{B} = \mathbf{0}$, one that has its origin in *magnetic* charge. If ρ_m is the density of such magnetic charge, the analogy with electrostatics suggests that

$$\boldsymbol{\nabla}\cdot\mathbf{B} = 4\pi\rho_m. \tag{2.5}$$

The implication of (2.5) is that a further source of magnetic fields, other than moving electric charges, could exist in magnetic charge. Whether this possibility is realized in nature still awaits experimental confirmation.

Further changes in Maxwell's equations are required if magnetic charge exists. Since then $\boldsymbol{\nabla}\cdot\mathbf{B} \neq 0$, the co-moving time derivative of $\mathbf{B}$ becomes [*cf.* (1.45a)]

$$\mathbf{0} = \frac{d}{dt}\mathbf{B} = \frac{\partial}{\partial t}\mathbf{B} + (\mathbf{v}\cdot\boldsymbol{\nabla})\mathbf{B} = \frac{\partial}{\partial t}\mathbf{B} - \boldsymbol{\nabla}\times(\mathbf{v}\times\mathbf{B}) + \mathbf{v}(\boldsymbol{\nabla}\cdot\mathbf{B}). \tag{2.6}$$

Then, using (2.5) and the magnetic current density $\mathbf{j}_m$, defined as

$$\mathbf{j}_m = \rho_m\mathbf{v}, \tag{2.7}$$

together with (2.2), we obtain the following modified Maxwell equation

$$-\boldsymbol{\nabla}\times\mathbf{E} = \frac{1}{c}\frac{\partial}{\partial t}\mathbf{B} + \frac{4\pi}{c}\mathbf{j}_m. \tag{2.8}$$

DOI: 10.1201/9781003057369-2

Notice that (2.8) implies the conservation of magnetic charge:

$$\frac{\partial}{\partial t}\rho_m + \boldsymbol{\nabla}\cdot\mathbf{j}_m = 0. \tag{2.9}$$

The complete set of Maxwell's equations, when magnetic charge is present, now reads

$$\begin{aligned}
\boldsymbol{\nabla}\times\mathbf{B} &= \frac{1}{c}\frac{\partial}{\partial t}\mathbf{E} + \frac{4\pi}{c}\mathbf{j}_e, \qquad &\boldsymbol{\nabla}\cdot\mathbf{E} = 4\pi\rho_e,\\
-\boldsymbol{\nabla}\times\mathbf{E} &= \frac{1}{c}\frac{\partial}{\partial t}\mathbf{B} + \frac{4\pi}{c}\mathbf{j}_m, \qquad &\boldsymbol{\nabla}\cdot\mathbf{B} = 4\pi\rho_m,
\end{aligned} \tag{2.10}$$

where we have consistently used the subscript e to denote densities for electric charge. Observe that these equations are invariant in form under the replacements (duality transformation)

$$\begin{aligned}
\rho_e &\to \rho_m, \qquad &\mathbf{E} &\to \mathbf{B}, \qquad &\mathbf{j}_e &\to \mathbf{j}_m,\\
\rho_m &\to -\rho_e, \qquad &\mathbf{B} &\to -\mathbf{E}, \qquad &\mathbf{j}_m &\to -\mathbf{j}_e.
\end{aligned} \tag{2.11}$$

The generalized Lorentz force law, suggested by this symmetry, is

$$\mathbf{F} = e\left(\mathbf{E} + \frac{\mathbf{v}}{c}\times\mathbf{B}\right) + g\left(\mathbf{B} - \frac{\mathbf{v}}{c}\times\mathbf{E}\right), \tag{2.12}$$

for a particle carrying both electric and magnetic charge, e and g, respectively.

2.1 A Very Brief History of Magnetic Charge

It is said that Peregrinus in 1269 observed that magnets (lodestones) always have two poles, which he called north and south. This was elevated to a "hypothesis" by Ampère in the early 19th Century. The first theoretical calculation of the motion of a charged particle in the presence of a single magnetic pole was performed by Poincaré in 1896 to explain recent observations.[1] A few years later, J. J. Thomson (1856–1940) showed[2] that a static system consisting of a magnetic pole and an electric charge possessed an angular momentum—see Problems 2.3, 2.4, 3.10, and 3.11. It was Dirac in 1931 who showed that magnetic charge was consistent with quantum mechanics only if electric and magnetic charges were quantized: For a system consisting of a pure magnetic charge g and a pure electric charge e, eg had to be an integral (or half-integral) multiple of $\hbar c$. Many people have contributed to the theory of magnetic charge subsequently; notable is the work of Schwinger in the 1960s and 1970s, especially his concept of dyons, particles which carry both electric and magnetic charge.

Although from time to time there have been spectacular reports of the discovery of magnetic charge [10, 12], these "discoveries" were never replicated, and serious objections were raised in each instance. Nevertheless, there are strong theoretical reasons to believe that magnetic charge exists in nature, and may have played an important role in the development of the universe. Modern unified theories of fundamental interactions typically imply the existence of magnetic monopoles, or of dyons, often at extremely high mass scales ($\sim 10^{16}$ GeV), but perhaps at accessible energies ($\sim$10 TeV). Moreover, there appears to be no reason why an elementary monopole or dyon of the Dirac-Schwinger type could not exist.

[1] Based on earlier work by Jean-Gaston Darboux (1878).
[2] Yale Lectures, 1904.

As Michael Faraday said, "Nothing is too wonderful to be true, if it be consistent with the laws of nature; and in such things as these, experiment is the best test of such consistency." [13] Searches for magnetic charge continue at the present time [14, 15], emphasizing that electromagnetism is very far from being a closed subject.

2.2 Problems for Chapter 2

1. Write Maxwell's equations with magnetic charge in terms of

$$\mathbf{F} = \mathbf{E} + i\mathbf{B}, \quad i = \sqrt{-1},$$

and related combinations of charge and current. Verify that these equations retain their form under the transformation illustrated by

$$\mathbf{F} \to e^{-i\phi}\mathbf{F},$$

where ϕ is an arbitrary constant. Express this as a transformation of **E**, **B**, and the charge-current quantities. What is the geometric interpretation? What is the particular form of this transformation when $\phi = \pi/2$?

2. Suppose every charged particle carried electric *and* magnetic charge in the universal ratio $g_k/e_k = \lambda$. Is there another way of looking at this situation in which we would be unaware of magnetic charge?

3. Consider the motion of an electron of mass m moving in the field of a magnetic monopole of mass M, $m \ll M$, where the charge of the electron is e and the magnetic charge of the monopole is g. Assume the motion of the electron remains nonrelativistic. Derive the equation of motion of the electron. From this equation, derive the equation of energy conservation. What does this tell you about the speed of the electron? Derive the equation for the time rate of change of the orbital angular momentum of the electron, $\mathbf{L} = m\mathbf{r}\times\mathbf{v}$. Deduce the presence of an extra angular momentum of the system. How does this contribution depend on the distance between the electron and the magnetic monopole? What follows if quantum ideas of the quantization of angular momentum are applied to the radial component of the angular momentum?

4. Consider the relative motion of two particles with masses m_1 and m_2 and electric and magnetic charges, e_1, g_1, and e_2, g_2, respectively. In deriving the equation of relative motion, which involves the reduced mass, remember that moving electric (magnetic) charges produce magnetic (electric) fields, but do not retain more than one factor of v_1/c or v_2/c. How do the combinations of electric and magnetic charges behave under the duality transformations (2.11)? What is the conserved angular momentum?

3

Conservation Laws

In order to check the physical consistency of the above set of equations governing Maxwell-Lorentz electrodynamics [(2.10) and (2.12) or (1.61) and (1.64)], we examine the action of, and reaction on, the sources of the electromagnetic fields. To be precise, we ask whether there is a correct balance in the exchange of energy, momentum, and angular momentum between the charged particles and the electromagnetic fields. As we shall see, the Maxwell-Lorentz system as it stands implies the conservation of these mechanical properties, no matter how rapidly the charges are moving.

3.1 Conservation of Energy

We start with a consideration of the rate at which work is done on the particles, that is, the rate of energy transfer, or the power absorbed by the particles. For one particle, we know that the rate at which work is done on it is

$$\mathbf{F}\cdot\mathbf{v} = e\mathbf{v}\cdot\mathbf{E} + g\mathbf{v}\cdot\mathbf{B} = \int (d\mathbf{r})\,(\mathbf{j}_e\cdot\mathbf{E} + \mathbf{j}_m\cdot\mathbf{B}), \tag{3.1}$$

where we have used the Lorentz force law, (2.12), and the expressions for the currents, (1.41) and (2.7), for a point particle. We interpret this equation as meaning, even for general current distributions, that $\mathbf{j}_e\cdot\mathbf{E} + \mathbf{j}_m\cdot\mathbf{B}$ is the rate of energy transfer from the field to the particles, per unit volume. Then, through elimination of the currents by use of Maxwell's equations, (2.10), this rate can be rewritten as

$$\begin{aligned}\mathbf{j}_e\cdot\mathbf{E} + \mathbf{j}_m\cdot\mathbf{B} &= \frac{c}{4\pi}\left(\boldsymbol{\nabla}\times\mathbf{B} - \frac{1}{c}\frac{\partial}{\partial t}\mathbf{E}\right)\cdot\mathbf{E} + \frac{c}{4\pi}\left(-\boldsymbol{\nabla}\times\mathbf{E} - \frac{1}{c}\frac{\partial}{\partial t}\mathbf{B}\right)\cdot\mathbf{B}\\ &= -\frac{\partial}{\partial t}\left(\frac{E^2+B^2}{8\pi}\right) - \boldsymbol{\nabla}\cdot\left(\frac{c}{4\pi}\mathbf{E}\times\mathbf{B}\right).\end{aligned} \tag{3.2}$$

The general form of any local conservation law, (1.42) or (1.43), suggests the following interpretations:

1. In the absence of charges ($\mathbf{j}_e = \mathbf{j}_m = \mathbf{0}$), this is the local energy conservation law:

$$\frac{\partial}{\partial t}\frac{E^2+B^2}{8\pi} + \boldsymbol{\nabla}\cdot\frac{c}{4\pi}\mathbf{E}\times\mathbf{B} = 0. \tag{3.3}$$

We label the two objects appearing here as

$$\text{energy density} = U = \frac{E^2+B^2}{8\pi}, \tag{3.4a}$$

$$\text{energy flux vector} = \mathbf{S} = \frac{c}{4\pi}\mathbf{E}\times\mathbf{B}. \tag{3.4b}$$

DOI: 10.1201/9781003057369-3

[The latter is usually called the Poynting vector, after John Henry Poynting (1852–1914).]

2. In the presence of charges, the relation (3.2) is

$$\frac{\partial}{\partial t}U + \boldsymbol{\nabla}\cdot\mathbf{S} + \mathbf{j}_e\cdot\mathbf{E} + \mathbf{j}_m\cdot\mathbf{B} = 0, \tag{3.5}$$

which, if we integrate over an arbitrary volume V, bounded by a surface S, becomes

$$\frac{d}{dt}\int_V (d\mathbf{r})\,U + \oint_S d\mathbf{S}\cdot\mathbf{S} + \int_V (d\mathbf{r})\,(\mathbf{j}_e\cdot\mathbf{E} + \mathbf{j}_m\cdot\mathbf{B}) = 0. \tag{3.6}$$

The three terms here are identified, respectively, as the rate of change of the electromagnetic field energy within the volume, the rate of flow of electromagnetic energy out of the volume, and the rate of transfer of electromagnetic energy to the charged particles. Thus, (3.5) gives a complete description of energy conservation.

3.2 Conservation of Momentum

Next we consider the force on a particle, (2.12), as the rate of change of momentum,

$$\begin{aligned}\mathbf{F} &= e\left(\mathbf{E} + \frac{\mathbf{v}}{c}\times\mathbf{B}\right) + g\left(\mathbf{B} - \frac{\mathbf{v}}{c}\times\mathbf{E}\right)\\ &= \int (d\mathbf{r})\left(\rho_e\mathbf{E} + \frac{1}{c}\mathbf{j}_e\times\mathbf{B} + \rho_m\mathbf{B} - \frac{1}{c}\mathbf{j}_m\times\mathbf{E}\right)\\ &\equiv \int (d\mathbf{r})\,\mathbf{f},\end{aligned} \tag{3.7}$$

where $\mathbf{f}$ is the force density. Removing reference to the (generalized) charge and current densities by use of Maxwell's equations, (2.10), we rewrite the force density $\mathbf{f}$ as

$$\begin{aligned}\mathbf{f} &= \frac{1}{4\pi}[\mathbf{E}(\boldsymbol{\nabla}\cdot\mathbf{E}) + \mathbf{B}(\boldsymbol{\nabla}\cdot\mathbf{B})]\\ &\quad + \frac{1}{4\pi}\left[\left(-\frac{1}{c}\frac{\partial}{\partial t}\mathbf{E} + \boldsymbol{\nabla}\times\mathbf{B}\right)\times\mathbf{B} + \mathbf{E}\times\left(-\frac{1}{c}\frac{\partial}{\partial t}\mathbf{B} - \boldsymbol{\nabla}\times\mathbf{E}\right)\right]\\ &= -\frac{\partial}{\partial t}\frac{\mathbf{E}\times\mathbf{B}}{4\pi c} + \frac{1}{4\pi}[-\mathbf{E}\times(\boldsymbol{\nabla}\times\mathbf{E}) + \mathbf{E}(\boldsymbol{\nabla}\cdot\mathbf{E}) - \mathbf{B}\times(\boldsymbol{\nabla}\times\mathbf{B}) + \mathbf{B}(\boldsymbol{\nabla}\cdot\mathbf{B})].\end{aligned} \tag{3.8}$$

The quadratic structure in $\mathbf{E}$ occurring here is

$$\begin{aligned}-\mathbf{E}\times(\boldsymbol{\nabla}\times\mathbf{E}) + \mathbf{E}(\boldsymbol{\nabla}\cdot\mathbf{E}) &= -\boldsymbol{\nabla}\frac{E^2}{2} + (\mathbf{E}\cdot\boldsymbol{\nabla})\mathbf{E} + \mathbf{E}(\boldsymbol{\nabla}\cdot\mathbf{E})\\ &= \boldsymbol{\nabla}\cdot\left(-\mathbf{1}\frac{E^2}{2} + \mathbf{E}\mathbf{E}\right),\end{aligned} \tag{3.9}$$

which introduces dyadic notation, including the unit dyadic $\mathbf{1}$, with components

$$\mathbf{1}_{kl} = \delta_{kl} = \begin{cases}1, & k = l,\\ 0, & k \neq l,\end{cases} \tag{3.10}$$

where δ_{kl} is the Kronecker δ symbol. (See Problem 3.1.) The analogous result holds for $\mathbf{B}$. Accordingly, the force density is

$$\mathbf{f} = -\frac{\partial}{\partial t}\frac{\mathbf{E}\times\mathbf{B}}{4\pi c} - \boldsymbol{\nabla}\cdot\left(\mathbf{1}\frac{E^2+B^2}{8\pi} - \frac{\mathbf{EE}+\mathbf{BB}}{4\pi}\right). \tag{3.11}$$

We interpret this equation physically by identifying

$$\text{momentum density} = \mathbf{G} = \frac{\mathbf{E}\times\mathbf{B}}{4\pi c}, \tag{3.12a}$$

and

$$\text{momentum flux (stress tensor)} = \mathbf{T} = \mathbf{1}\frac{E^2+B^2}{8\pi} - \frac{\mathbf{EE}+\mathbf{BB}}{4\pi}. \tag{3.12b}$$

When $\mathbf{f} = \mathbf{0}$, we obtain the local statement of the conservation of momentum of the electromagnetic field. A full account of momentum balance is contained in

$$\frac{\partial}{\partial t}\mathbf{G} + \boldsymbol{\nabla}\cdot\mathbf{T} + \mathbf{f} = \mathbf{0}. \tag{3.13}$$

The volume integral of this equation for electromagnetic momentum is interpreted analogously to the energy result, (3.6).

The components of the stress tensor are given by

$$T_{kl} = \delta_{kl}U - \frac{E_kE_l + B_kB_l}{4\pi}. \tag{3.14}$$

Notice that the stress tensor is symmetrical, $T_{kl} = T_{lk}$, which, as we shall see in the next section, is required in order to obtain a local conservation law for angular momentum. The trace of $\mathbf{T}$, the sum of the diagonal elements T_{kk}, is simply the energy density, (3.4a),

$$\operatorname{Tr}\mathbf{T} = \sum_k T_{kk} = U. \tag{3.15}$$

We also note that the Poynting vector, (3.4b), is proportional to the momentum density,

$$\mathbf{S} = c^2\mathbf{G}, \tag{3.16}$$

which has the structure of

$$\text{energy density} \times \text{velocity} = c^2(\text{mass density} \times \text{velocity}). \tag{3.17}$$

This is the first indication of the relativistic connection between energy and mass, $E = mc^2$.

3.3 Conservation of Angular Momentum. Virial Theorem

Having discussed (linear) momentum, we now turn to angular momentum. We will use tensor notation to write (3.13) in component form,

$$\frac{\partial}{\partial t}G_k + \nabla_l T_{lk} + f_k = 0, \tag{3.18}$$

where we have also used the summation convention: Whenever an index is repeated, a sum over all values of that index is assumed,

$$a_i b_i \equiv \sum_{i=1}^{3} a_i b_i = \mathbf{a}\cdot\mathbf{b}. \tag{3.19}$$

The rate of change of angular momentum is the torque $\boldsymbol{\tau}$, which, for one particle, is

$$\boldsymbol{\tau} = \mathbf{r}\times\mathbf{F} = \int (d\mathbf{r})\,\mathbf{r}\times\mathbf{f}, \tag{3.20}$$

where the volume-integrated form is no longer restricted to a single particle. The torque density, the moment of the force density, can be written in component form as

$$(\mathbf{r}\times\mathbf{f})_i = \epsilon_{ijk} x_j f_k, \tag{3.21}$$

where we have introduced the totally antisymmetric (Levi-Civita[1]) symbol ϵ_{ijk}, which changes sign under any interchange of two indices,

$$\epsilon_{ijk} = -\epsilon_{jik} = -\epsilon_{kji} = -\epsilon_{ikj} = +\epsilon_{kij} = +\epsilon_{jki}, \tag{3.22}$$

and is normalized by $\epsilon_{123} = 1$. In particular, then, it vanishes if any two indices are equal, $\epsilon_{112} = 0$, for example. The torque density may be obtained by first taking the moment of the force density equation (3.18),

$$\frac{\partial}{\partial t} x_j G_k + \nabla_l (x_j T_{lk}) - T_{jk} + x_j f_k = 0, \tag{3.23}$$

where we have noted that

$$\nabla_l x_j = \delta_{lj}. \tag{3.24}$$

When we now multiply (3.23) with ϵ_{ijk} and sum over repeated indices, we find that the terms involving spatial derivatives can be written as a divergence:

$$\frac{\partial}{\partial t}(\epsilon_{ijk} x_j G_k) + \nabla_l(\epsilon_{ijk} x_j T_{lk}) + \epsilon_{ijk} x_j f_k = 0. \tag{3.25}$$

This final step is justified *only* because T_{kl} is symmetrical (thus this symmetry is required for the existence of a local conservation law of angular momentum). We therefore identify the following electromagnetic angular momentum quantities:

$$\text{angular momentum density} = \boldsymbol{\mathcal{J}} = \mathbf{r}\times\mathbf{G}, \tag{3.26a}$$

$$\text{angular momentum flux tensor} = \boldsymbol{\mathcal{K}} = -\mathbf{T}\times\mathbf{r}, \quad \mathcal{K}_{ij} = \epsilon_{jkl} x_k T_{il}. \tag{3.26b}$$

The interpretation of (3.25) as a local account of angular momentum conservation for fields and particles proceeds as before:

$$\frac{\partial}{\partial t}\boldsymbol{\mathcal{J}} + \boldsymbol{\nabla}\cdot\boldsymbol{\mathcal{K}} + \mathbf{r}\times\mathbf{f} = \mathbf{0}, \tag{3.27}$$

Another important application of (3.23) results if we set $j = k$ and sum. With the aid of (3.15) this gives

$$\frac{\partial}{\partial t}(\mathbf{r}\cdot\mathbf{G}) + \boldsymbol{\nabla}\cdot(\mathbf{T}\cdot\mathbf{r}) - U + \mathbf{r}\cdot\mathbf{f} = 0, \tag{3.28}$$

which we call the electromagnetic virial theorem, in analogy with the mechanical virial theorem of Rudolf Clausius (1822–1888). (See Chapter 9.)

[1] Tullio Levi-Civita, 1873–1941.

3.4 Conservation Laws and the Speed of Light

In this section, we restrict our attention to electromagnetic fields in domains free of charged particles, specifically, moving, finite regions occupied by electromagnetic fields, which we will refer to as electromagnetic pulses. The total electromagnetic energy of such a pulse is constant in time:

$$\frac{d}{dt}E = \int_{\text{pulse}} (d\mathbf{r})\, \frac{\partial}{\partial t}U = -\int_{\text{pulse}} (d\mathbf{r})\, \boldsymbol{\nabla}\cdot\mathbf{S} = 0, \tag{3.29}$$

inasmuch as the resulting surface integral, conducted over an enclosing surface on which all fields vanish, equals zero. Similar considerations apply to the total electromagnetic linear and angular momentum,

$$\frac{d}{dt}\mathbf{P} = \int_{\text{pulse}} (d\mathbf{r})\, \frac{\partial}{\partial t}\mathbf{G} = -\int_{\text{pulse}} (d\mathbf{r})\, \boldsymbol{\nabla}\cdot\mathbf{T} = \mathbf{0}, \tag{3.30a}$$

$$\frac{d}{dt}\mathbf{J} = \int_{\text{pulse}} (d\mathbf{r})\, \frac{\partial}{\partial t}(\mathbf{r}\times\mathbf{G}) = -\int_{\text{pulse}} (d\mathbf{r})\, \boldsymbol{\nabla}\cdot(-\mathbf{T}\times\mathbf{r}) = \mathbf{0}. \tag{3.30b}$$

With an eye toward relativity, we consider the space and time moments of (3.5) and (3.13), respectively, combined as a single vector statement:

$$0 = x_k\left(\frac{\partial}{\partial t}U + \boldsymbol{\nabla}\cdot\mathbf{S}\right) - c^2 t\left(\frac{\partial}{\partial t}G_k + \nabla_l T_{lk}\right), \tag{3.31}$$

outside the charge and current distributions. Exploiting the connection between $\mathbf{S}$ and $\mathbf{G}$ [(3.16)], we can rewrite (3.31) as a local conservation law, much as the equality of T_{jk} and T_{kj} lead to the conservation of angular momentum:

$$\frac{\partial}{\partial t}(\mathbf{r}U - c^2 t\mathbf{G}) + \boldsymbol{\nabla}\cdot(\mathbf{S}\mathbf{r} - c^2 t\mathbf{T}) = \mathbf{0}. \tag{3.32}$$

When (3.32) is integrated over a volume enclosing the electromagnetic pulse, the surface term does not contribute, and we find

$$\frac{d}{dt}\int_{\text{pulse}} (d\mathbf{r})\, (\mathbf{r}U - c^2 t\mathbf{G}) = \mathbf{0}. \tag{3.33}$$

The volume integral of the momentum density is the total momentum $\mathbf{P}$,

$$\int_{\text{pulse}} (d\mathbf{r})\, \mathbf{G} = \mathbf{P}, \tag{3.34}$$

which as noted in (3.30a) is constant in time. Consequently, we can rewrite (3.33) as

$$\frac{d}{dt}\int_{\text{pulse}} (d\mathbf{r})\, \mathbf{r}U = c^2\mathbf{P}, \tag{3.35}$$

where the integral here provides an energy weighting of the position vector, at each instant of time,

$$\int_{\text{pulse}} (d\mathbf{r})\, \mathbf{r}U(\mathbf{r},t) = E\,\langle\mathbf{r}\rangle_E(t), \tag{3.36}$$

where, as in (3.29), the energy E is

$$E = \int_{\text{pulse}} (d\mathbf{r})\, U. \tag{3.37}$$

Thus, the motion of this energy-centroid vector is governed by

$$\frac{E}{c^2}\frac{d}{dt}\langle \mathbf{r}\rangle_E(t) = \mathbf{P}, \tag{3.38}$$

which is to say that the center of energy, $\langle \mathbf{r}\rangle_E(t)$, moves with constant velocity,

$$\frac{d}{dt}\langle \mathbf{r}\rangle_E(t) = \mathbf{v}_E, \tag{3.39}$$

the total momentum being that velocity multiplied by a mass,

$$m = E/c^2. \tag{3.40}$$

The application of the virial theorem, (3.28), to an electromagnetic pulse supplies another velocity. We infer that

$$\frac{d}{dt}\int_{\text{pulse}} (d\mathbf{r})\, \mathbf{r}\cdot\mathbf{G} = E. \tag{3.41}$$

By introducing a momentum weighting for the position vector,

$$\int_{\text{pulse}} (d\mathbf{r})\, \mathbf{r}\cdot\mathbf{G}(\mathbf{r},t) = \langle \mathbf{r}\rangle_P(t)\cdot\mathbf{P}, \tag{3.42}$$

we deduce that the center of momentum moves with velocity

$$\frac{d}{dt}\langle \mathbf{r}\rangle_P(t) = \mathbf{v}_P, \tag{3.43}$$

which is constant in the direction of the momentum,

$$\mathbf{v}_P\cdot\mathbf{P} = E. \tag{3.44}$$

We combine (3.44) with (3.38) to yield

$$\mathbf{v}_P\cdot\mathbf{v}_E = c^2. \tag{3.45}$$

If the flow of energy and momentum takes place in a single direction, it would be reasonable to expect that these mechanical properties are being transported with a common velocity,

$$\mathbf{v}_E = \mathbf{v}_P = \mathbf{v}, \tag{3.46}$$

which then has a definite magnitude,

$$\mathbf{v}\cdot\mathbf{v} = c^2, \quad v = c, \tag{3.47}$$

which supplies the physical identification of c as the speed of light. Of course, this identification was an input to our inference of Maxwell's equations. We here recover it from a consideration of energy and momentum, thus indicating the consistency of Maxwell's equations. The relation between the momentum and the energy of this electromagnetic pulse is then

$$E = \mathbf{v}\cdot\mathbf{P}, \quad \mathbf{P} = \frac{E}{c^2}\mathbf{v}, \tag{3.48}$$

so we learn that

$$E = Pc, \quad \mathbf{v} = c\frac{\mathbf{P}}{P}, \tag{3.49}$$

which results express the mechanical properties of a localized electromagnetic pulse carrying both energy and momentum at the speed of light, in the direction of the momentum.

There is another, somewhat more direct, mechanical proof that electromagnetic pulses propagate at speed c. When no charges or currents are present, the local equation of energy conservation, (3.3), implies

$$(r^2 - c^2t^2)\left[\frac{\partial}{\partial t}U + \boldsymbol{\nabla}\cdot\mathbf{S}\right] = 0, \tag{3.50}$$

which can be rewritten, using (3.16), as

$$\frac{\partial}{\partial t}[(r^2 - c^2t^2)U] + \boldsymbol{\nabla}\cdot[(r^2 - c^2t^2)\mathbf{S}] + 2c^2[tU - \mathbf{r}\cdot\mathbf{G}] = 0. \tag{3.51}$$

Integrating this over all space and using the idea of energy and momentum weighting to define averages, as before, we obtain

$$\frac{d}{dt}\left[\langle r^2\rangle_E(t) - c^2t^2\right]E = 2c^2\left[\langle\mathbf{r}\rangle_P(t)\cdot\mathbf{P} - tE\right]. \tag{3.52}$$

According to (3.44), the combination appearing on the right is a constant of the motion, which we can put equal to zero by identifying the coordinate origin with $\langle\mathbf{r}\rangle_P$ at $t = 0$. The time integral of this equation is then

$$\langle r^2\rangle_E = (ct)^2 + \text{constant}, \tag{3.53}$$

which implies, for large times,

$$\left(\langle r^2\rangle_E(t)\right)^{1/2} \sim ct; \tag{3.54}$$

the center of energy of the pulse moves away from the origin at the speed of light.

What are the fields doing to enforce the conditions (3.48) of simple mechanical flow in a single direction? The relation between momentum and energy, (3.49), $E = |\mathbf{P}|c$, can be expressed in terms of the fields as

$$\int(d\mathbf{r})\,\frac{E^2 + B^2}{8\pi} = \left|\int(d\mathbf{r})\,\frac{\mathbf{E}\times\mathbf{B}}{4\pi}\right|, \tag{3.55}$$

where the volume integrations are extended over the pulse. Now, a sum of vectors of given magnitudes is of maximum magnitude when all those vectors are parallel, which is to say here that

$$\int(d\mathbf{r})\,\frac{E^2 + B^2}{2} \leq \int(d\mathbf{r})\,|\mathbf{E}\times\mathbf{B}|, \tag{3.56}$$

where equality holds *only* when $\mathbf{E}\times\mathbf{B}$ everywhere points in the same direction, that of the pulse's total momentum or velocity. On the other hand, we note the inequality,

$$\begin{aligned}(\mathbf{E}\times\mathbf{B})^2 &= E^2B^2 - (\mathbf{E}\cdot\mathbf{B})^2\\ &= \left(\frac{E^2 + B^2}{2}\right)^2 - \left[\left(\frac{E^2 - B^2}{2}\right)^2 + (\mathbf{E}\cdot\mathbf{B})^2\right] \leq \left(\frac{E^2 + B^2}{2}\right)^2,\end{aligned} \tag{3.57}$$

where the equality holds *only* if both $\mathbf{E}\cdot\mathbf{B} = 0$ and $E^2 = B^2$. So we deduce the opposite inequality to (3.56),

$$\int(d\mathbf{r})\,|\mathbf{E}\times\mathbf{B}| \leq \int(d\mathbf{r})\,\frac{E^2 + B^2}{2}. \tag{3.58}$$

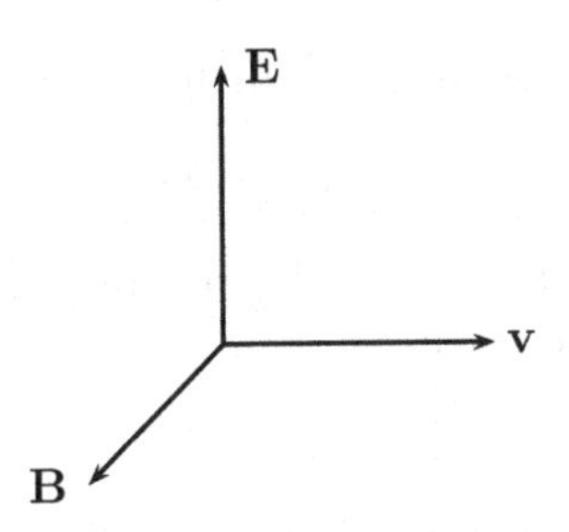

FIGURE 3.1
Electric and magnetic fields for an electromagnetic pulse propagating with velocity **v**.

Comparing (3.56) and (3.58), we see that both equalities must hold so that

$$\mathbf{E} \cdot \mathbf{B} = 0, \qquad E^2 = B^2, \tag{3.59}$$

and $\mathbf{E} \times \mathbf{B}$ is unidirectional, pointing in the direction of propagation. Accordingly, the electric and magnetic fields in a unidirectional pulse are, everywhere within the pulse, of equal magnitude, mutually perpendicular, and perpendicular to the direction of motion of the pulse. (See Fig. 3.1.) These are the familiar properties of the electromagnetic fields of a light wave, which are here derived without recourse to explicit solutions to Maxwell's equations.[2]

3.5 Problems for Chapter 3

1. The unit dyadic $\mathbf{1}$ is defined in terms of orthogonal unit vectors $\mathbf{i}$, $\mathbf{j}$, $\mathbf{k}$ by

$$\mathbf{1} = \mathbf{ii} + \mathbf{jj} + \mathbf{kk}. \tag{3.60}$$

   Verify that ($\mathbf{A}$ is an arbitrary vector)

$$\mathbf{A} \cdot \mathbf{1} = \mathbf{A}, \quad \mathbf{1} \cdot \mathbf{A} = \mathbf{A}, \quad \mathbf{1} \cdot \mathbf{1} = \mathbf{1}. \tag{3.61}$$

   Repeat, using components, i.e., $\mathbf{A} \cdot \mathbf{B} = A_i B_i$. Expand the following products of vectors with dyadics:

$$\mathbf{A} \cdot (\mathbf{BC}), \quad (\mathbf{AB}) \cdot \mathbf{C}, \quad \mathbf{A} \times (\mathbf{BC}), \quad (\mathbf{AB}) \times \mathbf{C}. \tag{3.62}$$

2. Let $\mathbf{A}(\mathbf{r})$ and $\mathbf{B}(\mathbf{r})$ be vector fields. Show that

$$\boldsymbol{\nabla} \cdot (\mathbf{AB}) = (\mathbf{A} \cdot \boldsymbol{\nabla})\mathbf{B} + \mathbf{B}(\boldsymbol{\nabla} \cdot \mathbf{A}). \tag{3.63}$$

   Let $\lambda(\mathbf{r})$ further be an arbitrary scalar function. Simplify

$$\boldsymbol{\nabla} \cdot (\lambda \mathbf{AB}), \quad \boldsymbol{\nabla} \cdot (\lambda \mathbf{A} \times \mathbf{B}). \tag{3.64}$$

[2]There is now an emerging literature about electromagnetic knots, linked closed electromagnetic field lines without sources. See, for example, Refs. [16, 17].

3. An infinitesimal rotation is described by its effect on an arbitrary vector $\mathbf{V}$ by

$$\delta\mathbf{V} = \delta\boldsymbol{\omega}\times\mathbf{V}, \tag{3.65}$$

where the direction of $\delta\boldsymbol{\omega}$ points in the direction of the rotation, and has magnitude equal to the (infinitesimal) amount of the rotation. Check that

$$\delta(\mathbf{V}^2) = 0. \tag{3.66}$$

The statement that, if $\mathbf{B}$ and $\mathbf{C}$ are vectors, so is $\mathbf{B}\times\mathbf{C}$, is expressed by

$$\delta(\mathbf{B}\times\mathbf{C}) = \delta\mathbf{B}\times\mathbf{C} + \mathbf{B}\times\delta\mathbf{C} = \delta\boldsymbol{\omega}\times(\mathbf{B}\times\mathbf{C}). \tag{3.67}$$

Verify directly the resulting relation among the arbitrary vectors.

4. Verify the following relations for the electromagnetic stress tensor:

(a)

$$\mathrm{Tr}\,\mathbf{T} = T_{kk} = U, \tag{3.68}$$

(b)

$$\mathrm{Tr}\,\mathbf{T}^2 = T_{kl}T_{lk} = 3U^2 - 2(c\mathbf{G})^2 \geq U^2, \tag{3.69}$$

and

(c)

$$\det\mathbf{T} = -U[U^2 - (c\mathbf{G})^2]. \tag{3.70}$$

Here the summation convention is employed, and the trace and determinant refer to $\mathbf{T}$ thought of as a 3×3 matrix.

5. A plane wave is described by the electric and magnetic fields of the form:

$$\mathbf{E} = \mathbf{e}_0\cos(i\mathbf{k}\cdot\mathbf{r} - i\omega t), \quad \mathbf{B} = \mathbf{b}_0\cos(i\mathbf{k}\cdot\mathbf{r} - i\omega t), \tag{3.71}$$

where $\mathbf{e_0}$, $\mathbf{b_0}$, $\mathbf{k}$, and ω are constants. From Maxwell's equation in free space (no charges or currents), determine the relation between $\mathbf{e_0}$, $\mathbf{b_0}$, and $\mathbf{k}$, and determine the relation between ω and $k = |\mathbf{k}|$.What is the momentum density $\mathbf{G}$ of this electromagnetic field? What is the energy flux vector $\mathbf{S}$? Compute the energy density U, and show that the energy is locally conserved in that

$$\frac{\partial U}{\partial t} + \boldsymbol{\nabla}\cdot\mathbf{S} = 0. \tag{3.72}$$

Calculate the stress tensor $\mathbf{T}$, and show that momentum is conserved,

$$\frac{\partial\mathbf{G}}{\partial t} + \boldsymbol{\nabla}\cdot\mathbf{T} = \mathbf{0}. \tag{3.73}$$

6. What if $\langle\mathbf{r}\rangle_P(0) \neq 0$ in (3.52)? Show that the integral of that equation can be interpreted in analogy with a group of particles that, at time $t = 0$, are set off with various positions and velocities, thereafter to move with those constant velocities,

$$\mathbf{r}(t) = \mathbf{r}(0) + \mathbf{v}t. \tag{3.74}$$

Square and average this position vector, and upon comparison with the solution of (3.52), identify $\langle\mathbf{r}(0)\cdot\mathbf{v}\rangle$ and v.

7. Consider a spherical shell of radius a with charge Q uniformly distributed on the surface. Choose a coordinate system with origin at the center of the sphere.

(a) Determine the Coulomb potential energy from

$$U = \frac{1}{2}\oint dS\, dS' \frac{\sigma(\mathbf{r})\sigma(\mathbf{r}')}{|\mathbf{r}-\mathbf{r}'|}, \tag{3.75}$$

where $\sigma(\mathbf{r})$ is the surface charge density, and the double integral extends over all points on the sphere. Give the answer as an explicit function of Q and a.

(b) Calculate, using Gauss' law to determine the electric field, the radial-radial component of the electric stress tensor, T_{rr}, just outside the surface of the sphere. This gives the flux of momentum flowing out of the sphere, and hence $-T_{rr}$ is the electrostatic force per unit area on the sphere. From T_{rr} compute the total stress F on the sphere which tends to expand it outward. Show that

$$F = -\frac{\partial}{\partial a}U, \tag{3.76}$$

where U is the electrostatic energy found in part a. This is an example of the *principle of virtual work.*

8. Suppose we define an energy density and energy flux vector for a system of charged particles by

$$U_{\text{ch}}(\mathbf{r},t) = \sum_a \delta(\mathbf{r}-\mathbf{r}_a(t))E_a(t), \tag{3.77a}$$

$$\mathbf{S}_{\text{ch}}(\mathbf{r},t) = \sum_a \delta(\mathbf{r}-\mathbf{r}_a(t))E_a(t)\mathbf{v}_a(t), \tag{3.77b}$$

where $E_a = \frac{1}{2}m_a v_a^2$. Show that energy is conserved, in that

$$\frac{\partial}{\partial t}U_{\text{ch}} + \boldsymbol{\nabla}\cdot\mathbf{S}_{\text{ch}} = \mathbf{j}\cdot\mathbf{E}, \tag{3.78}$$

where the right-hand side is the work done on the particles by the field. Thus, show that if the total energy density and flux vector are defined by the sum of the particle and field contributions,

$$U = U_{\text{ch}} + U_f, \quad \mathbf{S} = \mathbf{S}_{\text{ch}} + \mathbf{S}_f, \tag{3.79}$$

the total energy density satisfies the local conservation law:

$$\frac{\partial}{\partial t}U + \boldsymbol{\nabla}\cdot\mathbf{S} = 0. \tag{3.80}$$

If, similarly, we define the momentum density and stress tensor for the particles by

$$\mathbf{G}(\mathbf{r},t) = \sum_a \delta(\mathbf{r}-\mathbf{r}_a(t))m_a\mathbf{v}_a(t), \quad \mathbf{T}(\mathbf{r},t) = \sum_a \delta(\mathbf{r}-\mathbf{r}_a)(t)m_a\mathbf{v}_a(t)\mathbf{v}_a(t), \tag{3.81}$$

show that a statement for the conservation of total (particle plus field) momentum follows.

9. As in Problem 2.1, let

$$\mathbf{F} = \mathbf{E} + i\mathbf{B}, \quad \mathbf{F}^* = \mathbf{E} - i\mathbf{B}. \tag{3.82}$$

Identify the scalar

$$\frac{1}{8\pi}\mathbf{F}^*\cdot\mathbf{F}, \tag{3.83a}$$

the vector

$$\frac{1}{8\pi i}\mathbf{F}^*\times\mathbf{F}, \tag{3.83b}$$

and the dyadic

$$\frac{1}{8\pi}(\mathbf{F}\mathbf{F}^* + \mathbf{F}^*\mathbf{F}). \tag{3.83c}$$

What happens to these quantities if $\mathbf{F}$ is replaced by $e^{-i\phi}\mathbf{F}$, ϕ being a constant?

10. Electric charge e is located at the fixed point $\frac{1}{2}\mathbf{R}$. Magnetic charge g is stationed at the fixed point $-\frac{1}{2}\mathbf{R}$. What is the momentum density at the arbitrary point $\mathbf{r}$? Verify that it is divergenceless by writing it as a curl. Evaluate the electromagnetic angular momentum, the integrated moment of the momentum density. Recognize that it is a gradient with respect to $\mathbf{R}$. Continue the evaluation to discover that it depends only on the direction of $\mathbf{R}$, not its magnitude. This is the naive, semiclassical basis for the charge quantization condition of Dirac,

$$eg = \frac{n}{2}\hbar c. \tag{3.84}$$

11. As an alternative to the preceding problem, calculate the angular momentum of the above configuration by integrating the angular momentum density directly,

$$\mathbf{J} = \int (d\mathbf{r})\,\mathbf{r}\times\mathbf{G}, \tag{3.85}$$

as follows: Show that the above integral can be written as

$$\mathbf{J} = \frac{eg}{2c}\hat{\mathbf{R}}\int_{-1}^{1} d\mu(1-\mu^2)\int_0^\infty dt\,t\frac{1}{(1+t^2-2t\mu)^{3/2}}. \tag{3.86}$$

Do the t integral by noting that

$$\frac{t}{(1+t^2-2t\mu)^{3/2}} = \frac{\partial}{\partial\mu}\frac{1}{(1+t^2-2t\mu)^{1/2}}, \tag{3.87}$$

and use the indefinite integral

$$\int dx\frac{1}{(x^2+a^2)^{1/2}} = \ln\left(x+\sqrt{x^2+a^2}\right). \tag{3.88}$$

The μ integral is then trivial, and thus derive

$$\mathbf{J} = \frac{eg}{c}\hat{\mathbf{R}}. \tag{3.89}$$

4

Macroscopic Electrodynamics

4.1 Force on an Atom

The Maxwell-Lorentz system of equations, (1.61) and (1.64), provides a microscopic description of electromagnetic phenomena, at the classical level, ranging from the simplest two-particle system to the detailed behavior of all particles in a macroscopic system. However, for the latter case, we usually do not require such a complete description, since our measurements involve macroscopic quantities which are only indirectly related to the microscopic behavior of individual atoms. We must develop a theory that is directly applicable to the macroscopic situation with only an implicit reference back to the detailed characterization of the system. The resulting macroscopic electrodynamics is a *phenomenological* theory, by which is meant a theory that operates at the level of the phenomena being correlated and predicted, while maintaining the possibility of contact with a more fundamental theory—here, microscopic electrodynamics—that operates at a deeper level. That contact exists to the extent that the macroscopic measurements can be considered to be averages, over very many atoms, of the results of hypothetical microscopic measurements. Such an argument was apparently first made by Lorentz.

To begin, we consider an atom, an electrically neutral assembly of point charges,

$$\sum_a e_a = 0, \tag{4.1}$$

that are bound together in a small region. We want to study the response of such a system to external electric and magnetic fields that vary only slightly over the spatial extent of that system. We will first concentrate our attention on the net force on the system at a given time, the sum of the forces on its constituents, (1.64),

$$\mathbf{F} = \sum_a \left[e_a \mathbf{E}(\mathbf{r}_a) + e_a \frac{\mathbf{v}_a}{c} \times \mathbf{B}(\mathbf{r}_a) \right]. \tag{4.2}$$

Since the system is small, all the charges are near the center of mass of the charge distribution, which lies at the position $\mathbf{R}$. (For the purposes of the following expansion we could let $\mathbf{R}$ represent an arbitrary point inside the charge distribution; the use of the center of mass allows us to separate intrinsic properties from those due to the motion of the atom as a whole.) We can then expand the electric and magnetic fields about this reference point,

$$\mathbf{E}(\mathbf{r}_a) = \mathbf{E}(\mathbf{R}) + [(\mathbf{r}_a - \mathbf{R}) \cdot \boldsymbol{\nabla}]\mathbf{E}(\mathbf{R}) + \dots, \tag{4.3}$$

and likewise for $\mathbf{B}$, in which the subsequent terms are considered negligible. Here $\boldsymbol{\nabla}$ means the gradient with respect to $\mathbf{R}$. Now, the total force on the atom, (4.2), can be rewritten

DOI: 10.1201/9781003057369-4

in terms of this expansion as

$$\mathbf{F} = \left(\sum_a e_a\right)\mathbf{E}(\mathbf{R}) + \sum_a e_a[(\mathbf{r}_a - \mathbf{R})\cdot\boldsymbol{\nabla}]\mathbf{E}(\mathbf{R}) + \left(\sum_a e_a \frac{\mathbf{v}_a}{c}\right)\times\mathbf{B}(\mathbf{R}) + \sum_a e_a \frac{\mathbf{v}_a}{c}\times[(\mathbf{r}_a - \mathbf{R})\cdot\boldsymbol{\nabla}]\mathbf{B}(\mathbf{R}) + \dots. \tag{4.4}$$

Let us denote the four terms displayed here as I–IV, respectively. The first term I is zero because of the neutrality of the system, (4.1). In the second term II, we identify the electric dipole moment, $\mathbf{d}$,

$$\mathbf{d} = \sum_a e_a(\mathbf{r}_a - \mathbf{R}) = \sum_a e_a \mathbf{r}_a, \tag{4.5}$$

(which is independent of $\mathbf{R}$), while in III, we recognize its time derivative,

$$\sum_a e_a \mathbf{v}_a = \frac{d}{dt}\mathbf{d}. \tag{4.6}$$

Momentarily setting aside the fourth term IV, we find the force on the system to be

$$\mathbf{F}_{\text{I--III}} = (\mathbf{d}\cdot\boldsymbol{\nabla})\mathbf{E}(\mathbf{R}) + \frac{1}{c}\left(\frac{d}{dt}\mathbf{d}\right)\times\mathbf{B}(\mathbf{R}) + \dots. \tag{4.7}$$

For the second term here, we can transfer the time derivative,

$$\frac{1}{c}\left(\frac{d}{dt}\mathbf{d}\right)\times\mathbf{B}(\mathbf{R}) = \frac{1}{c}\frac{d}{dt}[\mathbf{d}\times\mathbf{B}(\mathbf{R})] - \frac{1}{c}\mathbf{d}\times\left(\frac{\partial}{\partial t} + \mathbf{V}\cdot\boldsymbol{\nabla}\right)\mathbf{B}(\mathbf{R}), \tag{4.8}$$

where $\mathbf{V} = d\mathbf{R}/dt$. Using (1.60) for $\partial\mathbf{B}/\partial t$, and rewriting the resulting double cross product according to

$$\mathbf{d}\times(\boldsymbol{\nabla}\times\mathbf{E}(\mathbf{R})) + (\mathbf{d}\cdot\boldsymbol{\nabla})\mathbf{E}(\mathbf{R}) = \boldsymbol{\nabla}(\mathbf{d}\cdot\mathbf{E}(\mathbf{R})), \tag{4.9}$$

we can present (4.7) as

$$\mathbf{F}_{\text{I--III}} = \boldsymbol{\nabla}[\mathbf{d}\cdot\mathbf{E}(\mathbf{R})] - \frac{1}{c}(\mathbf{V}\cdot\boldsymbol{\nabla})\mathbf{d}\times\mathbf{B}(\mathbf{R}) + \frac{1}{c}\frac{d}{dt}[\mathbf{d}\times\mathbf{B}(\mathbf{R})] + \dots. \tag{4.10}$$

Recalling that force is the time rate of change of momentum, we see that $(1/c)\mathbf{d}\times\mathbf{B}$ introduces a redefinition of the momentum of the system.

We now return to the fourth term IV of (4.4), which would seem to correspond to a small effect, since for atomic systems, $v_a/c \ll 1$. A rearrangement of it is

$$F_{\text{IV}} = \sum_a e_a\frac{\mathbf{v}_a}{c}\times[(\mathbf{r}_a - \mathbf{R})\cdot\boldsymbol{\nabla}]\mathbf{B}(\mathbf{R}) = \sum_a \frac{e_a}{c}\mathbf{V}\times[(\mathbf{r}_a - \mathbf{R})\cdot\boldsymbol{\nabla}]\mathbf{B}(\mathbf{R}) + \sum_a \frac{e_a}{c}(\mathbf{v}_a - \mathbf{V})\times[(\mathbf{r}_a - \mathbf{R})\cdot\boldsymbol{\nabla}]\mathbf{B}(\mathbf{R}), \tag{4.11}$$

where, recalling the definition of the electric dipole moment (4.5), we can express the first term on the right side as

$$F_{\text{IVa}} = \frac{1}{c}\mathbf{V}\times(\mathbf{d}\cdot\boldsymbol{\nabla})\mathbf{B}(\mathbf{R}). \tag{4.12}$$

Combining this contribution with the second term on the right side of (4.10), and using (1.50), we obtain

$$F_{\text{IVa}} + F_{(\text{I--III})b} = \frac{1}{c}[(\mathbf{d}\cdot\boldsymbol{\nabla})\mathbf{V} - (\mathbf{V}\cdot\boldsymbol{\nabla})\mathbf{d}]\times\mathbf{B}(\mathbf{R}) = \frac{1}{c}[(\mathbf{d}\times\mathbf{V})\times\boldsymbol{\nabla}]\times\mathbf{B}(\mathbf{R}) = \boldsymbol{\nabla}\left[\frac{1}{c}(\mathbf{d}\times\mathbf{V})\cdot\mathbf{B}(\mathbf{R})\right]. \tag{4.13}$$

Collecting the various results to this point, we can now rewrite the total force on the atom, (4.4), as

$$\mathbf{F} = \boldsymbol{\nabla}[\mathbf{d}\cdot\mathbf{E}(\mathbf{R})] + \boldsymbol{\nabla}\left[\frac{1}{c}(\mathbf{d}\times\mathbf{V})\cdot\mathbf{B}(\mathbf{R})\right] + \frac{d}{dt}\left[\frac{1}{c}\mathbf{d}\times\mathbf{B}(\mathbf{R})\right] + \mathbf{F}_{\text{IVb}}, \tag{4.14}$$

where $\mathbf{F}_{\text{IVb}}$ represents the second term on the right side of (4.11),

$$\mathbf{F}_{\text{IVb}} = \sum_a \frac{e_a}{c}(\mathbf{v}_a - \mathbf{V})\times[(\mathbf{r}_a - \mathbf{R})\cdot\boldsymbol{\nabla}]\mathbf{B}(\mathbf{R}), \tag{4.15}$$

which can be rearranged as follows:

$$\begin{aligned}\mathbf{F}_{\text{IVb}} = &\frac{d}{dt}\sum_a \frac{e_a}{c}(\mathbf{r}_a - \mathbf{R})\times[(\mathbf{r}_a - \mathbf{R})\cdot\boldsymbol{\nabla}]\mathbf{B}(\mathbf{R})\\ &-\sum_a \frac{e_a}{c}(\mathbf{r}_a - \mathbf{R})\times[(\mathbf{v}_a - \mathbf{V})\cdot\boldsymbol{\nabla}]\,\mathbf{B}(\mathbf{R})\\ &-\sum_a \frac{e_a}{c}(\mathbf{r}_a - \mathbf{R})\times[(\mathbf{r}_a - \mathbf{R})\cdot\boldsymbol{\nabla}]\,\frac{d}{dt}\mathbf{B}(\mathbf{R}).\end{aligned} \tag{4.16}$$

We must now recall the restricted nature of this description: The electric and magnetic fields change only slightly over the dimensions of the system. The first of the three terms on the right side of (4.16), F_{IVb1}, is a small correction to what is already present in (4.14) as $\frac{d}{dt}[(1/c)\mathbf{d}\times\mathbf{B}(\mathbf{R})]$ and is therefore to be neglected. Furthermore, the last term of (4.16), which is well approximated by

$$F_{\text{IVb3}} = \sum_a e_a(\mathbf{r}_a - \mathbf{R})\times[(\mathbf{r}_a - \mathbf{R})\cdot\boldsymbol{\nabla}]\,\boldsymbol{\nabla}\times\mathbf{E}(\mathbf{R}), \tag{4.17}$$

is of the same order of magnitude as the omitted terms in the expansion (4.4) and is therefore also to be neglected. An average of the initial form of $\mathbf{F}_{\text{IVb}}$, (4.15), with the single remaining contribution of (4.16), the second line there, called $\mathbf{F}_{\text{IVb2}}$, now gives

$$\begin{aligned}\mathbf{F}_{\text{IVb}} = &\frac{1}{2}\sum_a \frac{e_a}{c}(\mathbf{v}_a - \mathbf{V})\times[(\mathbf{r}_a - \mathbf{R})\cdot\boldsymbol{\nabla}]\mathbf{B}(\mathbf{R})\\ &-\frac{1}{2}\sum_a \frac{e_a}{c}(\mathbf{r}_a - \mathbf{R})\times[(\mathbf{v}_a - \mathbf{V})\cdot\boldsymbol{\nabla}]\,\mathbf{B}(\mathbf{R})\\ =&\frac{1}{2}\sum_a \frac{e_a}{c}\{[(\mathbf{r}_a - \mathbf{R})\times(\mathbf{v}_a - \mathbf{V})]\times\boldsymbol{\nabla}\}\times\mathbf{B}(\mathbf{R}).\end{aligned} \tag{4.18}$$

What has finally emerged here is the magnetic dipole moment of the system, $\boldsymbol{\mu}$,

$$\boldsymbol{\mu} = \frac{1}{2c}\sum_a e_a(\mathbf{r}_a - \mathbf{R})\times(\mathbf{v}_a - \mathbf{V}), \tag{4.19}$$

so (4.18) is equal to ($\boldsymbol{\mu}$ is constant in space)

$$\mathbf{F}_{\text{IVb}} = (\boldsymbol{\mu}\times\boldsymbol{\nabla})\times\mathbf{B}(\mathbf{R}) = \boldsymbol{\nabla}[\boldsymbol{\mu}\cdot\mathbf{B}(\mathbf{R})], \tag{4.20}$$

where we have used (1.50). (It is, of course, the similarity of this structure to $\boldsymbol{\nabla}[\mathbf{d}\cdot\mathbf{E}(\mathbf{R})]$ that justifies the identification of $\boldsymbol{\mu}$ as the magnetic analog of $\mathbf{d}$.) We also recognize that a contribution of this form already appears in the second term on the right side of (4.14), bearing the information that a moving electric dipole also acts as a magnetic dipole. The comparison of the two effects, characterized by $\frac{1}{c}\mathbf{d}\times\mathbf{V}$ and $\boldsymbol{\mu}$, is that of the typical speeds of the relatively heavy atoms, $|\mathbf{V}|$, and of the light electrons, $|\mathbf{v}_a|$, in the interior of atoms,

$$|\mathbf{V}| \ll |\mathbf{v}_a| \ll c. \tag{4.21}$$

Accordingly, we neglect the motional effects of the atoms, and finally write (4.14) as

$$\mathbf{F} = \boldsymbol{\nabla}[\mathbf{d}\cdot\mathbf{E}(\mathbf{R}) + \boldsymbol{\mu}\cdot\mathbf{B}(\mathbf{R})] + \frac{d}{dt}\left(\frac{1}{c}\mathbf{d}\times\mathbf{B}(\mathbf{R})\right). \tag{4.22}$$

In the absence of time variation, what remains is a force associated with the respective potential energies of a given electric dipole in an electric field,

$$U_E = -\mathbf{d}\cdot\mathbf{E}, \tag{4.23}$$

and of a given magnetic dipole in a magnetic field,

$$U_B = -\boldsymbol{\mu}\cdot\mathbf{B}. \tag{4.24}$$

The energy interpretation does more than supply the force components as negative gradients with respect to position coordinates. It also produces torques as negative gradients with respect to angles. Take the example of a magnetic dipole $\boldsymbol{\mu}$ in the presence of a magnetic field $\mathbf{B}$. If θ is the angle between $\boldsymbol{\mu}$ and $\mathbf{B}$, the magnetic potential energy is

$$U_B = -|\boldsymbol{\mu}||\mathbf{B}|\cos\theta. \tag{4.25}$$

The implied internal torque, that is, the torque on this individual dipole, and not the moment of the force on the dipole, is then

$$\frac{\partial}{\partial\theta}\left(|\boldsymbol{\mu}||\mathbf{B}|\cos\theta\right) = -|\boldsymbol{\mu}||\mathbf{B}|\sin\theta, \tag{4.26}$$

(the reference point of this torque is at the position of the dipole), which can be represented by a vector perpendicular to the plane formed by $\boldsymbol{\mu}$ and $\mathbf{B}$,

$$\boldsymbol{\tau} = \boldsymbol{\mu}\times\mathbf{B}. \tag{4.27}$$

We shall now derive this vectorial result directly, along with its electric counterpart; for simplicity, additional time derivative terms are omitted. The torque, the moment of the force about the center of the charge distribution at $\mathbf{R}$ is

$$\boldsymbol{\tau} = \sum_a (\mathbf{r}_a - \mathbf{R})\times\left(e_a\mathbf{E}(\mathbf{r}_a) + \frac{1}{c}e_a\mathbf{v}_a\times\mathbf{B}(\mathbf{r}_a)\right). \tag{4.28}$$

The part proportional to the electric field is, when we neglect the variation of $\mathbf{E}$ over the system, the electric torque

$$\boldsymbol{\tau}_E = \mathbf{d}\times\mathbf{E}(\mathbf{R}), \tag{4.29}$$

as expected in analogy with (4.27). In deriving the magnetic torque, we first make the unimportant change, $\mathbf{v}_a \to \mathbf{v}_a - \mathbf{V}$, using (4.21), and then transfer the time derivative to get

$$\begin{aligned}
\boldsymbol{\tau}_B &= \sum_a (\mathbf{r}_a - \mathbf{R})\times\left(\frac{1}{c}e_a(\mathbf{v}_a - \mathbf{V})\times\mathbf{B}(\mathbf{R})\right) \\
&\to -\sum_a (\mathbf{v}_a - \mathbf{V})\times\left[\frac{1}{c}e_a(\mathbf{r}_a - \mathbf{R})\times\mathbf{B}(\mathbf{R})\right] \\
&\to \frac{1}{2}\sum_a \frac{e_a}{c}\{(\mathbf{r}_a - \mathbf{R})\times[(\mathbf{v}_a - \mathbf{V})\times\mathbf{B}(\mathbf{R})] - (\mathbf{v}_a - \mathbf{V})\times[(\mathbf{r}_a - \mathbf{R})\times\mathbf{B}(\mathbf{R})]\} \\
&= \boldsymbol{\mu}\times\mathbf{B}(\mathbf{R}),
\end{aligned} \tag{4.30}$$

where in the second line we have omitted the $-\frac{1}{c}\frac{\partial}{\partial t}\mathbf{B} = \boldsymbol{\nabla}\times\mathbf{E}$ contribution as negligible in comparison with $\boldsymbol{\tau}_E$. (See Problem 4.2 for a justification of (4.30).) In the third line, we averaged the two preceding forms, and then used the identity (1.70a). Putting all this together, we find the torque on the system is given by

$$\boldsymbol{\tau} = \mathbf{d}\times\mathbf{E} + \boldsymbol{\mu}\times\mathbf{B}, \tag{4.31}$$

so that, as with the force, the result can be expressed in terms of the electric and magnetic dipole moments, $\mathbf{d}$ and $\boldsymbol{\mu}$.

4.2 Force on a Macroscopic Body

To this point, we have considered the response of a small system, an atom, for example, to external electric and magnetic fields, which vary smoothly over the system. Macroscopic materials are made up of large numbers of atoms. What is the total force on such a piece of material? We must sum up all the forces on the individual atoms. To the extent that the forces on the atoms vary but slightly from one atom to another, the summation can be replaced by a volume integration, weighted by the atomic density, $n(\mathbf{r})$, the number of atoms per unit volume at the macroscopic point $\mathbf{r}$:

$$\mathbf{F} = \int (d\mathbf{r})\, n(\mathbf{r}) \Big[\mathbf{d}\times(\boldsymbol{\nabla}\times\mathbf{E}) + (\mathbf{d}\cdot\boldsymbol{\nabla})\mathbf{E} + \boldsymbol{\mu}\times(\boldsymbol{\nabla}\times\mathbf{B}) + (\boldsymbol{\mu}\cdot\boldsymbol{\nabla})\mathbf{B} + \frac{d}{dt}\left(\frac{1}{c}\mathbf{d}\times\mathbf{B}\right)\Big]. \tag{4.32}$$

Notice that we have rewritten (4.22) with the aid of the identities

$$\boldsymbol{\nabla}(\mathbf{d}\cdot\mathbf{E}) = \mathbf{d}\times(\boldsymbol{\nabla}\times\mathbf{E}) + (\mathbf{d}\cdot\boldsymbol{\nabla})\mathbf{E}, \tag{4.33a}$$

$$\boldsymbol{\nabla}(\boldsymbol{\mu}\cdot\mathbf{B}) = \boldsymbol{\mu}\times(\boldsymbol{\nabla}\times\mathbf{B}) + (\boldsymbol{\mu}\cdot\boldsymbol{\nabla})\mathbf{B}. \tag{4.33b}$$

First a word about $\mathbf{d}$ and $\boldsymbol{\mu}$ in these expressions. In the single atom formula (4.22), the derivatives act only on $\mathbf{E}$ and $\mathbf{B}$, which is reflected in (4.32). For a many-atom system, the dipole moments could well vary from one location to another and so have macroscopic spatial dependence. Accordingly, $\mathbf{d}(\mathbf{r})$ and $\boldsymbol{\mu}(\mathbf{r})$ are the average dipole moments at the point $\mathbf{r}$. We now define the electric polarization, $\mathbf{P}$, and the magnetization, $\mathbf{M}$, by

$$\mathbf{P}(\mathbf{r},t) = n(\mathbf{r})\mathbf{d}(\mathbf{r},t), \tag{4.34a}$$

and

$$\mathbf{M}(\mathbf{r},t) = n(\mathbf{r})\boldsymbol{\mu}(\mathbf{r},t), \tag{4.34b}$$

respectively. The resulting macroscopic form of the total force at time t is

$$\mathbf{F}(t) = \int (d\mathbf{r}) \Big[\mathbf{P}(\mathbf{r},t)\times[\boldsymbol{\nabla}\times\mathbf{E}(\mathbf{r},t)] + [\mathbf{P}(\mathbf{r},t)\cdot\boldsymbol{\nabla}]\mathbf{E}(\mathbf{r},t) + \mathbf{M}(\mathbf{r},t)\times[\boldsymbol{\nabla}\times\mathbf{B}(\mathbf{r},t)] + [\mathbf{M}(\mathbf{r},t)\cdot\boldsymbol{\nabla}]\mathbf{B}(\mathbf{r},t) + \frac{\partial}{\partial t}\left(\frac{1}{c}\mathbf{P}(\mathbf{r},t)\times\mathbf{B}(\mathbf{r},t)\right)\Big]. \tag{4.35}$$

(Here, the distinction between $\frac{d}{dt}\mathbf{B}$ and $\frac{\partial}{\partial t}\mathbf{B}$ has been dropped, because the difference is of order of the small atomic velocity $\mathbf{V}$, which is averaged to zero in any case.)

We proceed to simplify this in various ways. First, we use Faraday's law to obtain

$$\mathbf{P}\times(\boldsymbol{\nabla}\times\mathbf{E}) + \frac{\partial}{\partial t}\left(\frac{1}{c}\mathbf{P}\times\mathbf{B}\right) = \left(\frac{1}{c}\frac{\partial}{\partial t}\mathbf{P}\right)\times\mathbf{B}, \tag{4.36}$$

and then we use the identity

$$\boldsymbol{\nabla}(\mathbf{M}\cdot\mathbf{B}) = \mathbf{M}\times(\boldsymbol{\nabla}\times\mathbf{B}) + (\mathbf{M}\cdot\boldsymbol{\nabla})\mathbf{B} + \mathbf{B}\times(\boldsymbol{\nabla}\times\mathbf{M}) + (\mathbf{B}\cdot\boldsymbol{\nabla})\mathbf{M}, \tag{4.37}$$

which is a generalization of (4.33b). All subsequent steps involve the statement that the integral is extended over a volume that includes the whole body so that, on the bounding surface of that volume, $n(\mathbf{r}) = 0$. This means that in performing partial integrations through the use of the divergence theorem, the surface integrals vanish. In effect, then,

$$(\mathbf{P}\cdot\boldsymbol{\nabla})\mathbf{E} \to -(\boldsymbol{\nabla}\cdot\mathbf{P})\mathbf{E}, \tag{4.38}$$

and similarly, using $\boldsymbol{\nabla}\cdot\mathbf{B} = 0$, (4.37) yields

$$\mathbf{M}\times(\boldsymbol{\nabla}\times\mathbf{B}) + (\mathbf{M}\cdot\boldsymbol{\nabla})\mathbf{B} \to (\boldsymbol{\nabla}\times\mathbf{M})\times\mathbf{B}. \tag{4.39}$$

The immediate result is

$$\mathbf{F} = \int (d\mathbf{r}) \left[-(\boldsymbol{\nabla}\cdot\mathbf{P})\mathbf{E} + \frac{1}{c}\left(\frac{\partial}{\partial t}\mathbf{P}\right)\times\mathbf{B} + (\boldsymbol{\nabla}\times\mathbf{M})\times\mathbf{B}\right]. \tag{4.40}$$

The comparison of this with the microscopic description of the force on charge and current densities, (3.7) for zero magnetic charge, suggests the definition of an effective charge density, ρ_{eff}, and an effective current density, $\mathbf{j}_{\text{eff}}$, as

$$\rho_{\text{eff}}(\mathbf{r},t) = -\boldsymbol{\nabla}\cdot\mathbf{P}(\mathbf{r},t), \tag{4.41a}$$

$$\mathbf{j}_{\text{eff}}(\mathbf{r},t) = \frac{\partial}{\partial t}\mathbf{P}(\mathbf{r},t) + c\boldsymbol{\nabla}\times\mathbf{M}(\mathbf{r},t). \tag{4.41b}$$

Notice that these effective densities satisfy the equation of charge conservation,

$$\frac{\partial}{\partial t}\rho_{\text{eff}} + \boldsymbol{\nabla}\cdot\mathbf{j}_{\text{eff}} = 0. \tag{4.42}$$

It is left to the reader to verify (Problem 4.3) that the total torque, $\boldsymbol{\tau}$, on the body, the sum over all atoms of the external torques:

$$\boldsymbol{\tau}_{\text{ext}} = \int (d\mathbf{r})\, n\mathbf{r}\times\left[\mathbf{d}\times(\boldsymbol{\nabla}\times\mathbf{E}) + (\mathbf{d}\cdot\boldsymbol{\nabla})\mathbf{E} + \boldsymbol{\mu}\times(\boldsymbol{\nabla}\times\mathbf{B}) + (\boldsymbol{\mu}\cdot\boldsymbol{\nabla})\mathbf{B} + \frac{1}{c}\frac{d}{dt}(\mathbf{d}\times\mathbf{B})\right], \tag{4.43}$$

and of the internal torques:

$$\boldsymbol{\tau}_{\text{int}} = \int (d\mathbf{r})\, n(\mathbf{d}\times\mathbf{E} + \boldsymbol{\mu}\times\mathbf{B}), \tag{4.44}$$

is properly reproduced as the integrated moment of the effective force density,

$$\boldsymbol{\tau} = \int (d\mathbf{r})\, \mathbf{r}\times\left[\rho_{\text{eff}}\mathbf{E} + \frac{1}{c}\mathbf{j}_{\text{eff}}\times\mathbf{B}\right]. \tag{4.45}$$

4.3 Macroscopic Electrodynamics

Now we construct a phenomenological macroscopic electrodynamics. And what is that? Nothing more than the form in which electrodynamics first arose, in the pre-atomic period, when only the properties of bulk matter were involved. But the challenge here is to derive the phenomenological theory from the microscopic Maxwell-Lorentz description. Both theories will employ concepts that are abstracted from the kinds of measurements that are appropriate to their level of description. The microscopic regime is characterized by rapid space-time variations unlike the macroscopic one, which is characterized by scales large compared to those of atoms. Laboratory instruments, being large, directly measure average quantities. Macroscopic fields are thus defined in terms of averages over space and time intervals, V and T, large on the atomic scale but small compared to typical macroscopic intervals. We adopt the convention that lower-case letters, like $f(\mathbf{r},t)$, represent microscopic quantities while capital letters, like $F(\mathbf{r},t)$, represent the corresponding macroscopic quantities. The connection between the two is

$$F(\mathbf{r},t)=\frac{1}{T}\int_T dt'\,\frac{1}{V}\int_V (d\mathbf{r}')\,f(\mathbf{r}+\mathbf{r}',t+t')=\overline{f(\mathbf{r},t)}. \tag{4.46}$$

This is a linear relation, in the sense that

$$\overline{f_1+f_2}=\overline{f_1}+\overline{f_2},\quad \overline{\lambda f}=\lambda\overline{f}, \tag{4.47}$$

where λ is a constant. From this follows the connection between derivatives of microscopic and macroscopic quantities, that is, that the averaged derivative of a function is the derivative of the average:

$$\begin{aligned}\frac{\partial}{\partial t}\overline{f(\mathbf{r},t)}&=\overline{\frac{\partial}{\partial t}f(\mathbf{r},t)},\\ \boldsymbol{\nabla}\overline{f(\mathbf{r},t)}&=\overline{\boldsymbol{\nabla}f(\mathbf{r},t)}.\end{aligned} \tag{4.48}$$

The microscopic charge distribution is composed of two parts. That which is confined to atoms is called bound charge. When the remaining, "free," microscopic charge distributions are appropriately averaged, we obtain the macroscopic densities

$$\rho=\overline{\rho_{\text{free}}},\qquad \mathbf{J}=\overline{\mathbf{j}_{\text{free}}}. \tag{4.49}$$

What is the macroscopic role of the bound charge distributions? It must be related to the effective charge and current densities given in terms of the polarization and the magnetization by (4.41a) and (4.41b),

$$\rho_{\text{eff}}=-\boldsymbol{\nabla}\cdot\mathbf{P}, \tag{4.50a}$$

$$\mathbf{j}_{\text{eff}}=\frac{\partial}{\partial t}\mathbf{P}+c\boldsymbol{\nabla}\times\mathbf{M}. \tag{4.50b}$$

As we have seen in the preceding section, these densities are examples of macroscopically measured quantities, disclosed by slowly varying electric and magnetic fields. The physical measurements necessary for the definitions of ρ_{eff} and $\mathbf{j}_{\text{eff}}$, since they employ slowly varying fields, should correspond to the mathematical process of averaging involved in the definitions of $\overline{\rho_{\text{bound}}}$ and $\overline{\mathbf{j}_{\text{bound}}}$, so we have the identifications

$$\begin{aligned}\overline{\rho_{\text{bound}}}&=\rho_{\text{eff}}\\ \overline{\mathbf{j}_{\text{bound}}}&=\mathbf{j}_{\text{eff}}.\end{aligned} \tag{4.51}$$

In view of (4.42), these two forms of macroscopic charge are separately conserved.

TABLE 4.1
Connection between microscopic and macroscopic quantities.

Electric field	Magnetic field	Charge density	Current density
$\mathbf{e}$	$\mathbf{b}$	$\rho_{\text{free}} + \rho_{\text{bound}}$	$\mathbf{j}_{\text{free}} + \mathbf{j}_{\text{bound}}$
$\mathbf{E}$	$\mathbf{B}$	$\rho - \boldsymbol{\nabla}\cdot\mathbf{P}$	$\mathbf{J} + \frac{\partial}{\partial t}\mathbf{P} + c\boldsymbol{\nabla}\times\mathbf{M}$

The correspondence between microscopic and macroscopic quantities is given by Table 4.1: The microscopic Maxwell equations now read

$$\begin{aligned} \boldsymbol{\nabla}\times\mathbf{b} &= \frac{1}{c}\frac{\partial}{\partial t}\mathbf{e} + \frac{4\pi}{c}(\mathbf{j}_{\text{free}} + \mathbf{j}_{\text{bound}}), & \boldsymbol{\nabla}\cdot\mathbf{e} &= 4\pi(\rho_{\text{free}} + \rho_{\text{bound}}), \\ -\boldsymbol{\nabla}\times\mathbf{e} &= \frac{1}{c}\frac{\partial}{\partial t}\mathbf{b}, & \boldsymbol{\nabla}\cdot\mathbf{b} &= 0. \end{aligned} \tag{4.52}$$

These are averaged to yield the macroscopic equations,

$$\begin{aligned} \boldsymbol{\nabla}\times\mathbf{B} &= \frac{1}{c}\frac{\partial}{\partial t}\mathbf{E} + \frac{4\pi}{c}\left(\mathbf{J} + \frac{\partial}{\partial t}\mathbf{P} + c\boldsymbol{\nabla}\times\mathbf{M}\right), & \boldsymbol{\nabla}\cdot\mathbf{E} &= 4\pi(\rho - \boldsymbol{\nabla}\cdot\mathbf{P}), \\ -\boldsymbol{\nabla}\times\mathbf{E} &= \frac{1}{c}\frac{\partial}{\partial t}\mathbf{B}, & \boldsymbol{\nabla}\cdot\mathbf{B} &= 0, \end{aligned} \tag{4.53}$$

which can be cast into the form of the microscopic equations if we define the displacement, $\mathbf{D}$,

$$\mathbf{D} = \mathbf{E} + 4\pi\mathbf{P}, \tag{4.54}$$

and the magnetic field, $\mathbf{H}$,

$$\mathbf{H} = \mathbf{B} - 4\pi\mathbf{M} \tag{4.55}$$

(recall that $\mathbf{B}$ is properly called the magnetic induction). The final form of the historical, macroscopic Maxwell equations is

$$\begin{aligned} \boldsymbol{\nabla}\times\mathbf{H} &= \frac{1}{c}\frac{\partial}{\partial t}\mathbf{D} + \frac{4\pi}{c}\mathbf{J}, & \boldsymbol{\nabla}\cdot\mathbf{D} &= 4\pi\rho, \\ -\boldsymbol{\nabla}\times\mathbf{E} &= \frac{1}{c}\frac{\partial}{\partial t}\mathbf{B}, & \boldsymbol{\nabla}\cdot\mathbf{B} &= 0. \end{aligned} \tag{4.56}$$

Note that the macroscopic charge is conserved,

$$\boldsymbol{\nabla}\cdot\mathbf{J} + \frac{\partial}{\partial t}\rho = 0, \tag{4.57}$$

which follows from the first pair of equations in (4.56). As microscopically smooth distributions, the density and flux of free charge will serve to measure the macroscopic fields $\mathbf{E}$ and $\mathbf{B}$. That is exhibited in the expression for the force on a macroscopic charge distribution,

$$\mathbf{F} = \int (d\mathbf{r}) \left(\rho\mathbf{E} + \frac{1}{c}\mathbf{J}\times\mathbf{B}\right). \tag{4.58}$$

[If bound charge is present, there is an additional contribution to the force coming from (4.40).]

For a complete description of the system, we require further relations between **D**, **E**, **P**, and **J**, expressing how material bodies respond to electric fields. Similar remarks hold for **H**, **B**, and **M**. These constitutive relations depend on the characteristics of the particular material under consideration. Simple classical models—which are not qualitatively misleading—will be considered in the following three chapters.

4.4 Problems for Chapter 4

1. Find the total charge and the dipole moment of the charge density

$$\rho(\mathbf{r}) = -\mathbf{d}\cdot\boldsymbol{\nabla}\delta(\mathbf{r}). \tag{4.59}$$

2. Justify the approximation leading to the final form of $\boldsymbol{\tau}$ in (4.30). In particular, show that the total time derivative omitted in going from the first to the second line of (4.30) leads to

$$\frac{d}{dt}\sum_a \mathbf{r}_a\times\mathbf{p}_a = \boldsymbol{\tau}, \tag{4.60}$$

where the "canonical momentum" $\mathbf{p}_a$ is defined by

$$\mathbf{p}_a = m_a\mathbf{v}_a + \frac{e_a}{c}\mathbf{A}(\mathbf{r}_a), \tag{4.61}$$

where the vector potential $\mathbf{A}$ for a constant magnetic field $\mathbf{B}$ is

$$\mathbf{A} = -\frac{1}{2}\mathbf{r}\times\mathbf{B}, \quad \boldsymbol{\nabla}\times\mathbf{A} = \mathbf{B}. \tag{4.62}$$

3. By summing the torque on an individual charge,

$$\boldsymbol{\tau}_a = \mathbf{r}_a\times\left(e_a\mathbf{E}(\mathbf{r}_a) + e_a\frac{\mathbf{v}_a}{c}\times\mathbf{B}(\mathbf{r}_a)\right), \tag{4.63}$$

first, over the charges in an individual atom, and thereby obtaining expressions in terms of $\mathbf{d}_a$ and $\boldsymbol{\mu}_a$, the dipole moments of the atom, and then over the atoms making up a macroscopic body, obtain the result that

$$\boldsymbol{\tau} = \boldsymbol{\tau}_{\text{ext}} + \boldsymbol{\tau}_{\text{int}}, \tag{4.64}$$

where the external and internal torques are given by (4.43) and (4.44), respectively. Then, verify that the torque acting on a macroscopic object in electric and magnetic fields is given in terms of ρ_{eff} and $\mathbf{j}_{\text{eff}}$ according to (4.45).

5

Simple Model for Constitutive Relations

5.1 Conductivity

We start by considering a simple model of a metal in which the current is linearly related to the electric field. The model is to be considered suggestive only, but it does lead to a qualitative understanding of the important phenomena of conduction. Of course, an accurate description requires quantum mechanics.

First consider a free electric charge (an electron) moving under the influence of an external electric field, and subject to collisions with the atoms of the substance. The electric field accelerates the charge, and the collisions slow it down. Our model represents the effects of the collisions by a frictional force that is proportional—and opposed—to the velocity. The equation of motion for the particle, having charge e and mass m, is

$$m\frac{d}{dt}\mathbf{v}(t) = -m\gamma\mathbf{v}(t) + e\mathbf{E}(t), \qquad \gamma > 0, \tag{5.1a}$$

or

$$\frac{d}{dt}\mathbf{v}(t) = -\gamma\mathbf{v}(t) + \frac{e}{m}\mathbf{E}(t). \tag{5.1b}$$

This model was discussed by Paul Drude in 1900. (The variation of the electric field with position is ignored here—the velocities of interest are of very small magnitude compared with c.) The frictional constant γ is given a physical interpretation by considering the situation for $\mathbf{E} = \mathbf{0}$:

$$\frac{d}{dt}\mathbf{v}(t) = -\gamma\mathbf{v}(t), \quad \mathbf{v}(t) = \mathbf{v}_0 e^{-\gamma t}; \tag{5.2}$$

any initial velocity decreases exponentially in time, due to collisions with atoms, with $1/\gamma$ supplying the characteristic decay time. The general solution to (5.1b) is found by first rewriting it as

$$\frac{d}{dt}\left[e^{\gamma t}\mathbf{v}(t)\right] = \frac{e}{m}e^{\gamma t}\mathbf{E}(t), \tag{5.3}$$

and then integrating from $t' = -\infty$ (a time before any field has been applied), to t (the time of observation),

$$\mathbf{v}(t) = \frac{e}{m}\int_{-\infty}^{t} dt'\, e^{-\gamma(t-t')}\mathbf{E}(t'). \tag{5.4}$$

Notice that the response of the system under the action of an external electric field is nonlocal in time. (That is, the velocity at a given time depends on the electric field at earlier times.) The main contribution to the integral comes from the region of time differences that are of the order of $1/\gamma$.

The current density for a single charge is proportional to its velocity. If n_f is the (constant) density of (free) conduction electrons, then $\mathbf{J}$ is

$$\mathbf{J}(t) = n_f e\mathbf{v}(t) = \frac{n_f e^2}{m}\int_{-\infty}^{t} dt'\, e^{-\gamma(t-t')}\mathbf{E}(t'). \tag{5.5}$$

DOI: 10.1201/9781003057369-5

For the particular example of a constant electric field, this reduces to

$$\mathbf{J} = \frac{n_f e^2}{m\gamma}\mathbf{E} = \sigma\mathbf{E}, \tag{5.6}$$

which is a statement of Ohm's law (Georg Simon Ohm, 1787–1854), σ being the *static conductivity*. [This, of course, can be more directly obtained by looking for the static solution of (5.1b).] A more general situation arises when the electric field exhibits harmonic variation (i.e., has a definite frequency),

$$E \sim \cos(\omega t + \phi) = \mathrm{Re}\left(e^{-i(\omega t+\phi)}\right), \tag{5.7}$$

or in terms of the complex amplitude $\mathbf{E}(\omega)$,

$$\mathbf{E}(t) = \mathrm{Re}(\mathbf{E}(\omega)e^{-i\omega t}). \tag{5.8}$$

Now, the current density, (5.5), becomes

$$\begin{aligned}\mathbf{J}(t) &= \mathrm{Re}\left(\frac{n_f e^2}{m}\mathbf{E}(\omega)\int_{-\infty}^{t} dt'\, e^{-\gamma(t-t')}e^{-i\omega t'}\right)\\ &= \mathrm{Re}\left[\frac{n_f e^2}{m}\frac{1}{\gamma - i\omega}\mathbf{E}(\omega)e^{-i\omega t}\right].\end{aligned} \tag{5.9}$$

Here displayed is a complex amplitude for $\mathbf{J}(t)$, $\mathbf{J}(\omega)$. In terms of it, the complex conductivity, $\sigma(\omega)$, is defined by

$$\mathbf{J}(\omega) = \sigma(\omega)\mathbf{E}(\omega), \tag{5.10}$$

where, from (5.9),

$$\sigma(\omega) = \frac{n_f e^2}{m}\frac{1}{\gamma - i\omega}. \tag{5.11}$$

For $\omega = 0$, we regain $\sigma(0) = \sigma$, the static conductivity given in (5.6).

The conductivity is a function of $i\omega$, which means that complex conjugation is equivalent to changing the sign of ω:

$$\sigma(\omega)^* = \sigma(-\omega). \tag{5.12}$$

This says that the real and imaginary parts of $\sigma(\omega)$,

$$\sigma(\omega) = \frac{n_f e^2}{m}\frac{\gamma + i\omega}{\gamma^2 + \omega^2} = \mathrm{Re}\,\sigma(\omega) + i\,\mathrm{Im}\,\sigma(\omega), \tag{5.13}$$

are, respectively, even and odd functions of ω, as shown in Fig. 5.1. Finally, we note that the integral of the conductivity over all frequencies is

$$\begin{aligned}\int_{-\infty}^{\infty} d\omega\,\sigma(\omega) &= \frac{n_f e^2}{m}2\int_0^{\infty} d\omega\frac{\gamma}{\gamma^2+\omega^2} = \frac{n_f e^2}{m}2\int_0^{\infty}\frac{dx}{1+x^2}\\ &= \frac{n_f e^2}{m}\pi,\end{aligned} \tag{5.14}$$

which is called a "sum rule." The significant feature of this sum rule is that the right-hand side is independent of the frictional force constant so that we could use it to determine n_f experimentally. What underlies this is the simplicity of the response to an electric field pulse that is localized at time $t = 0$. Without time to act, the frictional forces are effectively absent. (See Problem 5.1.)

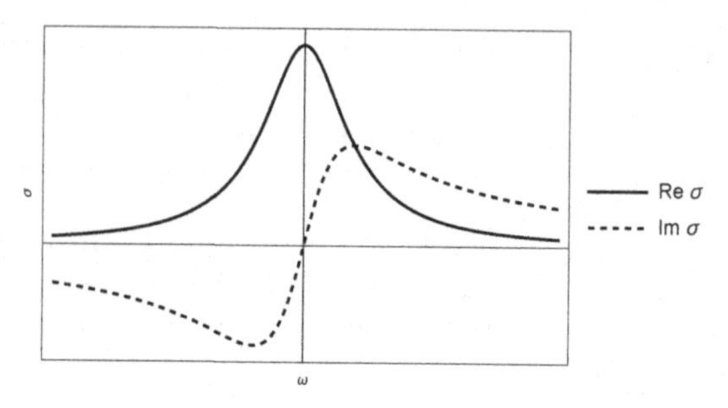

FIGURE 5.1
The real and imaginary parts of the conductivity, as given in (5.13).

5.2 Dielectric Constant

We now modify the above model in order to discuss bound charge by including an additional binding force term in (5.1a). We will take as the simplest model of such binding a harmonic oscillator force, which turns out, for the most part, to give qualitatively correct results. That is, we will adopt, taking the origin to be the center of the force,

$$m\frac{d}{dt}\mathbf{v} = -m\omega_0^2\mathbf{r} - m\gamma\mathbf{v} + e\mathbf{E}, \quad \mathbf{v} = \frac{d\mathbf{r}}{dt}, \tag{5.15}$$

as the new equation of motion. Here ω_0 is the natural (angular) frequency of the electron bound in the atom, while γ is a damping constant, primarily due to electromagnetic *radiation*. (More about this in Chapter 38.)

For a harmonic time dependence of the driving electric field, (5.8), the above force equation becomes

$$\frac{d^2}{dt^2}\mathbf{r} + \omega_0^2\mathbf{r} + \gamma\frac{d}{dt}\mathbf{r} = \frac{e}{m}\,\mathrm{Re}\left(\mathbf{E}(\omega)e^{-i\omega t}\right). \tag{5.16}$$

This implies that the steady-state solution for the position vector will also exhibit harmonic time variation, that is,

$$\mathbf{r}(t) = \frac{e}{m}\,\mathrm{Re}\left[\frac{\mathbf{E}(\omega)e^{-i\omega t}}{-\omega^2 + \omega_0^2 - i\gamma\omega}\right]. \tag{5.17}$$

Under the usual circumstance of $\gamma \ll \omega_0$, the amplitude of the induced oscillation becomes very large for $\omega = \omega_0$, the condition of resonance.

It is now immediate to calculate the polarization (4.34a) in terms of the induced electric dipole moment and the density of bound electrons, n_b,

$$\mathbf{P} = n_b e\mathbf{r}, \tag{5.18}$$

or, explicitly in terms of the electric field,

$$\begin{aligned}\mathbf{P}(t) &= \frac{n_b e^2}{m}\,\mathrm{Re}\left[\frac{\mathbf{E}(\omega)e^{-i\omega t}}{-\omega^2 + \omega_0^2 - i\gamma\omega}\right] \\ &= \mathrm{Re}[\chi_e(\omega)\mathbf{E}(\omega)e^{-i\omega t}],\end{aligned} \tag{5.19}$$

where χ_e is the (frequency-dependent) electric susceptibility,[1]

$$\chi_e(\omega) = \frac{n_b e^2}{m}\frac{1}{-\omega^2 - i\omega\gamma + \omega_0^2}, \tag{5.20}$$

[1]This is often called the Fermi model, after Enrico Fermi (1901–1954). However, it is more properly referred to as the Drude-Lorentz model.

which satisfies

$$\chi_e(\omega) = \chi_e(-\omega)^*. \tag{5.21}$$

The static susceptibility, the real value when $\omega = 0$, is

$$\chi_e = \frac{n_b e^2}{m\omega_0^2} > 0. \tag{5.22}$$

What is the order of magnitude of χ_e? Since ω_0 is identified with a characteristic atomic frequency,

$$\omega_0 \sim \frac{v}{l} \tag{5.23}$$

where v and l are representative atomic speeds and distances, respectively, we can estimate χ_e in terms of atomic quantities as

$$\frac{n_b e^2}{m\omega_0^2} \sim n_b \left(\frac{e^2/l}{mv^2}\right) l^3. \tag{5.24}$$

Since a typical value of the electron's kinetic energy is of the same order of magnitude as a typical Coulomb potential energy (this is the virial theorem—see Chapter 9),

$$mv^2 \sim \frac{e^2}{l}, \tag{5.25}$$

the value of the electric susceptibility is of order

$$\frac{n_b e^2}{m\omega_0^2} \sim n_b l^3, \tag{5.26}$$

the number of bound electrons in an atomic volume. For dense matter, where the atoms are tightly packed, we may have, therefore,

$$\chi_e \sim 1. \tag{5.27}$$

The displacement field, (4.54), is

$$\begin{aligned} \mathbf{D}(t) &= \mathbf{E}(t) + 4\pi\mathbf{P}(t) \\ &= \operatorname{Re}\left[(1 + 4\pi\chi_e(\omega))\mathbf{E}(\omega)e^{-i\omega t}\right], \end{aligned} \tag{5.28}$$

which defines the frequency-dependent dielectric "constant," or the permittivity, $\varepsilon(\omega)$, through the relation

$$\varepsilon(\omega) = 1 + 4\pi\chi_e(\omega) = \varepsilon(-\omega)^*. \tag{5.29}$$

Therefore, there is a linear relation between the field amplitudes,

$$\mathbf{D}(\omega) = \varepsilon(\omega)\mathbf{E}(\omega). \tag{5.30}$$

In particular, the real, static dielectric constant $\epsilon = \varepsilon(0)$ is

$$\epsilon = 1 + 4\pi\chi_e > 1, \tag{5.31}$$

where, as noted, the excess over unity can be significant in dense substances. For example, the static dielectric constant for mica is about 6, and for water, about 80.

We can also derive a sum rule for $\chi_e(\omega)$, by considering the integral

$$\int_{-\infty}^{\infty} d\omega\,(-i\omega)\chi_e(\omega) = -\frac{n_b e^2}{m}\int_{-\infty}^{\infty} d\omega \frac{-i\omega}{\left(\omega + \frac{i}{2}\gamma\right)^2 - \omega_0'^2}, \quad \omega_0'^2 = \omega_0^2 - \left(\frac{1}{2}\gamma\right)^2. \tag{5.32}$$

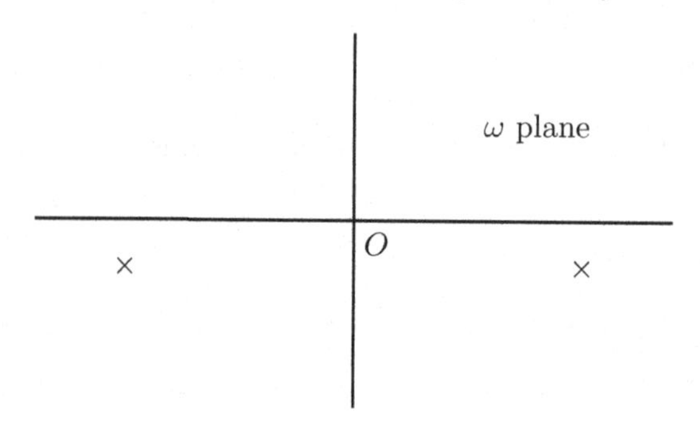

FIGURE 5.2
Singularities of the integrand in (5.32) in the complex ω plane.

(Although this integral might appear divergent, the leading part of the integrand for large $|\omega|$ is odd in ω.) For the following discussion, we assume that $\gamma < 2\omega_0$ so that $\omega_0'^2 > 0$. (This restriction is actually unnecessary—see Problem 5.3.) Now think of ω as a complex variable, and note that the integrand has singularities at

$$\omega = -\frac{i}{2}\gamma \pm \omega_0', \tag{5.33}$$

which lie in the lower half plane (see Fig. 5.2). Accordingly, the integrand is everywhere regular in the upper half plane, and the integral over a path that is closed by a large semicircle in the upper half plane equals zero. The integral of interest, (5.32), is therefore the negative of that over the semicircle of arbitrarily large radius, which is

$$\frac{n_b e^2}{m}\int_{\cap} d\omega \frac{-i}{\omega} = \pi n_b \frac{e^2}{m}. \tag{5.34}$$

Thus, the desired sum rule, independent of γ and ω_0, is

$$\int_{-\infty}^{\infty} d\omega\,(-i\omega)\chi_e(\omega) = \pi\frac{n_b e^2}{m}. \tag{5.35}$$

In retrospect, the same method could have been applied to the conductivity integral, (5.14), using (5.11), which integral is obtained from that in (5.32) by putting $\omega_0 = 0$. Physically, we should have expected this result on the basis of the σ relation, (5.14), since the two quantities ε and σ are just parts of a whole. The two phenomena being discussed are just the free electron and bound electron contributions to the total current, which have the following form for a definite frequency ω,

$$\mathbf{J} + \frac{\partial}{\partial t}\mathbf{P} \to \operatorname{Re}\left([\sigma(\omega) - i\omega\chi_e(\omega)]\mathbf{E}(\omega)e^{-i\omega t}\right), \tag{5.36}$$

so that the sum rule corresponding to this total current is proportional to the total electron density $n_b + n_f$,

$$\int_{-\infty}^{\infty} d\omega[\sigma(\omega) - i\omega\chi_e(\omega)] = (n_f + n_b)\pi\frac{e^2}{m}, \tag{5.37}$$

which expresses the fact that, in the response to an electric pulse localized at time zero, only the inertia of the electrons matter—frictional and binding forces have no time in which to act.

5.3 Plasma

Let us combine the results of the preceding two sections by considering the motion of free charge in a conducting dielectric material, for which the conduction current is

$$\mathbf{J} = \sigma\mathbf{E} = \frac{\sigma}{\epsilon}\mathbf{D}. \tag{5.38}$$

First suppose that both σ and ϵ are taken to be independent of frequency, an approximation which is valid for low frequencies. Then the local charge conservation equation,

$$\frac{\partial}{\partial t}\rho + \boldsymbol{\nabla}\cdot\mathbf{J} = 0, \tag{5.39}$$

becomes

$$\frac{\partial}{\partial t}\rho + \boldsymbol{\nabla}\cdot\frac{\sigma}{\epsilon}\mathbf{D} = 0. \tag{5.40}$$

In the interior of a homogeneous substance, the use of (4.56), $\boldsymbol{\nabla}\cdot\mathbf{D} = 4\pi\rho$, produces the differential equation

$$\frac{\partial}{\partial t}\rho + \frac{\sigma}{\epsilon}4\pi\rho = 0. \tag{5.41}$$

The solution to this equation, corresponding to an initial charge density $\rho(\mathbf{r}, 0)$, is

$$\rho(\mathbf{r}, t) = \rho(\mathbf{r}, 0)e^{-4\pi\sigma t/\epsilon}, \tag{5.42}$$

implying that the charge disappears from the interior of the conducting body at a rate measured by

$$\gamma' = \frac{4\pi\sigma}{\epsilon}. \tag{5.43}$$

The charge in the interior of the volume eventually all migrates to the surface. (See Problem 55.4.)

The use of the static conductivity here assumes that this decay rate, γ', is small compared to the frictional constant γ, $\gamma' \ll \gamma$, or, using (5.6), that

$$\frac{4\pi}{\epsilon}\frac{n_f e^2}{m} \ll \gamma^2. \tag{5.44}$$

This must be satisfied by some combination of small density of free charge and a high coefficient of friction. But what of the opposite situation, where there is a high density of free charge and little friction, as encountered in a plasma? For that, it may be clearer to return to the equation of motion (5.1b), presented as the differential equation

$$\left(\frac{\partial}{\partial t} + \gamma\right)\mathbf{J} = \frac{n_f e^2}{m}\mathbf{E} = \frac{n_f e^2}{m\epsilon}\mathbf{D} \tag{5.45}$$

for the current

$$\mathbf{J} = n_f e\mathbf{v}. \tag{5.46}$$

Now we can use the charge conservation equation, in the form

$$\left(\frac{\partial}{\partial t} + \gamma\right)\frac{\partial}{\partial t}\rho + \boldsymbol{\nabla}\cdot\left(\frac{\partial}{\partial t} + \gamma\right)\mathbf{J} = 0, \tag{5.47}$$

to get, when we again use $\boldsymbol{\nabla}\cdot\mathbf{D}=4\pi\rho$,

$$\left(\frac{\partial^2}{\partial t^2}+\gamma\frac{\partial}{\partial t}+\omega_p^2\right)\rho=0, \tag{5.48}$$

where the plasma frequency is defined by

$$\omega_p^2=\frac{4\pi n_f e^2}{m\epsilon}. \tag{5.49}$$

(Note that the usual form of the plasma frequency has $\epsilon=1$.) In general, the solutions to (5.48) are damped oscillations. If the rate of change is small compared to the scale set by γ, the equation (5.48) simplifies to

$$\left(\frac{\partial}{\partial t}+\gamma'\right)\rho=0,\quad \gamma'=\frac{\omega_p^2}{\gamma}\ll\gamma, \tag{5.50}$$

which will be recognized as the previous result (5.41), including the restriction (5.44). This limit corresponds to exponential decay. But in the other limit, where change occurs rapidly relative to γ, it is the γ term that can be approximately neglected,

$$\left(\frac{\partial^2}{\partial t^2}+\omega_p^2\right)\rho=0, \tag{5.51}$$

and the charge oscillates at the angular frequency ω_p.

The appearance of ϵ in this plasma frequency is a reminder of a restriction still in force, that ω_p be small compared to ω_0, the characteristic atomic frequency, near which the frequency dependence of $\varepsilon(\omega)$ can no longer be ignored. There is an extreme plasma circumstance in which ω_0 is not relevant. Let the physical conditions be such that all atoms are completely ionized, removing the distinction between free and bound charge. Then the entire charge density can be viewed as the result of polarization,

$$\rho=-\boldsymbol{\nabla}\cdot\mathbf{P}, \tag{5.52}$$

arising from the displacement $\mathbf{r}$ of the electrons relative to the oppositely charged heavy ions,

$$\mathbf{P}=ne\mathbf{r}. \tag{5.53}$$

That displacement changes in time, responding to the electric field as

$$m\frac{d^2\mathbf{r}}{dt^2}=e\mathbf{E}, \tag{5.54}$$

or

$$\frac{\partial^2}{\partial t^2}\mathbf{P}=\frac{ne^2}{m}\mathbf{E}. \tag{5.55}$$

From this, and the field equation

$$\boldsymbol{\nabla}\cdot\mathbf{E}=4\pi(-\boldsymbol{\nabla}\cdot\mathbf{P}), \tag{5.56}$$

we derive (5.51), describing the plasma oscillations, with $\epsilon=1$.

5.4 Polar Molecules

In the above model for the electric susceptibility, leading to (5.20), the dipole moments were induced by the applied electric field. But what about permanent electric dipole moments? Do individual atoms possess such static properties? With the exception of atomic hydrogen where the orbital motion respects a preferred direction in space (as in the classical elliptical orbits), atomic electric dipole moments change direction in space so rapidly in response to the fast electronic motion that no average effect survives. But things are different with molecules, specifically those of a polar nature. In the example of H^+Cl^-, the hydrogenic electron is transferred to form the chlorine ion, and a dipole moment is associated with the relative motion of the heavy ions. Other examples of polar molecules associated with familiar substances are H_2O, SO_2, NH_3, and CH_3Cl. For such molecules, in isolation, it is not misleading to think of a permanent electric dipole moment that changes its spatial orientation only in response to the slow rotation of the molecule.

But molecules are not ordinarily isolated; they exist in an environment in which other molecules collide with them at a rate determined by the temperature of the substance. The effect of these collisions is to remove any particular spatial orientation of the dipole moments; it still requires an electric field to provide a preferred direction. But now there is a competition between the organizing effect of the electric field, with its preference for lower values of the energy,

$$E = -\mathbf{d} \cdot \mathbf{E} = -|\mathbf{d}||\mathbf{E}| \cos\theta, \tag{5.57}$$

and the disorganizing effect of the ambient temperature T. For a static field, the net balance of that competition is expressed by the Boltzmann factor, which gives the probability of finding a configuration of energy E,

$$e^{-E/kT}, \tag{5.58}$$

where k, the constant of Ludwig Boltzmann (1844–1906), has the value

$$k = 1.381 \times 10^{-16} \text{erg/K}. \tag{5.59}$$

In thermal equilibrium, the fraction of dipole moments that are directed within the solid angle

$$d\Omega = \sin\theta \, d\theta \, d\phi \tag{5.60}$$

is proportional to the product of $d\Omega$ with the Boltzmann factor:

$$\frac{1}{Z}\frac{d\Omega}{4\pi} e^{\mathbf{d} \cdot \mathbf{E}/kT}, \tag{5.61}$$

where the choice of the normalization constant, the so-called "partition function"

$$Z = \int \frac{d\Omega}{4\pi} e^{\mathbf{d} \cdot \mathbf{E}/kT} \tag{5.62}$$

ensures that the totality of such fractions equals unity. Consequently, the average dipole moment is

$$\langle \mathbf{d} \rangle_T = \frac{1}{Z} \int \frac{d\Omega}{4\pi} \mathbf{d} \, e^{\mathbf{d} \cdot \mathbf{E}/kT} = \frac{1}{Z} kT \frac{\partial}{\partial \mathbf{E}} Z = kT \frac{\partial}{\partial \mathbf{E}} \ln Z. \tag{5.63}$$

Except for very low temperatures or very high fields, the condition $|\mathbf{d}|\,|\mathbf{E}| \ll kT$ holds true so that we may expand the exponential Boltzmann factor in the partition function Z:

$$Z \approx \int \frac{d\Omega}{4\pi} \left[1 + \frac{\mathbf{d} \cdot \mathbf{E}}{kT} + \frac{1}{2} \left(\frac{\mathbf{d} \cdot \mathbf{E}}{kT} \right)^2 + \dots \right] \approx 1 + \frac{1}{6} \left(\frac{|\mathbf{d}||\mathbf{E}|}{kT} \right)^2. \tag{5.64}$$

In this uniformly weighted average over all directions, where $\mathbf{d}$ and $-\mathbf{d}$ appear with equal weight, the average of $\mathbf{d}$ is zero, and that of the square of some component, say d_z^2, is the same as any other component so that

$$\langle d_x^2 \rangle = \langle d_y^2 \rangle = \langle d_z^2 \rangle = \frac{1}{3}(\mathbf{d})^2. \tag{5.65}$$

The leading contribution to (5.63) is produced by

$$\frac{\partial}{\partial \mathbf{E}} \ln Z = \frac{1}{3}\frac{d^2}{(kT)^2}\mathbf{E}, \tag{5.66}$$

so that in this approximation $\langle \mathbf{d} \rangle_T$ and $\mathbf{E}$ are linearly related,

$$\langle \mathbf{d} \rangle_T \approx \frac{1}{3}\frac{d^2}{kT}\mathbf{E}. \tag{5.67}$$

What kind of limit does our weak field condition impose on the magnitude of the electric field? If we recall that room temperature, $T = 300\text{K}$, corresponds to $kT \approx \frac{1}{40}$ eV, and typical atomic dimensions set the scale for the dipole moment, $|\mathbf{d}| \sim 10^{-8}\, e\,\text{cm}$ (where e represents the magnitude of the charge on the electron), then $|\mathbf{d}|\,|\mathbf{E}| \ll kT$ requires

$$|\mathbf{E}| \ll 3 \times 10^6\,\text{volts/cm}, \tag{5.68}$$

which is indeed a large electric field. (For the average dipole moment when this approximation is not valid, see Problem 5.5.) We note that one of the largest observed moments, that of potassium chloride, $d_{\text{KCl}} = 2.13 \times 10^{-8}\, e\,\text{cm}$, is slightly more than twice the value that was used in the estimate of (5.68).

With a density of n_{mol} polar molecules per unit volume, the (weak field) contribution to the polarization is

$$\mathbf{P} = n_{\text{mol}}\frac{d^2}{3kT}\mathbf{E}, \tag{5.69}$$

and the complete static susceptibility becomes

$$\chi_e = \chi_{e,\text{atom}} + n_{\text{mol}}\frac{d^2}{3kT}, \tag{5.70}$$

where $\chi_{e,\text{atom}}$ is the atomic susceptibility due to the induced dipole moments of the atoms [see (5.22)]. This result is accurate for a "polar gas," where the densities are low. Note that the first term in (5.70) is independent of T, while the second is proportional to $1/T$, allowing us to separate the two contributions experimentally, and thereby measure the dipole moment of a polar molecule. The expression (5.70) is known as the Langevin-Debye equation,[2] which has been of great importance in interpreting molecular structure.

The Boltzmann factor describes the situation of a static electric field. But suppose the electric field is vibrating, or rotating, with a definite frequency. Owing to the large inertia of the molecules, it takes a significant time for the dipole moment to readjust or relax into the configuration demanded by the new direction of the electric field. With increasing frequency the ability to readjust decreases and eventually the dipole moment ceases to follow the variations of the electric field. Experiments indicate that this occurs already for radio frequencies; at infrared and visible light frequencies, the dipole moments are effectively inert.

---

[2]Paul Langevin, 1872–1946, and Peter Debye, 1884–1966.

A simple quantitative version of this picture combines two ideas. First, in the absence of the electric field, any net dipole moment relaxes to zero with a characteristic decay rate, or inverse relaxation time, γ, and in the presence of a static field, the average moment is given by (5.67). These are united in the differential equation

$$\frac{d}{dt}\langle\mathbf{d}\rangle(t) = -\gamma\left[\langle\mathbf{d}\rangle(t) - \frac{d^2}{3kT}\mathbf{E}(t)\right]. \tag{5.71}$$

The steady-state solution, when

$$\mathbf{E}(t) = \mathrm{Re}\left(\mathbf{E}(\omega)e^{-i\omega t}\right), \tag{5.72}$$

is directly verified to be

$$\langle\mathbf{d}\rangle(t) = \mathrm{Re}\left[\frac{d^2}{3kT}\frac{\gamma}{\gamma - i\omega}\mathbf{E}(\omega)e^{-i\omega t}\right]. \tag{5.73}$$

The implication for the static situation, $\omega = 0$, and for the high-frequency limit, $\omega \gg \gamma$, are as already described. In particular, (5.73) expresses the fact that as ω increases, $\langle\mathbf{d}\rangle$ decreases eventually like γ/ω. In general, the static susceptibility contribution of the polar molecules is multiplied by

$$\frac{1}{1 - i\omega\tau}, \tag{5.74}$$

where

$$\tau = \frac{1}{\gamma} \tag{5.75}$$

is the relaxation time of the electric dipole moments. The molecular effect disappears at relatively low frequencies, leaving only the atomic contribution, since the atomic frequency, ω_0, is much greater than the molecular frequency, γ.

5.5 Clausius-Mossotti Equation

One further point needs discussion. In our atomic model for the dielectric constant in Section 5.2, we used the following equation for the electronic motion, (5.15):

$$m\frac{d^2}{dt^2}\mathbf{r} = -m\omega_0^2\mathbf{r} - m\gamma\frac{d}{dt}\mathbf{r} + e\mathbf{E}. \tag{5.76}$$

Here the driving force has been taken to be $e\mathbf{E}$ where the macroscopic field is $\mathbf{E}(\mathbf{r},t) = \overline{\mathbf{e}(\mathbf{r},t)}$. The same assumption for the driving field was made for the alignment of polar molecules in Section 5.4. This is incorrect since $\mathbf{e}$ includes the field of the atom (or molecule) itself, the effects of which are already represented in the harmonic restoring force. The correct driving field is the field acting on the electron due to all the other atoms (we use the word *atom* to stand, generically, for either atom or molecule),

$$\mathbf{E}_{\text{driving}} = \overline{\mathbf{e}_{\text{driving}}} = \overline{\mathbf{e} - \mathbf{e}_{\text{atom}}} = \mathbf{E} - \overline{\mathbf{e}_{\text{atom}}}, \tag{5.77}$$

where now the overbar represents a spatial average over a volume V which contains exactly one atom, that average volume per atom being the inverse of the density,

$$n = \frac{1}{V}. \tag{5.78}$$

[In (5.77), we assume that no significant contribution is produced by neighboring atoms so that $\mathbf{E}_{\text{driving}}$ does not differ appreciably from $\mathbf{e}_{\text{driving}}$, and also that the average over the atom is already sufficiently representative of a macroscopic average that the field $\mathbf{E}$ can be introduced. What we are doing should not be expected to apply for a strongly polar liquid or solid where the forces produced by neighboring molecules could be the dominant effect.] The field due to the atom in which the electron is located is

$$\mathbf{e}_{\text{atom}} = -\boldsymbol{\nabla} \sum_a \frac{e_a}{|\mathbf{r} - \mathbf{r}_a|} = \sum_a e_a \boldsymbol{\nabla}_a \frac{1}{|\mathbf{r} - \mathbf{r}_a|}, \tag{5.79}$$

where the summation extends over all charges in the atom. We can calculate $\overline{\mathbf{e}_{\text{atom}}}$ by averaging over a sphere (of radius a and volume $V = 4\pi a^3/3$) which is large enough to include the atom; the negative of the average field can be written as

$$-\overline{\mathbf{e}_{\text{atom}}} = \sum_a (-\boldsymbol{\nabla}_a) e_a \frac{1}{V} \int_V \frac{(d\mathbf{r})}{|\mathbf{r} - \mathbf{r}_a|}. \tag{5.80}$$

An individual term here,

$$\mathbf{e}_a(\mathbf{r}_a) = -\boldsymbol{\nabla}_a \int_V (d\mathbf{r}) \frac{e_a/V}{|\mathbf{r}_a - \mathbf{r}|}, \tag{5.81}$$

is the electric field, at the point $\mathbf{r}_a$ within the sphere, arising from a uniform charge distribution within the sphere of density e_a/V. As such, it obeys the differential equation (1.25)

$$\boldsymbol{\nabla}_a \cdot \mathbf{e}_a = 4\pi \frac{e_a}{V} = \boldsymbol{\nabla}_a \cdot \frac{4\pi}{3} \frac{e_a}{V} (\mathbf{r}_a - \mathbf{R}), \tag{5.82}$$

where $\mathbf{R}$ is the position vector of the center of the sphere. The last form merely uses (with subscript a) the fact that

$$\boldsymbol{\nabla} \cdot (\mathbf{r} - \mathbf{R}) = \boldsymbol{\nabla} \cdot \mathbf{r} = 3. \tag{5.83}$$

And now the spherical symmetry of the situation, telling us that the electric field at a point must be directed along the line from the center, immediately yields (see also Problem 5.7)

$$\mathbf{e}_a(\mathbf{r}_a) = \frac{4\pi}{3} n e_a (\mathbf{r}_a - \mathbf{R}), \tag{5.84}$$

and then

$$-\overline{\mathbf{e}_{\text{atom}}} = \frac{4\pi}{3} n \sum_a e_a (\mathbf{r}_a - \mathbf{R}) = \frac{4\pi}{3} n \mathbf{d} = \frac{4\pi}{3} \mathbf{P}. \tag{5.85}$$

We conclude that the correct driving field (subject to the caveats mentioned) is

$$\mathbf{E}_{\text{driving}} = \mathbf{E} + \frac{4\pi}{3} \mathbf{P}. \tag{5.86}$$

To appreciate what effect this has on our earlier results, let use denote by χ_e' what was previously, and incorrectly, called the susceptibility. For static fields, then,

$$\chi_e' = n_b \frac{e^2}{m\omega_0^2} + n_{\text{mol}} \frac{d^2}{3kT}, \tag{5.87}$$

and now we have

$$\mathbf{P} = \chi_e' \mathbf{E}_{\text{driving}} = \chi_e' \left(\mathbf{E} + \frac{4\pi}{3} \mathbf{P} \right), \tag{5.88}$$

or

$$\chi_e = \chi_e' \left(1 + \frac{4\pi}{3} \chi_e \right). \tag{5.89}$$

From this form, or the equivalent versions

$$\frac{\chi_e}{1+\frac{4\pi}{3}\chi_e} = \chi'_e, \quad \chi_e = \frac{\chi'_e}{1-\frac{4\pi}{3}\chi'_e}, \tag{5.90}$$

it is clear that the earlier identification of χ_e with χ'_e is valid only when $\chi_e \ll 1$, as in substances of low density. Yet another way of presenting matters, one that introduces the dielectric constant according to (5.31)

$$4\pi\chi_e = \epsilon - 1, \tag{5.91}$$

namely

$$\frac{\epsilon - 1}{\epsilon + 2} = \frac{4\pi}{3}\chi'_e, \tag{5.92}$$

is known, from its historical origins, as the Clausius-Mossotti formula [O. F. Mossotti (1850), R. Clausius (1879); the years cited indicate the dates of significant publication].

When one turns to fields of definite frequency, the names change, the relation

$$\frac{\varepsilon(\omega) - 1}{\varepsilon(\omega) + 2} = \frac{4\pi}{3}\chi'_e(\omega) \tag{5.93}$$

being called that of Lorenz and Lorentz [Ludvig Lorenz (1869) and Hendrik A. Lorentz (1880)]. For the particular example of a nonpolar substance, where [(5.20)]

$$\chi'_e(\omega) = \frac{n_b e^2/m}{\omega_0^2 - \omega^2 - i\omega\gamma}, \tag{5.94}$$

the formula (5.90),

$$4\pi\chi_e(\omega) = \frac{4\pi\chi'_e(\omega)}{1-\frac{4\pi}{3}\chi'_e(\omega)} = \frac{1}{(4\pi\chi'_e(\omega))^{-1} - \frac{1}{3}}, \tag{5.95}$$

immediately gives

$$\chi_e(\omega) = \frac{n_b e^2/m}{\omega_1^2 - \omega^2 - i\omega\gamma}, \tag{5.96}$$

with the shifted resonant frequency, ω_1, being given by

$$\omega_1^2 = \omega_0^2 - \frac{4\pi}{3}\frac{n_b e^2}{m} = \omega_0^2 \frac{3}{\epsilon + 2} < \omega_0^2, \tag{5.97}$$

where ϵ is the static dielectric constant given by $1 + 4\pi\chi_e(0)$. [We see here the appearance of something with the same structure as the plasma frequency, (5.49), but with the density of bound charges replacing that of free charges.] Clearly the sum rule (5.35) remains unchanged since it is independent of ω_0.

We conclude with an experimental example of a situation in which the modification of the driving field is both significant and accurate. The static dielectric constant of nonpolar nitrogen gas has been measured at low and high densities. At the density 0.06604 g/cm^3, the observed value of $\epsilon - 1$ is 0.03109. If $\epsilon - 1$ were equal to $4\pi\chi'_e$, which is proportional to the density, the value predicted for a density of 0.5780 g/cm^3—slightly more than half the density of water—would be $\epsilon - 1 = 0.2721$. The measured value is 9% higher: 0.29633. If, however, one uses the Clausius-Mossotti relation (5.92), it is $(\epsilon - 1)/(\epsilon + 2)$ that is proportional to the density. Now the value of $\epsilon - 1$ predicted for the larger density is 0.2959, which falls short of the measurement by only 0.1%.

5.6 Problems for Chapter 5

1. Use (5.1b) to find the related equation for the conduction current $\mathbf{J} = n_f e\mathbf{v}$. Solve this equation for $\mathbf{E}(t) = \mathbf{E}_0\delta(t)$ if $\mathbf{J}(t < 0) = \mathbf{0}$. What is $\mathbf{J}$ immediately after $t = 0$? Connect this with (5.14).

2. Recall a charged particle of mass m and charge e subject to a harmonic restoring force and a damping force, under the influence of an external electric field is described by the equation of motion

$$\left(\frac{d^2}{dt^2} + \omega_0^2 + \gamma\frac{d}{dt}\right)\mathbf{r}(t) = \frac{e}{m}\mathbf{E}(t). \tag{5.98}$$

If we multiply this equation by ne, where n is the number density of such charged particles, we obtain a differential equation for the electric polarization $\mathbf{P}$. Find a solution of this equation in the form

$$\mathbf{P}(t) = \int_{-\infty}^{\infty} dt'\, \chi(t - t')\mathbf{E}(t'), \tag{5.99}$$

and write down the differential equation satisfied by $\chi(t-t')$. Solve this equation by requiring that $\chi(t - t') = 0$ if $t - t' < 0$ (causality). Show that for $t - t' > 0$ the solution has the form

$$\chi(t - t') = ae^{i\lambda_+(t-t')} + be^{i\lambda_-(t-t')}, \tag{5.100}$$

where a and b are constants. What are the (constant) complex numbers $\lambda_\pm$? Determine the constants a and b by requiring χ be continuous at $t = t'$ but has a discontinuity in its first derivative dictated by the differential equation. Thus, give an explicit form for χ. Show that the same result follows from the Drude model result (5.20) by Fourier transforming it:

$$\chi(t - t') = \int_{-\infty}^{\infty} \frac{d\omega}{2\pi}\chi(\omega)e^{-i\omega(t-t')}. \tag{5.101}$$

3. Evaluate the integral in (5.32) in the case that $\gamma \geq 2\omega_0$, and, thereby, establish the sum rule (5.35) in general.

4. Show that the charge that disappears from the interior of a conductor according to (5.42) appears on the surface of that conductor.

5. Evaluate Z in (5.62) without the weak field approximation. Then deduce $\langle\mathbf{d}\rangle_T$. What happens in the limit $|\mathbf{d}||\mathbf{E}| \gg kT$?

6. In the text, we discussed the effect of thermal fluctuations on the orientation of an electric dipole in an external electric field. We used a classical Boltzmann distribution for this purpose. However, it is more accurate to recognize that angular momentum is quantized, and consequently, that the possible values of the electric dipole moment along the direction of the electric field take on only a discrete set of values. Consider the simplest case, that of a spin-1/2 molecule, where the possible values of $\mathbf{d}$ along the z direction, the direction of the electric field $\mathbf{E}$, are $\pm d$. (a) Calculate $\langle d_z\rangle$ as a function of d, E, and T. (b) What does $\langle d_z\rangle$ become when dE/kT is small? (c) Calculate the electric susceptibility corresponding to a gas of such molecules, with n molecules per unit volume.

7. Prove that the solution of (5.82) is (5.84), directly by the application of Gauss' law.

8. Another way to evaluate (5.80) is to compute the integral

$$\int_V \frac{(d\mathbf{r})}{|\mathbf{r}-\mathbf{r}_a|} = 4\pi\left[\frac{1}{2}a^2 - \frac{1}{6}r_a^2\right], \tag{5.102}$$

where the volume integral is over a sphere of radius a which contains the point $\mathbf{r}_a$, and the origin is taken to be the center of the sphere. [Hint: Take $\mathbf{r}_a$ to lie in the z direction.] Upon taking the gradient in (5.80), the result in (5.85) follows.

9. Verify the numerical improvement stated in the paragraph at the end of Section 5.5.

10. On the basis of the formula derived for the dielectric constant, (5.31) and (5.22), estimate as accurately as you can, based on simple physical arguments, the value of ϵ for air and water, in the regime $\omega_{\text{mol}} \ll \omega \ll \omega_{\text{atom}}$. What about the regime probed by capacitive measurements, $\omega \ll \omega_{\text{mol}}$, where, for water, $\epsilon \approx 80$? How closely do your estimates agree with the observed values? Discuss possible sources of error in your estimate. How are your results changed if the Clausius-Mossotti equation (5.92) is used instead?

11. A way of determining the sign of the charge carriers in a conductor is by means of the Hall effect, discovered by Edwin Hall in 1879. A magnetic field $\mathbf{B}$ is applied perpendicular to the direction of current flow in a conductor, and as a consequence a transverse voltage drop appears across the conductor. If d is the transverse length of the conductor, and v is the average drift speed of the charge carriers, show that the voltage, in magnitude, is

$$V = \frac{v}{c}dB. \tag{5.103}$$

What is the sense of this potential drop?

12. Consider the Hall effect discussed in the previous problem. If the magnetic field (perpendicular to the current flow) is in the z direction, the displacement current is neglected, and the resistivity $1/\sigma$ is very small, show that there exist waves of the form $\exp[i(kz-\omega t)]$ in $\mathbf{E}$ and $\mathbf{J}$, with the dispersion relation

$$\omega = k^2\frac{Bc}{2\pi ne}. \tag{5.104}$$

These are helicon waves, which provide a means of measuring $1/ne$.

13. (a) Show that Ohm's law for a neutral conducting fluid moving with velocity v is

$$\mathbf{J} = \sigma\left(\mathbf{E} + \frac{\mathbf{v}}{c}\times\mathbf{B}\right), \tag{5.105}$$

and thereby derive from Maxwell's equations

$$\frac{\partial\mathbf{B}}{\partial t} = \boldsymbol{\nabla}\times(\mathbf{v}\times\mathbf{B}) + \frac{c^2}{4\pi\sigma}\nabla^2\mathbf{B}. \tag{5.106}$$

(Argue that the displacement current, ignored here, gives only v^2/c^2 corrections.)

(b) For a fluid at rest this means that **B** satisfies the diffusion equation

$$\frac{\partial \mathbf{B}}{\partial t} = \frac{c^2}{4\pi\sigma}\nabla^2\mathbf{B}. \tag{5.107}$$

If **B** varies over a characteristic distance L, what is the characteristic time τ for the decay of the field? Estimate τ for the earth's core, where $L \sim 10^6$ m, $\sigma \sim 10^{17}$ esu. (See the Appendix for a discussion of units.)

(c) For times short compared to the diffusion time τ, **B** satisfies

$$\frac{\partial \mathbf{B}}{\partial t} = \boldsymbol{\nabla}\times(\mathbf{v}\times\mathbf{B}). \tag{5.108}$$

Show that this means that the magnetic flux through any closed loop moving with the local fluid velocity is constant in time.

14. Show that a compressible, nonviscous, perfectly conducting fluid in a magnetic field is governed by

$$\frac{\partial \rho}{\partial t} + \boldsymbol{\nabla}\cdot(\rho\mathbf{v}) = 0 \tag{5.109a}$$

(conservation of fluid),

$$\rho\frac{\partial}{\partial t}\mathbf{v} + \rho(\mathbf{v}\cdot\boldsymbol{\nabla})\mathbf{v} = -\boldsymbol{\nabla}p - \frac{1}{4\pi}\mathbf{B}\times(\boldsymbol{\nabla}\times\mathbf{B}) \tag{5.109b}$$

(Newton's law, where p is the pressure), and

$$\frac{\partial \mathbf{B}}{\partial t} = \boldsymbol{\nabla}\times(\mathbf{v}\times\mathbf{B}) \tag{5.109c}$$

(conservation of magnetic flux).

Linearize these equations by substituting

$$\mathbf{B} = \mathbf{B}_0 + \mathbf{B}_1(\mathbf{r}, t), \tag{5.110a}$$

$$\rho = \rho_0 + \rho_1(\mathbf{r}, t), \tag{5.110b}$$

$$\mathbf{v} = \mathbf{v}_1(\mathbf{r}, t), \tag{5.110c}$$

where quantities with 1 subscripts are supposed to be small. Show that $\mathbf{v}_1$ satisfies

$$\frac{\partial^2}{\partial t^2}\mathbf{v}_1 - s^2\boldsymbol{\nabla}(\boldsymbol{\nabla}\cdot\mathbf{v}_1) + \mathbf{v}_A\times(\boldsymbol{\nabla}\times[\boldsymbol{\nabla}\times(\mathbf{v}_1\times\mathbf{v}_A)]) = 0, \tag{5.111}$$

where s is the sound speed,

$$s^2 = \left(\frac{\partial p}{\partial \rho}\right)_0, \tag{5.112a}$$

and $\mathbf{v}_A$ is the Alfvén velocity,[3]

$$\mathbf{v}_A = \frac{\mathbf{B}_0}{\sqrt{4\pi\rho}}. \tag{5.112b}$$

If $\mathbf{v}_1(,\mathbf{r}, t)$ represents a plane wave propagating parallel to $\mathbf{B}_0$,

$$\mathbf{v}_1(\mathbf{r}, t) = \mathbf{v}_1 e^{i\mathbf{k}\cdot\mathbf{r} - i\omega t}, \quad \mathbf{k} \parallel \mathbf{B}_0, \tag{5.113}$$

show that there are two modes:

[3] Hannes Alfvén, 1908–1995.

(a) A longitudinal wave, $\mathbf{v}_1 \parallel \mathbf{k}$, with phase velocity s. (This is an ordinary sound wave).

(b) A transverse wave, $\mathbf{v}_1 \perp \mathbf{k}$, with phase velocity v_A. (This is the Alfvén wave.)

What are the modes and wave velocities if $\mathbf{k} \perp \mathbf{B}_0$?

15. Show that for a collisionless plasma, for which the charge density satisfies

$$\left(\frac{\partial^2}{\partial t^2} + \omega_p^2\right) \rho = 0, \tag{5.114}$$

the electric field satisfies the same equation provided $\mathbf{B} = \mathbf{0}$. (Why is this consistent?) Derive the linearized equation then satisfied by the fluid velocity $\mathbf{v}(\mathbf{r}, t)$.

6

Dispersion Relations for the Susceptibility

We obtained a dispersion relation for the dielectric constant, or for the electric susceptibility, that is, an expression for ϵ or χ as a function of ω, in Chapter 5 [see (5.20), for example], by assuming a particular model for the motion of the charges in the medium. Here we will, starting from general physical principles, derive the form of the dispersion relation satisfied by $\epsilon(\omega)$ or $\chi(\omega)$, of which (5.20) is a particular example. For simplicity, we will assume the medium to be nonmagnetic, that is

$$\mu \approx 1 \quad \text{or} \quad \mathbf{B} \approx \mathbf{H}, \tag{6.1}$$

which is the usual situation. In a linear, isotropic, dielectric medium, the displacement, the polarization, and the electric field are related by (4.54) and (5.30),

$$\mathbf{D} = \mathbf{E} + 4\pi\mathbf{P} = \epsilon\mathbf{E}, \tag{6.2}$$

which implies that

$$\mathbf{P} = \frac{\epsilon - 1}{4\pi}\mathbf{E} = \chi\mathbf{E}, \tag{6.3}$$

where ϵ, or χ, is a complex function of ω.

Such a relation between **P** and **E** must be consistent with the principle of causality, which here implies that the effects produced by $\mathbf{E}(t)$ occur only at a later time. Bearing this in mind, we may write the most general linear, time-translationally invariant, causal relation between **P** and **E** as

$$\mathbf{P}(t) = \int_{-\infty}^{\infty} dt'\, \eta(t-t') f(t-t')\mathbf{E}(t'), \tag{6.4}$$

which is valid for most materials for sufficiently weak fields. The unit step function,[1] $\eta(t-t')$,

$$\eta(t-t') = \begin{cases} 1, \text{if } t > t', \\ 0, \text{if } t < t', \end{cases} \tag{6.5}$$

is introduced to satisfy the causality requirement; it has the integral representation

$$\eta(t-t') = \frac{i}{2\pi}\int_{-\infty}^{\infty} d\nu\, \frac{e^{-i\nu(t-t')}}{\nu + i\epsilon}, \qquad \epsilon \to +0. \tag{6.6}$$

The function $f(t)$ may be represented by a Fourier integral

$$f(t) = \int_{-\infty}^{\infty} \frac{d\omega'}{2\pi} e^{-i\omega' t} f(\omega'), \tag{6.7}$$

which, for $f(t)$ real, implies for the Fourier transform

$$f^*(\omega') = f(-\omega'). \tag{6.8}$$

[1] Usually denoted by θ, and attributed to Heaviside; but we prefer η whose capital form is written as H.

DOI: 10.1201/9781003057369-6

Substituting the representation (6.6) into (6.4), and using the Fourier transform (6.7), we identify

$$\mathbf{P}(\omega) = \left[i \int_{-\infty}^{\infty} \frac{d\omega'}{2\pi} \frac{1}{\omega - \omega' + i\epsilon} f(\omega') \right] \mathbf{E}(\omega). \tag{6.9}$$

Consequently, causality demands that the following mathematical representation for the susceptibility is always possible:

$$\chi(\omega) = \frac{\epsilon(\omega) - 1}{4\pi} = i \int_{-\infty}^{\infty} \frac{d\omega'}{2\pi} \frac{f(\omega')}{\omega + i\epsilon - \omega'}. \tag{6.10}$$

In order to simplify this relation, we note that $f(t - t')$, defined in (6.4), is arbitrary for $t - t' < 0$. A convenient assumption is that the function is odd,

$$f(t - t') = -f(t' - t). \tag{6.11}$$

This, together with (6.8), means that we can write

$$f(\omega') = \frac{1}{2} i\omega' \Phi(\omega'), \tag{6.12}$$

where $\Phi(\omega')$ is a real even function. Consequently, an alternative expression for (6.10) is

$$\chi(\omega) = \int_0^{\infty} \frac{d\omega'}{2\pi} \frac{\omega'^2 \Phi(\omega')}{\omega'^2 - (\omega + i\epsilon)^2}. \tag{6.13}$$

There is a physical requirement which will impose a further condition on $\epsilon(\omega)$. As a charged particle moves through a medium, it loses energy to the atoms, not the other way around. The origin of this energy loss is that the electric field, $\mathbf{E}$, produced by the particle, does work on the induced currents, $\frac{\partial}{\partial t}\mathbf{P}$, the total work per unit volume being

$$\int dt \left(\frac{\partial}{\partial t} \mathbf{P} \right) \cdot \mathbf{E} > 0. \tag{6.14}$$

If the Fourier representation for the electric field is introduced, (6.14) becomes

$$\begin{aligned} \int dt \frac{\partial}{\partial t} \mathbf{P}(t) \cdot \int_{-\infty}^{\infty} \frac{d\omega}{2\pi} e^{-i\omega t} \mathbf{E}(\omega) &= \int_{-\infty}^{\infty} \frac{d\omega}{2\pi} i\omega \mathbf{E}(\omega) \cdot \mathbf{P}(-\omega) \\ &= \int_{-\infty}^{\infty} \frac{d\omega}{2\pi} i\omega \chi(-\omega) |\mathbf{E}(\omega)|^2 > 0. \end{aligned} \tag{6.15}$$

On the other hand, since $|\mathbf{E}(\omega)|^2$ is even in ω, (6.15) can also be written

$$\int_{-\infty}^{\infty} \frac{d\omega}{2\pi} (-i\omega) \chi(\omega) |\mathbf{E}(\omega)|^2 > 0. \tag{6.16}$$

The sum of these two equations yields

$$\int_{-\infty}^{\infty} \frac{d\omega}{2\pi} \omega \, \operatorname{Im} \epsilon(\omega) |\mathbf{E}(\omega)|^2 = \int_0^{\infty} \frac{d\omega}{\pi} \omega \, \operatorname{Im} \epsilon(\omega) |\mathbf{E}(\omega)|^2 > 0, \tag{6.17}$$

where we have used the analog of (6.8) for $\epsilon(\omega)$. Therefore, since this holds for any electric field, $\mathbf{E}(\omega)$, we obtain the positivity condition

$$\operatorname{Im} \epsilon(\omega) \geq 0 \quad \text{for} \quad \omega \geq 0. \tag{6.18}$$

When this condition is applied to (6.13), we find, for $\omega \geq 0$,

$$\operatorname{Im}\chi(\omega) = \operatorname{Im}\left(\frac{\epsilon(\omega)-1}{4\pi}\right) = \operatorname{Im}\int_0^\infty \frac{d\omega'}{2\pi}\frac{\omega'^2\Phi(\omega')}{\omega'^2-(\omega+i\epsilon)^2}$$
$$= \int_0^\infty \frac{d\omega'}{2\pi}\omega'^2\Phi(\omega')\pi\delta(\omega'^2-\omega^2) = \frac{\omega}{4}\Phi(\omega) \geq 0. \qquad (6.19)$$

In deriving (6.19), we have used the identity

$$\operatorname{Im}\frac{1}{x-i\epsilon} = \pi\delta(x). \qquad (6.20)$$

An additional property of the dielectric constant may be inferred by considering the high-frequency limit, where ω is much larger than a characteristic excitation frequency of the atoms making up the medium, ω_{atomic}. In this situation, the effect of binding on the electrons can be neglected, leaving us with the response of free electrons to the applied electric field. This is given by the high frequency limit of (5.20),

$$\chi(\omega) \approx -\frac{ne^2}{m\omega^2}, \quad \omega \gg \omega_{\text{atomic}}. \qquad (6.21)$$

(Of course, we assume that $\hbar\omega \ll mc^2$ so that the electron remains nonrelativistic.) On the other hand, the dispersion relation (6.13) becomes, in this limit,

$$\chi(\omega) = -\frac{1}{\omega^2}\int_0^\infty \frac{d\omega'}{2\pi}\omega'^2\Phi(\omega'), \quad \omega \gg \omega_{\text{atomic}}, \qquad (6.22)$$

if the integral exists. In this way, we arrive at a sum rule for Φ:

$$\int_0^\infty \frac{d\omega'}{2\pi}\omega'^2\Phi(\omega') = \frac{ne^2}{m}. \qquad (6.23)$$

Consequently, it is convenient to define $p(\omega')$ by

$$\frac{\omega'^2}{2\pi}\Phi(\omega') = \frac{ne^2}{m}p(\omega'), \qquad (6.24)$$

which has the following properties,

$$p(\omega') \geq 0, \quad \omega' > 0, \qquad (6.25)$$

and

$$\int_0^\infty p(\omega')d\omega' = 1. \qquad (6.26)$$

The dispersion relation, (6.13), now appears as

$$\chi(\omega) = \frac{ne^2}{m}\int_0^\infty d\omega'\frac{p(\omega')}{\omega'^2-(\omega+i\epsilon)^2}. \qquad (6.27)$$

An alternative, and perhaps more familiar, way of writing this relation, according to (6.19), is as

$$\chi(\omega) = \frac{2}{\pi}\int_0^\infty d\nu\frac{\nu\operatorname{Im}\chi(\nu)}{\nu^2-(\omega+i\epsilon)^2}, \quad \operatorname{Im}\chi(\nu) \geq 0. \qquad (6.28)$$

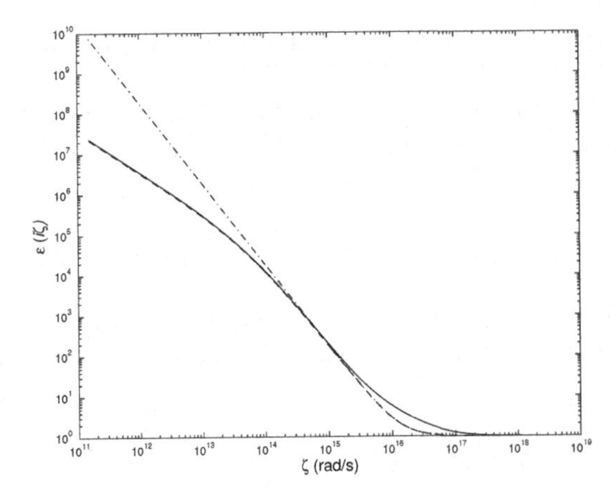

FIGURE 6.1
Solid line: Permittivity $\epsilon(i\zeta)$ as function of imaginary frequency ζ for gold. The curve is calculated on the basis of experimental data. Courtesy of Astrid Lambrecht and Serge Reynaud [19]. Dashed lines: $\epsilon(i\zeta)$ versus ζ with T as parameter, based upon the temperature-dependent Drude model; *cf.* Appendix D of Ref. [18]. This graph first appeared in that reference. The upper curve is for $T = 10$ K; the lower is for $T = 300$ K, which for energies below 1 eV (1.5×10^{15} rad/s) nicely fits the experimental data. Both curves are below the experimental one for $\zeta > 2 \times 10^{15}$ rad/s.

This determines the real part of the susceptibility in terms of the imaginary part; the susceptibility is necessarily complex. This relation is attributed to Hendrik Kramers (1927) and Ralph Kronig (1926). Because $\chi(\omega)$ is regular in the upper half-plane, this dispersion relation gives the following expression for purely imaginary frequencies:

$$\chi(i\zeta) = \frac{2}{\pi}\int_0^\infty d\nu \frac{\nu \operatorname{Im}\chi(\nu)}{\nu^2 + \zeta^2}, \tag{6.29}$$

which is useful for many practical application, for it avoids the oscillations of resonances. A graph of this form for real data for gold, compared to the Drude model, (6.30), (with $\omega_0 = 0$, $\omega_p = 9$ eV, and the room-temperature value of $\gamma = 0.035$ eV) is shown in Fig. 6.1. (In the comparison, the temperature-dependence of the damping is modeled by a Debye model, as explained in Ref. [18].)

As a simple example, recall the situation when the system has only one excitation frequency, ω_0, with a damping constant γ. This is the Fermi model (or Drude-Lorentz model), which we considered in Chapter 5. In particular, we found there the dielectric constant to be [see (5.20)]

$$\epsilon(\omega) = 1 + \frac{\omega_p^2}{\omega_0^2 - \omega^2 - i\gamma\omega}, \tag{6.30}$$

where the plasma frequency of the medium is defined by

$$\omega_p^2 = \frac{4\pi n e^2}{m}. \tag{6.31}$$

The spectral weight function $p(\omega')$ in (6.27) can be obtained from the imaginary part of ϵ:

$$p(\omega') = \frac{2}{\pi}\frac{\omega'^2\gamma}{(\omega'^2 - \omega_0^2)^2 + (\omega'\gamma)^2}. \tag{6.32}$$

For small γ ($\gamma \ll \omega_0$), this exhibits the structure of the Lorentzian line shape. (See Section 38.3.)

It will be useful in the following to have a dispersion relation for the inverse of $\epsilon(\omega)$ as well. Its form is suggested by the Fermi model. The reciprocal of (6.30) is

$$\frac{1}{\epsilon(\omega)} = \frac{\omega_0^2 - i\omega\gamma - \omega^2}{(\omega_0^2 + \omega_p^2) - i\omega\gamma - \omega^2} = 1 - \omega_p^2 \frac{1}{(\omega_0^2 + \omega_p^2) - i\omega\gamma - \omega^2}, \tag{6.33}$$

which differs from (6.30) by the sign of the second term, and the replacement of ω_0^2 by $\omega_0^2 + \omega_p^2$. Therefore, we anticipate that $1/\epsilon(\omega)$ satisfies a dispersion relation of the form

$$\frac{1}{\epsilon(\omega)} = 1 - \omega_p^2 \int_0^\infty d\omega' \frac{q(\omega')}{\omega'^2 - (\omega + i\epsilon)^2}, \tag{6.34}$$

where the spectral function $q(\omega')$ satisfies the conditions

$$q(\omega') > 0, \tag{6.35a}$$

$$\int_0^\infty q(\omega')\, d\omega' = 1. \tag{6.35b}$$

To show that (6.34) is the correct dispersion relation for $1/\epsilon(\omega)$, we may proceed in the same manner as we did in deriving the dispersion relation for $\epsilon(\omega)$. We apply the causality condition to write a general relationship between **P** and **D**, leading to a dispersion relation for $1/\epsilon(\omega)$. The positivity of the spectral function, (6.35a), follows from the physical requirement (6.14), expressed in terms of $|\mathbf{D}|^2$, while the sum rule (6.35b), results from the expression for $1/\epsilon(\omega)$ at high frequencies, $\omega \gg \omega_{\text{atomic}}, \omega_p$, which follows from (6.21)

$$\frac{1}{\epsilon(\omega)} \approx 1 + \frac{\omega_p^2}{\omega^2}. \tag{6.36}$$

6.1 Problems for Chapter 6

1. Derive the form of the dispersion relation (6.34) by following the steps outlined above.
2. Show that the sum rule (5.35) follows from the dispersion relation (6.27) and the sum rule (6.26).

7

Magnetic Properties of Matter

7.1 Canonical Equations of Motion in Electromagnetic Fields

Since magnetic effects are more subtle than electric ones, it is helpful to first develop the formalism describing a charge moving in the presence of electric and magnetic fields. We start with the equation of motion for such a charge,

$$m\frac{d}{dt}\mathbf{v} = e\left(\mathbf{E} + \frac{\mathbf{v}}{c}\times\mathbf{B}\right). \tag{7.1}$$

Although, unlike an electric field, a magnetic field does no work on a point charge, there is a magnetic term in the energy, because the act of turning on a magnetic field produces an electric field, according to Faraday's law, (1.60). To see this, it is convenient to recast (7.1) so that potentials appear instead of field strengths. Since the magnetic field has zero divergence,

$$\boldsymbol{\nabla}\cdot\mathbf{B} = 0, \tag{7.2}$$

we may introduce a vector potential, $\mathbf{A}$, such that

$$\mathbf{B} = \boldsymbol{\nabla}\times\mathbf{A}. \tag{7.3}$$

When we substitute (7.3) into Faraday's law

$$-\frac{1}{c}\frac{\partial}{\partial t}\mathbf{B} = \boldsymbol{\nabla}\times\mathbf{E}, \tag{7.4}$$

we observe that the quantity $\mathbf{E} + \frac{1}{c}\frac{\partial}{\partial t}\mathbf{A}$ is irrotational,

$$\boldsymbol{\nabla}\times\left(\mathbf{E} + \frac{1}{c}\frac{\partial}{\partial t}\mathbf{A}\right) = \mathbf{0}, \tag{7.5}$$

which means that a corresponding scalar potential ϕ exists,

$$\mathbf{E} + \frac{1}{c}\frac{\partial}{\partial t}\mathbf{A} = -\boldsymbol{\nabla}\phi. \tag{7.6}$$

This is the generalization of (1.12) to incorporate time dependence. In terms of the potentials $\mathbf{A}$ and ϕ, the equation of motion (7.1) reads

$$m\frac{d}{dt}\mathbf{v} = -e\boldsymbol{\nabla}\phi - \frac{e}{c}\frac{\partial}{\partial t}\mathbf{A} + \frac{e}{c}\mathbf{v}\times(\boldsymbol{\nabla}\times\mathbf{A}), \tag{7.7}$$

which can be rewritten as

$$m\frac{d}{dt}\mathbf{v} = -\frac{e}{c}\frac{d}{dt}\mathbf{A} - e\boldsymbol{\nabla}\left(\phi - \frac{1}{c}\mathbf{v}\cdot\mathbf{A}\right), \tag{7.8}$$

DOI: 10.1201/9781003057369-7

since (∇ operates only on $\mathbf{A}$),

$$\mathbf{v}\times(\nabla\times\mathbf{A}) = \nabla(\mathbf{v}\cdot\mathbf{A}) - (\mathbf{v}\cdot\nabla)\mathbf{A}, \tag{7.9}$$

and

$$\frac{\partial}{\partial t}\mathbf{A}(\mathbf{r},t) + (\mathbf{v}\cdot\nabla)\mathbf{A}(\mathbf{r},t) = \frac{d}{dt}\mathbf{A}(\mathbf{r}(t),t). \tag{7.10}$$

This suggests that we define the canonical momentum of the particle by

$$\mathbf{p} = m\mathbf{v} + \frac{e}{c}\mathbf{A}, \tag{7.11}$$

which supplements the kinetic momentum $m\mathbf{v}$ by the potential momentum $\frac{e}{c}\mathbf{A}$. In terms of $\mathbf{p}$ the equation of motion reads

$$\frac{d}{dt}\mathbf{p} = -\nabla\left(e\phi - \frac{e}{c}\mathbf{v}\cdot\mathbf{A}\right). \tag{7.12}$$

Note that if the potentials are spatially constant, $\mathbf{p}$ is conserved. (The astute reader will note that the potentials ϕ and $\mathbf{A}$ are not uniquely defined. We will discuss the freedom to redefine them by so-called gauge transformations later, see Section 10.5.)

Next, let us derive the corresponding equation for the energy of the system. Taking the scalar product of $\mathbf{v}$ with (7.1), we obtain

$$\frac{d}{dt}\left(\frac{1}{2}mv^2\right) = e\mathbf{E}\cdot\mathbf{v} = -e\frac{d}{dt}\phi + e\frac{\partial}{\partial t}\left(\phi - \frac{\mathbf{v}}{c}\cdot\mathbf{A}\right), \tag{7.13}$$

where we have used (7.6), as well as (7.10) for ϕ instead of $\mathbf{A}$. [In (7.13), $\mathbf{v}$ is an implicit function of t, so $\partial/\partial t$ does not act on it; that is, the trajectory of the particle, $\mathbf{r}(t)$ is not affected by the $\partial/\partial t$ operation.] Rewriting this equation as

$$\frac{d}{dt}\left(\frac{1}{2}mv^2 + e\phi\right) = \frac{\partial}{\partial t}\left(e\phi - \frac{e}{c}\mathbf{v}\cdot\mathbf{A}\right), \tag{7.14}$$

we see that the energy, E,

$$E = \frac{1}{2}mv^2 + e\phi, \tag{7.15}$$

is conserved if there is no explicit time dependence in the potentials. Since we have shifted our emphasis from $\mathbf{v}$ to $\mathbf{p}$, it is convenient to rewrite E in terms of this new variable,

$$E = \frac{1}{2m}\left(\mathbf{p} - \frac{e}{c}\mathbf{A}\right)^2 + e\phi. \tag{7.16}$$

We now summarize the system of equations we have derived,

$$\mathbf{v} \equiv \frac{d}{dt}\mathbf{r} = \frac{1}{m}\left(\mathbf{p} - \frac{e}{c}\mathbf{A}\right), \tag{7.17a}$$

$$\frac{d}{dt}\mathbf{p} = -\nabla\left(e\phi - \frac{e}{c}\mathbf{v}\cdot\mathbf{A}\right), \tag{7.17b}$$

$$\frac{d}{dt}E = \frac{\partial}{\partial t}\left(e\phi - \frac{e}{c}\mathbf{v}\cdot\mathbf{A}\right). \tag{7.17c}$$

Observe that this set has the form of the equations of motion in the Hamiltonian[1] formulation of mechanics. If the Hamiltonian function is $H(q,p,t)$, the Hamiltonian equations of motion are

$$\frac{d}{dt}q = \frac{\partial}{\partial p}H, \qquad \frac{d}{dt}p = -\frac{\partial}{\partial q}H, \qquad \frac{d}{dt}H = \frac{\partial}{\partial t}H, \tag{7.18}$$

[1]William Rowan Hamilton, 1805–1865.

where, in the case we are considering here, the variables are identified as

$$q \to \mathbf{r}, \qquad p \to \mathbf{p}, \tag{7.19}$$

while the Hamiltonian is identified with (7.16).

7.2 Diamagnetism

We now apply this formalism to a simple mutually interacting system (an atom) immersed in a homogeneous magnetic field, $\mathbf{B}$, for which the vector potential can be taken to be

$$\mathbf{A} = \frac{1}{2}\mathbf{B}\times\mathbf{r}. \tag{7.20}$$

The energy of this system, as a simple generalization of (7.16), is

$$E = \sum_a \left[\frac{1}{2m_a}\left(\mathbf{p}_a - \frac{e_a}{2c}\mathbf{B}\times\mathbf{r}_a\right)^2\right] + U, \tag{7.21}$$

where the last term, U, represents the potential energy of atomic forces that keep the atom together. The energy can be rewritten as

$$E = \left(\sum_a \frac{p_a^2}{2m_a} + U\right) - \mathbf{B}\cdot\sum_a \frac{e_a}{2m_a c}\mathbf{r}_a\times\mathbf{p}_a + \sum_a \frac{e_a^2}{8m_a c^2}(\mathbf{B}\times\mathbf{r}_a)^2, \tag{7.22}$$

where the first term is the ordinary expression for the energy of the atom. The second term in (7.22) involves the intrinsic magnetic moment of the atom, $\boldsymbol{\mu}_0$, defined by

$$\boldsymbol{\mu}_0 = \sum_a \frac{e_a}{2m_a c}\mathbf{r}_a\times\mathbf{p}_a, \tag{7.23}$$

when $\mathbf{r}_a$ is the position of the ath particle relative to the center of mass of the atom. This is very similar to the definition of the magnetic moment given in (4.19),

$$\boldsymbol{\mu} = \sum_a \frac{e_a}{2c}\mathbf{r}_a\times\mathbf{v}_a, \tag{7.24}$$

but is not the same since $\mathbf{p} \neq m\mathbf{v}$. To see that $\boldsymbol{\mu}$ is the actual magnetic moment, which is required to satisfy [see (4.24)]

$$-\frac{\partial E}{\partial \mathbf{B}} = \boldsymbol{\mu}, \tag{7.25}$$

we note that the change in the energy, $\delta_{\mathbf{B}}E$, caused by a change in the magnetic field, $\delta\mathbf{B}$, is

$$\begin{aligned}
\delta_{\mathbf{B}}E &= \sum_a \frac{\mathbf{p}_a - \frac{e_a}{2c}\mathbf{B}\times\mathbf{r}_a}{m_a}\cdot\frac{e_a}{2c}\mathbf{r}_a\times\delta\mathbf{B} \\
&= -\left(\sum_a \frac{e_a}{2c}\mathbf{r}_a\times\mathbf{v}_a\right)\cdot\delta\mathbf{B} = -\boldsymbol{\mu}\cdot\delta\mathbf{B}.
\end{aligned} \tag{7.26}$$

The actual field-dependent magnetic moment can then be read off from (7.26) to be given by (7.24) or explicitly by

$$\boldsymbol{\mu} = \boldsymbol{\mu}_0 - \sum_a \frac{e_a^2}{4m_a c^2} \mathbf{r}_a \times (\mathbf{B} \times \mathbf{r}_a), \tag{7.27}$$

which can be derived from (7.24) by use of (7.11), the second term being an induced effect. [Note that this argument provides a demonstration that the intrinsic magnetic moment $\boldsymbol{\mu}_0$ is independent of the magnetic field $\mathbf{B}$.]

We first consider an atom which has zero intrinsic magnetic moment,

$$\boldsymbol{\mu}_0 = \mathbf{0}. \tag{7.28}$$

This will be true for a spherically symmetric system or, since for a single particle

$$\boldsymbol{\mu}_0 \sim \mathbf{r} \times \mathbf{p} = \boldsymbol{\ell} = \text{angular momentum}, \tag{7.29}$$

for a system with zero total angular momentum as long as all the particles involved have the same e/m ratio (which is very nearly true for an atom). For such a spherically symmetric system, which has no preferred direction, we see that

$$\langle \mathbf{r}\mathbf{r} \rangle = \frac{1}{3} r^2 \mathbf{1}, \tag{7.30}$$

and hence

$$\langle \mathbf{r} \times (\mathbf{B} \times \mathbf{r}) \rangle = \langle [r^2 \mathbf{B} - (\mathbf{r} \cdot \mathbf{B})\mathbf{r}] \rangle = \frac{2}{3} r^2 \mathbf{B}, \tag{7.31}$$

so that the time average of $\boldsymbol{\mu}$ is

$$\boldsymbol{\mu} = -\sum_a \frac{e_a^2}{6m_a c^2} r_a^2 \mathbf{B}. \tag{7.32}$$

Note that the induced magnetic moment is directed *oppositely* to the magnetic field. The proportionality constant between $\boldsymbol{\mu}$ and $\mathbf{B}$ is called the (static) magnetic polarizability, β,

$$\boldsymbol{\mu} = \beta \mathbf{B}, \quad \beta = -\sum_a \frac{e_a^2}{6m_a c^2} r_a^2. \tag{7.33}$$

If n is the density of such atoms, the magnetization (4.34b) is

$$\mathbf{M} = n\beta \mathbf{B} = \frac{\mathbf{B} - \mathbf{H}}{4\pi} \equiv \chi_m \mathbf{H}, \tag{7.34}$$

where the magnetic susceptibility,

$$\chi_m = n\beta \left[1 - 4\pi n\beta\right]^{-1}, \tag{7.35}$$

is negative. If we define the permeability, μ, as

$$\mathbf{B} = \mu \mathbf{H}, \tag{7.36}$$

we have

$$\mu = 1 + 4\pi \chi_m = \frac{1}{1 - 4\pi n\beta} \approx 1 + 4\pi n\beta < 1, \tag{7.37}$$

where the last step is justified since $-\chi_m \ll 1$. The relative sizes of χ_m and χ_e, from (5.22), are estimated by the ratio

$$-\frac{\chi_m}{\chi_e} \sim \frac{1}{6}\left(\frac{\omega_0 r}{c}\right)^2 \sim \frac{1}{6}\left(\frac{v}{c}\right)^2 \sim 10^{-5}\text{–}10^{-7}, \tag{7.38}$$

if we use typical atomic dimensions r and speeds v. The effect we have discussed here is called diamagnetism, which is universally exhibited by all matter.

That the induced magnetization in atoms is opposite in sign to the inducing field is a microscopic example of what is called Lenz's law [Heinrich Lenz (1804-1865)]. Bodies for which this effect is the dominant one are called diamagnetic. They are repelled by regions of strong magnetic field, where the energy is increased. (See Chapter 28.) The characterization of diamagnetism, $\mu < 1$, is seen to be analogous to the usual situation for dielectrics, $\epsilon > 1$, since the parallel relationship between magnetic and electric quantities is $\mathbf{H} = \frac{1}{\mu}\mathbf{B}$ and $\mathbf{D} = \epsilon\mathbf{E}$.

We close this section by remarking that the simple formula (7.37) suggests that the ratio of susceptibility to density for a given substance is independent of temperature. This is almost universally valid experimentally. In the example of water, where the mass density is practically constant, the measured susceptibility is close to -0.7×10^{-6}. The major exception is the metal—with the largest diamagnetic susceptibility—bismuth. Its susceptibility at room temperature is about -1.3×10^{-6}, but decreases significantly with rising temperature. Here the quantum insights of the modern theory of metals are indispensable.

7.3 Paramagnetism

What happens if there is a permanent intrinsic moment,

$$\boldsymbol{\mu}_0 \neq \mathbf{0}? \tag{7.39}$$

This situation is analogous to that of permanent electric dipole moments, discussed in Section 5.4. Due to thermal motion, the average magnetization will be zero if there is no external field. In the presence of a magnetic field, we obtain the thermal averaged magnetic moment from (5.63) by replacing $\mathbf{d}$ by $\boldsymbol{\mu}_0$ and $\mathbf{E}$ by $\mathbf{B}$. The obvious high temperature limit, from (5.67), is

$$\langle\boldsymbol{\mu}_0\rangle_T = \frac{\mu_0^2}{3kT}\mathbf{B}, \tag{7.40}$$

corresponding to the magnetization

$$\mathbf{M} = \frac{n\mu_0^2}{3kT}\mathbf{B}. \tag{7.41}$$

This is appropriate to the weak field circumstance

$$\mu_0 B \ll kT. \tag{7.42}$$

Inasmuch as the typical magnitudes of μ_0/d are of order $v/c \sim 10^{-2}$–10^{-3}, the upper limit to B, at room temperature, for (7.41) to be valid is in the range of millions of gauss, or hundreds of Teslas. Note that unlike in diamagnetism, the magnetization here is *parallel* to the magnetic field. The permeability is [*cf.* (7.37)]

$$\mu = \frac{1}{1 - 4\pi n\frac{\mu_0^2}{3kT}} \approx 1 + 4\pi n\frac{\mu_0^2}{3kT} > 1, \quad \chi_m = n\frac{\mu_0^2}{3kT}, \tag{7.43}$$

since, again, the magnetization is small. Substances with positive magnetic susceptibilities are called *paramagnetic*. For this class of materials, the permeability is greater than one. The simple models indicate that the ratio of paramagnetic to diamagnetic susceptibilities is of the order

$$\frac{\chi_{m,\text{para}}}{\chi_{m,\text{dia}}} \sim \frac{mv^2}{kT} \sim 100 \quad \text{at room temperature,} \tag{7.44}$$

where mv^2 is related to the magnitude of energies in the atom. The estimate in (7.44) is in general agreement with the observation that paramagnetic gaseous oxygen at standard pressure and room temperature has a positive susceptibility about one fifth the susceptibility of water, although the molecular density of the oxygen is less than a thousandth of that of water. The susceptibilities of paramagnetic substances are still so small compared with unity (for liquid oxygen, $\chi_m = 3 \times 10^{-4}$) that the approximation of neglecting the distinction between $\mathbf{B}$ and $\mathbf{H}$ in (7.43) is well justified. The inverse dependence on temperature displayed there was discovered experimentally by Pierre Curie (1859–1906).

Again, we have persisted in an error. In the electric case (see Section 5.5), the correct driving electric field, $\mathbf{E}_{\text{driving}}$ was obtained by removing the field of the atom itself. Exactly the same arguments apply here. The effect is one of no practical importance but is conceptually significant. Analogously to the previous arguments, we define the driving magnetic field by

$$\mathbf{B}_{\text{driving}} = \mathbf{B} - \overline{\mathbf{b}_{\text{atom}}}. \tag{7.45}$$

Recalling from Section 1.2 that the magnetic field, which is present in a coordinate system moving relative to a static electric field, is

$$\mathbf{B} = \frac{\mathbf{v}}{c} \times \mathbf{E}, \tag{7.46}$$

we can approximate the atomic magnetic field by

$$\mathbf{b}_{\text{atom}} = \sum_a \frac{\mathbf{v}_a}{c} \times \mathbf{e}_a, \tag{7.47}$$

where now $\mathbf{e}_a$ is an individual term in the second sum in (5.79). The average value of this field is, according to (5.84) [$\overline{\mathbf{e}_a} = -\mathbf{e}_a(\mathbf{r}_a)$ there],

$$\begin{aligned} \overline{\mathbf{b}_{\text{atom}}} &= \sum_a \frac{\mathbf{v}_a}{c} \times \overline{\mathbf{e}_a} = \sum_a \frac{\mathbf{v}_a}{c} \times \frac{-\frac{4\pi}{3} e_a \mathbf{r}_a}{V} \\ &= \frac{4\pi}{3} \frac{1}{V} \sum_a \frac{e_a}{c} \mathbf{r}_a \times \mathbf{v}_a = \frac{8\pi}{3} n \boldsymbol{\mu} = \frac{8\pi}{3} \mathbf{M}, \end{aligned} \tag{7.48}$$

where we recall (7.24). Hence, the driving magnetic field is

$$\mathbf{B}_{\text{driving}} = \mathbf{B} - \frac{8\pi}{3} \mathbf{M} = \mathbf{H} + \frac{4\pi}{3} \mathbf{M}, \tag{7.49}$$

which is analogous to the electric case, (5.86). Since the magnetic susceptibility is defined by

$$\mathbf{M} = \chi_m \mathbf{H}, \tag{7.50}$$

the correct driving field is negligibly different from $\mathbf{B}$,

$$\mathbf{B}_{\text{driving}} = \left(1 + \frac{4\pi}{3} \chi_m\right) \mathbf{H} \approx \mathbf{B}, \tag{7.51}$$

because $\chi_m \ll 1$.

7.4 Ferromagnetism

The history of magnetism did not begin with the phenomena of paramagnetism and diamagnetism, which were first recognized by Faraday in 1845. The ancients were familiar with the remarkable properties of Magnesian stone, the iron oxide Fe_3O_4. The term ferromagnetism refers to the property of such substances, primarily members of the iron group, of exhibiting permanent magnetization. A simple model of this effect was introduced by Pierre-Ernest Weiss (1865–1940), who effectively postulated that the driving magnetic field within ferromagnets is not (7.49), but rather

$$\mathbf{B}_{\text{driving}} = \mathbf{H} + \lambda \mathbf{M}, \tag{7.52}$$

where $\lambda \gg 1$. In terms of $\mathbf{B}_{\text{driving}}$ we wish to calculate the thermal average of the intrinsic magnetic moment, $\langle \boldsymbol{\mu}_0 \rangle_T$. Rather than use a classical distribution (but see Problem 7.3), it is simpler and more accurate quantum mechanically to suppose that the atomic magnetic moment $\boldsymbol{\mu}_0$ is either lined up parallel or anti-parallel to $\mathbf{B}_{\text{driving}}$, which defines the z axis. Since the interaction energies, for the two possibilities, are

$$-\boldsymbol{\mu}_0 \cdot \mathbf{B}_{\text{driving}} = \mp \mu_0 B_{\text{driving}}, \tag{7.53}$$

the Boltzmann weighting of states yields

$$\langle \mu_{0z} \rangle_T = \frac{\mu_0 e^x - \mu_0 e^{-x}}{e^x + e^{-x}} = \mu_0 \tanh x, \tag{7.54}$$

with

$$x = \frac{\mu_0}{kT}(H + \lambda M). \tag{7.55}$$

The resulting magnetization has magnitude

$$M = n\mu_0 \tanh \frac{\mu_0}{kT}(H + \lambda M). \tag{7.56}$$

The possible existence of a magnetization in the absence of the field H is implied by the equation

$$\frac{M}{n\mu_0} = \tanh\left(\frac{T_c}{T}\frac{M}{n\mu_0}\right), \tag{7.57}$$

in which

$$T_c \equiv \frac{n\mu_0^2}{k}\lambda. \tag{7.58}$$

In Fig. 7.1 there is plotted the left side of this equation, and also the right side with examples of the two situations $T > T_c$ and $T < T_c$. For $T > T_c$ the curves corresponding to the two sides of (7.57) intersect only at $M = 0$; there is no magnetization. But, for $T < T_c$, there is also an intersection at a positive value of $M/n\mu_0$ that is less than unity; a permanent magnetization can exist. The critical temperature T_c above which no permanent magnetization is possible is called the Curie temperature.[2] Its value is of the order $T_c \sim 1000$ K. If $T \ll T_c$ all the magnetic moments are lined up,

$$M = n\mu_0, \tag{7.59}$$

which corresponds to a permanent magnet.

[2] After Pierre Curie.

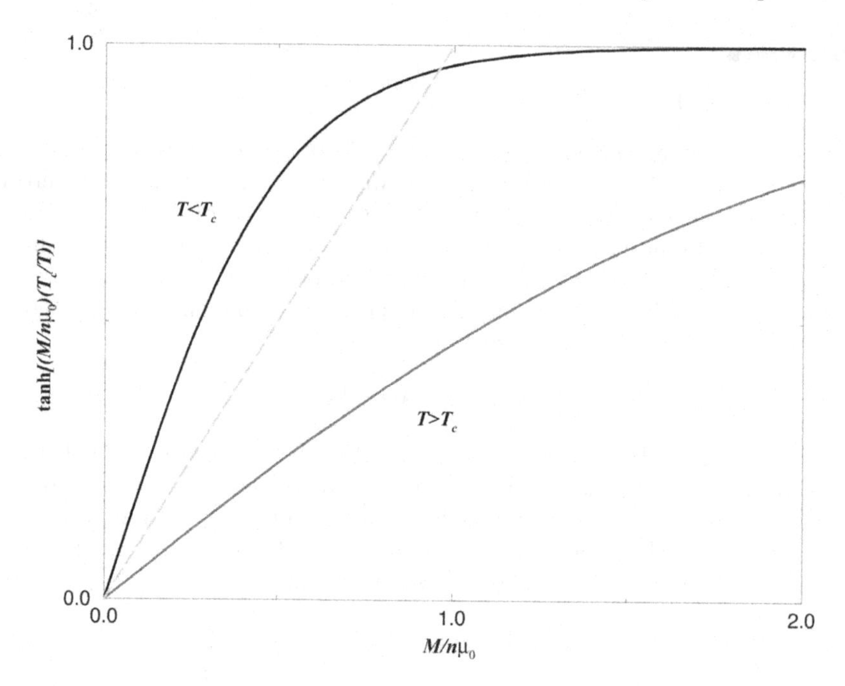

FIGURE 7.1
Solutions to (7.57) for temperatures above and below the Curie temperature. The solid curves are the plots of the right side of (7.57) for $T < T_c$ and $T > T_c$, respectively, while the dashed line is the left side of that equation.

Above the Curie temperature, an external field is required to produce magnetization. Suppose we are sufficiently above the Curie temperature so we are allowed to take x in (7.55) as a small quantity. Then (7.56) reduces to

$$\frac{M}{n\mu_0} = \frac{\mu_0 H}{kT} + \frac{T_c}{T}\frac{M}{n\mu_0}, \tag{7.60}$$

implying the magnetic susceptibility

$$\chi_m = \frac{M}{H} = \frac{n\mu_0^2/k}{T - T_c} = \frac{1}{\lambda}\frac{T_c}{T - T_c}, \tag{7.61}$$

which is valid providing

$$\frac{T - T_c}{T_c} \gg \frac{\mu_0 H}{kT_c}. \tag{7.62}$$

The result (7.61), which is called the Curie-Weiss law, exhibits the characteristic paramagnetic behavior seen in (7.43).

Finally, we estimate the phenomenological parameter λ, by inserting $\mu_0 \sim er(v/c)$ into (7.58):

$$kT_c \sim \lambda \left(\frac{v}{c}\right)^2 \frac{e^2}{r}(nr^3). \tag{7.63}$$

In a solid at ordinary densities, the product of n ($n \sim 10^{23}$ cm^{-3}) with r^3 (10^{-8} cm)^3, the fraction of the solid occupied by the atoms, is of the order of 1/10. And a tenth of the atomic energy magnitude e^2/r is not far from the value kT_c. We conclude that

$$\lambda \sim \frac{c^2}{v^2}, \tag{7.64}$$

which is the inverse of the typical factor that relates magnetic energy to electric energy. The clear suggestion is that the underlying mechanism of ferromagnetism is not magnetic, but electrical, in origin. The quantum theory of ferromagnetism initiated by Werner Heisenberg (1901–1976) vindicates this conclusion.

7.5 Problems for Chapter 7

1. (a) Verify that a uniform electric field is described by

$$\phi = -\mathbf{E}\cdot\mathbf{r}, \qquad \mathbf{A} = \mathbf{0} \qquad (\mathbf{E} = \text{constant}). \tag{7.65}$$

   (b) Show that for such a uniform field acting on a system of charged particles $\{e_a\}$, the electric dipole moment is given by the analog of (7.25),

$$\frac{\partial E}{\partial \mathbf{E}} = -\mathbf{d}, \tag{7.66}$$

   where E is the energy.

   (c) Taking this differential equation as defining $\mathbf{d}$ generally, find the $\mathbf{E}$ dependence of E in the case that $\mathbf{d} = \mathbf{d}_0 + \alpha\mathbf{E}$, where α is the electric polarizability.

2. What electromagnetic fields do the following potentials describe? Here the fields are functions of $\mathbf{r} = (x, y, z)$ and t.

   (a)

$$\phi = 0, \quad A_x = A_z = 0, \quad A_y = Kx, \quad (K = \text{constant}), \tag{7.67a}$$

   (b)

$$\phi = 0, \quad A_x = A_y = 0, \quad A_z = Kt, \tag{7.67b}$$

   (c)

$$\phi = 0, \quad \mathbf{A} = \mathbf{a}(\mathbf{a}\cdot\mathbf{r}), \quad (\mathbf{a} = \text{constant}), \tag{7.67c}$$

   (d)

$$\phi = \frac{\mathbf{a}\cdot\mathbf{r}}{r^3}, \quad \mathbf{A} = \mathbf{0}; \quad r > 0, \tag{7.67d}$$

   (e)

$$\phi = 0, \quad \mathbf{A} = \frac{\mathbf{a}\times\mathbf{r}}{r^3}; \quad r > 0. \tag{7.67e}$$

3. Suppose we used the classical canonical (Boltzmann) distribution instead of the quantum mechanical one treated in Section 7.4 for ferromagnetism, that is,

$$\langle \boldsymbol{\mu}_0 \rangle_T = \frac{1}{Z}\int d\Omega\, \boldsymbol{\mu}_0 \exp[\boldsymbol{\mu}_0\cdot(\mathbf{H} + \lambda\mathbf{M})/kT], \tag{7.68a}$$

   where the partition function is

$$Z = \int d\Omega \exp[\boldsymbol{\mu}_0\cdot(\mathbf{H} + \lambda\mathbf{M})/kT]. \tag{7.68b}$$

   Repeat the analysis given in Section 7.4 for this distribution and discuss similarities and differences, particularly in the low and high temperature limits.

4. Generalize (5.15) by including a time-dependent magnetic field $\mathbf{B}$. Find the equivalent of (5.17) in the case of a weak $\mathbf{B}$. Specify the meaning of *weak*. Show that the polarization vector is linearly related to the electric field,

$$\mathbf{P}(\omega) = \chi_e(\omega) \cdot \mathbf{E}(\omega), \tag{7.69}$$

and find the electric susceptibility dyadic $\chi_e(\omega)$, whenever the magnetic field is constant in time.

5. A charge e is bound harmonically to a center that is moving nonrelativistically ($v/c \ll 1$) along the trajectory $\mathbf{R}(t)$. Show that the equation of relative motion, including a damping force, is

$$m\frac{d^2}{dt^2}\mathbf{r} = -m\omega_0^2\mathbf{r} - m\gamma\frac{d\mathbf{r}}{dt} - m\frac{d^2}{dt^2}\mathbf{R} + e\left(\mathbf{E} + \frac{1}{c}\frac{d\mathbf{R}}{dt}\times\mathbf{B} + \frac{1}{c}\frac{d\mathbf{r}}{dt}\times\mathbf{B}\right). \tag{7.70}$$

For $\mathbf{E} = \mathbf{0}$, and $\mathbf{R}(t)$ a circular motion with constant angular velocity $\mathbf{\Omega}$, show that to first order in Ω there is an effective electric field acting on the charge. What would thus be the polarization of a dielectric body rotating in an external constant magnetic field?

6. A circular cylinder carries an axially symmetric charge density, and rotates around its axis of symmetry. Write the resulting current as a curl, and by inspection infer the induced magnetic field.

7. Use the two preceding problems to find the magnetic susceptibility of a circular cylinder with axially symmetric dielectric constant, that is rotating around its axis $\mathbf{n}$ in a constant magnetic field $\mathbf{B} = B\mathbf{n}$. Can you confirm the general rule $\mathbf{M} = -\frac{\mathbf{v}}{c}\times\mathbf{P}$? For a homogeneous cylinder of radius a and length l find the total magnetization.

8

Macroscopic Energy and Momentum

8.1 General Discussion

We now turn to the consideration of the distribution of energy and momentum of electromagnetic fields within material media, following closely the development of Chapter 3 (however, here we will assume no magnetic charge is present). We will base our discussion on the macroscopic form of Maxwell's equations, (4.56). Accordingly, the rate at which the electric field does work on the free charges is

$$\mathbf{J}\cdot\mathbf{E} = \mathbf{E}\cdot\left[\frac{c}{4\pi}\boldsymbol{\nabla}\times\mathbf{H} - \frac{1}{4\pi}\frac{\partial}{\partial t}\mathbf{D}\right]. \tag{8.1a}$$

If we add to this the parallel equation, appropriate to the absence of magnetic charge,

$$0 = \mathbf{H}\cdot\left[-\frac{c}{4\pi}\boldsymbol{\nabla}\times\mathbf{E} - \frac{1}{4\pi}\frac{\partial}{\partial t}\mathbf{B}\right], \tag{8.1b}$$

we obtain the suggestive form

$$\mathbf{J}\cdot\mathbf{E} = -\boldsymbol{\nabla}\cdot\left(\frac{c}{4\pi}\mathbf{E}\times\mathbf{H}\right) - \frac{1}{4\pi}\left(\mathbf{E}\cdot\frac{\partial}{\partial t}\mathbf{D} + \mathbf{H}\cdot\frac{\partial}{\partial t}\mathbf{B}\right). \tag{8.2}$$

[Recall that if there were free magnetic currents, (8.1b) would represent the work done on the magnetic charges.] Our aim is to write this result as a local energy conservation law. We immediately identify, from the divergence term, the energy flux or Poynting's vector, $\mathbf{S}$, to be

$$\mathbf{S} = \frac{c}{4\pi}\mathbf{E}\times\mathbf{H}, \tag{8.3}$$

which has the same form as that of the microscopic flux, (3.4b), except that here $\mathbf{B}$ is replaced by $\mathbf{H}$. More intractable is the identification of the last term in (8.2). To what extent is it the negative time derivative of an energy density, $-\partial U/\partial t$? If there does exist some quantity U such that

$$\frac{\partial}{\partial t}U \stackrel{?}{=} \frac{1}{4\pi}\left(\mathbf{E}\cdot\frac{\partial}{\partial t}\mathbf{D} + \mathbf{H}\cdot\frac{\partial}{\partial t}\mathbf{B}\right), \tag{8.4}$$

we would have a local statement of energy conservation,

$$\frac{\partial}{\partial t}U + \boldsymbol{\nabla}\cdot\mathbf{S} + \mathbf{J}\cdot\mathbf{E} = 0. \tag{8.5}$$

Similarly, we consider the rate at which momentum is transferred to the charges, or equivalently, the force density, $\mathbf{f}$,

$$\mathbf{f} = \rho\mathbf{E} + \frac{1}{c}\mathbf{J}\times\mathbf{B}. \tag{8.6}$$

DOI: 10.1201/9781003057369-8

The relevant pair of equations here is

$$\rho\mathbf{E} + \frac{1}{c}\mathbf{J}\times\mathbf{B} = \frac{1}{4\pi}(\boldsymbol{\nabla}\cdot\mathbf{D})\mathbf{E} + \frac{1}{4\pi}\left(\boldsymbol{\nabla}\times\mathbf{H} - \frac{1}{c}\frac{\partial}{\partial t}\mathbf{D}\right)\times\mathbf{B}, \tag{8.7a}$$

$$\mathbf{0} = \frac{1}{4\pi}(\boldsymbol{\nabla}\cdot\mathbf{B})\mathbf{H} + \frac{1}{4\pi}\left(\boldsymbol{\nabla}\times\mathbf{E} + \frac{1}{c}\frac{\partial}{\partial t}\mathbf{B}\right)\times\mathbf{D}, \tag{8.7b}$$

which is added to yield

$$\rho\mathbf{E} + \frac{1}{c}\mathbf{J}\times\mathbf{B} = -\frac{\partial}{\partial t}\left(\frac{1}{4\pi c}\mathbf{D}\times\mathbf{B}\right) - \frac{1}{4\pi}[D_i\boldsymbol{\nabla}E_i - \boldsymbol{\nabla}\cdot(\mathbf{DE})]$$
$$-\frac{1}{4\pi}[B_i\boldsymbol{\nabla}H_i - \boldsymbol{\nabla}\cdot(\mathbf{BH})]. \tag{8.8}$$

From the time derivative, we identify the momentum density, $\mathbf{G}$, to be

$$\mathbf{G} = \frac{1}{4\pi c}\mathbf{D}\times\mathbf{B}, \tag{8.9}$$

which is analogous to (3.12a) except the correct electric field here is $\mathbf{D}$.[1] If a local momentum conservation law holds, the last two terms of (8.8) should be the divergence of a tensor, representing the flow of momentum,

$$\frac{1}{4\pi}[D_i\boldsymbol{\nabla}E_i - \boldsymbol{\nabla}\cdot(\mathbf{DE}) + B_i\boldsymbol{\nabla}H_i - \boldsymbol{\nabla}\cdot(\mathbf{BH})] \stackrel{?}{=} \boldsymbol{\nabla}\cdot\mathbf{T}. \tag{8.10}$$

Then, a local law of momentum conservation would hold,

$$\frac{\partial}{\partial t}\mathbf{G} + \boldsymbol{\nabla}\cdot\mathbf{T} + \rho\mathbf{E} + \frac{1}{c}\mathbf{J}\times\mathbf{B} = \mathbf{0}. \tag{8.11}$$

8.2 Nondispersive Medium

We cannot proceed further without specific assumptions about the properties of the material medium. The simplest hypothesis is that of a homogeneous, isotropic, nondispersive medium,

$$\mathbf{D} = \epsilon\mathbf{E}, \qquad \mathbf{B} = \mu\mathbf{H}, \tag{8.12}$$

where ϵ and μ are constants. This is not an unrealistic situation for many substances over a sufficiently limited frequency range. For this case, the energy density and the stress tensor exist, and have the following forms:

$$U = \frac{\epsilon E^2 + \mu H^2}{8\pi}, \qquad \mathbf{T} = \mathbf{1}\frac{\epsilon E^2 + \mu H^2}{8\pi} - \frac{\epsilon\mathbf{EE} + \mu\mathbf{HH}}{4\pi}, \tag{8.13}$$

while we recall that the energy flux and the momentum density are given by

$$\mathbf{S} = \frac{c}{4\pi}\mathbf{E}\times\mathbf{H}, \qquad \mathbf{G} = \frac{\epsilon\mu}{4\pi c}\mathbf{E}\times\mathbf{H} = \frac{\epsilon\mu}{c^2}\mathbf{S}. \tag{8.14}$$

[1] We are adopting the Minkowski definition of the field momentum rather than the Abraham form, constructed in the same way from $\mathbf{E}$ and $\mathbf{H}$. (Hermann Minkowski, 1864–1909; Max Abraham, 1875–1922.) Surprisingly, this controversy over the form of field momentum in a medium persists to the present day. For a recent discussion see Ref. [20].

It is interesting that we can transform these expressions, as well as Maxwell's equations, to look like those in vacuum, by redefining the fields, the charges, and the speed of light as follows:

$$E' = \sqrt{\epsilon}E, \quad H' = \sqrt{\mu}H, \quad c' = \frac{c}{\sqrt{\epsilon\mu}}, \quad \rho' = \frac{1}{\sqrt{\epsilon}}\rho, \quad J' = \frac{1}{\sqrt{\epsilon}}J. \tag{8.15}$$

(See Problem 8.1.) The ratio of c to c' is the index of refraction for the medium,

$$\frac{c}{c'} = n = \sqrt{\epsilon\mu}. \tag{8.16}$$

By this transformation we see that the speed of propagation of electromagnetic waves in the medium is c'; for propagation in a definite direction, the transcription from the vacuum statement (see Section 3.4) that $\mathbf{E}'$ and $\mathbf{B}'$ are mutually perpendicular and equal in magnitude is

$$\epsilon E^2 = \mu H^2, \qquad \mathbf{E}\cdot\mathbf{H} = 0. \tag{8.17}$$

In the usual situation at low frequencies, where $\mu \approx 1$, $\epsilon > 1$, the speed of light in the medium, c', is less than that in vacuum, c, and the electric field is smaller than the magnetic field, by the same factor:

$$c' = \frac{c}{n}, \quad |\mathbf{E}| = \frac{|\mathbf{H}|}{n}, \quad n > 1. \tag{8.18}$$

8.3 Dispersive Medium

The naive way to incorporate the effects of a frequency dependence in the dielectric constant (a dispersive medium) is to modify (8.13) and (8.16) by replacing the constants ϵ and μ with frequency dependent functions, $\epsilon(\omega)$ and $\mu(\omega)$, which for the simple model leading to (5.96), are

$$\mu = 1, \qquad \epsilon(\omega) = 1 + \frac{4\pi ne^2}{m}\frac{1}{\omega_1^2 - \omega^2 - i\omega\gamma}. \tag{8.19}$$

However, this replacement cannot be correct, because if we consider a frequency large compared to ω_1, $\omega \gg \omega_1$, the dielectric constant becomes less than unity,

$$\epsilon(\omega) \to 1 - \frac{\omega_p^2}{\omega^2} < 1, \tag{8.20}$$

where ω_p is the plasma frequency, (5.49) with $\epsilon = 1$. Substituting this into (8.16), we see that the speed of light in the medium, c', is larger than the speed of light in the vacuum: $c' = \frac{c}{n} > c$. If c' is the speed of energy flow, as is true for a nondispersive medium, this is impossible, since it manifestly violates causality. (Causality is the concept that information cannot travel faster than the speed of light in vacuum.) We must now return to our starting point and carefully re-examine energy flow. A realistic electromagnetic wave is characterized by a finite spatial extent, so correspondingly contains a finite range of frequencies. This is relevant because the identification of the energy density according to (8.4) involves a time integration, the building up of the field. But a field with a definite frequency has no such transient behavior; a range of frequencies is required. Since the properties of the medium are frequency dependent, the spread in frequency has very important consequences, no matter how small the spread is.

We begin again by representing the time behavior of the electric field as a superposition of frequencies (Fourier transform)

$$\mathbf{E}(t) = \int_{-\infty}^{\infty} d\omega\, \mathbf{E}(\omega) e^{-i\omega t} = \int_{-\infty}^{\infty} d\omega\, \mathbf{E}(\omega)^* e^{i\omega t}, \tag{8.21}$$

where we have used the fact that $\mathbf{E}(t)$ is a real function, which implies that

$$\mathbf{E}(-\omega)^* = \mathbf{E}(\omega). \tag{8.22}$$

Likewise, the displacement vector can be written as

$$\mathbf{D}(t) = \int_{-\infty}^{\infty} d\omega\, \mathbf{D}(\omega) e^{-i\omega t}, \tag{8.23}$$

where [see (5.30)]

$$\mathbf{D}(\omega) = \epsilon(\omega)\mathbf{E}(\omega). \tag{8.24}$$

Again, since $\mathbf{D}(t)$ is real, the dielectric constant obeys

$$\epsilon(-\omega)^* = \epsilon(\omega), \tag{8.25}$$

just as we saw in the model (8.19)—see (5.29). For simplicity, we will assume $\epsilon(\omega)$ is real (that is, we stay away from the absorption regions) so that it is an even function of ω. Since ordinary materials are non-magnetic, $\mu = 1$, we will take

$$\mathbf{B} = \mathbf{H}. \tag{8.26}$$

With this preparation we now return to our general discussion of energy flow, in particular, to the purported equality (8.4). Since (8.26) holds, the magnetic term is simple:

$$\frac{\partial}{\partial t}\frac{H^2}{8\pi}. \tag{8.27}$$

The electric part can be written in terms of the product of two Fourier integrals, (8.23) and the second form of (8.21) with $\omega \to \omega'$:

$$\begin{aligned}\frac{1}{4\pi}\mathbf{E}\cdot\frac{\partial}{\partial t}\mathbf{D} &= \frac{1}{4\pi}\int d\omega\, d\omega'\, e^{i\omega' t}\mathbf{E}(-\omega')\cdot\epsilon(\omega)\mathbf{E}(\omega)(-i\omega)e^{-i\omega t}\\ &= \frac{1}{8\pi}\int d\omega\, d\omega' e^{-i(\omega-\omega')t}(-i)[\omega\epsilon(\omega) - \omega'\epsilon(\omega')]\mathbf{E}(\omega)\cdot\mathbf{E}(-\omega').\end{aligned} \tag{8.28}$$

In the last form the symmetry $\omega \leftrightarrow -\omega'$ has been made manifest. This can be easily seen to be a time derivative by inserting the factor $(\omega-\omega')/(\omega-\omega')$ so that (8.28) becomes

$$\frac{\partial}{\partial t}\frac{1}{8\pi}\int d\omega\, d\omega'\, e^{-i(\omega-\omega')t}\frac{\omega\epsilon(\omega) - \omega'\epsilon(\omega')}{\omega-\omega'}\mathbf{E}(\omega)\cdot\mathbf{E}(-\omega'). \tag{8.29}$$

[A check of this result is to note that if we let $\epsilon(\omega) = \epsilon$, we recover the result for the nondispersive energy density, $(\epsilon/8\pi)(\mathbf{E}(t))^2$.] This provides a general expression for the electric part of the energy density.

Now suppose that the turning on of the field takes place so slowly that only a very small band of frequencies about a central frequency $\overline{\omega}$ occurs (the precise relation between turning-on time and band width is not required here—but see Problem 8.5). Now with $\omega \approx \omega' \approx \overline{\omega}$ or $-\overline{\omega}$, we have, since $\epsilon(-\omega) = \epsilon(\omega)$,

$$\frac{\omega\epsilon(\omega) - \omega'\epsilon(\omega')}{\omega - \omega'} \to \frac{d}{d\omega}[\omega\epsilon(\omega)]\big|_{\omega=\overline{\omega}}. \tag{8.30}$$

On the other hand, for $\omega \approx -\omega'$, the above quantity is simply $\epsilon(\overline{\omega})$. However, this contribution to the energy density oscillates in time with the frequency $2\overline{\omega}$ and will not contribute significantly when averaged over one or more periods. In addition, the spatial dependence will also be rapidly oscillating, suppressing the contribution in spatial averages. In the sense of such averages, then, the final form for the energy density in this situation is inferred from (8.27) and (8.29) to be[2]

$$U = \frac{H^2}{8\pi} + \frac{E^2}{8\pi}\left[\frac{d}{d\omega}(\omega\epsilon(\omega))\right]_{\omega=\overline{\omega}}. \tag{8.31}$$

To find the speed at which energy is propagated in an electromagnetic pulse, we proceed as in Section 3.4. In regions without charges, ρ, $\mathbf{J} = 0$, local energy conservation implies

$$\mathbf{r}\left[\frac{\partial}{\partial t}U + \boldsymbol{\nabla}\cdot\mathbf{S}\right] = \mathbf{0}, \tag{8.32}$$

which, when integrated over the pulse, implies

$$\frac{d}{dt}\int(d\mathbf{r})\,\mathbf{r}U = \int(d\mathbf{r})\,\mathbf{S}, \tag{8.33}$$

if we discard the surface term. Recalling the notion of energy weightings, we write the magnitude of (8.33) as

$$|\mathbf{v}_E|\int(d\mathbf{r})\left\{\frac{H^2}{8\pi} + \left[\frac{d}{d\omega}(\omega\epsilon)\right]\frac{E^2}{8\pi}\right\} = \frac{c}{4\pi}\left|\int(d\mathbf{r})\,\mathbf{E}\times\mathbf{H}\right|, \tag{8.34}$$

where [(3.36) and (3.39)]

$$\mathbf{v}_E = \frac{d}{dt}\langle\mathbf{r}\rangle_E. \tag{8.35}$$

Since we are considering a nearly monochromatic pulse, with the effects of dispersion adequately summarized in (8.30), the relations given at the end of Section 8.2 between **E** and **H** are still true, at frequency $\overline{\omega}$:

$$|\mathbf{E}\times\mathbf{H}| = EH = \frac{H^2}{\sqrt{\epsilon}}, \quad E^2 = \frac{H^2}{\epsilon}. \tag{8.36}$$

We therefore conclude from (8.34) that the speed of propagation of energy at frequency ω is

$$|\mathbf{v}_E| = \frac{2\sqrt{\epsilon}}{\left[\frac{d}{d\omega}(\omega\epsilon) + \epsilon\right]}c. \tag{8.37}$$

(For the situation when $\mu \neq 1$, see Problem 8.3.)

For the specific model where the dielectric constant is given by (8.20),

$$\frac{d}{d\omega}[\omega\epsilon(\omega)] = 1 + \frac{\omega_p^2}{\omega^2} > 1. \tag{8.38}$$

[2] Although this classical argument might seem a bit unsatisfactory, it is greatly strengthened by the fact that quantum field theory yields precisely the same dispersive factor in the energy when the field products are understood as quantum-mechanical expectation values.

The corresponding speed of energy flow is

$$|\mathbf{v}_E| = \sqrt{1 - \omega_p^2/\omega^2}\, c < c, \quad \omega > \omega_p. \tag{8.39}$$

The speed of propagation of energy is less than the speed of light in vacuum, c, in accord with the general idea that information cannot travel faster than c.

8.4 Problems for Chapter 8

1. Verify that Maxwell's equations transform to the corresponding equations in vacuum (where $D' = E'$, $B' = H'$) under the transformation (8.15) when ϵ and μ are constant.

2. Show that (8.37) can be presented as

$$|\mathbf{v}_E| = c\left(\frac{d}{d\omega}n\omega\right)^{-1}, \tag{8.40}$$

where $n = \sqrt{\epsilon}$ is the index of refraction

3. Repeat the discussion in Section 8.3 when $\epsilon(\omega) = 1$ and $\mu(\omega) \neq 1$. What happens if both ϵ and μ are frequency dependent functions? In such a case, the energy density is generalized from (8.31) to

$$U = \frac{1}{8\pi}\left[\frac{d(\omega\epsilon)}{d\omega}|\mathbf{E}|^2 + \frac{d(\omega\mu)}{d\omega}|\mathbf{H}|^2\right]. \tag{8.41}$$

For a unidirectional light pulse, with the properties (8.17), show that the speed of energy propagation is

$$v = \frac{c}{\frac{d}{d\omega}n\omega}, \quad n = \sqrt{\epsilon\mu}. \tag{8.42}$$

4. Use the model for the dielectric constant, (8.19), for $|\omega_1^2 - \omega^2| \gg \omega\gamma$ so that $\epsilon(\omega)$ is real, to calculate the speed of energy flow, $\mathbf{v}_E$, from (8.37) in the two cases $\omega < \omega_1$ and $\omega > \omega_1$. In both cases, show that the speed of energy flow is less than c. (While this conclusion is not altered, the story is more complicated when $|\omega - \omega_1| \sim \gamma$. Treatment of this question by Arnold Sommerfeld (1868–1951) and Léon Brillouin (1889–1969) can be found in Ref. [21] See Problems 7 and 8 below.)

5. Imagine an electric field $\mathbf{E}(t)$ is slowly built up from a value of zero in the distant past to a value $\mathrm{Re}\left(\mathbf{E}(\omega)e^{-i\omega t}\right)$ in the present. Let T be the time scale over which this turning on of the field occurs. Compute the band width of frequencies, $\Delta\omega$, occurring in the Fourier transform of $\mathbf{E}(t)$.

6. A metal may be thought of, roughly, as a plasma in which electrons are free to move. The dielectric constant is therefore approximately given by

$$\epsilon(\omega) = 1 - \frac{\omega_p^2}{\omega^2}.$$

The metal becomes transparent when the frequency ω is large enough so that $n = \sqrt{\epsilon}$ is real. Estimate the frequency at which copper ($N_c = 8.5 \times 10^{22}$ electrons/cm^3) becomes transparent.

7. In Section 8.2, we considered the electric part of the energy density in the absence of absorption. We will here consider absorption within the framework of the dielectric model discussed in Section 5.2. There, a simple damping model is employed, and, it is clear that this damping will remove a certain amount of power per unit volume, $\mathcal{P}$.

   (a) Calculate $\mathcal{P}$ and write it in the form of (8.28).

   (b) We might expect that (8.4) needs to be modified in the presence of absorption to read
   $$\frac{1}{4\pi}\left(\mathbf{E}\cdot\frac{\partial}{\partial t}\mathbf{D}+\mathbf{H}\cdot\frac{\partial}{\partial t}\mathbf{B}\right)=\frac{\partial}{\partial t}U+\mathcal{P}.$$
   Determine the electric part of U.

   (c) Find the narrow band analog of (8.31).

8. Use the above formulation to calculate a propagation speed in the presence of absorption. In particular, for the oscillator model of Section 5.2, derive the speed of energy flow found by Brillouin (suitably corrected),
   $$v_E = c\frac{\text{Re } n}{\epsilon_1},$$
   where
   $$\epsilon_1=\frac{1}{2}\left[1+\frac{\omega_p^2(\omega_0^2+\omega^2)}{(\omega_0^2-\omega^2)^2+\omega^2\gamma^2}+|\epsilon|\right].$$

9. Consider a plane electromagnetic wave impinging normally from vacuum onto a flat dielectric surface with permittivity ϵ. Let the direction of propagation of this wave be in the $+z$ direction and the interface lie in the x-y plane at $z=0$.

   (a) If the incident plane wave has the form for $z<0$
   $$\mathbf{E}(\mathbf{r},t)=\mathbf{E}_0 e^{ikz-i\omega t},\quad \mathbf{B}(\mathbf{r},t)=\mathbf{B}_0 e^{ikz-i\omega t}, \tag{8.43}$$
   give the relation between the amplitudes $\mathbf{E}_0$ and $\mathbf{B}_0$, and between their directions.

   (b) At the interface, some of the wave is reflected and some is transmitted. Because of the normal incidence of the incoming wave, the transmitted wave propagates along the $+z$ direction, while the reflected wave propagates in the $-z$ direction. If the reflected wave has magnetic amplitude $r\mathbf{B}_0$, where r is the reflection coefficient to be determined below, what is the amplitude of the transmitted magnetic field? (Hint: the tangential magnetic field is continuous at the interface between two dielectric bodies.)

   (c) What are the amplitudes of the transmitted and reflected electric fields?

   (d) By adding the electric and the magnetic fields on the two sides of the interface, calculate the energy flux vector on both sides in terms of the initial amplitude $\mathbf{B}_0$. (Assume r is real, and omit rapidly oscillating terms.)

   (e) Use energy conservation at the interface to obtain an equation for r, the reflection coefficient, in terms of ϵ. Solve this equation. Is this result physically reasonable? What happens as $\epsilon\to 1$, $\epsilon\to\infty$? (The latter result should correspond to the $z>0$ region being a perfect conductor.)

10. (a) The Pauli[3] exclusion principle asserts that only one electron can occupy a given state. The number of states per element of phase space is given by

$$2\frac{(d\mathbf{r})(d\mathbf{p})}{h^3}$$

where h is Planck's[4] constant, and the factor of 2 arises from the two possible spin orientations of the electron. At zero temperature the lowest energy states are completely filled. Show that the energy of the highest occupied state, the Fermi energy, is

$$E_F = \frac{\hbar^2}{2m}(3\pi^2 n)^{2/3},$$

where $\hbar = \frac{h}{2\pi} = 1.05 \times 10^{-27}$ erg-s, and n is the number density.

(b) In a metal, the conduction electrons are essentially free. Let $n(\mathbf{r}) - n_0$ be the (small) deviation from a uniform electron concentration. The chemical potential

$$\mu = E_F(\mathbf{r}) - e\phi(\mathbf{r}),$$

where ϕ is the electrostatic potential, is constant is space in equilibrium. Show then that ϕ satisfies

$$\nabla^2\phi = \lambda^2\phi,$$

and compute λ in terms of n_0. What value does λ have for copper? Show that the spherically symmetric solution of this equation is the screened Coulomb potential,

$$\phi(\mathbf{r}) = q\frac{e^{-\lambda r}}{r}.$$

(c) From the analysis of (b), we see that placing an external charge distribution ρ_{ext} in a metal leads to an induced electronic charge density ρ_e. Derive the following relation between the Fourier transforms,

$$\rho_e(\mathbf{k}) = \frac{\lambda^2}{k^2}\left(\rho_{\text{ext}}(\mathbf{k}) + \rho_e(\mathbf{k})\right),$$

and hence show that this situation corresponds to a wavevector-dependent dielectric constant,

$$\epsilon(\mathbf{k}) = 1 + \frac{\lambda^2}{k^2}.$$

(d) An alternative derivation of this last result follows by noting that

$$\epsilon(\mathbf{k}) = \frac{\phi_{\text{ext}}(\mathbf{k})}{\phi(\mathbf{k})},$$

ϕ being the total potential, and ϕ_{ext} that due to the external charge distribution. From (b) show that

$$\phi_{\text{ext}}(\mathbf{k}) = \frac{4\pi\rho_{\text{ext}}(\mathbf{k})}{k^2}, \quad \phi(\mathbf{k}) = \frac{4\pi\rho_{\text{ext}}(\mathbf{k})}{k^2 + \lambda^2},$$

from which the result follows.

---

[3]Wolfgang Pauli, 1900–1958.
[4]Max Planck, 1858–1947.

9

Review of Action Principles

We have already mentioned the Hamiltonian formulation for the particle equations of motion [see (7.17a)–(7.17c)]. We now want to show that the whole system of particles and fields is a mechanical system derivable from a Hamilton action principle.

We start by reviewing and generalizing the Lagrange-Hamilton[1] principle for a single particle. The action, W_{12}, is defined as the time integral of the Lagrangian, L, where the integration extends from an initial configuration or state at time t_2 to a final state at time t_1:

$$W_{12} = \int_{t_2}^{t_1} dt\, L. \tag{9.1}$$

The integral refers to any path, any line of time development, from the initial to the final state, as shown in Fig. 9.1. The actual time evolution of the system is selected by the principle of stationary action: In response to infinitesimal variations of the integration path, the action W_{12} is stationary—does not have a corresponding infinitesimal change—for variations about the correct path, provided the initial and final configurations are held fixed,

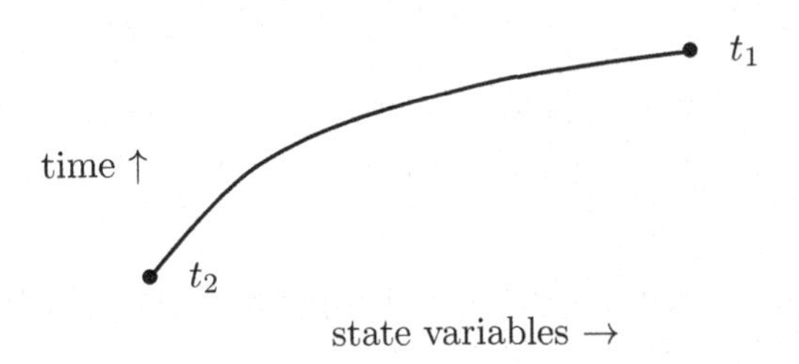

FIGURE 9.1
A possible path from initial state to final state.

$$\delta W_{12} = 0. \tag{9.2}$$

This means that, if we allow infinitesimal changes at the initial and final times, including alterations of those times, the only contribution to δW_{12} then comes from the endpoint variations, or

$$\delta W_{12} = G_1 - G_2, \tag{9.3}$$

where the "generator," G_a, $a = 1$ or 2, is a function depending on dynamical variables only at time t_a. In the following, we will consider three different realizations of the action principle, where, for simplicity, we will restrict our attention to a single particle.

[1] Joseph-Louis Lagrange, 1736–1813

DOI: 10.1201/9781003057369-9

9.1 Lagrangian Viewpoint

The nonrelativistic motion of a particle of mass m moving in a potential $V(\mathbf{r},t)$ is described by the Lagrangian

$$L=\frac{1}{2}m\left(\frac{d\mathbf{r}}{dt}\right)^2-V(\mathbf{r},t). \tag{9.4}$$

Here, the independent variables are $\mathbf{r}$ and t so that two kinds of variations can be considered. First, a particular motion is altered infinitesimally, that is, the path is changed by an amount $\delta\mathbf{r}$:

$$\mathbf{r}(t)\to\mathbf{r}(t)+\delta\mathbf{r}(t). \tag{9.5}$$

Second, the final and initial times can be altered infinitesimally, by δt_1 and δt_2, respectively. It is more convenient, however, to think of these time displacements as produced by a continuous variation of the time parameter, $\delta t(t)$,

$$t\to t+\delta t(t), \tag{9.6}$$

so chosen that, at the endpoints,

$$\delta t(t_1)=\delta t_1,\qquad \delta t(t_2)=\delta t_2. \tag{9.7}$$

The corresponding change in the time differential is

$$dt\to d(t+\delta t)=\left(1+\frac{d\delta t}{dt}\right)dt, \tag{9.8}$$

which implies the transformation of the time derivative,

$$\frac{d}{dt}\to\left(1-\frac{d\delta t}{dt}\right)\frac{d}{dt}. \tag{9.9}$$

Because of this redefinition of the time variable, the limits of integration in the action,

$$W_{12}=\int_2^1\left[\frac{1}{2}m\frac{(d\mathbf{r})^2}{dt}-dt\,V\right], \tag{9.10}$$

are *not* changed, the time displacement being produced through $\delta t(t)$ subject to (9.7). The resulting variation in the action is now

$$\begin{aligned}\delta W_{12}&=\int_2^1 dt\left\{m\frac{d\mathbf{r}}{dt}\cdot\frac{d}{dt}\delta\mathbf{r}-\delta\mathbf{r}\cdot\boldsymbol{\nabla}V-\frac{d\delta t}{dt}\left[\frac{1}{2}m\left(\frac{d\mathbf{r}}{dt}\right)^2+V\right]-\delta t\frac{\partial}{\partial t}V\right\}\\&=\int_2^1 dt\left\{\frac{d}{dt}\left[m\frac{d\mathbf{r}}{dt}\cdot\delta\mathbf{r}-\left(\frac{1}{2}m\left(\frac{d\mathbf{r}}{dt}\right)^2+V\right)\delta t\right]\right.\\&\quad\left.+\delta\mathbf{r}\cdot\left[-m\frac{d^2}{dt^2}\mathbf{r}-\boldsymbol{\nabla}V\right]+\delta t\left(\frac{d}{dt}\left[\frac{1}{2}m\left(\frac{d\mathbf{r}}{dt}\right)^2+V\right]-\frac{\partial}{\partial t}V\right)\right\},\end{aligned} \tag{9.11}$$

where, in the last form, we have shifted the time derivatives in order to isolate $\delta\mathbf{r}$ and δt.

Because $\delta\mathbf{r}$ and δt are independent variations, the principle of stationary action implies that the actual motion is governed by

$$m\frac{d^2}{dt^2}\mathbf{r}=-\boldsymbol{\nabla}V, \tag{9.12a}$$

$$\frac{d}{dt}\left[\frac{1}{2}m\left(\frac{d\mathbf{r}}{dt}\right)^2+V\right]=\frac{\partial}{\partial t}V, \tag{9.12b}$$

while the total time derivative gives the change at the endpoints,

$$G = \mathbf{p} \cdot \delta\mathbf{r} - E\delta t, \tag{9.13}$$

with

$$\text{momentum} = \mathbf{p} = m\frac{d\mathbf{r}}{dt}, \qquad \text{energy} = E = \frac{1}{2}m\left(\frac{d\mathbf{r}}{dt}\right)^2 + V. \tag{9.14}$$

Therefore, we have derived Newton's second law [the equation of motion in second-order form], (9.12a), and, for a static potential, $\partial V/\partial t = 0$, the conservation of energy, (9.12b). The significance of (9.13) will be discussed later in Section 9.4.

9.2 Hamiltonian Viewpoint

Using the above definition of the momentum, we can rewrite the Lagrangian as

$$L = \mathbf{p} \cdot \frac{d\mathbf{r}}{dt} - H(\mathbf{r}, \mathbf{p}, t), \tag{9.15}$$

where we have introduced the Hamiltonian

$$H = \frac{p^2}{2m} + V(\mathbf{r}, t). \tag{9.16}$$

This conversion from the Lagrangian to the Hamiltonian is called a Legendre transformation (Adrien-Marie Legendre, 1752–1833). We are here to regard $\mathbf{r}$, $\mathbf{p}$, and t as independent variables in

$$W_{12} = \int_2^1 [\mathbf{p} \cdot d\mathbf{r} - dt\, H]. \tag{9.17}$$

The change in the action, when $\mathbf{r}$, $\mathbf{p}$, and t are all varied, is

$$\begin{aligned}
\delta W_{12} &= \int_2^1 dt \left[\mathbf{p} \cdot \frac{d}{dt}\delta\mathbf{r} - \delta\mathbf{r} \cdot \frac{\partial H}{\partial \mathbf{r}} + \delta\mathbf{p} \cdot \frac{d\mathbf{r}}{dt} - \delta\mathbf{p} \cdot \frac{\partial H}{\partial \mathbf{p}} - \frac{d\delta t}{dt}H - \delta t\frac{\partial H}{\partial t}\right] \\
&= \int_2^1 dt \left[\frac{d}{dt}(\mathbf{p} \cdot \delta\mathbf{r} - H\delta t) + \delta\mathbf{r} \cdot \left(-\frac{d\mathbf{p}}{dt} - \frac{\partial H}{\partial \mathbf{r}}\right)\right. \\
&\qquad \left. + \delta\mathbf{p} \cdot \left(\frac{d\mathbf{r}}{dt} - \frac{\partial H}{\partial \mathbf{p}}\right) + \delta t\left(\frac{dH}{dt} - \frac{\partial H}{\partial t}\right)\right].
\end{aligned} \tag{9.18}$$

The action principle then implies

$$\frac{d\mathbf{r}}{dt} = \frac{\partial H}{\partial \mathbf{p}} = \frac{\mathbf{p}}{m}, \tag{9.19a}$$

$$\frac{d\mathbf{p}}{dt} = -\frac{\partial H}{\partial \mathbf{r}} = -\boldsymbol{\nabla} V, \tag{9.19b}$$

$$\frac{dH}{dt} = \frac{\partial H}{\partial t}, \tag{9.19c}$$

$$G = \mathbf{p} \cdot \delta\mathbf{r} - H\delta t. \tag{9.19d}$$

In contrast with the Lagrangian differential equations of motion, which involve second derivatives, these Hamiltonian equations contain only first derivatives; they are called first-order equations. They describe the same physical system, because when (9.19a) is substituted into (9.19b), we recover the Lagrangian-Newtonian equation (9.12a). Furthermore, if we insert (9.19a) into the Hamiltonian (9.16), we identify H with E. The third equation (9.19c) is then identical with (9.12b). We also note the equivalence of the two versions of G.

But probably the most direct way of seeing that the same physical system is involved comes by writing the Lagrangian in the Hamiltonian viewpoint as

$$L = \frac{m}{2}\left(\frac{d\mathbf{r}}{dt}\right)^2 - V - \frac{1}{2m}\left(\mathbf{p} - m\frac{d\mathbf{r}}{dt}\right)^2. \tag{9.20}$$

The result of varying $\mathbf{p}$ in the stationary action principle is to produce

$$\mathbf{p} = m\frac{d\mathbf{r}}{dt}. \tag{9.21}$$

But, if we accept this as the *definition* of $\mathbf{p}$, the corresponding term in L disappears and we explicitly regain the Lagrangian description. We are justified in completely omitting the last term on the right side of (9.20), despite its dependence on the variables $\mathbf{r}$ and t, because of its quadratic structure. Its explicit contribution to δL is

$$-\frac{1}{m}\left(\mathbf{p} - m\frac{d\mathbf{r}}{dt}\right)\cdot\left(\delta\mathbf{p} - m\frac{d}{dt}\delta\mathbf{r} + m\frac{d\mathbf{r}}{dt}\frac{d\delta t}{dt}\right), \tag{9.22}$$

and the equation supplied by the stationary action principle for $\mathbf{p}$ variations, (9.21), also guarantees that there is no contribution here to the results of $\mathbf{r}$ and t variations.

9.3 A Third Viewpoint

Here we take $\mathbf{r}$, $\mathbf{p}$, and the velocity, $\mathbf{v}$, as independent variables so that the Lagrangian is written in the form

$$L = \mathbf{p}\cdot\left(\frac{d\mathbf{r}}{dt} - \mathbf{v}\right) + \frac{1}{2}mv^2 - V(\mathbf{r},t) \equiv \mathbf{p}\cdot\frac{d\mathbf{r}}{dt} - H(\mathbf{r},\mathbf{p},\mathbf{v},t), \tag{9.23}$$

where

$$H(\mathbf{r},\mathbf{p},\mathbf{v},t) = \mathbf{p}\cdot\mathbf{v} - \frac{1}{2}mv^2 + V(\mathbf{r},t). \tag{9.24}$$

The variation of the action is now

$$\begin{aligned}
\delta W_{12} &= \delta\int_2^1 [\mathbf{p}\cdot d\mathbf{r} - H\,dt] \\
&= \int_2^1 dt\left[\delta\mathbf{p}\cdot\frac{d\mathbf{r}}{dt} + \mathbf{p}\cdot\frac{d}{dt}\delta\mathbf{r} - \delta\mathbf{r}\cdot\frac{\partial H}{\partial\mathbf{r}} - \delta\mathbf{p}\cdot\frac{\partial H}{\partial\mathbf{p}} - \delta\mathbf{v}\cdot\frac{\partial H}{\partial\mathbf{v}}\right. \\
&\qquad\left. - \delta t\frac{\partial H}{\partial t} - H\frac{d\delta t}{dt}\right] \\
&= \int_2^1 dt\left[\frac{d}{dt}(\mathbf{p}\cdot\delta\mathbf{r} - H\delta t) - \delta\mathbf{r}\cdot\left(\frac{d\mathbf{p}}{dt} + \frac{\partial H}{\partial\mathbf{r}}\right)\right. \\
&\qquad\left. + \delta\mathbf{p}\cdot\left(\frac{d\mathbf{r}}{dt} - \frac{\partial H}{\partial\mathbf{p}}\right) - \delta\mathbf{v}\cdot\frac{\partial H}{\partial\mathbf{v}} + \delta t\left(\frac{dH}{dt} - \frac{\partial H}{\partial t}\right)\right],
\end{aligned} \tag{9.25}$$

so that the action principle implies

$$\frac{d\mathbf{p}}{dt} = -\frac{\partial H}{\partial \mathbf{r}} = -\boldsymbol{\nabla} V, \tag{9.26a}$$

$$\frac{d\mathbf{r}}{dt} = \frac{\partial H}{\partial \mathbf{p}} = \mathbf{v}, \tag{9.26b}$$

$$\mathbf{0} = -\frac{\partial H}{\partial \mathbf{v}} = -\mathbf{p} + m\mathbf{v}, \tag{9.26c}$$

$$\frac{dH}{dt} = \frac{\partial H}{\partial t}, \tag{9.26d}$$

$$G = \mathbf{p} \cdot \delta\mathbf{r} - H\delta t. \tag{9.26e}$$

Notice that there is no equation of motion for $\mathbf{v}$ since $d\mathbf{v}/dt$ does not occur in the Lagrangian, nor is it multiplied by a time derivative. Consequently, (9.26c) refers to a single time and is an equation of constraint.

From this third approach, we have the option of returning to either of the other two viewpoints by imposing an appropriate restriction. Thus, if we write (9.24) as

$$H(\mathbf{r}, \mathbf{p}, \mathbf{v}, t) = \frac{p^2}{2m} + V(\mathbf{r}, t) - \frac{1}{2m}(\mathbf{p} - m\mathbf{v})^2, \tag{9.27}$$

and we adopt

$$\mathbf{v} = \frac{1}{m}\mathbf{p} \tag{9.28}$$

as the *definition* of $\mathbf{v}$, we recover the Hamiltonian description, (9.15) and (9.16). Alternatively, we can present the Lagrangian (9.23) as

$$L = \frac{m}{2}\left(\frac{d\mathbf{r}}{dt}\right)^2 - V + (\mathbf{p} - m\mathbf{v}) \cdot \left(\frac{d\mathbf{r}}{dt} - \mathbf{v}\right) - \frac{m}{2}\left(\frac{d\mathbf{r}}{dt} - \mathbf{v}\right)^2. \tag{9.29}$$

Then, if we adopt the following as *definitions*,

$$\mathbf{v} = \frac{d\mathbf{r}}{dt}, \quad \mathbf{p} = m\mathbf{v}, \tag{9.30}$$

the resultant form of L is that of the Lagrangian viewpoint, (9.4). It might seem that only the definition $\mathbf{v} = d\mathbf{r}/dt$, inserted in (9.29), suffices to regain the Lagrangian description. But then the next to last term in (9.29) would give the following additional contribution to δL, associated with the variation $\delta\mathbf{r}$:

$$(\mathbf{p} - m\mathbf{v}) \cdot \frac{d}{dt}\delta\mathbf{r}. \tag{9.31}$$

In the next chapter, where the action formulation of electrodynamics is considered, we will see the advantage of adopting this third approach, which is characterized by the introduction of additional variables, similar to $\mathbf{v}$, for which there are no equations of motion.

9.4 Invariance and Conservation Laws

There is more content to the principle of stationary action than equations of motion. Suppose one considers a variation such that

$$\delta W_{12} = 0, \tag{9.32}$$

independently of the choice of initial and final times. We say that the action, which is left unchanged, is *invariant* under this alteration of path. Then the stationary action principle (9.3) asserts that

$$\delta W_{12} = G_1 - G_2 = 0, \tag{9.33}$$

or, there is a quantity $G(t)$ that has the same value for any choice of time t; it is conserved in time. A differential statement of that is

$$\frac{d}{dt}G(t) = 0. \tag{9.34}$$

The G functions, which are usually referred to as generators, express the interrelation between conservation laws and invariances of the system.

Invariance implies conservation, and vice versa. A more precise statement is the following:

> If there is a conservation law, the action is stationary under an infinitesimal transformation in an appropriate variable.

The converse of this statement is also true.

> If the action W is invariant under an infinitesimal transformation (that is, $\delta W = 0$), then there is a corresponding conservation law.

This is the celebrated theorem proved by Amalie Emmy Noether (1882–1935).

Here are some examples. Suppose the Hamiltonian of (9.15) does not depend explicitly on time, or

$$W_{12} = \int_2^1 [\mathbf{p}\cdot d\mathbf{r} - H(\mathbf{r},\mathbf{p})dt]. \tag{9.35}$$

Then the variation (which as a rigid displacement in time, amounts to a shift in the time origin)

$$\delta t = \text{constant} \tag{9.36}$$

will give $\delta W_{12} = 0$ [see the first line of (9.18), with $\delta\mathbf{r} = 0$, $\delta\mathbf{p} = 0$, $d\delta t/dt = 0$, $\partial H/\partial t = 0$]. The conclusion is that G in (9.19d), which here is just

$$G_t = -H\delta t, \tag{9.37}$$

is a conserved quantity, or that

$$\frac{dH}{dt} = 0. \tag{9.38}$$

This inference, that the Hamiltonian—the energy—is conserved, if there is no explicit time dependence in H, is already present in (9.19c). But now a more general principle is at work.

Next, consider an infinitesimal, rigid rotation, one that maintains the lengths and scalar products of all vectors. Written explicitly for the position vector $\mathbf{r}$, it is

$$\delta\mathbf{r} = \delta\boldsymbol{\omega}\times\mathbf{r}, \tag{9.39}$$

where the constant vector $\delta\boldsymbol{\omega}$ gives the direction and magnitude of the rotation (see Fig. 9.2). Now specialize (9.16) to

$$H = \frac{p^2}{2m} + V(r), \tag{9.40}$$

where $r = |\mathbf{r}|$, a rotationally invariant structure. Then

$$W_{12} = \int_2^1 [\mathbf{p}\cdot d\mathbf{r} - H\,dt] \tag{9.41}$$

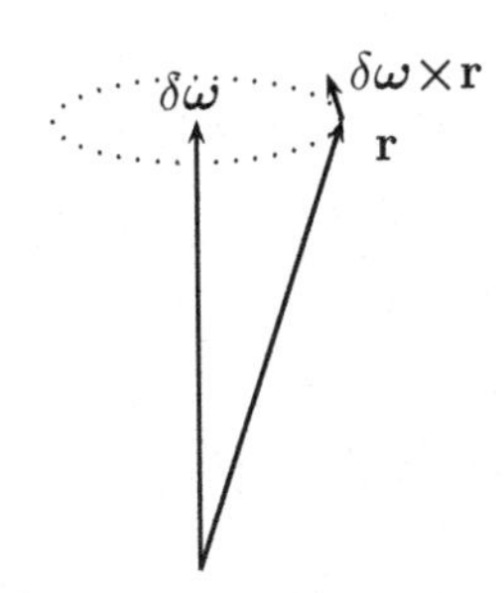

FIGURE 9.2
$\delta\boldsymbol{\omega}\times\mathbf{r}$ is perpendicular to $\delta\boldsymbol{\omega}$ and $\mathbf{r}$, and represents an infinitesimal rotation of $\mathbf{r}$ about the $\delta\boldsymbol{\omega}$ axis.

is also invariant under the rigid rotation, implying the conservation of

$$G_{\delta\boldsymbol{\omega}} = \mathbf{p}\cdot\delta\mathbf{r} = \delta\boldsymbol{\omega}\cdot\mathbf{r}\times\mathbf{p}. \tag{9.42}$$

This is the conservation of angular momentum,

$$\mathbf{L} = \mathbf{r}\times\mathbf{p}, \quad \frac{d}{dt}\mathbf{L} = \mathbf{0}. \tag{9.43}$$

Of course, this is also contained within the equation of motion,

$$\frac{d}{dt}\mathbf{L} = -\mathbf{r}\times\boldsymbol{\nabla}V = -\mathbf{r}\times\hat{\mathbf{r}}\frac{\partial V}{\partial r} = \mathbf{0}, \tag{9.44}$$

since V depends only on $|\mathbf{r}|$.

Conservation of linear momentum appears analogously when there is invariance under a rigid translation. For a single particle, (9.19b) tells us immediately that $\mathbf{p}$ is conserved if V is a constant, say zero. Then, indeed, the action

$$W_{12} = \int_2^1 \left[\mathbf{p}\cdot d\mathbf{r} - \frac{p^2}{2m}dt\right] \tag{9.45}$$

is invariant under the displacement

$$\delta\mathbf{r} = \delta\boldsymbol{\epsilon} = \text{constant}, \tag{9.46}$$

and

$$G_{\delta\boldsymbol{\epsilon}} = \mathbf{p}\cdot\delta\boldsymbol{\epsilon} \tag{9.47}$$

is conserved. But the general principle acts just as easily for, say, a system of two particles, a and b, interaction through a potential that depends only on the relative position, with Hamiltonian

$$H = \frac{p_a^2}{2m_a} + \frac{p_b^2}{2m_b} + V(\mathbf{r}_a - \mathbf{r}_b). \tag{9.48}$$

This Hamiltonian and the associated action

$$W_{12} = \int_2^1 [\mathbf{p}_a\cdot d\mathbf{r}_a + \mathbf{p}_b\cdot d\mathbf{r}_b - H\,dt] \tag{9.49}$$

are invariant under the rigid translation

$$\delta\mathbf{r}_a = \delta\mathbf{r}_b = \delta\boldsymbol{\epsilon}, \tag{9.50}$$

with the implication that

$$G_{\delta\epsilon} = \mathbf{p}_a \cdot \delta\mathbf{r}_a + \mathbf{p}_b \cdot \delta\mathbf{r}_b = (\mathbf{p}_a + \mathbf{p}_b) \cdot \delta\boldsymbol{\epsilon} \tag{9.51}$$

is conserved. This is the conservation of the total linear momentum,

$$\mathbf{P} = \mathbf{p}_a + \mathbf{p}_b, \quad \frac{d}{dt}\mathbf{P} = \mathbf{0}. \tag{9.52}$$

Something a bit more general appears when we consider a rigid translation that grows linearly in time:

$$\delta\mathbf{r}_a = \delta\mathbf{r}_b = \delta\mathbf{v}\, t, \quad \text{with} \quad \delta\mathbf{v} = \text{constant}, \tag{9.53}$$

using the example of two particles. This gives each particle the common additional velocity $\delta\mathbf{v}$, and therefore must also change their momenta,

$$\delta\mathbf{p}_a = m_a \delta\mathbf{v}, \quad \delta\mathbf{p}_b = m_b \delta\mathbf{v}. \tag{9.54}$$

The response of the action (9.49) to this variation is

$$\begin{aligned} \delta W_{12} &= \int_2^1 \left[(\mathbf{p}_a + \mathbf{p}_b) \cdot \delta\mathbf{v}\, dt + \delta\mathbf{v} \cdot (m_a d\mathbf{r}_a + m_b d\mathbf{r}_b) - (\mathbf{p}_a + \mathbf{p}_b) \cdot \delta\mathbf{v}\, dt\right] \\ &= \int_2^1 d[(m_a\mathbf{r}_a + m_b\mathbf{r}_b) \cdot \delta\mathbf{v}]. \end{aligned} \tag{9.55}$$

The action is *not* invariant; its variation has end-point contributions. But there is still a conservation law, not of $G = \mathbf{P} \cdot \delta\mathbf{v}t$, but of $\mathbf{N} \cdot \delta\mathbf{v}$, where

$$\mathbf{N} = \mathbf{P}t - (m_a\mathbf{r}_a + m_b\mathbf{r}_b). \tag{9.56}$$

Written in terms of the center-of-mass position vector

$$\mathbf{R} = \frac{m_a\mathbf{r}_a + m_b\mathbf{r}_b}{M}, \quad M = m_a + m_b, \tag{9.57}$$

the statement of conservation of

$$\mathbf{N} = \mathbf{P}t - M\mathbf{R}, \tag{9.58}$$

namely

$$\mathbf{0} = \frac{d\mathbf{N}}{dt} = \mathbf{P} - M\frac{d\mathbf{R}}{dt}, \tag{9.59}$$

is the familiar fact that the center of mass of an isolated system moves at the constant velocity given by the ratio of the total momentum to the total mass of that system.

9.5 Nonconservation Laws. The Virial Theorem

The action principle also supplies useful nonconservation laws. Consider, for constant $\delta\lambda$,

$$\delta\mathbf{r} = \delta\lambda\mathbf{r}, \quad \delta\mathbf{p} = -\delta\lambda\mathbf{p}, \tag{9.60}$$

which leaves $\mathbf{p}\cdot d\mathbf{r}$ invariant,

$$\delta(\mathbf{p}\cdot d\mathbf{r}) = (-\delta\lambda\mathbf{p})\cdot d\mathbf{r} + \mathbf{p}\cdot(\delta\lambda d\mathbf{r}) = 0. \tag{9.61}$$

But the response of the Hamiltonian

$$H = T(p) + V(\mathbf{r}), \quad T(p) = \frac{p^2}{2m}, \tag{9.62}$$

is given by the noninvariant form

$$\delta H = \delta\lambda(-2T + \mathbf{r}\cdot\boldsymbol{\nabla}V). \tag{9.63}$$

Therefore, we have, for an arbitrary time interval, for the variation of the action (9.17),

$$\delta W_{12} = \int_2^1 dt[\delta\lambda(2T - \mathbf{r}\cdot\boldsymbol{\nabla}V)] = G_1 - G_2 = \int_2^1 dt\frac{d}{dt}(\mathbf{p}\cdot\delta\lambda\mathbf{r}) \tag{9.64}$$

or, the theorem

$$\frac{d}{dt}\mathbf{r}\cdot\mathbf{p} = 2T - \mathbf{r}\cdot\boldsymbol{\nabla}V. \tag{9.65}$$

This is an example of the mechanical virial theorem to which we referred at the end of Section 3.3.

For the particular situation of the Coulomb potential between charges, $V = \text{constant}/r$, where

$$\mathbf{r}\cdot\boldsymbol{\nabla}V = r\frac{d}{dr}V = -V, \tag{9.66}$$

the virial theorem asserts that

$$\frac{d}{dt}(\mathbf{r}\cdot\mathbf{p}) = 2T + V. \tag{9.67}$$

We apply this to a *bound* system produced by a force of attraction. On taking the time average of (9.67) the time derivative term disappears. That is because, over an arbitrarily long time interval $\tau = t_1 - t_2$, the value of $\mathbf{r}\cdot\mathbf{p}(t_1)$ can differ by only a finite amount from $\mathbf{r}\cdot\mathbf{p}(t_2)$, and

$$\overline{\frac{d}{dt}(\mathbf{r}\cdot\mathbf{p})} = \frac{1}{\tau}\int_{t_2}^{t_1} dt\frac{d}{dt}\mathbf{r}\cdot\mathbf{p} = \frac{\mathbf{r}\cdot\mathbf{p}(t_1) - \mathbf{r}\cdot\mathbf{p}(t_2)}{\tau} \to 0, \tag{9.68}$$

as $\tau \to \infty$. The conclusion,

$$2\overline{T} = -\overline{V}, \tag{9.69}$$

has been used qualitatively in Section 5.2.

Here is one more example of a nonconservation law: Consider the variations

$$\delta\mathbf{r} = \delta\lambda\frac{\mathbf{r}}{r}, \tag{9.70a}$$

$$\delta\mathbf{p} = -\delta\lambda\left(\frac{\mathbf{p}}{r} - \frac{\mathbf{r}\,\mathbf{p}\cdot\mathbf{r}}{r^3}\right) = \delta\lambda\frac{\mathbf{r}\times(\mathbf{r}\times\mathbf{p})}{r^3}. \tag{9.70b}$$

Again $\mathbf{p}\cdot d\mathbf{r}$ is invariant:

$$\delta(\mathbf{p}\cdot d\mathbf{r}) = -\delta\lambda\left(\frac{\mathbf{p}}{r} - \frac{\mathbf{r}\,\mathbf{p}\cdot\mathbf{r}}{r^3}\right)\cdot d\mathbf{r} + \mathbf{p}\cdot\left(\delta\lambda\frac{d\mathbf{r}}{r} - \delta\lambda\mathbf{r}\frac{\mathbf{r}\cdot d\mathbf{r}}{r^3}\right) = 0, \tag{9.71}$$

and the change of the Hamiltonian (9.62) is now

$$\delta H = \delta\lambda\left[-\frac{\mathbf{L}^2}{mr^3} + \frac{\mathbf{r}}{r}\cdot\boldsymbol{\nabla}V\right]. \tag{9.72}$$

The resulting theorem, for $V = V(r)$, is

$$\frac{d}{dt}\left(\frac{\mathbf{r}}{r}\cdot\mathbf{p}\right) = \frac{\mathbf{L}^2}{mr^3} - \frac{dV}{dr}, \tag{9.73}$$

which, when applied to the Coulomb potential, gives the bound-state time average relation

$$\frac{L^2}{m}\overline{\left(\frac{1}{r^3}\right)} = -\overline{\left(\frac{V}{r}\right)}. \tag{9.74}$$

This relation is significant in hydrogen fine-structure calculations.

9.6 Problems for Chapter 9

1. Suppose the system consists of N particles interacting through a pairwise potential $V(\mathbf{r}_a - \mathbf{r}_b)$. Write down the Lagrangian and obtain the equations of motion. What is the Hamiltonian, $H(\mathbf{r}_a, \mathbf{p}_a)$? Show that energy and total momentum are conserved. What is required for angular momentum to be conserved?

2. For a free relativistic particle of rest mass m_0, the energy is

$$E = \sqrt{p^2c^2 + m_0^2c^4}. \tag{9.75}$$

   Use this as the Hamiltonian H, and from the Lagrangian

$$L = \mathbf{p}\cdot\frac{d\mathbf{r}}{dt} - H, \tag{9.76}$$

   determine the relationship between the velocity $\mathbf{v} = d\mathbf{r}/dt$ and the momentum. Compute the energy in terms of the velocity. Write the Lagrangian in terms of $\mathbf{v}$.

3. Consider a particle bound by a potential of the form

$$V = ar^b. \tag{9.77}$$

   Derive the time-averaged virial theorem relating $\overline{T}$ to $\overline{V}$. What is the smallest value of b for which a bound state can occur?

10

Action Principle for Electrodynamics

10.1 Action of Particle in Field

It was stated in our review of mechanical action principles that the third viewpoint, which employs the variables $\mathbf{r}$, $\mathbf{p}$, and $\mathbf{v}$, was particularly convenient for describing electromagnetic forces on charged particles. With the explicit, and linear, appearance of $\mathbf{v}$ in what plays the role of the potential function in (7.12), we begin to see the basis for that remark. Indeed, we have only to consult (9.23) to find the appropriate Lagrangian:

$$L = \mathbf{p} \cdot \left(\frac{d\mathbf{r}}{dt} - \mathbf{v} \right) + \frac{1}{2} m v^2 - e\phi + \frac{e}{c} \mathbf{v} \cdot \mathbf{A}. \tag{10.1}$$

To recapitulate, the equations resulting from variations of $\mathbf{p}$, $\mathbf{r}$, and $\mathbf{v}$ are, respectively,

$$\frac{d\mathbf{r}}{dt} = \mathbf{v}, \tag{10.2a}$$

$$\frac{d}{dt}\mathbf{p} = -e\boldsymbol{\nabla} \left[\phi - \frac{1}{c} \mathbf{v} \cdot \mathbf{A} \right], \tag{10.2b}$$

$$\mathbf{p} = m\mathbf{v} + \frac{e}{c}\mathbf{A}. \tag{10.2c}$$

We can now move to either the Lagrangian or the Hamiltonian formulation. For the first, we simply adopt $\mathbf{v} = d\mathbf{r}/dt$ as a definition (but see the discussion in Sec. 9.3) and get

$$L = \frac{1}{2} m \left(\frac{d\mathbf{r}}{dt} \right)^2 - e\phi + \frac{e}{c} \frac{d\mathbf{r}}{dt} \cdot \mathbf{A}. \tag{10.3}$$

Alternatively, we use (10.2c) to define

$$\mathbf{v} = \frac{1}{m} \left(\mathbf{p} - \frac{e}{c}\mathbf{A} \right), \tag{10.4}$$

and find

$$L = \mathbf{p} \cdot \frac{d\mathbf{r}}{dt} - H, \tag{10.5a}$$

$$H = \frac{1}{2m} \left(\mathbf{p} - \frac{e}{c}\mathbf{A} \right)^2 + e\phi. \tag{10.5b}$$

Here we make contact with the energy considerations of Sec. 7.1; in particular, H coincides with the form of the energy given in (7.16).

DOI: 10.1201/9781003057369-10

10.2 Electrodynamic Action

The electromagnetic field is a mechanical system. It contributes its variables to the action, to the Lagrangian of the whole system of charges and fields. In contrast with the point charges, the field is distributed in space. Its Lagrangian should therefore be, not a summation over discrete points, but an integration over all spatial volume elements,

$$L_{\text{field}} = \int (d\mathbf{r})\, \mathcal{L}_{\text{field}}; \tag{10.6}$$

this introduces the Lagrange function, or Lagrangian density, $\mathcal{L}$. The total Lagrangian must be the sum of the particle part, (10.1), and the field part, (10.6), where the latter must be chosen so as to give the Maxwell equations (1.61b):

$$\begin{aligned} \boldsymbol{\nabla}\times\mathbf{B} &= \frac{1}{c}\frac{\partial}{\partial t}\mathbf{E} + \frac{4\pi}{c}\mathbf{j}, & \boldsymbol{\nabla}\cdot\mathbf{E} &= 4\pi\rho, \\ -\boldsymbol{\nabla}\times\mathbf{E} &= \frac{1}{c}\frac{\partial}{\partial t}\mathbf{B}, & \boldsymbol{\nabla}\cdot\mathbf{B} &= 0. \end{aligned} \tag{10.7}$$

The homogeneous equations here are equivalent to (7.6) and (7.3),

$$\frac{1}{c}\frac{\partial}{\partial t}\mathbf{A} = -\mathbf{E} - \boldsymbol{\nabla}\phi, \tag{10.8a}$$

$$\mathbf{B} = \boldsymbol{\nabla}\times\mathbf{A}. \tag{10.8b}$$

Thus, we recognize that $\mathbf{A}(\mathbf{r},t)$, $\mathbf{E}(\mathbf{r},t)$, in analogy with $\mathbf{r}(t)$, $\mathbf{p}(t)$, obey equations of motion while $\phi(\mathbf{r},t)$, $\mathbf{B}(\mathbf{r},t)$, as analogs of $\mathbf{v}(t)$, do not. There are enough clues here to give the structure of $\mathcal{L}_{\text{field}}$, apart from an overall factor. The anticipated complete Lagrangian for microscopic electrodynamics is

$$\begin{aligned} L = {} & \sum_a \left[\mathbf{p}_a\cdot\left(\frac{d\mathbf{r}_a}{dt} - \mathbf{v}_a\right) + \frac{1}{2}m_a v_a^2 - e_a\phi(\mathbf{r}_a) + \frac{e_a}{c}\mathbf{v}_a\cdot\mathbf{A}(\mathbf{r}_a)\right] \\ & + \frac{1}{4\pi}\int (d\mathbf{r}) \left[\mathbf{E}\cdot\left(-\frac{1}{c}\frac{\partial}{\partial t}\mathbf{A} - \boldsymbol{\nabla}\phi\right) - \mathbf{B}\cdot\boldsymbol{\nabla}\times\mathbf{A} + \frac{1}{2}(B^2 - E^2)\right]. \end{aligned} \tag{10.9}$$

The terms that are summed in (10.9) describe the behavior of charged particles under the influence of the fields, while the terms that are integrated describe the field behavior. The independent variables are

$$\mathbf{r}_a(t),\quad \mathbf{v}_a(t),\quad \mathbf{p}_a(t),\quad \phi(\mathbf{r},t),\quad \mathbf{A}(\mathbf{r},t),\quad \mathbf{E}(\mathbf{r},t),\quad \mathbf{B}(\mathbf{r},t),\quad t. \tag{10.10}$$

We now look at the response of the Lagrangian to variations in each of these variables separately, starting with the particle part:

$$\delta\mathbf{r}_a:\quad \delta L = \frac{d}{dt}(\delta\mathbf{r}_a\cdot\mathbf{p}_a) + \delta\mathbf{r}_a\cdot\left[-\frac{d\mathbf{p}_a}{dt} - \boldsymbol{\nabla}_a e_a\left(\phi(\mathbf{r}_a) - \frac{\mathbf{v}_a}{c}\cdot\mathbf{A}(\mathbf{r}_a)\right)\right], \tag{10.11a}$$

$$\delta\mathbf{v}_a:\quad \delta L = \delta\mathbf{v}_a\cdot\left[-\mathbf{p}_a + m_a\mathbf{v}_a + \frac{e_a}{c}\mathbf{A}(\mathbf{r}_a)\right], \tag{10.11b}$$

$$\delta\mathbf{p}_a:\quad \delta L = \delta\mathbf{p}_a\cdot\left(\frac{d\mathbf{r}_a}{dt} - \mathbf{v}_a\right). \tag{10.11c}$$

The stationary action principle now implies the equations of motion

$$\frac{d\mathbf{p}_a}{dt} = -e_a\boldsymbol{\nabla}_a\left(\phi(\mathbf{r}_a) - \frac{\mathbf{v}_a}{c}\cdot\mathbf{A}(\mathbf{r}_a)\right), \tag{10.12a}$$

$$m_a\mathbf{v}_a = \mathbf{p}_a - \frac{e_a}{c}\mathbf{A}(\mathbf{r}_a), \tag{10.12b}$$

$$\mathbf{v}_a = \frac{d\mathbf{r}_a}{dt}, \tag{10.12c}$$

which are the known results, (10.2a)–(10.2c).

The real work now lies in deriving the equations of motion for the fields. In order to cast all the field-dependent terms into integral form, we introduce charge and current densities,

$$\rho(\mathbf{r},t) = \sum_a e_a\delta(\mathbf{r} - \mathbf{r}_a(t)), \tag{10.13a}$$

$$\mathbf{j}(\mathbf{r},t) = \sum_a e_a\mathbf{v}_a(t)\delta(\mathbf{r} - \mathbf{r}_a(t)), \tag{10.13b}$$

so that

$$\sum_a\left[-e_a\phi(\mathbf{r}_a) + \frac{e_a}{c}\mathbf{v}_a\cdot\mathbf{A}(\mathbf{r}_a)\right] = \int(d\mathbf{r})\left[-\rho\phi + \frac{1}{c}\mathbf{j}\cdot\mathbf{A}\right]. \tag{10.14}$$

The volume integrals extend over sufficiently large regions to contain all the fields of interest. Consequently, we can integrate by parts and ignore the surface terms. The responses of the Lagrangian (10.9) to field variations, and the corresponding equations of motion deduced from the action principle are

$$\delta\phi: \quad \delta L = \frac{1}{4\pi}\int(d\mathbf{r})\,\delta\phi(\boldsymbol{\nabla}\cdot\mathbf{E} - 4\pi\rho), \tag{10.15a}$$

$$\boldsymbol{\nabla}\cdot\mathbf{E} = 4\pi\rho, \tag{10.15b}$$

$$\begin{aligned}\delta\mathbf{A}: \quad \delta L = &-\frac{1}{4\pi c}\frac{d}{dt}\int(d\mathbf{r})\,\delta\mathbf{A}\cdot\mathbf{E} \\ &+ \frac{1}{4\pi}\int(d\mathbf{r})\,\delta\mathbf{A}\cdot\left(\frac{1}{c}\frac{\partial\mathbf{E}}{\partial t} + \frac{4\pi}{c}\mathbf{j} - \boldsymbol{\nabla}\times\mathbf{B}\right),\end{aligned} \tag{10.15c}$$

$$\boldsymbol{\nabla}\times\mathbf{B} = \frac{1}{c}\frac{\partial}{\partial t}\mathbf{E} + \frac{4\pi}{c}\mathbf{j}, \tag{10.15d}$$

$$\delta\mathbf{E}: \quad \delta L = \frac{1}{4\pi}\int(d\mathbf{r})\,\delta\mathbf{E}\cdot\left(-\frac{1}{c}\frac{\partial}{\partial t}\mathbf{A} - \boldsymbol{\nabla}\phi - \mathbf{E}\right), \tag{10.15e}$$

$$\mathbf{E} = -\frac{1}{c}\frac{\partial}{\partial t}\mathbf{A} - \boldsymbol{\nabla}\phi, \tag{10.15f}$$

$$\delta\mathbf{B}: \quad \delta L = \frac{1}{4\pi}\int(d\mathbf{r})\,\delta\mathbf{B}\cdot(-\boldsymbol{\nabla}\times\mathbf{A} + \mathbf{B}), \tag{10.15g}$$

$$\mathbf{B} = \boldsymbol{\nabla}\times\mathbf{A}. \tag{10.15h}$$

We therefore recover Maxwell's equations, two of which are implicit in the construction of $\mathbf{E}$ and $\mathbf{B}$ in terms of potentials. By making a time variation of the action [variations due to the time dependence of the fields vanish by virtue of the stationary action principle—that is, they are already subsumed in (10.15a)–(10.15h)],

$$\delta t: \quad \delta W = \int dt\left[\frac{d}{dt}(-H\delta t) + \delta t\frac{dH}{dt}\right], \tag{10.16}$$

we identify the Hamiltonian of the system to be

$$H = \sum_a \left[\left(\mathbf{p}_a - \frac{e_a}{c}\mathbf{A}(\mathbf{r}_a) \right) \cdot \mathbf{v}_a - \frac{1}{2}m_a v_a^2 + e_a\phi(\mathbf{r}_a) \right] + \frac{1}{4\pi}\int (d\mathbf{r}) \left[\mathbf{E}\cdot\nabla\phi + \mathbf{B}\cdot\nabla\times\mathbf{A} + \frac{1}{2}(E^2 - B^2) \right], \tag{10.17}$$

which is a constant of the motion, $dH/dt = 0$. The generators are inferred from the total time derivative terms in (10.11a), (10.15c), and (10.16),

$$\delta W_{12} = G_1 - G_2, \tag{10.18}$$

to be

$$G = \sum_a \delta\mathbf{r}_a \cdot \mathbf{p}_a - \frac{1}{4\pi c}\int (d\mathbf{r})\, \mathbf{E}\cdot\delta\mathbf{A} - H\delta t. \tag{10.19}$$

10.3 Energy

Notice that the total Lagrangian (10.9) can be presented as

$$L = \sum_a \mathbf{p}_a \cdot \frac{d\mathbf{r}_a}{dt} - \frac{1}{4\pi c}\int (d\mathbf{r})\, \mathbf{E}\cdot\frac{\partial}{\partial t}\mathbf{A} - H, \tag{10.20}$$

where the Hamiltonian is given by (10.17). The narrower, Hamiltonian, description is reached by eliminating all variables that do not obey equations of motion, and, correspondingly, do not appear in G. Those "superfluous" variables are the $\mathbf{v}_a$ and the fields ϕ and $\mathbf{B}$, which are determined by using (10.12b), (10.15b), and (10.15h), the equations without time derivatives, resulting, first, in the intermediate form

$$H = \sum_a \left(\frac{1}{2m_a}\left(\mathbf{p}_a - \frac{e_a}{c}\mathbf{A}(\mathbf{r}_a) \right)^2 + e_a\phi(\mathbf{r}_a) \right) + \int (d\mathbf{r}) \left[\frac{E^2 + B^2}{8\pi} - \rho\phi \right]. \tag{10.21}$$

The first term here is the energy of the particles moving in the field [particle energy—see (10.5b)], so we might call the second term the field energy. The ambiguity of these terms (whether the potential energy of particles is attributed to them or to the fields, or to both) is evident from the existence of a simpler form of the Hamiltonian

$$H = \sum_a \frac{1}{2m_a}\left(\mathbf{p}_a - \frac{e_a}{c}\mathbf{A}(\mathbf{r}_a) \right)^2 + \int (d\mathbf{r})\, \frac{E^2 + B^2}{8\pi}, \tag{10.22}$$

where we have used the equivalence of the two terms involving ϕ, given in (10.14).

This apparently startling result suggests that the scalar potential has disappeared from the dynamical description. But, in fact, it has not. If we vary the Lagrangian (10.20), where H is given by (10.22), with respect to $\mathbf{E}$, we find

$$\delta L = -\frac{1}{4\pi}\int (d\mathbf{r})\, \delta\mathbf{E}\cdot\left(\frac{1}{c}\frac{\partial}{\partial t}\mathbf{A} + \mathbf{E} \right) = 0. \tag{10.23}$$

Do we conclude that $\frac{1}{c}\frac{\partial}{\partial t}\mathbf{A} + \mathbf{E} = \mathbf{0}$? That would be true if the $\delta\mathbf{E}(\mathbf{r}, t)$ were arbitrary. They are not; $\mathbf{E}$ is subject to the restriction—the constraint—(10.15b), which means that any change in $\mathbf{E}$ must obey

$$\boldsymbol{\nabla} \cdot \delta\mathbf{E} = 0. \tag{10.24}$$

The proper conclusion is that the vector multiplying $\delta\mathbf{E}$ in (10.23) is the gradient of a scalar function, just as in (10.15f),

$$\frac{1}{c}\frac{\partial}{\partial t}\mathbf{A} + \mathbf{E} = -\boldsymbol{\nabla}\phi, \tag{10.25}$$

for that leads to

$$\delta L = -\frac{1}{4\pi}\int (d\mathbf{r})\,(\boldsymbol{\nabla} \cdot \delta\mathbf{E})\phi = 0, \tag{10.26}$$

as required.

The fact that the energy is conserved,

$$\frac{dH}{dt} = 0, \tag{10.27}$$

where

$$H = \sum_a \frac{1}{2} m_a v_a^2 + \int (d\mathbf{r})\,U, \quad U = \frac{E^2 + B^2}{8\pi}, \tag{10.28}$$

is a simple sum of particle kinetic energy and integrated field energy density, can be verified directly by taking the time derivative of (10.21). The time rate of change of the particle energy was computed in (7.14),

$$\frac{d}{dt}\sum_a \left(\frac{1}{2} m_a v_a^2 + e_a\phi(\mathbf{r}_a)\right) = \sum_a \frac{\partial}{\partial t}\left(e_a\phi(\mathbf{r}_a) - \frac{e_a}{c}\mathbf{v}_a \cdot \mathbf{A}(\mathbf{r}_a)\right). \tag{10.29}$$

We can compute the time derivative of the field energy by using the equation of energy conservation, (3.6),

$$\frac{d}{dt}\int (d\mathbf{r})\,U = -\int (d\mathbf{r})\,\mathbf{j} \cdot \mathbf{E}, \tag{10.30}$$

to be

$$\begin{aligned} \frac{d}{dt}\int (d\mathbf{r})\left(\frac{E^2 + B^2}{8\pi} - \rho\phi\right) &= \int (d\mathbf{r})\left[-\mathbf{j} \cdot \mathbf{E} - \phi\frac{\partial}{\partial t}\rho - \rho\frac{\partial}{\partial t}\phi\right] \\ &= -\int (d\mathbf{r})\left[\rho\frac{\partial}{\partial t}\phi - \frac{1}{c}\mathbf{j} \cdot \frac{\partial}{\partial t}\mathbf{A}\right] \\ &= -\sum_a e_a\left(\frac{\partial}{\partial t}\phi(\mathbf{r}_a) - \frac{1}{c}\mathbf{v}_a \cdot \frac{\partial}{\partial t}\mathbf{A}(\mathbf{r}_a)\right). \end{aligned} \tag{10.31}$$

Here we have used (10.15f), and have noted that

$$\int (d\mathbf{r})\left[\mathbf{j} \cdot \boldsymbol{\nabla}\phi - \phi\frac{\partial}{\partial t}\rho\right] = 0 \tag{10.32}$$

by charge conservation. Observe that (10.29) and (10.31) are equal in magnitude and opposite in sign so that their sum is zero. This proves the statement of energy conservation (10.27).

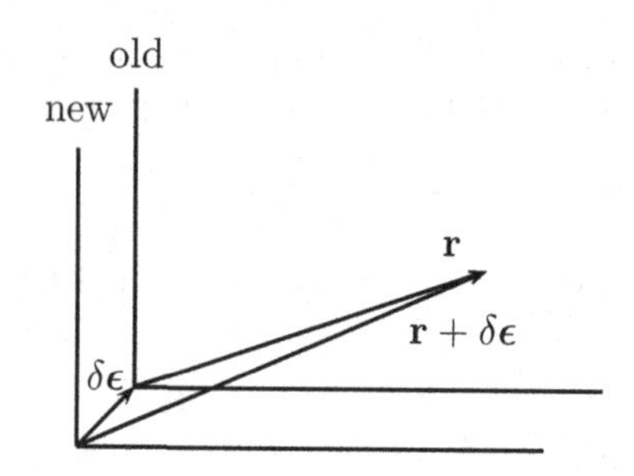

FIGURE 10.1
Rigid coordinate displacement.

10.4 Momentum and Angular Momentum Conservation

The action principle not only provides us with the field equations, particle equations of motion, and expressions for the energy, but also with the generators (10.19). The generators provide a connection between conservation laws and invariances of the action (recall Section 9.4). Here we will further illustrate this connection by deriving momentum and angular momentum conservation from the invariance of the action under rigid coordinate translations and rotations, respectively. [In a similar way we could derive energy conservation, (10.27), from the invariance under time displacements—see also Section 10.6.]

Under an infinitesimal rigid coordinate displacement, $\delta\boldsymbol{\epsilon}$, a given point which is described by $\mathbf{r}$ in the old coordinate system is described by $\mathbf{r}+\delta\boldsymbol{\epsilon}$ in the new one. (See Fig. 10.1.) The response of the particle term in (10.19) is simple: $\delta\boldsymbol{\epsilon}\cdot\sum_a \mathbf{p}_a$; for the field part, we require the change, $\delta\mathbf{A}$, of the vector potential induced by the rigid coordinate displacement. The value of a field $\mathcal{F}$ at a physical point P is unchanged under such a displacement so that if $\mathbf{r}$ and $\mathbf{r}+\delta\boldsymbol{\epsilon}$ are the coordinates of P in the two frames, there are corresponding functions F and $\overline{F}$ such that

$$\mathcal{F}(P) = F(\mathbf{r}) = \overline{F}(\mathbf{r}+\delta\boldsymbol{\epsilon}), \tag{10.33}$$

that is, the new function $\overline{F}$ of the new coordinate equals the old function F of the old coordinate. The change in the function F at the *same coordinate* is given by

$$\overline{F}(\mathbf{r}) = F(\mathbf{r}) + \delta F(\mathbf{r}), \tag{10.34}$$

so that

$$\delta F(\mathbf{r}) = F(\mathbf{r}-\delta\boldsymbol{\epsilon}) - F(\mathbf{r}) = -\delta\boldsymbol{\epsilon}\cdot\boldsymbol{\nabla}F(\mathbf{r}), \tag{10.35}$$

for a rigid translation (not a rotation).

As an example, consider the charge density

$$\rho(\mathbf{r}) = \sum_a e_a\delta(\mathbf{r}-\mathbf{r}_a). \tag{10.36}$$

If the positions of all the particles, the $\mathbf{r}_a$, are displaced by $\delta\boldsymbol{\epsilon}$, the charge density changes to

$$\rho(\mathbf{r}) + \delta\rho(\mathbf{r}) = \sum_a e_a\delta(\mathbf{r}-\mathbf{r}_a-\delta\boldsymbol{\epsilon}), \tag{10.37}$$

where

$$\delta(\mathbf{r}-\mathbf{r}_a-\delta\boldsymbol{\epsilon}) = \delta(\mathbf{r}-\mathbf{r}_a) - \delta\boldsymbol{\epsilon}\cdot\boldsymbol{\nabla}_r\delta(\mathbf{r}-\mathbf{r}_a), \tag{10.38}$$

and therefore

$$\delta\rho(\mathbf{r}) = -\delta\boldsymbol{\epsilon}\cdot\boldsymbol{\nabla}\rho(\mathbf{r}), \tag{10.39}$$

in agreement with (10.35).

So the field part of G in (10.19) is

$$\begin{aligned}-\int(d\mathbf{r})\,\frac{1}{4\pi c}\mathbf{E}\cdot\delta\mathbf{A} &= \frac{1}{4\pi c}\int(d\mathbf{r})\,E_i(\delta\boldsymbol{\epsilon}\cdot\boldsymbol{\nabla})A_i\\ &= -\frac{1}{c}\sum_a e_a\delta\boldsymbol{\epsilon}\cdot\mathbf{A}(\mathbf{r}_a) + \frac{1}{4\pi c}\int(d\mathbf{r})\,(\mathbf{E}\times\mathbf{B})\cdot\delta\boldsymbol{\epsilon},\end{aligned} \tag{10.40}$$

where the last rearrangement makes use of (10.15b) and (10.15h), and the vector identity

$$\delta\boldsymbol{\epsilon}\times(\boldsymbol{\nabla}\times\mathbf{A}) = \boldsymbol{\nabla}(\delta\boldsymbol{\epsilon}\cdot\mathbf{A}) - (\delta\boldsymbol{\epsilon}\cdot\boldsymbol{\nabla})\mathbf{A}. \tag{10.41}$$

Including the particle part from (10.19), we find the generator corresponding to a rigid coordinate displacement can be written as

$$G = \delta\boldsymbol{\epsilon}\cdot\mathbf{P}, \tag{10.42}$$

where

$$\mathbf{P} = \sum_a\left(\mathbf{p}_a - \frac{e_a}{c}\mathbf{A}(\mathbf{r}_a)\right) + \frac{1}{4\pi c}\int(d\mathbf{r})\,\mathbf{E}\times\mathbf{B} = \sum_a m_a\mathbf{v}_a + \int(d\mathbf{r})\,\mathbf{G}, \tag{10.43}$$

with $\mathbf{G}$ the momentum density, (3.12a). Since the action is invariant under a rigid displacement,

$$0 = \delta W = G_1 - G_2 = (\mathbf{P}_1 - \mathbf{P}_2)\cdot\delta\mathbf{r}, \tag{10.44}$$

we see that

$$\mathbf{P}_1 = \mathbf{P}_2, \tag{10.45}$$

that is, the total momentum, $\mathbf{P}$, is conserved. This, of course, can also be verified by explicit calculation:

$$\begin{aligned}\frac{d}{dt}\int(d\mathbf{r})\,\frac{1}{4\pi c}\mathbf{E}\times\mathbf{B} &= -\int(d\mathbf{r})\,\left[\rho\mathbf{E} + \frac{1}{c}\mathbf{j}\times\mathbf{B}\right]\\ &= -\sum_a e_a\left(\mathbf{E}(\mathbf{r}_a) + \frac{1}{c}\mathbf{v}_a\times\mathbf{B}(\mathbf{r}_a)\right),\end{aligned} \tag{10.46}$$

which restates (3.13), from which the constancy of $\mathbf{P}$ follows from (7.1).

Similar arguments can be carried out for a rigid rotation for which the change in the coordinate vector is

$$\delta\mathbf{r} = \delta\boldsymbol{\omega}\times\mathbf{r}, \tag{10.47}$$

with $\delta\boldsymbol{\omega}$ constant. The corresponding change in a vector function is

$$\overline{\mathbf{A}}(\mathbf{r}+\delta\mathbf{r}) = \mathbf{A}(\mathbf{r}) + \delta\boldsymbol{\omega}\times\mathbf{A}(\mathbf{r}) \tag{10.48}$$

since a vector transforms in the same way as $\mathbf{r}$, so the new function at the initial numerical values of the coordinates is

$$\overline{\mathbf{A}}(\mathbf{r}) = \mathbf{A}(\mathbf{r}) - (\delta\mathbf{r}\cdot\boldsymbol{\nabla})\mathbf{A}(\mathbf{r}) + \delta\boldsymbol{\omega}\times\mathbf{A}(\mathbf{r}). \tag{10.49}$$

(See Problem 10.3.) The change in the vector potential is

$$\delta\mathbf{A} = -(\delta\mathbf{r}\cdot\boldsymbol{\nabla})\mathbf{A} + \delta\boldsymbol{\omega}\times\mathbf{A}. \tag{10.50}$$

The generator can now be written in the form

$$G = \delta\boldsymbol{\omega}\cdot\mathbf{J}, \tag{10.51}$$

where the total angular momentum, $\mathbf{J}$, is found to be (see Problem 10.5)

$$\mathbf{J} = \sum_a \mathbf{r}_a\times m_a\mathbf{v}_a + \int (d\mathbf{r})\,\mathbf{r}\times\mathbf{G}, \tag{10.52}$$

which again is a constant of the motion.

10.5 Gauge Invariance and the Conservation of Charge

An electromagnetic system possesses a conservation law, that of electric charge, which has no place in the usual mechanical framework. It is connected to a further invariance of the electromagnetic fields—the potentials are not uniquely defined in that if we let

$$\mathbf{A}\to\mathbf{A}+\boldsymbol{\nabla}\lambda, \qquad \phi\to\phi-\frac{1}{c}\frac{\partial}{\partial t}\lambda, \tag{10.53}$$

the electric and magnetic fields defined by (10.15f) and (10.15h) remain unaltered, for an arbitrary function λ. This is called gauge invariance; the corresponding substitution (10.53) is a gauge transformation. [The term has its origin in a now obsolete theory (1918) of Hermann Weyl (1885–1955).]

This invariance of the action must imply a corresponding conservation law. To determine what is conserved, we compute the change in the Lagrangian, (10.9), explicitly. Trivially, the field part of L remains unchanged. In considering the change of the particle part, we recognize that (10.53) is incomplete; since $\mathbf{v}$ is a physical quantity, $\mathbf{p}-(e/c)\mathbf{A}$ must be invariant under a gauge transformation, which will only be true if (10.53) is supplemented by

$$\mathbf{p}\to\mathbf{p}+\frac{e}{c}\boldsymbol{\nabla}\lambda. \tag{10.54}$$

Under the transformation (10.53) and (10.54), the Lagrangian becomes

$$\begin{aligned} L\to\overline{L} &\equiv L+\sum_a\left[\frac{e_a}{c}\boldsymbol{\nabla}\lambda\cdot\left(\frac{d\mathbf{r}_a}{dt}-\mathbf{v}_a\right)+\frac{e_a}{c}\frac{\partial}{\partial t}\lambda+\frac{e_a}{c}\mathbf{v}_a\cdot\boldsymbol{\nabla}\lambda\right]\\ &= L+\sum_a\frac{e_a}{c}\left(\frac{\partial}{\partial t}\lambda+\frac{d\mathbf{r}_a}{dt}\cdot\boldsymbol{\nabla}\lambda\right) \qquad (10.55\text{a})\\ &= L+\frac{d}{dt}w, \qquad (10.55\text{b}) \end{aligned}$$

where

$$w=\sum_a\frac{e_a}{c}\lambda(\mathbf{r}_a,t). \tag{10.56}$$

What is the physical consequence of adding a total time derivative to a Lagrangian? It does not change the equations of motion, so the system is unaltered. Since the entire change is in the end point behavior,

$$\overline{W}_{12}=W_{12}+(w_1-w_2), \tag{10.57}$$

the whole effect is a redefinition of the generators, G,

$$\overline{G} = G + \delta w. \tag{10.58}$$

This alteration reflects the fact that the Lagrangian itself is ambiguous up to a total time derivative term.

To ascertain the implication of gauge invariance, we rewrite the change in the Lagrangian given in (10.55a) by use of (10.13),

$$\overline{L} - L = \frac{1}{c}\int (d\mathbf{r})\left[\rho\frac{\partial}{\partial t}\lambda + \mathbf{j}\cdot\boldsymbol{\nabla}\lambda\right], \tag{10.59}$$

and apply this result to an infinitesimal gauge transformation, $\lambda \to \delta\lambda$. The change in the action is then

$$\delta W_{12} = G_{\delta\lambda_1} - G_{\delta\lambda_2} - \int_{t_2}^{t_1} dt \int (d\mathbf{r})\,\frac{1}{c}\delta\lambda\left(\frac{\partial}{\partial t}\rho + \boldsymbol{\nabla}\cdot\mathbf{j}\right), \tag{10.60}$$

with the generator being

$$G_{\delta\lambda} = \int (d\mathbf{r})\,\frac{1}{c}\rho\,\delta\lambda. \tag{10.61}$$

In view of the arbitrary nature of $\delta\lambda(\mathbf{r},t)$, the stationary action principle now demands that, at every point,

$$\frac{\partial}{\partial t}\rho + \boldsymbol{\nabla}\cdot\mathbf{j} = 0, \tag{10.62}$$

that is, gauge invariance implies local charge conservation. (Of course, this same result follows from Maxwell's equations.) Then, the special situation $\delta\lambda =$ constant, where $\delta\mathbf{A} = \delta\phi = 0$, and W_{12} is certainly invariant, implies a conservation law, that of

$$G_{\delta\lambda} = \frac{1}{c}\delta\lambda\, Q, \tag{10.63}$$

in which

$$Q = \int (d\mathbf{r})\,\rho \tag{10.64}$$

is the conserved total charge.

10.6 Gauge Invariance and Local Conservation Laws

We have just derived the local conservation law of electric charge. Electric charge is a property carried only by the particles, not by the electromagnetic field. In contrast, the mechanical properties of energy, linear, and angular momentum are attributes of both particles and fields. For these we have conservation laws of total quantities. What about local conservation laws? Early in this development (Chapter 3), we produced local *non*-conservation laws; they concentrated on the fields and characterized the charged particles as sources (or sinks) of field mechanical properties. It is natural to ask for a more even-handed treatment of both charges and fields. We shall supply it, in the framework of a particular example. The property of gauge invariance will be both a valuable guide, and an aid to simplifying the calculations.

The time displacement of a complete physical system identifies its total energy. This suggests that time displacement of a part of the system provides energetic information about that portion. The ultimate limit of this spatial subdivision, a local description, should appear in response to an (infinitesimal) time displacement that varies arbitrarily in space as well as in time, $\delta t(\mathbf{r}, t)$.

Now we need a clue. How do fields, and potentials, respond to such coordinate-dependent displacements? This is where the freedom of gauge transformations enters: The change of the vector and scalar potentials, by $\boldsymbol{\nabla}\lambda(\mathbf{r}, t)$, $-(1/c)(\partial/\partial t)\lambda(\mathbf{r}, t)$, respectively, serves as a model for the potentials themselves. The advantage here is that the response of the scalar $\lambda(\mathbf{r}, t)$ to the time displacement can be reasonably taken to be

$$(\lambda + \delta\lambda)(\mathbf{r}, t + \delta t) = \lambda(\mathbf{r}, t), \tag{10.65}$$

or

$$\delta\lambda(\mathbf{r}, t) = -\delta t(\mathbf{r}, t)\frac{\partial}{\partial t}\lambda(\mathbf{r}, t). \tag{10.66}$$

Then we derive

$$\delta(\boldsymbol{\nabla}\lambda) = -\delta t\frac{\partial}{\partial t}(\boldsymbol{\nabla}\lambda) + \left(-\frac{1}{c}\frac{\partial}{\partial t}\lambda\right)c\boldsymbol{\nabla}\delta t, \tag{10.67a}$$

$$\delta\left(-\frac{1}{c}\frac{\partial}{\partial t}\lambda\right) = -\delta t\left(-\frac{1}{c}\frac{\partial^2}{\partial t^2}\lambda\right) - \left(-\frac{1}{c}\frac{\partial}{\partial t}\lambda\right)\frac{\partial}{\partial t}\delta t, \tag{10.67b}$$

which is immediately generalized to

$$\delta\mathbf{A} = -\delta t\frac{\partial}{\partial t}\mathbf{A} + \phi c\boldsymbol{\nabla}\delta t, \tag{10.68a}$$

$$\delta\phi = -\delta t\frac{\partial}{\partial t}\phi - \phi\frac{\partial}{\partial t}\delta t, \tag{10.68b}$$

or, equivalently,

$$\delta\mathbf{A} = c\delta t\mathbf{E} + \boldsymbol{\nabla}(\phi c\delta t), \tag{10.69a}$$

$$\delta\phi = -\frac{1}{c}\frac{\partial}{\partial t}(\phi c\delta t). \tag{10.69b}$$

In the latter form we recognize a gauge transformation, produced by the scalar $\phi c\delta t$, which will not contribute to the changes of field strengths. Accordingly, for that calculation we have, effectively, $\delta\mathbf{A} = c\delta t\mathbf{E}$, $\delta\phi = 0$, leading to

$$\delta\mathbf{E} = -\frac{1}{c}\frac{\partial}{\partial t}(c\delta t\mathbf{E}) = -\delta t\frac{\partial}{\partial t}\mathbf{E} - \mathbf{E}\frac{\partial}{\partial t}\delta t, \tag{10.70a}$$

$$\delta\mathbf{B} = \boldsymbol{\nabla}\times(c\delta t\mathbf{E}) = -\delta t\frac{\partial}{\partial t}\mathbf{B} - \mathbf{E}\times\boldsymbol{\nabla}c\delta t; \tag{10.70b}$$

the last line employs the field equation $\boldsymbol{\nabla}\times\mathbf{E} = -(1/c)(\partial\mathbf{B}/\partial t)$.

In the following we adopt a viewpoint in which such homogeneous field equations are accepted as consequences of the definition of the fields in terms of potentials. That permits the field Lagrange function (10.9) to be simplified:

$$\mathcal{L}_{\text{field}} = \frac{1}{8\pi}(E^2 - B^2). \tag{10.71}$$

Then we can apply the field variation (10.70b) directly, and get

$$\begin{aligned}\delta\mathcal{L}_{\text{field}} &= -\delta t\frac{\partial}{\partial t}\mathcal{L}_{\text{field}} - \frac{1}{4\pi}E^2\frac{\partial}{\partial t}\delta t - \frac{c}{4\pi}\mathbf{E}\times\mathbf{B}\cdot\boldsymbol{\nabla}\delta t\\ &= -\frac{\partial}{\partial t}(\delta t\mathcal{L}_{\text{field}}) - \frac{1}{8\pi}(E^2 + B^2)\frac{\partial}{\partial t}\delta t - \frac{c}{4\pi}\mathbf{E}\times\mathbf{B}\cdot\boldsymbol{\nabla}\delta t.\end{aligned} \tag{10.72}$$

Before commenting on these last, not unfamiliar, field structures, we turn to the charged particles and put them on a somewhat similar footing in terms of a continuous, rather than a discrete, description.

We therefore present the Lagrangian of the charges in (10.9) in terms of a corresponding Lagrange function,

$$L_{\text{charges}} = \int (d\mathbf{r})\, \mathcal{L}_{\text{charges}}, \tag{10.73a}$$

where

$$\mathcal{L}_{\text{charges}} = \sum_a \mathcal{L}_a \tag{10.73b}$$

and

$$\mathcal{L}_a = \delta(\mathbf{r} - \mathbf{r}_a(t)) \left[\frac{1}{2} m_a v_a(t)^2 - e_a \phi(\mathbf{r}_a, t) + \frac{e_a}{c} \mathbf{v}_a(t) \cdot \mathbf{A}(\mathbf{r}_a, t)\right]; \tag{10.73c}$$

the latter adopts the Lagrangian viewpoint, with $\mathbf{v}_a = d\mathbf{r}_a/dt$ accepted as a definition. Then, the effect of the time displacement on the variables $\mathbf{r}_a(t)$, taken as

$$(\mathbf{r}_a + \delta\mathbf{r}_a)(t + \delta t) = \mathbf{r}_a(t), \tag{10.74a}$$

$$\delta\mathbf{r}_a(t) = -\delta t(\mathbf{r}_a, t)\mathbf{v}_a(t), \tag{10.74b}$$

implies the velocity variation

$$\delta\mathbf{v}_a(t) = -\delta t(\mathbf{r}_a, t)\frac{d}{dt}\mathbf{v}_a(t) - \mathbf{v}_a(t)\left[\frac{\partial}{\partial t}\delta t + \mathbf{v}_a \cdot \boldsymbol{\nabla}\delta t\right]; \tag{10.75}$$

the last step exhibits both the explicit and the implicit dependences of $\delta t(\mathbf{r}_a, t)$ on t. In computing the variation of $\phi(\mathbf{r}_a, t)$, for example, we combine the potential variation given in (10.68b) with the effect of $\delta\mathbf{r}_a$:

$$\delta\phi(\mathbf{r}_a(t), t) = -\delta t\frac{\partial}{\partial t}\phi - \phi\frac{\partial}{\partial t}\delta t - \delta t\mathbf{v}_a \cdot \boldsymbol{\nabla}_a\phi = -\delta t\frac{d}{dt}\phi - \phi\frac{\partial}{\partial t}\delta t, \tag{10.76a}$$

and, similarly,

$$\delta\mathbf{A}(\mathbf{r}_a(t), t) = -\delta t\frac{\partial}{\partial t}\mathbf{A} + \phi c\boldsymbol{\nabla}\delta t - \delta t\mathbf{v}_a \cdot \boldsymbol{\nabla}_a\mathbf{A} = -\delta t\frac{d}{dt}\mathbf{A} + \phi c\boldsymbol{\nabla}\delta t. \tag{10.76b}$$

The total effect of these variations on $\mathcal{L}_a$ is thus

$$\delta\mathcal{L}_a = -\delta t\frac{d}{dt}\mathcal{L}_a + \delta(\mathbf{r} - \mathbf{r}_a(t))\left(-m_a v_a^2 - \frac{e_a}{c}\mathbf{A} \cdot \mathbf{v}_a + e_a\phi\right)\left(\frac{\partial}{\partial t}\delta t + \mathbf{v}_a \cdot \boldsymbol{\nabla}\delta t\right), \tag{10.77}$$

or

$$\delta\mathcal{L}_a = -\frac{d}{dt}(\delta t\mathcal{L}_a) - \delta(\mathbf{r} - \mathbf{r}_a(t))E_a\left(\frac{\partial}{\partial t}\delta t + \mathbf{v}_a \cdot \boldsymbol{\nabla}\delta t\right), \tag{10.78}$$

where we see the kinetic energy of the charged particle,

$$E_a = \frac{1}{2}m_a v_a^2. \tag{10.79}$$

We have retained the particle symbol d/dt to the last, but now, being firmly back in the field, space-time viewpoint, it should be written as $\partial/\partial t$, referring to all t dependence, with $\mathbf{r}$ being held fixed. The union of these various contributions to the variation of the total Lagrange function is

$$\delta\mathcal{L}_{\text{tot}} = -\frac{\partial}{\partial t}(\delta t\mathcal{L}_{\text{tot}}) - U_{\text{tot}}\frac{\partial}{\partial t}\delta t - \mathbf{S}_{\text{tot}} \cdot \boldsymbol{\nabla}\delta t, \tag{10.80}$$

where, from (10.72) and (10.78),

$$U_{\text{tot}} = \frac{1}{8\pi}(E^2 + B^2) + \sum_a \delta(\mathbf{r} - \mathbf{r}_a(t))E_a \tag{10.81}$$

and

$$\mathbf{S}_{\text{tot}} = \frac{c}{4\pi}\mathbf{E}\times\mathbf{B} + \sum_a \delta(\mathbf{r} - \mathbf{r}_a(t))E_a\mathbf{v}_a, \tag{10.82}$$

are physically transparent forms for the total energy density and total energy flux vector.

To focus on what is new in this development, we ignore boundary effects in the stationary action principle, by setting the otherwise arbitrary $\delta t(\mathbf{r}, t)$ equal to zero at t_1 and t_2. Then, through partial integration, we conclude that

$$\delta W_{12} = \int_{t_2}^{t_1} dt \int (d\mathbf{r})\, \delta t \left(\frac{\partial}{\partial t}U_{\text{tot}} + \boldsymbol{\nabla}\cdot\mathbf{S}_{\text{tot}}\right) = 0, \tag{10.83}$$

from which follows the local statement of total energy conservation,

$$\frac{\partial}{\partial t}U_{\text{tot}} + \boldsymbol{\nabla}\cdot\mathbf{S}_{\text{tot}} = 0, \tag{10.84}$$

which generalizes (3.3).

10.7 Problems for Chapter 10

1. Consider the Lagrangian of a particle in a given electromagnetic field, (10.1):

$$L(\mathbf{r}, \mathbf{p}, \mathbf{v}, t) = \mathbf{p}\cdot\left(\frac{d\mathbf{r}}{dt} - \mathbf{v}\right) + \frac{1}{2}mv^2 - e\phi(\mathbf{r}, t) + \frac{e}{c}\mathbf{v}\cdot\mathbf{A}(\mathbf{r}, t). \tag{10.85}$$

   (a) Reduce to Lagrangian form and then derive the equations of motion.
   (b) Reduce to Hamiltonian form and then derive the equations of motion.
   (c) Show the equivalence between a) and b).

2. A nonrelativistic particle of charge e and mass m moves in a homogeneous magnetic field $\mathbf{B}$. Let the direction of the magnetic field be the z axis. What is a vector potential corresponding to such a field? Write down the Hamiltonian for the system. Obtain the Hamiltonian equations of motion for the particles, that is, the equations for $\dot{\mathbf{r}}$ and $\dot{\mathbf{p}}$, expressed in terms of $\mathbf{B}$. Show that these equations can be expressed as a constant precession of the velocity vector $\dot{\mathbf{r}}$ around the direction of $\mathbf{B}$. What is the frequency of this precession?

3. Verify the transformation laws under rigid rotations for scalars and vectors,

$$\delta S(\mathbf{r}) = -(\delta\boldsymbol{\omega}\times\mathbf{r})\cdot\boldsymbol{\nabla}S(\mathbf{r}), \tag{10.86a}$$
$$\delta\mathbf{V}(\mathbf{r}) = -(\delta\boldsymbol{\omega}\times\mathbf{r})\cdot\boldsymbol{\nabla}\mathbf{V}(\mathbf{r}) + \delta\boldsymbol{\omega}\times\mathbf{V}(\mathbf{r}), \tag{10.86b}$$

   by considering the transformations of the charge and current densities (10.13b).

4. Derive the behavior of a vector field $\mathbf{V}(\mathbf{r})$ under infinitesimal rotations by considering the example of the gradient of a scalar field $S(\mathbf{r})$.

5. Fill in the details to derive (10.52).
6. Prove directly from the equations of motion that the total angular momentum (10.52) is conserved.
7. Verify directly the local conservation law obeyed by U_{tot} and $\mathbf{S}_{\text{tot}}$, (10.84).
8. By considering $\delta\mathbf{r}(\mathbf{r}, t)$, analogous to $\delta t(\mathbf{r}, t)$ in Section 10.6, in its effect on $\mathcal{L}_{\text{field}}$, derive the field momentum density and stress dyadic.
9. Consider a scalar field, $\phi(\mathbf{r}, t)$, described by the Lagrangian density (here units are chosen so that $c = 1$)

$$\mathcal{L} = \frac{1}{2}\left(\frac{\partial\phi}{\partial t}\right)^2 - \frac{1}{2}(\boldsymbol{\nabla}\phi)^2 - \frac{1}{2}\mu^2\phi^2. \tag{10.87}$$

This describes a relativistic spinless particle of mass μ in quantum field theory. Consider the action

$$W_{12} = \int_2^1 dt \int (d\mathbf{r})\mathcal{L}, \tag{10.88}$$

where the spatial integration is carried out over all space.

(a) By requiring $\delta W = 0$ for a ϕ variation in the interior of the integration region, determine the field equation satisfied by ϕ.

(b) By requiring $\delta W = 0$ for a time variation other than at the endpoints, describe the energy or Hamiltonian H, and find the Hamiltonian density given by

$$H = \int (d\mathbf{r})\mathcal{H}. \tag{10.89}$$

The latter should be expressed in terms of ϕ and the canonical momentum $\pi = \partial\mathcal{L}/\partial\dot{\phi}$, where the dot denotes a partial derivative with respect to time.

(c) Now consider variations at the starting and ending times, t_2 and t_1, respectively. Calculate the generators, and again identify the canonical momentum π and the Hamiltonian density $\mathcal{H}$.

(d) Show that the Lagrangian and Hamiltonian densities are related by

$$\mathcal{L}(\phi, \dot{\phi}) = \pi\dot{\phi} - \mathcal{H}(\phi, \pi). \tag{10.90}$$

Show that requiring that the right-hand side of this equation to be stationary with respect to ϕ and π variations gives a set of first-order equations for $\dot{\phi}$, $\dot{\pi}$, which are equivalent to the second-order equation found in part 9a.

11

Einsteinian Relativity

11.1 Relativistic Modification

After the discussion in the previous section one might well ask whether a unified dynamics of charges and fields has now been attained. The answer is no—there is still a major flaw. An electromagnetic pulse is a mechanical system that travels at the speed of light, carrying a mass proportional to the total energy content, as in (3.40),

$$m = \frac{E}{c^2}. \tag{11.1}$$

In contrast, the masses of the charged particles are fixed quantities that have no reference to the particle's state of motion and its associated energy. Another way of expressing this lack of mechanical unity between fields and particles comes from the physically evident expression for the total momentum density (Problem 10.8)

$$\mathbf{G}_{\text{tot}} = \frac{1}{4\pi c}\mathbf{E}\times\mathbf{B} + \sum_a \delta(\mathbf{r} - \mathbf{r}_a(t)) m_a \mathbf{v}_a. \tag{11.2}$$

The relation

$$\mathbf{G}_{\text{tot}} = \frac{1}{c^2}\mathbf{S}_{\text{tot}}, \tag{11.3}$$

which is valid for the field terms, does not hold for the particle contribution [see (10.82)]. Could it be that Newtonian mechanics is mistaken, and that the correct expressions for particle inertia and energy do satisfy $m = E/c^2$? We now follow this unifying suggestion—that the relation between inertia and energy, which the electromagnetic field has disclosed, is, in fact, universally valid.

Consider a single particle in the absence of applied electromagnetic fields. What we are proposing is that the connection between momentum and velocity is actually

$$\mathbf{p} = \frac{E}{c^2}\mathbf{v}. \tag{11.4}$$

To this we add the relation of Hamiltonian mechanics,

$$\mathbf{v} = \frac{\partial E}{\partial \mathbf{p}}, \tag{11.5}$$

and deduce that

$$c^2\mathbf{p}\cdot d\mathbf{p} = E\,dE, \tag{11.6}$$

which is integrated to

$$E^2 = c^2p^2 + \text{constant}. \tag{11.7}$$

DOI: 10.1201/9781003057369-11

We already know (Section 3.4) that the added constant is zero for an electromagnetic pulse, moving at the speed c. What is its value for an ordinary particle? The energy (11.7) is smallest for $\mathbf{p} = \mathbf{0}$, when the particle is at rest. Then we write

$$\mathbf{p} = \mathbf{0}: \quad E = m_0 c^2, \tag{11.8}$$

where m_0 is the mass appropriate to zero velocity—the rest mass. Therefore, we have in general,

$$E^2 = p^2c^2 + m_0^2 c^4, \quad E = \sqrt{p^2c^2 + m_0^2 c^4}. \tag{11.9}$$

From (11.4), we have

$$p^2c^2 = E^2 \frac{v^2}{c^2} \tag{11.10}$$

implying that the energy and momentum are explicitly

$$E = \frac{m_0 c^2}{\sqrt{1 - v^2/c^2}}, \quad \mathbf{p} = \frac{m_0 \mathbf{v}}{\sqrt{1 - v^2/c^2}}. \tag{11.11}$$

The last momentum construction exhibits the relation to, and the limitation of, the initial Newtonian formulation of particle mechanics. For speeds small in comparison to that of light, $|\mathbf{v}| \ll c$, the momentum is $\mathbf{p} = m_0 \mathbf{v}$ and the particle inertia is constant. This is the domain of Newtonian mechanics. But even here something is different, as we see from the energy derived from the approximation

$$\frac{v}{c} \ll 1: \quad \left(1 - \frac{v^2}{c^2}\right)^{-1/2} \approx 1 + \frac{1}{2}\frac{v^2}{c^2}, \tag{11.12}$$

namely,

$$E \approx m_0 c^2 + \frac{1}{2} m_0 v^2. \tag{11.13}$$

In addition to the Newtonian kinetic energy $\frac{1}{2} m_0 v^2$ there is a constant, the rest energy $m_0 c^2$, displacing the Newtonian origin of energy. The same thing appears in the momentum form of E, (11.9), as

$$p \ll m_0 c^2: \quad E \approx m_0 c^2 + \frac{p^2}{2m_0}. \tag{11.14}$$

For speeds approaching the speed of light we enter a new physical domain, one where the speed of light is an impassable barrier. This we can see from the particle velocity, exhibited as

$$\mathbf{v} = c \frac{\mathbf{p}}{(p^2 + m_0^2 c^2)^{1/2}}; \tag{11.15}$$

it is such that

$$v \leq c, \tag{11.16}$$

with the equality sign occurring only for $m_0 = 0$. As for the last conclusion, it is not unreasonable that a system, such as an electromagnetic pulse, which can never be at rest, has no rest mass.

Now we must reconstruct the Lagrangian-Hamiltonian dynamics of particles. If we omit the potential from the Lagrangian of (9.23), we have

$$L = \mathbf{p} \cdot \left(\frac{d\mathbf{r}}{dt} - \mathbf{v}\right) + \frac{1}{2} m v^2. \tag{11.17}$$

Clearly, the Newtonian term $\frac{1}{2}mv^2$ must be replaced by a new function of $\mathbf{v}$, $L(\mathbf{v})$:

$$L = \mathbf{p} \cdot \left(\frac{d\mathbf{r}}{dt} - \mathbf{v} \right) + L(\mathbf{v}), \tag{11.18}$$

that will reproduce the new forms. We can find $L(\mathbf{v})$ by using the velocity construction of the energy,

$$E = \mathbf{p} \cdot \mathbf{v} - L(\mathbf{v}), \tag{11.19}$$

and of the momentum. It is

$$L(\mathbf{v}) = \frac{m_0 \mathbf{v}}{(1 - v^2/c^2)^{1/2}} \cdot \mathbf{v} - \frac{m_0 c^2}{(1 - v^2/c^2)^{1/2}} = -m_0 c^2 \left(1 - \frac{v^2}{c^2} \right)^{1/2}. \tag{11.20}$$

In the Newtonian regime, the $L(\mathbf{v})$ reproduces the original form to within a constant,

$$\frac{v}{c} \ll 1: \quad L(\mathbf{v}) = -m_0 c^2 + \frac{1}{2} m_0 v^2. \tag{11.21}$$

The consistency of the action principle is verified on noting the consequence of a $\mathbf{v}$ variation:

$$\mathbf{p} = \frac{\partial L(\mathbf{v})}{\partial \mathbf{v}} = \frac{m_0 \mathbf{v}}{\sqrt{1 - v^2/c^2}}. \tag{11.22}$$

The elimination of $\mathbf{v}$ produces the Hamiltonian version

$$L = \mathbf{p} \cdot \frac{d\mathbf{r}}{dt} - H, \quad H = c(p^2 + m_0^2 c^2)^{1/2} = E. \tag{11.23}$$

11.2 Lorentz Transformations

We shall find it especially rewarding to use this new particle dynamics in re-examining a subject previously discussed in the Newtonian framework. The topic is the coordinate translation that grows linearly in time, (9.53), or equivalently, the introduction of a new coordinate system with a constant relative velocity. Here we consider only a single particle. The displacement

$$\delta \mathbf{r}(t) = \delta \mathbf{v}\, t, \tag{11.24}$$

with $\delta \mathbf{v}$ constant, combined with the momentum change

$$\delta \mathbf{p}(t) = \frac{E}{c^2} \delta \mathbf{v}, \tag{11.25}$$

induces the following alteration of the action element $dt\, L$:

$$\delta_{\mathbf{r},\mathbf{p}}[dt\, L] = \delta_{\mathbf{r},\mathbf{p}}[\mathbf{p} \cdot d\mathbf{r} - H\, dt] = \mathbf{p} \cdot \delta\mathbf{v} dt + \frac{H}{c^2} \delta\mathbf{v} \cdot d\mathbf{r} - \mathbf{p} \cdot \delta\mathbf{v} dt = H d \left(\frac{1}{c^2} \delta\mathbf{v} \cdot \mathbf{r} \right), \tag{11.26}$$

where we have used (11.6), or $\partial H / \partial \mathbf{p} = \mathbf{p} c^2 / H$. At the analogous Newtonian stage, (9.55), m $(= m_0)$ appeared in place of H/c^2, and we concluded that the action was not invariant, but changed by a differential. Now a totally new situation presents itself. If we also vary the *time* by

$$\delta t = \frac{1}{c^2}\delta\mathbf{v}\cdot\mathbf{r}, \tag{11.27}$$

the additional contribution, $-Hd(\delta t)$, will cancel (11.26), and the action *is* invariant under the combined space and time transformations

$$\delta\mathbf{r} = \delta\mathbf{v}\,t, \quad \delta t = \frac{1}{c^2}\delta\mathbf{v}\cdot\mathbf{r}. \tag{11.28}$$

In view of the invariance of the action, the implied conservation law should now follow directly. Indeed

$$G = \mathbf{p}\cdot\delta\mathbf{r} - H\delta t = \delta\mathbf{v}\cdot\mathbf{N}, \tag{11.29}$$

where $(H = E)$

$$\mathbf{N} = \mathbf{p}t - \frac{E}{c^2}\mathbf{r} \tag{11.30}$$

is conserved:

$$\frac{d\mathbf{N}}{dt} = \mathbf{p} - \frac{E}{c^2}\frac{d\mathbf{r}}{dt} = \mathbf{0}, \tag{11.31}$$

and we have recovered our starting point, $m = E/c^2$.

But, from this initial dynamical modification of Newtonian dynamics has now emerged a change in Newtonian kinematics: The absolute distinction between time and space has been removed. That is emphasized by the fact that neither r^2 nor $(ct)^2$ is left unchanged by the transformation (11.28), whereas the difference $r^2 - (ct)^2$ is invariant:

$$\delta[r^2 - (ct)^2] = 2[\mathbf{r}\cdot\delta\mathbf{v}\,t - t\,\delta\mathbf{v}\cdot\mathbf{r}] = 0. \tag{11.32}$$

The physical meaning of this invariance appears on considering an electromagnetic pulse, that, at time $t = 0$, is emitted from the origin, $\mathbf{r} = \mathbf{0}$. Moving at the speed of light, c, the pulse, at time t, will have traveled the distance $r = ct$ so that

$$r^2 - c^2t^2 = 0. \tag{11.33}$$

The fact that an observer in uniform relative motion will assign different values to the elapsed time, and to the distance traversed, but agree that (11.33) is still valid, means that he also measures the speed of light as c. This is Einsteinian relativity.

One might object that it could all be true only for infinitesimal transformations. But, from infinitesimal transformations, finite transformations grow. We make this explicit by letting $\delta\mathbf{v}$ point along the z-axis so that

$$\delta x = 0, \quad \delta y = 0, \quad \delta z = \frac{\delta v}{c}ct, \quad \delta ct = \frac{\delta v}{c}z. \tag{11.34}$$

Let us regard this infinitesimal transformation as the result of changing a parameter θ by the infinitesimal amount

$$\delta\theta = \frac{\delta v}{c}. \tag{11.35}$$

The implied differential equations in the variable θ are

$$\frac{dx(\theta)}{d\theta} = 0, \quad \frac{dy(\theta)}{d\theta} = 0, \quad \frac{dz(\theta)}{d\theta} = ct(\theta), \quad \frac{dct(\theta)}{d\theta} = z(\theta). \tag{11.36}$$

From the latter we derive

$$\frac{d^2z(\theta)}{d\theta^2} = z(\theta), \quad \frac{d^2ct(\theta)}{d\theta^2} = ct(\theta), \tag{11.37}$$

which are solved by the hyperbolic functions, $\cosh\theta$ and $\sinh\theta$. The explicit solutions of these equations that obey the initial conditions

$$z(0) = z, \quad ct(0) = ct,$$
$$\frac{dz}{d\theta}(0) = ct, \quad \frac{dct}{d\theta}(0) = z \tag{11.38}$$

are

$$z(\theta) = z\cosh\theta + ct\sinh\theta,$$
$$ct(\theta) = z\sinh\theta + ct\cosh\theta. \tag{11.39}$$

Physical interpretation is facilitated by focusing on $\tanh\theta$, the ratio of $\sinh\theta$ and $\cosh\theta$, which cannot exceed unity in magnitude. We now write

$$\tanh\theta = \frac{v}{c}, \tag{11.40}$$

which reduces to (11.35) for infinitesimal values of these parameters. Then, the constructions

$$\cosh\theta = \frac{1}{(1 - v^2/c^2)^{1/2}}, \quad \sinh\theta = \frac{v/c}{(1 - v^2/c^2)^{1/2}} \tag{11.41}$$

satisfy (11.40) as well as the hyperbolic relation

$$\cosh^2\theta - \sinh^2\theta = 1. \tag{11.42}$$

(The quantity θ is called the rapidity, with the idea of a Lorentz transformation being an imaginary rotation tracing back to Minkowski in 1908.) If we distinguish the transformed values of the coordinates by a prime, the transformation equations read

$$z' = \frac{1}{(1 - v^2/c^2)^{1/2}}(z + vt),$$
$$t' = \frac{1}{(1 - v^2/c^2)^{1/2}}\left(t + \frac{v}{c^2}z\right), \tag{11.43a}$$

along with

$$x' = x, \quad y' = y. \tag{11.43b}$$

We see that the point with coordinates

$$x = 0, \quad y = 0, \quad z = -vt \tag{11.44a}$$

is represented by the transformed coordinates

$$x' = 0, \quad y' = 0, \quad z' = 0. \tag{11.44b}$$

It is the origin of the new reference frame which therefore moves with velocity $-\mathbf{v}$ relative to the initial one. See Fig. 11.1.

To see that $r^2 - (ct)^2$ is left invariant by these finite transformations, it helps to present (11.43a) as

$$z' + ct' = \left(\frac{1 + v/c}{1 - v/c}\right)^{1/2}(z + ct),$$
$$z' - ct' = \left(\frac{1 - v/c}{1 + v/c}\right)^{1/2}(z - ct), \tag{11.45}$$

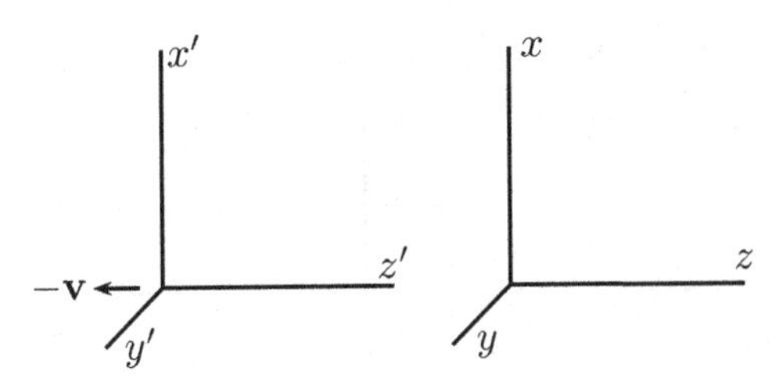

FIGURE 11.1
The transformation (11.43a), (11.43b) carries us to a new coordinate frame moving relative to the original one with a velocity $-v$ along the z-axis.

for it is immediately apparent, on multiplication, that

$$z'^2 - (ct')^2 = z^2 - (ct)^2 \tag{11.46}$$

and then [(11.43b)]

$$r'^2 - (ct')^2 = r^2 - (ct)^2. \tag{11.47}$$

The space-time transformations of the new kinematics are called Lorentz transformations, although it was Albert Einstein who, in 1905, first understood their significance as describing the full physical equivalence of reference frames in uniform relative motion. As an aspect of that equivalence, we note the following. The original reference frame moves with velocity $+v$ relative to the new one (Fig. 11.2):

$$x = 0, \quad y = 0, \quad z = 0 \tag{11.48a}$$

implies

$$x' = 0, \quad y' = 0, \quad z' = vt'. \tag{11.48b}$$

Then, should not the transformation that produces the unprimed coordinates from the primed ones be of precisely the same form as (11.43a), (11.43b) but with the sign of v reversed? Indeed it is, as is evident on rewriting (11.45) as

$$\begin{aligned} z + ct &= \left(\frac{1 - v/c}{1 + v/c}\right)^{1/2} (z' + ct'), \\ z - ct &= \left(\frac{1 + v/c}{1 - v/c}\right)^{1/2} (z' - ct'), \end{aligned} \tag{11.49}$$

together with $x = x'$, $y = y'$. More generally, suppose the coordinate transformation $z, t \to z', t'$, produced by relative velocity $-v_1$ along the common z-z' axis, is followed by the transformation $z', t' \to z'', t''$, produced by a relative velocity $-v_2$ along the common z'-z'' axis. Is the net result a transformation $z, t \to z'', t''$ that is produced by some relative velocity $-v$? Yes. It suffices to consider just one of the pairs of equations analogous to (11.45), say

$$z' + ct' = \left(\frac{1 + v_1/c}{1 - v_1/c}\right)^{1/2} (z + ct), \tag{11.50}$$

and, similarly

$$z'' + ct'' = \left(\frac{1 + v_2/c}{1 - v_2/c}\right)^{1/2} (z' + ct'), \tag{11.51}$$

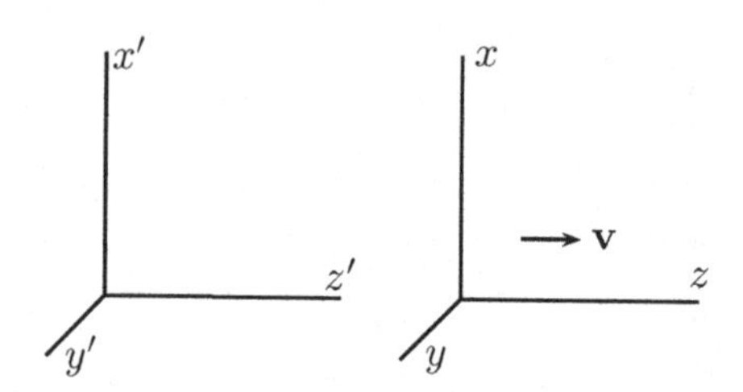

FIGURE 11.2
Motion of original frame relative to the new frame.

(the other set emerges by the systematic substitution of $-c$ for c) which immediately yields

$$z'' + ct'' = \left(\frac{1+v/c}{1-v/c}\right)^{1/2} (z+ct), \tag{11.52}$$

with

$$\left(\frac{1+v/c}{1-v/c}\right)^{1/2} = \left(\frac{1+v_1/c}{1-v_1/c}\right)^{1/2} \left(\frac{1+v_2/c}{1-v_2/c}\right)^{1/2}. \tag{11.53}$$

For any of the square root factors, the variation of the value of the appropriate v/c from -1 to $+1$ changes the square root from 0 to ∞; it is a positive number, and the product of two positive numbers is again a positive number. In other words, no succession of Lorentz transformations can produce a net transformation with $|v| > c$. The specific value of v in (11.53) is identified by writing this relation as

$$\frac{1+v/c}{1-v/c} = \frac{1+v_1v_2/c^2 + (v_1+v_2)/c}{1+v_1v_2/c^2 - (v_1+v_2)/c}, \tag{11.54}$$

or

$$v = \frac{v_1+v_2}{1+v_1v_2/c^2}. \tag{11.55}$$

Simple addition of the velocities occurs only in the Newtonian regime, $|v_{1,2}| \ll c$.

11.3 Transformation of Fields

We cannot end this chapter without showing that the kinematical space-time transformations of (11.43a), (11.43b) do indeed produce a dynamical unification of charged particles and electromagnetic fields. That requires a study of the behavior of fields and potentials under the infinitesimal Lorentz transformations. That has already been touched on in Problem 1.10, but we prefer to apply our recently developed methods here, beginning with the analog of (10.66) for Lorentz transformations (11.28):

$$\delta\lambda(\mathbf{r},t) = -\left(\delta\mathbf{v}t\cdot\boldsymbol{\nabla} + \frac{1}{c^2}\delta\mathbf{v}\cdot\mathbf{r}\frac{\partial}{\partial t}\right)\lambda(\mathbf{r},t) \equiv -\delta_{\text{coor}}\lambda(\mathbf{r},t). \tag{11.56}$$

Differentiation supplies the model for potential variations,

$$\begin{aligned}\delta\mathbf{A}(\mathbf{r},t) &= -\delta_{\text{coor}}\mathbf{A}(\mathbf{r},t) + \frac{1}{c}\delta\mathbf{v}\phi(\mathbf{r},t),\\ \delta\phi(\mathbf{r},t) &= -\delta_{\text{coor}}\phi(\mathbf{r},t) + \frac{1}{c}\delta\mathbf{v}\cdot\mathbf{A}(\mathbf{r},t),\end{aligned} \tag{11.57}$$

which are alternatively presented as (see Problem 11.6)

$$\delta \mathbf{A} = \delta \mathbf{v} t \times \mathbf{B} + \frac{1}{c}\delta \mathbf{v} \cdot \mathbf{r} \mathbf{E} + \boldsymbol{\nabla}\left(-\delta \mathbf{v} t \cdot \mathbf{A} + \frac{1}{c}\delta \mathbf{v} \cdot \mathbf{r}\phi\right),$$
$$\delta\phi = \delta \mathbf{v} t \cdot \mathbf{E} - \frac{1}{c}\frac{\partial}{\partial t}\left(-\delta \mathbf{v} t \cdot \mathbf{A} + \frac{1}{c}\delta \mathbf{v} \cdot \mathbf{r}\phi\right). \tag{11.58}$$

The use of the latter simplifies the calculation of the field variations; they emerge as

$$\delta \mathbf{E} = -\delta_{\text{coor}} \mathbf{E} - \frac{1}{c}\delta \mathbf{v} \times \mathbf{B},$$
$$\delta \mathbf{B} = -\delta_{\text{coor}} \mathbf{B} + \frac{1}{c}\delta \mathbf{v} \times \mathbf{E}. \tag{11.59}$$

(These induced electric and magnetic fields were already encountered in Chapter 1.) Then further differentiation in accordance with the field equations,

$$\rho = \frac{1}{4\pi}\boldsymbol{\nabla} \cdot \mathbf{E},$$
$$\mathbf{j} = \frac{c}{4\pi}\left(\boldsymbol{\nabla} \times \mathbf{B} - \frac{1}{c}\frac{\partial}{\partial t}\mathbf{E}\right), \tag{11.60}$$

yields

$$\delta\rho = -\delta_{\text{coor}}\rho + \frac{1}{c^2}\delta \mathbf{v} \cdot \mathbf{j},$$
$$\delta \mathbf{j} = -\delta_{\text{coor}} \mathbf{j} + \delta \mathbf{v}\rho. \tag{11.61}$$

Notice that the Lorentz transformation properties of $\mathbf{j}/c$, ρ are the same as those for $\mathbf{A}$, ϕ. If we introduce a new symbol 'δ' that involves both changes of fields (δ) and of coordinates (δ_{coor}), then its meaning as applied to any field $F(\mathbf{r}, t)$ is

$$\text{`}\delta\text{'} F(\mathbf{r}, t) = (F + \delta F)(\mathbf{r} + \delta \mathbf{r}, t + \delta t) - F(\mathbf{r}, t) = \delta F(\mathbf{r}, t) + \delta_{\text{coor}} F(\mathbf{r}, t). \tag{11.62}$$

Note that the use of δ_{coor} in (11.56) is consistent with this, since $\lambda(\mathbf{r}, t)$ is a scalar, 'δ'$\lambda(\mathbf{r}, t) = 0$.

We must now examine the response of the various parts of the total Lagrange function. First consider

$$\mathcal{L}_{\text{field}} = \frac{1}{8\pi}(E^2 - B^2). \tag{11.63}$$

It is immediately apparent that the contributions of the last terms in the transformation equations (11.59) just cancel and

$$\delta\mathcal{L}_{\text{field}} = -\delta_{\text{coor}}\mathcal{L}_{\text{field}}. \tag{11.64}$$

Then we consider the interaction contribution to the Lagrange function [see (10.14)]

$$\mathcal{L}_{\text{int}} = -\rho\phi + \frac{1}{c}\mathbf{j} \cdot \mathbf{A}. \tag{11.65}$$

Again, the last terms in the transformation (11.57) and (11.61) have no net effect, and

$$\delta\mathcal{L}_{\text{int}} = -\delta_{\text{coor}}\mathcal{L}_{\text{int}}. \tag{11.66}$$

Finally, we come to the Lagrange function of the individual particles. For one particle, with the Lagrangian description adopted ($\mathbf{v} = d\mathbf{r}/dt$), the Lagrangian (11.20) is

$$L_a = -m_{0a}c^2\left(1 - \frac{v_a^2}{c^2}\right)^{1/2}. \tag{11.67}$$

In contrast with the procedure of (10.73a), our introduction of the Lagrange function now is dictated by the impossibility of maintaining a common time for particles at different spatial points—that Newtonian concept has disappeared in the world of Einsteinian relativistic kinematics. Accordingly, we give each particle its individual time coordinate, and present its contribution to the action as

$$\int dt_a L_a = \int (-m_{0a}c)[(cdt_a)^2 - (d\mathbf{r}_a)^2]^{1/2} = \int dt\,(d\mathbf{r})\mathcal{L}_a, \tag{11.68}$$

where

$$\mathcal{L}_a(\mathbf{r},t) = -m_{0a}c\int \delta(\mathbf{r}-\mathbf{r}_a)\delta(t-t_a)[(cdt_a)^2 - (d\mathbf{r}_a)^2]^{1/2}. \tag{11.69}$$

[Unlike (10.73c), interaction terms are not included here, because they are already present in the coupling of charges and currents to the potentials.] The last integral is extended over the trajectory of the particle, with $\mathbf{r}_a$ varying as a function of t_a, or, better, with $\mathbf{r}_a$ and t_a given as functions of some parameter (the proper time) that is not changed by Lorentz transformations. Apart from a sign change and the considerations of infinitesimals, the space-time structure of the square root has just the invariant form in (11.32). Then, in response to

$$\delta\mathbf{r}_a = \delta\mathbf{v}t_a, \quad \delta t_a = \frac{1}{c}\delta\mathbf{v}\cdot\mathbf{r}_a, \tag{11.70}$$

the delta function product becomes

$$\delta(\mathbf{r}-\mathbf{r}_a-\delta\mathbf{v}t_a)\delta(t-t_a-\frac{1}{c^2}\delta\mathbf{v}\cdot\mathbf{r}_a) = \delta(\mathbf{r}-\delta\mathbf{v}t-\mathbf{r}_a)\delta(t-\frac{1}{c^2}\delta\mathbf{v}\cdot\mathbf{r}-t_a), \tag{11.71}$$

with the last form following from the delta function property, and the resulting change is just

$$-\delta_{\text{coor}}[\delta(\mathbf{r}-\mathbf{r}_a)\delta(t-t_a)]. \tag{11.72}$$

We have now verified, for every individual constituent of $\mathcal{L}_{\text{tot}}$, that

$$\delta\mathcal{L}_{\text{tot}} = -\delta_{\text{coor}}\mathcal{L}_{\text{tot}}, \tag{11.73}$$

so that the total variation in the sense of (11.62) is then zero. More generally this can be expressed as follows: On introducing the transformed particle and field variables associated with the transformed coordinates,

$$\mathcal{L}'(\mathbf{r}',t') = \mathcal{L}(\mathbf{r},t); \tag{11.74}$$

the Lagrange function of the system of interacting charges and fields is invariant under the Lorentz transformations of Einsteinian relativity.

The action

$$W_{12} = \int_2^1 dt(d\mathbf{r})\,\mathcal{L} \tag{11.75}$$

is also Lorentz invariant because the space-time, four dimensional, element of volume has that property. To prove this it suffices to examine the Jacobian determinant of the transformation (11.39), for example,

$$\cosh^2\theta - \sinh^2\theta = 1. \tag{11.76}$$

We shall not repeat the discussion of the various conservation laws in the light of these relativistic modifications. It should be sufficiently evident that such expressions as the total energy density and energy flux vector, (10.81) and (10.82), will be regained with E_a replaced by the relativistic energy (11.11). And, certainly (11.3) is now satisfied!

Finally, we supply the finite version of the Lorentz transformations for the various fields. Proceeding in analogy with (11.36), we write (11.57), for a Lorentz transformation along the z axis, as

$$\frac{dA_x(\theta)}{d\theta} = 0, \quad \frac{dA_y(\theta)}{d\theta} = 0, \quad \frac{dA_z(\theta)}{d\theta} = \phi(\theta), \quad \frac{d\phi(\theta)}{d\theta} = A_z(\theta), \tag{11.77}$$

where d denotes the total change in the sense of (11.62). The finite transformation equations are then

$$\begin{aligned} A'_x(\mathbf{r}', t') &= A_x(\mathbf{r}, t), \\ A'_y(\mathbf{r}', t') &= A_y(\mathbf{r}, t), \\ A'_z(\mathbf{r}', t') &= \frac{1}{(1 - v^2/c^2)^{1/2}}\left(A_z(\mathbf{r}, t) + \frac{v}{c}\phi(\mathbf{r}, t)\right), \\ \phi'(\mathbf{r}', t') &= \frac{1}{(1 - v^2/c^2)^{1/2}}\left(\phi(\mathbf{r}, t) + \frac{v}{c}A_z(\mathbf{r}, t)\right). \end{aligned} \tag{11.78}$$

The same forms apply with $\mathbf{A}$, ϕ replaced by $\mathbf{j}/c$, ρ. The transformations given in (11.78) identify these quantities as four-vectors (*cf.* Problem 1.10 and Chapter 12). The electric and magnetic field equations supplied by (11.59) are

$$\begin{aligned} &\frac{dE_z(\theta)}{d\theta} = 0, \quad \frac{dB_z(\theta)}{d\theta} = 0, \\ &\frac{dE_x(\theta)}{d\theta} = B_y(\theta), \quad \frac{dB_y(\theta)}{d\theta} = E_x(\theta), \\ &\frac{dE_y(\theta)}{d\theta} = -B_x(\theta), \quad \frac{dB_x(\theta)}{d\theta} = -E_y(\theta), \end{aligned} \tag{11.79}$$

and therefore

$$\begin{aligned} E'_z(\mathbf{r}', t') &= E_z(\mathbf{r}, t), \quad B'_z(\mathbf{r}', t') = B_z(\mathbf{r}, t), \\ E'_x(\mathbf{r}', t') &= \frac{1}{(1 - v^2/c^2)^{1/2}}[E_x(\mathbf{r}, t) + \frac{v}{c}B_y(\mathbf{r}, t)], \\ B'_y(\mathbf{r}', t') &= \frac{1}{(1 - v^2/c^2)^{1/2}}[B_y(\mathbf{r}, t) + \frac{v}{c}E_x(\mathbf{r}, t)], \end{aligned} \tag{11.80}$$

and

$$\begin{aligned} E'_y(\mathbf{r}', t') &= \frac{1}{(1 - v^2/c^2)^{1/2}}[E_y(\mathbf{r}, t) - \frac{v}{c}B_x(\mathbf{r}, t)], \\ B'_x(\mathbf{r}', t') &= \frac{1}{(1 - v^2/c^2)^{1/2}}[B_x(\mathbf{r}, t) - \frac{v}{c}E_y(\mathbf{r}, t)]. \end{aligned} \tag{11.81}$$

The invariance, under finite transformations, of $E^2 - B^2$ is readily apparent, as is the invariance property

$$\mathbf{E}'(\mathbf{r}', t') \cdot \mathbf{B}'(\mathbf{r}', t') = \mathbf{E}(\mathbf{r}, t) \cdot \mathbf{B}(\mathbf{r}, t). \tag{11.82}$$

11.4 Problems for Chapter 11

1. The relativistic modification of the action principle involves the replacement

$$\frac{1}{2}mv^2 \to -m_0c^2\sqrt{1-\frac{v^2}{c^2}} \tag{11.83}$$

in $L(\mathbf{r}, \mathbf{p}, \mathbf{v}, t)$ in Problem 10.1. Use the Lagrangian viewpoint and find the equations of motion.

2. Repeat the above problem using the Hamiltonian viewpoint.

3. A relativistic particle of rest mass m_0 and electric charge e moves under the influence of prescribed electromagnetic fields, described in terms of vector and scalar potentials, $\mathbf{A}$ and ϕ. The Lagrangian is

$$L = -m_0c^2\sqrt{1-\frac{v^2}{c^2}} - e\phi(\mathbf{r}) + \frac{e}{c}\mathbf{v}\cdot\mathbf{A}(\mathbf{r}), \quad \mathbf{v} = \frac{d\mathbf{r}}{dt}. \tag{11.84}$$

   (a) Consider an infinitesimal variation in the particle trajectory, $\mathbf{r} \to \mathbf{r} + \delta\mathbf{r}$, and calculate the change in the action,

$$\delta\int_{t_2}^{t_1} dt\, L(\mathbf{r}(t)) = G_1 - G_2, \tag{11.85}$$

   where $G = \mathbf{p}\cdot\delta\mathbf{r}$, and give an expression for $\mathbf{p}$.

   (b) From the variation of the Lagrangian between the two endpoints, determine the equation of motion for $\mathbf{p}$.

   (c) Rewrite this equation in terms of the rate of change of mechanical momentum, and the Lorentz force $\mathbf{F} = e\mathbf{E} + e\frac{\mathbf{v}}{c}\times\mathbf{B}$.

4. Show, very simply, that the magnetic field $\mathbf{B}$ of a uniformly moving charge is related to its electric field by $\frac{1}{c}\mathbf{v}\times\mathbf{E}$. Then, consider two charges, moving with common velocity $\mathbf{v}$ along parallel tracks, and show that the magnetic force between them is *opposite* to the electric force, and smaller by a factor of v^2/c^2. (The first property is an example of the rule that like charges *repel* while like current *attract.*) Can you derive the same result by Lorentz transforming the equation of motion in the common rest frame of the particles?

5. Find the explicit form of θ in terms of v/c as a logarithm. How does the composition law of velocities appear in terms of θ?

6. Verify (11.58) and use the results to derive the statements of (11.59).

7. Supply the proof of (11.61)

8. Given the infinitesimal coordinate variations

$$\delta\mathbf{r} = \frac{1}{c}\delta\mathbf{v}ct, \tag{11.86a}$$

$$\delta(ct) = \frac{1}{c}\delta\mathbf{v}\cdot\mathbf{r}, \tag{11.86b}$$

   what is

$$\delta(\boldsymbol{\nabla}), \quad \delta\left(\frac{\partial}{\partial t}\right)? \tag{11.87}$$

   Do you recognize another four-dimensional vector?

9. Use these variations of derivatives, and the total variations of fields defined in (11.62), to confirm that Maxwell's equations are maintained under the combined variations.

10. Show that the finite Lorentz transformations have the vectorial form $[\gamma = (1 - v^2/c^2)^{-1/2}]$

$$\mathbf{r}' = \mathbf{r} + \frac{\gamma^2}{1+\gamma}\frac{1}{c^2}\mathbf{v}\mathbf{v}\cdot\mathbf{r} + \gamma\mathbf{v}t, \tag{11.88a}$$

$$t' = \gamma\left(t + \frac{1}{c^2}\mathbf{v}\cdot\mathbf{r}\right). \tag{11.88b}$$

Check the invariance of $\mathbf{r}^2 - (ct)^2$.

11. Derive the expressions analogous to (10.81) and (10.82) for relativistic particles.

12. In vacuum, show that the electromagnetic field energy U,

$$U = \int (d\mathbf{r})\frac{E^2 + B^2}{8\pi}, \tag{11.89a}$$

and the field momentum P,

$$\mathbf{P} = \int (d\mathbf{r})\frac{\mathbf{E}\times\mathbf{B}}{4\pi c}, \tag{11.89b}$$

transform under an infinitesimal Lorentz transformation as

$$\delta U = \frac{\delta\mathbf{v}}{c}\cdot\mathbf{P}c, \quad \delta\mathbf{P}c = \frac{\delta\mathbf{v}}{c}U, \tag{11.90}$$

where $\delta\mathbf{v}$ is a constant velocity boost. This is to say that $(U/c, \mathbf{P})$ transform as a four-vector.

12

Relativistic Formulation

In the previous chapter, we employed three-dimensional notation of electric and magnetic fields, even though the natural arena of Einstein's relativity is four-dimensional spacetime. We will continue to use the former in the rest of the book, because it is convenient for most practical purposes. However, theoretically it is often more appropriate to use the explicit four-dimensional language of special relativity. Since the latter was discovered by Einstein from the unmodified Maxwell theory, it is only Newtonian mechanics that is modified by special relativity, not electrodynamics.

12.1 Four-Dimensional Notation

A space-time coordinate of an event can be represented by a contravariant vector,

$$x^\mu \ : \quad x^0 = ct \ , \quad x^1 = x \ , \quad x^2 = y \ , \quad x^3 = z, \tag{12.1}$$

where μ is an index which takes on the values 0, 1, 2, 3. The corresponding covariant vector is

$$x_\mu \ : \quad x_0 = -ct \ , \quad x_1 = x \ , \quad x_2 = y \ , \quad x_3 = z. \tag{12.2}$$

The contravariant and covariant vector components are related by the metric tensor $g_{\mu\nu}$,

$$x_\mu = g_{\mu\nu} x^\nu, \tag{12.3}$$

which uses the Einstein summation convention of summing over repeated covariant and contravariant indices,

$$g_{\mu\nu} x^\nu = \sum_{\nu=0}^{3} g_{\mu\nu} x^\nu . \tag{12.4}$$

From the above explicit forms for x_μ and x^ν we read off, in matrix form (here the first index labels the rows, the second the columns, both enumerated from 0 to 3)[1]

$$g_{\mu\nu} = \begin{pmatrix} -1 & 0 & 0 & 0 \\ 0 & 1 & 0 & 0 \\ 0 & 0 & 1 & 0 \\ 0 & 0 & 0 & 1 \end{pmatrix}, \tag{12.5}$$

where evidently $g_{\mu\nu}$ is symmetric,

$$g_{\mu\nu} = g_{\nu\mu}. \tag{12.6}$$

Similarly,

$$x^\mu = g^{\mu\nu} x_\nu, \tag{12.7}$$

[1] In this book, we use what could be referred to as the democratic metric (formerly the East-coast metric), in which the signature is dictated by the larger number of entries.

DOI: 10.1201/9781003057369-12

where

$$g^{\mu\nu} = \begin{pmatrix} -1\,0\,0\,0 \\ 0\ \ 1\,0\,0 \\ 0\ \ 0\,1\,0 \\ 0\ \ 0\,0\,1 \end{pmatrix}, \quad g^{\mu\nu} = g^{\nu\mu}. \tag{12.8}$$

The four-dimensional analog of a rotationally invariant length is the proper length s or the proper time τ:

$$s^2 = -c^2\tau^2 = x^\mu x_\mu = x^\mu g_{\mu\nu} x^\nu = x_\mu g^{\mu\nu} x_\nu = \mathbf{r}\cdot\mathbf{r} - (ct)^2. \tag{12.9}$$

Recall the transformation of a scalar field under a coordinate displacement, as in (10.35),

$$\delta\phi(\mathbf{r}) = -\delta\mathbf{r}\cdot\boldsymbol{\nabla}\phi(\mathbf{r}), \quad \boldsymbol{\nabla} = \frac{\partial}{\partial\mathbf{r}}. \tag{12.10}$$

The corresponding four-dimensional statement is

$$\delta\phi(x) = -\delta x^\mu \partial_\mu \phi(x) \quad \partial_\mu = \frac{\partial}{\partial x^\mu}, \tag{12.11}$$

which shows the definition of the covariant gradient operator, so defined in order that $\partial_\mu x^\mu$ be invariant. The corresponding contravariant gradient is

$$\partial^\mu = g^{\mu\nu}\partial_\nu = \frac{\partial}{\partial x_\mu}. \tag{12.12}$$

Using these operators we can write the equation of electric current conservation, (1.42),

$$\boldsymbol{\nabla}\cdot\mathbf{j} + \frac{\partial}{\partial t}\rho = 0, \tag{12.13}$$

in the four-dimensional form

$$\partial_\mu j^\mu = 0, \tag{12.14}$$

where we define the components of the electric current four-vector as

$$j^\mu : \quad j^0 = c\rho, \quad \{j^i\} = \mathbf{j}, \tag{12.15}$$

where we have adopted the convention that Latin indices run over the values 1, 2, 3, corresponding to the three spatial directions. Note that (12.15) is quite analogous to the construction of the position four-vector, (12.1).

The invariant interaction term (11.65)

$$\mathcal{L}_{\text{int}} = -\rho\phi + \frac{1}{c}\mathbf{j}\cdot\mathbf{A} \tag{12.16}$$

has the four-dimensional form

$$\mathcal{L}_{\text{int}} = \frac{1}{c} j^\mu A_\mu, \tag{12.17}$$

where

$$A_\mu = g_{\mu\nu}A^\nu, \quad A^0 = \phi, \quad \{A^i\} = \mathbf{A}. \tag{12.18}$$

The four-dimensional generalization of

$$\mathbf{B} = \boldsymbol{\nabla}\times\mathbf{A} \tag{12.19}$$

is the tensor construction

$$F_{\mu\nu} = \partial_\mu A_\nu - \partial_\nu A_\mu, \tag{12.20}$$

where the antisymmetric field strength tensor

$$F_{\mu\nu} = -F_{\nu\mu} \tag{12.21}$$

contains the magnetic field components as

$$F_{23} = B_1 , \quad F_{31} = B_2 , \quad F_{12} = B_3, \tag{12.22}$$

which may be presented more succinctly as

$$F_{ij} = \epsilon_{ijk} B^k, \tag{12.23}$$

which uses the totally antisymmetric Levi-Civita symbol in (3.22). The construction (12.20) includes the other potential statement (10.8a),

$$\mathbf{E} = -\boldsymbol{\nabla}\phi - \frac{1}{c}\frac{\partial}{\partial t}\mathbf{A}, \tag{12.24}$$

provided

$$F_{0i} = -E_i. \tag{12.25}$$

Alternatively, with

$$F^\mu{}_\nu = g^{\mu\lambda} F_{\lambda\nu}, \tag{12.26}$$

we have

$$F^0{}_i = E_i. \tag{12.27}$$

Maxwell's equations with only electric currents present are summarized by

$$\partial_\nu F^{\mu\nu} = \frac{4\pi}{c} j^\mu, \tag{12.28a}$$

$$\partial_\lambda F_{\mu\nu} + \partial_\mu F_{\nu\lambda} + \partial_\nu F_{\lambda\mu} = 0, \tag{12.28b}$$

where

$$F^{\mu\nu} = F^\mu{}_\lambda g^{\lambda\nu} = g^{\mu\kappa} F_{\kappa\lambda} g^{\lambda\nu}. \tag{12.29}$$

It is convenient to define a dual field-strength tensor by

$${}^*F^{\mu\nu} = \frac{1}{2}\epsilon^{\mu\nu\kappa\lambda} F_{\kappa\lambda} = -{}^*F^{\nu\mu}, \tag{12.30}$$

where $\epsilon^{\mu\nu\kappa\lambda}$ is the four-dimensional totally antisymmetric Levi-Civita symbol, which therefore vanishes if any two of the indices are equal, normalized by

$$\epsilon^{0123} = +1 . \tag{12.31}$$

We now have

$${}^*F^{01} = F_{23} = B_1 , \quad {}^*F^{02} = F_{31} = B_2 , {}^*F^{03} = F_{12} = B_3, \tag{12.32}$$

and

$${}^*F^{23} = F_{01} = -E_1 , \quad {}^*F^{31} = F_{02} = -E_2 , {}^*F^{12} = F_{03} = -E_3, \tag{12.33}$$

so indeed the dual transformation corresponds to the replacement

$$\mathbf{E} \to \mathbf{B} , \quad \mathbf{B} \to -\mathbf{E}. \tag{12.34}$$

[This is the duality relation we saw in (2.11).] Note that two dual operations brings you back to the beginning:

$$^*(^*F^{\mu\nu}) = -F^{\mu\nu}. \tag{12.35}$$

We summarize the field strength tensor, and its dual, by writing them in matrix form:

$$F^{\mu\nu} = \begin{pmatrix} 0 & E_x & E_y & E_z \\ -E_x & 0 & B_z & -B_y \\ -E_y & -B_z & 0 & B_x \\ -E_z & B_y & -B_x & 0 \end{pmatrix}, \quad {}^*F^{\mu\nu} = \begin{pmatrix} 0 & B_x & B_y & B_z \\ -B_x & 0 & -E_z & E_y \\ -B_y & E_z & 0 & -E_x \\ -B_z & -E_y & E_x & 0 \end{pmatrix}. \tag{12.36}$$

Using the dual, Maxwell's equations including both the electric (j^μ) and the magnetic ($^*j^\mu$) currents (called j_e and j_m in Chapter 2) are given by

$$\partial_\nu F^{\mu\nu} = \frac{4\pi}{c} j^\mu, \quad \partial_\nu {}^*F^{\mu\nu} = \frac{4\pi}{c} {}^*j^\mu, \tag{12.37}$$

where both currents must be conserved,

$$4\pi \partial_\mu j^\mu = c\partial_\mu \partial_\nu F^{\mu\nu} = 0, \tag{12.38a}$$

$$4\pi \partial_\mu {}^*j^\mu = c\partial_\mu \partial_\nu {}^*F^{\mu\nu} = 0, \tag{12.38b}$$

because of the symmetry in μ and ν of $\partial_\mu \partial_\nu$ and the antisymmetry of $F^{\mu\nu}$ and $^*F^{\mu\nu}$.

We had earlier in (12.9) introduced the proper time. The corresponding differential statement is

$$d\tau = \frac{1}{c}\sqrt{-dx^\mu dx_\mu} = dt\sqrt{1 - \frac{v^2}{c^2}}, \tag{12.39}$$

which is an invariant time interval. The particle equations of motion using τ as the time parameter read (see Problem 12.1)

$$m_0 \frac{d^2 x^\mu}{d\tau^2} = \frac{e}{c} F^\mu{}_\nu \frac{dx^\nu}{d\tau}, \tag{12.40}$$

to which is to be added $\frac{g}{c} {}^*F^\mu{}_\nu dx^\nu/d\tau$ if the particle possesses magnetic charge g. We can write down three alternative forms for the action of the particle:

$$W_{12} = \int_2^1 \left(-m_0 c^2 d\tau + \frac{e}{c} A_\mu dx^\mu\right) \tag{12.41a}$$

$$= \int_2^1 d\tau \left[\frac{1}{2} m_0 \left(\frac{dx^\mu}{d\tau}\frac{dx_\mu}{d\tau} - c^2\right) + \frac{e}{c} A_\mu \frac{dx^\mu}{d\tau}\right] \tag{12.41b}$$

$$= \int_2^1 d\tau \left[p_\mu \left(\frac{dx^\mu}{d\tau} - v^\mu\right) + \frac{1}{2} m_0 \left(v^\mu v_\mu - c^2\right) + \frac{e}{c} A_\mu v^\mu\right]. \tag{12.41c}$$

In the last two forms, τ is an independent parameter, with the added requirement that each generator G [recall the action principle states $\delta W_{12} = G_1 - G_2$] is independent of $\delta\tau$. In the third version, where x^μ, v^μ, and p^μ are independent dynamical variables, it is a consequence of the action principle that

$$v^\mu = \frac{dx^\mu}{d\tau}, \quad p^\mu = m_0 v^\mu + \frac{e}{c} A^\mu, \quad v^\mu v_\mu = -c^2, \tag{12.42a}$$

$$\frac{dp^\mu}{d\tau} = \frac{e}{c} \partial_\mu A_\lambda v^\lambda. \tag{12.42b}$$

The invariant Lagrange function for the electromagnetic field (10.71) is

$$\mathcal{L}_f = -\frac{1}{16\pi}F^{\mu\nu}F_{\mu\nu} = \frac{E^2 - B^2}{8\pi}. \tag{12.43}$$

The energy-momentum, or stress tensor, subsumes the energy density, the momentum density (or energy flux vector) and the three-dimensional stress tensor:

$$T^{\mu\nu} = T^{\nu\mu} = \frac{1}{4\pi}\left(F^{\mu\lambda}F^{\nu}{}_{\lambda} - g^{\mu\nu}\frac{1}{4}F^{\kappa\lambda}F_{\kappa\lambda}\right); \tag{12.44}$$

It has the property of being traceless:

$$T^{\mu}{}_{\mu} = g_{\mu\nu}T^{\mu\nu} = 0, \tag{12.45}$$

and has the following explicit components:

$$T^{00} = U\,, \quad T^{0}{}_{k} = \frac{1}{c}S_k = cG_k, \quad T_{ij} = \mathbf{T}_{ij}, \tag{12.46}$$

in terms of the energy density, (3.4a), the energy flux vector (3.4b) or momentum density (3.12a), and the stress tensor (3.12b). It satisfies the equation

$$\partial_\nu T^{\mu\nu} = -F^{\mu\nu}\frac{1}{c}j_\nu, \tag{12.47}$$

which restates the energy and momentum conservation laws (3.5) and (3.13).

12.2 Field Transformations

A Lorentz transformation, or more properly a *boost,* is a transformation that mixes the time and space coordinates without changing the invariant distance s^2. An infinitesimal transformation of this class is that given in (11.28):

$$\delta\mathbf{r} = \delta\mathbf{v}\,t\,, \quad \delta t = \frac{1}{c^2}\delta\mathbf{v}\cdot\mathbf{r}, \tag{12.48}$$

where $-\delta\mathbf{v}$ is the velocity with which the new coordinate frame moves relative to the old one. (It is assumed that the two coordinate frames coincide at $t = 0$.) In terms of the four-vector position, $x^\mu = (ct, \mathbf{r})$, we can write this result compactly as

$$\delta x^\mu = \delta\omega^{\mu\nu}x_\nu, \tag{12.49}$$

where the only nonzero components of the transformation parameter $\delta\omega^{\mu\nu}$ are

$$\delta\omega^{0i} = -\delta\omega^{i0} = \frac{\delta v^i}{c}. \tag{12.50}$$

Ordinary rotations of course also preserve s^2, so they must be included in the transformations (12.49), and they are, corresponding to $\delta\omega^{\mu\nu}$ having no time components, and spatial components

$$\delta\omega^{ij} = -\epsilon^{ijk}\delta\omega_k, \tag{12.51}$$

so, as in (9.39),

$$\delta \mathbf{r} = \delta \boldsymbol{\omega} \times \mathbf{r}. \tag{12.52}$$

In fact, the only property $\delta\omega^{\mu\nu}$ must have in order to preserve the invariant length s^2 is antisymmetry:

$$\delta\omega^{\mu\nu} = -\delta\omega^{\nu\mu}, \tag{12.53}$$

for

$$\delta(x^\mu x_\mu) = 2\delta\omega^{\mu\nu} x_\mu x_\nu = 0, \tag{12.54}$$

and a scalar product, such as that in $j^\mu A_\mu$ is similarly invariant. Any infinitesimal transformation with this property we will dub a Lorentz transformation.

Now consider the transformation of a four-vector field, such as the vector potential, $A^\mu = (\phi, \mathbf{A})$. This field undergoes the same transformation as given by the coordinate four-vector, but one must also transform to the new coordinate representing the same physical point. That is, under a Lorentz transformation,

$$A^\mu(x) \to \overline{A}^\mu(\overline{x}) = A^\mu(x) + \delta\omega^{\mu\nu} A_\nu(x), \tag{12.55}$$

where

$$\overline{x}^\mu = x^\mu + \delta x^\mu = x^\mu + \delta\omega^{\mu\nu} x_\nu. \tag{12.56}$$

So that the transformation may be considered a field variation only, we define the change in the field at the same coordinate value (which refers to different physical points in the two frames):

$$\begin{aligned} \delta A^\mu(x) &= \overline{A}^\mu(x) - A^\mu(x) \\ &= A^\mu(x - \delta x) + \delta\omega^{\mu\nu} A_\nu(x) - A^\mu(x) \\ &= -\delta x^\nu \partial_\nu A^\mu(x) + \delta\omega^{\mu\nu} A_\nu(x). \end{aligned} \tag{12.57}$$

This is equivalent to (11.57). The four-vector current $j^\mu = (c\rho, \mathbf{j})$ must transform in the same way:

$$\delta j^\mu = -\delta x^\nu \partial_\nu j^\mu(x) + \delta\omega^{\mu\nu} j_\nu(x). \tag{12.58}$$

A scalar field, $\lambda(x)$, on the other hand, only undergoes the coordinate transformation:

$$\lambda(x) \to \overline{\lambda}(\overline{x}) = \lambda(x), \tag{12.59}$$

so

$$\delta\lambda(x) = -\delta x^\nu \partial_\nu \lambda(x). \tag{12.60}$$

Because a vector potential can be changed by a gauge transformation,

$$A^\mu \to A^\mu + \partial^\mu \lambda, \tag{12.61}$$

without altering any physical quantity, in particular the field strength tensor $F^{\mu\nu}$, the transformation law for the vector potential must follow by differentiating that of λ, and indeed it does.

What about the transformation property of the field strength tensor? Again, it follows by direct differentiation:

$$\begin{aligned} \delta F_{\mu\nu} &= \delta(\partial_\mu A_\nu - \partial_\nu A_\mu) \\ &= -\delta x^\lambda \partial_\lambda F_{\mu\nu} - (\partial_\mu \delta x^\lambda)\partial_\lambda A_\nu + (\partial_\nu \delta x^\lambda)\partial_\lambda A_\mu + \delta\omega_{\nu\lambda}\partial_\mu A^\lambda - \delta\omega_{\mu\lambda}\partial_\nu A^\lambda \\ &= -\delta x^\lambda \partial_\lambda F_{\mu\nu} + \delta\omega_\mu{}^\lambda F_{\lambda\nu} + \delta\omega_\nu{}^\lambda F_{\mu\lambda}. \end{aligned} \tag{12.62}$$

So we see that each index of a tensor transforms like that of a vector. From this it is easy to work out how the components of the electric and magnetic fields transform under a boost (12.50). Apart from the coordinate change—which just says we are evaluating fields at the same physical point— we see [cf. (11.59)]

$$`\delta\text{'}\mathbf{E} = -\frac{\delta\mathbf{v}}{c}\times\mathbf{B}, \tag{12.63a}$$

$$`\delta\text{'}\mathbf{B} = \frac{\delta\mathbf{v}}{c}\times\mathbf{E}, \tag{12.63b}$$

The proof of the Lorentz invariance the relativistic Lagrangian is now immediate. That is,

$$\delta\mathcal{L} = -\delta x^\lambda \partial_\lambda \mathcal{L}, \tag{12.64}$$

which just says that $\overline{\mathcal{L}}(\overline{x}) = \mathcal{L}(x)$, implying that $\delta W = \delta\int(dx)\mathcal{L}(x) = 0$. We have already remarked that $`\delta\text{'}\mathcal{L}_{\text{int}} = 0$. The invariance of the field Lagrangian (12.43) is simply the statement

$$4\pi`\delta\text{'}\mathcal{L}_f = \delta\omega_{\mu\nu}F^{\mu\lambda}F^\nu{}_\lambda = 0, \tag{12.65}$$

and the particle action in (12.41a)–(12.41c) is manifestly invariant.

12.3 Problems for Chapter 12

1. Show that the time and space components of (12.40) are equivalent to the particle equations of motion

$$\frac{dE_a}{dt} = e_a\mathbf{v}_a\cdot\mathbf{E}(\mathbf{r}_a), \tag{12.66a}$$

$$\frac{d\mathbf{p}_a}{dt} = e_a\left(\mathbf{E}(\mathbf{r}_a) + \frac{1}{c}\mathbf{v}_a\times\mathbf{B}(\mathbf{r}_a)\right) = \mathbf{F}_a, \tag{12.66b}$$

provided the relativistic forms of the particle kinetic energy and momentum, (11.11), are employed.

2. Derive the first form of the particle action (12.41a) from the relativistic particle Lagrangian $-m_0c^2\sqrt{1-v^2/c^2}$, (11.20), and the interaction (12.16).

3. Obtain the equations resulting from variations of the second form of the particle action (12.41b) with respect to both x^μ and τ variations, and verify that these are as expected.

4. A covariant form for the current vector of a moving point charge e is the proper-time integral

$$\frac{1}{c}j^\mu(x) = \int_{-\infty}^{\infty} d\tau\, e\,\frac{dx^\mu(\tau)}{d\tau}\delta(x - x(\tau)). \tag{12.67}$$

Verify that

$$\partial_\mu j^\mu = 0, \tag{12.68}$$

under the assumption that the charge is infinitely remote at $\tau = \pm\infty$. Show that

$$\int(d\mathbf{r})\frac{1}{c}j^0(x) = e, \tag{12.69}$$

provided $dx^0(\tau)/d\tau$ is always positive. The stress tensor $T^{\mu\nu}$ for a mass point is given analogously by

$$T^{\mu\nu}(x) = \int_{-\infty}^{\infty} d\tau\, m_0 c \frac{dx^\mu(\tau)}{d\tau}\frac{dx^\nu(\tau)}{d\tau}\delta(x - x(\tau)). \tag{12.70}$$

Verify that

$$\partial_\nu T^{\mu\nu}(x) = 0\,, \tag{12.71}$$

provided the particle is unaccelerated ($d^2x^\mu(\tau)/d\tau^2 = 0$). Then show that

$$\int (d\mathbf{r})\, T^{0\nu} = m_0 c \frac{dx^\nu(\tau)}{d\tau}. \tag{12.72}$$

Does this comprise the expected values for the energy and momentum (multiplied by c) of a uniformly moving particle?

5. Suppose the particle of the previous problem is accelerated—it carries charge e and moves in an electromagnetic field. Use the covariant equations of motion (12.40) to show that

$$\partial_\nu T^{\mu\nu}_{\text{part}} = \frac{1}{c}F^\mu{}_\nu j^\nu. \tag{12.73}$$

What do you conclude by comparison with the corresponding divergence of the electromagnetic stress tensor, (12.47)?

6. The next several problems refer to a purely electromagnetic model of the electron [22]. A spherically symmetrical distribution of charge e at rest has the potentials $\phi = ef(r^2)$, $\mathbf{A} = \mathbf{0}$, where, at distances large compared with its size, $f(r^2) \sim 1/\sqrt{r^2}$. As observed in a frame in uniform relative motion, the potentials are

$$A^\mu(x) = \frac{e}{c}v^\mu f(\xi^2)\,, \quad \xi^\mu = x^\mu + \frac{v^\mu}{c}\left(\frac{v^\lambda}{c}x_\lambda\right), \tag{12.74}$$

where

$$v^\lambda \xi_\lambda = 0\,, \quad \xi^2 = x^2 + \left(\frac{v^\lambda}{c}x_\lambda\right)^2. \tag{12.75}$$

Check that for motion along the z axis with velocity v,

$$\xi^2 = x^2 + y^2 + \frac{(z - vt)^2}{1 - v^2/c^2}, \tag{12.76}$$

which anticipates Problem 34.1. Compute the field strengths $F^{\mu\nu}$ and evaluate the electromagnetic field stress tensor (12.44).

7. From the previous problem, use the field equation (12.28a) to produce $j^\mu(x)$. Check that $\partial_\mu j^\mu = 0$. Construct $F^{\mu\nu}\frac{1}{c}j_\nu$ and note that its vector nature lets one write

$$F^{\mu\nu}\frac{1}{c}j_\nu = -\partial^\mu t(\xi^2). \tag{12.77}$$

Exhibit $t(\xi^2)$ for the example $f(\xi^2) = (\xi^2 + a^2)^{-1/2}$. Inasmuch as the field tensor obeys (12.47)

$$\partial_\nu T^{\mu\nu}_f = -F^{\mu\nu}\frac{1}{c}j_\nu = \partial^\mu t, \tag{12.78}$$

one has realized a divergenceless electromagnetic tensor:

$$T^{\mu\nu} = T^{\mu\nu}_f - g^{\mu\nu}t\,, \quad \partial_\nu T^{\mu\nu} = 0. \tag{12.79}$$

It is the basis of a purely electromagnetic relativistic model of mass. There is, however, an ambiguity, because from (12.75)

$$\partial_\nu \left(\frac{v^\mu}{c} \frac{v^\nu}{c} t(\xi^2) \right) = 0. \tag{12.80}$$

Therefore,

$$T^{\mu\nu} = T_f^{\mu\nu} - \left(g^{\mu\nu} + \frac{v^\mu}{c} \frac{v^\nu}{c} \right) t, \tag{12.81}$$

for example, is also a possible electromagnetic tensor. Choice (12.79) has the property that the momentum density of the moving system (multiplied by c) is just that of the field,

$$T^{0k} = T_f^{0k} = \frac{1}{4\pi} (\mathbf{E} \times \mathbf{B})_k. \tag{12.82}$$

Choice (12.81) is such that the energy density of the system at rest is just that of the field,

$$\mathbf{v} = \mathbf{0}: \quad T^{00} = T_f^{00} = \frac{E^2}{8\pi}. \tag{12.83}$$

One cannot have both. That requires $t = 0$; that is, no charge. The system then is an electromagnetic pulse – it moves at the speed c.

8. Without specializing $f(\xi^2)$, integrate over all space (by introducing the variable $z' = (z - vt)/\sqrt{1 - v^2/c^2}$) to show that, whether one uses tensor (12.79) or (12.81),

$$E = \int (d\mathbf{r})\, T^{00} = \frac{mc^2}{\sqrt{1 - v^2/c^2}}, \quad p_k = \frac{1}{c} \int (d\mathbf{r})\, T^0{}_k = \frac{mv_k}{\sqrt{1 - v^2/c^2}}. \tag{12.84}$$

What numerical factor relates m is scheme (12.79) to that in scheme (12.81)?

9. Repeat the action discussion following from (12.41c) with $m_0 = 0$ and unspecified $f(\xi^2)$. What mass emerges?

10. Verify that the Maxwell equations involving magnetic currents, the second set in (12.37), can also be given by

$$\partial_\lambda F_{\mu\nu} + \partial_\mu F_{\nu\lambda} + \partial_\nu F_{\lambda\mu} = \epsilon_{\mu\nu\lambda\kappa} \frac{4\pi}{c} {}^*j^\kappa. \tag{12.85}$$

11. A particle has velocity components $v_x = \frac{dx}{dt}$ and $v_z = \frac{dz}{dt}$ in one coordinate frame. There is a second frame with relative velocity $\mathbf{V}$ along the z axis. What are the velocity components $v_x' = \frac{dx'}{dt'}$ and $v_z' = \frac{dz'}{dt'}$ in this frame? Give a simple interpretation of the v_x' result for $v_z = 0$.

12. Let the motion referred to in the previous problem be that of light, moving at angle θ with respect to the z axis. Find $\cos\theta'$ and $\sin\theta'$ in terms of $\cos\theta$ and $\sin\theta$. Check that $\cos^2\theta' + \sin^2\theta' = 1$. Exhibit θ' explicitly when $\beta = v/c \ll 1$.

13. The infinitesimal transformation contained in (12.50)

$$\delta\mathbf{p} = \frac{\delta\mathbf{v}}{c} \frac{E}{c}, \qquad \frac{\delta E}{c} = \frac{\delta\mathbf{v}}{c} \cdot \mathbf{p}, \tag{12.86}$$

identify the four-vector of momentum $p^\mu = (E/c, \mathbf{p})$ What is the value of the invariant $p^\mu p_\mu$ for a particle of rest mass m_0? Apply the analog of the space-time transformation equations

$$t' = \frac{t + \mathbf{v}\cdot\mathbf{r}/c^2}{\sqrt{1 - v^2/c^2}}, \qquad \mathbf{v}\cdot\mathbf{r}' = \mathbf{v}\cdot\frac{\mathbf{r} + \mathbf{v}t}{\sqrt{1 - v^2/c^2}} \tag{12.87}$$

to find the energy and momentum of a moving particle from their values when the particle is at rest.

14. A body of mass M is at rest relative to one observer. Two photons, each of energy ϵ, moving in opposite directions along the x-axis, fall on the body and are absorbed. Since the photons carry equal and opposite momenta, no net momentum is transferred to the body, and it remains at rest. Another observer is moving slowly along the y axis. Relative to him, the two photons and the body, both before and after the absorption act, have a common velocity $\mathbf{v}$ ($|\mathbf{v}| \ll c$) along the y axis, Reconcile conservation of the y-component of momentum with the fact that the velocity of the body does *not* change when the photons are absorbed.

15. Show, very simply, that $\mathbf{B}$, the magnetic field of a uniformly moving charge is $\frac{1}{c}\mathbf{v}\times\mathbf{E}$. Then consider two charges, moving with a common velocity $\mathbf{v}$ along parallel tracks, and show that the magnetic force between them is *opposite* to the electric force, and smaller by a factor of v^2/c^2. (This is an example of the rule that like *charges* repel, like *currents* attract.) Can you derive the same result by Lorentz transforming the equation of motion in the common rest frame of the two charges? (Hint: Coordinates perpendicular to the line of relative motion are unaffected by the transformation.)

16. From the response of a particle momentum to an infinitesimal Lorentz transformation (12.86), find the infinitesimal change of the particle velocity $\mathbf{V}$ when $\mathbf{V}$ and $\delta\mathbf{v}$ are in the same direction. Compare your result with the implication of the formula for the relativistic addition of velocities.

17. Light travels at the speed c/n in a stationary, nondispersive medium. What is the speed of light when this medium is moving at speed v parallel or antiparallel to the direction of the light? To what does this simplify when $v/c \ll 1$?

18. The frequency ω and the propagation vector $\mathbf{k}$ of a plane wave form a four-vector: $k^\mu = (\omega/c, \mathbf{k})$. Check that $k^\mu k_\mu = 0$ and that $\exp(ik_\mu x^\mu) = \exp[i(\mathbf{k}\cdot\mathbf{r} - \omega t)]$. Use Lorentz transformations to show that radiation, of frequency ω, propagating at an angle θ with respect to the z axis, will, to an observer moving with relative velocity $v = \beta c$ along the z axis, have the frequency

$$\omega' = \frac{1}{\sqrt{1-\beta^2}}\omega(1 - \beta\cos\theta) \tag{12.88}$$

(this is the Doppler[2] effect) and an angle relative to the z axis given by

$$\cos\theta' = \frac{\cos\theta - \beta}{1 - \beta\cos\theta} \tag{12.89}$$

(this is aberration). Find θ' explicitly for $|\beta| \ll 1$.

[2]Christian Doppler, 1803–1853.

19. By writing the angle relation (12.89) as

$$\cos\theta - \cos\theta' = \beta(1 - \cos\theta\cos\theta'), \tag{12.90}$$

show that

$$\tan\frac{1}{2}\theta' = \sqrt{\frac{1+\beta}{1-\beta}}\tan\frac{1}{2}\theta, \tag{12.91}$$

or, replacing the angle θ for the direction of travel by the angle $\alpha = \pi - \theta$ for the direction of arrival,

$$\tan\frac{1}{2}\alpha' = \sqrt{\frac{1-\beta}{1+\beta}}\tan\frac{1}{2}\alpha. \tag{12.92}$$

20. Integrate the invariant $(dk)2\delta(k^2)$, where $(dk) = dk^0(d\mathbf{k})$, over all $k^0 > 0$ to arrive at the invariant

$$\frac{(d\mathbf{k})}{|\mathbf{k}|} = \frac{\omega\, d\omega}{c^2}d\Omega\,, \quad \omega = kc, \tag{12.93}$$

in which $d\Omega$ is an element of solid angle. Use the Doppler effect formula (12.88) to deduce the solid angle transformation law,

$$d\Omega' = \frac{1-\beta^2}{(1-\beta\cos\theta)^2}d\Omega\,. \tag{12.94}$$

Then get it directly from the aberration formula (12.89). What did you assume about the azimuthal angle ϕ, and why? Check that the above relation is consistent with the requirement that $\int d\Omega' = 4\pi$.

21. In covariant notation, the Lagrange function for the electromagnetic field interacting with a prescribed current $j^\mu = (\rho, \mathbf{j})$ is (we set $c = 1$ for convenience)

$$\mathcal{L} = \frac{1}{4\pi}\left[-\frac{1}{2}F^{\mu\nu}(\partial_\mu A_\nu - \partial_\nu A_\mu) + \frac{1}{4}F^{\mu\nu}F_{\mu\nu}\right] + j^\mu A_\mu, \quad \partial_\mu j^\mu = 0. \tag{12.95}$$

The action is obtained from the Lagrange function by integrating over all space and time, $W = \int(dx)\mathcal{L}$, where $(dx) = dt(d\mathbf{r})$. In $\mathcal{L}$ the vector potential, A_μ, and the field strength tensor, $F_{\mu\nu}$, are regarded as independent variables.

(a) What equations are obtained by requiring that W be stationary under independent variations in A_μ and $F_{\mu\nu}$?

(b) Consider a coordinate displacement δx_μ. A scalar field ϕ changes under such a displacement by $\delta\phi(x) = \phi(x-\delta x) - \phi(x) = -\delta x_\nu \partial^\nu \phi$. Because $\mathcal{L}$ is invariant under a gauge transformation, $A_\mu \to A_\mu + \partial_\mu\lambda$, while $F_{\mu\nu} \to F_{\mu\nu}$, where λ is an arbitrary function of spacetime, conclude that A_μ must respond to a coordinate displacement by $\delta A_\mu = -\delta x_\lambda \partial^\lambda A_\mu - A^\lambda \partial_\mu \delta x_\lambda$.

(c) Now consider a source-free region, where $j_\mu = 0$. Assume now that $F_{\mu\nu} = \partial_\mu A_\nu - \partial_\nu A_\mu$. Show that δW is zero under a rigid coordinate displacement, $\delta x_\lambda =$ constant, provided the fields vanish outside the space-time region in question.

(d) Now suppose $\delta x_\lambda \neq$ constant. Show that

$$\delta W = \int (dx)\partial_\mu \delta x_\nu T^{\mu\nu}, \tag{12.96}$$

where $T^{\mu\nu}$ is the energy-momentum, or stress, tensor,

$$4\pi T^{\mu\nu} = F^{\mu\lambda}F^\nu{}_\lambda - \frac{1}{4}g^{\mu\nu}F^{\alpha\beta}F_{\alpha\beta}. \tag{12.97}$$

(e) Use the action principle to show that $t^{\mu\nu}$ is conserved, $\partial_\mu T^{\mu\nu} = 0$.

(f) Verify that T^{00} is the energy density, T^{0i} is the energy flux vector (or the momentum density), and that T^{ij} is the three-dimensional stress dyadic discussed in Chapter 3.

(g) What is the trace of $T^{\mu\nu}$? What is the significance of that result?

22. How are the above considerations leading to a conserved energy-momentum tensor modified if a fixed current source $j^\mu(x)$ is present? Show that (12.47) follows.

23. Show that the infinitesimal coordinate transformations considered in Problem 21 imply the infinitesimal Lorentz transformations of the fields, (12.57) and (12.62), when $\delta x^\mu = \delta\omega^{\mu\nu} x_\nu$.

24. The photon (quantum of light) is usually assumed to be massless (rest mass zero) so that it travels at speed c in all inertial coordinate frames. However, this is an experimental question. The best current limits are that the rest mass of the photon satisfies [23]

$$m_0 c^2 < 10^{-18}\,\text{eV}, \tag{12.98}$$

or even 8 orders of magnitude smaller than that. In this problem we will explore what happens if the photon rest mass is nonzero.

(a) The Lagrangian density for a spin-1 massive field coupled to a specified current j^μ is

$$\mathcal{L} = -\frac{1}{4}F^{\mu\nu}F_{\mu\nu} - \frac{1}{2}m_0^2 A^\mu A_\mu + j^\mu A_\mu. \tag{12.99}$$

Here, for simplicity, we have adopted rationalized natural units (see Appendix), with $c = 1$. The field strength $F^{\mu\nu}$ is defined in terms of the 4-vector potential to be

$$F^{\mu\nu} = \partial^\mu A^\nu - \partial^\nu A^\mu. \tag{12.100}$$

The action is defined by

$$W_{12} = \int_{t_2}^{t_1} dt \int (d\mathbf{r})\,\mathcal{L}, \tag{12.101}$$

where the spatial integration is over all space. Using the stationary action principle, varying the action with respect to A^μ, determine the equation of motion for A^μ.

(b) Is the resulting theory gauge invariant? Why or why not?

(c) Is the current necessarily conserved? Work out $\partial_\mu A^\mu$.

(d) In working out the equation of motion for A^μ you had to integrate by parts. Since the time integration is over a finite range, there is a boundary term. That boundary term corresponds to a generator. What is that generator? Express this in terms of an integral over the electric field and the 3-vector potential $\mathbf{A}$.

(e) Suppose $\partial_\mu A^\mu = 0$. Then, obtain the equation satisfied by the scalar potential $\phi = A^0$ in the presence of a static point charge at the origin, $j^0 = e\delta(\mathbf{r})$.

Part II

Electrostatics

13

Stationary Principles for Electrostatics

An overview of electrodynamics—the mechanics of charged particles and electromagnetic fields in interaction—is now before us. In this, and in the following several chapters, we specialize to more restricted situations—by considering prescribed distributions of charges and currents, without inquiring how those distributions were established and maintained; and by discussing only arrangements in which there is no time dependence in charge and field quantities—a static regime. We do this both because such situations are of physical interest, and because of the opportunity offered to develop general mathematical methods in simpler contexts.

Let us look back at the Lagrangian (10.9) and isolate those terms that make reference to the electromagnetic field, including the interaction with charges. With the further omission of the time derivative terms, what remains (apart from a minus sign) is the energy of the field in the presence of prescribed charges and currents: [see also the Hamiltonian (10.17)]

$$E = E_{\text{elec}} + E_{\text{mag}}, \tag{13.1}$$

with

$$E_{\text{elec}} = \int (d\mathbf{r}) \left[\rho\phi + \frac{1}{4\pi}\left(\mathbf{E}\cdot\boldsymbol{\nabla}\phi + \frac{1}{2}\mathbf{E}^2\right)\right] \tag{13.2a}$$

and

$$E_{\text{mag}} = \int (d\mathbf{r}) \left[-\frac{1}{c}\mathbf{j}\cdot\mathbf{A} + \frac{1}{4\pi}\left(\mathbf{B}\cdot\boldsymbol{\nabla}\times\mathbf{A} - \frac{1}{2}\mathbf{B}^2\right)\right]. \tag{13.2b}$$

In the latter, $\mathbf{j}$ is restricted by

$$\boldsymbol{\nabla}\cdot\mathbf{j} = 0, \tag{13.3}$$

as the time independent version of local charge conservation. The two energy expressions are entirely independent. We will consider only the situation governed by (13.2a) in the next several chapters and will defer consideration of magnetostatics, governed by (13.2b), until Chapter 28.

13.1 Stationary Principles for the Energy

In this chapter, we consider only the electrostatic energy (omitting the subscript)

$$E[\phi, \mathbf{E}] = \int (d\mathbf{r}) \left[\rho\phi + \frac{1}{4\pi}\mathbf{E}\cdot\boldsymbol{\nabla}\phi + \frac{1}{8\pi}\mathbf{E}^2\right]. \tag{13.4}$$

Derived as it is, by a specialization of the action, this energy must also have the property of being stationary for variations about the solutions of the field equations. Indeed, independent variations of $\phi(\mathbf{r})$ and $\mathbf{E}(\mathbf{r})$ gives

$$\delta E[\phi, \mathbf{E}] = \int (d\mathbf{r}) \left[\rho\delta\phi + \frac{1}{4\pi}\mathbf{E}\cdot\boldsymbol{\nabla}\delta\phi + \frac{1}{4\pi}\delta\mathbf{E}\cdot(\boldsymbol{\nabla}\phi + \mathbf{E})\right], \tag{13.5}$$

DOI: 10.1201/9781003057369-13

which may be simplified by use of the identity

$$\mathbf{E} \cdot \boldsymbol{\nabla} \delta\phi = \boldsymbol{\nabla} \cdot (\mathbf{E}\delta\phi) - (\boldsymbol{\nabla} \cdot \mathbf{E})\delta\phi, \tag{13.6}$$

to read

$$\delta E[\phi, \mathbf{E}] = \int (d\mathbf{r}) \left[\delta\phi \left(\rho - \frac{\boldsymbol{\nabla} \cdot \mathbf{E}}{4\pi}\right) + \frac{1}{4\pi}\delta\mathbf{E} \cdot (\boldsymbol{\nabla}\phi + \mathbf{E})\right], \tag{13.7}$$

where we have ignored the surface term since we assume the integral extends over the entire region where the fields are nonzero. (In particular, we explicitly assume that ϕ retains a zero value at infinity: $\delta\phi = 0$.) The stationary requirement on the energy E, $\delta E = 0$, now implies the two basic equations of electrostatics,

$$\boldsymbol{\nabla} \cdot \mathbf{E} = 4\pi\rho, \quad \mathbf{E} = -\boldsymbol{\nabla}\phi. \tag{13.8}$$

Of course, the connection between $\mathbf{E}$ and ϕ implies

$$\boldsymbol{\nabla} \times \mathbf{E} = \mathbf{0}. \tag{13.9}$$

Having seen this, we can immediately modify the energy expression, (13.4), to incorporate the effects of a dielectric medium, where the permittivity may vary in space, $\epsilon = \epsilon(\mathbf{r})$,

$$E[\phi, \mathbf{D}] = \int (d\mathbf{r}) \left[\rho\phi + \frac{\epsilon\mathbf{E} \cdot \boldsymbol{\nabla}\phi}{4\pi} + \frac{\epsilon\mathbf{E}^2}{8\pi}\right]. \tag{13.10}$$

The validity of this form is indicated by noting that if it is required to be stationary under variations in ϕ and $\mathbf{D} = \epsilon\mathbf{E}$, we recover the equations of electrostatics in a dielectric. That is, the variation in the energy,

$$\delta E[\phi, \mathbf{D}] = \int (d\mathbf{r}) \left[\delta\phi \left(\rho - \frac{1}{4\pi}\boldsymbol{\nabla} \cdot \mathbf{D}\right) + \frac{1}{4\pi}\delta\mathbf{D} \cdot (\boldsymbol{\nabla}\phi + \mathbf{E})\right] = 0, \tag{13.11}$$

implies

$$\boldsymbol{\nabla} \cdot \mathbf{D} = 4\pi\rho, \quad \mathbf{E} = -\boldsymbol{\nabla}\phi. \tag{13.12}$$

(As before, in writing the variation in the form of (13.11), we have integrated by parts and omitted the surface term.)

The discussion of action forms has accustomed us to the idea of restricting the choice of variables by adopting one or more of the (field) equations as definitions. Here we can use two restrictive versions of the above energy functional, (13.10).

13.1.1 The Scalar Field Form

We adopt, as the definition of $\mathbf{E}$, its construction in terms of the scalar potential,

$$\mathbf{E} = -\boldsymbol{\nabla}\phi, \tag{13.13}$$

the curl of which is zero. The energy, (13.10), as a functional of the potential, is then

$$E[\phi] = \int (d\mathbf{r}) \left[\rho\phi - \frac{1}{8\pi}\epsilon(\boldsymbol{\nabla}\phi)^2\right]. \tag{13.14}$$

The requirement that E be stationary under the variation of ϕ yields Maxwell's equation

$$\boldsymbol{\nabla} \cdot \mathbf{D} = 4\pi\rho, \tag{13.15}$$

where $\mathbf{D}$ is defined by

$$\mathbf{D} = -\epsilon\boldsymbol{\nabla}\phi. \tag{13.16}$$

An infinitesimal variation in the scale of the function $\phi(\mathbf{r})$,

$$\delta\phi(\mathbf{r}) = \delta\lambda\phi(\mathbf{r}), \tag{13.17}$$

for which

$$\delta E[\phi] = \delta\lambda \int (d\mathbf{r}) \left[\rho\phi - \frac{1}{4\pi}\epsilon\mathbf{E}^2\right] = 0, \tag{13.18}$$

supplies information about E, the value of $E[\phi]$ for the correct field. It is

$$E = \int (d\mathbf{r})\, \frac{1}{8\pi}\epsilon\mathbf{E}^2, \tag{13.19}$$

which is indeed the total energy of the system, according to (8.13).

The energy functional, (13.14), contains yet further information. At the stationary point ϕ_0, $E[\phi_0] = E$ is an absolute maximum, as is seen by making a finite variation in the potential, $\Delta\phi$, provided $\epsilon > 0$,

$$E[\phi_0 + \Delta\phi] = E - \int (d\mathbf{r})\, \frac{\epsilon}{8\pi}(\boldsymbol{\nabla}\Delta\phi)^2 \leq E, \tag{13.20}$$

since the linear term in $\Delta\phi$ vanishes by the stationary principle. The correct energy, given by the physical ϕ_0, is a maximum of the functional (13.14). Evaluating the energy functional for an arbitrary potential bounds the energy from below,

$$E \geq E[\phi]. \tag{13.21}$$

13.1.2 The Vector Field Form

Now we use the vector $\mathbf{D}$ as the independent variable, but restrict its choice by insisting on the validity of the field equation

$$\boldsymbol{\nabla}\cdot\mathbf{D} = 4\pi\rho. \tag{13.22}$$

Then, if we replace $\epsilon\mathbf{E}$ by $\mathbf{D}$ in the expression for the energy, (13.10), and integrate by parts on the $\mathbf{D}\cdot\boldsymbol{\nabla}\phi$ term, we obtain

$$E[\mathbf{D}] = \int (d\mathbf{r})\, \frac{D^2}{8\pi\epsilon}, \tag{13.23}$$

which is a functional of $\mathbf{D}$. How does the stationary principle work here? The variation of (13.23) is

$$\delta E = \int (d\mathbf{r})\, \frac{1}{4\pi\epsilon}\mathbf{D}\cdot\delta\mathbf{D}, \tag{13.24}$$

where the displacement vector is varied subject to the restriction that (13.22) be satisfied:

$$\boldsymbol{\nabla}\cdot(\mathbf{D} + \delta\mathbf{D}) = 4\pi\rho, \tag{13.25a}$$

or

$$\boldsymbol{\nabla}\cdot\delta\mathbf{D} = 0. \tag{13.25b}$$

Therefore, any variation in $\mathbf{D}$ must be a curl,

$$\delta\mathbf{D} \equiv \boldsymbol{\nabla}\times\delta\boldsymbol{\mathcal{A}}, \tag{13.26}$$

where $\delta\boldsymbol{\mathcal{A}}$ is an arbitrary, infinitesimal vector, enabling us to write the variation in the energy as

$$\delta E = \int (d\mathbf{r})\, \frac{\mathbf{E}}{4\pi} \cdot \boldsymbol{\nabla}\times\delta\boldsymbol{\mathcal{A}}. \tag{13.27a}$$

An integration by parts then implies the irrotational property of $\mathbf{E}$,

$$\boldsymbol{\nabla}\times\mathbf{E} = \mathbf{0}. \tag{13.27b}$$

A second way of getting this result is based on the replacement of the local restriction (13.25b) by the equivalent integral statement

$$0 = \int (d\mathbf{r})\, \frac{1}{4\pi}\phi(\mathbf{r})\boldsymbol{\nabla}\cdot\delta\mathbf{D}(\mathbf{r}) = -\int (d\mathbf{r})\, \frac{1}{4\pi}\boldsymbol{\nabla}\phi\cdot\delta\mathbf{D}, \tag{13.28}$$

where $\phi(\mathbf{r})$ is an arbitrary scalar function, a Lagrange multiplier. Then the comparison of

$$\delta E[\mathbf{D}] = \int (d\mathbf{r})\, \frac{1}{4\pi}\mathbf{E}\cdot\delta\mathbf{D} = 0 \tag{13.29}$$

with the restriction (13.28) yields the condition $\mathbf{E} = -\boldsymbol{\nabla}\phi$, which is equivalent to (13.27b).

The advantage to this functional form of the energy, (13.23), is seen by again considering finite variations (called $\Delta\mathbf{D}$). If $\mathbf{D}_0$ is the correct physical solution, then $E[\mathbf{D}_0] = E$ is an absolute minimum,

$$E[\mathbf{D}_0 + \Delta\mathbf{D}] = E + \int (d\mathbf{r})\, \frac{(\Delta\mathbf{D})^2}{8\pi\epsilon} \geq E, \tag{13.30}$$

(since the linear term in $\Delta\mathbf{D}$ is zero due to the stationary principle). Therefore, the correct energy is the minimum value of (13.23) while an arbitrary $\mathbf{D}$ [compatible with (13.22)] will give an upper bound to E. These bounds, (13.20) and (13.30),

$$E[\phi] \leq E \leq E[\mathbf{D}], \tag{13.31}$$

are useful for finding approximate solutions when exact solutions are difficult or impossible to obtain.

13.2 Force on Dielectrics

First, we present an essentially trivial example of the effect on the energy of a finite change in dielectric constant. It requires that ϵ be independent of position. Let us consider the scalar field form (13.14) and write

$$\phi = \frac{1}{\epsilon}\phi_0. \tag{13.32}$$

The consequence for the energy is

$$E = \frac{1}{\epsilon}E_0, \tag{13.33}$$

where

$$E_0 = \int (d\mathbf{r})\left[\phi_0\rho - \frac{1}{8\pi}(\boldsymbol{\nabla}\phi_0)^2\right] \tag{13.34}$$

will be recognized as the energy expression for $\epsilon = 1$, the vacuum. The same energy relation follows immediately from (13.23), inasmuch as $\mathbf{D}$ retains its meaning in order to satisfy

(13.22). That the presence of a uniform dielectric medium reduces the energy by a factor of ϵ is familiar in elementary presentations of electrostatics as the corresponding reduction in the strength of the force between charges.

Now we turn to a position-dependent dielectric constant, $\epsilon(\mathbf{r})$, and examine the effect of an infinitesimal alteration,

$$\epsilon(\mathbf{r}) \to \epsilon(\mathbf{r}) + \delta\epsilon(\mathbf{r}). \tag{13.35}$$

This will change the fields, $\phi(\mathbf{r})$ or $\mathbf{D}(\mathbf{r})$, but these changes have no first-order effect on the energy, which is stationary. Accordingly, the only contribution to the energy change is produced by the explicit appearance of $\epsilon(\mathbf{r})$ in the respective energy expressions. Using (13.14), we get

$$\delta E[\phi] = -\int (d\mathbf{r})\, \frac{\delta\epsilon(\boldsymbol{\nabla}\phi)^2}{8\pi} = -\int (d\mathbf{r})\, \frac{\delta\epsilon \mathbf{E}^2}{8\pi}. \tag{13.36a}$$

Equivalently, from the second form, (13.23), the first-order variation in the energy is

$$\delta E[\mathbf{D}] = -\int (d\mathbf{r})\, \frac{1}{8\pi} \left(\frac{\mathbf{D}}{\epsilon}\right)^2 \delta\epsilon = -\int (d\mathbf{r})\, \frac{\delta\epsilon \mathbf{E}^2}{8\pi}, \tag{13.36b}$$

in agreement with (13.36a).

An example is provided by the infinitesimal displacement of an uncharged inhomogeneous dielectric. If the material is displaced by $\delta\mathbf{r}$, the new dielectric constant at $\mathbf{r}$ is the old dielectric constant at $\mathbf{r} - \delta\mathbf{r}$:

$$\epsilon(\mathbf{r}) \to \epsilon(\mathbf{r}) + \delta\epsilon(\mathbf{r}) = \epsilon(\mathbf{r} - \delta\mathbf{r}) = \epsilon(\mathbf{r}) - \delta\mathbf{r} \cdot \boldsymbol{\nabla}\epsilon(\mathbf{r}), \tag{13.37}$$

or

$$\delta\epsilon(\mathbf{r}) = -\delta\mathbf{r} \cdot \boldsymbol{\nabla}\epsilon(\mathbf{r}). \tag{13.38}$$

Then, the change in energy is

$$\delta E = \delta\mathbf{r} \cdot \int (d\mathbf{r})\, (\boldsymbol{\nabla}\epsilon) \frac{E^2}{8\pi} = -\mathbf{F} \cdot \delta\mathbf{r}. \tag{13.39}$$

We therefore identify the force, $\mathbf{F}$, on the dielectric due to the inhomogeneity of the medium to be

$$\mathbf{F} = -\int (d\mathbf{r})\, \frac{E^2}{8\pi} \boldsymbol{\nabla}\epsilon. \tag{13.40}$$

This result, (13.40), also shows the continued relevance of the electric field part of the stress tensor, (8.13) with $\mathbf{H} = \mathbf{0}$, even though $\epsilon(\mathbf{r})$ is no longer constant,

$$\mathbf{T} = \mathbf{1}\frac{\epsilon E^2}{8\pi} - \frac{\epsilon \mathbf{E}\mathbf{E}}{4\pi}, \tag{13.41}$$

for spatially inhomogeneous dielectrics. Since the stress tensor (dyadic) describes the outward flow of momentum across a directed unit area, the total force on a body bounded by a closed surface S, the net flow of momentum *into* that body, is

$$\mathbf{F} = -\oint_S d\mathbf{S} \cdot \mathbf{T} = -\int (d\mathbf{r})\, \boldsymbol{\nabla} \cdot \mathbf{T}, \tag{13.42}$$

where the volume integral extends over the body. The divergence of the stress tensor is

$$\boldsymbol{\nabla} \cdot \mathbf{T} = (\boldsymbol{\nabla}\epsilon)\frac{E^2}{8\pi} + \frac{1}{4\pi} D_i \boldsymbol{\nabla} E_i - \frac{\boldsymbol{\nabla} \cdot \mathbf{D}}{4\pi}\mathbf{E} - \frac{(\mathbf{D} \cdot \boldsymbol{\nabla})\mathbf{E}}{4\pi}, \tag{13.43}$$

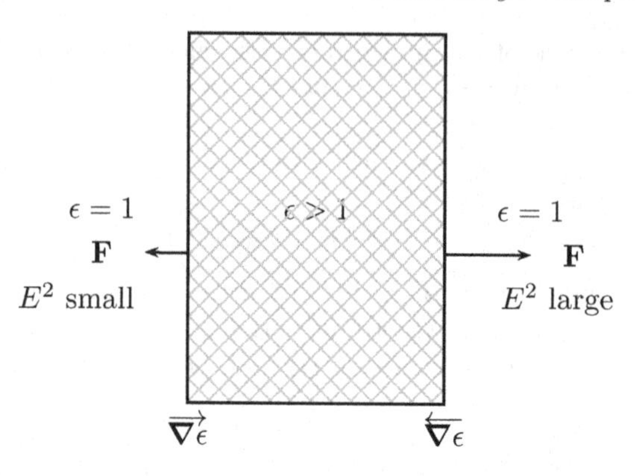

FIGURE 13.1
Dielectric slab immersed in an inhomogeneous electric field. The forces **F** shown are those on each surface due to (13.42), (13.46).

which is the generalization of (8.10). Since for electrostatics,

$$D_i \boldsymbol{\nabla} E_i - (\mathbf{D} \cdot \boldsymbol{\nabla})\mathbf{E} = \mathbf{D} \times (\boldsymbol{\nabla} \times \mathbf{E}) = \mathbf{0}, \tag{13.44}$$

and when no free charge is present,

$$\boldsymbol{\nabla} \cdot \mathbf{D} = 4\pi\rho = 0, \tag{13.45}$$

the divergence reduces to

$$\boldsymbol{\nabla} \cdot \mathbf{T} = (\boldsymbol{\nabla}\epsilon)\frac{E^2}{8\pi}, \tag{13.46}$$

so that the force calculated from (13.42) is identical with (13.40).

As an application of the above result, consider a slab of dielectric material, with $\epsilon =$ constant, immersed in an inhomogeneous electric field. The gradient of ϵ arises from the discontinuity of the dielectric constant between vacuum and medium. The situation might be described by Fig. 13.1. If E^2 is small on the left and large on the right, the dielectric material will be pulled to the right, that is, into the region of strong field. [Note that this depends on the slab having $\epsilon > 1$ so that the directions of $\boldsymbol{\nabla}\epsilon$ are as shown in the figure.]

Incidentally, if there is a distribution of free charge in the body, an additional term survives in (13.43), $-\rho\mathbf{E}$, which leads to the expected force contribution

$$\int (d\mathbf{r})\, \rho\mathbf{E}. \tag{13.47}$$

The same result emerges from the energy form (13.14) if we introduce the analog of (13.38) for the charge density,

$$\delta_\rho E = \int (d\mathbf{r})\, (-\delta\mathbf{r} \cdot \boldsymbol{\nabla}\rho)\phi = -\delta\mathbf{r} \cdot \int (d\mathbf{r})\, \rho\mathbf{E}. \tag{13.48}$$

In using the vector field form (13.23), we must recognize that the displacement of the charge distribution requires a corresponding displacement of the field **D**, in order to preserve (13.22):

$$\delta_\rho E = \int (d\mathbf{r})\, \frac{1}{4\pi\epsilon}\mathbf{D} \cdot \delta\mathbf{D} = \int (d\mathbf{r})\, \frac{1}{4\pi}\mathbf{E} \cdot (-\delta\mathbf{r} \cdot \boldsymbol{\nabla})\mathbf{D}. \tag{13.49}$$

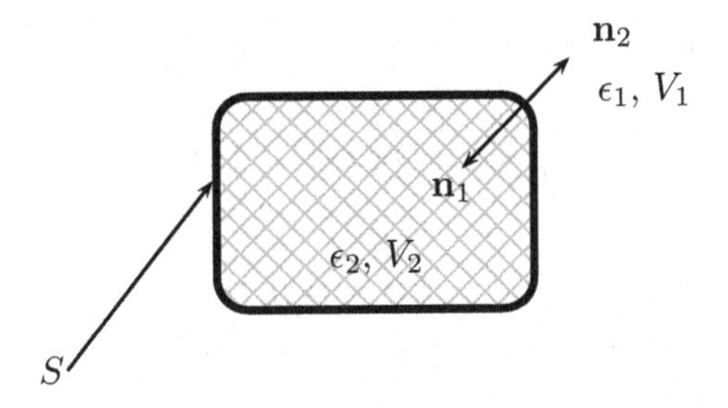

FIGURE 13.2
Boundary between two regions with different dielectric constants.

Then, the identity

$$-(\delta\mathbf{r}\cdot\boldsymbol{\nabla})\mathbf{D}+\delta\mathbf{r}\boldsymbol{\nabla}\cdot\mathbf{D}=\boldsymbol{\nabla}\times(\delta\mathbf{r}\times\mathbf{D}), \tag{13.50}$$

and use of both electrostatic field equations then gives the anticipated result (13.47).

13.3 Boundary Conditions

An arrangement of objects with different values of ϵ that are in contact with each other provides the simplest example of an inhomogeneous dielectric constant. Imagine we have two volumes, V_1 and V_2, with dielectric constants ϵ_1 and ϵ_2, respectively, sharing a common surface S, as shown in Fig. 13.2. Because of this discontinuity of ϵ on the surface, when the energy expression (13.10) is varied and expressed in the form (13.11), there is an additional contribution from the surface term previously omitted. In the interior of both V_1 and V_2 the same arguments as those given in Section 13.1 still apply so that we have the same field equations,

$$\boldsymbol{\nabla}\cdot\mathbf{D}=4\pi\rho;\qquad \boldsymbol{\nabla}\times\mathbf{E}=\mathbf{0}. \tag{13.51}$$

The additional surface term, which is now the total variation in the energy, is

$$\begin{aligned}\delta E&=\int_{V_1}(d\mathbf{r})\,\boldsymbol{\nabla}\cdot\left(\frac{\mathbf{D}}{4\pi}\delta\phi\right)+\int_{V_2}(d\mathbf{r})\,\boldsymbol{\nabla}\cdot\left(\frac{\mathbf{D}}{4\pi}\delta\phi\right)\\&=\int_S dS\left(\mathbf{n}_1\cdot\frac{\mathbf{D}_1}{4\pi}\delta\phi_1+\mathbf{n}_2\cdot\frac{\mathbf{D}_2}{4\pi}\delta\phi_2\right),\end{aligned} \tag{13.52}$$

where S is the common surface and $\mathbf{n}_1$ and $\mathbf{n}_2$ are the oppositely directed outward unit normal vectors for V_1 and V_2, respectively:

$$\mathbf{n}_1=-\mathbf{n}_2. \tag{13.53}$$

At the atomic level, the transition between two macroscopically different substances is reasonably smooth. And certainly the averaged microscopic field that is $\mathbf{E}(\mathbf{r})$ remains finite in that transition region. In the static regime now under consideration, where $\mathbf{E}(\mathbf{r})=-\boldsymbol{\nabla}\phi(\mathbf{r})$, this implies continuity of the scalar potential across the contact surface, or

$$\phi_1=\phi_2,\qquad \delta\phi_1=\delta\phi_2, \tag{13.54}$$

at a common point on the surface S. The variation of the energy, (13.52), at the stationary point is therefore

$$\delta E = \int_S dS \frac{1}{4\pi}[\mathbf{n}_2 \cdot (\mathbf{D}_2 - \mathbf{D}_1)]\delta\phi = 0 \tag{13.55a}$$

or, since $\delta\phi$ is arbitrary,

$$\mathbf{n}_2 \cdot (\mathbf{D}_2 - \mathbf{D}_1) = 0. \tag{13.55b}$$

The normal component of $\mathbf{D}$ is continuous across the boundary between the two media because of our implicit assumption that there is no free surface charge density ($\sigma = 0$).

There is, however, the possibility of forming a distribution of free charge along the contact surface S. Then the volume integral involving the charge density ρ in (13.10) is supplemented by a surface integral referring to the surface charge density σ (which symbol is not likely to be confused with that for conductivity in these static circumstances), namely

$$\int_S dS\, \sigma\phi. \tag{13.56}$$

Thus, there is a further term in the variation of the energy (13.10),

$$\int_S dS\, \sigma\delta\phi. \tag{13.57}$$

The resulting generalization of (13.55a) is

$$\delta E = \int_S dS \frac{1}{4\pi}[\mathbf{n}_2 \cdot (\mathbf{D}_2 - \mathbf{D}_1) + 4\pi\sigma]\delta\phi = 0, \tag{13.58a}$$

so correspondingly the stationary energy principle now asserts that

$$\mathbf{n}_2 \cdot (\mathbf{D}_1 - \mathbf{D}_2) = 4\pi\sigma \tag{13.58b}$$

is the general boundary condition on the normal component of $\mathbf{D}$. We may regard (13.58b) as the surface version of the volume statement

$$\boldsymbol{\nabla} \cdot \mathbf{D} = 4\pi\rho. \tag{13.59}$$

Likewise, there is a surface statement corresponding to $\boldsymbol{\nabla}\times\mathbf{E} = \mathbf{0}$. This follows from the continuity of ϕ, (13.54), which implies that of the tangential derivative (that is, the component of the gradient parallel to the surface):

$$(\mathbf{n}_2\times\boldsymbol{\nabla})(\phi_1 - \phi_2) = \mathbf{0} \tag{13.60a}$$

or

$$\mathbf{n}_2\times(\mathbf{E}_1 - \mathbf{E}_2) = \mathbf{0}. \tag{13.60b}$$

Equation (13.60b) states that the tangential component of the electric field is continuous. This result can also be derived directly from the vector field form by considering the surface terms that follow from (13.27a).

13.4 Conductors

The surface of a conducting body provides an important example of these boundary conditions. Within a conductor electric current flows in response to the presence of an electric field. Since in the static situation there is no flow of charge, $\mathbf{E} = \mathbf{0}$ everywhere inside the conductor; the interior of a conductor is a region of constant potential ϕ. The continuity of the tangential component of $\mathbf{E}$, (13.60b), then implies

$$\mathbf{n} \times \mathbf{E} = \mathbf{0} \tag{13.61}$$

on the surface of the conductor. Indeed, if it were otherwise, charge would flow on the surface, which must also be a region of constant potential, of value equal to that in the interior. Moreover, since $\mathbf{E}$ and $\mathbf{D}$ vanish inside, the other boundary condition, (13.58b), implies

$$\mathbf{n} \cdot \mathbf{D} = 4\pi\sigma \tag{13.62}$$

just outside the conductor, where $\mathbf{n}$ is the outward normal to the surface, and again σ is the surface charge density.

We use these properties of the field to examine the electrical forces exerted on the conducting surface, as described by the stress dyadic (13.41)

$$\mathbf{T} = \mathbf{1}\frac{\epsilon E^2}{8\pi} - \frac{\epsilon \mathbf{E}\mathbf{E}}{4\pi}, \tag{13.63}$$

where ϵ is the dielectric constant of the medium surrounding the conductor. The force acting on a unit element of area, with its outwardly directed normal vector $\mathbf{n}$, is

$$-\mathbf{n} \cdot \mathbf{T} = -\frac{\epsilon E^2}{8\pi}\mathbf{n} + \frac{\mathbf{n} \cdot \mathbf{D}}{4\pi}\mathbf{E}, \tag{13.64}$$

where

$$\epsilon E^2 = \epsilon(\mathbf{n} \times \mathbf{E})^2 + \frac{1}{\epsilon}(\mathbf{n} \cdot \mathbf{D})^2 \tag{13.65}$$

is a helpful decomposition into tangential and normal field components. We now recognize that

$$-\mathbf{n} \cdot \mathbf{T} = -\frac{2\pi\sigma^2}{\epsilon}\mathbf{n} + \sigma\mathbf{E}, \tag{13.66}$$

which force has no tangential components; there is only a normal component of force per unit area:

$$(-\mathbf{n}) \cdot \mathbf{T} \cdot (-\mathbf{n}) = \frac{2\pi\sigma^2}{\epsilon} - \frac{\sigma}{\epsilon}\mathbf{n} \cdot \mathbf{D} = -\frac{2\pi}{\epsilon}\sigma^2. \tag{13.67}$$

The negative sign attached to this normal force per unit area tells us that it is not a pressure but a traction or tension, drawing the surface element into the region occupied by the electric field.

Here is a simple application of this result. Two identical conducting plates of cross sectional area A and negligible thickness are placed in parallel proximity, just short of total contact. Equal and opposite charges are bought up and placed on the respective conductors, resulting in the final charges Q and $-Q$. In view of the almost complete cancellation of the electric fields produced by the opposite charges, essentially no work is performed during the act of charging up the plates. Over most of the area of each plate, the surface charge density is uniform, of magnitude

$$\sigma = \frac{Q}{A}. \tag{13.68}$$

Therefore, the total force on each plate, drawing it toward its partner, has the strength

$$F = \frac{2\pi}{\epsilon}\left(\frac{Q}{A}\right)^2 A, \tag{13.69}$$

where, again, ϵ is the dielectric constant of the medium in which the plates are inserted (only the material between the plates is physically significant).

Now holding one plate fixed, move the other away to a distance a, which distance is very small on the linear scale provided by $A^{1/2}$. The work required, the energy of the resulting parallel plate configuration, is just Fa, or

$$E = \frac{1}{2}\frac{Q^2}{C}, \tag{13.70}$$

where

$$C = \frac{\epsilon A}{4\pi a} \tag{13.71}$$

is identified as the capacitance of the electric capacitor. We will discuss capacitance in general in Sec. 26.6.

13.5 Problems for Chapter 13

1. An uncharged pith ball is placed in the vicinity of a positive charge. Which way will the pith ball move? What if the charge is negative? Get your answer (1) by using the electrostatic stationary principle, and (2) by direct physical argument.
2. Consider the boundary conditions between two different media with different dielectric constants. Prove that the continuity of the potential implies that the tangential derivative is also continuous, hence that the tangential component of the electric field is continuous.
3. Derive (13.60b) from the vector field form of the energy functional (13.23).
4. Derive (13.40) from the force given in (4.40), namely

$$\mathbf{F} = \int (d\mathbf{r})\,[-(\boldsymbol{\nabla}\cdot\mathbf{P})\mathbf{E}]. \tag{13.72}$$

(Hint: Integrate by parts and use $\boldsymbol{\nabla}\times\mathbf{E} = \mathbf{0}$.)

5. Consider two parallel nonconducting plates with uniform charge densities $+\sigma$ and $-\sigma$. As shown in Fig. 13.3, half the region between the plates is filled with a dielectric with dielectric constant $\epsilon > 1$, and half is filled with vacuum ($\epsilon = 1$). Ignoring fringing effects at the ends of the plates, calculate the electric field $\mathbf{E}$ in both the dielectric and the vacuum. Is there a force on the dielectric? Which way does it point? (Does it matter if the dielectric extends in the region outside the plates?) What happens if the plates are conductors, so the charge is free to redistribute itself on them?
6. By applying a scale transformation $\phi(\mathbf{r}) \to \lambda\phi(\mathbf{r})$ in $E[\phi]$, (13.14), derive the stationary, scale-independent energy form

$$E = \frac{1}{4}\frac{\left[\int (d\mathbf{r})\,\rho\phi\right]^2}{\int (d\mathbf{r})\,\epsilon(\boldsymbol{\nabla}\phi)^2/8\pi}. \tag{13.73}$$

Verify that this is indeed stationary.

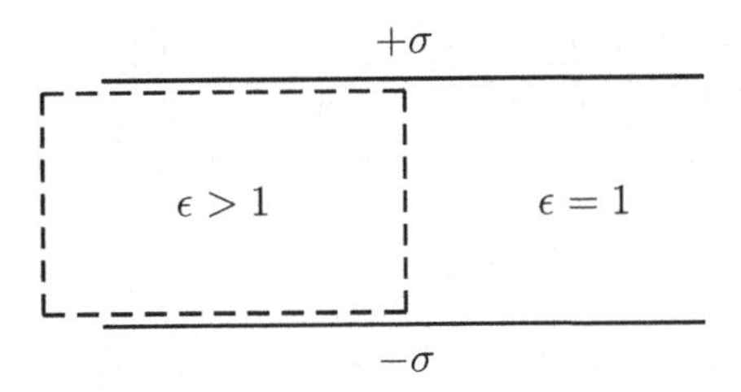

FIGURE 13.3
Dielectric between parallel plates, as discussed in Problem 11.5.

7. Recall that for homogeneous ϵ, $\phi = \phi_0/\epsilon$. Now use $E[\phi]$ and $E[\mathbf{D}]$ to find bounds for the energy of a charge distribution $\rho(\mathbf{r})$ in the dielectric medium $\epsilon(\mathbf{r})$ where ϵ is *slowly* varying:
$$\frac{|\boldsymbol{\nabla}\epsilon(\mathbf{r})|}{\epsilon(\mathbf{r})} \ll \frac{|\boldsymbol{\nabla}\phi(\mathbf{r})|}{\phi(\mathbf{r})} \approx \frac{|\boldsymbol{\nabla}\phi_0(\mathbf{r})|}{\phi_0(\mathbf{r})}. \tag{13.74}$$
Here ϕ_0 is the solution to Poisson's equation in the absence of the dielectric,
$$-\nabla^2\phi_0 = 4\pi\rho,$$
and the bounds should be expressed in terms of integrals involving ϵ, ϕ_0, $\boldsymbol{\nabla}\phi_0$, and $\boldsymbol{\nabla}\epsilon$.

8. Use Gauss' law to show that the electric field outside a static spherically symmetric distribution of charge is that same as if all the charge were concentrated at its center. Then consider the interaction between two such spherically symmetric charge distributions, which are completely non-overlapping. If the electric field of the first charge distribution is $\mathbf{E}_1$, and the charge density of the second charge distribution is ρ_2, show that the force on the second charge distribution is
$$\mathbf{F}_{12} = \int_{V_2} (d\mathbf{r})\rho_2(\mathbf{r})\mathbf{E}_1(\mathbf{r}), \tag{13.75}$$
where the integral is over the volume V_2 of the second charge distribution. Argue that this integral can only be of the form
$$\mathbf{F}_{12} = Q_1 Q_2 \frac{\mathbf{R}}{R^3} f\left(\frac{b}{R}\right), \tag{13.76}$$
where f is some function of the ratio of b to R, where b is the radius of the second charge distribution and $\mathbf{R}$ is the vector displacement of the two centers of the charge distributions from each other. (Otherwise, f is a function of parameters that describe the internal distribution of charge.) Here, Q_i is the total charge of the ith distribution,
$$Q_i = \int_{V_i} (d\mathbf{r})\rho_i(\mathbf{r}). \tag{13.77}$$
But, you could equally well interchange the roles of the two charge distributions, and then you would have for the force on distribution 1 due to distribution 2
$$\mathbf{F}_{21} = -Q_1 Q_2 \frac{\mathbf{R}}{R^3} g\left(\frac{a}{R}\right), \tag{13.78}$$

where a is the radius of the second charge distribution. The two functions, f and g might be different because the two distributions are different. (That is, the internal distribution of charge might differ.) However, the only consistent conclusion is that

$$f\left(\frac{b}{R}\right) = g\left(\frac{a}{R}\right) = 1, \tag{13.79}$$

that is, the force due to the interaction between two spherically symmetric charge distributions is as though both charges were concentrated at the center of their respective spherical distributions.

9. An alternative way to prove the above theorem is to use the electric stress tensor. Let the line connected the two spherically symmetric, non-overlapping charge distributions lie along the z axis. Consider T_{zz} integrated over a plane perpendicular to the z-axis bisecting that line connecting the centers, but between and outside both distributions. Argue that the force on one of the spherical distributions must be given by

$$F = \int dx\, dy\, T_{zz}(x, y, z), \tag{13.80}$$

integrated over that plane. The electric field is the vector sum of the fields from each of the charge distributions. Argue that since, outside the distributions, each field is that of a point charge, located at the center of the corresponding distribution, the force between the bodies must be identical to that between the corresponding point charges.

10. A point charge e is placed at the center of a spherical cavity of radius a embedded in a uniform dielectric. That is, with origin chosen at the center of the sphere, the permittivity satisfies

$$\epsilon(\mathbf{r}) = \begin{cases} 1, r < a, \\ \varepsilon, r > a, \end{cases} \tag{13.81}$$

with ε a constant. Suppose that besides the point charge at the origin, there is otherwise no free charge.

(a) Calculate the electric field $\mathbf{E}$ at each point in space.

(b) Show that the pressure exerted by the charge on the walls of the cavity, the radial integral of the radial components of the force density,

$$\mathbf{f} = -\frac{E^2}{8\pi}\boldsymbol{\nabla}\epsilon, \tag{13.82}$$

is

$$P = -\frac{\epsilon - 1}{8\pi}\left(E_\perp^2 + \frac{1}{\varepsilon}E_r^2\right)\Big|_{r=a-}, \tag{13.83}$$

using the boundary conditions of the electric field $\mathbf{E}$ and the electric displacement $\mathbf{D}$ across a dielectric interface.

(c) Insert the value the electric field found in part 10a to find the total stress on the spherical wall of the cavity, S:

$$\mathcal{F} = \oint_S dS P. \tag{13.84}$$

(d) Compute the radial-radial component of the stress tensor T_{rr} and compute the pressure on the surface from

$$T_{rr}\Big|_{r=a-} - T_{rr}\Big|_{r=a+}. \tag{13.85}$$

Does the result agree with that found in part 10c?

(e) Show in general (arbitrary $\mathbf{E}$) that the discontinuity of the stress tensor found in part 10d coincides with that given in (13.83).

11. Repeat the above problem, but for a point charge at the center of a dielectric ball, where the permittivity is described by

$$\epsilon(\mathbf{r}) = \begin{cases} \varepsilon, r < a, \\ 1, r > a, \end{cases}. \tag{13.86}$$

12. (a) Prove Earnshaw's[1] theorem [24], that a charged particle cannot be held in stable equilibrium by an electrostatic field, by showing that any Cartesian component of $\mathbf{E}$ cannot have a maximum or a minimum in any region where the charge density is zero.

    (b) However, a neutral particle, such as a molecule, can be trapped by the Stark effect,[2] in which the energy depends on $|\mathbf{E}|$. Show from the equations of electrostatics that $\mathbf{E}^2$ can have a minimum but not a maximum. This allows for electrostatic trapping of molecules. However, this theorem does mean that trapping of a dielectric nanoparticle by an inhomogeneous electric field by the gradient force (13.40) in which a dielectric is attracted to the strongest electric field strength, is not possible, A minimum-seeking trap is possible in the quantum mechanical Stark situation.

[1] Samuel Earnshaw, 1805-1888.

[2] This quantum effect was discovered by Johannes Stark in 1913.

14

Introduction to Green's Functions

The differential equation governing the scalar potential of a given charge distribution in a dielectric medium, the combination of

$$\boldsymbol{\nabla}\cdot\mathbf{D} = 4\pi\rho, \quad \mathbf{D} = \epsilon\mathbf{E}, \quad \mathbf{E} = -\boldsymbol{\nabla}\phi, \tag{14.1}$$

is

$$-\boldsymbol{\nabla}\cdot(\epsilon\boldsymbol{\nabla}\phi) = 4\pi\rho. \tag{14.2}$$

In the special situation of the vacuum, $\epsilon = 1$ (or, effectively, for a spatially homogeneous dielectric constant), this becomes

$$-\nabla^2\phi = 4\pi\rho, \tag{14.3}$$

the equation of Siméon Denis Poisson (1781–1840); in regions devoid of charge, it reduces to the equation of Pierre Simon Laplace (1749–1827),

$$\nabla^2\phi = 0. \tag{14.4}$$

Our task is to find the solution of (14.2) for ϕ, for a given charge distribution. Since (14.2) is a linear differential equation relating ρ and ϕ, the potential at a point $\mathbf{r}$ is the additive contribution of all individual charge elements $(d\mathbf{r}')\rho(\mathbf{r}')$:

$$\phi(\mathbf{r}) = \int (d\mathbf{r}')\, G(\mathbf{r},\mathbf{r}')\rho(\mathbf{r}'). \tag{14.5}$$

It is evident that $G(\mathbf{r},\mathbf{r}')$ is the potential at $\mathbf{r}$ arising from a unit point charge at $\mathbf{r}'$ so that it satisfies the differential equation

$$-\boldsymbol{\nabla}\cdot[\epsilon(\mathbf{r})\boldsymbol{\nabla}G(\mathbf{r},\mathbf{r}')] = 4\pi\delta(\mathbf{r}-\mathbf{r}'). \tag{14.6}$$

Equation (14.5) expresses the fact that the potential due to a charge distribution is simply the sum of the contributions of each of the charges. $G(\mathbf{r},\mathbf{r}')$ is an example of a class of functions introduced by George Green. Once we have Green's function, the solution for any charge distribution in a given dielectric arrangement is a matter of integration.

14.1 Reciprocity Relation

The most striking property of Green's function is the reciprocity between the position of the unit charge and the point at which the potential is evaluated. This is the symmetry

$$G(\mathbf{r},\mathbf{r}') = G(\mathbf{r}',\mathbf{r}). \tag{14.7}$$

DOI: 10.1201/9781003057369-14

A first derivation of this follows from a consideration of the energy of the system. From the stationary property of the energy functional (13.14), we notice that an arbitrary change in the charge density, $\delta\rho(\mathbf{r})$, produces the following change in the energy

$$\delta E = \int (d\mathbf{r})\, \delta\rho(\mathbf{r})\phi(\mathbf{r}); \tag{14.8}$$

only the explicit dependence of E on ρ contributes. Of course, this energy relation merely restates the significance of $\phi(\mathbf{r})$ as the interaction energy of a unit point charge at $\mathbf{r}$ with the given charge distribution. But, its importance for us here is in its emphasis that the potential function, and thereby Green's function, is uniquely determined by a knowledge of the energy for an arbitrary charge distribution.

Now, from (13.18) and (13.19), we recognize that the energy of the physical configuration can be written in two alternative forms:

$$E = \int (d\mathbf{r})\, \epsilon \frac{\mathbf{E}^2}{8\pi} = \frac{1}{2}\int (d\mathbf{r})\, \rho\phi. \tag{14.9}$$

[See also (13.23).] If we use the second of these expressions together with (14.5),

$$\begin{aligned} E &= \frac{1}{2}\int (d\mathbf{r})\, \rho(\mathbf{r})\phi(\mathbf{r}) \\ &= \frac{1}{2}\int (d\mathbf{r})\,(d\mathbf{r}')\, \rho(\mathbf{r})G(\mathbf{r},\mathbf{r}')\rho(\mathbf{r}') = \frac{1}{2}\int (d\mathbf{r})\,(d\mathbf{r}')\, \rho(\mathbf{r})G(\mathbf{r}',\mathbf{r})\rho(\mathbf{r}') \end{aligned} \tag{14.10}$$

(where the last form has used $\mathbf{r} \to \mathbf{r}'$), we see that only the symmetrical part of $G(\mathbf{r},\mathbf{r}')$ contributes to the energy. This proves (14.7) because the energy completely determines Green's function. In other words, the reciprocity relation states that Green's function is the interaction energy of a pair of unit charges at specified points and is a property of that pair of points, not of their individual labels. [Note that (14.8) and (14.10) imply the construction (14.5).]

A clumsier, but more conventional proof of the symmetry of Green's function uses the differential equation that it obeys. The equations satisfied by the Green's function due to point charges at $\mathbf{r}'$ and $\mathbf{r}''$, respectively, are

$$-\boldsymbol{\nabla}\cdot[\epsilon(\mathbf{r})\boldsymbol{\nabla}G(\mathbf{r},\mathbf{r}')] = 4\pi\delta(\mathbf{r}-\mathbf{r}'), \tag{14.11a}$$

$$-\boldsymbol{\nabla}\cdot[\epsilon(\mathbf{r})\boldsymbol{\nabla}G(\mathbf{r},\mathbf{r}'')] = 4\pi\delta(\mathbf{r}-\mathbf{r}''). \tag{14.11b}$$

We now multiply (14.11a) by $G(\mathbf{r},\mathbf{r}'')$ and (14.11b) by $G(\mathbf{r},\mathbf{r}')$ and subtract the resulting equations:

$$\begin{aligned} &4\pi[\delta(\mathbf{r}-\mathbf{r}')G(\mathbf{r}',\mathbf{r}'') - \delta(\mathbf{r}-\mathbf{r}'')G(\mathbf{r}'',\mathbf{r}')] \\ &\quad = \boldsymbol{\nabla}\cdot[\epsilon(\mathbf{r})\{G(\mathbf{r},\mathbf{r}')\boldsymbol{\nabla}G(\mathbf{r},\mathbf{r}'') - G(\mathbf{r},\mathbf{r}'')\boldsymbol{\nabla}G(\mathbf{r},\mathbf{r}')\}], \end{aligned} \tag{14.12}$$

in which we have also applied the following identity for three arbitrary scalar functions,

$$\phi\boldsymbol{\nabla}\cdot(\lambda\boldsymbol{\nabla}\psi) - \psi\boldsymbol{\nabla}\cdot(\lambda\boldsymbol{\nabla}\phi) = \boldsymbol{\nabla}\cdot[\lambda(\phi\boldsymbol{\nabla}\psi - \psi\boldsymbol{\nabla}\phi)], \tag{14.13}$$

which is a slight generalization of what is known as Green's second identity. On integrating (14.12) over a volume that is bounded by a remote surface S, we learn that

$$G(\mathbf{r}',\mathbf{r}'') - G(\mathbf{r}'',\mathbf{r}') = \frac{1}{4\pi}\oint_S d\mathbf{S}\cdot[\epsilon(\mathbf{r})\{G(\mathbf{r},\mathbf{r}')\boldsymbol{\nabla}G(\mathbf{r},\mathbf{r}'') - G(\mathbf{r},\mathbf{r}'')\boldsymbol{\nabla}G(\mathbf{r},\mathbf{r}')\}]. \tag{14.14}$$

Let us stop for a moment and appreciate that this same volume integration, applied to the differential equation (14.6), gives

$$1 = \frac{1}{4\pi}\oint_S d\mathbf{S}\cdot\epsilon(\mathbf{r})\{-\boldsymbol{\nabla}G(\mathbf{r},\mathbf{r}')\}. \tag{14.15}$$

If we suppose, for definiteness, that the remote surface S is a sphere of radius R drawn in the vacuum ($\epsilon = 1$), it is clear that the magnitude of the surface area, $4\pi R^2$, is balanced by the magnitude of the electric field derived from the potential $1/R$, appropriate to unit charge. The same elements are present on the right-hand side of (14.14), with the additional factors of $G(\mathbf{r},\mathbf{r}')$ or $G(\mathbf{r},\mathbf{r}'')$, which are of order $1/R$. Accordingly, the surface integral in (14.14) certainly vanishes in the limit $R\to\infty$, and again we conclude that Green's function depends symmetrically upon its two variables.

14.2 Problems for Chapter 14

1. Derive (14.9) from the equations of electrostatics directly.
2. Verify the identity (14.13).
3. Show that if ϕ is a solution to Poisson's equation,

$$-\nabla^2\phi = 4\pi\rho \tag{14.16}$$

and G is the corresponding Green's function

$$-\nabla^2 G(\mathbf{r},\mathbf{r}') = 4\pi\delta(\mathbf{r}-\mathbf{r}'), \tag{14.17}$$

ϕ is expressed in terms of the charge density and the boundary values by

$$\begin{aligned}&-\oint_S \frac{d\mathbf{S}'}{4\pi}\cdot[G(\mathbf{r},\mathbf{r}')\boldsymbol{\nabla}'\phi(\mathbf{r}') - \phi(\mathbf{r}')\boldsymbol{\nabla}'G(\mathbf{r},\mathbf{r}')]\\ &\quad+\int_V (d\mathbf{r}')\,G(\mathbf{r},\mathbf{r}')\rho(\mathbf{r}') = \begin{cases}\phi(\mathbf{r}), & \mathbf{r}\text{ in }V,\\ 0, & \mathbf{r}\text{ outside }V,\end{cases}\end{aligned} \tag{14.18}$$

where S is the closed surface bounding the volume V, and $d\mathbf{S}'$ points in the direction of the inward normal. If ϕ assumes specified values on S, and G vanishes there, then inside V

$$\phi(\mathbf{r}) = \int_V (d\mathbf{r}')\,G(\mathbf{r},\mathbf{r}')\rho(\mathbf{r}') + \oint_S \frac{d\mathbf{S}'}{4\pi}\cdot[\boldsymbol{\nabla}'G(\mathbf{r},\mathbf{r}')]\phi(\mathbf{r}'). \tag{14.19}$$

If, instead, we choose a Green's function that has vanishing normal derivative on the surface,

$$\hat{\mathbf{n}}\cdot\boldsymbol{\nabla}G(\mathbf{r},\mathbf{r}')\big|_{\mathbf{r}\in S} = 0, \tag{14.20}$$

we get an expression for ϕ in terms of this Green's function and a specified normal derivative of ϕ on the surface. The two situations contemplated here, specifying boundary values of the potential or of its normal derivative, are referred to as Dirichlet and Neumann boundary conditions, respectively.[1]

[1] Johann Peter Gustav Lejeune Dirichlet (1805–1859) and Carl Gottfried Neumann (1832–1925).

4. This problem concerns Green's function in an inhomogeneous medium, which is governed by the differential equation (14.6).

   (a) Write

$$G(\mathbf{r},\mathbf{r}') = \frac{1}{\sqrt{\epsilon(\mathbf{r})}}\overline{G}(\mathbf{r},\mathbf{r}')\frac{1}{\sqrt{\epsilon(\mathbf{r}')}}, \tag{14.21}$$

   and then establish

$$\epsilon^{1/2}\boldsymbol{\nabla}\epsilon^{-1/2} = \boldsymbol{\nabla} - \boldsymbol{\nabla}\ln\epsilon^{1/2}. \tag{14.22}$$

   Show that as a consequence, $\overline{G}$ satisfies

$$\left[-\nabla^2 + \left(\boldsymbol{\nabla}\ln\epsilon^{1/2}\right)^2 + \nabla^2\ln\epsilon^{1/2}\right]\overline{G}(\mathbf{r},\mathbf{r}') = 4\pi\delta(\mathbf{r}-\mathbf{r}'). \tag{14.23}$$

   (b) What does this differential equation become when

$$\epsilon(\mathbf{r}) = e^{2\gamma z}, \quad \gamma = \text{constant}, \tag{14.24}$$

   everywhere.

   (c) Show that the solution to the differential equation in part 4b is of the form

$$\overline{G}(\mathbf{r},\mathbf{r}') = \frac{1}{|\mathbf{r}-\mathbf{r}'|}e^{-\alpha|\mathbf{r}-\mathbf{r}'|}, \tag{14.25}$$

   and establish the relation between α and γ. (Recall that the Laplacian operator in spherical polar coordinates is

$$\nabla^2 = \frac{1}{r^2}\frac{\partial}{\partial r}r^2\frac{\partial}{\partial r} + \frac{1}{r^2}\frac{1}{\sin\theta}\frac{\partial}{\partial\theta}\sin\theta\frac{\partial}{\partial\theta} + \frac{1}{r^2\sin^2\theta}\frac{\partial^2}{\partial\phi^2}.\bigg] \tag{14.26}$$

   (d) What is now the original Green's function, $G(\mathbf{r},\mathbf{r}')$? Is it bounded as $|z|$, $|z'|$ go to infinity?

5. A dielectric body of arbitrary $\epsilon(\mathbf{r})$ is completely enclosed by a single conducting surface, which is assumed to lie completely outside the dielectric body, so $\epsilon = 1$ on S. Assume that there is no free charge lying within the hollow conductor. Prove that the electric field vanishes everywhere in that internal region, as follows:

   (a) From the differential equation satisfied by Green's function, chosen to vanish on S, determine the value of the surface integral

$$\oint_S d\mathbf{S}\cdot\boldsymbol{\nabla}G(\mathbf{r},\mathbf{r}'), \tag{14.27}$$

   for $\mathbf{r}$ inside S.

   (b) for $\rho = 0$ establish the relation between the value of $\phi(\mathbf{r})$ at any interior point and the constant potential ϕ_0 on the surface S.

   (c) Then, verify that the corresponding electric field in the region interior to S is zero.

6. (a) Charge q is distributed in the interior of a region enclosed by a conducting surface. Use Gauss' law to prove that the total charge induced on the surface is $-q$.

   (b) Explain why this is also true for a point charge q placed in front of an infinite plane conducting surface, although, now, the charge q is not enclosed by conductors.

15

Electrostatics in Free Space

The simplest electrostatic situation is for the vacuum, $\epsilon = 1$. The differential equation for the potential is then Poisson's equation,

$$-\nabla^2\phi(\mathbf{r}) = 4\pi\rho(\mathbf{r}), \tag{15.1}$$

so that the corresponding Green's function equation is

$$-\nabla^2 G(\mathbf{r}, \mathbf{r}') = 4\pi\delta(\mathbf{r} - \mathbf{r}'). \tag{15.2}$$

The solution to (15.2) is the well-known Coulomb potential, since it is the potential at $\mathbf{r}$ produced by a unit point source at $\mathbf{r}'$:

$$G(\mathbf{r}, \mathbf{r}') = \frac{1}{|\mathbf{r} - \mathbf{r}'|}. \tag{15.3}$$

We will now derive (15.3) from the differential equation (15.2). In order to do this, we require an integral representation for the delta function, the charge density of a unit point charge. The relevant properties are that it vanish everywhere except at a single point, while possessing a unit integrated value. In one spatial dimension, these properties appear as

$$\delta(x - x') = 0 \quad \text{if} \quad x \neq x', \tag{15.4}$$

while $\delta(x - x')$ is so singular at $x = x'$ that

$$\int_{-\infty}^{\infty} dx\, \delta(x - x') = 1. \tag{15.5}$$

No conventional function can satisfy these requirements. They can, however, be realized in a limit. Consider, for example, the function

$$\delta_\epsilon(x - x') = \frac{1}{\pi}\frac{\epsilon}{(x - x')^2 + \epsilon^2}, \quad (\epsilon > 0). \tag{15.6}$$

First, the limiting values of the function for $\epsilon \to +0$ are

$$\lim_{\epsilon \to +0} \delta_\epsilon(x - x') = \begin{cases} 0, \; x \neq x', \\ \infty, x = x', \end{cases} \tag{15.7}$$

while the value of the integral of the function over the whole range of x is unity, independent of the choice of ϵ:

$$\int_{-\infty}^{\infty} d(x - x')\, \frac{1}{\pi}\frac{\epsilon}{(x - x')^2 + \epsilon^2} = \frac{1}{\pi}\int_{-\infty}^{\infty} dt \frac{1}{t^2 + 1} = 1, \tag{15.8}$$

which uses the substitution $x - x' = \epsilon t$.

DOI: 10.1201/9781003057369-15

To arrive at a more useful form of the function, we observe that

$$\delta_\epsilon(x-x') = \mathrm{Re}\,\frac{1}{\pi}\frac{i}{x-x'+i\epsilon} = \mathrm{Re}\,\frac{1}{\pi}\int_0^\infty dk\, e^{ik(x-x'+i\epsilon)}, \tag{15.9a}$$

or

$$\delta_\epsilon(x-x') = \int_{-\infty}^{\infty}\frac{dk}{2\pi}\,e^{ik(x-x')}e^{-\epsilon|k|}. \tag{15.9b}$$

Can we now set ϵ equal to zero? No; the resulting integral would not exist—it would not converge at infinity. But we can think of giving ϵ an arbitrarily small positive value, in order to make the integral mathematically meaningful, without significantly altering thereby the physical properties being represented. After all, when we speak of a point charge, we mean no more than one of size that is very small on the scale set by all the other significant lengths in the given physical situation. In the following we shall keep in mind the necessary presence of the convergence factor, $\exp(-\epsilon|k|)$, but not write it explicitly. Accordingly, the desired expression for the delta function, in one dimension, is

$$\delta(x-x') = \int_{-\infty}^{\infty}\frac{dk}{2\pi}e^{ik(x-x')}. \tag{15.10}$$

The delta function in three dimensions is now realized by

$$\delta(\mathbf{r}-\mathbf{r}') = \delta(x-x')\delta(y-y')\delta(z-z'). \tag{15.11}$$

Indeed, on integrating over a volume V,

$$\int_V (d\mathbf{r})\,\delta(\mathbf{r}-\mathbf{r}') = \int dx\,\delta(x-x')\int dy\,\delta(y-y')\int dz\,\delta(z-z'), \tag{15.12}$$

the possible outcomes for the right-hand side are just zero and unity, the latter being realized only if the range of x-integration includes x', that of y-integration includes y', and the range of z-integration includes z'—which is just to say that the point $\mathbf{r}'$ is within the volume V. The use of the integral representation (15.10) for the three individual delta functions, with the integration variables labeled k_x, k_y, and k_z, respectively, then produces the three-dimensional representation

$$\delta(\mathbf{r}-\mathbf{r}') = \int\frac{(d\mathbf{k})}{(2\pi)^3}e^{i\mathbf{k}\cdot(\mathbf{r}-\mathbf{r}')}. \tag{15.13}$$

We now employ the above representation of the delta function to solve (15.2) for Green's function. This can be very simply accomplished if we note that

$$\boldsymbol{\nabla}e^{i\mathbf{k}\cdot(\mathbf{r}-\mathbf{r}')} = i\mathbf{k}e^{i\mathbf{k}\cdot(\mathbf{r}-\mathbf{r}')}, \tag{15.14a}$$

or, effectively,

$$\boldsymbol{\nabla} \to i\mathbf{k}, \tag{15.14b}$$

so that we can read off *a* solution to (15.2),

$$G(\mathbf{r},\mathbf{r}') = 4\pi\int\frac{(d\mathbf{k})}{(2\pi)^3}\frac{e^{i\mathbf{k}\cdot(\mathbf{r}-\mathbf{r}')}}{k^2}. \tag{15.15}$$

We can verify that this is in fact the known result, (15.3), by using spherical coordinates for $\mathbf{k}$, with $\mathbf{r}-\mathbf{r}'$ pointing along the z-axis,

$$\mathbf{k}\cdot(\mathbf{r}-\mathbf{r}') = kR\cos\theta, \quad R = |\mathbf{r}-\mathbf{r}'|, \quad (d\mathbf{k}) = k^2dk\,2\pi d(\cos\theta), \tag{15.16}$$

so that

$$G(\mathbf{r},\mathbf{r}') = 4\pi \int \frac{k^2 dk\, 2\pi d(\cos\theta)}{8\pi^3} \frac{e^{ikR\cos\theta}}{k^2}$$
$$= \frac{1}{\pi}\int_0^\infty dk \frac{1}{ikR}\left(e^{ikR} - e^{-ikR}\right) = \frac{2}{\pi}\frac{1}{R}\int_0^\infty \frac{\sin x}{x} dx = \frac{1}{R}. \quad (15.17)$$

The convergence of the final integral here shows that only values of k of order $1/R$ are significant, which also applies to the individual components of $\mathbf{k}$. Accordingly, the implicit convergence factors will indeed effectively equal unity if the individual value of ϵ are restricted by $\epsilon \ll R$. This is entirely consistent with the physical context in which a point charge has a linear extent small in comparison to the distance R.

15.1 2 + 1 Dimensions

In the above, we have treated all three directions of space on an equal footing. However, we need not do this. We can separate out one direction, say that of z, and treat it differently. The reason this is useful is because there are geometries in which physically interesting quantities vary in only a single direction. Singling out the z direction, we can write Green's function, (15.15), as

$$G(\mathbf{r},\mathbf{r}') = 4\pi \int \frac{dk_x dk_y}{(2\pi)^2} e^{i[k_x(x-x')+k_y(y-y')]} \int \frac{dk_z}{2\pi} \frac{e^{ik_z(z-z')}}{k_x^2+k_y^2+k_z^2}. \quad (15.18)$$

Adopting the nomenclature that the two-dimensional space of x and y is transverse ($\perp$) to the selected direction, we write the first part of (15.18) as

$$\int \frac{(d\mathbf{k}_\perp)}{(2\pi)^2} e^{i\mathbf{k}_\perp \cdot (\mathbf{r}-\mathbf{r}')_\perp}. \quad (15.19)$$

The remaining integration over k_z ($k_\perp^2 = k_x^2 + k_y^2$ and $k_\perp \geq 0$)

$$\int_{-\infty}^\infty \frac{dk_z}{2\pi} \frac{e^{ik_z(z-z')}}{k_\perp^2 + k_z^2} = \frac{1}{2k_\perp} e^{-k_\perp|z-z'|}, \quad (15.20)$$

is performed by doing a contour integration as indicated in Fig. 15.1, which indicates the simple poles occurring in the integrand of (15.20) at $k_z = \pm ik_\perp$, and with the contour being closed by infinitely remote semicircles which give no contribution. We have therefore recast Green's function into the form

$$G(\mathbf{r},\mathbf{r}') = 4\pi \int \frac{(d\mathbf{k}_\perp)}{(2\pi)^2} e^{i\mathbf{k}_\perp \cdot (\mathbf{r}-\mathbf{r}')_\perp} g(z, z'; k_\perp), \quad (15.21)$$

where, for free space, the "reduced" Green's function is

$$g(z, z'; k_\perp) = \frac{1}{2k_\perp} e^{-k_\perp|z-z'|}. \quad (15.22)$$

The form (15.21) applies to any problem which is translationally invariant in x and y but not necessarily in z. The representation is particularly adapted to the situation in which

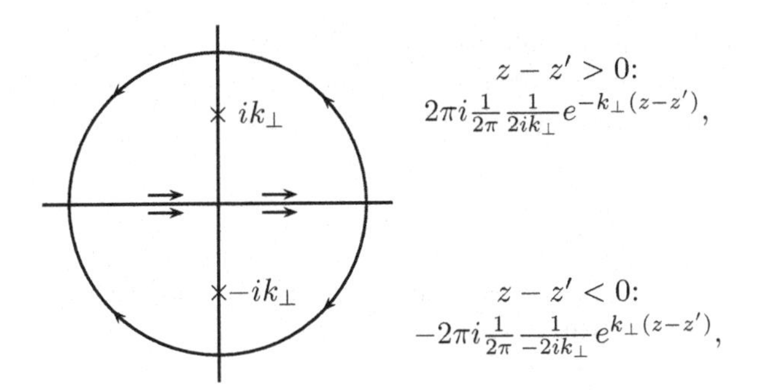

FIGURE 15.1
Contour used to evaluate (15.20).

the dielectric constant is a function only of z. For the case at hand, where Green's function satisfies (15.2), we can easily derive the differential equation for $g(z, z'; k_\perp)$ as follows:

$$\begin{aligned} -\nabla^2 G &= 4\pi \int \frac{(d\mathbf{k}_\perp)}{(2\pi)^2} e^{i\mathbf{k}_\perp \cdot (\mathbf{r}-\mathbf{r}')_\perp} \left[k_\perp^2 - \frac{\partial^2}{\partial z^2} \right] g(z, z'; k_\perp) \\ &= 4\pi \int \frac{(d\mathbf{k}_\perp)}{(2\pi)^2} e^{i\mathbf{k}_\perp \cdot (\mathbf{r}-\mathbf{r}')_\perp} \delta(z - z'), \end{aligned} \tag{15.23}$$

which implies

$$\left[k_\perp^2 - \frac{\partial^2}{\partial z^2} \right] g(z, z'; k_\perp) = \delta(z - z'). \tag{15.24}$$

Of course, we could also have derived this equation directly from the left-hand side of (15.20), which is only to say that the latter expression is an obvious solution of the one-dimensional differential equation. Can one also arrive directly at the form of the solution given in (15.22) by solving the differential equation (15.24)? Yes.

This differential equation is solved by noting that for $z \neq z'$,

$$\left[k_\perp^2 - \frac{\partial^2}{\partial z^2} \right] g(z, z'; k_\perp) = 0, \tag{15.25}$$

while the derivative of g at $z = z'$ is discontinuous. Indeed, if we merely accept that g is bounded when z is in the infinitesimal neighborhood of z', integration of (15.24) over the interval between $z' - 0$ and $z' + 0$ yields

$$-\frac{\partial}{\partial z} g(z, z'; k_\perp) \Big|_{z'-0}^{z'+0} = 1. \tag{15.26}$$

Then, the continuity of g at $z = z'$ follows. To prove that last statement, we multiply (15.24) by z, and present the result as

$$-\frac{\partial}{\partial z} \left[z \frac{\partial}{\partial z} g - g \right] + k_\perp^2 z g = z' \delta(z - z'), \tag{15.27}$$

and integrate between $z' - 0$ and $z' + 0$. The result,

$$z' \left(-\frac{\partial}{\partial z} g \right) \Big|_{z'-0}^{z'+0} + g \Big|_{z'-0}^{z'+0} = z', \tag{15.28}$$

taken together with (15.26), proves that g is continuous.

The two independent solutions of the differential equation (15.25) are given by $\exp(\pm k_\perp z)$. The choice between them is dictated by the physical requirement that g be bounded as z recedes to infinity in either direction:

$$g = \begin{cases} A(z')e^{-k_\perp z}, & z > z', \\ B(z')e^{k_\perp z}, & z < z', \end{cases} \tag{15.29a}$$

where $A(z')$ and $B(z')$ are independent of z. Imposing continuity of g at $z = z'$, as well discontinuity of the derivative there, according to (15.26), gives the system of equations

$$A(z')e^{-k_\perp z'} - B(z')e^{k_\perp z'} = 0, \tag{15.30}$$

$$A(z')k_\perp e^{-k_\perp z'} + B(z')k_\perp e^{k_\perp z'} = 1, \tag{15.31}$$

which when solved yields

$$A = \frac{1}{2k_\perp}e^{k_\perp z'}, \quad B = \frac{1}{2k_\perp}e^{-k_\perp z'}, \tag{15.32}$$

which implies (15.22).

In summary, we have found two alternative expressions for Green's function in empty space:

$$\frac{1}{|\mathbf{r}-\mathbf{r}'|} = 4\pi \int \frac{(d\mathbf{k}_\perp)}{(2\pi)^2} e^{i\mathbf{k}_\perp \cdot (\mathbf{r}-\mathbf{r}')_\perp} \frac{1}{2k_\perp} e^{-k_\perp|z-z'|}. \tag{15.33}$$

We shall have more to say in Chapter 18 about the equivalence of these two forms.

15.2 Problems for Chapter 15

1. Consider

$$\delta_\eta(x-x') = Ae^{-(x-x')^2/2\eta^2} \tag{15.34}$$

as another model of the delta function, in the limit $\eta \to 0$. Find the normalization factor A and verify the counterpart of (15.7). Write $\delta_\eta(x-x')$ in a form analogous to (15.9b).

2. Show that

$$\bar{\delta}_\epsilon(x) = -x\frac{d}{dx}\delta_\epsilon(x) \tag{15.35}$$

is also a model of the delta function. Can you support otherwise the inference that

$$-x\frac{d}{dx}\delta(x) = \delta(x)? \tag{15.36}$$

3. Express

$$\delta(x^2 - a^2), \quad a > 0 \tag{15.37}$$

in terms of $\delta(x \pm a)$. What can you say about $\delta(f(x))$, where f is a real function with a simple zero at $x = a$?

4. Perform the integral in (15.8) and the last integral in (15.17) by contour integration.

16

Semi-Infinite Dielectric

16.1 Green's Function for Charge Outside Dielectric

We now apply the above representation, (15.21), to find Green's function for the simplest situation involving a nonuniform dielectric constant, that of a body of uniform dielectric constant occupying the semi-infinite region $z < 0$, while the region $z > 0$ is a vacuum—see Fig. 16.1. The inhomogeneity in $\epsilon(\mathbf{r})$ is limited to a discontinuity across the surface $z = 0$. This physical arrangement has the required translational invariance in the transverse plane, that is, in x and y. The Green's function equation, (14.6),

$$-\boldsymbol{\nabla} \cdot [\epsilon(z)\boldsymbol{\nabla} G(\mathbf{r}, \mathbf{r}')] = 4\pi\delta(\mathbf{r} - \mathbf{r}'), \tag{16.1}$$

becomes, in the two regions,

$$z > 0: \quad -\nabla^2 G = 4\pi\delta(\mathbf{r} - \mathbf{r}'), \tag{16.2a}$$

$$z < 0: \quad -\epsilon\nabla^2 G = 4\pi\delta(\mathbf{r} - \mathbf{r}'). \tag{16.2b}$$

To solve these equations, we must impose appropriate boundary conditions. Across the interface between the vacuum and the medium, the normal component of the displacement vector $\mathbf{D}$ must be continuous [see (13.55b)], assuming no surface charge,

$$\left(-\frac{\partial}{\partial z}\right) G\bigg|_{z=+0} = \epsilon \left(-\frac{\partial}{\partial z}\right) G\bigg|_{z=-0}. \tag{16.3}$$

Of course, G must be continuous across the interface because it represents the potential of a point charge, recalling (13.54):

$$G(\mathbf{r}, \mathbf{r}')\bigg|_{z=+0} = G(\mathbf{r}, \mathbf{r}')\bigg|_{z=-0}. \tag{16.4}$$

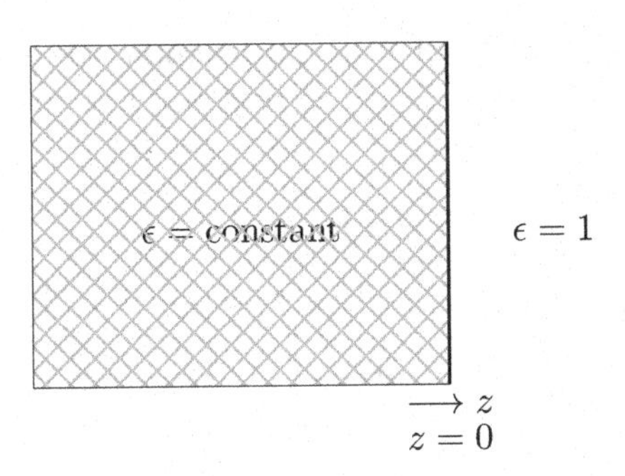

FIGURE 16.1
Geometry of semi-infinite dielectric region.

DOI: 10.1201/9781003057369-16

The reduced Green's function, $g(z, z'; k_\perp)$, introduced in the 2+1 dimensional representation (15.21),

$$G(\mathbf{r}, \mathbf{r}') = 4\pi \int \frac{(d\mathbf{k}_\perp)}{(2\pi)^2} e^{i\mathbf{k}_\perp \cdot (\mathbf{r}-\mathbf{r}')_\perp} g(z, z'; k_\perp), \tag{16.5}$$

satisfies the differential equations [see (15.24)]

$$z > 0: \quad \left(-\frac{\partial^2}{\partial z^2} + k_\perp^2\right) g = \delta(z - z'), \tag{16.6a}$$

$$z < 0: \quad \epsilon\left(-\frac{\partial^2}{\partial z^2} + k_\perp^2\right) g = \delta(z - z'), \tag{16.6b}$$

subject to the boundary conditions

$$g\Big|_{z=-0} = g\Big|_{z=+0}, \tag{16.7a}$$

$$\epsilon\frac{\partial}{\partial z} g\Big|_{z=-0} = \frac{\partial}{\partial z} g\Big|_{z=+0}. \tag{16.7b}$$

In the following we will first solve this problem by assuming that $z' > 0$ (that is, the unit charge lies in the vacuum and not in the dielectric). [For the converse situation see Problem 16.1 and (16.44).] The solutions to (16.6a), (16.6b), (16.7a), and (16.7b) can be expressed in terms of the solutions, $e^{k_\perp z}$ and $e^{-k_\perp z}$, of the corresponding homogeneous equation. The forms of the solution in the three regions are as follows:

$$z < 0: \quad g = Ae^{k_\perp z}, \tag{16.8a}$$

$$z > z': \quad g = Be^{-k_\perp z}, \tag{16.8b}$$

$$0 < z < z': \quad g = Ce^{k_\perp z} + De^{-k_\perp z}, \tag{16.8c}$$

where the single exponentials in (16.8a) and (16.8b) are required by the boundary condition that g remain finite for $|z| \to \infty$. The boundary conditions at $z = 0$, (16.7a) and (16.7b), require

$$A = C + D, \tag{16.9a}$$

$$\epsilon k_\perp A = k_\perp (C - D), \tag{16.9b}$$

from which we infer

$$C = \frac{\epsilon + 1}{2} A, \quad D = -\frac{\epsilon - 1}{2} A. \tag{16.10}$$

Just as in the situation mentioned at the end of the last chapter [see (15.26)], the singularity in the differential equation requires, at $z = z'$, that g be continuous, while $-\partial g/\partial z$ be discontinuous,

$$g\Big|_{z=z'-0}^{z=z'+0} = 0, \tag{16.11a}$$

$$-\frac{\partial}{\partial z} g\Big|_{z=z'-0}^{z=z'+0} = 1, \tag{16.11b}$$

which imply, explicitly ,

$$Be^{-k_\perp z'} - \left(Ce^{k_\perp z'} + De^{-k_\perp z'}\right) = 0, \tag{16.12a}$$

$$k_\perp Be^{-k_\perp z'} + k_\perp (Ce^{k_\perp z'} - De^{-k_\perp z'}) = 1. \tag{16.12b}$$

The elimination of B and D between (16.12a) and (16.12b) immediately gives

$$C = \frac{1}{2k_\perp} e^{-k_\perp z'}, \tag{16.13}$$

and then, from (16.10), we obtain

$$A = \frac{2}{\epsilon+1} \frac{1}{2k_\perp} e^{-k_\perp z'}, \quad D = -\frac{\epsilon-1}{\epsilon+1} \frac{1}{2k_\perp} e^{-k_\perp z'}. \tag{16.14}$$

Finally, (16.12a) supplies

$$B = -\frac{\epsilon-1}{\epsilon+1} \frac{1}{2k_\perp} e^{-k_\perp z'} + \frac{1}{2k_\perp} e^{k_\perp z'}. \tag{16.15}$$

Inserting these coefficients back into (16.8a), (16.8b), and (16.8c), we find for the reduced Green's function, g, in the two media,

$$z > 0: \quad g = \frac{1}{2k_\perp} \left[e^{-k_\perp |z-z'|} - \frac{\epsilon-1}{\epsilon+1} e^{-k_\perp (z+z')} \right], \tag{16.16a}$$

$$\begin{aligned} z < 0: \quad g &= \frac{1}{2k_\perp} \frac{2}{\epsilon+1} e^{-k_\perp (z'-z)} \\ &= \frac{1}{2k_\perp} \left[e^{-k_\perp |z-z'|} - \frac{\epsilon-1}{\epsilon+1} e^{-k_\perp (z'-z)} \right], \end{aligned} \tag{16.16b}$$

or, taken together,

$$z' > 0: \quad g(z, z'; k_\perp) = \frac{1}{2k_\perp} e^{-k_\perp |z-z'|} - \frac{\epsilon-1}{\epsilon+1} \frac{1}{2k_\perp} e^{-k_\perp |z|} e^{-k_\perp |z'|}. \tag{16.17}$$

Of course, if we set $\epsilon = 1$, we recover the vacuum result, (15.22), as we must.

It is helpful to analyze the Green's function we have found in terms of primary and secondary fields. The primary field results from the point charge at z' and so is represented by the Green's function (15.22). The secondary field is due to the bound charge built up on the interface and is given by

$$g_S = -\frac{\epsilon-1}{\epsilon+1} \frac{1}{2k_\perp} e^{-k_\perp z'} e^{-k_\perp |z|}. \tag{16.18}$$

The situation is illustrated by Fig. 16.2, for $\epsilon > 1$.

16.2 Derivation in Terms of Bound Charge

Let us derive this result (16.17) more directly. Recall that macroscopic electrostatics begins by recognizing that the charge bound in atoms contributes the polarization charge density, see (4.50a),

$$\rho_{\text{eff}} = \overline{\rho_{\text{bound}}} = -\boldsymbol{\nabla} \cdot \mathbf{P} = -\boldsymbol{\nabla} \cdot \left(\frac{\epsilon(\mathbf{r}) - 1}{4\pi} \mathbf{E} \right), \tag{16.19a}$$

or

$$\rho_{\text{eff}} = -\boldsymbol{\nabla} \cdot \left(\frac{\epsilon(\mathbf{r}) - 1}{4\pi\epsilon(\mathbf{r})} \mathbf{D} \right) = -\frac{\epsilon(\mathbf{r}) - 1}{\epsilon(\mathbf{r})} \rho - \frac{1}{4\pi} \mathbf{D} \cdot \boldsymbol{\nabla} \left(\frac{\epsilon(\mathbf{r}) - 1}{\epsilon(\mathbf{r})} \right), \tag{16.19b}$$

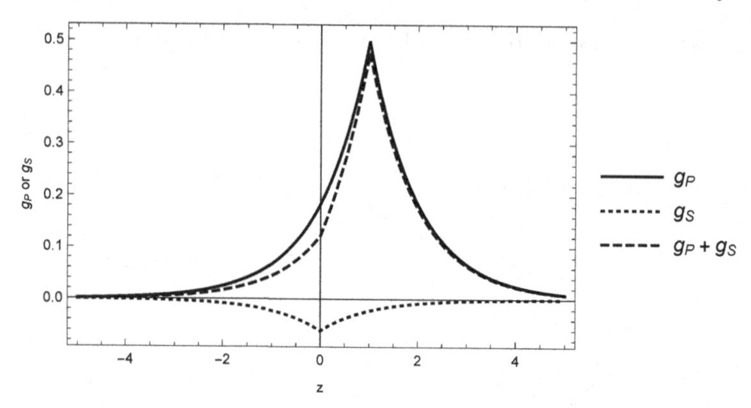

FIGURE 16.2
The primary field, P, and the secondary field, S, contributing to Green's function $g_P + g_S$ (16.17). Here we have illustrated the behavior by setting $k_\perp = 1$ and $\epsilon = 2$.

which introduces the average free charge density ρ. The problem of finding the potential that is produced in a dielectric medium by the free charge density is therefore equivalent to determining the potential that is produced in the vacuum by the total charge density

$$\rho_{\text{tot}}(\mathbf{r}) = \rho(\mathbf{r}) + \rho_{\text{eff}}(\mathbf{r}) = \frac{1}{\epsilon(\mathbf{r})}\rho(\mathbf{r}) + \frac{1}{4\pi}\epsilon(\mathbf{r})\mathbf{E}(\mathbf{r}) \cdot \boldsymbol{\nabla}\frac{1}{\epsilon(\mathbf{r})}. \tag{16.20}$$

If ϵ were completely uniform, the total charge density, and the potential it produces, would be reduced below the vacuum values by the factor ϵ, as already given in (13.32).

In the situation before us, $\epsilon(\mathbf{r})$ is discontinuous across the surface $z = 0$ so that

$$\frac{\partial}{\partial z}\frac{1}{\epsilon(\mathbf{r})} = \left(1 - \frac{1}{\epsilon}\right)\delta(z), \tag{16.21}$$

as one verifies by integrating over an interval that includes $z = 0$. Note that the multiplier of this derivative in (16.20), $\epsilon(\mathbf{r})E_z(\mathbf{r})$, is continuous at $z = 0$ and can be evaluated as E_z at $z = +0$, the vacuum side of the boundary. Accordingly, the bound charge density has a surface charge component,

$$\rho'_{\text{bound}} = \delta(z)\sigma(\mathbf{r}_\perp), \tag{16.22}$$

where the surface charge density of bound charge is

$$\sigma(\mathbf{r}_\perp) = \frac{1}{4\pi}\frac{\epsilon - 1}{\epsilon}E_z(\mathbf{r}_\perp, z = +0). \tag{16.23}$$

Green's function is the potential of a unit point charge,

$$\rho(\mathbf{r}) = \delta(\mathbf{r} - \mathbf{r}'), \tag{16.24}$$

and therefore the multiplier of ρ in (16.20) is $1/\epsilon(\mathbf{r}')$, which is unity for the present arrangement ($z' > 0$). Accordingly, G for $z' > 0$ is the potential produced in the vacuum by the combination of the point and surface charge distributions:

$$\rho_{\text{tot}}(\mathbf{r}) = \delta(\mathbf{r} - \mathbf{r}') + \delta(z)\sigma(\mathbf{r}_\perp), \tag{16.25a}$$

$$\sigma(\mathbf{r}_\perp) = \frac{1}{4\pi}\frac{\epsilon - 1}{\epsilon}\left(-\frac{\partial}{\partial z}\right)G(\mathbf{r}, \mathbf{r}')\bigg|_{z=+0}. \tag{16.25b}$$

That potential is given by

$$G(\mathbf{r},\mathbf{r}') = \frac{1}{|\mathbf{r}-\mathbf{r}'|} + \int_S dS_1 \frac{\sigma(\mathbf{r}_{1\perp})}{|\mathbf{r}-\mathbf{r}_{1\perp}|}, \tag{16.26}$$

where the element of area on the boundary surface S, which is the interface at $z_1 = 0$, is

$$dS_1 = (d\mathbf{r}_{1\perp}). \tag{16.27}$$

This construction of G in terms of itself (through σ) is an integral equation. That we are able to solve it with ease is directly attributable to the two-dimensional translational invariance embodied in the representation (16.5), along with the one for $1/|\mathbf{r}-\mathbf{r}'|$, (15.33), which we now write as

$$\frac{1}{|\mathbf{r}-\mathbf{r}'|} = 4\pi \int \frac{(d\mathbf{k}_\perp)}{(2\pi)^2} e^{i\mathbf{k}_\perp \cdot (\mathbf{r}-\mathbf{r}')_\perp} g_0(z-z';k_\perp), \quad g_0(z-z';k_\perp) = \frac{1}{2k_\perp} e^{-k_\perp|z-z'|}. \tag{16.28}$$

Let us multiply the terms of (16.26) by $\exp(-i\mathbf{k}_\perp \cdot \mathbf{r}_\perp)$ and integrate over $\mathbf{r}_\perp$. We encounter the delta function

$$\delta(\mathbf{k}_\perp - \mathbf{k}'_\perp) = \int \frac{(d\mathbf{r}_\perp)}{(2\pi)^2} e^{-i\mathbf{k}_\perp \cdot \mathbf{r}_\perp} e^{i\mathbf{k}'_\perp \cdot \mathbf{r}_\perp}, \tag{16.29}$$

which produces (apart from a common factor of 4π)

$$\begin{aligned} e^{-i\mathbf{k}_\perp \cdot \mathbf{r}'_\perp} g(z,z';k_\perp) &= e^{-i\mathbf{k}_\perp \cdot \mathbf{r}'_\perp} g_0(z-z';k_\perp) \\ &\quad + g_0(z;k_\perp) \int (d\mathbf{r}_{1\perp}) e^{-i\mathbf{k}_\perp \cdot \mathbf{r}_{1\perp}} \sigma(\mathbf{r}_{1\perp}), \end{aligned} \tag{16.30}$$

where, in turn, from (16.25b),

$$\int (d\mathbf{r}_{1\perp}) e^{-i\mathbf{k}_\perp \cdot \mathbf{r}_{1\perp}} \sigma(\mathbf{r}_{1\perp}) = e^{-i\mathbf{k}_\perp \cdot \mathbf{r}'_\perp} \frac{\epsilon - 1}{\epsilon} \left(-\frac{\partial}{\partial z_1}\right) g(z_1, z'; k_\perp)\bigg|_{z_1=+0}. \tag{16.31}$$

Thus, the one-dimensional reduction of the integral equation (16.26) is the algebraic relation

$$g(z,z';k_\perp) = g_0(z-z';k_\perp) + g_0(z;k_\perp) \frac{\epsilon-1}{\epsilon} \left(-\frac{\partial}{\partial z_1}\right) g(z_1,z';k_\perp)\bigg|_{z_1=+0}. \tag{16.32}$$

Now we have only to compute the negative z-derivative at $z = +0$,

$$-\frac{\partial}{\partial z} g(z,z';k_\perp)\bigg|_{z=+0} = -\frac{1}{2} e^{-k_\perp z'} + \frac{1}{2}\frac{\epsilon-1}{\epsilon} \left(-\frac{\partial}{\partial z}\right) g(z,z';k_\perp)\bigg|_{z=+0}, \tag{16.33}$$

to learn that

$$-\frac{\partial}{\partial z} g(z,z';k_\perp)\bigg|_{z=+0} = -\frac{\epsilon}{\epsilon+1} e^{-k_\perp z'}. \tag{16.34}$$

Therefore, the solution to (16.32) is

$$g(z,z';k_\perp) = g_0(z-z';k_\perp) - \frac{\epsilon-1}{\epsilon+1} g_0(z;k_\perp) e^{-k_\perp z'}, \tag{16.35}$$

which is just the result stated in (16.17).

16.3 Green's Function for Charge Within Dielectric

In the preceding discussion we chose to return to fundamentals [(16.19a)] and then encountered an integral equation [(16.26)] that was easily reduced to a one-dimensional statement for the 2+1-dimensional geometry under consideration. It is, then, not altogether surprising that, in the pursuit of the same outcome, one can also avoid any reference to an integral equation. To this end, let us consider the differential equation (16.1), but without specializing to the two regions $z > 0$ and $z < 0$, as in (16.2a) and (16.2b). The left-hand side of (16.1) is also expressed by

$$-\epsilon(z)\nabla^2 G - \frac{\partial\epsilon(z)}{\partial z}\frac{\partial}{\partial z}G, \tag{16.36}$$

and, after dividing by $\epsilon(z)$, we can write

$$-\nabla^2 G(\mathbf{r},\mathbf{r}') = \frac{4\pi}{\epsilon(z)}\delta(\mathbf{r}-\mathbf{r}') + \left(\frac{\partial}{\partial z}\frac{1}{\epsilon(z)}\right)\epsilon(z)\left(-\frac{\partial}{\partial z}\right)G(\mathbf{r},\mathbf{r}'). \tag{16.37}$$

Then, we have only to insert the representation (16.5) to obtain

$$\left(-\frac{\partial^2}{\partial z^2}+k_\perp^2\right)g(z,z';k_\perp) = \delta(z-z')\frac{1}{\epsilon(z')} + \left(\frac{\partial}{\partial z}\frac{1}{\epsilon(z)}\right)\epsilon(z)\left(-\frac{\partial}{\partial z}\right)g(z,z';k_\perp). \tag{16.38}$$

Now if we specialize to the homogeneous dielectric in contact with the vacuum, this becomes [(16.21)]

$$\left(-\frac{\partial^2}{\partial z^2}+k_\perp^2\right)g(z,z';k_\perp) = \delta(z-z')\frac{1}{\epsilon(z')} + \delta(z)\left(1-\frac{1}{\epsilon}\right)\left(-\frac{\partial}{\partial z_1}\right)g(z_1,z';k_\perp)\bigg|_{z_1=+0}. \tag{16.39}$$

The above equation describes the z-dependence of the potential produced in the vacuum by the combination of two localized charge distributions: one at $z = z'$, the other at $z = 0$. Expressed in terms of the vacuum function g_0, which obeys

$$\left(-\frac{\partial^2}{\partial z^2}+k_\perp^2\right)g_0(z-z';k_\perp) = \delta(z-z'), \tag{16.40}$$

that potential is given by

$$g(z,z';k_\perp) = g_0(z-z';k_\perp)\frac{1}{\epsilon(z')} + g_0(z;k_\perp)\frac{\epsilon-1}{\epsilon}\left(-\frac{\partial}{\partial z_1}\right)g(z_1,z';k_\perp)\bigg|_{z_1=+0}. \tag{16.41}$$

For $z' > 0$, $\epsilon(z') = 1$, this is (16.32).

Having left the choice of z' open in (16.41), we can now proceed to the arrangement in which the unit point charge is embedded in the dielectric: $z' < 0$. With $\epsilon(z') = \epsilon$, we have

$$-\frac{\partial}{\partial z}g(z,z';k_\perp)\bigg|_{z=+0} = \frac{1}{\epsilon}\frac{1}{2}e^{k_\perp z'} + \frac{1}{2}\frac{\epsilon-1}{\epsilon}\left(-\frac{\partial}{\partial z}\right)g(z,z';k_\perp)\bigg|_{z=+0}, \tag{16.42}$$

and therefore

$$-\frac{\partial}{\partial z}g(z,z';k_\perp) = \frac{1}{\epsilon+1}e^{-k_\perp|z'|}. \tag{16.43}$$

The outcome is

$$z' < 0: \quad g(z, z'; k_\perp) = \frac{1}{\epsilon} g_0(z - z'; k_\perp) + \frac{1}{\epsilon}\frac{\epsilon - 1}{\epsilon + 1} g_0(z; k_\perp) e^{-k_\perp |z'|}$$
$$= \frac{1}{\epsilon}\frac{1}{2k_\perp} e^{-k_\perp |z - z'|} + \frac{1}{\epsilon}\frac{\epsilon - 1}{\epsilon + 1}\frac{1}{2k_\perp} e^{-k_\perp |z|} e^{-k_\perp |z'|}. \tag{16.44}$$

It was really unnecessary to do the last calculation, however. This situation in which the point charge is in the dielectric medium can be derived from the one with the charge in the vacuum. First note that the use of vacuum, $\epsilon = 1$, is a matter of convenience. What is physically significant is the ratio of the two dielectric constants. To change all dielectric constants by a common constant factor only modifies the potential—Green's function—by the inverse of that factor. [Again, (13.32) is an example.] Now, begin with Green's function for $z' > 0$ and replace ϵ by $1/\epsilon$. That converts the region containing the point charge into the one with the larger dielectric constant, in the ratio $\epsilon : 1$. On restoring the value unity to the smaller dielectric constant—raising the overall scale by the factor ϵ—Green's function becomes divided by ϵ. And finally, to have the region of higher dielectric constant in the half-space $z < 0$, one replaces all z-coordinates by their negatives. Doing this in (16.17) yields (16.44).

16.4 Full Green's Function and Image Charge

A slightly different way of presenting these results leads to the full three-dimensional Green's function, $G(\mathbf{r}, \mathbf{r}')$. Let us return to (16.16a) and (16.16b), referring to $z' > 0$, and write

$$z > 0: \quad g(z, z'; k_\perp) = g_0(z - z'; k_\perp) - \frac{\epsilon - 1}{\epsilon + 1} g_0(z - \overline{z'}), \tag{16.45a}$$

$$z < 0: \quad g(z, z'; k_\perp) = \frac{2}{\epsilon + 1} g_0(z - z'; k_\perp), \tag{16.45b}$$

where we have introduced $\overline{z'}$, the image point of z', defined by

$$\overline{z'} = -z'. \tag{16.46}$$

Because (16.45a) and (16.45b) involve only the vacuum form of the one-dimensional function, we can interpret (16.45a) and (16.45b) as follows. For $z > 0$, the Green's function appears to describe the potential due to two point charges, one, of strength unity, at $\mathbf{r}' = (x', y', z')$, and another, the image charge, of strength $-\frac{\epsilon-1}{\epsilon+1}$, at the image point $\overline{\mathbf{r}'} = (x', y', \overline{z'})$. For $z < 0$, only one point charge appears, of strength $\frac{2}{\epsilon+1}$, located at $\mathbf{r}'$. In either medium, the total effective charge is the same. With this interpretation, the full Green's function may be written down immediately, ($z' > 0$)

$$z > 0: \quad G(\mathbf{r}, \mathbf{r}') = \frac{1}{|\mathbf{r} - \mathbf{r}'|} - \frac{\epsilon - 1}{\epsilon + 1}\frac{1}{|\mathbf{r} - \overline{\mathbf{r}'}|}, \tag{16.47a}$$

$$z < 0: \quad G(\mathbf{r}, \mathbf{r}') = \frac{2}{\epsilon + 1}\frac{1}{|\mathbf{r} - \mathbf{r}'|}. \tag{16.47b}$$

The related forms for $z' < 0$, as produced, for example, by the substitution $\epsilon \to 1/\epsilon$ followed by the procedure in the preceding paragraph, are

$$z < 0: \quad G(\mathbf{r}, \mathbf{r}') = \frac{1}{\epsilon} \frac{1}{|\mathbf{r} - \mathbf{r}'|} + \frac{1}{\epsilon} \frac{\epsilon - 1}{\epsilon + 1} \frac{1}{|\mathbf{r} - \overline{\mathbf{r}'}|}, \tag{16.48a}$$

$$z > 0: \quad G(\mathbf{r}, \mathbf{r}') = \frac{2}{\epsilon + 1} \frac{1}{|\mathbf{r} - \mathbf{r}'|}. \tag{16.48b}$$

One sees directly, particularly on comparing the $z < 0$, $z' > 0$ and $z > 0$, $z' < 0$ forms, that the symmetry of Green's function is realized in these results.

Applications of these Green's functions will be treated in the next chapter.

16.5 Problems for Chapter 16

1. Consider the half-infinite dielectric region discussed in this chapter. Repeat the calculation for $z' < 0$ by the discontinuity method which led to (16.16a) and (16.16b), and reproduce the result stated in (16.44). Give the primary and secondary field picture. Check the symmetry of the Green's function (both G and g) for all z and z'.
2. A homogeneous dielectric medium, with dielectric constant ϵ, occupies not only the region $z < 0$ as discussed in this chapter but also the region $z > a > 0$. A unit point charge is stationed in the vacuum between the dielectrics, $a > z' > 0$. Use any method to find the function g in the three regions. What happens as $a \to \infty$, and as $a \to 0$?
3. Now suppose a dielectric slab occupies the region $0 < z < a$. Assume the unit point charge is in the vacuum region $z' > a$. Compute the reduced Green's function in this case. Can you derive from this result additional information about the arrangement of Problem 16.2?

17

Application of Green's Function

17.1 Force between Charge and Dielectric

Armed with Green's function for the semi-infinite dielectric arrangement, we can now answer mechanical questions about interaction energy and force. The total energy of a charge distribution in the presence of the dielectric medium, (14.10),

$$E = \frac{1}{2}\int (d\mathbf{r})(d\mathbf{r}')\,\rho(\mathbf{r})G(\mathbf{r},\mathbf{r}')\rho(\mathbf{r}'), \tag{17.1}$$

includes the mutual interactions of the charges. We are not interested in this but rather in the *change* of the energy due to the introduction of the dielectric. To calculate this change, we let G_0 be Green's function in vacuum,

$$G_0(\mathbf{r}-\mathbf{r}') = \frac{1}{|\mathbf{r}-\mathbf{r}'|}, \tag{17.2}$$

while G is Green's function in the presence of the dielectric, as found above. Therefore, the interaction energy between the charge distribution and the dielectric is

$$E_{\text{int}} = \frac{1}{2}\int (d\mathbf{r})(d\mathbf{r}')\rho(\mathbf{r})[G(\mathbf{r},\mathbf{r}') - G_0(\mathbf{r},\mathbf{r}')]\rho(\mathbf{r}'). \tag{17.3}$$

Evaluating this for a point charge at position $\mathbf{r}_0$, with $z_0 > 0$,

$$\rho(\mathbf{r}) = e\delta(\mathbf{r}-\mathbf{r}_0), \tag{17.4}$$

and making use of (16.47a), we find the energy of interaction to be

$$E_{\text{int}} = \frac{e^2}{2}[G(\mathbf{r},\mathbf{r}') - G_0(\mathbf{r},\mathbf{r}')]_{\mathbf{r},\mathbf{r}'=\mathbf{r}_0} = -\frac{e^2}{2}\frac{\epsilon-1}{\epsilon+1}\frac{1}{2z_0}. \tag{17.5}$$

Is this a physically meaningful result? If $\epsilon > 1$, $E < 0$ so that there is a force of attraction pulling the dielectric toward the charge and into the region of higher fields, in agreement with the earlier discussion of Section 13.2. The magnitude of this force is

$$F = -\frac{\partial E_{\text{int}}}{\partial(-z_0)} = \frac{\epsilon-1}{\epsilon+1}\frac{e^2}{4z_0^2} = \frac{\left|-\frac{\epsilon-1}{\epsilon+1}e\right|\,|e|}{(2z_0)^2}, \tag{17.6}$$

which can be interpreted as the force between the charge and the image charge.

The field point of view, as opposed to that of action-at-a-distance, provides an alternate derivation of this result. To calculate the force on the dielectric, we calculate the normal component of the flow of momentum into the dielectric. In terms of the stress tensor, this force is

$$F = -\int dx\,dy\,T_{zz} \tag{17.7}$$

DOI: 10.1201/9781003057369-17

where the integration is over a surface just outside the dielectric, at $z = +0$. Correspondingly, we use the vacuum form of T_{zz}, (3.12b),

$$T_{zz} = \frac{1}{8\pi}(E^2 - 2E_z^2) = \frac{1}{8\pi}(E_\perp^2 - E_z^2). \tag{17.8}$$

Since Green's function is the potential of a unit point charge, the electric field is

$$\mathbf{E}(\mathbf{r}) = -\boldsymbol{\nabla} eG(\mathbf{r},\mathbf{r}_0) = e\left[\frac{\mathbf{r}-\mathbf{r}_0}{|\mathbf{r}-\mathbf{r}_0|^3} - \frac{\epsilon-1}{\epsilon+1}\frac{\mathbf{r}-\overline{\mathbf{r}_0}}{|\mathbf{r}-\overline{\mathbf{r}_0}|^3}\right], \tag{17.9}$$

according to the Green's function form for $z > 0$ in (16.47a). Then, on writing

$$\rho = |(\mathbf{r}-\mathbf{r}_0)_\perp|, \tag{17.10}$$

the field components on the surface are found to be

$$\begin{aligned} z = +0: \quad \mathbf{E}_\perp &= e\frac{2}{\epsilon+1}\frac{(\mathbf{r}-\mathbf{r}_0)_\perp}{(\rho^2+z_0^2)^{3/2}}, \\ E_z &= -e\frac{2\epsilon}{\epsilon+1}\frac{z_0}{(\rho^2+z_0^2)^{3/2}}, \end{aligned} \tag{17.11}$$

giving us this expression for the force:

$$\begin{aligned} F &= \frac{e^2}{8\pi}\left(\frac{2}{\epsilon+1}\right)^2 \int_0^\infty 2\pi\,\rho\,d\rho\,\frac{\epsilon^2 z_0^2 - \rho^2}{(\rho^2+z_0^2)^3} \\ &= \left(\frac{1}{\epsilon+1}\right)^2 \frac{e^2}{2z_0^2}\int_0^\infty dt\frac{\epsilon^2-1-(t-1)}{(t+1)^3} = \frac{\epsilon-1}{\epsilon+1}\frac{e^2}{4z_0^2}, \end{aligned} \tag{17.12}$$

where we have introduced the variable $t = \rho^2/z_0^2$, in agreement with (17.6).

But let us not hurry on without examining, in the context of this simple example, the connection between these two modes of computation. We know that, in regions of homogeneous dielectric constant, the stress dyadic (13.41) obeys the electrostatic specialization of (8.11),

$$\boldsymbol{\nabla}\cdot\mathbf{T} + \rho\mathbf{E} = \mathbf{0}. \tag{17.13}$$

Now integrate the z-component of this equation over the entire volume of the semi-infinite region $z > 0$ to get (infinitely remote surfaces do not contribute):

$$F = -\int dS\,T_{zz}(z=+0) = -\int (d\mathbf{r})\,\rho E_z = -e\left[-\frac{\partial}{\partial z}eG(\mathbf{r},\mathbf{r}_0)\right]_{\mathbf{r}\to\mathbf{r}_0}. \tag{17.14}$$

The electric field required here is (17.9). The force on the dielectric body is now found as the negative of that acting on the charge e, as produced by the latter's own Coulomb field (which force we know to be zero), and by the Coulomb field of the image charge. This derivation of (17.6) is in the spirit of the first such calculation, but lacks the reference to the energy.

As far as the field on the surface $z = +0$ is concerned, the construction given in (17.9) could as well apply everywhere. If we accept this fiction, the volume integral of the z-component of (17.13), extended over the semi-infinite region $z < +0$, gives

$$F = -\int dS\,T_{zz}(z=+0) = \int (d\mathbf{r})\,\rho_{\text{image}}E_z = -e\frac{\epsilon-1}{\epsilon+1}\left[-\frac{\partial}{\partial z}eG(\mathbf{r},\mathbf{r}_0)\right]_{\mathbf{r}\to\overline{\mathbf{r}_0}}, \tag{17.15}$$

comprising the zero self-force of the image charge, and the force exerted on the image charge by the charge e. Quite correct.

But suppose we want to use the *physical* field produced in the dielectric region by the point charge in the vacuum. First we move across the boundary by writing

$$F = -\int dS\,[T_{zz}(z=+0) - T_{zz}(z=-0)] - \int dS\,T_{zz}(z=-0). \tag{17.16}$$

Then, integration of the z-component of (17.13) over the region $z < -0$ relates the last integral above to the force on the free charge within the dielectric. There isn't any! That effectively reduces the calculation to the one already carried out in (17.12) or (17.14). Nevertheless, let us evaluate F by means of the first part of (17.16), involving the discontinuity of T_{zz} across the interface between dielectric and vacuum:

$$F = -\int dS\,[T_{zz}(z=+0) - T_{zz}(z=-0)]. \tag{17.17}$$

Written in terms of the electric field in the vacuum ($z = +0$), we have

$$T_{zz}(z=-0) = \frac{1}{8\pi}[\epsilon E_\perp^2 - \frac{1}{\epsilon}E_z^2], \tag{17.18}$$

and then

$$T_{zz}(z=+0) - T_{zz}(z=-0) = -\frac{1}{8\pi}(\epsilon-1)[E_\perp^2 + \frac{1}{\epsilon}E_z^2]. \tag{17.19}$$

Incidentally, the same expression for the force emerges from (13.40), with its z-component presented as

$$F = -\int dS \int_{-0}^{+0} dz \frac{d\epsilon(z)}{dz}\frac{1}{8\pi}\left[E_\perp^2 + \left(\frac{1}{\epsilon(z)}\right)^2 D_z^2\right], \tag{17.20}$$

which involves the continuous transverse electric field and normal component of $\mathbf{D}$. With these field components given their $z = +0$ values, using (16.21), the resulting z-integration over the discontinuity region yields

$$F = \int dS \frac{1}{8\pi}(\epsilon-1)[E_\perp^2 + \frac{1}{\epsilon}E_z^2], \tag{17.21}$$

in agreement with (17.17), (17.19). The insertion of the fields displayed in (17.11) then produces

$$\begin{aligned} F &= \frac{e^2}{8\pi}\left(\frac{2}{\epsilon+1}\right)^2(\epsilon-1)\int_0^\infty d\rho\,\rho\,2\pi\frac{\epsilon z_0^2+\rho^2}{(\rho^2+z_0^2)^3} \\ &= \frac{\epsilon-1}{(\epsilon+1)^2}\frac{e^2}{2z_0^2}\int_0^\infty dt\frac{\epsilon+1+(t-1)}{(t+1)^3}, \end{aligned} \tag{17.22}$$

which is indeed equal to (17.12), in consequence of the null value of the integral appearing in both (17.12) and (17.22),

$$\int_0^\infty dt\frac{t-1}{(t+1)^3} = 0, \tag{17.23}$$

as can be seen from the substitution $t \to 1/t$. This property just expresses the zero value possessed by the surface integral of $T_{zz}(z=-0)$.

The version given in (17.17) directs attention to the physical interpretation of F as the force on the bound charge that is localized on the interface between the two regions. As such, the force should also equal

$$F = \int dS\,\sigma\,E_z(z=0). \tag{17.24}$$

But what is the value of $E_z(z=0)$? One way to answer that question has us returning to (16.21), which describes how a surface distribution of charge emerges in the limit where $\epsilon(z)$ is discontinuous. If we regress to the continuous description, through the replacement [the other factors in σ, (16.23), and the reference to a particular point on the surface, are understood]

$$\left(1-\frac{1}{\epsilon}\right)E_z(z=0) \to \int_{-0}^{+0} dz\left(\frac{d}{dz}\frac{1}{\epsilon(z)}\right)E_z(z), \tag{17.25}$$

and use the continuity of D_z, we find this integral to be

$$\int_{-0}^{+0} dz\left(\frac{d}{dz}\frac{1}{\epsilon(z)}\right)\frac{1}{\epsilon(z)}D_z(z=0) = \frac{1}{2}\left(1-\frac{1}{\epsilon^2}\right)D_z(z=0). \tag{17.26}$$

That identifies $E_z(z=0)$ with

$$E_z(z=0) = \frac{1}{2}\left(1+\frac{1}{\epsilon}\right)D_z(z=0) = \frac{1}{2}[E_z(z=+0)+E_z(z=-0)]. \tag{17.27}$$

According to the form given for $E_z(z=+0)$ in (17.11), this is

$$E_z(z=0) = -e\frac{z_0}{(\rho^2+z_0^2)^{3/2}}, \tag{17.28}$$

which is just the field of the point charge. But we could have anticipated that; the self-force of the surface charge distribution is zero. The above field is now combined with (16.23), (17.11),

$$\sigma = -\frac{e}{2\pi}\frac{\epsilon-1}{\epsilon+1}\frac{z_0}{(\rho^2+z_0^2)^{3/2}} \tag{17.29}$$

to produce

$$F = \int dS\,\sigma\,E_z(z=0) = \frac{e^2}{2\pi}\frac{\epsilon-1}{\epsilon+1}\int_0^\infty d\rho\,\rho\,2\pi\frac{z_0^2}{(\rho^2+z_0^2)^3} = \frac{\epsilon-1}{\epsilon+1}\frac{e^2}{(2z_0)^2}. \tag{17.30}$$

Six derivations of the same quantity? Time to move on. Yes, but there is one little question about the surface charge distribution that has not been answered. What is the total amount of that charge? Obviously, one can compute it by integration—and so we shall, for a particular example. But a more immediate answer comes from Green's function ($z'>0$) as displayed in (16.47a), when this is regarded as the potential produced in the vacuum by the unit point charge and the surface charge. At very large distances, on the scale set by z' so that the finite dimensions of the surface charge distribution are negligible, the behavior of G in both half-spaces is simply

$$G(\mathbf{r},\mathbf{r}') \sim \left(1-\frac{\epsilon-1}{\epsilon+1}\right)\frac{1}{r}; \tag{17.31}$$

the total surface charge equals the image charge.

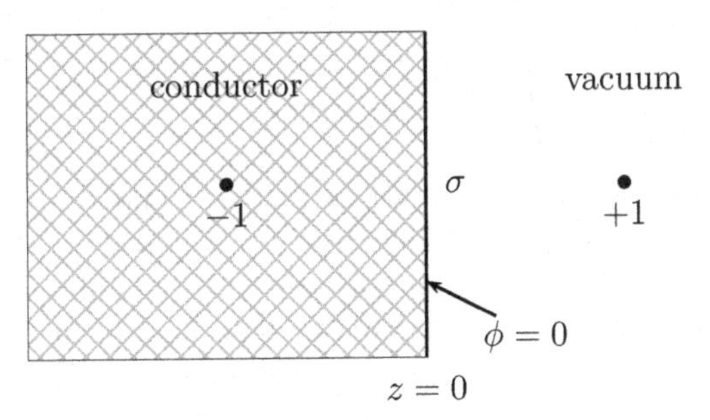

FIGURE 17.1
Image charge for a grounded conducting plane.

17.2 Infinite Conducting Plate

The apparently unphysical process of letting ϵ become arbitrarily large, in fact, produces a new physical situation, one in which the dielectric body acts like a conductor. Indeed, we see from (16.47b) that $G(z' > 0)$ then vanishes in the region $z < 0$; the latter has become a region of constant—zero—potential. In the limit that $\epsilon \to \infty$, (16.47a) and (16.47b) become

$$G(\mathbf{r},\mathbf{r}') = \begin{cases} \frac{1}{|\mathbf{r}-\mathbf{r}'|} - \frac{1}{|\mathbf{r}-\overline{\mathbf{r}'}|}, & z > 0, \\ 0, & z < 0, \end{cases} \tag{17.32}$$

which is obviously Green's function for a grounded conductor. For $z > 0$, Green's function can be interpreted as the potential of a unit point charge at $\mathbf{r}' = (0, 0, z')$ and an image charge of strength -1 at $\mathbf{r}' = (0, 0, -z')$, as shown in Fig. 17.1. The image charge is exactly opposite to the charge placed in the vacuum, and G vanishes on the surface, all of its points being at the same distance from the equal and opposite charges.

For such a unit point charge, we now calculate the *free* surface charge density, σ, induced on the conductor. We know, from (13.62), that

$$4\pi\sigma(\mathbf{r}_\perp) = E_z(\mathbf{r})\Big|_{z=+0} = -\frac{\partial}{\partial z}G(\mathbf{r},\mathbf{r}')\Big|_{z=+0}, \tag{17.33}$$

which is just the $\epsilon \to \infty$ limit of what is displayed in (16.25b). Alternative and equivalent forms for Green's function are, for $z > 0$, [recall (16.28)]

$$\begin{aligned} G(\mathbf{r},\mathbf{r}') &= \frac{1}{|\mathbf{r}-\mathbf{r}'|} - \frac{1}{|\mathbf{r}-\overline{\mathbf{r}'}|} \\ &= 4\pi \int \frac{(d\mathbf{k}_\perp)}{(2\pi)^2} e^{i\mathbf{k}_\perp \cdot (\mathbf{r}-\mathbf{r}')_\perp} \frac{1}{2k_\perp} \left[e^{-k_\perp|z-z'|} - e^{-k_\perp(z+z')} \right]. \end{aligned} \tag{17.34}$$

This is the $\epsilon \to \infty$ limit of (16.17). The outcome for the form of σ computed from the first form can be read off from (17.29) with $\epsilon \to \infty$, $e = 1$, and $z_0 = z'$:

$$\sigma = -\frac{1}{4\pi}\frac{2z'}{(\rho^2 + z'^2)^{3/2}}. \tag{17.35}$$

The second form gives

$$\sigma = -\int \frac{(d\mathbf{k}_\perp)}{(2\pi)^2} e^{i\mathbf{k}_\perp \cdot (\mathbf{r}-\mathbf{r}')_\perp} e^{-k_\perp z'}, \tag{17.36}$$

which is also available from the $\epsilon \to \infty$ limit of (16.34). The equivalence of (17.35) and (17.36) will be exploited in the next chapter.

For now, let us check that in both cases the total induced charge on the surface of the conductor is -1, the strength of the image charge. The first form, (17.35), yields for the total charge

$$\begin{aligned} Q_{\text{ind}} &= \int dS\,\sigma = \int 2\pi\,\rho\,d\rho \left(-\frac{1}{2\pi}\right) \frac{z'}{(\rho^2 + z'^2)^{3/2}} \\ &= -\frac{1}{2}\int_0^\infty dt \frac{1}{(t+1)^{3/2}} = -1. \end{aligned} \tag{17.37}$$

In using the second form, (17.36), one encounters the two-dimensional delta function $[dS = (d\mathbf{r}_\perp)]$, as already noted in (16.29),

$$\int \frac{(d\mathbf{r}_\perp)}{(2\pi)^2} e^{i\mathbf{k}_\perp \cdot (\mathbf{r}-\mathbf{r}')_\perp} = \delta(\mathbf{k}_\perp). \tag{17.38}$$

That gives

$$Q_{\text{ind}} = -\int (d\mathbf{k}_\perp)\delta(\mathbf{k}_\perp) e^{-k_\perp z'} = -1. \tag{17.39}$$

In another exercise of this kind we give two derivations of the force pulling the conductor toward the unit point charge, using the force per unit area presented in (13.67), but with $\epsilon = 1$ and omitting the minus sign:

$$F = \int dS\, 2\pi\,\sigma^2. \tag{17.40}$$

We need not trouble with the calculation that employs (17.35) for σ; it is just what is exhibited in (17.30), with $\epsilon \to \infty$, $e = 1$, and z_0 replaced by z':

$$F = \frac{1}{(2z')^2}. \tag{17.41}$$

The calculation that uses (17.36) begins with

$$F = 2\pi \int (d\mathbf{r}_\perp) \int \frac{(d\mathbf{k}_\perp)}{(2\pi)^2} \frac{(d\mathbf{k}'_\perp)}{(2\pi)^2} e^{i(\mathbf{k}-\mathbf{k}')_\perp \cdot (\mathbf{r}-\mathbf{r}')_\perp} e^{-(k_\perp + k'_\perp)z'}, \tag{17.42}$$

in which it is convenient to write $-\mathbf{k}'_\perp$ as the integration variable for the second σ. Performing the $\mathbf{r}_\perp$ integration with the aid of (17.38), we get the same result,

$$\begin{aligned} F &= 2\pi \int \frac{(d\mathbf{k}_\perp)}{(2\pi)^2} \frac{(d\mathbf{k}'_\perp)}{(2\pi)^2} (2\pi)^2 \delta(\mathbf{k}_\perp - \mathbf{k}'_\perp) e^{-(k_\perp + k'_\perp)z'} \\ &= \int \frac{(d\mathbf{k}_\perp)}{2\pi} e^{-2k_\perp z'} = \int_0^\infty dk_\perp\, k_\perp e^{-2k_\perp z'} = \frac{1}{(2z')^2}. \end{aligned} \tag{17.43}$$

17.3 Problems for Chapter 17

1. Look at the answer to Problem 16.2 in the limit $\epsilon \to \infty$, that is, Green's function for a cavity, of width a, between two conducting plates, and bring it into the form, for $0 < z' < z < a$,

$$g = \frac{\sinh k(a - z) \sinh kz'}{k \sinh ka}. \tag{17.44}$$

What is it for $0 < z < z' < a$? Show directly that this satisfies the correct differential equation with the proper boundary conditions.

2. Prove the identity (17.23) and show that it expresses the fact that the last integral in (17.16) vanishes.

3. In Section 17.2 we considered a charge in front of a conducting plane. Here we consider a different situation, one in which the surface at $z = 0$ has a specified potential, $V(\mathbf{r}_\perp)$. Use either (a) the explicit coordinate representation of the Green's function, or (b) the transverse Fourier representation (17.34), together with the construction (14.19), to obtain a formula for $\phi(\mathbf{r})$, $z > 0$, in terms of the boundary values, $V(\mathbf{r}_\perp)$ on the $z = 0$ surface. What happens when the potential on the surface is constant, $V(\mathbf{r}_\perp) = V$?

18

Bessel Functions

Useful mathematical identities can be obtained if we solve a physical problem by using different representations for Green's function. In particular, through the consideration of situations where physical quantities vary only in a single direction, we learn of the properties of the important class of functions called Bessel functions. An illustration of this was encountered in the last chapter, where we obtained two forms for the surface charge density, (17.35) and (17.36). If we let $\rho = |(\mathbf{r} - \mathbf{r}')_\perp|$ and introduce polar coordinates for $\mathbf{k}_\perp$, that is, the magnitude $k = |\mathbf{k}_\perp|$, and the angle ϕ, which may be taken to be the angle between $\mathbf{k}_\perp$ and $(\mathbf{r} - \mathbf{r}')_\perp$, we have for the exponent in (17.36)

$$\mathbf{k}_\perp \cdot (\mathbf{r} - \mathbf{r}')_\perp = k\rho\cos\phi, \tag{18.1}$$

and for the element of integration

$$(d\mathbf{k}_\perp) = k\,dk\,d\phi. \tag{18.2}$$

Accordingly, the identity may be written as (for $z > 0$)

$$-\frac{2z}{(\rho^2 + z^2)^{3/2}} = -4\pi \int \frac{k\,dk\,d\phi}{(2\pi)^2} e^{ik\rho\cos\phi} e^{-kz}. \tag{18.3}$$

The ϕ integral in (18.3) is defined as the Bessel function of zeroth order, J_0,

$$\int_0^{2\pi} \frac{d\phi}{2\pi} e^{ik\rho\cos\phi} = J_0(k\rho); \tag{18.4}$$

it is the first member of the class of functions that are named for Friedrich Wilhelm Bessel (1784–1846), although the infinite series that represents these functions had appeared more than half a century earlier, in a work by Leonard Euler (1707–1783). Now expressed in terms of this function, the equivalence (18.3) appears as

$$\frac{z}{(\rho^2 + z^2)^{3/2}} = \int_0^\infty k\,dk\,J_0(k\rho)e^{-kz}, \quad z > 0. \tag{18.5}$$

Since this result was obtained by equating z derivatives [*cf.* (17.33)], it may be immediately integrated to yield

$$\frac{1}{\sqrt{\rho^2 + z^2}} = \int_0^\infty dk\,J_0(k\rho)e^{-k|z|}. \tag{18.6}$$

(The integration constant vanishes since both sides go to zero as $|z| \to \infty$.) Note that (18.6) may be directly derived from the equality of the two forms of the Coulomb potential in (15.33), which with $\mathbf{r}' = \mathbf{0}$ reads

$$\frac{1}{r} = 4\pi \int \frac{(d\mathbf{k}_\perp)}{(2\pi)^2} e^{i\mathbf{k}_\perp \cdot \mathbf{r}_\perp} \frac{1}{2k} e^{-k|z|}. \tag{18.7}$$

DOI: 10.1201/9781003057369-18

Here in (18.6), we have recognized two different forms for the potential of a unit point charge at the origin, which satisfies Laplace's equation,

$$\nabla^2 \phi = 0, \tag{18.8}$$

except at the origin. Actually, the integrand of the right-hand side of (18.6) is also a solution of Laplace's equation, that is, for each positive value of k,

$$\nabla^2 [J_0(k\rho)e^{-k|z|}] = 0, \tag{18.9}$$

provided $z \neq 0$. This is most easily proved by returning to (18.7) and noting that

$$z \neq 0: \quad \nabla^2 \left(e^{i\mathbf{k}_\perp \cdot \mathbf{r}_\perp} e^{-k|z|}\right) = (-\mathbf{k}_\perp^2 + k^2) e^{i\mathbf{k}_\perp \cdot \mathbf{r}_\perp} e^{-k|z|} = 0. \tag{18.10}$$

Integration over the direction angle of $\mathbf{k}_\perp$ according to (18.4) then produces (18.9).

We can convert (18.9) into the differential equation that J_0 obeys by writing ∇^2, the Laplacian differential operator, in terms of circular cylindrical coordinates. Let us first recall that in using orthogonal curvilinear coordinates, u_1, u_2, u_3, such that the distance represented by the infinitesimal coordinate change du_k $(k = 1, 2, 3)$ is $h_k(u)du_k$, the element of volume becomes

$$h_1 du_1\, h_2 du_2\, h_3 du_3 \equiv h\, du_1\, du_2\, du_3, \tag{18.11}$$

while the element of area perpendicular to the u_1 direction, for example, is of magnitude

$$h_2 du_2\, h_3 du_3. \tag{18.12}$$

Then, the translation into curvilinear coordinates of the divergence theorem,

$$\int_V (d\mathbf{r})\, \nabla^2 \phi(\mathbf{r}) = \oint_S d\mathbf{S} \cdot \boldsymbol{\nabla} \phi(\mathbf{r}), \tag{18.13}$$

is as indicated by

$$\int_V du_1 du_2 du_3\, h\, \nabla^2 \phi(u) = \oint_S \left(du_2 du_3 \frac{h_2 h_3}{h_1} \frac{\partial}{\partial u_1} \phi(u) + \dots\right). \tag{18.14}$$

Rewriting the latter surface integral in terms of a curvilinear coordinate volume integral (with due attention to the sense of the normals) yields the desired Laplacian form (see Problem 18.1)

$$\nabla^2 = \frac{1}{h} \sum_{k=1}^{3} \frac{\partial}{\partial u_k} \left(\frac{h}{h_k^2} \frac{\partial}{\partial u_k}\right). \tag{18.15}$$

In the example of circular cylindrical coordinates, where infinitesimal distances in the three orthogonal directions are produced by $d\rho$, $\rho d\phi$, dz, and $h = \rho$, the Laplacian is

$$\nabla^2 = \frac{1}{\rho} \frac{\partial}{\partial \rho} \left(\rho \frac{\partial}{\partial \rho}\right) + \frac{1}{\rho^2} \frac{\partial^2}{\partial \phi^2} + \frac{\partial^2}{\partial z^2}, \tag{18.16}$$

leading to the reduced form for (18.9),

$$\left[\frac{1}{\rho} \frac{d}{d\rho} \left(\rho \frac{d}{d\rho}\right) + k^2\right] J_0(k\rho) = 0, \tag{18.17a}$$

or, with $t = k\rho$,

$$\left[\frac{1}{t} \frac{d}{dt} \left(t \frac{d}{dt}\right) + 1\right] J_0(t) = \left(\frac{d^2}{dt^2} + \frac{1}{t} \frac{d}{dt} + 1\right) J_0(t) = 0. \tag{18.17b}$$

Inserting here the integral representation of the Bessel function, (18.4),

$$J_0(t) = \int_0^{2\pi} \frac{d\phi}{2\pi} e^{it\cos\phi}, \tag{18.18}$$

and multiplying by t, we have

$$\int_0^{2\pi} \frac{d\phi}{2\pi}[-t\cos^2\phi + i\cos\phi + t]e^{it\cos\phi} = \int_0^{2\pi} \frac{d\phi}{2\pi}\frac{d}{d\phi}[i\sin\phi e^{it\cos\phi}] = 0, \tag{18.19}$$

thereby again proving (18.9), and establishing the equation satisfied by the Bessel function of order zero, (18.17b).

To exploit the more general solution of Laplace's equation exhibited in (18.10), we take

$$e^{i\mathbf{k}_\perp \cdot \mathbf{r}_\perp} = e^{ik\rho\cos\phi} = \exp\left[it\frac{1}{2}\left(e^{i\phi} + e^{-i\phi}\right)\right], \tag{18.20}$$

introduce the symbol $u = ie^{i\phi}$, and then encounter the *generating function* for the Bessel functions of integer order,

$$e^{\frac{1}{2}t(u-1/u)} = \sum_{m=-\infty}^{\infty} u^m J_m(t), \tag{18.21}$$

which serves as a definition of all the Bessel functions of integer order m. Since the generating function is invariant under the substitution

$$u \to -\frac{1}{u}, \tag{18.22}$$

that is,

$$\sum_{m=-\infty}^{\infty} u^m J_m(t) = \sum_{m=-\infty}^{\infty} (-1)^m u^{-m} J_m(t) = \sum_{m=-\infty}^{\infty} (-1)^m u^m J_{-m}(t), \tag{18.23}$$

we learn that Bessel functions of positive and negative integer orders are related by

$$J_{-m}(t) = (-1)^m J_m(t). \tag{18.24}$$

Now let us regard the generating function as the product of two exponentials, with arguments proportional to u and $-1/u$, respectively. The insertion of the appropriate infinite power series for these exponentials,

$$\sum \frac{\left(\frac{1}{2}t\right)^{m+n} u^{m+n}}{(m+n)!} \sum \frac{\left(-\frac{1}{2}t\right)^n u^{-n}}{n!}, \tag{18.25}$$

which summations extend over all non-negative integer values of $m+n$ and n, then supplies the infinite power series representation for J_m,

$$J_m(t) = \sum_{n=0}^{\infty} (-1)^n \frac{\left(\frac{1}{2}t\right)^{m+2n}}{(m+n)!\,n!}, \tag{18.26}$$

that (to within a constant factor) Euler first encountered in 1764.

By writing (18.21) in terms of ϕ,

$$e^{it\cos\phi} = \sum_{m=-\infty}^{\infty} i^m e^{im\phi} J_m(t), \tag{18.27}$$

and using the orthogonality condition

$$\int_0^{2\pi} \frac{d\phi}{2\pi} e^{-im'\phi} e^{im\phi} = \delta_{mm'}, \tag{18.28}$$

we obtain an integral representation for J_m,

$$i^m J_m(t) = \int_0^{2\pi} \frac{d\phi}{2\pi} e^{i(t\cos\phi - m\phi)}, \tag{18.29}$$

which contains the result for J_0, (18.4). Incidentally, changing the sign of m, and replacing ϕ by $2\pi - \phi$, leaves the integral in (18.29) intact, or

$$i^{-m} J_{-m}(t) = i^m J_m(t), \tag{18.30}$$

which is the content of (18.24).

Since (18.29) is the Fourier coefficient of $e^{ik\rho\cos\phi}$, we see from (18.10) that another set of solutions to Laplace's equation is

$$e^{im\phi} J_m(k\rho) e^{-k|z|}. \tag{18.31}$$

The Laplacian, (18.16), acting on this solution yields the differential equation satisfied by J_m,

$$\left[\frac{1}{\rho}\frac{d}{d\rho}\left(\rho\frac{d}{d\rho}\right) - \frac{m^2}{\rho^2} + k^2\right] J_m(k\rho) = 0, \tag{18.32a}$$

or

$$\left[\frac{d^2}{dt^2} + \frac{1}{t}\frac{d}{dt} - \frac{m^2}{t^2} + 1\right] J_m(t) = 0. \tag{18.32b}$$

One can, of course, verify that $J_m(t)$, as represented by the integral (18.29), obeys this differential equation. (See Problems 18.4 and 18.5.)

Another way to present the differential equation uses the differential operator relation (an operand is understood)

$$t^{-1/2}\frac{d}{dt}t^{1/2} = \frac{d}{dt} + \frac{1}{2t}, \tag{18.33}$$

and its square

$$\left(t^{-1/2}\frac{d}{dt}t^{1/2}\right)^2 = t^{-1/2}\frac{d^2}{dt^2}t^{1/2} = \frac{d^2}{dt^2} + \frac{1}{t}\frac{d}{dt} - \frac{1}{4t^2}. \tag{18.34}$$

The latter leads to the following differential equation for $t^{1/2}J_m(t)$:

$$\left(\frac{d^2}{dt^2} - \frac{m^2 - 1/4}{t^2} + 1\right) t^{1/2} J_m(t) = 0. \tag{18.35}$$

This version has the advantage of suggesting more clearly the behavior of the Bessel functions at extreme values of t. For sufficiently small values of t, such that unity is negligible in comparison with $(m^2 - 1/4)/t^2$, the differential equation indicates that $t^{1/2}J_m(t) \sim t^{|m|+1/2}$, in accordance with the leading term of (18.26),

$$m \geq 0: \quad J_m(t) = \frac{\left(\frac{1}{2}t\right)^m}{m!} + \dots, \quad (t \to 0). \tag{18.36}$$

And, at sufficiently large values of t, such that $(m^2 - 1/4)/t^2$ is negligible compared to unity, $t^{1/2}J_m(t)$ should be a trigonometric function of t. Indeed, the leading asymptotic term for $J_m(t)$ is

$$J_m(t) \sim \left(\frac{2}{\pi t}\right)^{1/2} \cos\left(t - (m + 1/2)\frac{\pi}{2}\right). \tag{18.37}$$

18.1 Delta Functions and Completeness

We have spoken of Coulomb's potential as a solution of Laplace's equation, which it is when the position of the charge is excluded. If all of space is considered, however, we are dealing with a solution of Poisson's inhomogeneous equation, where the delta function appears as the charge density. That brings us to the question: How is the delta function represented in curvilinear coordinates? To answer this, we refer to the basic integration property

$$\int_V (d\mathbf{r})\,\delta(\mathbf{r}-\mathbf{r}') = \begin{cases} 1, & \mathbf{r}'\,\text{in}\,V \\ 0, & \mathbf{r}'\,\text{not in}\,V \end{cases}, \tag{18.38}$$

and recall that [(18.11)]

$$(d\mathbf{r}) = h(u)\,du_1du_2du_3. \tag{18.39}$$

Accordingly, we must have

$$\delta(\mathbf{r}-\mathbf{r}') = \frac{1}{h(u)}\delta(u_1-u_1')\,\delta(u_2-u_2')\,\delta(u_3-u_3'). \tag{18.40}$$

This is illustrated in circular cylindrical coordinates, where

$$(d\mathbf{r}) = d\rho\,\rho\,d\phi\,dz, \tag{18.41}$$

by the construction

$$\delta(\mathbf{r}-\mathbf{r}') = \frac{1}{\rho}\delta(\rho-\rho')\,\delta(\phi-\phi')\,\delta(z-z'). \tag{18.42}$$

A particular example of a curvilinear coordinate delta function appears when we return to (18.6), or

$$\frac{1}{r} = \int_0^\infty dk\,J_0(k\rho)e^{-k|z|}, \tag{18.43}$$

and apply the Laplacian differential operator, without excluding $z=0$. On recalling from (15.22) and (15.24) that

$$\left(-\frac{d^2}{dz^2}+k^2\right)\frac{e^{-k|z|}}{2k} = \delta(z), \tag{18.44}$$

we arrive at

$$-\nabla^2\frac{1}{r} = 4\pi\delta(z)\frac{1}{2\pi}\int_0^\infty dk\,k\,J_0(k\rho). \tag{18.45}$$

What can we say about the latter integral? A glance at (18.5) shows that

$$\frac{1}{2\pi}\int_0^\infty dk\,k\,J_0(k\rho) = \lim_{z\to+0}\frac{1}{2\pi}\frac{z}{(\rho^2+z^2)^{3/2}} = \begin{cases} 0 & \text{if } \rho\neq 0 \\ \infty & \text{if } \rho=0 \end{cases}. \tag{18.46}$$

To these limiting results we append the value of the polar coordinate surface integral [this is (17.37) again]

$$\int_0^\infty d\rho\,\rho\,2\pi\frac{1}{2\pi}\frac{z}{(\rho^2+z^2)^{3/2}} = \frac{1}{2}\int_0^\infty dt\frac{1}{(t+1)^{3/2}} = 1, \tag{18.47}$$

independent of the particular choice of $z>0$. Here, then, is the realization, in polar coordinates, of $\delta(x)\delta(y)$, as required to complete the structure of $\delta(\mathbf{r})$ in (18.45):

$$\frac{1}{2\pi}\int_0^\infty dk\,k\,J_0(k\rho) = \frac{1}{\rho}\delta(\rho)\frac{1}{2\pi}. \tag{18.48}$$

Notice that there is no reference here to ϕ; that angle is indeterminate at $\rho = 0$. [One can regard the factor $1/(2\pi)$ as the average of $\delta(\phi - \phi')$ over all values of ϕ'; it also helps to interpret $\delta(\rho)$ as the limit of $\delta(\rho - \rho')$ as ρ' approaches zero. See also (18.66).]

The characteristics of any delta function—say the one-dimensional $\delta(x - x')$—are expressed in terms of an arbitrary function $f(x)$ by the integration property

$$\int_{-\infty}^{\infty} dx'\, \delta(x - x')\, f(x') = f(x). \tag{18.49}$$

Then, the insertion of the integral construction of the delta function

$$\delta(x - x') = \int_{-\infty}^{\infty} \frac{dk}{2\pi} e^{ik(x-x')}, \tag{18.50}$$

produces

$$f(x) = \int_{-\infty}^{\infty} \frac{dk}{2\pi} e^{ikx} \left[\int_{-\infty}^{\infty} dx' e^{-ikx'} f(x')\right]. \tag{18.51}$$

This exhibits the completeness of the set of functions $\exp(ikx)$ in which k ranges continuously from $-\infty$ to ∞; any function of x can be constructed as a linear combination of that set, with the coefficients as displayed in (18.51). Such integral representations of functions were introduced by Jean Baptiste Joseph Fourier (1768–1830), although the particular complex form appearing in (18.51) can be attributed to the independent investigations of Augustin Louis Cauchy (1789–1857). The analogous two- and three-dimensional completeness statements are similar consequences of the appropriate delta function constructions. We now find the polar coordinate version of this completeness in two dimensions.

The starting point is the two-dimensional analog of (15.13) and (18.50),

$$\int \frac{(d\mathbf{k}_\perp)}{(2\pi)^2} e^{i\mathbf{k}_\perp \cdot (\mathbf{r}-\mathbf{r}')_\perp} = \delta((\mathbf{r} - \mathbf{r}')_\perp) = \frac{1}{\rho}\delta(\rho - \rho')\delta(\phi - \phi'), \tag{18.52}$$

where we have introduced polar coordinates $\mathbf{r}_\perp(\rho, \phi)$ and $\mathbf{r}'_\perp(\rho', \phi')$. Correspondingly, if we use polar coordinates for $\mathbf{k}_\perp(k, \alpha)$, (18.52) becomes

$$\int_0^{\infty} \frac{k\, dk}{2\pi} \int_0^{2\pi} \frac{d\alpha}{2\pi} e^{ik\rho\cos(\phi-\alpha)} e^{-ik\rho'\cos(\phi'-\alpha)} = \frac{1}{\rho}\delta(\rho - \rho')\delta(\phi - \phi'). \tag{18.53}$$

We next expand the exponentials here by use of the generating function expression, (18.27) [with $\phi \to \phi - \alpha$] together with its complex conjugate [with $\phi \to \phi' - \alpha$] and perform the α integration by means of (18.28):

$$\begin{aligned}
&\int_0^{2\pi} \frac{d\alpha}{2\pi} \left[\sum_m i^m e^{im(\phi-\alpha)} J_m(k\rho)\right] \left[\sum_{m'} (-i)^{m'} e^{-im'(\phi'-\alpha)} J_{m'}(k\rho')\right] \\
&= \sum_{m,m'} i^m e^{im\phi} J_m(k\rho) \delta_{mm'} (-i)^{m'} e^{-im'\phi'} J_{m'}(k\rho') \\
&= \sum_{m=-\infty}^{\infty} e^{im\phi} J_m(k\rho) e^{-im\phi'} J_m(k\rho').
\end{aligned} \tag{18.54}$$

The result of these operations is the identity

$$\int_0^{\infty} \frac{kdk}{2\pi} \sum_{m=-\infty}^{\infty} e^{im\phi} J_m(k\rho) e^{-im\phi'} J_m(k\rho') = \frac{1}{\rho}\delta(\rho - \rho')\delta(\phi - \phi'). \tag{18.55}$$

Again this is a completeness statement, this time for the functions $e^{im\phi}J_m(k\rho)$, where k ranges continuously from 0 to ∞, and m assumes all integer values: 0, ±1, ±2, ... That is made explicit on multiplying (18.55) by an arbitrary function $f(\rho', \phi')$, and by the area element $\rho'\,d\rho'\,d\phi'$, followed by integration:

$$f(\rho,\phi) = \int_0^\infty \frac{kdk}{2\pi} \sum_{m=-\infty}^{\infty} e^{im\phi} J_m(k\rho) \int_0^\infty \rho' d\rho' \int_0^{2\pi} d\phi' e^{-im\phi'} J_m(k\rho') f(\rho',\phi'). \quad (18.56)$$

We may now easily isolate the individual ρ and ϕ dependencies of (18.55). If we multiply (18.55) by $e^{-i\overline{m}(\phi-\phi')}$ and integrate over ϕ, we select a particular value of m, according to (18.28) so that we obtain

$$\int_0^\infty k\,dk\, J_m(k\rho)J_m(k\rho') = \frac{1}{\rho}\delta(\rho-\rho'). \quad (18.57a)$$

This states that $J_m(k\rho)$ is a complete set of functions of ρ for *any* integer m:

$$f(\rho) = \int_0^\infty k\,dk\, J_m(k\rho) \left[\int_0^\infty \rho' d\rho' J_m(k\rho') f(\rho')\right]. \quad (18.57b)$$

Putting this information back into (18.55), we determine the completeness relation for the functions $e^{im\phi}$:

$$\sum_{m=-\infty}^{\infty} \frac{1}{2\pi} e^{im(\phi-\phi')} = \delta(\phi-\phi'), \quad (18.58a)$$

and correspondingly

$$f(\phi) = \sum_{m=-\infty}^{\infty} e^{im\phi} \int_0^{2\pi} \frac{d\phi'}{2\pi} e^{-im\phi'} f(\phi'), \quad (18.58b)$$

which is the Fourier series expansion for $f(\phi)$, that is, the statement of completeness, over a 2π interval, of the set of functions $\exp(im\phi)$, m integral.

Perhaps we should remark here that the validity of (18.58a) depends on the restricted 2π range of the variables ϕ and ϕ'. But the restriction can be lifted if the right-hand side of (18.58a) is made periodic in its variable to match that property of the left-hand side. With this unlimited extension of $\phi - \phi'$ now called x, that is displayed by

$$\frac{1}{2\pi} \sum_{m=-\infty}^{\infty} e^{imx} = \sum_{\nu=-\infty}^{\infty} \delta(x - 2\pi\nu), \quad (18.59)$$

and, indeed, when x is restricted to the interval between -2π and 2π, only $\nu = 0$ can contribute to the right-hand side. It's interesting to check the correctness of (18.59) by introducing the integral representation for the delta function, (18.50). That produces, for the right-hand side of (18.59),

$$\int_{-\infty}^\infty \frac{dk}{2\pi} e^{ikx} \sum_{\nu=-\infty}^{\infty} e^{-i2\pi\nu k}, \quad (18.60)$$

which can equal the summation on the left-hand side of (18.59) only if

$$\sum_{\nu=-\infty}^{\infty} e^{-i2\pi\nu k} = \sum_{m=-\infty}^{\infty} \delta(k-m). \quad (18.61)$$

After appropriate redefinitions ($k \to x/2\pi$, $m \to \nu$, $\nu \to -m$), this becomes just the original relation (18.59). [The identity (18.59) is called the Poisson summation formula.]

A further property of the Bessel functions may be obtained by expanding $e^{i\mathbf{k}_\perp \cdot (\mathbf{r}-\mathbf{r}')_\perp}$ using (18.27), and integrating over α [as we did in (18.54)]:

$$\int_0^{2\pi} \frac{d\alpha}{2\pi} e^{i\mathbf{k}_\perp \cdot (\mathbf{r}-\mathbf{r}')_\perp} = \sum_{m=-\infty}^{\infty} e^{im\phi} J_m(k\rho) e^{-im\phi'} J_m(k\rho'). \tag{18.62}$$

On the other hand, we could also have specified α as the angle between the two vectors $\mathbf{k}_\perp$ and $(\mathbf{r}-\mathbf{r}')_\perp$. That would identify the left-hand side of (18.62) as the Bessel function of zeroth order, (18.4),

$$\int_0^{2\pi} \frac{d\alpha}{2\pi} e^{ik|(\mathbf{r}-\mathbf{r}')_\perp|\cos\alpha} = J_0(k|(\mathbf{r}-\mathbf{r}')_\perp|), \tag{18.63}$$

where

$$|(\mathbf{r}-\mathbf{r}')_\perp| = \sqrt{\rho^2 + \rho'^2 - 2\rho\rho' \cos(\phi - \phi')} = P \tag{18.64}$$

(read P as capital rho). Therefore, we have derived the *addition theorem* for the Bessel functions of integer order:

$$J_0(kP) = \sum_{m=-\infty}^{\infty} e^{im\phi} J_m(k\rho) e^{-im\phi'} J_m(k\rho'). \tag{18.65}$$

Now, if we return to the completeness relation (18.55), and recall the integral (18.48), the implication of (18.65) is that

$$\frac{1}{\rho}\delta(\rho - \rho')\delta(\phi - \phi') = \frac{1}{P}\delta(P)\frac{1}{2\pi}, \tag{18.66}$$

which is a quantitative version of the remark that coincident points have zero spatial separation.

18.2 Problems for Chapter 18

1. Fill in the steps leading to (18.15).
2. Using the divergence theorem, show that the divergence in orthogonal curvilinear coordinates is

$$\boldsymbol{\nabla} \cdot \mathbf{V} = \frac{1}{h} \sum_k \frac{\partial}{\partial u_k} \left(\frac{h}{h_k} V_k \right). \tag{18.67}$$

   Use Stokes' theorem (1.78) to show that the curl is

$$(\boldsymbol{\nabla} \times \mathbf{V})_i = \frac{h_i}{h} \left(\frac{\partial}{\partial u_j}(h_k V_k) - \frac{\partial}{\partial u_k}(h_j V_j) \right), \tag{18.68}$$

   where i, j, and k are in cyclic order.
3. Derive the differential equation (18.32b) satisfied by J_m, directly from the generating function (18.21).

4. Use the integral representation of J_m, (18.29), to prove the recurrence relations

$$2\frac{d}{dt}J_m(t) = J_{m-1}(t) - J_{m+1}(t), \tag{18.69a}$$
$$2\frac{m}{t}J_m(t) = J_{m-1}(t) + J_{m+1}(t). \tag{18.69b}$$

5. Using the results of the previous problem, show that

$$\left(-\frac{d}{dt}+\frac{m-1}{t}\right)\left(\frac{d}{dt}+\frac{m}{t}\right)J_m(t) = \left(\frac{d}{dt}+\frac{m+1}{t}\right)\left(-\frac{d}{dt}+\frac{m}{t}\right)J_m(t) = J_m(t) \tag{18.70}$$

and from this derive the differential equation satisfied by J_m.

6. Using the integral representation of J_m, (18.29), for $m > 0$, develop the power series for J_m, (18.26).

7. Bessel's equation of order ν,

$$u'' + \frac{1}{x}u' + \left(1 - \frac{\nu^2}{x^2}\right)u = 0, \tag{18.71}$$

possesses two solutions. One of them is Bessel's function of the first kind, $J_\nu(x)$, which is regular at $x = 0$. Show that a series solution of Bessel's equation yields, up to a multiplicative constant

$$J_\nu(x) = \left(\frac{x}{2}\right)^\nu \sum_{m=0}^{\infty} \frac{(-1)^m}{m!\,\Gamma(m+\nu+1)}\left(\frac{x}{2}\right)^{2m}, \tag{18.72}$$

where $\Gamma(z)$ is the gamma function, which generalizes the factorial,

$$\Gamma(z+1) = z\Gamma(z), \quad \Gamma(n+1) = n!. \tag{18.73}$$

Note that the series expansion of $J_\nu(x)$ agrees with (18.26) if $\nu = n$, an integer. Show that this series converges uniformly and absolutely for all finite x, but possesses a branch line for $\nu \neq$ integer.

8. If ν is not an integer, $J_\nu(x)$ and $J_{-\nu}(x)$ are independent solutions to Bessel's equation. However, from (18.24),

$$J_n(x) = (-1)^n J_{-n}(x). \tag{18.74}$$

Prove this directly from the series representation. (Note that $1/\Gamma(-k) = 0$ for $k = 0, 1, 2, \ldots$.) In this case we need a second solution, called Bessel's function of the second kind (or the Neumann function, named for Carl Gottfried Neumann, 1832–1925)

$$N_\nu(x) = \frac{\cos\nu\pi\, J_\nu(x) - J_{-\nu}(x)}{\sin\nu\pi}. \tag{18.75}$$

Show that

$$N_n = \lim_{\nu\to n} N_\nu(x) = \frac{1}{\pi}\left\{\left.\frac{\partial J_\nu(x)}{\partial\nu}\right|_{\nu=n} - (-1)^n \left.\frac{\partial J_{-\nu}(x)}{\partial\nu}\right|_{\nu=n}\right\} \tag{18.76}$$

is a linearly independent solution.

9. Show that the recursion relations in Problem 18.4 hold for non-integral order. [Use the series representation.]

10. By substitution, show that the asymptotic solution to Bessel's equation (18.71) has the form ($x \to \infty$)

$$u(x) = w_1(x)x^{-1/2}\cos\left(x - \frac{1}{2}\nu\pi - \frac{\pi}{4}\right) + w_2(x)x^{-1/2}\sin\left(x - \frac{1}{2}\nu\pi - \frac{\pi}{4}\right), \tag{18.77}$$

where

$$w_1(x) \sim a_0 \sum_{n=0}^{\infty}(-1)^n c_{2n}x^{-2n} + b_0 \sum_{n=0}^{\infty}(-1)^n c_{2n+1}x^{-2n-1} \tag{18.78a}$$

$$w_2(x) \sim b_0 \sum_{n=0}^{\infty}(-1)^n c_{2n}x^{-2n} - a_0 \sum_{n=0}^{\infty}(-1)^n c_{2n+1}x^{-2n-1}, \tag{18.78b}$$

with a_0, b_0 arbitrary. Compute c_n,

$$c_0 = 1, \quad c_n = \frac{(4\nu^2 - 1)(4\nu^2 - 3^2)\cdots(4\nu^2 - (2n-1)^2)}{n!\,8^n}, \quad n = 1, 2, \ldots. \tag{18.79}$$

Note that the series terminates when $\nu = n + 1/2$ and then these asymptotic relations become exact. $J_\nu(x)$ is represented by the above with $a_0 = (2/\pi)^{1/2}$, $b_0 = 0$, and N_ν by the above with $a_0 = 0$, $b_0 = (2/\pi)^{1/2}$.

11. Coulomb's potential in circular cylindrical coordinates is given by the two alternative representations in (18.6). Insert the integral representation for J_0, (18.4), and integrate over k. Express in simplest terms the integral you have thereby evaluated.

19

Parallel Conducting Plates

19.1 Reduced Green's Function

Having developed some mathematical machinery, let us now turn to another essentially one-dimensional problem, that of the potential due to a point charge between two parallel grounded conducting plates, as illustrated in Fig. 19.1. Green's function—the potential of a unit point charge—is defined by the differential equation

$$-\nabla^2 G(\mathbf{r},\mathbf{r}') = 4\pi\delta(\mathbf{r}-\mathbf{r}'), \tag{19.1}$$

together with the boundary conditions

$$G = 0 \quad \text{at} \quad z = 0,\ a. \tag{19.2}$$

Since the geometry depends only on the z coordinate, Green's function can be written in the $(2+1)$-dimensional form (15.21),

$$G(\mathbf{r},\mathbf{r}') = 4\pi \int \frac{(d\mathbf{k}_\perp)}{(2\pi)^2} e^{i\mathbf{k}_\perp \cdot (\mathbf{r}-\mathbf{r}')_\perp} g(z,z';k_\perp), \tag{19.3}$$

where the reduced Green's function $g(z,z';k_\perp)$ satisfies (15.24),

$$\left(-\frac{\partial^2}{\partial z^2} + k_\perp^2\right) g(z,z';k_\perp) = \delta(z-z'), \tag{19.4}$$

subject to the boundary conditions

$$g(z,z';k_\perp) = 0 \quad \text{at} \quad z = 0, a. \tag{19.5}$$

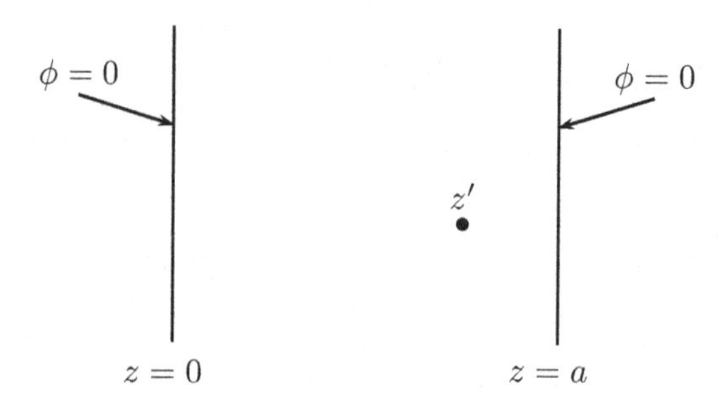

FIGURE 19.1
Geometry of grounded, parallel, conducting plates.

DOI: 10.1201/9781003057369-19

Our first response is in the spirit of the straightforward development used in Chapters 15 and 16, based on solving the homogeneous equation appropriate to $z \neq z'$, (15.25), with the connections provided by the continuity of g and the discontinuity condition of (15.26). As before, the basic solutions of the homogeneous equation are still of the form $e^{\pm kz}$ (with $k = |\mathbf{k}_\perp|$); the linear combinations that satisfy the boundary conditions (19.5) are expressed in terms of the hyperbolic function $\sinh x$:

$$0 < z < z' : \quad g = A \sinh kz, \tag{19.6a}$$

$$a > z > z' : \quad g = B \sinh k(a - z). \tag{19.6b}$$

The constants A and B are to be determined by the conditions on g in the neighborhood of $z = z'$ [recall (16.11a) and (16.11b)],

$$g \text{ continuous}, \qquad -\frac{\partial}{\partial z} g \Big|_{z=z'-0}^{z=z'+0} = 1, \tag{19.7}$$

which leads to the equations

$$A \sinh kz' = B \sinh k(a - z'), \tag{19.8a}$$

$$kB \cosh k(a - z') + kA \cosh kz' = 1. \tag{19.8b}$$

It is convenient to satisfy (19.8a) by letting

$$A = C \sinh k(a - z'), \quad B = C \sinh kz', \tag{19.9}$$

which, when substituted into (19.8b) yields

$$kC \sinh ka = 1. \tag{19.10}$$

The reduced Green's function is thus found to be

$$g(z, z'; k) = \frac{\sinh kz_< \sinh k(a - z_>)}{k \sinh ka}, \tag{19.11}$$

where $z_>$ ($z_<$) is the greater (lesser) of z and z'. Note that the reciprocity condition, (14.7), is satisfied because $g(z, z'; k)$ is symmetrical in z and z'.

It is worth noting what happens as the two plates are withdrawn from the neighborhood of the points specified by z and z'. This means that both $kz_<$ and $k(a - z_>)$, and, of course, ka, are considered to become very large. Then every $\sinh x$ appearing in (19.11) is dominated by $\frac{1}{2}e^x$, and

$$g(z, z'; k) \sim \frac{1}{2k} e^{-k(z_> - z_<)} = \frac{1}{2k} e^{-k|z - z'|}, \tag{19.12}$$

the vacuum form displayed in (15.22). Alternatively, we can withdraw just the right-hand wall, for example, as accomplished by having ka become very large without reference to z and z'. The outcome

$$g(z, z'; k) \sim \frac{1}{k} \sinh kz_< e^{-kz_>} = \frac{1}{2k} e^{-k|z - z'|} - \frac{1}{2k} e^{-k(z+z')}, \tag{19.13}$$

is just what is contained in (17.34), appropriate to a single conducting plate at $z = 0$.

With a knowledge of Green's function, we can now answer questions about the charge induced on the conducting plates, and about the interaction energy and force between the plates and the point charge.

19.2 Induced Charge

One application of this result lies in calculating the charge densities on the conducting plates induced by a unit point charge at z'. According to (13.62), these charge densities are

$$z = 0: \quad 4\pi\sigma = E_z = -\frac{\partial}{\partial z}G, \tag{19.14a}$$

$$z = a: \quad 4\pi\sigma = -E_z = \frac{\partial}{\partial z}G. \tag{19.14b}$$

As in Chapter 17, it is simplest to calculate the total charge induced on each plate:

$$Q_{\text{ind}}(z=0) = -\frac{1}{4\pi}4\pi \int (d\mathbf{r}_\perp) \int \frac{(d\mathbf{k}_\perp)}{(2\pi)^2} e^{i\mathbf{k}_\perp \cdot (\mathbf{r}-\mathbf{r}')_\perp} \frac{\partial}{\partial z} g(z,z';k)\bigg|_{z=0}, \tag{19.15a}$$

$$Q_{\text{ind}}(z=a) = \frac{1}{4\pi}4\pi \int (d\mathbf{r}_\perp) \int \frac{(d\mathbf{k}_\perp)}{(2\pi)^2} e^{i\mathbf{k}_\perp \cdot (\mathbf{r}-\mathbf{r}')_\perp} \frac{\partial}{\partial z} g(z,z';k)\bigg|_{z=a}. \tag{19.15b}$$

We have seen such integrals previously [see (17.39) and (17.38)]. The spatial integrations over $\mathbf{r}_\perp$ yields $(2\pi)^2\delta(\mathbf{k}_\perp)$, while the subsequent $\mathbf{k}_\perp$ integration sets $k = 0$ so that

$$g(z,z';0) = \frac{1}{a} z_<(a - z_>). \tag{19.16}$$

The total induced charges are therefore

$$Q_{\text{ind}}(z=0) = -\left(1 - \frac{z'}{a}\right), \tag{19.17a}$$

$$Q_{\text{ind}}(z=a) = -\frac{z'}{a}, \tag{19.17b}$$

with the total induced charge on both plates being -1, of course. The total induced charge is divided between the two plates in inverse proportion to their distances from the point charge:

$$z' Q_{\text{ind}}(z=0) = (a - z') Q_{\text{ind}}(z=a). \tag{19.18}$$

19.3 Energy

The interaction energy between the conducting plates and a point charge e, at the location $\mathbf{r}_0$, is given by [*cf.* (17.5)]

$$\begin{aligned} E_{\text{int}} &= \frac{1}{2}e^2[G(\mathbf{r},\mathbf{r}') - G_0(\mathbf{r},\mathbf{r}')]_{\mathbf{r},\mathbf{r}'\to\mathbf{r}_0} \\ &= e^2 2\pi \int \frac{(d\mathbf{k}_\perp)}{(2\pi)^2}\left[\frac{\sinh kz_0 \sinh k(a-z_0)}{k \sinh ka} - \frac{1}{2k}\right], \end{aligned} \tag{19.19a}$$

or

$$E_{\text{int}} = \frac{1}{2}e^2 \int_0^\infty dk \left[\frac{\cosh ka}{\sinh ka} - 1 - \frac{\cosh k(a - 2z_0)}{\sinh ka}\right]. \tag{19.19b}$$

Then, if we introduce the quantities

$$x = e^{-ka}, \quad \alpha = 1 - \frac{2z_0}{a}, \tag{19.20}$$

where α ranges between $+1$ and -1, this reads

$$E_{\text{int}} = -\frac{1}{2}\frac{e^2}{a}\int_0^1 dx \frac{x^\alpha + x^{-\alpha} - 2x}{1 - x^2}. \tag{19.21}$$

One situation is particularly simple: $\alpha = 0$ or $z_0 = \frac{1}{2}a$; the point charge is equidistant between the two plates. Then, we have

$$z_0 = \frac{1}{2}a: \quad E_{\text{int}} = -\frac{e^2}{a}\int_0^1 dx \frac{1}{1+x} = -\frac{e^2}{a}\ln 2. \tag{19.22}$$

To proceed with the general situation, we introduce the power series expansion for the denominator in (19.21), and get

$$\begin{aligned} -\frac{E_{\text{int}}}{(e^2/a)} &= \frac{1}{2}\int_0^1 dx (x^\alpha + x^{-\alpha} - 2x)\sum_{n=0}^{\infty} x^{2n} \\ &= \frac{1}{4}\sum_{n=0}^{\infty}\left[\frac{1}{n + \frac{1+\alpha}{2}} + \frac{1}{n + \frac{1-\alpha}{2}} - \frac{2}{n+1}\right]. \end{aligned} \tag{19.23}$$

But before presenting this summation in terms of a known transcendental function, we use it as it stands to isolate the leading terms for the two situations, $1 - \alpha \ll 1$ ($z_0 \ll a$) and $1 + \alpha \ll 1$ ($a - z_0 \ll a$), which position the point charge close to one of the conducting plates. To focus on the circumstance $z_0/a \ll 1$, we first rewrite the summation (without the factor $1/4$) as

$$\begin{aligned} &\sum_{n=0}^{\infty}\left[\frac{1}{n + (z_0/a)} + \frac{1}{n + 1 - (z_0/a)} - \frac{2}{n+1}\right] \\ &= \frac{1}{(z_0/a)} + \sum_{n=1}^{\infty}\left[\frac{1}{n + (z_0/a)} + \frac{1}{n - (z_0/a)} - \frac{2}{n}\right], \end{aligned} \tag{19.24}$$

which makes clear that the two leading terms are

$$z_0/a \ll 1: \quad \frac{1}{(z_0/a)} + 2\left(\frac{z_0}{a}\right)^2 \sum_{n=1}^{\infty}\frac{1}{n^3}. \tag{19.25}$$

The summation encountered here is an example of the zeta function introduced by Georg Friedrich Bernhard Riemann (1826–1866),

$$\zeta(s) = \sum_{n=1}^{\infty}\frac{1}{n^s}; \tag{19.26}$$

the relevant numerical value is

$$\zeta(3) = 1.202\ldots. \tag{19.27}$$

Thus, we arrive at the interaction energy

$$z_0 \ll a: \quad E_{\text{int}} = -\frac{1}{2}\frac{e^2}{2z_0} - \frac{(ez_0)^2}{2a^3}\zeta(3), \tag{19.28}$$

where the first term is recognized as that associated with the nearby conductor [(17.5) with $\epsilon = \infty$], thereby identifying the next term as the leading contribution of the second, distant conductor (more about this later). The analogous result for the other situation, $a - z_0 \ll a$, requires only the substitution $z_0 \to a - z_0$.

The gamma function, first studied by Euler, is conveniently defined in the manner of Karl Wilhelm Theodor Weierstrass (1815–1897), by stating its reciprocal as an infinite product:

$$\frac{1}{\Gamma(t)} = te^{\gamma t} \prod_{n=1}^{\infty} \left[\left(1 + \frac{t}{n}\right) e^{-t/n} \right]; \tag{19.29}$$

here, γ is Euler's constant,

$$\gamma = \lim_{N \to \infty} \left[\sum_{n=1}^{N} \frac{1}{n} - \ln N \right] = 0.5772\ldots. \tag{19.30}$$

Of more immediate interest to us, however, is the related logarithmic derivative function, the digamma function,

$$\psi(t) = \frac{d}{dt} \ln \Gamma(t) = -\gamma - \frac{1}{t} - \sum_{n=1}^{\infty} \left(\frac{1}{n+t} - \frac{1}{n} \right). \tag{19.31}$$

Indeed, on introducing the relation

$$\sum_{n=1}^{\infty} \left(\frac{1}{n} - \frac{1}{n+1} \right) = 1, \tag{19.32}$$

we can present (19.31) as

$$\psi(t) = -\gamma - \sum_{n=0}^{\infty} \left(\frac{1}{n+t} - \frac{1}{n+1} \right), \tag{19.33}$$

which allows us to evaluate (19.23) generally as

$$E_{\text{int}} = \frac{e^2}{4a} \left[\psi\left(\frac{1+\alpha}{2}\right) + \psi\left(\frac{1-\alpha}{2}\right) + 2\gamma \right]. \tag{19.34}$$

The comparison of this form with the particular example of (19.22), corresponding to $\alpha = 0$, supplies the value

$$\psi\left(\frac{1}{2}\right) = -\gamma - 2\ln 2; \tag{19.35}$$

this can be checked directly through the following evaluation of the summation:

$$\lim_{x \to 1} \sum_{n=0}^{\infty} x^{2n} \left(\frac{1}{n+1/2} - \frac{1}{n+1} \right) = \lim_{x \to 1} \left(\frac{1}{x} \ln \frac{1+x}{1-x} - \frac{1}{x^2} \ln \frac{1}{1-x^2} \right) = 2\ln 2, \tag{19.36}$$

which uses the series expansion of the logarithm,

$$\ln(1+y) = -\sum_{n=1}^{\infty} (-1)^n \frac{y^n}{n}. \tag{19.37}$$

Figure 19.2 shows how the interaction energy (19.34) varies with the position of the charge between the plates.

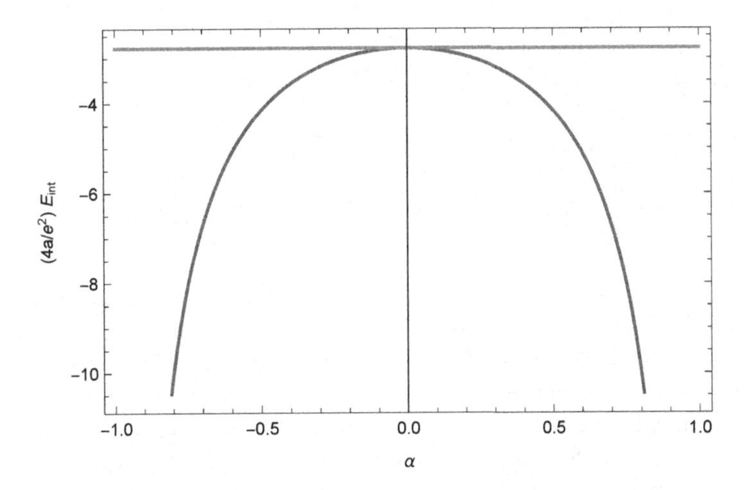

FIGURE 19.2
The energy of the charge shown in Fig. 19.1 as a function of the distance from the left-hand plate, $z' = z_0$, in terms of $\alpha = 1 - 2z_0/a$. The energy diverges as the charge approaches either plate, and when equidistant has the value $-(e^2/a)\ln 2$, (19.22), as shown by the horizontal line.

19.4 Force

The force exerted on the point charge is given by the negative derivative of the interaction energy with respect to the coordinate z_0. Expressed in terms of the derivative of the digamma function,

$$\psi'(t) = \frac{d}{dt}\psi(t) = \sum_{n=0}^{\infty} \frac{1}{(n+t)^2}, \tag{19.38}$$

that force is

$$\begin{aligned} F &= \frac{e^2}{4a^2}\left[\psi'\left(\frac{1+\alpha}{2}\right) - \psi'\left(\frac{1-\alpha}{2}\right)\right] \\ &= \frac{e^2}{4a^2}\left[\psi'\left(1-\frac{z_0}{a}\right) - \psi'\left(\frac{z_0}{a}\right)\right]. \end{aligned} \tag{19.39}$$

As one would expect, this force vanishes when the charge is equidistant between the plates, $z_0 = \frac{1}{2}a$. One also expects that a displacement from the equilibrium point should produce a larger force of attraction toward the nearer conducting plate, resulting in a net force that acts to increase that displacement—the equilibrium at the midpoint is unstable. That is made explicit on expanding (19.39) for small values of α,

$$|\alpha| \ll 1: \quad F \approx \frac{e^2}{4a^2}\alpha\psi''(1/2) = \frac{e^2}{a^3}\left[-\frac{1}{2}\psi''(1/2)\right](z_0 - a/2), \tag{19.40}$$

where (see Problem 19.2)

$$-\frac{1}{2}\psi''(1/2) = \sum_{n=0}^{\infty} \frac{1}{(n+1/2)^3} = 7\zeta(3), \tag{19.41}$$

although the qualitatively important point here is that the latter is a positive number; the force is proportional to a small displacement from the equilibrium point, and acts in the same sense.

The explicit expression for the force,

$$F = -\sum_{n=0}^{\infty} \frac{e^2}{(2z_0 + 2na)^2} + \sum_{n=0}^{\infty} \frac{e^2}{(2(a - z_0) + 2na)^2}, \tag{19.42}$$

clearly suggests an interpretation in terms of an equivalent distribution of point charges—image charges. It is not possible, however, just from a knowledge of the force to assign uniquely the values and locations of these charges. For example, a positive contribution to the force on charge e could equally well arise from an image charge e located to the left, or from an image charge $-e$ positioned to its right. And, one would be unaware of the possible compensating force contributions of identical charges arrayed equidistantly on opposite sides of charge e. Accordingly, having thus been led to the question of an image charge description, we make a direct attack on this problem.

19.5 Images

Let us look back at Green's function in the physical region to the right of a single conducting plate, as it is expressed by (17.34),

$$z > 0: \quad g(z, z'; k) = \frac{1}{2k} \left[e^{-k|z-z'|} - e^{-k(z+z')} \right]. \tag{19.43}$$

This function vanishes at $z = 0$, satisfying the boundary condition for a conductor at zero potential, and it obeys

$$z > 0: \quad \left(-\frac{\partial^2}{\partial z^2} + k^2 \right) g(z, z'; k) = \delta(z - z'), \tag{19.44}$$

which makes explicit the unit charge at the point $z' > 0$. Now we get a *bounded* extrapolation of (19.43) into a fictitious vacuum region $z < 0$ by writing g as

$$g(z, z'; k) = \frac{1}{2k} \left[e^{-k|z-z'|} - e^{-k|z+z'|} \right], \tag{19.45}$$

which function obeys the differential equation

$$\left(-\frac{\partial^2}{\partial z^2} + k^2 \right) g(z, z'; k) = \delta(z - z') - \delta(z + z'); \tag{19.46}$$

the image charge appears here on the same footing as the physical charge.

The effect of the coordinate reflection $z \to -z$ is to reverse the sign of the right-hand side of (19.46). Then the function g, being determined by the right-hand side of its differential equation, will also reverse sign. Accordingly, the continuous function g necessarily vanishes at $z = 0$. Here is the recognition that symmetry considerations in extended regions can occasionally be applied to satisfy boundary conditions. With this lesson in mind we turn back to the pair of conducting plates.

In this circumstance the function g must vanish at $z = a$, as well as at $z = 0$. Let us begin at the latter point and follow g to its zero at $z = a$. Now we extrapolate beyond $z = a$,

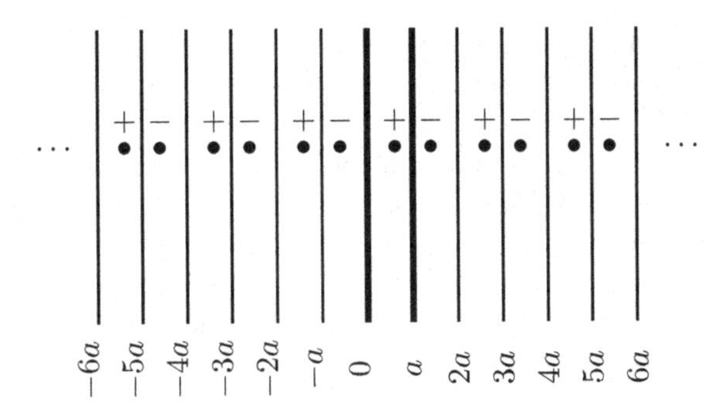

FIGURE 19.3
Positions of image charges for two parallel conductors.

where the continuous function g will start out with reversed sign. Suppose that the course of the function from $z = a$ to $z = 2a$ duplicates that from $z = a$ to $z = 0$, but with the opposite sign. Then the function g will vanish at $z = 2a$. If we continue in the same way, with another sign reversal at this zero, the behavior of g from $z = 2a$ to $z = 3a$ duplicates exactly that from $z = 0$ to $z = a$. In short, we are led to extrapolate the function over the entire range of z, $-\infty$ to ∞, by imposing the requirement of periodicity, with period $2a$. That periodicity of g must be matched by a like periodicity of the point charges so that the initial unit positive charge at z' is supplemented by unit positive charges at $z' \pm 2a$, $z' \pm 4a$, ... Doing the same thing with the unit negative image charge at $-z'$ then replaces (19.44) with

$$\left(-\frac{\partial^2}{\partial z^2} + k^2\right) g(z, z'; k) = \sum_{\nu=-\infty}^{\infty} [\delta(z - z' - 2\nu a) - \delta(z + z' - 2\nu a)]. \tag{19.47}$$

Fine, but does it work? Will this charge structure reverse sign under reflection, both at $z = 0$: $z \to -z$, and at $z = a$: $z - a \to -(z - a)$? Without question for $z = 0$: The replacement $\nu \to -\nu$ leaves the charge structure intact. And then we recognize that $z \to -z + 2a$ differs only in the additional displacement of $2a$, which is just the periodicity interval of the charges. The solution of (19.47) will indeed obey the boundary conditions for the pair of conducting plates. See Fig. 19.3.

But before we construct that solution, let us look back at the force and energy in light of the now known pattern of image charges. For the circumstance of charge e stationed at z_0, the image charges comprise charges e at the positions $z_0 + 2\nu a$, $\nu = \pm 1, \pm 2, \ldots$, and charges $-e$ located at $-z_0 + 2\nu a$, $\nu = 0, \pm 1, \pm 2, \ldots$. The image charges of the same sign as the physical charge are disposed symmetrically about the latter, resulting in no net force. The distances between the physical charge and the image charges of opposite sign that stand on its right, and on its left, are $2(a - z_0)$, $2(a - z_0) + 2a$, ..., and $2z_0$, $2z_0 + 2a$, ..., respectively. Inasmuch as forces of attraction from the right and left are counted as positive and negative, respectively, the total force is precisely as given in (19.42). Turning to the interaction energy given in (19.23), we first rewrite it as

$$E_{\text{int}} = \frac{1}{2} \sum_{n=1}^{\infty} \left[2\frac{e^2}{2na} - \frac{e^2}{2na - 2z_0} - \frac{e^2}{2(n-1)a + 2z_0} \right]. \tag{19.48}$$

Then we can recognize in the successive terms of the summation the electrostatic energy of interaction between the physical charge and the image charges; first, those of like sign, and then the oppositely signed charges, positioned to the right and to the left, respectively. Very good, but why the additional factor of 1/2? [This question might well have been raised at our first encounter with the interaction energy that was interpreted in terms of an image charge, (17.5).]

The electrostatic energy of the physical point charge can be identified with the work performed in assembling successive infinitesimal multiples, $\delta\lambda e$, of the final charge, with λ increasing from zero to unity. At a stage where the assembled charge is λe, the image charges that it induces are also reduced by the fraction λ of their final values, and so also is the electrostatic potential of the image charges. If ϕ is the final value of that potential, the total work performed in raising λ from 0 to 1 is

$$\int_0^1 d\lambda e\,\lambda\phi = \frac{1}{2}e\phi; \tag{19.49}$$

just one half of the electrostatic energy associated with the final charges, regarded as independently assignable.

19.6 Linear Lattices

The latter remark directs us to the solution of a different physical problem. Consider a one-dimensional lattice of charges $-e$, with uniform separation $2a$, and a similar lattice composed of charges $+e$, which is displaced to the right (along the z-axis), relative to the first lattice, by the distance $0 < 2z_0 < 2a$. In particular, for $z_0 = \frac{1}{2}a$, the complete one-dimensional lattice is composed of alternatingly signed charges with uniform spacing a. What is the electrostatic energy of any individual charge? The charge pattern here is just that of a physical charge e and its images in the parallel conducting plates, as shown in Fig. 19.3. Accordingly, the electrostatic energy per unit charge is twice that given in (19.34); in the example with all charge spacings equal to a, the energy per charge, twice that presented in (19.22), is $-(e^2/a)2\ln 2$.

19.7 Periodic Green's Function

We shall now solve the differential equation (19.47) with the aid of the relation (18.59). First we rewrite the delta function by supplying the variables with a constant multiplicative factor in accordance with the simple relation

$$\kappa\delta(\kappa x) = \delta(x), \quad \kappa > 0, \tag{19.50}$$

which has already been used without comment in connection with (18.61). Thus we have

$$\begin{aligned}
&\sum_{\nu=-\infty}^{\infty} [\delta(z - z' - 2\nu a) - \delta(z + z' - 2\nu a)] \\
&= \frac{\pi}{a}\sum_{\nu=-\infty}^{\infty} \left[\delta\left(\frac{\pi}{a}(z - z') - 2\pi\nu\right) - \delta\left(\frac{\pi}{a}(z + z') - 2\pi\nu\right)\right] \\
&= \frac{1}{2a}\sum_{m=-\infty}^{\infty} \left[e^{im\pi(z-z')/a} - e^{im\pi(z+z')/a}\right],
\end{aligned} \tag{19.51}$$

which, in turn, can be rearranged as

$$\frac{1}{a}\sum_{n=1}^{\infty}\left[\cos\frac{n\pi}{a}(z-z')-\cos\frac{n\pi}{a}(z+z')\right]=\frac{2}{a}\sum_{n=1}^{\infty}\sin\frac{n\pi}{a}z\sin\frac{n\pi}{a}z'. \tag{19.52}$$

Should we now return to the physical range of the coordinates z and z' so that

$$-a<z-z'<a,\quad 0<z+z'<2a, \tag{19.53}$$

only the physical charge contributes to (19.51) and this relation becomes

$$\frac{2}{a}\sum_{n=1}^{\infty}\sin\frac{n\pi}{a}z\sin\frac{n\pi}{a}z'=\delta(z-z'). \tag{19.54a}$$

This is a statement of completeness of the functions $\sin n\pi z/a$, $n=1,2,\ldots$, over the interval between 0 and a. That is, for any function $f(z)$ defined over this interval,

$$f(z)=\sum_{n=1}^{\infty}\sin\frac{n\pi z}{a}\left[\frac{2}{a}\int_0^a dz'\sin\frac{n\pi z'}{a}f(z')\right], \tag{19.54b}$$

which is a (sine) Fourier series.

With the right-hand side of (19.47) replaced by (19.52), the solution of the differential equation is evident:

$$g(z,z';k)=\frac{2}{a}\sum_{n=1}^{\infty}\frac{\sin\frac{n\pi}{a}z\sin\frac{n\pi}{a}z'}{k^2+\left(\frac{n\pi}{a}\right)^2}. \tag{19.55}$$

And this function obviously obeys the boundary conditions, inasmuch as all $\sin(n\pi/a)z$ vanish at $z=0$ and $z=a$. Or is it really that cut and dried? Let's construct the Fourier series for the function $1-(z/a)$:

$$1-\frac{z}{a}=\sum_{n=1}^{\infty}\sin\frac{n\pi}{a}z\left[\frac{2}{a}\int_0^a dz'\sin\frac{n\pi}{a}z'\left(1-\frac{z'}{a}\right)\right], \tag{19.56}$$

where the latter integral is evaluated as

$$\int_0^a d\left[-\frac{a}{n\pi}\cos\frac{n\pi}{a}z'\left(1-\frac{z'}{a}\right)-\frac{a}{(n\pi)^2}\sin\frac{n\pi}{a}z'\right]=\frac{a}{n\pi}. \tag{19.57}$$

The result is

$$1-\frac{z}{a}=\sum_{n=1}^{\infty}\frac{2}{n\pi}\sin\frac{n\pi}{a}z. \tag{19.58}$$

Certainly both sides of this equation vanish for $z=a$. But suppose we put $z=0$?

In order to refute the possible inference that the sine Fourier series applies only to functions that vanish at $z=0$ and a, let us evaluate explicitly the right-hand side of (19.58):

$$\sum_{n=1}^{\infty}\frac{2}{n\pi}\sin\frac{n\pi}{a}z=\operatorname{Im}\frac{2}{\pi}\sum_{n=1}^{\infty}\frac{1}{n}\left(e^{i\pi z/a}\right)^n=\operatorname{Im}\frac{2}{\pi}\ln\frac{1}{1-e^{i\pi z/a}}, \tag{19.59a}$$

which, in turn, is $[0<z<a]$

$$\operatorname{Im}\frac{2}{\pi}\ln\left[\frac{e^{-i(\pi z/2a)}}{2\sin(\pi z/2a)}e^{\pi i/2}\right]=1-\frac{z}{a}. \tag{19.59b}$$

The point here is that, while individual terms of a sine Fourier series vanish as $z \to 0$, the series as a whole can converge to a nonzero value in this limit. Thus, suppose we were to assume that, with sufficiently small z, $\sin(n\pi/a)z \approx (n\pi/a)z$, for all values of n that contribute significantly to the summation in (19.58). Then the latter would be approximated by

$$\sum_{n=1}^{\infty} \frac{2}{n\pi}\frac{n\pi}{a}z = \frac{2z}{a}\sum_{n=1}^{\infty} 1, \tag{19.60}$$

which does *not* exist. We are being told that unlimitedly large values of n do contribute, contradicting the assumption involved in replacing $\sin(n\pi/a)z$ by $(n\pi/a)z$. But if the dominant contribution comes from very large values of n, the summation can effectively be replaced by an integral,

$$\sum_{n=1}^{\infty} \frac{2}{n\pi}\sin\frac{n\pi}{a}z \approx \frac{2}{\pi}\int_0^{\infty}\frac{dn}{n}\sin\frac{n\pi}{a}z = \frac{2}{\pi}\int_0^{\infty}\frac{dt}{t}\sin t, \tag{19.61}$$

yielding a finite limit as $z \to 0$. As to its specific value [already quoted in (15.17)—see also Problem 15.4], note that, through the option of redefining the integration variable, we have, for arbitrary x,

$$\frac{2}{\pi}\int_0^{\infty}\frac{dt}{t}\sin(xt) = \frac{2}{\pi}\epsilon(x)\int_0^{\infty}\frac{dt}{t}\sin t, \tag{19.62}$$

where

$$\epsilon(x) = \begin{cases} 1, & x > 0 \\ -1, & x < 0 \end{cases}. \tag{19.63}$$

Now we differentiate with respect to x and observe that the derivative of the step function $\epsilon(x)$ is $2\delta(x)$, reflecting the value of the discontinuity at $x = 0$:

$$\frac{2}{\pi}\int_0^{\infty} dt\cos(xt) = 2\delta(x)\frac{2}{\pi}\int_0^{\infty}\frac{dt}{t}\sin t, \tag{19.64}$$

where the left-hand side is recognized to be

$$\frac{1}{\pi}\int_{-\infty}^{\infty} dt\, e^{ixt} = 2\delta(x); \tag{19.65}$$

as expected, we learn that

$$\frac{2}{\pi}\int_0^{\infty}\frac{dt}{t}\sin t = 1. \tag{19.66}$$

It is now clear that $g(z, z'; k)$ will indeed vanish at $z = 0$, say, if the infinite series that results from the replacement of $\sin(n\pi/a)z$ by $(n\pi/a)z$ is a convergent one. This being a question referring to very large values of n, it suffices to make the test for $k = 0$. Then, with $z/a \ll 1$, we get just the sum given in (19.58),

$$\frac{z}{a} \ll 1: \quad g(z, z'; 0) = z\sum_{n=1}^{\infty}\frac{2}{n\pi}\sin\frac{n\pi}{a}z' = \frac{z(a - z')}{a}. \tag{19.67}$$

Yes, $g(z, z'; k)$ does vanish at $z = 0$ and at $z = a$. Incidentally, if one looks at the $k = 0$ limit of (19.11),

$$g(z, z'; 0) = \frac{z_<(a - z_>)}{a}, \tag{19.68}$$

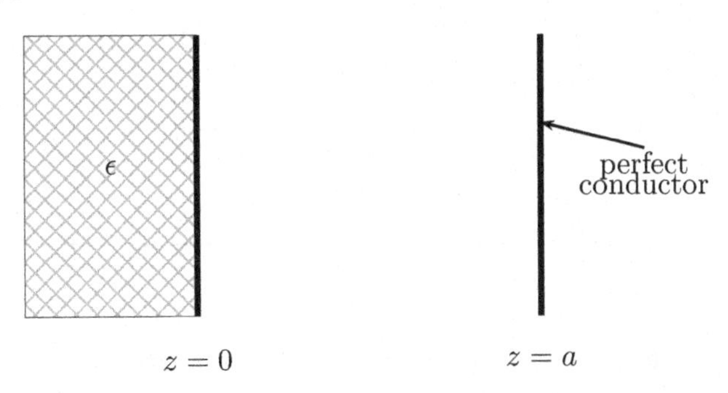

FIGURE 19.4
Geometry of parallel dielectric and conducting surfaces for Problem 19.1.

it is evident that (19.67) holds, not only for $z \ll a$, but for all $z < z'$. More generally, the equivalence of the two forms for $g(z, z'; k)$, (19.11) and (19.55),

$$\frac{\sinh kz_< \sinh k(a - z_>)}{k \sinh ka} = \frac{2}{a} \sum_{n=1}^{\infty} \frac{\sin(n\pi z/a) \sin(n\pi z'/a)}{(n\pi/a)^2 + k^2}, \tag{19.69}$$

can be checked by carrying out the integrations involved in representing the first version as a Fourier series in z, or in z'. (See Problem 19.4.)

19.8 Problems for Chapter 19

1. A semi-infinite dielectric slab, with dielectric constant ϵ, fills all space with $z < 0$. A perfect conducting plate occupies the plane $z = a > 0$. The geometry is shown in Fig. 19.4. Calculate the reduced Green's function, $g(z, z'; k_\perp)$, for $0 < z, z' < a$, and the full Green's function $G(\mathbf{r}, \mathbf{r}')$ in the same region.

2. Prove that
$$\sum_{n=0}^{\infty} \frac{1}{(n + 1/2)^s} = (2^s - 1)\zeta(s), \tag{19.70}$$
and hence establish (19.41).

3. We now wish to investigate the properties of $g(z, z'; k)$, (19.11), considered as a function of a complex variable k. First notice that g is even in k and is finite at $k = 0$. The behavior of g when the real part of k is large and positive is given by (19.12). This limiting form is the reduced Green's function for empty space, (15.22), which is evidently bounded (in fact, goes to zero) as $\text{Re}\, k \to \infty$. The singularities of $g(z, z'; k)$ occur on the imaginary axis where
$$\sinh ka = 0, \tag{19.71}$$
that is, at the points where
$$ka = in\pi, \quad n = \pm 1, \pm 2, \pm 3, \ldots. \tag{19.72}$$

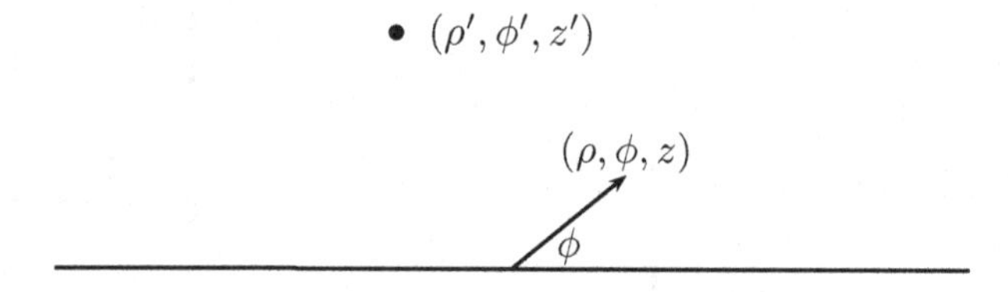

FIGURE 19.5
Infinite conducting sheet, described in cylindrical coordinates by $\phi = 0, \pi$. The coordinate z is perpendicular to the page. A unit point charge is located at the point (ρ', ϕ', z').

By examining the behavior of g in the neighborhood of these singularities, where g has simple poles, and noting that apart from these poles, g is a bounded, analytic function of k, express g entirely as a sum over these pole contributions,

$$g(z, z'; k) = \sum_{n\neq 0} \frac{1}{n\pi} \frac{\sin \frac{n\pi}{a} z \sin \frac{n\pi}{a} z'}{\frac{n\pi}{a} + ik}. \tag{19.73}$$

Finally, by combining the contributions for n and $-n$, obtain the result (19.55).

4. Alternatively, show the equivalence of the two forms of g in (19.69) by expanding the left-hand side in a sine Fourier series as suggested.
5. Consider an infinite conducting sheet, with a unit point charge above it. Use circular cylindrical coordinates so that the point charge is located at (ρ', ϕ', z'), as shown in Fig. 19.5. Find Green's function by using appropriate representations of $\delta(z - z')$ and $\delta(\phi - \phi')$. Get the differential equation for the radial functions $g(\rho, \rho')$. Now consider the potential of an infinite line charge of unit density,

$$\Gamma(\rho, \phi; \rho', \phi') = \int dz'\, G(\rho, \phi, z; \rho', \phi', z'), \tag{19.74}$$

and the associated radial functions $\gamma(\rho, \rho')$. Construct the latter from the differential equation and boundary conditions. Show that the resulting function can be expressed in terms of the complex quantities

$$\zeta = \rho e^{i\phi}, \quad \zeta' = \rho' e^{i\phi'} \tag{19.75}$$

as

$$\Gamma = 2 \operatorname{Re} \ln \frac{\zeta - \zeta'^*}{\zeta - \zeta'}. \tag{19.76}$$

6. Do the same for a semi-infinite conducting sheet, defined by $\phi = 0$. Under the circumstance where ϕ and ϕ' are small positive angles while ρ and ρ' are large, in the sense that $|\rho - \rho'| \ll \rho, \rho'$, the physical situation is effectively that of Problem 5 (the edge is far away). Show the correspondence.
7. Consider the same z-dependent permittivity considered in Problem 14.4, but now confined between two conducting plates, at $z = 0$ and $z = a$. Solve the differential equation for $\overline{G}$, written in the two-dimensional transverse Fourier representation, for the model (14.24). Find the total charge induced on each plate by a unit charge at $z' \in [0, a]$, and the sum of these charges. Show that, when the parameter γ that characterizes the z-dependence of ϵ is zero, the induced charges for parallel conducting plates in vacuum (19.17) are recovered.

20

Modified Bessel Functions

An eigenfunction expansion for Green's function, $G(\mathbf{r},\mathbf{r}')$, for the geometry shown in Fig. 19.1 can now be obtained by substituting the corresponding form for the reduced Green's function, (19.55), into (19.3). The Bessel function of zeroth order, J_0, (18.63), is introduced upon performing the angular integration associated with $\mathbf{k}_\perp$ so that Green's function becomes

$$G(\mathbf{r},\mathbf{r}') = 2\int_0^\infty dk\, k\, J_0(kP)\frac{2}{a}\sum_{n=1}^{\infty}\frac{\sin\frac{n\pi}{a}z\sin\frac{n\pi}{a}z'}{k^2+\left(\frac{n\pi}{a}\right)^2}, \tag{20.1}$$

where

$$P = |(\mathbf{r}-\mathbf{r}')_\perp| = [\rho^2+\rho'^2-2\rho\rho'\cos(\phi-\phi')]^{1/2}. \tag{20.2}$$

A complete *eigenfunction decomposition* of G could be obtained by using the addition theorem (18.65). In (20.1), we encounter a new type of Bessel function, K_0, the modified Bessel function of zeroth order, defined by the integral representation

$$K_0(\lambda P) = \int_0^\infty dk\, k\frac{J_0(kP)}{k^2+\lambda^2}. \tag{20.3a}$$

It suffices to make the change of variable $kP = s$ to recognize that the integral depends only on the combination $\lambda P = t$:

$$K_0(t) = \int_0^\infty ds\, s\frac{J_0(s)}{s^2+t^2}. \tag{20.3b}$$

In terms of this new function, the Green's function in the region between two parallel plates is

$$G(\mathbf{r},\mathbf{r}') = \frac{4}{a}\sum_{n=1}^{\infty}\sin\left(\frac{n\pi}{a}z\right)\sin\left(\frac{n\pi}{a}z'\right)K_0\left(\frac{n\pi}{a}P\right). \tag{20.4}$$

If we had not performed the angular integration in (20.1), we would have encountered the modified Bessel function as

$$K_0(\lambda P) = 2\pi\int\frac{(d\mathbf{k}_\perp)}{(2\pi)^2}\frac{e^{i\mathbf{k}_\perp\cdot(\mathbf{r}-\mathbf{r}')_\perp}}{k^2+\lambda^2}. \tag{20.5}$$

Then, we see immediately that K_0 is a two-dimensional Green's function,

$$(-\nabla_\perp^2+\lambda^2)K_0(\lambda P) = 2\pi\delta((\mathbf{r}-\mathbf{r}')_\perp). \tag{20.6}$$

As always, that statement comprises two bits of information. The first one is the homogeneous equation applicable for $\mathbf{r}_\perp \neq \mathbf{r}'_\perp$, which is

$$P>0: \quad (-\nabla_\perp^2+\lambda^2)K_0(\lambda P) = \left(-\frac{1}{P}\frac{d}{dP}P\frac{d}{dP}+\lambda^2\right)K_0(\lambda P) = 0, \tag{20.7a}$$

DOI: 10.1201/9781003057369-20

or

$$t > 0: \quad \left(-\frac{1}{t}\frac{d}{dt}t\frac{d}{dt} + 1\right) K_0(t) = 0. \tag{20.7b}$$

Notice that the latter equation differs from the one obeyed by $J_0(t)$, (18.17b), only in a sign change, such as would be produced by the substitution $t \to it$. Accordingly, $K_0(t)$ is also sometimes called a Bessel function of imaginary argument. Incidentally, the relationship between the two kinds of differential equation is emphasized on using the construction (20.3a),

$$\begin{aligned}\left(-\frac{1}{P}\frac{d}{dP}P\frac{d}{dP} + \lambda^2\right) K_0(\lambda P) &= \int_0^\infty dk \frac{k}{k^2 + \lambda^2}\left(-\frac{1}{P}\frac{d}{dP}P\frac{d}{dP} + \lambda^2\right) J_0(kP) \\ &= \int_0^\infty dk\, k\, J_0(kP), \end{aligned} \tag{20.8}$$

which is

$$\frac{1}{P}\delta(P) = 0, \quad P > 0, \tag{20.9}$$

according to (18.48).

The second bit of information, that carried by the inhomogeneous term in (20.6), is extracted on integrating the latter equation [or (20.8), (20.9)] over an arbitrarily small circle surrounding the point $\mathbf{r}'_\perp$, using the two-dimensional form of the divergence theorem:

$$\lambda P \ll 1: \quad -2\pi P \frac{d}{dP} K_0(\lambda P) = 2\pi. \tag{20.10}$$

We have assumed that the integral of K_0 is arbitrarily small, which is justified by the implication of (20.10),

$$t \ll 1: \quad K_0(t) \approx \ln\frac{1}{t} + \text{constant}. \tag{20.11}$$

The value of the constant appearing here, along with other useful results, can be obtained from alternative representations of K_0. For these we return to (20.5) and evaluate this rotationally invariant integral by using the direction of $(\mathbf{r} - \mathbf{r}')_\perp$ as the x-axis:

$$K_0(\lambda P) = \frac{1}{2\pi}\int dk_x\, dk_y \frac{e^{ik_x P}}{k_x^2 + k_y^2 + \lambda^2}. \tag{20.12}$$

We now have the options of integrating first with respect to k_x, using the integral of (15.20), which results in ($k_y = k$)

$$K_0(\lambda P) = \frac{1}{2}\int_{-\infty}^\infty dk \frac{e^{-\sqrt{k^2+\lambda^2}P}}{\sqrt{k^2 + \lambda^2}}, \tag{20.13a}$$

or, of applying a particular example of the cited integral to carry out the integration with respect to k_y (here $k = k_x$)

$$K_0(\lambda P) = \frac{1}{2}\int_{-\infty}^\infty dk \frac{e^{ikP}}{\sqrt{k^2 + \lambda^2}}. \tag{20.13b}$$

Equivalent versions of these two forms are

$$K_0(t) = \int_0^\infty ds \frac{e^{-\sqrt{s^2+t^2}}}{\sqrt{s^2 + t^2}} = \int_0^\infty d\theta\, e^{-t\cosh\theta} \tag{20.14a}$$

(with the latter produced by $s = t\sinh\theta$), and

$$K_0(t) = \int_0^\infty ds \frac{\cos s}{\sqrt{s^2+t^2}} = \int_0^\infty d\theta \cos(t\sinh\theta), \tag{20.14b}$$

respectively.

We apply the exponential version (20.14a), for small t, by introducing a transitional value of the integration variable s, S, such that $t \ll S \ll 1$. Then the integral can be divided into two parts, with appropriate approximations, as

$$K_0(t) \approx \int_0^S ds \frac{1}{\sqrt{s^2+t^2}} + \int_S^\infty ds \frac{e^{-s}}{s}. \tag{20.15}$$

The first of these two parts is equal to

$$\ln\left(s+\sqrt{s^2+t^2}\right)\Big|_0^S \approx \ln\frac{2S}{t}, \tag{20.16a}$$

while the second one, rearranged by partial integration, is

$$\ln s\, e^{-s}\Big|_S^\infty + \int_S^\infty ds \ln s\, e^{-s} \approx -\ln S + \int_0^\infty ds \ln s\, e^{-s}. \tag{20.16b}$$

This gives

$$t \ll 1: \quad K_0(t) \approx \ln\frac{2}{t} + \int_0^\infty ds \ln s\, e^{-s}, \tag{20.17}$$

which is indeed of the form (20.11).

The integral that appears above can be recognized as something known if we look at the following [perhaps more familiar than that appearing in (19.29)] representation of the gamma function,

$$\Gamma(t) = \int_0^\infty ds\, s^{t-1} e^{-s}, \tag{20.18}$$

or, more precisely, its derivative:

$$\Gamma'(t) = \frac{d}{dt}\Gamma(t) = \int_0^\infty ds \ln s\, s^{t-1} e^{-s}. \tag{20.19}$$

We now see that the integral appearing in (20.17) is $\Gamma'(1)$, which is also $\psi(1)$, inasmuch as $\Gamma(1) = 1$. Then on consulting (19.33), we learn that $\Gamma'(1)$ is just the negative of Euler's constant γ:

$$t \ll 1: \quad K_0(t) \approx \ln(2/t) - \gamma = \ln(1/t) + 0.1159\ldots. \tag{20.20}$$

The exponential representation (20.14a) is also particularly useful in finding the leading asymptotic form of $K_0(t)$ for $t \gg 1$. In this circumstance, contributions to the integral have already begun to decrease significantly when the integration variable attains values $s \sim t^{1/2} \ll t$, which leads to the asymptotic approximation

$$K_0(t) \sim \int_0^\infty ds \frac{1}{t} \exp\left[-\left(t + \frac{1}{2}\frac{s^2}{t}\right)\right] = \frac{e^{-t}}{t}\left(\frac{\pi t}{2}\right)^{1/2} \tag{20.21a}$$

or

$$K_0(t) \sim \sqrt{\frac{\pi}{2t}} e^{-t}, \quad t \to \infty. \tag{20.21b}$$

Of course, apart from the specific numerical factor, this asymptotic form is immediately apparent in the version of the differential equation for $K_0(t)$ that is analogous ($m = 0$) to (18.35),

$$\left(\frac{d^2}{dt^2} + \frac{1}{4t^2} - 1\right) t^{1/2} K_0(t) = 0. \tag{20.22}$$

An application of the asymptotic behavior of $K_0(t)$ appears on returning to the three-dimensional Green's function (20.4) and considering the situation $P \gg a$, where the distance between $\mathbf{r}$ and $\mathbf{r}'$ in the transverse plane is large compared with the separation of the conducting plates. Then, of the sequence of decreasing exponentials, $\exp(-n\pi(P/a))$, it is the first one that dominates,

$$P \gg a, \quad G \sim \sqrt{\frac{8}{Pa}} \sin\frac{\pi}{a}z \sin\frac{\pi}{a}z' e^{-\pi P/a}. \tag{20.23}$$

The parallel conducting plates suppress most effectively the Coulomb field of the point charge at large distances, on the scale set by the distance a. That is qualitatively clear in the picture wherein the conductors are replaced by the infinite set of image charges; the potentials of the two lattices composed of positive and negative charges, respectively, will cancel almost completely at large distances.

The image charge picture also makes evident that $G(\mathbf{r}, \mathbf{r}')$ approaches Green's function for the vacuum—Coulomb's potential—as $|\mathbf{r} - \mathbf{r}'|$ becomes very small on the scale set by a (except when the unit charge is quite close to one of the conductors). Indeed we can recognize in (19.19a) a statement about the approach to that limit:

$$|\mathbf{r} - \mathbf{r}'| \ll a: \quad G(\mathbf{r}, \mathbf{r}') \approx \frac{1}{|\mathbf{r} - \mathbf{r}'|} + 2E_{\text{int}}, \tag{20.24}$$

where E_{int} is the energy of interaction, with the conductors, of a unit charge that is stationed at a point $\mathbf{r} \approx \mathbf{r}'$.

20.1 More Bessel Functions

First, let us summarize what we have learned about $(2+1)$-dimensional representations of Coulomb's potential. Let's begin with (15.15), presented as

$$\begin{aligned} \frac{1}{|\mathbf{r} - \mathbf{r}'|} &= 4\pi \int \frac{(d\mathbf{k}_\perp)}{(2\pi)^2} e^{i\mathbf{k}_\perp \cdot (\mathbf{r}-\mathbf{r}')_\perp} \int_{-\infty}^{\infty} \frac{dk_z}{2\pi} \frac{e^{ik_z(z-z')}}{k_\perp^2 + k_z^2} \\ &= \frac{1}{\pi} \int_0^\infty dk\, k \int_{-\infty}^{\infty} dk_z \frac{J_0(kP) e^{ik_z Z}}{k^2 + k_z^2}; \end{aligned} \tag{20.25}$$

in the latter version $k = k_\perp$ and $Z = z - z'$. Now we have the options of integrating first over k_z, using (15.20):

$$\frac{1}{|\mathbf{r} - \mathbf{r}'|} = \frac{1}{\sqrt{P^2 + Z^2}} = \int_0^\infty dk\, J_0(kP) e^{-k|Z|}, \tag{20.26a}$$

[this is (18.6)] or, of integrating first over k, using (20.3a):

$$\frac{1}{\sqrt{P^2 + Z^2}} = \frac{1}{\pi} \int_{-\infty}^{\infty} dk_z K_0(|k_z|P) e^{ik_z Z} = \frac{2}{\pi} \int_0^\infty dk\, K_0(kP) \cos kZ, \tag{20.26b}$$

where k replaces k_z.

Observe that in one representation an exponentially decreasing function of Z is combined with (for sufficiently large kP) an oscillating function of P, while in the other representation it is an oscillating function of Z that combines with (for sufficiently large kP) an exponentially decreasing function of P. The necessity for this mixing in the functional character of the multiplicative constituents stems from the significance of $1/|\mathbf{r}-\mathbf{r}'|$, $\mathbf{r}\neq\mathbf{r}'$, as a solution of Laplace's equation:

$$\left(\frac{\partial^2}{\partial P^2}+\frac{1}{P}\frac{\partial}{\partial P}+\frac{\partial^2}{\partial Z^2}\right)\frac{1}{\sqrt{P^2+Z^2}}=0. \tag{20.27}$$

To the extent that the first derivative term is relatively negligible (P large), the two second derivatives must be of opposite sign, requiring that a convex function of one variable be combined with a concave function of the other variable.

Perhaps we should also point to the close relationship, indeed, equivalence of (20.14b) and (20.26b). First notice that the statement of completeness in (18.51) reduces, for an even function of x, to

$$f(x)=\frac{2}{\pi}\int_0^\infty dk\cos kx\left[\int_0^\infty dx'\cos kx'\,f(x')\right]. \tag{20.28}$$

Now regard (20.26b), with Z playing the role of x, as such a Fourier integral representation of $(P^2+Z^2)^{-1/2}$ and conclude that

$$K_0(kP)=\int_0^\infty dZ'\cos kZ'\frac{1}{\sqrt{P^2+Z'^2}}; \tag{20.29}$$

the substitution $kZ'=s$, $kP=t$ then yields (20.14b). In a similar application of the completeness of the Bessel functions $J_0(kP)$, (18.57b), we infer from (20.26a), written as ($Z>0$)

$$\frac{1}{\sqrt{P^2+Z^2}}=\int_0^\infty dk\,k\,J_0(kP)\frac{e^{-kZ}}{k}, \tag{20.30}$$

that

$$\frac{e^{-kZ}}{k}=\int_0^\infty dP'P'J_0(kP')\frac{1}{\sqrt{P'^2+Z^2}}, \tag{20.31a}$$

or

$$\int_0^\infty ds\,s\,J_0(s)\frac{1}{\sqrt{s^2+t^2}}=e^{-t}. \tag{20.31b}$$

The latter formula lends itself to the derivation of additional mathematical relations. For example, one can integrate (20.31b) with respect to t from 0 to ∞, using the integral

$$\int_0^\Upsilon dt\frac{1}{\sqrt{s^2+t^2}}=\sinh^{-1}\left(\frac{\Upsilon}{s}\right)\approx\ln\frac{2\Upsilon}{s},\quad \Upsilon\gg s, \tag{20.32}$$

and the null value recorded in (18.46),

$$\int_0^\infty ds\,s\,J_0(s)=0, \tag{20.33}$$

to arrive at [a convergence factor, $\exp(-\epsilon s)$, $\epsilon\to+0$, is understood]

$$\int_0^\infty ds\,s\ln\frac{1}{s}\,J_0(s)=1. \tag{20.34}$$

Or, we might multiply (20.31b) by $\cos\lambda t$ before integrating with respect to t from 0 to ∞. Then the following version of (20.14b),

$$K_0(\lambda s) = \int_0^\infty dt \frac{\cos\lambda t}{\sqrt{s^2+t^2}}, \tag{20.35}$$

yields

$$\int_0^\infty ds\, s\, J_0(s)\, K_0(\lambda s) = \int_0^\infty ds\, s\, J_0(\lambda s)\, K_0(s) = \frac{1}{1+\lambda^2}, \tag{20.36}$$

where the alternative integral is produced by first setting $\lambda s = t$, and then replacing λ by λ^{-1}. [This formula is attributed to Oliver Heaviside.] Note the connection between (20.34) and (20.36) that is provided by the initial behavior of K_0, as presented in (20.20). But enough mathematical doodling. Let's ask this question: The addition theorem relating $J_0(kP)$ to the $J_m(k\rho)$ and $J_m(k\rho')$, (18.65), is of obvious physical importance. Does $K_0(kP)$, as it enters (20.4), for example, also possess an addition theorem?

To set the stage for our affirmative response, we first present an alternative derivation of the addition theorem for $J_0(kP)$. And for that we need relations among the Bessel functions. Let's begin with another version of (18.29),

$$\int_0^{2\pi} \frac{d\alpha}{2\pi} e^{im\alpha} e^{i\mathbf{k}_\perp \cdot \mathbf{r}_\perp} = i^m e^{im\phi} J_m(k\rho), \tag{20.37}$$

in which α and ϕ are the polar angles of $\mathbf{k}_\perp$ and $\mathbf{r}_\perp$, respectively. (One returns to the earlier form by using $\alpha - \phi$, or its negative, as the integration variable while noting that the resulting integral is independent of the algebraic sign of m.) Now observe that

$$\frac{1}{ik}\left(\frac{\partial}{\partial x} \pm i\frac{\partial}{\partial y}\right) e^{i\mathbf{k}_\perp \cdot \mathbf{r}_\perp} = \frac{1}{k}(k_x \pm ik_y) e^{i\mathbf{k}_\perp \cdot \mathbf{r}_\perp} = e^{\pm i\alpha} e^{i\mathbf{k}_\perp \cdot \mathbf{r}_\perp}, \tag{20.38}$$

from which follows

$$\frac{1}{ik}\left(\frac{\partial}{\partial x} \pm i\frac{\partial}{\partial y}\right)\left(i^m e^{im\phi} J_m(k\rho)\right) = i^{m\pm 1} e^{i(m\pm 1)\phi} J_{m\pm 1}(k\rho). \tag{20.39}$$

Although this is the preferred form for our present purposes, it is also possible to remove the reference to ϕ and present these just as relations among the Bessel functions. (See Problem 20.1.) The two differential operators appearing here behave as inverses when acting on the functions $\exp(im\phi)J_m(k\rho)$:

$$\left[\frac{1}{ik}\left(\frac{\partial}{\partial x} \mp i\frac{\partial}{\partial y}\right)\right]\left[\frac{1}{ik}\left(\frac{\partial}{\partial x} \pm i\frac{\partial}{\partial y}\right)\right] \to 1. \tag{20.40}$$

It is in this sense that we begin with $J_0(k\rho)$ and derive from it the following constructions for both positive and negative values of m:

$$i^m e^{im\phi} J_m(k\rho) = \left[\frac{1}{ik}\left(\frac{\partial}{\partial x} + i\frac{\partial}{\partial y}\right)\right]^m J_0(k\rho). \tag{20.41}$$

Now let us consider

$$J_0(kP) = J_0(k|(\mathbf{r}-\mathbf{r}')_\perp|) = e^{-\mathbf{r}'_\perp \cdot \boldsymbol{\nabla}_\perp} J_0(k\rho), \tag{20.42}$$

where the latter form uses a symbolic expression of a Taylor series expansion in powers of $\mathbf{r}'_\perp$. It's time to recognize that (20.40), or

$$-\frac{1}{k^2}\nabla_\perp^2 \to 1, \tag{20.43}$$

which is a statement of the differential equation [see (18.32a)] obeyed by the functions $\exp(im\phi)J_m(k\rho)$, asserts that $(i/k)\boldsymbol{\nabla}_\perp$ behaves like a unit vector, when $J_0(k\rho)$ is the operand. We can also assign to this unit vector a polar angle Φ, in the sense that

$$e^{i\Phi} = \cos\Phi + i\sin\Phi = \frac{i}{k}\left(\frac{\partial}{\partial x} + i\frac{\partial}{\partial y}\right). \tag{20.44}$$

In consequence, we have from (18.27)

$$e^{-\mathbf{r}'_\perp \cdot \boldsymbol{\nabla}_\perp} = e^{ik\mathbf{r}'_\perp \cdot (i/k)\boldsymbol{\nabla}_\perp} \to e^{ik\rho'\cos(\Phi-\phi')} = \sum_{m=-\infty}^{\infty} i^m e^{im(\Phi-\phi')} J_m(k\rho'), \tag{20.45}$$

where, with $J_0(k\rho)$ made explicit,

$$i^m e^{im\Phi} J_0(k\rho) = i^{-m}\left[\frac{1}{ik}\left(\frac{\partial}{\partial x} + i\frac{\partial}{\partial y}\right)\right]^m J_0(k\rho) = e^{im\phi} J_m(k\rho), \tag{20.46}$$

so that (20.45) is the addition theorem (18.65),

$$J_0(kP) = \sum_{m=-\infty}^{\infty} e^{im\phi} J_m(k\rho) e^{-im\phi'} J_m(k\rho'). \tag{20.47}$$

And so, with this example before us, we now write

$$K_0(kP) = K_0(k|(\mathbf{r}-\mathbf{r}')_\perp|) = e^{-\mathbf{r}'_\perp \cdot \boldsymbol{\nabla}_\perp} K_0(k\rho). \tag{20.48}$$

But we see immediately that matters are somewhat different here. First, the function $K_0(kP)$ has a singularity at $P = 0$ [(20.20)] and the power series expansion cannot converge outside the domain $|\mathbf{r}'_\perp| = \rho' < \rho$. This fits in well, however, with another, related difference. Unlike $J_0(k\rho)$, $K_0(k\rho)$ obeys a homogeneous differential equation only if $\rho = 0$ is excluded. And that brings us to the fact that the homogeneous differential equation obeyed by K_0 [(20.7a)] is not the same as the one for J_0. Indeed, now we have, from (20.6),

$$\frac{1}{k^2}\nabla_\perp^2 \to 1, \tag{20.49}$$

and it is $(-1/k)\boldsymbol{\nabla}_\perp$ that behaves like a unit vector.

That is one way to recognize that the Bessel functions entering the addition theorem of K_0 must be those of imaginary argument, the ones that obey the same homogeneous differential equation that governs K_0, corresponding to (20.49). The real function $I_m(t)$ is conventionally defined by [see (18.24)]

$$i^{-m} J_m(it) = I_m(t) = I_{-m}(t), \tag{20.50}$$

where either sign of i can be adopted on the left-hand side. Indeed, we get the form we want by replacing t with $-it$ in (18.27), thereby arriving at

$$\begin{aligned} e^{t\cos\phi} &= \sum_{m=-\infty}^{\infty} e^{im\phi} I_m(t) \\ &= I_0(t) + 2\sum_{m=1}^{\infty} \cos m\phi\, I_m(t), \end{aligned} \tag{20.51}$$

which also supplies the integral representation

$$I_m(t) = \int_0^\pi \frac{d\phi}{\pi} \cos(m\phi) e^{t\cos\phi}. \tag{20.52}$$

We also note the appropriate modification of (20.41), as produced by the substitution $k \to -ik$:

$$e^{im\phi} I_m(k\rho) = \left[\frac{1}{k}\left(\frac{\partial}{\partial x} + i\frac{\partial}{\partial y}\right)\right]^m I_0(k\rho). \tag{20.53}$$

Accordingly, this time we have

$$e^{-\mathbf{r}'_\perp \cdot \boldsymbol{\nabla}_\perp} = e^{k\mathbf{r}'_\perp \cdot (-1/k)\boldsymbol{\nabla}_\perp} \to e^{k\rho' \cos(\Phi-\phi')} = \sum_{m=-\infty}^{\infty} e^{im(\Phi-\phi')} I_m(k\rho'), \tag{20.54}$$

where

$$e^{i\Phi} = -\frac{1}{k}\left(\frac{\partial}{\partial x} + i\frac{\partial}{\partial y}\right), \tag{20.55}$$

and the outcome is

$$\rho > \rho': \quad K_0(kP) = \sum_{m=-\infty}^{\infty} e^{im\phi} K_m(k\rho) e^{-im\phi'} I_m(k\rho'), \tag{20.56}$$

where we have defined

$$e^{im\phi} K_m(k\rho) = \left[-\frac{1}{k}\left(\frac{\partial}{\partial x} + i\frac{\partial}{\partial y}\right)\right]^m K_0(k\rho). \tag{20.57}$$

This definition is analogous to (20.53), but differs in the additional sign factor $(-1)^m$. That flaw is accepted in order to achieve the positiveness of both $I_m(t>0)$ and $K_m(t>0)$, which property of the I_m is evident in the power series inferred from (18.26),

$$I_m(t) = \sum_{n=0}^{\infty} \frac{\left(\frac{1}{2}t\right)^{m+2n}}{(m+n)!\,n!}. \tag{20.58}$$

To discuss the positivity of the $K_m(t)$, as defined by (20.57), we return to the initial definition of K_0, (20.5), which is now presented as

$$K_0(k\rho) = 2\pi \int \frac{(d\mathbf{k}'_\perp)}{(2\pi)^2} \frac{e^{i\mathbf{k}'_\perp \cdot \mathbf{r}_\perp}}{k'^2 + k^2}. \tag{20.59}$$

The effect of the differential operator in (20.57) on the exponential function appearing here is given by

$$\left[-\frac{1}{k}\left(\frac{\partial}{\partial x} + i\frac{\partial}{\partial y}\right)\right]^m e^{i\mathbf{k}'_\perp \cdot \mathbf{r}_\perp} = \left[-\frac{i}{k}(k'_x + ik'_y)\right]^m e^{i\mathbf{k}'_\perp \cdot \mathbf{r}_\perp}. \tag{20.60}$$

It is clear that integration over the polar angle of $\mathbf{k}'_\perp$ $[k'_x + ik'_y = k'e^{i\alpha}]$ will produce the factor of $\exp(im\phi)$ that stands on the left-hand side of (20.57). Accordingly, we are free to specialize to $\phi = 0$, which is to say, $x = \rho$, $y = 0$, and get

$$\begin{aligned} K_m(k\rho) &= \int \frac{dk'_x dk'_y}{2\pi} \left[\frac{1}{k}(-ik'_x + k'_y)\right]^m \frac{e^{ik'_x\rho}}{k'^2_x + k'^2_y + k^2} \\ &= \int_{-\infty}^{\infty} dk'_y \left[\frac{1}{k}\left(-\frac{\partial}{\partial\rho} + k'_y\right)\right]^m \int_{-\infty}^{\infty} \frac{dk'_x}{2\pi} \frac{e^{ik'_x\rho}}{k'^2_x + k'^2_y + k^2} \\ &= \int_{-\infty}^{\infty} dk'_y \left[\frac{1}{k}\left(-\frac{\partial}{\partial\rho} + k'_y\right)\right]^m \frac{1}{2} \frac{e^{-\sqrt{k'^2_y + k^2}\,\rho}}{\sqrt{k'^2_y + k^2}}, \end{aligned} \tag{20.61}$$

which uses (15.20), or, with $k_y' = k\sinh\theta$,

$$K_m(k\rho) = \frac{1}{2}\int_{-\infty}^{\infty} d\theta\,(\cosh\theta + \sinh\theta)^m e^{-k\rho\cosh\theta}. \tag{20.62a}$$

We are thus led to the integral representation

$$K_m(t) = \int_0^\infty d\theta\cosh m\theta\, e^{-t\cosh\theta}, \tag{20.62b}$$

which is indeed positive, and also displays the property

$$K_{-m}(t) = K_m(t). \tag{20.63}$$

We present one application of the K_0 addition theorem that is related to the integral (20.36). First, let's use the integral representation of $K_0(t)$ given in (20.14a) to evaluate

$$\int_0^\infty dt\,t\,K_0(t) = \int_0^\infty ds\int_0^\infty d\left[-e^{-(s^2+t^2)^{1/2}}\right] = \int_0^\infty ds\,e^{-s} = 1. \tag{20.64}$$

Now observe from (20.56) that ($\phi' = 0$)

$$\int_0^{2\pi}\frac{d\phi}{2\pi}K_0(kP) = K_0(k\rho)I_0(k\rho'), \quad \rho > \rho', \tag{20.65}$$

from which follows

$$\int_0^\infty dk\,k\,K_0(k\rho)I_0(k\rho') = \int_0^{2\pi}\frac{d\phi}{2\pi}\int_0^\infty dk\,k\,K_0(kP) = \int_0^{2\pi}\frac{d\phi}{2\pi}\frac{1}{P^2}; \tag{20.66}$$

the latter integral is

$$\int_0^{2\pi}\frac{d\phi}{2\pi}\frac{1}{\rho^2+\rho'^2-2\rho\rho'\cos\phi} = \frac{1}{[(\rho^2+\rho'^2)^2-(2\rho\rho')^2]^{1/2}} = \frac{1}{\rho^2-\rho'^2}. \tag{20.67}$$

With a change of variables, this result appears as

$$\int_0^\infty ds\,s\,I_0(\lambda s)K_0(s) = \frac{1}{1-\lambda^2}, \quad \lambda < 1, \tag{20.68}$$

which is just what emerges from the second form of (20.36) after $\lambda \to i\lambda$.

Finally, we comment on the asymptotic form of $K_m(t)$ for any given m and sufficiently large t. Approximations analogous to those in (20.21a) give the leading term from (20.62b),

$$K_m(t) \sim \int_0^\infty d\theta\,e^{-t(1+\theta^2/2)} = \sqrt{\frac{\pi}{2t}}e^{-t}, \quad t \gg 1, \tag{20.69}$$

independently of m. And, a related treatment of the integral representation for $I_m(t)$, (20.52), yields

$$I_m(t) \sim \int_0^\infty \frac{d\phi}{\pi}e^{t(1-\phi^2/2)} = \frac{1}{\sqrt{2\pi t}}e^t, \quad t \gg 1, \tag{20.70}$$

which, of course, coincides with the implication of (18.37) for $i^{-m}J_m(it)$.

20.2 Problems for Chapter 20

1. Derive the recursion relations in Problem 18.4 [(18.69)] from (20.39)
2. Show that, for $m > 0$, the differential operator in (20.41) acting on $J_0(k\rho)$ satisfies

$$\left[\frac{1}{ik}\left(\frac{\partial}{\partial x}+i\frac{\partial}{\partial y}\right)\right]^m = i^m e^{im\phi} F\left(k\rho, \frac{\partial}{\partial k\rho}\right), \tag{20.71}$$

where F is a differential operator constructed from $k\rho$ and $\partial/\partial k\rho$.

3. The modified Bessel functions I_ν, K_ν satisfy the differential equation

$$u'' + \frac{1}{x}u' - \left(1+\frac{\nu^2}{x^2}\right)u = 0. \tag{20.72}$$

Determine the leading asymptotic behavior, given in (20.70) and (20.69), directly from this differential equation. Because

$$I_\nu(x) = i^{-\nu} J_\nu(ix), \tag{20.73a}$$

$$K_\nu(x) = \frac{\pi}{2} i^{\nu+1}(J_\nu(ix) + iN_\nu(ix)), \tag{20.73b}$$

show that this asymptotic behavior is consistent with that found in Problem 18.10, which holds for complex x, $|\arg x| < \pi$.

4. From (20.53) and (20.57) derive recursion relations similar to those in Problem 18.4 for I_m and $(-1)^m K_m$.
5. Show that $K_m(t)$ defined by (20.62b) continues to obey the appropriate differential equation for nonintegral values of m.
6. Starting from the integral representation for $K_9(t)$, which can be written as

$$K_0(t) = \frac{1}{2}\int_{-\infty}^{\infty} d\lambda \frac{1}{\sqrt{\lambda^2+1}} e^{i\lambda t}, \tag{20.74}$$

distort the contour of integration along the real λ axis to one encircling the branch line along the positive imaginary axis. To do so, you must show that the contributions along the quarter circles at "infinitely" large radius in the first and second quadrants of the complex λ plane vanish. By adding the contributions to the left and right of the branch line, show that an alternate integral representation for K_0 is

$$K_0(t) = \int_1^{\infty} d\mu \frac{1}{\sqrt{\mu^2-1}} e^{-\mu t}, \tag{20.75}$$

which is equivalent to (20.62b) for $m = 0$. Show that this form yields the asymptotic behavior (20.69) rather immediately.

21

Cylindrical Conductors

21.1 Rectangle

In a step beyond the parallel conducting plates of Chapter 19, we consider another set of plates that intersect the first set at right angles, thereby producing a cylindrical region with a rectangular cross section. The latter is displayed in Fig. 21.1 along with a convenient coordinate system. Again, we seek Green's function for a vacuum region,

$$-\nabla^2 G(\mathbf{r},\mathbf{r}') = 4\pi\delta(\mathbf{r}-\mathbf{r}'), \qquad \left.\begin{array}{c} x=0,a \\ y=0,b \end{array}\right\}: \quad G(\mathbf{r},\mathbf{r}')=0. \tag{21.1}$$

Now we begin with both (19.54a)

$$\delta(x-x') = \frac{2}{a}\sum_{l=1}^{\infty} \sin\frac{l\pi}{a}x \,\sin\frac{l\pi}{a}x', \tag{21.2a}$$

and

$$\delta(y-y') = \frac{2}{b}\sum_{m=1}^{\infty} \sin\frac{m\pi}{b}y \,\sin\frac{m\pi}{b}y', \tag{21.2b}$$

combining them into the two-dimensional delta function

$$\delta(x-x')\delta(y-y') = \sum_{lm} \phi_{lm}(x,y)\phi_{lm}(x',y'). \tag{21.3}$$

The complete set of functions

$$\phi_{lm}(x,y) = \sqrt{\frac{2}{a}}\sin\frac{l\pi}{a}x\,\sqrt{\frac{2}{b}}\sin\frac{m\pi}{b}y, \quad l,m=1,2,3,\ldots, \tag{21.4}$$

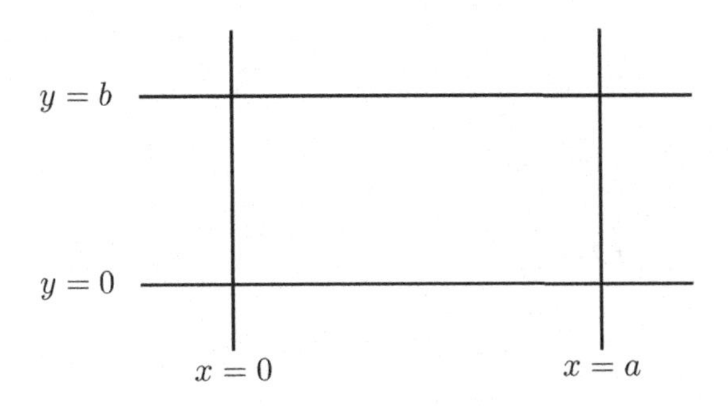

FIGURE 21.1
Cylindrical region created by intersecting perpendicular conducting planes.

DOI: 10.1201/9781003057369-21

possesses the orthonormality property:

$$\int_0^a dx \int_0^b dy\, \phi_{lm}(x,y)\phi_{l',m'}(x,y) = \delta_{ll'}\delta_{mm'}. \tag{21.5}$$

The related construction of Green's function,

$$G(\mathbf{r},\mathbf{r}') = 4\pi \sum_{lm} \phi_{lm}(x,y)\phi_{lm}(x',y')g_{lm}(z,z'), \tag{21.6}$$

then supplies the one-dimensional differential equation

$$\left(-\frac{\partial^2}{\partial z^2} + \gamma_{lm}^2\right) g_{lm}(z,z') = \delta(z-z'), \tag{21.7}$$

in which

$$\gamma_{lm}^2 = \left(\frac{l\pi}{a}\right)^2 + \left(\frac{m\pi}{b}\right)^2 \geq \left(\frac{\pi}{a}\right)^2 + \left(\frac{\pi}{b}\right)^2. \tag{21.8}$$

We know the solution of (21.7) for a cylinder of unlimited length [see (15.22)],

$$g_{lm}(z,z') = \frac{1}{2\gamma_{lm}} e^{-\gamma_{lm}|z-z'|}. \tag{21.9}$$

When $|z-z'|$ is large, on the scale set by the greater of the rectangular dimensions, it is the $l = m = 1$ term that dominates (21.6):

$$G(\mathbf{r},\mathbf{r}') \sim 4\pi\phi_{11}(x,y)\phi_{11}(x',y')\frac{1}{2\gamma_{11}} e^{-\gamma_{11}|z-z'|}. \tag{21.10}$$

Suppose, now, that conducting walls at $z = 0$ and c truncate the infinite cylinder, thereby encasing the unit point charge in a rectangular conducting enclosure or cavity. The solution of (21.7) under such circumstances is given in Chapter 19. But there is a more symmetrical procedure here; begin with the three-dimensional delta function construction

$$\delta(\mathbf{r}-\mathbf{r}') = \sum_{lmn} \phi_{lmn}(\mathbf{r})\phi_{lmn}(\mathbf{r}'), \tag{21.11}$$

where the complete set of functions

$$\phi_{lmn}(\mathbf{r}) = \sqrt{\frac{2}{a}} \sin\frac{l\pi}{a}x \sqrt{\frac{2}{b}} \sin\frac{m\pi}{b}y \sqrt{\frac{2}{c}} \sin\frac{l\pi}{c}z, \tag{21.12}$$

obeys

$$\int (d\mathbf{r})\, \phi_{lmn}(\mathbf{r})\phi_{l'm'n'}(\mathbf{r}') = \delta_{ll'}\delta_{mm'}\delta_{nn'}. \tag{21.13}$$

Green's function is then given by

$$G(\mathbf{r},\mathbf{r}') = 4\pi \sum_{lmn} \frac{\phi_{lmn}(\mathbf{r})\phi_{lmn}(\mathbf{r}')}{\gamma_{lmn}^2}, \tag{21.14}$$

in which

$$\gamma_{lmn}^2 = \left(\frac{l\pi}{a}\right)^2 + \left(\frac{m\pi}{b}\right)^2 + \left(\frac{n\pi}{c}\right)^2 \geq \left(\frac{\pi}{a}\right)^2 + \left(\frac{\pi}{b}\right)^2 + \left(\frac{\pi}{c}\right)^2. \tag{21.15}$$

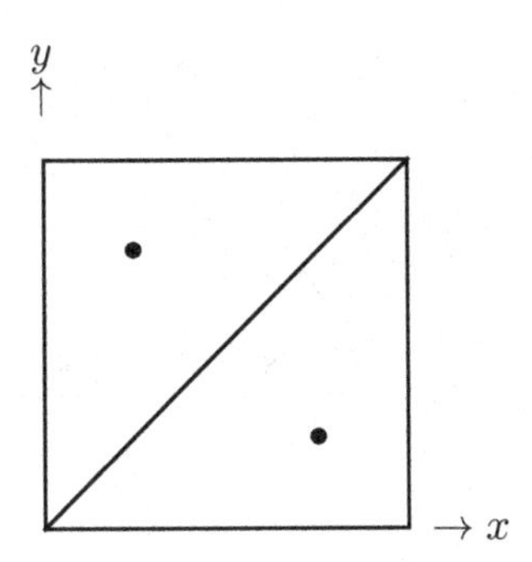

FIGURE 21.2
Bisection of a cylinder with square cross section.

21.2 Isosceles Right Angle Triangle

The situation in which the rectangular cross section of the cylinder is a square ($b = a$) has only one special feature, the symmetry

$$\gamma_{lm}^2 = \left(\frac{\pi}{a}\right)^2 (l^2 + m^2) = \gamma_{ml}^2. \tag{21.16}$$

Yet this is at the heart of the possibility of constructing a new cylindrical shape by applying the sort of symmetry consideration developed in Chapter 19. As a glance at Fig. 21.2 indicates, Green's function for a cylindrical conductor with the cross section of an isosceles right angle triangle will be produced if we satisfy the boundary condition of vanishing on the diagonal line $y = x$. And that will be accomplished by introducing an image charge through reflection in this line.

If the positive charge is stationed at the point with coordinates x', y', the negative image charge produced by the interchange of the x- and y-axes is found at y', x'. Thus the two-dimensional charge structure is

$$\begin{aligned}
&\delta(x - x')\delta(y - y') - \delta(x - y')(y - x') = \\
&= \left(\frac{2}{a}\right)^2 \sum_{lm} \sin\frac{l\pi}{a}x \sin\frac{m\pi}{a}y \left[\sin\frac{l\pi}{a}x' \sin\frac{m\pi}{a}y' - \sin\frac{l\pi}{a}y' \sin\frac{m\pi}{a}x'\right] \\
&= \sum_{l<m} \phi_{lm}(x,y)\phi_{lm}(x',y'),
\end{aligned} \tag{21.17}$$

where now

$$\phi_{lm}(x,y) = \frac{2}{a}\left[\sin\frac{l\pi}{a}x \sin\frac{m\pi}{a}y - \sin\frac{m\pi}{a}x \sin\frac{l\pi}{a}y\right], \tag{21.18}$$

a function that reverses sign on interchanging l and m, and on interchanging x and y. With $l < m$, these functions constitute a complete orthonormal set over the area of the isosceles triangle. And the significance of the symmetry (21.16) is now seen; both parts of (21.18) are associated with the same value of γ, which is therefore assigned to the complete function,

$$-\left(\frac{\partial^2}{\partial x^2} + \frac{\partial^2}{\partial y^2}\right)\phi_{lm}(x,y) = \gamma_{lm}^2 \phi_{lm}(x,y), \tag{21.19}$$

where ($l < m$)

$$\gamma_{lm}^2 = \left(\frac{\pi}{a}\right)^2 (l^2 + m^2) \geq 5\left(\frac{\pi}{a}\right)^2. \tag{21.20}$$

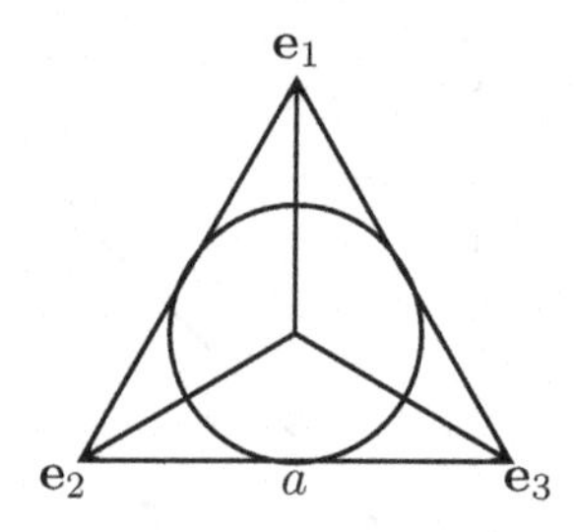

FIGURE 21.3
Cross section of equilateral triangular cylinder.

21.3 Equilateral Triangle

Some of the geometry of an equilateral triangle is displayed in Fig. 21.3; a is the common length of the sides, the length of the perpendicular from any apex to the opposite side is

$$h = a \sin \frac{\pi}{3} = \frac{\sqrt{3}}{2} a, \tag{21.21a}$$

and the radius of the inscribed circle equals

$$r = \frac{1}{2} a \tan \frac{\pi}{6} = \frac{1}{2\sqrt{3}} a = \frac{1}{3} h. \tag{21.21b}$$

We employ two coordinate systems in the transverse plane perpendicular to the axis of the triangular cylinder. One is the conventional rectangular frame—coordinates x and y—with its origin placed at the center of the inscribed circle. The second is a trilinear coordinate system based on the three unit vectors directed from the center of the inscribed circle to the three apexes. These three vectors are presented in terms of the unit orthogonal vectors $\mathbf{i}$ and $\mathbf{j}$ by

$$\mathbf{e}_1 = \mathbf{j}, \quad \mathbf{e}_2 = -\frac{\sqrt{3}}{2}\mathbf{i} - \frac{1}{2}\mathbf{j}, \quad \mathbf{e}_3 = \frac{\sqrt{3}}{2}\mathbf{i} - \frac{1}{2}\mathbf{j}, \tag{21.22}$$

where

$$\sum_{a=1}^{3} \mathbf{e}_a = \mathbf{0}. \tag{21.23}$$

One notes that

$$\sum_a \mathbf{e}_a \mathbf{e}_a = \frac{3}{2}(\mathbf{ii} + \mathbf{jj}) = \frac{3}{2}\mathbf{1}_\perp, \tag{21.24}$$

from which the trilinear coordinate representation of $\mathbf{r}_\perp$ is derived as

$$\mathbf{r}_\perp = \mathbf{1}_\perp \cdot \mathbf{r}_\perp = \frac{2}{3}\sum_{a=1}^{3} \mathbf{e}_a \mu_a, \tag{21.25}$$

where

$$\mu_a = \mathbf{e}_a \cdot \mathbf{r}_\perp. \tag{21.26a}$$

These coordinates, which obey

$$\sum_{a=1}^{3} \mu_a = 0, \tag{21.26b}$$

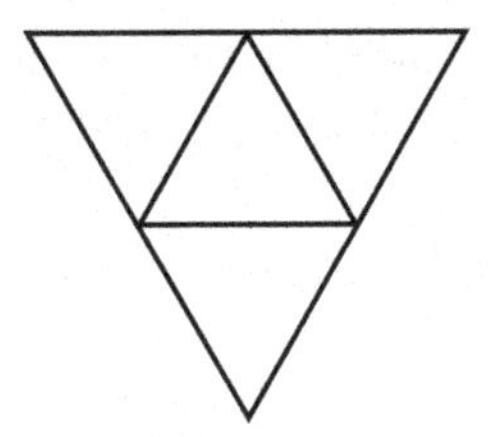

FIGURE 21.4
Reflected equilateral triangle.

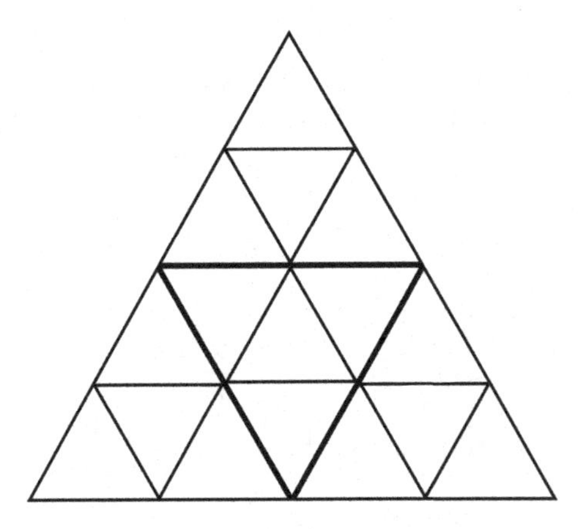

FIGURE 21.5
Repeated reflection of equilateral triangle.

are given individually by

$$\mu_1 = y, \quad \mu_2 = -\frac{\sqrt{3}}{2}x - \frac{1}{2}y, \quad \mu_3 = \frac{\sqrt{3}}{2}x - \frac{1}{2}y. \tag{21.26c}$$

The trilinear coordinates of a point are the perpendicular distances from the origin to the three lines, drawn through the point, parallel to the sides of the triangle. A coordinate is negative if the associated line lies between the origin and the related side. Thus, the three sides of the triangle are represented in trilinear coordinates as $\mu_1 = -r$, $\mu_2 = -r$, and $\mu_3 = -r$, respectively.

The potential that is Green's function can be extrapolated in a way similar to that of the parallel plates, by reflection in the three sides of the equilateral triangle. It is thereby initially defined in a larger region (see Fig. 21.4), one that is an inverted triangle of side $2a$. The repetition of this process for the bigger triangle produces an enlarged copy of the initial triangle, with side $4a$ [Fig. 21.5]. Having been generated by two reflections, the values of Green's function within the three triangles, of side a, that are produced by a displacement of $2h$ in each of the directions $\mathbf{e}_a$, $a = 1, 2, 3$, duplicate those in the initial triangle. In short, the indefinite repetition of this reflection process supplies a function defined over the entire two-dimensional space, a function that is periodic in each of the three directions, with period $2h$.

We now construct a set of point charges with the same periodicity. A first impulse might be to locate them at positions along the three trilinear directions according to

$$\mathbf{r}'_\perp + 2h\sum_{a=1}^{3}\nu_a\mathbf{e}_a, \quad \nu_a = 0, \pm 1, \ldots. \tag{21.27}$$

That would be redundant, however, in view of the relations given in (21.23). Indeed,

$$\begin{aligned}\sum_{a=1}^{3}\nu_a\mathbf{e}_a &= (\nu_1-\nu_3)\mathbf{e}_1 + (\nu_2-\nu_3)\mathbf{e}_2 = (\nu_2-\nu_1)\mathbf{e}_2 + (\nu_3-\nu_1)\mathbf{e}_3\\ &= (\nu_3-\nu_2)\mathbf{e}_3 + (\nu_1-\nu_2)\mathbf{e}_1,\end{aligned} \tag{21.28}$$

which means that any pair of trilinear basis vectors will do (no surprise in a two-dimensional space). In actuality, then, we extend $\delta(\mathbf{r}_\perp - \mathbf{r}'_\perp)$, the initial point charge density (in the transverse plane) to

$$\begin{aligned}&\sum_{\nu_2,\nu_3=-\infty}^{\infty}\delta(\mathbf{r}_\perp - \mathbf{r}'_\perp - 2h(\nu_2\mathbf{e}_2+\nu_3\mathbf{e}_3))\\ &\quad = \int\frac{(d\mathbf{k}_\perp)}{(2\pi)^2}e^{i\mathbf{k}_\perp\cdot(\mathbf{r}-\mathbf{r}')_\perp}\sum_{\nu_2\nu_3}e^{-i2h\nu_2k_2}e^{-i2h\nu_3k_3},\end{aligned} \tag{21.29}$$

which introduces some of the trilinear coordinates of $\mathbf{k}_\perp$,

$$k_a = \mathbf{e}_a\cdot\mathbf{k}_\perp, \quad \sum_a k_a = 0. \tag{21.30}$$

All of these coordinates are employed in writing

$$\mathbf{k}_\perp\cdot(\mathbf{r}-\mathbf{r}')_\perp = \mathbf{k}_\perp\cdot\mathbf{1}_\perp\cdot(\mathbf{r}-\mathbf{r}')_\perp = \frac{2}{3}\sum_a k_a(\mu_a - \mu'_a). \tag{21.31}$$

The ν_2 and ν_3 summations in (21.29) are evaluated, using (18.59) and (19.50), as

$$\begin{aligned}&\frac{1}{(2\pi)^2}\sum_{\nu_2=-\infty}^{\infty}e^{-i2h\nu_2k_2}\sum_{\nu_3=-\infty}^{\infty}e^{-i2h\nu_3k_3}\\ &\quad = \frac{1}{(2h)^2}\sum_{l_2=-\infty}^{\infty}\delta(k_2-\frac{\pi}{h}l_2)\sum_{l_3=-\infty}^{\infty}\delta(k_3-\frac{\pi}{h}l_3),\end{aligned} \tag{21.32}$$

which tells us that the k_a have the discrete values

$$k_a = \frac{\pi}{h}l_a, \quad \sum_{a=1}^{3}l_a = 0. \tag{21.33}$$

Let us also record the discrete values of

$$(\mathbf{k}_\perp)^2 = \frac{2}{3}\sum_a k_a^2 = \frac{2}{3}\left(\frac{\pi}{h}\right)^2\sum_a l_a^2. \tag{21.34}$$

Now we must compute the integral

$$\int(d\mathbf{k}_\perp)\,\delta(k_2-\pi l_2/h)\delta(k_3-\pi l_3/h), \tag{21.35a}$$

which is

$$\frac{2}{\sqrt{3}}\int dk_2 dk_3 \delta(k_2 - \pi l_2/h)\delta(k_3 - \pi l_3/h) = \frac{2}{\sqrt{3}}, \tag{21.35b}$$

according to the value of the Jacobian determinant [Karl Gustav Jacob Jacobi (1804-1851)] that relates the integration variables k_2, k_3 and k_x, k_y:

$$\frac{\partial(k_2, k_3)}{\partial(k_x, k_y)} = \begin{vmatrix} -\sqrt{3}/2 & -1/2 \\ \sqrt{3}/2 & -1/2 \end{vmatrix} = \frac{\sqrt{3}}{2}. \tag{21.36}$$

The outcome for (21.29) is [recall that $a = 2h/\sqrt{3}$]

$$\sum_{\nu_2\nu_3} \delta(\mathbf{r}_\perp - \mathbf{r}'_\perp - 2h(\nu_2\mathbf{e}_2 + \nu_3\mathbf{e}_3))$$
$$= \frac{1}{3ha}\sum_l \exp\left[i\frac{2\pi}{3h}\sum_a l_a\mu_a\right]\exp\left[-i\frac{2\pi}{3h}\sum_a l_a\mu'_a\right], \tag{21.37}$$

in which the l-summation extends over all integral values of the l_a that satisfy the restrictive condition (21.33). It is worth noting here that a displacement of $\mathbf{r}_\perp$, for example, by $2h\mathbf{e}_b$, $b = 1, 2, 3$, alters the trilinear coordinates μ_a by

$$\mu_a \to \mu_a + \mathbf{e}_a \cdot \mathbf{e}_b 2h, \tag{21.38}$$

where, through inspection of (21.22) or otherwise,

$$\mathbf{e}_a \cdot \mathbf{e}_b = \frac{3}{2}\delta_{ab} - \frac{1}{2} = \begin{cases} a = b: & 1 \\ a \neq b: & -1/2 \end{cases}. \tag{21.39}$$

Accordingly, any exponential function of the μ_a in (21.37) changes by the factor

$$\exp\left[i\frac{2\pi}{3h}\left(l_b 2h + \sum_{a\neq b}(-l_a h)\right)\right] = \exp(i2\pi l_b) = 1. \tag{21.40}$$

The next step supplements the lattice of positive charges with a negative charge distribution that is designed to satisfy the specific boundary conditions on the three sides of the triangle. For that purpose we first observe that the image of $\mathbf{r}'_\perp$ in the triangle side perpendicular to $\mathbf{e}_1$ has the rectangular coordinates

$$\overline{x}' = x', \quad \overline{y}' = -y' - 2r, \tag{21.41}$$

the latter being more transparently written as

$$\overline{y}' + r = -(y' + r). \tag{21.42}$$

And more generally, the image in the side perpendicular to $\mathbf{e}_b$ is

$$\overline{\mathbf{r}}'_\perp = \mathbf{r}'_\perp - 2\mathbf{e}_b(\mathbf{e}_b \cdot \mathbf{r}'_\perp + r), \tag{21.43}$$

which appears in trilinear coordinates as

$$\overline{\mu}'_a = \mu'_a + \mu'_b + r - 3\delta_{ab}(\mu'_b + r) = \begin{cases} a = b: -\mu'_b - 2r \\ a \neq b: \ -\mu'_c + r \end{cases}, \tag{21.44}$$

where c designates the trilinear axis that is neither a nor b.

Let us select a particular exponential function of the μ'_a in (21.37), supply it, for convenience, with the numerical factor $\exp(i2\pi l_1/3)$, and then subtract from it the result of replacing μ'_a by $\overline{\mu}'_a$, using the example of reflection in the side perpendicular to $\mathbf{e}_1$. This yields the combination (recall $3r = h$)

$$\exp\left[-i\frac{2\pi}{3h}(l_1\mu'_1 + l_2\mu'_2 + l_3\mu'_3 - l_1 h)\right] - \exp\left[i\frac{2\pi}{3h}(l_1\mu'_1 + l_2\mu'_3 + l_3\mu'_2 - l_1 h)\right], \quad (21.45a)$$

which, from its construction, vanishes when $\mu'_1 = -r$, for that is the line in which reflection takes place. (See also Problem 21.1.) This property, being independent of the particular choice of the l_a, continues to hold when the indices of the l_a are cyclically permuted: $1 \to 2$, $2 \to 3$, $3 \to 1$, thereby producing

$$\exp\left[-i\frac{2\pi}{3h}(l_2\mu'_1 + l_3\mu'_2 + l_1\mu'_3 - l_2 h)\right] - \exp\left[i\frac{2\pi}{3h}(l_2\mu'_1 + l_3\mu'_3 + l_1\mu'_2 - l_2 h)\right], \quad (21.45b)$$

and

$$\exp\left[-i\frac{2\pi}{3h}(l_3\mu'_1 + l_1\mu'_2 + l_2\mu'_3 - l_3 h)\right] - \exp\left[i\frac{2\pi}{3h}(l_3\mu'_1 + l_1\mu'_3 + l_2\mu'_2 - l_3 h)\right]. \quad (21.45c)$$

Furthermore, the sum of the three pairs, which of course also vanishes for $\mu'_1 = -r$, responds in a simple way to cyclic permutations of the μ'_a indices. One can check that the substitution $\mu'_1 \to \mu'_2$, $\mu'_2 \to \mu'_3$, $\mu'_3 \to \mu'_1$ reproduces the set of six terms, multiplied by the common factor

$$\exp\left[-i\frac{2\pi}{3}(l_1 - l_2)\right] = \exp\left[-i\frac{2\pi}{3}(l_2 - l_3)\right] = \exp\left[-i\frac{2\pi}{3}(l_3 - l_1)\right]; \quad (21.46)$$

the equivalence of these forms follows from the relations

$$(l_1 - l_2) - (l_2 - l_3) = -3l_2, \quad (l_2 - l_3) - (l_3 - l_1) = -3l_2. \quad (21.47)$$

And the implication of this property is that as the complete structure of six terms vanishes for $\mu'_1 = -r$, so does it vanish for $\mu'_2 = -r$, and for $\mu'_3 = -r$; it obeys the boundary conditions.

In an extension of what could be done without comment in (21.17), we remark that the six term structure is either unaltered or reverses sign under the six operations of cyclic permutation of the l_a (three in number, including the identity), and anticyclic permutations combined with sign reversal (three in number), as illustrated by $l_1 \to -l_1$, $l_2 \to -l_3$, $l_3 \to -l_2$. These operations, applied to an exponential function of the μ_a in (21.37), produce an analogous six term structure, thereby presenting us with the final generalization of the initial transverse delta function:

$$\delta(\mathbf{r}_\perp - \mathbf{r}'_\perp) \to \sum_l \phi_l(\mu)\phi_l(\mu')^*, \quad (21.48)$$

where

$$\begin{aligned}
&(3ha)^{1/2}\phi_l(\mu)\\
&= \exp\left[i\frac{2\pi}{3h}(l_1\mu_1 + l_2\mu_2 + l_3\mu_3 - l_1 h)\right] - \exp\left[-i\frac{2\pi}{3h}(l_1\mu_1 + l_3\mu_2 + l_2\mu_3 - l_1 h)\right]\\
&+ \exp\left[i\frac{2\pi}{3h}(l_2\mu_1 + l_3\mu_2 + l_1\mu_3 - l_2 h)\right] - \exp\left[-i\frac{2\pi}{3h}(l_2\mu_1 + l_1\mu_2 + l_3\mu_3 - l_2 h)\right]\\
&+ \exp\left[i\frac{2\pi}{3h}(l_3\mu_1 + l_1\mu_2 + l_2\mu_3 - l_3 h)\right] - \exp\left[-i\frac{2\pi}{3h}(l_3\mu_1 + l_2\mu_2 + l_1\mu_3 - l_3 h)\right].
\end{aligned} \quad (21.49)$$

The l-summation is extended over all sets of the l_a that are distinct with respect to cyclic and anticyclic-reflection permutations. But one must not overlook the fact that $\phi_l(\mu)$ vanishes if any $l_a = 0$ (this is verified in Problem 21.2); the possible values of l_a are $\pm 1, \pm 2, \ldots$. Such $\phi_l(\mu)$ are complete, and orthonormal over the equilateral triangle, in the sense [analogous to (18.28)] that

$$\int (d\mathbf{r}_\perp)\, \phi_l(\mathbf{r}_\perp)^* \phi_{l'}(\mathbf{r}_\perp) = \delta_{ll'}. \tag{21.50}$$

Inasmuch as the functions $\phi_l(\mathbf{r}_\perp)$ have their origin [(21.29)] in the exponentials $\exp[i\mathbf{k}_\perp \cdot \mathbf{r}_\perp]$, they obey the differential equation

$$-\nabla_\perp^2 \phi_l = \gamma_l^2 \phi_l, \tag{21.51}$$

where [(21.34)]

$$\gamma_l^2 = \frac{2}{3}\left(\frac{\pi}{h}\right)^2 \sum_a l_a^2 \tag{21.52}$$

is indeed a quantity associated with the set l, being unaltered by cyclic and anticyclic-reflection permutations of the l_a. The smallest value of γ_l^2 is realized for the l-set indicated by $\{1, 1, -2\}$,

$$\gamma_l^2 \geq \left(\frac{2\pi}{h}\right)^2. \tag{21.53}$$

The associated function, multiplied by i [which leaves intact its contribution to (21.48)] is the real function

$$\phi_0(\mu) = \frac{2}{(3ha)^{1/2}} \sum_{a=1}^{3} \sin \frac{2\pi}{h}(\mu_a + r). \tag{21.54}$$

Its complete symmetry in the three trilinear coordinates is consistent with the general observation [(21.46)] that $\phi_l(\mu)$, with any two l_a equal, is unchanged by cyclic permutations of the μ_a. [Note that the two other possibilities, multiplication by $\exp(\pm i2\pi/3)$, are not options for a real function.]

It has been remarked that the $\phi_l(\mu)$ are normalized functions in the sense of (21.50). We rise (partly) to the challenge of verifying this for the simplest function, $\phi_0(\mu)$. That presents us with the integral

$$\begin{aligned}\int (d\mathbf{r}_\perp)\, \phi_0^2 &= \frac{4}{3ha} \int (d\mathbf{r}_\perp) \bigg\{ \sum_a \left[\sin \frac{2\pi}{h}(\mu_a + r)\right]^2 \\ &\qquad + \sum_{a \neq b} \sin \frac{2\pi}{h}(\mu_a + r) \sin \frac{2\pi}{h}(\mu_b + r) \bigg\} \\ &= \frac{4}{ha} \int (d\mathbf{r}_\perp) \bigg\{ \left[\sin \frac{2\pi}{h}(\mu_1 + r)\right]^2 \\ &\qquad + 2 \sin \frac{2\pi}{h}(\mu_1 + r) \sin \frac{2\pi}{h}(\mu_2 + r) \bigg\}.\end{aligned} \tag{21.55}$$

The second version exploits the threefold rotational symmetry of the equilateral triangle; the three integrals involving μ_a are independent of the particular choice of direction, and the six integrals containing μ_a and μ_b are the same for all pairs of directions. Now the latter trigonometric functions within braces can be presented as

$$\frac{1}{2} - \frac{1}{2} \cos \frac{4\pi}{h}(\mu_1 + r) - \cos \frac{2\pi}{h}(\mu_1 + r) + \cos \frac{2\pi}{h}(\mu_1 - \mu_2), \tag{21.56}$$

which also uses the equivalence under integration of $-\mu_1 - \mu_2 = \mu_3$ and μ_1. It is left to the reader [Problem 21.3] to prove the null value that the area integrals of these cosine functions possess. Then the outcome is just

$$\int (d\mathbf{r}_\perp)\phi_0^2 = \frac{2}{ha}\int (d\mathbf{r}_\perp) = 1, \tag{21.57}$$

which finally involves the area of the equilateral triangle. (See Chapter 27 for a simpler treatment.)

21.4 Circle

Parallel plate, rectangle, triangle—we come to the circle, which is to say, the circular cylinder. Let it be of radius a, with its axis identified as the z-axis of a circular cylindrical coordinate system. The potential of a unit point charge in vacuum within a conducting cylinder at zero potential—Green's function—is the sum of the Coulomb potential of the point charge and the potential of the surface charge distribution:

$$\begin{aligned} G(\mathbf{r},\mathbf{r}') &= \frac{1}{|\mathbf{r}-\mathbf{r}'|} + \int_S dS_1 \frac{\sigma(\mathbf{r}_1)}{|\mathbf{r}-\mathbf{r}_1|}, \\ \mathbf{r}\,\text{on}\,S: \quad & G(\mathbf{r},\mathbf{r}') = 0. \end{aligned} \tag{21.58}$$

We make use of the Coulomb potential construction (20.26b) and the addition theorem of (20.56) in writing

$$\frac{1}{|\mathbf{r}-\mathbf{r}'|} = \frac{1}{\pi}\int_{-\infty}^{\infty} dk_z e^{ik_z(z-z')} \sum_{m=-\infty}^{\infty} e^{im(\phi-\phi')} I_m(|k_z|\rho_<) K_m(|k_z|\rho_>). \tag{21.59}$$

Then the integration over the cylindrical surface in (21.58), where

$$dS_1 = dz_1\, a\, d\phi_1, \tag{21.60}$$

introduces the following measure of the surface charge density,

$$\sigma_m(k_z) = \int_{-\infty}^{\infty} dz_1\, a \int_0^{2\pi} d\phi_1 e^{-ik_z z_1} e^{-im\phi_1}\sigma(\mathbf{r}_1). \tag{21.61}$$

With this notation, the potential produced by the surface charge density is ($k = |k_z|$)

$$\frac{1}{\pi}\int_{-\infty}^{\infty} dk_z e^{ik_z z} \sum_{m=-\infty}^{\infty} e^{im\phi} I_m(k\rho) K_m(ka)\sigma_m(k_z). \tag{21.62}$$

The boundary condition on Green's function requires that the surface charge potential annul the point charge potential on the surface S ($\rho = a$, $z: -\infty \to \infty$, $\phi: 0 \to 2\pi$). In view of the completeness of the functions $\exp(ik_z z)\exp(im\phi)$ over this domain, that null statement must apply to every value of k_z and m (a null function is uniquely represented by zero coefficients). Thus, we have

$$0 = e^{-ik_z z'} e^{-im\phi'} I_m(k\rho') K_m(ka) + I_m(ka) K_m(ka)\sigma_m(k_z), \tag{21.63a}$$

or

$$\sigma_m(k_z) = -\frac{I_m(k\rho')}{I_m(ka)} e^{-ik_z z'} e^{-im\phi'}. \tag{21.63b}$$

The result is presented, in the notation

$$G(\mathbf{r}, \mathbf{r}') = 4\pi \int_{-\infty}^{\infty} \frac{dk_z}{2\pi} e^{ik_z(z-z')} \frac{1}{2\pi} \sum_{m=-\infty}^{\infty} e^{im(\phi-\phi')} g_m(\rho, \rho'; k), \tag{21.64}$$

where the reduced Green's function is

$$g_m(\rho, \rho'; k) = I_m(k\rho_<) \left[K_m(k\rho_>) - I_m(k\rho_>) \frac{K_m(ka)}{I_m(ka)} \right]. \tag{21.65}$$

Another derivation employs the differential equation

$$-\nabla^2 G(\mathbf{r}, \mathbf{r}') = 4\pi\delta(\mathbf{r} - \mathbf{r}') = 4\pi \frac{1}{\rho} \delta(\rho - \rho')\delta(\phi - \phi')\delta(z - z'). \tag{21.66}$$

The introduction of the construction (21.64) then yields the radial differential equation satisfied by g_m,

$$\left[-\frac{1}{\rho} \frac{\partial}{\partial \rho} \left(\rho \frac{\partial}{\partial \rho} \right) + k^2 + \frac{m^2}{\rho^2} \right] g_m(\rho, \rho'; k) = \frac{1}{\rho} \delta(\rho - \rho'). \tag{21.67}$$

It is to be solved under the general requirements of finiteness and continuity for g_m; the discontinuity of the derivative of g_m implied by the delta function (multiply by ρ before integrating):

$$-\rho' \frac{\partial}{\partial \rho} g_m(\rho, \rho'; k) \Big|_{\rho=\rho'-0}^{\rho=\rho'+0} = 1; \tag{21.68a}$$

and the boundary condition

$$g_m(a, \rho'; k) = 0. \tag{21.68b}$$

The homogeneous differential equation that applies for $\rho \neq \rho'$ identifies the solutions as Bessel functions of order m and imaginary argument, the functions $I_m(k\rho)$ and $K_m(k\rho)$. First, we recognize, from the small t behavior of $K_0(t)$ [(20.20)] and the general construction of (20.57), that the $K_m(t)$ are singular at $t = 0$. Accordingly, for $\rho < \rho'$, a domain that includes $\rho = 0$, only $I_m(k\rho)$ is acceptable:

$$\rho < \rho' : \quad g_m(\rho, \rho'; k) = A I_m(k\rho). \tag{21.69a}$$

As for $\rho > \rho'$, where both I_m and K_m can enter, the requirement that g_m vanish at $\rho = a$ picks out a specific linear combination [we see it in (21.65)]:

$$\rho > \rho' : \quad g_m(\rho, \rho'; k) = B \left[K_m(k\rho) - I_m(k\rho) \frac{K_m(ka)}{I_m(ka)} \right]. \tag{21.69b}$$

Continuity at $\rho = \rho'$ is satisfied by writing

$$A = C \left[K_m(k\rho') - I_m(k\rho') \frac{K_m(ka)}{I_m(ka)} \right],$$

$$B = C I_m(k\rho'). \tag{21.70a}$$

And, finally, the discontinuity condition (21.68a) tells us that

$$C\,k\rho'[K_m(k\rho')I'_m(k\rho') - I_m(k\rho')K'_m(k\rho')] = 1, \tag{21.70b}$$

where the prime on the function denotes a derivative, for example,

$$I'_m(t) \equiv \frac{d}{dt} I_m(t). \tag{21.71}$$

We meet here an example of the Wronskian [Hoëné Wronski (1778-1853)], characterizing two independent solutions of a differential equation. It is sufficient to write that equation as

$$\left[\frac{d}{dt}\left(t\frac{d}{dt}\right) + f(t)\right]\left\{\begin{matrix} I_m(t) \\ K_m(t) \end{matrix}\right\} = 0, \tag{21.72}$$

from which it follows that

$$\frac{d}{dt}\left[K_m(t)\,t\frac{d}{dt}I_m(t) - I_m(t)\,t\frac{d}{dt}K_m(t)\right] = 0, \tag{21.73a}$$

or

$$t[K_m(t)I'_m(t) - I_m(t)K'_m(t)] = 1; \tag{21.73b}$$

the value of the constant on the right-hand side is supplied by an inspection of the asymptotic forms of the functions, (20.69), (20.70). Therefore, C is unity. When the forms for $\rho < \rho'$ and $\rho > \rho'$ are united we regain (21.65).

Incidentally, the Wronskian appears explicitly in verifying that the surface charge density described by (21.63b) is what it should be in terms of the normal component of the electric field:

$$\sigma(\mathbf{r}_1) = \frac{1}{4\pi}\frac{\partial}{\partial\rho}G(\mathbf{r}, \mathbf{r}')\bigg|_{\rho=a}. \tag{21.74}$$

The introduction of the representation (21.64) for G, along with the definition of $\sigma_m(k_z)$, (21.61), converts the above relation into

$$\sigma_m(k_z) = e^{-ikz'}e^{-im\phi'}a\frac{\partial}{\partial\rho}g_m(\rho, \rho'; k)\bigg|_{\rho=a}. \tag{21.75}$$

Now, according to (21.65),

$$a\frac{\partial}{\partial\rho}g_m(\rho,\rho';k)\bigg|_{\rho=a} = I_m(k\rho')\,ka\left[K'_m(ka) - I'_m(ka)\frac{K_m(ka)}{I_m(ka)}\right] = -\frac{I_m(k\rho')}{I_m(ka)}, \tag{21.76}$$

on applying the Wronskian relation (21.73b). The result for $\sigma_m(k_z)$ reproduces (21.63b).

This reference to the surface density of induced charge naturally raises the question of its total amount. That is

$$Q_{\text{ind}} = \int_{-\infty}^{\infty} dz_1\, a\int_0^{2\pi} d\phi_1\,\sigma(\mathbf{r}_1) = \sigma_0(0) = -1, \tag{21.77}$$

according to (21.61) and (21.63b), along with the fact that $I_0(t)$ is unity at $t = 0$. We ask another question: What is the potential generated outside the radius $\rho = a$ by the combination of the internal point charge and the surface charge distributions? A glance back at the forms appropriate to $\rho > a$ will show that the significant combination of the two constituents is just what appears in (21.63a), with the common factor $K_m(ka)$ replaced by $K_m(k\rho)$. In short, the potential for all $\rho > a$ is zero; the world outside the cylinder is unaware of the balanced distribution of charge within.

Now let us turn back to the parallel conducting plates in order to develop a useful correspondence with the circular cylinder. We recall from (19.11) that the reduced Green's function for the former case is

$$g(z, z'; k) = \frac{\sinh kz_< \sinh k(a - z_>)}{k \sinh ka}, \tag{21.78a}$$

or, equivalently,

$$g(z, z'; k) = \frac{1}{k} \sinh kz_< \left[e^{-kz_>} - \sinh kz_> \frac{e^{-ka}}{\sinh ka} \right]. \tag{21.78b}$$

Notice, upon comparison with (21.65), the close analogy between the pairs $\sinh t$, e^{-t} and $I_m(t)$, $K_m(t)$, respectively. There is also an analogous Wronskian relation, appropriate to the simpler differential equation,

$$e^{-t} \frac{d}{dt} \sinh t - \sinh t \frac{d}{dt} e^{-t} = 1. \tag{21.79}$$

All this leads us to search for an analog of the alternative form of $g(z, z'; k)$ that is presented in (19.69). We stress that the statement of equivalence in the latter is an identity, holding for all values of k, real and complex. As shown in Problem 19.3, both sides of this equation have the same singularity structure, poles along the imaginary axis at $k = \pm i(n\pi/a)$ (n a positive integer), with the same residues, and vanish asymptotically as $k \to \infty$. The proof of equivalence of the two sides of (19.69) follows from the observation that, as a function of the complex variable k, the difference between the two sides is everywhere regular and bounded, which identifies it as a constant, according to the theorem of Cauchy, ascribed to Joseph Liouville (1809–1882), said constant being zero in virtue of the asymptotic behavior.

With this analogy before us, we now look back at $g_m(\rho, \rho'; k)$, (21.65), and recognize that it must be given equivalently as the sum over all the pole terms that are associated with the zeros of $I_m(ka)$. Again, as with $\sinh ka$, these occur for imaginary values of k [recall (20.50)]:

$$ka = \pm i\gamma_{mn}, \quad J_m(\gamma_{mn}) = 0, \quad n = 1, 2, \ldots, \tag{21.80}$$

in which n numbers the successive positive zeros of $J_m(t)$. In order, however, not to rely entirely on analogy, we supply an explicit proof that the zeros of $I_m(ka)$ occur only for imaginary values of k. Let us consider complex values of k and begin with the differential equations obeyed by $I_m(k\rho)$ and its complex conjugate, $I_m(k^*\rho)$:

$$\begin{aligned} \left[\frac{1}{\rho} \frac{d}{d\rho} \rho \frac{d}{d\rho} - k^2 - \frac{m^2}{\rho^2} \right] I_m(k\rho) &= 0, \\ \left[\frac{1}{\rho} \frac{d}{d\rho} \rho \frac{d}{d\rho} - k^{*2} - \frac{m^2}{\rho^2} \right] I_m(k^*\rho) &= 0. \end{aligned} \tag{21.81}$$

Now, as a generalization of the Wronskian relation, we multiply each equation by the other function, along with a factor of ρ, and subtract the two, finding that

$$\begin{aligned} &\frac{d}{d\rho} \left[I_m(k^*\rho) \rho \frac{d}{d\rho} I_m(k\rho) - I_m(k\rho) \rho \frac{d}{d\rho} I_m(k^*\rho) \right] \\ &\quad = (k^2 - k^{*2}) \rho I_m(k^*\rho) I_m(k\rho). \end{aligned} \tag{21.82}$$

Then we integrate over ρ, from 0 to a, under the assumption that

$$I_m(ka) = 0, \quad I_m(k^*a) = 0, \tag{21.83}$$

these being mutually complex conjugate statements. That yields

$$(k^2 - k^{*2}) \int_0^a d\rho\, \rho\, |I_m(k\rho)|^2 = 0, \tag{21.84}$$

which permits only one inference, $k^2 - k^{*2} = 0$, or

$$k^* = k; \quad k^* = -k. \tag{21.85}$$

The option $k^* = k$ is not realized, however, for I_m of real positive argument is everywhere positive—it has no real zeros except possibly for zero, as may be seen from the power series expression (20.58). Thus, k is imaginary, as stated in (21.80).

With k in the infinitesimal neighborhood of $\pm i\gamma_{mn}/a$, we have

$$I_m(ka) \sim [ka - (\pm i\gamma_{mn})]\, I'_m(\pm i\gamma_{mn}). \tag{21.86}$$

Then we refer to the Wronskian relation (21.73b), for $I_m(t) = 0$, to learn that

$$K_m(\pm i\gamma_{mn}) = \frac{1}{\pm i\gamma_{mn}} \frac{1}{I'_m(\pm i\gamma_{mn})}. \tag{21.87}$$

Finally, with due attention to the various powers of i [(20.50)], we arrive at

$$k \sim \pm i\gamma_{mn}/a: \quad g_m(\rho, \rho', k) \sim \frac{J_m(\gamma_{mn}\rho/a) J_m(\gamma_{mn}\rho'/a)}{[J'_m(\gamma_{mn})]^2} \frac{1}{\pm i\gamma_{mn}} \frac{1}{ka \mp i\gamma_{mn}}, \tag{21.88}$$

where the sum of the two complex conjugate pole contributions results in the following multiple for the Bessel function factor:

$$\frac{2}{k^2a^2 + \gamma_{mn}^2}. \tag{21.89}$$

The sum over all such pairs of poles then yields the desired construction

$$g_m(\rho, \rho'; k) = \sum_{n=1}^{\infty} \frac{P_{mn}(\rho) P_{mn}(\rho')}{k^2 + \gamma_{mn}^2/a^2}, \tag{21.90}$$

in which

$$P_{mn}(\rho) = \frac{\sqrt{2}}{a} \frac{J_m(\gamma_{mn}\rho/a)}{J'_m(\gamma_{mn})}. \tag{21.91}$$

The latter function obeys the Bessel differential equation

$$\left(\frac{1}{\rho}\frac{d}{d\rho}\rho\frac{d}{d\rho} + \frac{\gamma_{mn}^2}{a^2} - \frac{m^2}{\rho^2}\right) P_{mn}(\rho) = 0, \tag{21.92a}$$

and the boundary condition

$$P_{mn}(a) = 0. \tag{21.92b}$$

On applying the differential operator in (21.67) to the construction of (21.90), the denominator is canceled, leaving us with

$$\frac{1}{\rho}\delta(\rho - \rho') = \sum_{n=1}^{\infty} P_{mn}(\rho) P_{mn}(\rho'), \tag{21.93}$$

the statement of completeness of the functions $P_{mn}(\rho)$ over the interval from $\rho = 0$ to $\rho = a$. These functions are also orthonormal in the sense that

$$\int_0^a d\rho\, \rho\, P_{mn}(\rho) P_{mn'}(\rho) = \delta_{nn'}. \tag{21.94}$$

Although this property is quite analogous to those encountered in earlier examples of orthonormality, we provide a proof, one that begins with the explicit statement of completeness,

$$f(\rho) = \sum_{n=1}^{\infty} P_{mn}(\rho) \int_0^a d\rho'\, \rho' P_{mn}(\rho') f(\rho'). \tag{21.95}$$

We choose $f(\rho)$ to be $P_{mn}(\rho)$ and present the result as

$$\sum_n P_{mn}(\rho) \left[\int_0^a d\rho' \rho'\, P_{mn}(\rho') P_{mn'}(\rho') - \delta_{nn'} \right] = 0. \tag{21.96}$$

Then we assert that there is no such linear relationship among the $P_{mn}(\rho)$, except the one with completely null coefficients, which indeed yields the orthonormality relation (21.94). To provide a basis for that assertion, consider such a null combination of N different $P_{mn}(\rho)$ functions, where N can be arbitrarily large, but is finite in that we can identify a particular $P_{mn}(\rho)$ as possessing the largest value of γ_{mn} in the set of N such numbers. Thus, we accept that the members of the set can be ordered in the sense of increasing values of γ, and thereby labeled from 1 to N. Now the functions $P_{mn}(\rho)$ obey the differential equation given in (21.92a). Let us apply, to the sum of N terms, the differential operator associated with the largest value of γ. That will remove the Nth term and multiply all the other coefficients by positive numbers. Proceed similarly for the $(N-1)$th term in the sum of $N-1$ terms, and so on until one reaches a statement containing only $N = 1$, which declares its coefficient to be zero. Then read back along the list of relations with 2, 3, ... terms to conclude that all coefficients are zero—nontrivial linear relations do not exist. Of course, the orthonormality property can also be derived using the differential equation. That is left to the reader in Problems 21.4, 21.5.

What are the values of γ_{mn}? The starting point is the differential equation (18.35),

$$\left(\frac{d^2}{dt^2} + 1 - \frac{m^2 - 1/4}{t^2} \right) t^{1/2} J_m(t) = 0, \tag{21.97}$$

and the asymptotic behavior for large t, (18.37),

$$J_m(t) \sim \left(\frac{2}{\pi t} \right)^{1/2} \cos\left(t - (m + 1/2)\frac{\pi}{2} \right). \tag{21.98}$$

We use the latter as a guide to writing, generally,

$$\left(\frac{\pi t}{2} \right)^{1/2} J_m(t) = A(t) \cos \Phi(t). \tag{21.99}$$

Introducing this into the differential equation, and setting the coefficients of $\cos\Phi$ and $\sin\Phi$ equal to zero, yields

$$A'' - (\Phi')^2 A + \left(1 - \frac{m^2 - 1/4}{t^2} \right) A = 0, \tag{21.100a}$$

$$2A'\Phi' + A\Phi'' = 0. \tag{21.100b}$$

Equation (21.100b), multiplied by A, then supplies the information:

$$\frac{d}{dt}(A^2\Phi') = 0, \quad A^2\Phi' = 1, \tag{21.101}$$

where the stated value of the integration constant is provided by the asymptotic form,

$$t \gg |m| : \quad A(t) \sim 1, \quad \Phi(t) \sim t - \left(m + \frac{1}{2}\right)\frac{\pi}{2}. \tag{21.102}$$

We now use the relation of (21.101), along with (21.100a),

$$\Phi'^2 = 1 - \frac{m^2 - \frac{1}{4}}{t^2} + \frac{A''}{A}, \tag{21.103}$$

in a sequence of approximations that begins with (21.102) as the first approximation. For the second approximation we use $A \approx 1$ and $t \gg |m|$ to get

$$\Phi' \approx 1 - \frac{1}{2}\frac{m^2 - \frac{1}{4}}{t^2}, \tag{21.104a}$$

and

$$\Phi(t) \approx t - \left(m + \frac{1}{2}\right)\frac{\pi}{2} + \frac{1}{2}\frac{m^2 - \frac{1}{4}}{t}. \tag{21.104b}$$

Then (21.101) supplies the second approximation to A:

$$A(t) \approx 1 + \frac{1}{4}\frac{m^2 - \frac{1}{4}}{t^2}. \tag{21.105}$$

To go on to the third approximation, note that

$$\frac{A''}{A} \approx \frac{3}{2}\frac{m^2 - \frac{1}{4}}{t^4} \tag{21.106}$$

and then

$$\Phi' \approx 1 - \frac{1}{2}\frac{m^2 - \frac{1}{4}}{t^2} - \frac{1}{8}\frac{(m^2 - \frac{1}{4})^2}{t^4} + \frac{3}{4}\frac{m^2 - \frac{1}{4}}{t^4}, \tag{21.107}$$

leading to

$$\Phi(t) \approx t - \left(m + \frac{1}{2}\right)\frac{\pi}{2} + \frac{1}{2}\frac{m^2 - \frac{1}{4}}{t} + \frac{1}{24}\frac{\left(m^2 - \frac{1}{4}\right)^2}{t^3} - \frac{1}{4}\frac{m^2 - \frac{1}{4}}{t^3} + \dots. \tag{21.108}$$

The roots of the Bessel function $J_m(t)$, the zeros of $\cos\Phi(t)$, are inferred from

$$\Phi(\gamma) = \left(n - \frac{1}{2}\right)\pi, \quad n = 1, 2, \dots. \tag{21.109}$$

Accordingly, the successive approximations to $\Phi(t)$ provide approximations to γ as follows,

$$\begin{aligned}
1: \quad \gamma_{mn} &= \left(m + 2n - \frac{1}{2}\right)\frac{\pi}{2}, \\
2: \quad \gamma_{mn} &= \left(m + 2n - \frac{1}{2}\right)\frac{\pi}{2} - \frac{1}{\pi}\frac{m^2 - \frac{1}{4}}{m + 2n - \frac{1}{2}}, \\
3: \quad \gamma_{mn} &= \left(m + 2n - \frac{1}{2}\right)\frac{\pi}{2} - \frac{1}{\pi}\frac{m^2 - \frac{1}{4}}{m + 2n - \frac{1}{2}} \\
&\quad - \frac{1}{3\pi^3}\frac{m^2 - \frac{1}{4}}{(m + 2n - 1/2)^3}\left[7\left(m^2 - \frac{1}{4}\right) - 6\right].
\end{aligned} \tag{21.110}$$

(In effect, then, this is an expansion in powers of $1/\pi^2$.)

The smallest value of γ appears for $m = 0$, $n = 1$. The values produced by the successive approximations are

$$\begin{aligned} 1: &\quad \frac{3\pi}{4} = 2.3562, \\ 2: &\quad \frac{3\pi}{4} + \frac{1}{6\pi} = 2.4092, \\ 3: &\quad \frac{3\pi}{4} + \frac{1}{6\pi} - \frac{31}{162\pi^3} = 2.4031. \end{aligned} \tag{21.111a}$$

Compared to the actual value

$$\gamma_{01} = 2.4048255577, \tag{21.111b}$$

these are in error by -2.0%, 0.18%, and -0.07%, respectively. Turning to $m = 0$, $n = 2$ we get

$$\begin{aligned} 1: &\quad \frac{7\pi}{4} = 5.4978, \\ 2: &\quad \frac{7\pi}{4} + \frac{1}{14\pi} = 5.5205, \\ 3: &\quad \frac{7\pi}{4} + \frac{1}{14\pi} - \frac{31}{2058\pi^3} = 5.52004, \end{aligned} \tag{21.112a}$$

which are much more accurate approximations as compared with

$$\gamma_{02} = 5.5200781103, \tag{21.112b}$$

which represents errors of -0.4%, 0.008%, and -0.0007%, respectively. As a last numerical example, consider $m = 1$, $n = 1$, where

$$\begin{aligned} 1: &\quad \frac{5\pi}{4} = 3.9270, \\ 2: &\quad \frac{5\pi}{4} - \frac{3}{10\pi} = 3.8315, \\ 3: &\quad \frac{5\pi}{4} - \frac{3}{10\pi} + \frac{3}{250\pi^3} = 3.8319. \end{aligned} \tag{21.113a}$$

Relative to

$$\gamma_{11} = 3.8317059702, \tag{21.113b}$$

these are in error by 2.5%, -0.005%, and 0.005%, respectively.

We have met two statements of completeness for functions of the variable ρ, (18.57a) and (21.93):

$$\begin{aligned} \frac{1}{\rho}\delta(\rho - \rho') &= \int_0^\infty dk\, k\, J_m(k\rho) J_m(k\rho') \\ &= \sum_{n=1}^\infty \frac{2}{a^2} \frac{J_m(\gamma_{mn}\rho/a) J_m(\gamma_{mn}\rho'/a)}{[J'_m(\gamma_{mn})]^2}. \end{aligned} \tag{21.114}$$

The first of these is valid for all $\rho, \rho' > 0$; the second one applies with $0 < \rho, \rho' < a$. Does the latter yield the first version in the limit $a \to \infty$? Indeed. We begin by remarking that, for any appreciable value of γ_{mn}/a, which becomes k in the limit, γ_{mn} is arbitrarily large. That, in turn, implies correspondingly large values of the integer n, which is to say that the range of values—the spectrum—of $\gamma_{mn}/a \to k$ is effectively a continuum. In view of these

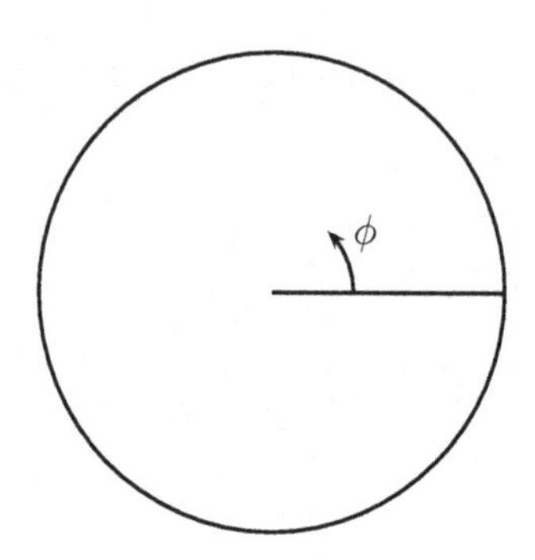

FIGURE 21.6
Cylinder with septum.

very large values of n, the first approximation of (21.110) suffices, and the asymptotic form (21.98) can be applied in evaluating $J'_m(\gamma_{mn})$, where

$$\gamma_{mn} = n\pi + \left(m - \frac{1}{2}\right)\frac{\pi}{2}. \tag{21.115}$$

Accordingly, we have

$$J'_m(\gamma_{mn}) = -\left(\frac{2}{\pi\gamma_{mn}}\right)^{1/2} \sin\left(n\pi - \frac{\pi}{2}\right) = (-1)^n \left(\frac{2}{\pi\gamma_{mn}}\right)^{1/2}, \tag{21.116}$$

and the summation in (21.114) becomes

$$\sum_n \frac{\pi}{a^2}\gamma_{mn} J_{mn}(\gamma_{mn}\rho/a) J_m(\gamma_{mn}\rho'/a). \tag{21.117}$$

Now we have only to recognize [(21.115)] that π/a is the interval between successive n-values of γ_{mn}/a, for then, with the replacements

$$\frac{\gamma_{mn}}{a} \to k, \quad \frac{\pi}{a} \to dk, \tag{21.118}$$

we realize the integral form of (21.114).

21.5 Circle and Septum

To illustrate a class of cylindrical conductors related to the circular cylinder, we introduce into the latter a radial septum, a conducting partition of negligible thickness, as illustrated in Fig. 21.6. Now Green's function must vanish at $\phi = 0$, the top face of the partition, and at $\phi = 2\pi$, the bottom face. That kind of situation is already present in parallel plates, and we have only to transcribe (19.54a) with $z, z' \to \phi, \phi'$ and $a \to 2\pi$ (we also change n into ν), to arrive at the proper replacement for $\delta(\phi - \phi')$:

$$\delta(\phi - \phi') = \frac{1}{\pi} \sum_{\nu=1}^{\infty} \sin\frac{\nu}{2}\phi \sin\frac{\nu}{2}\phi'. \tag{21.119}$$

Indeed, the functions $\sin m\phi$, which vanish at $\phi = 0$, will also vanish at $\phi = 2\pi$ provided

$$2m = \nu = 1, 2, 3, 4, \ldots, \quad \text{or} \quad m = \frac{1}{2}\nu = \frac{1}{2}, 1, \frac{3}{2}, 2, \ldots. \tag{21.120}$$

Thus, in addition to integral values of m, excluding zero, we need integer plus 1/2 values, beginning with $m = 1/2$.

The appropriate modification of the Green's function construction (21.64) is

$$G(\mathbf{r}, \mathbf{r}') = 4\pi \int_{-\infty}^{\infty} \frac{dk_z}{2\pi} e^{ik_z(z-z')} \frac{1}{\pi} \sum_{m=\nu/2} \sin m\phi \sin m\phi' g_m(\rho, \rho'; k), \tag{21.121}$$

and the differential equation obeyed by g_m is just (21.67), with m assuming the range of values in (21.120). It is to be emphasized here that in writing, for example,

$$(\nabla_\perp^2 + k^2) e^{im\phi} J_m(k\rho), \tag{21.122}$$

there is no inherent requirement that m be an integer—that is a matter of the boundary conditions. Which is to say that the constructions of g_m for the circle and septum, whether in terms of Bessel functions of imaginary argument or, alternatively, by means of Bessel functions and their zeros, have exactly the same appearance as for the unobstructed circle, only the implicit reference to the spectrum of m indicating the change. That directs our attention to what is new, the values $m = 1/2, 3/2, \ldots$.

As we know, the differential equation for $J_m(t)$ can be presented as (21.97). A striking consequence emerges for $m = 1/2$; we immediately get exact solutions: $\sin t$, $\cos t$. The latter possibility is rejected in order to achieve finiteness of $J_{1/2}(t)$ at $t = 0$,

$$J_{1/2}(t) = \left(\frac{2}{\pi t}\right)^{1/2} \sin t = \left(\frac{2}{\pi t}\right)^{1/2} \cos(t - \pi/2), \tag{21.123}$$

where the numerical factor is chosen to conform with the here everywhere valid asymptotic form (21.98). All the roots of this Bessel function are given exactly by

$$\gamma_n = n\pi, \quad n = 1, 2, \ldots, \tag{21.124}$$

which is also displayed in the three identical statements of (21.110) for $m = 1/2$.

Beginning with this simple form for $J_{1/2}$, we now proceed to construct $J_{3/2}$, $J_{5/2}$, ... In doing this, we apply (20.39), which continues to hold for non-integral values of m. As we have mentioned, $\exp(im\phi)J_m(k\rho)$ is a solution of the differential equation in (21.122), irrespective of the choice of m, and any derivative of this function is also a solution of the differential equation. Furthermore, the combination of derivatives employed in (20.39) is such as to introduce the additional $\exp(\pm i\phi)$ dependence:

$$\frac{\partial}{\partial x} \pm i\frac{\partial}{\partial y} = e^{\pm i\phi}\left(\frac{\partial}{\partial \rho} \pm \frac{i}{\rho}\frac{\partial}{\partial \phi}\right). \tag{21.125}$$

In using the relation (20.39), we shall find it convenient to write

$$e^{im\phi} = (e^{i\phi})^m = \left(\frac{x + iy}{\rho}\right)^m, \tag{21.126}$$

while noting that

$$\left(\frac{\partial}{\partial x} \pm i\frac{\partial}{\partial y}\right) f(x \pm iy) = f'(x \pm iy) + (\pm i)^2 f'(x \pm iy) = 0. \tag{21.127}$$

Accordingly, on choosing the upper sign of i, we get

$$-(x+iy)^m \frac{1}{k}\left(\frac{\partial}{\partial x}+i\frac{\partial}{\partial y}\right)\frac{J_m(k\rho)}{\rho^m} = (x+iy)^{m+1}\frac{J_{m+1}(k\rho)}{\rho^{m+1}}, \tag{21.128}$$

where, in view of (21.125) and (21.126),

$$\left(\frac{\partial}{\partial x}+i\frac{\partial}{\partial y}\right)\frac{J_m(k\rho)}{\rho^m} = \frac{x+iy}{\rho}\frac{d}{d\rho}\frac{J_m(k\rho)}{\rho^m}. \tag{21.129}$$

This produces the relation

$$-\frac{1}{t}\frac{d}{dt}\frac{J_m(t)}{t^m} = \frac{J_{m+1}(t)}{t^{m+1}}. \tag{21.130}$$

Its repeated use then yields (l is a positive integer)

$$\frac{J_{m+l}(t)}{t^{m+l}} = \left(-\frac{1}{t}\frac{d}{dt}\right)^l \frac{J_m(t)}{t^m}. \tag{21.131}$$

We apply the latter, with $m = 1/2$, to get

$$J_{l+1/2}(t) = \left(\frac{2}{\pi}\right)^{1/2} t^{l+1/2}\left(-\frac{1}{t}\frac{d}{dt}\right)^l \frac{\sin t}{t}. \tag{21.132}$$

This general construction is illustrated by

$$J_{3/2}(t) = \left(\frac{2}{\pi t}\right)^{1/2}\left(-\cos t + \frac{\sin t}{t}\right) \sim \left(\frac{2}{\pi t}\right)^{1/2}\cos(t-\pi), \tag{21.133}$$

which also displays the asymptotic form. For arbitrary l, the function

$$\sqrt{\frac{2\pi}{t}}\cos\left(t-(m+1/2)\frac{\pi}{2}\right) \tag{21.134a}$$

is multiplied by a polynomial in $1/t$, of even degree, that begins with unity, while

$$\sqrt{\frac{2\pi}{t}}\sin\left(t-(m+1/2)\frac{\pi}{2}\right) \tag{21.134b}$$

is multiplied by a odd polynomial in $1/t$.

We can give the construction (21.132) another form by observing that

$$\frac{\sin t}{t} = \frac{1}{2}\int_{-1}^{1} d\mu\, e^{i\mu t}. \tag{21.135}$$

Then we write

$$\begin{aligned}-\frac{1}{t}\frac{d}{dt}\frac{\sin t}{t} &= -\frac{i}{t}\frac{1}{2}\int_{-1}^{1} d\mu\,\mu\, e^{i\mu t}\\ &= \frac{i}{t}\frac{1}{2}\int_{-1}^{1} d\left(\frac{1-\mu^2}{2}\right) e^{i\mu t}\\ &= \frac{1}{2}\int_{-1}^{1} d\mu \frac{1}{2}(1-\mu^2)e^{i\mu t}.\end{aligned} \tag{21.136}$$

The repetition of this operation produces

$$\left(-\frac{1}{t}\frac{d}{dt}\right)^l \frac{\sin t}{t} = \frac{1}{2}\int_{-1}^{1} d\mu \frac{(1-\mu^2)^l}{2^l\, l!} e^{i\mu t}, \tag{21.137}$$

as we can check by doing it once more: $l \to l+1$. As a result we see that

$$\begin{aligned} J_{l+1/2}(t) &= \left(\frac{2}{\pi}\right)^{1/2} t^{l+1/2} \frac{1}{2}\int_{-1}^{1} d\mu \frac{(1-\mu^2)^l}{2^l\, l!} e^{i\mu t} \\ &= \left(\frac{2}{\pi}\right)^{1/2} t^{l+1/2} \frac{1}{2^l\, l!}\left(1+\frac{d^2}{dt^2}\right)^l \frac{\sin t}{t}. \end{aligned} \tag{21.138}$$

The equivalence of the two forms for $l = 1$ tells us that

$$\left(\frac{d^2}{dt^2} + \frac{2}{t}\frac{d}{dt} + 1\right)\frac{\sin t}{t} = 0. \tag{21.139}$$

More generally, the connection between $J_{l+3/2}$ and $J_{l+1/2}$ that is inferred from (21.138),

$$\frac{J_{l+3/2}(t)}{t^{l+3/2}} = \frac{1}{2l+2}\left(1+\frac{d^2}{dt^2}\right)\frac{J_{l+1/2}(t)}{t^{l+1/2}}, \tag{21.140}$$

combines with the implication of (21.131),

$$\frac{J_{l+3/2}(t)}{t^{l+3/2}} = -\frac{1}{t}\frac{d}{dt}\frac{J_{l+1/2}(t)}{t^{l+1/2}}, \tag{21.141}$$

to supply the differential equation

$$\left(\frac{d^2}{dt^2} + \frac{2l+2}{t}\frac{d}{dt} + 1\right)\frac{J_{l+1/2}(t)}{t^{l+1/2}} = 0. \tag{21.142}$$

This is, of course, a version of the Bessel differential equation for $m = l + 1/2$. One can rewrite it by noting the symbolic relations

$$t^l \frac{d}{dt} t^{-l} = \frac{d}{dt} - \frac{l}{t}, \tag{21.143a}$$

$$t^l \frac{d^2}{dt^2} t^{-l} = \left(\frac{d}{dt} - \frac{l}{t}\right)^2 = \frac{d^2}{dt^2} - \frac{2l}{t}\frac{d}{dt} + \frac{l(l+1)}{t^2}. \tag{21.143b}$$

They lead to the differential equation

$$\left(\frac{d^2}{dt^2} + \frac{2}{t}\frac{d}{dt} + 1 - \frac{l(l+1)}{t^2}\right) t^{-1/2} J_{l+1/2}(t) = 0; \tag{21.144}$$

we shall meet it again in Chapter 23.

21.6 Problems for Chapter 21

1. Show, explicitly, that the combination (21.45a) vanishes on the line of reflection, $\mu_1' = -r$. [Hint: remember the constraints (21.26b) and (21.33).

2. Prove that $\phi_l(\mu)$ defined by (21.49) vanishes if any $l_a = 0$.
3. Prove that the area integrals of the cosine functions in (21.56) vanish.
4. Show that

$$\frac{d}{d\rho}\left[\rho J_m(k\rho)\frac{d}{d\rho}J_m(k'\rho) - \rho J_m(k'\rho)\frac{d}{d\rho}J_m(k\rho)\right]$$
$$= (k^2 - k'^2)\rho J_m(k\rho)J_m(k'\rho). \tag{21.145}$$

Use this to verify that

$$n \neq n': \quad \int_0^a d\rho\, \rho\, P_{mn}(\rho)P_{mn'}(\rho) = 0. \tag{21.146}$$

5. Employ the first relation in Problem 4 to derive

$$\frac{d}{d\rho}\left[\rho^2 J'_m(k\rho)kJ'_m(k\rho) - \rho J_m(k\rho)\frac{d}{d\rho}\rho J'_m(k\rho)\right] = 2k\rho[J_m(k\rho)]^2. \tag{21.147}$$

Use this to verify that

$$\int_0^a d\rho\, \rho\, [P_{mn}(\rho)]^2 = 1. \tag{21.148}$$

6. Again employ the first relation of Problem 4, with $k' = k^*$, to prove that the roots of J_m are real.
7. Check the Wronskian relation between $I_m(t)$ and $K_m(t)$, (21.73b), using the known $t \to 0$ behavior for $m = 0$. Then, for $m > 0$, use the known $t \to 0$ behavior of $I_m(t)$ to deduce that of $K_m(t)$.
8. Apply an image method to find Green's function for the interior of an infinite conducting cylinder with cross section in the form of a semicircle of radius a. Get an approximation to the smallest value of γ in this geometry [the correct value is 3.8317...].
9. A point charge e is situated on the axis of an infinite conducting circular cylinder of radius a. Prove that the interaction energy with the cylinder is

$$E_{\text{int}} = -\frac{1}{\pi}\frac{e^2}{a}\int_0^\infty dt\, \frac{K_0(t)}{I_0(t)}, \tag{21.149a}$$

or, equivalently (think of the Wronskian)

$$E_{\text{int}} = -\frac{1}{\pi}\frac{e^2}{a}\int_0^\infty \frac{dt}{(I_0(t))^2}. \tag{21.149b}$$

10. Find Green's function for a unit point charge in vacuum outside an infinite conducting circular cylinder of radius a.
11. Having introduced the Wronskian for K_m and I_m,

$$W(K_m, I_m) = K_m(t)I'_m(t) - I_m(t)K'_m(t), \tag{21.150a}$$

in (21.73b), we can consider the corresponding quantity for the solutions of Bessel's equation of order ν,

$$W(J_\nu, N_\nu) = J_\nu(t)N'_\nu(t) - N_\nu(t)J'_\nu(t), \tag{21.150b}$$

where N_ν was introduced in Problem 18.8.

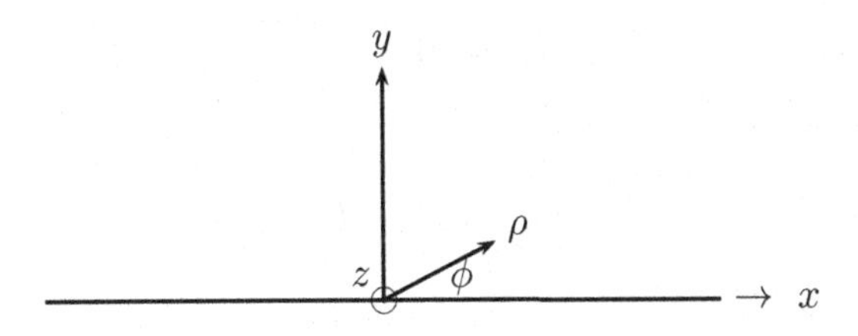

FIGURE 21.7
A perfectly conducting planar surface lying in the x-z plane. Shown are the Cartesian (x, y, z) and cylindrical polar (ρ, ϕ, z) coordinate systems used.

(a) From the differential equation satisfied by J_ν and N_ν (see Problem 18.7), show that

$$t\,W(J_\nu, N_\nu) = \text{constant.} \tag{21.151}$$

(b) Show that the constant here is $2/\pi$.

12. Calculate Green's function for Laplace's equation,

$$-\nabla^2 G(\mathbf{r}, \mathbf{r}') = 4\pi\delta(\mathbf{r} - \mathbf{r}') \tag{21.152}$$

in the interior of a hollow conducting circular cylinder of radius a and length L with conducting endcaps. Introduce a Fourier series in the direction of the axis. Give an image interpretation of the result. Compute the charge density on the surface.

13. Show that the orthonormality relations established in Problems 4 and 5 hold for nonintegral m. What further is required to prove that any function $f(\rho)$ defined on the interval $0 \leq \rho \leq a$ can be expanded in a "Fourier-Bessel" series,

$$f(\rho) = \sum_{n=1}^{\infty} c_n J_\nu\left(\gamma_{\nu n}\frac{\rho}{a}\right), \tag{21.153a}$$

$$c_n = \frac{2}{a^2 J_{\nu+1}^2(\gamma_{\nu n})}\int_0^a d\rho'\,\rho'\,f(\rho')\,J_\nu\left(\gamma_{\nu n}\frac{\rho'}{a}\right)? \tag{21.153b}$$

14. Consider a perfectly conducting surface lying in the $y = 0$ plane, as shown in Fig. 21.7. Describe the potential produced by a unit point charge, *i.e.*, the Green's function, using cylindrical coordinates, in which the z axis is perpendicular to the page, lying in the plane of the conducting surface, the x, z plane, and ρ and ϕ are the polar coordinates in the plane of the page so that

$$x = \rho\cos\phi, \quad y = \rho\sin\phi, \quad z = z. \tag{21.154}$$

The unit point charge is located in the region above the conductor, at

$$\mathbf{r}' = (\rho', \phi', z'), \quad 0 < \phi' < \pi. \tag{21.155}$$

Green's function satisfies

$$-\nabla^2 G(\mathbf{r}, \mathbf{r}') = 4\pi\delta(\mathbf{r} - \mathbf{r}'), \tag{21.156}$$

with the boundary condition

$$G\big|_{y=0} = G\big|_{\phi=0,\pi} = 0. \tag{21.157}$$

Solve for G in the upper region, $0 \leq \phi \leq \pi$, by first introducing a complete set of functions in ϕ and z satisfying the appropriate boundary conditions. Thus we may write

$$\delta(z - z') = \int_{-\infty}^{\infty} \frac{dk}{2\pi} e^{ik(z-z')}. \tag{21.158}$$

However, we cannot use

$$\delta(\phi - \phi') = \frac{1}{2\pi} \sum_{m=-\infty}^{\infty} e^{im(\phi-\phi')}. \tag{21.159}$$

Why?

(a) Show that we can write

$$G(\mathbf{r}, \mathbf{r}') = \frac{2}{\pi} \sum_{m=1}^{\infty} \sin m\phi \sin m\phi' \int_{-\infty}^{\infty} \frac{dk}{2\pi} e^{ik(z-z')} g_m(\rho, \rho'; k), \tag{21.160}$$

and derive the differential equation satisfied by g_m.

(b) Solve this differential equation for g_m in terms of Bessel functions of imaginary argument, defined by (20.73), where J_ν and N_ν are the two solutions of Bessel's equation (20.72). The function I_m is finite at the origin, unlike K_m, and, conversely, K_m goes to zero (exponentially fast) at infinity, while I_m diverges there. Use the discontinuity method; that is, consider the two regions $\rho < \rho'$ and $\rho > \rho'$, and apply the appropriate continuity and discontinuity conditions at $\rho = \rho'$. (What are the boundary conditions at $\rho = 0$, $\rho = \infty$?) In solving this equation you will need the Wronskian (21.73b). (The answer is very simple.)

(c) Now, put the result for g_m back into (21.160). First, carry out the sum over m using the addition theorem (20.56), which is valid for $\rho > \rho'$. What is the meaning of P?

(d) Then, do the k integral using the Fourier transform (20.26b). You should obtain a familiar result.

22

Spherical Harmonics

22.1 Solutions to Laplace's Equation

The fundamental solution to Laplace's equation in unbounded space is $1/r$,

$$\nabla^2 \frac{1}{r} = 0, \quad r > 0. \tag{22.1}$$

In terms of this solution, we can generate a large number of others. For example, taking $\mathbf{a}$ to be a constant vector,

$$\nabla^2 (\mathbf{a} \cdot \boldsymbol{\nabla}) \frac{1}{r} = 0, \tag{22.2}$$

we find

$$\mathbf{a} \cdot \boldsymbol{\nabla} \frac{1}{r} = -\frac{\mathbf{a} \cdot \mathbf{r}}{r^3} \tag{22.3a}$$

is also a solution for $\mathbf{r} \neq \mathbf{0}$. Continuing this operation, we see that

$$(\mathbf{a}_1 \cdot \boldsymbol{\nabla})(\mathbf{a}_2 \cdot \boldsymbol{\nabla}) \frac{1}{r} = \frac{3(\mathbf{a}_1 \cdot \mathbf{r})(\mathbf{a}_2 \cdot \mathbf{r}) - (\mathbf{a}_1 \cdot \mathbf{a}_2) r^2}{r^5} \tag{22.3b}$$

is yet a third solution. This process can be repeated an indefinite number of times, to yield the following solution to Laplace's equation,

$$[(\mathbf{a}_1 \cdot \boldsymbol{\nabla})(\mathbf{a}_2 \cdot \boldsymbol{\nabla}) \dots (\mathbf{a}_l \cdot \boldsymbol{\nabla})] \frac{1}{r} = \frac{1}{r^{2l+1}} f_l(\mathbf{r}), \tag{22.3c}$$

where $f_l(\mathbf{r})$ is a homogeneous function of $\mathbf{r}$ of degree l. We also observe that $f_l(\mathbf{r})$ itself is a solution to Laplace's equation. [This follows from the inversion theorem, that if $\phi(\mathbf{r})$ is a solution to Laplace's equation, so is $\frac{1}{r}\phi(\mathbf{r}/r^2)$. See Problem 22.1.]

Thus, our attention is directed to solutions that are homogeneous polynomials of degree l. The above construction provides examples for $l = 0, 1, 2$, which are summarized in Table 22.1. [Here we denote (x, y, z) by (x_1, x_2, x_3).] Why are there only five independent solutions for $l = 2$? A symmetrical tensor has six independent components but because of the constraint that the tensor satisfies Laplace's equation, it must be traceless, leaving but five independent components.

TABLE 22.1
Polynomial solutions of Laplace's equation.

l	f_l	Number of independent solutions
0	1	1
1	$x_1,\ x_2,\ x_3$	3
2	$3x_m x_n - \delta_{mn} r^2$	5

DOI: 10.1201/9781003057369-22

The general polynomial of degree l can be constructed from the monomials

$$x_1^{k_1} x_2^{k_2} x_3^{k_3}, \quad k_1 + k_2 + k_3 = l. \tag{22.4}$$

How many of these monomials are there? To answer this, we first ask the analogous question in two dimensions: how many monomials of the form

$$x_1^{k_1} x_2^{k_2} \tag{22.5}$$

are there with $k_1 + k_2 = n$? The answer to this question is simple since if k_1 goes from 0 to n, k_2 must go from n to 0, giving $n+1$ possibilities. Thus to answer our three-dimensional question, we first assign a definite value to k_3,

$$k_1 + k_2 = l - k_3. \tag{22.6}$$

The number of monomials with this value of k_3 is

$$l - k_3 + 1, \tag{22.7}$$

so the number of homogeneous polynomials of degree l is

$$\sum_{k_3=0}^{l} (l - k_3 + 1) = \frac{1}{2}(l + l)(l + 2). \tag{22.8}$$

From this set of polynomials, we wish to find those combinations which are solutions to Laplace's equation. Since ∇^2 acting on a homogeneous polynomial of degree l produces a homogeneous polynomial of degree $l-2$, of which there are

$$\frac{1}{2}(l - 2 + 1)(l - 2 + 2) = \frac{1}{2}l(l - 1) \tag{22.9}$$

independent ones, there are $\frac{1}{2}l(l-1)$ restrictions on the polynomials, that is, the number of independent solutions to Laplace's equation of degree l is

$$\frac{1}{2}(l + 1)(l + 2) - \frac{1}{2}l(l - 1) = 2l + 1. \tag{22.10}$$

For the cases $l = 0, 1, 2$, this agrees with what we found above. The solutions we find in this way are called *solid harmonics*, $Y_l(\mathbf{r})$. To emphasize the fact that they are homogeneous polynomials of degree l, the solid harmonic may be written in terms of a surface (or spherical) harmonic, $Y_l(\mathbf{r}/r)$:

$$Y_l(\mathbf{r}) = r^l Y_l\left(\frac{\mathbf{r}}{r}\right) \tag{22.11a}$$

$$\to \frac{1}{r^{l+1}} Y_l\left(\frac{\mathbf{r}}{r}\right), \tag{22.11b}$$

where the latter form, also a solid harmonic, results from inversion and is the solution constructed in (22.3c).

22.2 Spherical Harmonics

Our next task is to devise a systematic and convenient way to generate the spherical harmonics as functions of the spherical angles θ and ϕ. We first ask under what condition is the polynomial,

$$(\mathbf{a} \cdot \mathbf{r})^l, \tag{22.12}$$

a solution to Laplace's equation? Since

$$\boldsymbol{\nabla}(\mathbf{a}\cdot\mathbf{r})^l = l(\mathbf{a}\cdot\mathbf{r})^{l-1}\mathbf{a}, \tag{22.13}$$

we see that the Laplacian acting on this polynomial is

$$\nabla^2(\mathbf{a}\cdot\mathbf{r})^l = l(l-1)(\mathbf{a}\cdot\mathbf{r})^{l-2}\mathbf{a}^2, \tag{22.14}$$

which is Laplace's equation if $\mathbf{a}^2$ is zero (necessarily, $\mathbf{a}$ must then be complex). A convenient way to write $\mathbf{a}^2$ is

$$\mathbf{a}^2 = (a_1 - ia_2)(a_1 + ia_2) + a_3^2, \tag{22.15}$$

(again, note that the components a_i are complex), suggesting that the condition that $\mathbf{a}^2$ be zero can be automatically satisfied if we write

$$\begin{aligned} a_1 + ia_2 &= \xi_-^2, \\ a_1 - ia_2 &= -\xi_+^2, \\ a_3 &= \xi_+\xi_-, \end{aligned} \tag{22.16}$$

where $\xi_\pm$ are two arbitrary complex numbers. Then we have

$$\begin{aligned} \mathbf{a}\cdot\frac{\mathbf{r}}{r} &= \frac{1}{2}(a_1 - ia_2)\frac{x+iy}{r} + \frac{1}{2}(a_1+ia_2)\frac{x-iy}{r} + a_3\frac{z}{r} \\ &= \frac{1}{2}\left[-\xi_+^2 \sin\theta e^{i\phi} + \xi_-^2 \sin\theta e^{-i\phi} + 2\xi_+\xi_- \cos\theta\right], \end{aligned} \tag{22.17}$$

and the polynomial (22.12),

$$(\mathbf{a}\cdot\mathbf{r})^l = r^l\left(\frac{\mathbf{a}\cdot\mathbf{r}}{r}\right)^l, \tag{22.18}$$

can be rewritten in terms of $\xi_\pm$ as

$$\begin{aligned} \left(\mathbf{a}\cdot\frac{\mathbf{r}}{r}\right)^l &= \left(\frac{\xi_+^2 e^{i\phi}}{2\sin\theta}\right)^l \left[\left(\frac{\xi_-}{\xi_+}\sin\theta e^{-i\phi}\right)^2 + 2\frac{\xi_-}{\xi_+}\sin\theta e^{-i\phi}\cos\theta - \sin^2\theta\right]^l \\ &= \left(\frac{\xi_+^2 e^{i\phi}}{2\sin\theta}\right)^l \left[\left(\frac{\xi_-}{\xi_+}\sin\theta e^{-i\phi} + \cos\theta\right)^2 - 1\right]^l \\ &= \frac{\xi_+^{2l} e^{il\phi}}{2^l \sin^l\theta} \sum_{m=-l}^{l} \frac{\left(\frac{\xi_-}{\xi_+}\sin\theta e^{-i\phi}\right)^{l-m}}{(l-m)!} \left[\frac{d}{d\cos\theta}\right]^{l-m} (\cos^2\theta - 1)^l, \end{aligned} \tag{22.19}$$

where, in the last line, we have employed a convenient form of a Taylor expansion. In this way, we have constructed, from polynomials of degree l, $2l+1$ independent functions which are solutions to Laplace's equation, the coefficients of the powers of $\xi_\pm$ in the expansion

$$\begin{aligned} \left(\mathbf{a}\cdot\frac{\mathbf{r}}{r}\right)^l = l! \sum_{m=-l}^{l} & \frac{\xi_+^{l+m}\xi_-^{l-m}}{\sqrt{(l+m)!\,(l-m)!}} \sqrt{\frac{(l+m)!}{(l-m)!}} (\sin\theta)^{-m} e^{im\phi} \\ & \times \left[\frac{d}{d\cos\theta}\right]^{l-m} \frac{(\cos^2\theta-1)^l}{2^l l!}. \end{aligned} \tag{22.20}$$

All we need is a normalization factor in order to define the spherical harmonics. Employing the notation for the monomials

$$\frac{\xi_+^{l+m}\xi_-^{l-m}}{\sqrt{(l+m)!\,(l-m)!}} = \psi_{lm}, \tag{22.21}$$

we obtain the *generating function* for the spherical harmonics, $Y_{lm}(\theta,\phi)$,

$$\frac{(\mathbf{a}\cdot\mathbf{r})^l}{l!} = r^l \sum_{m=-l}^{l} \psi_{lm}\sqrt{\frac{4\pi}{2l+1}}Y_{lm}(\theta,\phi), \tag{22.22}$$

where, according to (22.20),

$$Y_{lm}(\theta,\phi) = \sqrt{\frac{2l+1}{4\pi}}\sqrt{\frac{(l+m)!}{(l-m)!}}e^{im\phi}(\sin\theta)^{-m}\left[\frac{d}{d\cos\theta}\right]^{l-m}\frac{(\cos^2\theta-1)^l}{2^l l!}, \tag{22.23}$$

in which $-l \leq m \leq l$, with $l = 0, 1, 2, \ldots$. An alternative form can be derived by noting that the left-hand side of (22.22) is unaltered by the transformation [see (22.17)]

$$\xi_+ \leftrightarrow \xi_-, \quad \theta \to -\theta, \quad \phi \to -\phi, \tag{22.24a}$$

which implies that the spherical harmonics must remain unchanged under the substitutions

$$m \to -m, \quad \theta \to -\theta, \quad \phi \to -\phi. \tag{22.24b}$$

In this way, we learn that

$$Y_{lm}(\theta,\phi) = Y_{l,-m}(-\theta,-\phi), \tag{22.25}$$

or, using the explicit form (22.23), we obtain the alternate version

$$Y_{lm}(\theta,\phi) = \sqrt{\frac{2l+1}{4\pi}}\sqrt{\frac{(l-m)!}{(l+m)!}}e^{im\phi}(-\sin\theta)^{m}\left[\frac{d}{d\cos\theta}\right]^{l+m}\frac{(\cos^2\theta-1)^l}{2^l l!}. \tag{22.26}$$

Sometimes it is convenient to separate Y_{lm} into its θ and ϕ dependences,

$$Y_{lm}(\theta,\phi) = \frac{e^{im\phi}}{\sqrt{2\pi}}\Theta_{lm}(\theta), \tag{22.27}$$

where[1]

$$\Theta_{lm}(\theta) = \sqrt{\frac{2l+1}{2}}\sqrt{\frac{(l\pm m)!}{(l\mp m)!}}(\pm\sin\theta)^{\mp m}\left[\frac{d}{d\cos\theta}\right]^{l\mp m}\frac{(\cos^2\theta-1)^l}{2^l l!}. \tag{22.29}$$

The functions $Y_{lm}(\theta,\phi)$ are called surface spherical harmonics, or simply spherical harmonics, a term introduced by Lord Kelvin, William Thomson (1824–1907). Here are the explicit forms of the harmonics of degree $l = 0, 1, 2$:

$$l = 0: \quad Y_{00} = \frac{1}{\sqrt{4\pi}}, \tag{22.30a}$$

$$\begin{aligned} l = 1: \quad Y_{11} &= -\sqrt{\frac{3}{8\pi}}\frac{x+iy}{r} = -\sqrt{\frac{3}{8\pi}}\sin\theta e^{i\phi}, \\ Y_{10} &= \sqrt{\frac{3}{4\pi}}\frac{z}{r} = \sqrt{\frac{3}{4\pi}}\cos\theta, \\ Y_{1,-1} &= \sqrt{\frac{3}{8\pi}}\frac{x-iy}{r} = \sqrt{\frac{3}{8\pi}}\sin\theta e^{-i\phi}, \end{aligned} \tag{22.30b}$$

[1] Apart from a normalization factor, this is the associated Legendre function

$$\Theta_{lm}(\theta) = \sqrt{\frac{2l+1}{2}}\sqrt{\frac{(l-m)!}{(l+m)!}}P_l^m(\mu), \quad \mu = \cos\theta. \tag{22.28}$$

$$\begin{aligned} l=2: \quad Y_{22} &= \sqrt{\frac{15}{32\pi}}\frac{(x+iy)^2}{r^2} = \sqrt{\frac{15}{32\pi}}\sin^2\theta e^{2i\phi}, \\ Y_{21} &= -\sqrt{\frac{15}{8\pi}}\frac{z(x+iy)}{r^2} = -\sqrt{\frac{15}{8\pi}}\cos\theta\sin\theta e^{i\phi}, \\ Y_{20} &= \sqrt{\frac{5}{16\pi}}\frac{(3z^2-r^2)}{r^2} = \sqrt{\frac{5}{16\pi}}(3\cos^2\theta-1), \\ Y_{2,-1} &= \sqrt{\frac{15}{8\pi}}\frac{z(x-iy)}{r^2} = \sqrt{\frac{15}{8\pi}}\cos\theta\sin\theta e^{-i\phi}, \\ Y_{2,-2} &= \sqrt{\frac{15}{32\pi}}\frac{(x-iy)^2}{r^2} = \sqrt{\frac{15}{32\pi}}\sin^2\theta e^{-2i\phi}. \end{aligned} \tag{22.30c}$$

22.3 Orthonormality Condition

The particular factors that occur in the definition of Y_{lm} are such as to make the spherical harmonics an orthonormal set of functions. To see this, consider the product of two generating functions, with parameters $\mathbf{a}$ and $\mathbf{a}^*$, integrated over all angles:

$$\int d\Omega\left(\mathbf{a}^*\cdot\frac{\mathbf{r}}{r}\right)^l\left(\mathbf{a}\cdot\frac{\mathbf{r}}{r}\right)^{l'} \tag{22.31a}$$

with

$$d\Omega = \sin\theta\, d\theta\, d\phi. \tag{22.31b}$$

This integral can only contain rotationally invariant combinations, that is, it has to be a function of scalars constructed from $\mathbf{a}$ and $\mathbf{a}^*$. Since $\mathbf{a}^2 = \mathbf{a}^{*2} = 0$, the only such scalar is $\mathbf{a}\cdot\mathbf{a}^*$. Therefore, there must be an equal number of factors of $\mathbf{a}$ and $\mathbf{a}^*$, which means that the integral (22.31a) is zero unless $l = l'$; we have

$$\int d\Omega\left(\mathbf{a}^*\cdot\frac{\mathbf{r}}{r}\right)^l\left(\mathbf{a}\cdot\frac{\mathbf{r}}{r}\right)^{l'} = \delta_{ll'}C_l(\mathbf{a}^*\cdot\mathbf{a})^l. \tag{22.32}$$

To calculate C_l we consider a particular form of $\mathbf{a}$:

$$\mathbf{a} = (1, i, 0), \quad \mathbf{a}^* = (1, -i, 0). \tag{22.33}$$

The quantities appearing in (22.32) are then

$$\mathbf{a}^*\cdot\frac{\mathbf{r}}{r} = \sin\theta e^{-i\phi}, \quad \mathbf{a}\cdot\frac{\mathbf{r}}{r} = \sin\theta e^{i\phi}, \quad \mathbf{a}^*\cdot\mathbf{a} = 2, \tag{22.34}$$

implying that the integral, (22.32), for $l = l'$, is

$$\begin{aligned} C_l 2^l &= \int_0^\pi \sin\theta\, d\theta \int_0^{2\pi} d\phi\,(\sin\theta e^{-i\phi})^l(\sin\theta e^{i\phi})^l \\ &= 2\pi\int_{-1}^1 d(\cos\theta)(1-\cos^2\theta)^l = 4\pi\int_0^1 d\mu\,(1-\mu^2)^l \\ &= 4\pi\frac{(2^l l!)^2}{(2l+1)!}. \end{aligned} \tag{22.35}$$

The final integral in (22.35) is evaluated as follows. Defining

$$I_l = \int_0^1 d\mu\,(1-\mu^2)^l, \quad I_0 = 1, \tag{22.36}$$

we integrate by parts, for $l > 0$, to derive the recursion formula

$$I_l = 2lI_{l-1} - 2lI_l, \tag{22.37}$$

which implies that

$$\begin{aligned} I_l &= \frac{2l}{2l+1} I_{l-1} \\ &= \frac{2l}{2l+1}\frac{2l-2}{2l-1}\frac{2l-4}{2l-3}\cdots I_0 \\ &= \frac{(2^l l!)^2}{(2l+1)!}, \end{aligned} \tag{22.38a}$$

the last form being valid for $l \geq 0$. Alternatively, we could evaluate I_l in terms of the beta function, $B(m,n)$,

$$\begin{aligned} I_l &= \frac{1}{2}\int_{-1}^1 d\mu\,(1-\mu)^l(1+\mu)^l \\ &= 2^{2l}\int_0^1 dt\,t^l(1-t)^l \qquad \left[t = \frac{1-\mu}{2}\right] \\ &= 2^{2l}B(l+1,l+1) \\ &= 2^{2l}\frac{l!\,l!}{(2l+1)!}, \end{aligned} \tag{22.38b}$$

where we have noted that for integer m and n,

$$B(m,n) = \int_0^1 dt\,t^{m-1}(1-t)^{n-1} = \frac{(m-1)!(n-1)!}{(m+n-1)!}. \tag{22.39}$$

We have therefore learned that

$$\begin{aligned} &\int d\Omega \frac{(\mathbf{a}^*\cdot\mathbf{r}/r)^l}{l!}\frac{(\mathbf{a}\cdot\mathbf{r}/r)^{l'}}{l'!} = \delta_{ll'}4\pi\frac{2^l}{(2l+1)!}(\mathbf{a}^*\cdot\mathbf{a})^l \\ &= \sum_{m=-l}^{l}\sum_{m'=-l'}^{l'}\psi^*_{lm}\psi_{l'm'}\sqrt{\frac{4\pi}{2l+1}}\sqrt{\frac{4\pi}{2l'+1}}\int d\Omega\, Y^*_{lm}(\theta,\phi)Y_{l'm'}(\theta,\phi), \end{aligned} \tag{22.40}$$

where we have used the generating function (22.22). What we now must do is extract the coefficient of $\psi^*_{lm}\psi_{l'm'}$ from $(\mathbf{a}^*\cdot\mathbf{a})^l$, which is achieved as follows:

$$\begin{aligned} \frac{2^l(\mathbf{a}^*\cdot\mathbf{a})^l}{(2l+1)!} &= \frac{(\xi^*_+\xi_+ + \xi^*_-\xi_-)^{2l}}{(2l+1)!} \\ &= \frac{1}{(2l+1)!}\sum_{m=-l}^{l}\frac{(2l)!}{(l+m)!(l-m)!}(\xi^*_+\xi_+)^{l+m}(\xi^*_-\xi_-)^{l-m} \\ &= \frac{1}{2l+1}\sum_{m=-l}^{l}\psi^*_{lm}\psi_{lm}, \end{aligned} \tag{22.41}$$

where we have used (22.16), (22.21), and the binomial expansion. Comparing this with (22.40), we obtain the orthonormality condition for the spherical harmonics:

$$\int d\Omega\, Y_{lm}^{*}(\theta,\phi) Y_{l'm'}(\theta,\phi) = \delta_{ll'}\delta_{mm'}. \tag{22.42}$$

When Y_{lm} is separated as in (22.27), the orthonormality condition reads

$$\int_0^{\pi} \sin\theta\, d\theta\, \Theta_{lm}(\theta)\Theta_{l'm}(\theta) = \delta_{ll'}. \tag{22.43}$$

22.4 Legendre's Polynomials

A few special cases of Θ_{lm} can be easily extracted from (22.29):

$$\Theta_{l,-l}(\theta) = \sqrt{\frac{(2l+l)!}{2}}\frac{(\sin\theta)^l}{2^l l!}, \tag{22.44a}$$

$$\Theta_{ll}(\theta) = (-1)^l \Theta_{l,-l}(\theta), \tag{22.44b}$$

$$\begin{aligned}\Theta_{l0}(\theta) &= \sqrt{\frac{2l+1}{2}}\left[\frac{d}{d(\cos\theta)}\right]^l \frac{(\cos^2\theta-1)^l}{2^l l!}\\ &= \sqrt{\frac{2l+1}{2}} P_l(\cos\theta).\end{aligned} \tag{22.44c}$$

Occurring in (22.44c) is Legendre's polynomial of order l,

$$P_l(\cos\theta) = \left[\frac{d}{d(\cos\theta)}\right]^l \frac{(\cos^2\theta-1)^l}{2^l l!}, \tag{22.45}$$

named for Adrien Marie Legendre (1752–1833), which is so normalized that

$$P_l(1) = 1. \tag{22.46}$$

According to (22.43), Legendre's polynomials satisfy the following orthonormality condition:

$$\int_{-1}^{1} d(\cos\theta) P_l(\cos\theta) P_{l'}(\cos\theta) = \frac{2}{2l+1}\delta_{ll'}. \tag{22.47}$$

We will begin to explore the physical significance of Legendre's polynomials in the next chapter.

22.5 Problems for Chapter 22

1. Assume that $\phi(\mathbf{r})$ is a solution of Laplace's equation. Show that for $r > 0$,

$$\frac{1}{r}\phi\left(\frac{\mathbf{r}}{r^2}\right) \tag{22.48}$$

is also a solution.

2. Check that

$$\xi_+ \frac{\partial}{\partial \xi_-} \psi_{lm} = \sqrt{(l-m)(l+m+1)}\psi_{lm+1}, \tag{22.49a}$$

$$\xi_- \frac{\partial}{\partial \xi_+} \psi_{lm} = \sqrt{(l+m)(l-m+1)}\psi_{lm-1}. \tag{22.49b}$$

Note that these are related by $\xi_+ \leftrightarrow \xi_-$, $\psi_{lm} \to \psi_{l,-m}$. Show that

$$e^{i\phi}\left(\frac{\partial}{\partial\theta} - \cot\theta \frac{1}{i}\frac{\partial}{\partial\phi}\right)(\mathbf{a}\cdot\mathbf{r})^l = \xi_- \frac{\partial}{\partial \xi_+}(\mathbf{a}\cdot\mathbf{r})^l, \tag{22.50a}$$

$$e^{-i\phi}\left(-\frac{\partial}{\partial\theta} - \cot\theta \frac{1}{i}\frac{\partial}{\partial\phi}\right)(\mathbf{a}\cdot\mathbf{r})^l = \xi_+ \frac{\partial}{\partial \xi_-}(\mathbf{a}\cdot\mathbf{r})^l, \tag{22.50b}$$

which are related by $\xi_+ \leftrightarrow \xi_-$, $\theta \to -\theta$, $\phi \to -\phi$. The generating function (22.22) shows that these are equivalent to

$$e^{i\phi}\left(\frac{\partial}{\partial\theta} - \cot\theta \frac{1}{i}\frac{\partial}{\partial\phi}\right)Y_{lm} = \sqrt{(l-m)(l+m+1)}Y_{lm+1}, \tag{22.51a}$$

$$e^{-i\phi}\left(-\frac{\partial}{\partial\theta} - \cot\theta \frac{1}{i}\frac{\partial}{\partial\phi}\right)Y_{lm} = \sqrt{(l+m)(l-m+1)}Y_{lm-1}. \tag{22.51b}$$

These two equations are related by $\theta \to -\theta$, $\phi \to -\phi$, $Y_{lm} \to Y_{l-m}$.

3. Rewrite the conclusion of the previous problem for $\Theta_{lm}(\theta)$. Derive the differential equations

$$\left(-\frac{\partial}{\partial\theta} - (m+1)\cot\theta\right)\left(\frac{\partial}{\partial\theta} - m\cot\theta\right)\Theta_{lm} = [l(l+1) - m(m+1)]\Theta_{lm}, \tag{22.52a}$$

$$\left(\frac{\partial}{\partial\theta} - (m-1)\cot\theta\right)\left(-\frac{\partial}{\partial\theta} - m\cot\theta\right)\Theta_{lm} = [l(l+1) - m(m-1)]\Theta_{lm}. \tag{22.52b}$$

4. Prove

$$Y^*_{lm} = (-1)^m Y_{l-m}, \tag{22.53}$$

by noting that $(\mathbf{a}\cdot\hat{\mathbf{r}})^*$ is obtained from $(\mathbf{a}\cdot\hat{\mathbf{r}})$ by the replacement $\xi_+ \to i\xi^*_-$, $\xi_- \to -i\xi^*_+$. Show the consistency of this result with the conclusion of Problem 2.

5. Assume $\mathbf{a}_1$, $\mathbf{a}_2$, $\mathbf{a}_3$ are all null vectors. Then

$$\int d\Omega\,(\mathbf{a}_1\cdot\hat{\mathbf{r}})^{l_1}(\mathbf{a}_2\cdot\hat{\mathbf{r}})^{l_2}(\mathbf{a}_3\cdot\hat{\mathbf{r}})^{l_3} = C(\mathbf{a}_2\cdot\mathbf{a}_3)^{n_1}(\mathbf{a}_3\cdot\mathbf{a}_1)^{n_2}(\mathbf{a}_1\cdot\mathbf{a}_2)^{n_3}. \tag{22.54}$$

Find the condition on l_1, l_2, l_3 such that

$$\int d\Omega\, Y_{l_1 m_1} Y_{l_2 m_2} Y_{l_3, m_3} \neq 0. \tag{22.55}$$

This is equivalent to

$$Y_{l_1 m_1} Y_{l_2 m_2} = \sum (\text{coefficients}) Y_{l_3 m_3}. \tag{22.56}$$

6. An example of a null vector is

$$\mathbf{a} = (-i\cos\alpha, -i\sin\alpha, 1), \tag{22.57}$$

that is,

$$\mathbf{a}\cdot\mathbf{r} = -i(x\cos\alpha + y\sin\alpha) + z. \tag{22.58}$$

Show that, for this **a**,

$$\int_0^{2\pi}\frac{d\alpha}{2\pi}(\mathbf{a}\cdot\mathbf{r})^l e^{im\alpha} = i^m \frac{l!}{\sqrt{(l+m)!\,(l-m)!}}\sqrt{\frac{4\pi}{2l+1}}r^l Y_{lm}(\theta,\phi), \tag{22.59}$$

and, in particular, that

$$P_l(\cos\theta) = \int_0^{\pi}\frac{d\alpha}{\pi}(\cos\theta - i\sin\theta\cos\alpha)^l. \tag{22.60}$$

This is called Laplace's first integral representation.

7. Apply the last formula to evaluate, for $|t| < 1$,

$$\sum_{l=0}^{\infty} t^l P_l(\cos\theta) \tag{22.61}$$

and arrive at a known result. Would you expect that $P_l(\cos\theta)$ is also given by

$$\int_0^{\pi}\frac{d\alpha}{\pi}(\cos\theta - i\sin\theta\cos\alpha)^{-l-1}? \tag{22.62}$$

(This is Laplace's second integral representation, presented by Jacobi in 1843. It requires $\cos\theta > 0$.) In any event, again work out $\sum_{l=0}^{\infty} t^l P_l(\cos\theta)$, using the latter formula.

8. Apply the integral representation of J_0, and the result in Problem 6, to show that

$$P_l(\cos\theta) = \left(\cos\theta - \sin\theta\frac{d}{dt}\right)^l J_0(t)\bigg|_{t=0}. \tag{22.63}$$

Verify this for $l = 0, 1, 2$. Now let $\theta = x/l$ and, for fixed x, consider the limit $l \to \infty$. You should get

$$\lim_{l\to\infty} P_l\left(\cos\frac{x}{l}\right) = J_0(x), \tag{22.64}$$

which is often used in the approximate form

$$\theta \ll 1,\ l \gg 1: \quad P_l(\cos\theta) \sim J_0(l\theta). \tag{22.65}$$

9. For what geometrical reason does one expect an asymptotic connection between spherical and cylindrical coordinate functions? Use this insight in comparing the differential equations for $P_l(\cos\theta)$ and $J_0(t)$ to arrive again at the result of the previous problem.

10. $\left(\frac{\partial}{\partial z}\right)^l \frac{1}{r}$ is a solution of Laplace's equation provided $r > 0$. (Why?) It is homogeneous in **r** of degree $-l-1$ and is independent of the angle ϕ. Therefore,

$$\left(\frac{\partial}{\partial z}\right)^l \frac{1}{r} = C_l \frac{1}{r^{l+1}} P_l(\cos\theta). \tag{22.66}$$

Find the constant C_l. Check the result for $l = 0, 1, 2$.

11. This is a generalization of Problem 8. Show that

$$\lim_{l\to\infty}\sqrt{\frac{2}{2l+1}}\Theta_{lm}\left(\frac{x}{l}\right)=(-1)^m J_m(x). \tag{22.67}$$

12. Show that

$$\sum_{l=0}^{\infty}\frac{(-1)^l}{l!}(kr)^l P_l(\cos\theta)=e^{-kz}J_0(k\rho). \tag{22.68}$$

13. This is a generalization of Problem 10. Indicate the analogous reasoning that leads to

$$\left(\frac{\partial}{\partial x}+i\frac{\partial}{\partial y}\right)^m\left(\frac{\partial}{\partial z}\right)^{l-m}\frac{1}{r}=C_{lm}\frac{1}{r^{l+1}}Y_{lm}(\theta,\phi), \tag{22.69}$$

which also holds for $m<0$, in the sense that

$$\left(\frac{\partial}{\partial x}+i\frac{\partial}{\partial y}\right)^{-1}\frac{\partial^2}{\partial z^2}\equiv-\left(\frac{\partial}{\partial x}-i\frac{\partial}{\partial y}\right), \tag{22.70}$$

understood as an operator statement acting on $1/r$. Find C_{lm}.

14. Spherical harmonics are defined by (22.22), so show that

$$\frac{\mathbf{a}\cdot\mathbf{r}}{l+1}\sum_{m=-l}^{l}\psi_{lm}\sqrt{\frac{4\pi}{2l+1}}r^l Y_{lm}(\theta,\phi)=\sum_{m=-l-1}^{l+1}\sqrt{\frac{4\pi}{2(l+1)+1}}r^{l+1}Y_{l+1,m}(\theta,\phi), \tag{22.71}$$

where $\mathbf{a}\cdot\mathbf{r}$ is given by (22.17). Write out the explicit construction this gives for the spherical harmonics of degree $l+1$ in terms of those of degree l:

$$\begin{aligned}Y_{l+1,m}(\theta,\phi)=\frac{1}{2(l+1)}\sqrt{\frac{2l+3}{2l+1}}\Big[&-\sqrt{(l+m)(l+m+1)}\sin\theta e^{i\phi}Y_{l,m-1}(\theta,\phi)\\&+\sqrt{(l-m)(l-m+1)}\sin\theta e^{-i\phi}Y_{l,m+1}(\theta,\phi)\\&+2\sqrt{(l+m+1)(l-m+1)}\cos\theta Y_{lm}(\theta,\phi)\Big].\end{aligned} \tag{22.72}$$

Begin with $Y_{00}=1/\sqrt{4\pi}$ and construct the three Y_{1m} and then find one or more of the five Y_{2m}. You should reproduce the results in (22.30b) and (22.30c).

23

Coulomb's Potential

The motivation for constructing the solid harmonics was that they formed, in terms of homogeneous functions, a particular set of solutions to Laplace's equation. Since these harmonics are functions of the spherical angles θ and ϕ, Laplace's equation should be expressed in spherical coordinates, where the Laplacian has the form

$$\nabla^2 = \frac{1}{r^2}\frac{\partial}{\partial r}\left(r^2\frac{\partial}{\partial r}\right) + \frac{1}{r^2}\left[\frac{1}{\sin\theta}\frac{\partial}{\partial\theta}\left(\sin\theta\frac{\partial}{\partial\theta}\right) + \frac{1}{\sin^2\theta}\frac{\partial^2}{\partial\phi^2}\right]. \tag{23.1}$$

(This may be immediately inferred from (18.15) by noting that the distances corresponding to infinitesimal changes of the spherical coordinates r, θ, and ϕ are dr, $r\,d\theta$, and $r\sin\theta\,d\phi$, respectively.) Thus, since the solid harmonics, (22.11a) and (22.11b),

$$Y_{lm}(\mathbf{r}) = \left\{\begin{matrix} r^l \\ r^{-l-1}\end{matrix}\right\} Y_{lm}(\theta,\phi), \tag{23.2}$$

are solutions to

$$\nabla^2 Y_{lm}(\mathbf{r}) = 0, \quad r \neq 0, \tag{23.3}$$

and since

$$\frac{d}{dr}\left(r^2\frac{d}{dr}\right)\left\{\begin{matrix} r^l \\ r^{-l-1}\end{matrix}\right\} = l(l+1)\left\{\begin{matrix} r^l \\ r^{-l-1}\end{matrix}\right\}, \tag{23.4}$$

the differential equation satisfied by the spherical harmonics, Y_{lm}, is

$$\left[\frac{1}{\sin\theta}\frac{\partial}{\partial\theta}\left(\sin\theta\frac{\partial}{\partial\theta}\right) + \frac{1}{\sin^2\theta}\frac{\partial^2}{\partial\phi^2} + l(l+1)\right] Y_{lm}(\theta,\phi) = 0. \tag{23.5}$$

When the θ and ϕ dependence of Y_{lm} is separated as in (22.27), the differential equation for Θ_{lm} is

$$\left[\frac{1}{\sin\theta}\frac{\partial}{\partial\theta}\left(\sin\theta\frac{\partial}{\partial\theta}\right) + l(l+1) - \frac{m^2}{\sin^2\theta}\right]\Theta_{lm}(\theta) = 0. \tag{23.6}$$

23.1 Legendre's Polynomials

The fundamental solution of Laplace's equation is Coulomb's potential, (15.3), for $\mathbf{r} \neq \mathbf{r}'$, which, written in spherical coordinates, is

$$\frac{1}{|\mathbf{r}-\mathbf{r}'|} = \frac{1}{\sqrt{r^2 + r'^2 - 2rr'\cos\gamma}}, \tag{23.7}$$

where γ is the angle between $\mathbf{r}$ and $\mathbf{r}'$, explicitly,

$$\cos\gamma = \cos\theta\cos\theta' + \sin\theta\sin\theta'\cos(\phi-\phi'). \tag{23.8}$$

DOI: 10.1201/9781003057369-23

We now expand (23.7) as

$$\frac{1}{\sqrt{r_>^2 + r_<^2 - 2r_> r_< \cos\gamma}} = \sum_{l=0}^{\infty} \left(\frac{r_<^l}{r_>^{l+1}}\right) (\text{polynomial of degree } l \text{ in } \cos\gamma), \tag{23.9}$$

where $r_>$ ($r_<$) is the greater (lesser) of r and r'. (The recognition that the square root is also the absolute value $r_>|1 - (r_</r_>)e^{i\gamma}|$ makes it evident that this expansion converges.) The polynomial of degree l appearing here is a solution to (23.5), and so must be a linear combination of $Y_{lm}(\theta, \phi)$'s, $-l < m < l$. On the other hand, as we will show below, this is just Legendre's polynomial in $\cos\gamma$, (22.45), that is

$$\frac{1}{|\mathbf{r} - \mathbf{r}'|} = \sum_{l=0}^{\infty} \frac{r_<^l}{r_>^{l+1}} P_l(\cos\gamma). \tag{23.10}$$

(In fact, this is how Legendre first introduced his polynomial in 1784.) For $\gamma = 0$, this expansion is trivially

$$\frac{1}{r_> - r_<} = \sum_{l=0}^{\infty} \frac{r_<^l}{r_>^{l+1}} \tag{23.11}$$

which supplies the normalization condition

$$P_l(1) = 1, \tag{23.12}$$

as is required for Legendre's polynomials [see (22.46)].

The demonstration that the polynomial introduced here coincides with Legendre's polynomial follows immediately if we take $\mathbf{r}'$ to lie on the z-axis so that γ and θ are identical. Then, with $r_< = r$, we recognize in (23.10) that $r^l P_l(\cos\theta)$ is a solution to Laplace's equation, a solid harmonic of degree l, in which the surface harmonic factor is independent of ϕ. In short, $P_l(\cos\theta)$, as defined in (23.10), is proportional to $\Theta_{l0}(\theta)$, and in virtue of the property (23.12), is identical with what is defined in (22.44c).

A more direct proof of this identity, and further properties of the spherical harmonics, follow if we expand Coulomb's potential

$$G(\mathbf{r}, \mathbf{r}') = \frac{1}{|\mathbf{r} - \mathbf{r}'|}, \tag{23.13}$$

which satisfies the inhomogeneous Green's function equation

$$-\nabla^2 G(\mathbf{r}, \mathbf{r}') = 4\pi\delta(\mathbf{r} - \mathbf{r}'), \tag{23.14}$$

in terms of the solutions to the homogeneous Laplace's equation, (23.3). In spherical coordinates, the delta function is [see (18.40)]

$$\delta(\mathbf{r} - \mathbf{r}') = \frac{1}{r^2}\delta(r - r')\frac{1}{\sin\theta}\delta(\theta - \theta')\delta(\phi - \phi'), \tag{23.15}$$

while the Laplacian is given by (23.1). For $r < r'$, (23.10) shows that the solution to (23.14) can be expanded in powers of r, [cf. (23.2)]

$$r < r': \quad G = \sum_{lm} A_{lm} r^l Y_{lm}(\theta, \phi), \tag{23.16a}$$

while for $r > r'$ the expansion is in terms of powers of $1/r$,

$$r > r': \quad G = \sum_{lm} B_{lm} r^{-l-1} Y_{lm}(\theta, \phi). \tag{23.16b}$$

(This guarantees that G is bounded both at $r = 0$ and as $r \to \infty$.) The expansion coefficients, A_{lm} and B_{lm}, depending on r', θ' and ϕ', are to be determined by the conditions on Green's function near the source:

$$G \text{ is continuous at } r = r'; \tag{23.17a}$$

and

$$\left[-r^2 \frac{\partial}{\partial r} G\right]_{r'-0}^{r'+0} = 4\pi \frac{1}{\sin\theta} \delta(\theta - \theta')\delta(\phi - \phi'). \tag{23.17b}$$

These two conditions imply, respectively,

$$r'^l A_{lm} = \frac{1}{r'^{l+1}} B_{lm}, \tag{23.18a}$$

$$\sum_{lm} \left[(l+1)\frac{1}{r'^l} B_{lm} + l r'^{l+1} A_{lm}\right] Y_{lm}(\theta, \phi) = 4\pi \frac{1}{\sin\theta} \delta(\theta - \theta')\delta(\phi - \phi'). \tag{23.18b}$$

If we write

$$A_{lm} = r'^{-l-1} C_{lm}, \quad B_{lm} = r'^l C_{lm}, \tag{23.19}$$

(23.18a) is satisfied automatically, while (23.18b) reads

$$\sum_{lm} (2l+1) C_{lm} Y_{lm}(\theta, \phi) = 4\pi \frac{1}{\sin\theta} \delta(\theta - \theta')\delta(\phi - \phi'). \tag{23.20}$$

The use of the orthonormality condition, (22.42), now yields

$$C_{lm} = \frac{4\pi}{2l+1} Y_{lm}^*(\theta', \phi'). \tag{23.21}$$

By substituting this into (23.20), we obtain the completeness statement for the spherical harmonics,

$$\sum_{lm} Y_{lm}(\theta, \phi) Y_{lm}^*(\theta', \phi') = \frac{1}{\sin\theta} \delta(\theta - \theta')\delta(\phi - \phi'), \tag{23.22}$$

which allows us to expand any function of θ and ϕ in terms of spherical harmonics. We therefore have obtained such an expansion for Green's function

$$G(\mathbf{r}, \mathbf{r}') = \frac{1}{|\mathbf{r} - \mathbf{r}'|} = \sum_{lm} \frac{r_<^l}{r_>^{l+1}} \frac{4\pi}{2l+1} Y_{lm}(\theta, \phi) Y_{lm}^*(\theta', \phi'). \tag{23.23}$$

Comparing this with the alternative representation, (23.10), we obtain the relation

$$P_l(\cos\gamma) = \frac{4\pi}{2l+1} \sum_m Y_{lm}(\theta, \phi) Y_{lm}^*(\theta', \phi'). \tag{23.24}$$

The relation (23.24) is called the addition theorem for spherical harmonics.

Let us finally show explicitly that this function of $\cos\gamma$ actually is Legendre's polynomial, (22.45). As noted above, this is easily done by considering a particular coordinate system, where

$$\theta' = 0 \Rightarrow \gamma = \theta. \tag{23.25}$$

From (22.29), we learn that

$$Y_{lm}(\theta', \phi') \propto (\sin\theta')^{|m|}, \tag{23.26}$$

implying [see (22.44c)]

$$Y_{lm}(0,\phi') = \delta_{m0}\sqrt{\frac{2l+1}{4\pi}}, \tag{23.27}$$

so that only the $m = 0$ term contributes to the right-hand side of (23.24), which is therefore, by (22.44c),

$$\frac{4\pi}{2l+1}\sqrt{\frac{2l+1}{4\pi}}P_l(\cos\theta)\sqrt{\frac{2l+1}{4\pi}} = P_l(\cos\theta). \tag{23.28}$$

Thus we have proved that the function of $\cos\gamma$ occurring in (23.10) is indeed Legendre's polynomial.

23.2 Infinitesimal Rotations

So far we have not explicitly used the differential equation satisfied by the spherical harmonics, or by $\Theta_{lm}(\theta)$, (23.6). We confine ourselves to $m = 0$ in noting that the introduction of the variable

$$\mu = \cos\theta \tag{23.29}$$

provides the following form for the differential equation obeyed by Legendre's polynomial:

$$\left[\frac{d}{d\mu}\left((1-\mu^2)\frac{d}{d\mu}\right) + l(l+1)\right] P_l(\mu) = 0. \tag{23.30}$$

One easily checks this equation for the first few polynomials,

$$\begin{aligned} P_0(\mu) &= 1, \qquad P_1(\mu) = \mu, \\ P_2(\mu) &= \frac{1}{2}(3\mu^2-1), \quad P_3(\mu) = \frac{1}{2}(5\mu^3-3\mu). \end{aligned} \tag{23.31}$$

While the relevance of Laplace's equation cannot be minimized, the differential properties of the harmonics are fundamentally statements about their responses to infinitesimal rotations of the coordinate system, or, equivalently, of the vectors $\mathbf{r}$ and $\mathbf{a}$. (Recall Sec. 22.2.) An infinitesimal rotation is described by a vector $\delta\boldsymbol{\omega}$, specifying the axis and angle of rotation. The common rotation of $\mathbf{r}$ and $\mathbf{a}$ is represented by the infinitesimal changes

$$\delta\mathbf{r} = \delta\boldsymbol{\omega}\times\mathbf{r}, \qquad \delta\mathbf{a} = \delta\boldsymbol{\omega}\times\mathbf{a}, \tag{23.32}$$

or by equivalent infinitesimal variations of the parameters that specify these vectors: θ and ϕ for $\mathbf{r}$ (r is invariant), ξ_+ and ξ_- for $\mathbf{a}$. We exhibit these for the three orthogonal axes of rotation as (see Problem 23.2)

$$\begin{aligned} (\delta\theta,\delta\phi,\delta\xi_+,\delta\xi_-) &= \delta\omega_x(-\sin\phi, -\cot\theta\cos\phi, -\frac{1}{2}i\xi_-, -\frac{1}{2}i\xi_+) \\ &+ \delta\omega_y(\cos\phi, -\cot\theta\sin\phi, -\frac{1}{2}\xi_-, \frac{1}{2}\xi_+) + \delta\omega_z(0,1,-\frac{1}{2}i\xi_+, \frac{1}{2}i\xi_-). \end{aligned} \tag{23.33}$$

We shall express the induced changes in $Y_{lm}(\theta,\phi)$ and ψ_{lm} by a vector rotational operator, the angular momentum, $\mathbf{L}$, which is defined initially by its action on $\mathbf{r}$:

$$\delta\mathbf{r} = (\delta\boldsymbol{\omega}\cdot i\mathbf{L})\mathbf{r}; \quad \mathbf{L} = \mathbf{r}\times\frac{1}{i}\boldsymbol{\nabla}. \tag{23.34}$$

Correspondingly, we write

$$\delta Y_{lm}(\theta,\phi) = (\delta\boldsymbol{\omega}\cdot i\mathbf{L})Y_{lm}(\theta,\phi), \quad \delta\psi_{lm} = (\delta\boldsymbol{\omega}\cdot i\mathbf{L})\psi_{lm}. \tag{23.35}$$

Thus, for the relatively simple situation of a rotation about the z-axis, given by the coefficient of $\delta\omega_z$, we have, respectively, from (23.33),

$$L_z = \frac{1}{i}\frac{\partial}{\partial\phi}, \quad L_z = -\frac{1}{2}\left(\xi_+\frac{\partial}{\partial\xi_+} - \xi_-\frac{\partial}{\partial\xi_-}\right). \tag{23.36a}$$

For the rotations about the other axes, we see that

$$iL_x = -\sin\phi\frac{\partial}{\partial\theta} - \cos\phi\cot\theta\frac{\partial}{\partial\phi}, \quad L_x = -\frac{1}{2}\left(\xi_-\frac{\partial}{\partial\xi_+} + \xi_+\frac{\partial}{\partial\xi_-}\right), \tag{23.36b}$$

and

$$iL_y = \cos\phi\frac{\partial}{\partial\theta} - \sin\phi\cot\theta\frac{\partial}{\partial\phi}, \quad L_y = \frac{1}{2i}\left(-\xi_-\frac{\partial}{\partial\xi_+} + \xi_+\frac{\partial}{\partial\xi_-}\right), \tag{23.36c}$$

although more convenient combinations are

$$L_x \pm iL_y = e^{\pm i\phi}\left(\pm\frac{\partial}{\partial\theta} - \cot\theta\frac{1}{i}\frac{\partial}{\partial\phi}\right), \quad L_x \pm iL_y = -\xi_\mp\frac{\partial}{\partial\xi_\pm}. \tag{23.37}$$

Let's begin with the direct effect of these operators on

$$\psi_{lm} = \frac{\xi_+^{l+m}\xi_-^{l-m}}{\sqrt{(l+m)!(l-m)!}}. \tag{23.38}$$

For example, we have

$$\xi_\pm\frac{\partial}{\partial\xi_\pm}\psi_{lm} = (l\pm m)\psi_{lm}, \tag{23.39}$$

from which follows

$$L_z\psi_{lm} = -m\psi_{lm}. \tag{23.40}$$

Next, we learn that

$$(L_x \pm iL_y)\psi_{lm} = -\sqrt{(l\pm m)(l\mp m+1)}\psi_{l,m\mp 1}. \tag{23.41}$$

We can also apply these two operations in either order:

$$(L_x \mp iL_y)(L_x \pm iL_y)\psi_{lm} = (l\pm m)(l\mp m+1)\psi_{lm}. \tag{23.42}$$

If we take half the sum of these, from which L_xL_y and L_yL_x cancel, we get

$$(L_x^2 + L_y^2)\psi_{lm} = [l(l+1) - m^2]\psi_{lm}. \tag{23.43}$$

Then, in view of (23.40), there emerges

$$\mathbf{L}^2\psi_{lm} = l(l+1)\psi_{lm}. \tag{23.44}$$

We have already made use of the fact that, as a scalar, $\mathbf{a}\cdot\mathbf{r}$ is unaltered by a common rotation of the two vectors. But now we point to the inference that the right-hand side of (22.22) is therefore unaltered by such a rotation. And then it takes only a glance at the addition theorem (23.24), where $P_l(\cos\gamma)$, $\cos\gamma = (\mathbf{r}/r)\cdot(\mathbf{r}'/r')$, is surely a rotational invariant, to conclude that the Y_{lm}^* and ψ_{lm} behave in the same manner under rotations.

The immediate conclusion is that every statement about the effect of $\mathbf{L}$ on ψ_{lm} has its precise counterpart in the action of $\mathbf{L}$ on Y_{lm}. (We may relate Y and Y^* by the result of Problem 22.4.) Thus, we infer from (23.40) that

$$\frac{\partial}{\partial\phi}Y_{lm}(\theta,\phi) = imY_{lm}(\theta,\phi), \tag{23.45}$$

which is certainly true. Then (23.41) leads analogously to (we divide by i)

$$e^{\pm i\phi}\left(\pm\frac{\partial}{\partial\theta} + i\cot\theta\frac{\partial}{\partial\phi}\right)Y_{lm}(\theta,\phi) = \sqrt{(l\mp m)(l\pm m+1)}Y_{l,m\pm 1}(\theta,\phi). \tag{23.46}$$

And it will be clear that the analog of (23.44), still written compactly as

$$[-\mathbf{L}^2 + l(l+1)]Y_{lm}(\theta,\phi) = 0, \tag{23.47}$$

can be no other than (23.5). [Note that the result (23.46) is derived by equivalent means, without the geometrical interpretation, in Problem 22.2.] Thus, we see that Y_{lm} is an eigenfunction of L_z and $\mathbf{L}^2$ with eigenvalues m and $l(l+1)$, respectively.

On making the ϕ-dependence explicit in (23.46), we are left with

$$\left(\pm\frac{d}{d\theta} - m\cot\theta\right)\Theta_{lm}(\theta) = \sqrt{(l\mp m)(l\pm m+1)}\Theta_{l,m\pm 1}(\theta), \tag{23.48}$$

where it should be noted that

$$\pm\frac{d}{d\theta} - m\cot\theta = \pm(\sin\theta)^{\pm m}\frac{d}{d\theta}(\sin\theta)^{\mp m}. \tag{23.49}$$

Immediate consequences appear if we set $m = \pm l$ in (23.48). Then, the right-hand side vanishes, which, on referring to (23.49), tells us that $\Theta_{l,\pm l}$ are proportional to $\sin^l\theta$ [see (22.44b), (22.44a)]. Then one can start with these extreme values of m and work up or down in m to construct all the other $\Theta_{lm}(\theta)$ functions. Specifically, we rewrite (23.48) with the upper sign as

$$-\frac{d}{d\cos\theta}(\sin\theta)^{-m}\Theta_{lm}(\theta) = \sqrt{(l-m)(l+m+1)}(\sin\theta)^{-m-1}\Theta_{l,m+1}(\theta). \tag{23.50}$$

Then the k-fold repetition of this operation gives

$$\begin{aligned}&\left(-\frac{d}{d\cos\theta}\right)^k(\sin\theta)^{-m}\Theta_{lm}(\theta)\\ &= \sqrt{\frac{(l-m)!}{(l-m-k)!}\frac{(l+m+k)!}{(l+m)!}}(\sin\theta)^{-m-k}\Theta_{l,m+k}(\theta).\end{aligned} \tag{23.51}$$

If we here set $m = -l$, where [(22.44a)]

$$(\sin\theta)^l\Theta_{l,-l}(\theta) = \sqrt{\frac{(2l+1)!}{2}}(-1)^l\frac{(\cos^2\theta-1)^l}{2^l\,l!}, \tag{23.52}$$

and replace the positive integer k by $l+m$, we get

$$\Theta_{lm}(\theta) = \sqrt{\frac{2l+1}{2}}\sqrt{\frac{(l-m)!}{(l+m)!}}(-\sin\theta)^m\left(\frac{d}{d\cos\theta}\right)^{l+m}\frac{(\cos^2\theta-1)^l}{2^l\,l!}, \tag{23.53}$$

in agreement with the lower sign choice in (22.29). Working down from $m = l$ produces the equivalent version.

23.3 Spherical Bessel Functions

We return to the discussion of completeness, in a manner that extends to three dimensions what has been studied for two dimensions in Chapter 18. The starting point is

$$\delta(\mathbf{r}-\mathbf{r}') = \int \frac{(d\mathbf{k})}{(2\pi)^3} e^{i\mathbf{k}\cdot(\mathbf{r}-\mathbf{r}')}, \tag{23.54}$$

which directs attention to an individual exponential,

$$e^{i\mathbf{k}\cdot\mathbf{r}} = e^{ikr\cos\theta}, \tag{23.55}$$

the latter form appearing when the direction of $\mathbf{k}$ is adopted as the z-axis. Now, for functions that depend only on the angle θ, the completeness statement of (23.22) reduces to [recall (22.27) and (22.44c)]

$$f(\theta) = \sum_{l=0}^{\infty}(2l+1)P_l(\cos\theta)\left[\frac{1}{2}\int_0^{\pi} d\theta'\,\sin\theta' P_l(\cos\theta')f(\theta')\right]. \tag{23.56}$$

The application of this expansion to (23.55) yields

$$e^{ikr\cos\theta} = \sum_{l=0}^{\infty}(2l+1)P_l(\cos\theta)i^l j_l(kr), \tag{23.57}$$

where

$$i^l j_l(kr) = \frac{1}{2}\int_{-1}^{1} d\mu\, P_l(\mu)e^{ikr\mu}. \tag{23.58}$$

To explain the presence of the factor i^l in this definition we must recall the nature of solid harmonics as being homogeneous in the components of the vector $\mathbf{r}$. In particular, on using the following notation to relate solid and surface harmonics,

$$Y_{lm}(\mathbf{r}) = r^l Y_{lm}(\theta,\phi), \tag{23.59}$$

we have the simple relations,

$$\begin{aligned} Y_{lm}(-\mathbf{r}) &= (-1)^l Y_{lm}(\mathbf{r}),\\ Y_{lm}(\pi-\theta,\pi+\phi) &= (-1)^l Y_{lm}(\theta,\phi), \end{aligned} \tag{23.60}$$

which includes

$$P_l(-\cos\theta) = (-1)^l P_l(\cos\theta). \tag{23.61}$$

This tells us what happens when the complex conjugation of the right-hand side of (23.58) is combined with the transformation $\mu \to -\mu$. The appearance of a factor of $(-1)^l$ is compensated by the response of i^l. In short, the function $j_l(kr)$ defined by (23.58) is a real function.

The introduction of the Legendre polynomial construction (22.45) into this definition yields

$$j_l(t) = i^{-l}\frac{1}{2}\int_{-1}^{1} d\mu \left[\left(\frac{d}{d\mu}\right)^l \frac{(\mu^2-1)^l}{2^l\, l!}\right] e^{i\mu t} = t^l \frac{1}{2}\int_{-1}^{1} d\mu \frac{(1-\mu^2)^l}{2^l\, l!} e^{i\mu t}, \tag{23.62}$$

the latter being produced by l-fold partial integration. [This can be turned into a differential formula. See Problem 23.3.] But we recognize this! According to (21.138), we have

$$j_l(t) = \left(\frac{\pi}{2t}\right)^{1/2} J_{l+1/2}(t), \tag{23.63}$$

which accounts for the name "spherical Bessel function" that is applied to j_l. We note the translation into this notation of (21.140), (21.141):

$$\frac{j_{l+1}(t)}{t^{l+1}} = \frac{1}{2l+2}\left(1+\frac{d^2}{dt^2}\right)\frac{j_l(t)}{t^l} = -\frac{1}{t}\frac{d}{dt}\frac{j_l(t)}{t^l}. \tag{23.64}$$

As for the differential version of (21.144), or

$$\left[\frac{d^2}{dt^2} + \frac{2}{t}\frac{d}{dt} + 1 - \frac{l(l+1)}{t^2}\right] j_l(t) = 0, \tag{23.65}$$

we have only to remark that

$$(\nabla^2 + k^2)e^{i\mathbf{k}\cdot\mathbf{r}} = 0; \tag{23.66}$$

the spherical coordinate version of this equation, valid for each l in the expansion of (23.57) is [see (23.1) and (23.5)]

$$\left[\frac{1}{r^2}\frac{d}{dr}\left(r^2\frac{d}{dr}\right) - \frac{l(l+1)}{r^2} + k^2\right] j_l(kr) = 0, \tag{23.67}$$

which is the content of (23.65).

The statement of the expansion (23.57) in an arbitrary coordinate frame is, using the addition theorem, (23.24),

$$e^{i\mathbf{k}\cdot\mathbf{r}} = \sum_{l=0}^{\infty}\sum_{m=-l}^{l} 4\pi i^l j_l(kr) Y_{lm}(\theta,\phi) Y_{lm}^*(\alpha,\beta), \tag{23.68a}$$

where α and β are the spherical coordinate angles for the vector $\mathbf{k}$. The complex conjugate version is, with $\mathbf{r} \to \mathbf{r}'$,

$$e^{-i\mathbf{k}\cdot\mathbf{r}'} = \sum_{l=0}^{\infty}\sum_{m=-l}^{l} 4\pi(-i)^l j_l(kr') Y_{lm}(\alpha,\beta) Y_{lm}^*(\theta',\phi'). \tag{23.68b}$$

Now we carry out a solid angle integration over the directions of $\mathbf{k}$,

$$(d\mathbf{k}) = k^2 dk\, d\Omega_k, \tag{23.69}$$

using the orthonormality of the $Y_{lm}(\alpha,\beta)$, to arrive at

$$\int d\Omega_k \frac{1}{(2\pi)^3} e^{i\mathbf{k}\cdot\mathbf{r}} e^{-i\mathbf{k}\cdot\mathbf{r}'} = \frac{2}{\pi}\sum_l j_l(kr) j_l(kr') \sum_m Y_{lm}(\theta,\phi) Y_{lm}^*(\theta',\phi'). \tag{23.70}$$

Accordingly, the spherical coordinate transcription of (23.54) is

$$\begin{aligned}&\frac{1}{r^2}\delta(r-r')\frac{1}{\sin\theta}\delta(\theta-\theta')\delta(\phi-\phi')\\ &\quad = \frac{2}{\pi}\sum_{l=0}^{\infty}\int_0^{\infty} dk\, k^2 j_l(kr) j_l(kr') \sum_{m=-l}^{l} Y_{lm}(\theta,\phi) Y_{lm}^*(\theta',\phi').\end{aligned} \tag{23.71}$$

Then, if we multiply by a particular $Y_{lm}(\theta', \phi')$, and integrate over solid angle in the primed coordinates, what emerges is

$$\frac{1}{r^2}\delta(r-r') = \frac{2}{\pi}\int_0^\infty dk\, k^2\, j_l(kr) j_l(kr'), \tag{23.72a}$$

the completeness property of the radial functions $j_l(kr)$, $0 < k < \infty$. It is not a new statement, for on introducing (23.63), we get

$$\frac{1}{r}\delta(r-r') = \int_0^\infty dk\, k\, J_{l+1/2}(kr) J_{l+1/2}(kr'), \tag{23.72b}$$

the completeness statement for the Bessel functions J_m, with $m = l + 1/2$ [*cf.* (18.57a)]. Incidentally, putting (23.72a) back into (23.71), yields the completeness relation for the spherical harmonics, (23.22).

The reverse side of the coin of completeness carries the image of orthonormality. We look to (23.71) for the basic functions,

$$\delta(\mathbf{r}-\mathbf{r}') = \sum_{lm}\int_0^\infty dk\, k^2 \left[\left(\frac{2}{\pi}\right)^{1/2} j_l(kr) Y_{lm}(\theta,\phi)\right]\left[\left(\frac{2}{\pi}\right)^{1/2} j_l(kr') Y_{lm}(\theta',\phi')\right]^*, \tag{23.73}$$

which combines summation over l and m with integration over k. The corresponding orthonormality statement is contained in the values of

$$\begin{aligned}
&\int_0^\infty dr\, r^2 \int d\Omega \left[\left(\frac{2}{\pi}\right)^{1/2} j_l(kr) Y_{lm}(\theta,\phi)\right]^* \left[\left(\frac{2}{\pi}\right)^{1/2} j_{l'}(k'r) Y_{l'm'}(\theta,\phi)\right] \\
&= \delta_{ll'}\delta_{mm'}\frac{2}{\pi}\int_0^\infty dr\, r^2\, j_l(kr) j_l(k'r) \\
&= \delta_{ll'}\delta_{mm'}\frac{1}{k^2}\delta(k-k'),
\end{aligned} \tag{23.74}$$

where the radial integral is just (23.72a), with the substitutions $k \to r$, $r \to k$, $r' \to k'$.

Let us not hurry on without noting the form taken by Coulomb's potential, (15.15):

$$\begin{aligned}
\frac{1}{|\mathbf{r}-\mathbf{r}'|} &= 4\pi \int \frac{(d\mathbf{k})}{(2\pi)^3}\frac{1}{k^2} e^{i\mathbf{k}\cdot(\mathbf{r}-\mathbf{r}')} \\
&= \sum_{l=0}^{\infty}(2l+1)P_l(\cos\gamma)\frac{2}{\pi}\int_0^\infty dk\, j_l(kr)\, j_l(kr'),
\end{aligned} \tag{23.75}$$

which produces the identification, from (23.10),

$$\frac{2}{\pi}\int_0^\infty dk\, j_l(kr) j_l(kr') = \frac{1}{2l+1}\frac{r_<^l}{r_>^{l+1}}, \tag{23.76a}$$

or, equivalently,

$$\lambda < 1: \quad \frac{2}{\pi}\int_0^\infty dt\, j_l(\lambda t)\, j_l(t) = \frac{1}{2l+1}\lambda^l. \tag{23.76b}$$

This is illustrated for $l = 0$ by (see Problem 23.3)

$$\begin{aligned}
\frac{2}{\pi}\frac{1}{\lambda}\int_0^\infty \frac{dt}{t^2}\sin\lambda t \sin t &= \frac{1}{\pi\lambda}\int_0^\infty d\left(\frac{1}{t}\right)[\cos(1+\lambda)t - \cos(1-\lambda)t] \\
&= \frac{1}{\pi\lambda}\left[(1+\lambda)\int_0^\infty \frac{dt}{t}\sin(1+\lambda)t - (1-\lambda)\int_0^\infty \frac{dt}{t}\sin(1-\lambda)t\right] \\
&= \frac{1}{\pi\lambda}\left[(1+\lambda)\frac{\pi}{2} - (1-\lambda)\frac{\pi}{2}\right] = 1.
\end{aligned} \tag{23.77}$$

23.4 Problems for Chapter 23

1. Another proof of the addition theorem, (23.24), proceeds as follows. By considering $r_< = r$, argue that $r^l P_l(\cos\gamma)$ is a solution of Laplace's equation that is homogeneous of degree l in the components of $\mathbf{r}$. As such, $P_l(\cos\gamma)$ must be a linear combination of the $2l+1$ spherical harmonics $Y_{lm}(\theta,\phi)$. A similar consideration of the situation $r' = r_<$ tells us that $P_l(\cos\gamma)$ is also a linear combination of the spherical harmonics $Y_{lm}(\theta',\phi')$. Then, show that because $\cos\gamma$ depends only on the difference of the azimuthal angles ϕ and ϕ' we can write
$$P_l(\cos\gamma) = \sum_{m=-l}^{l} a_{lm} Y_{lm}(\theta,\phi) Y^*_{lm}(\theta',\phi'), \tag{23.78}$$
in terms of $2l+1$ *real* coefficients a_{lm}. Now set θ', ϕ' equal to θ, ϕ, respectively, so that $\gamma = 0$, and deduce by integrating over all angles θ and ϕ, that
$$4\pi = \sum_{m=-l}^{l} a_{lm}. \tag{23.79}$$
Then, multiply the above expansion for $P_l(\cos\gamma)$ by its complex conjugate, and integrate over θ, ϕ and over θ', ϕ' separately, to deduce
$$\frac{(4\pi)^2}{2l+1} = \sum_{m=-l}^{l} a_{lm}^2. \tag{23.80}$$
Infer from these two relations that
$$a_{lm} = \frac{4\pi}{2l+1}, \tag{23.81}$$
hence proving (23.24).

2. Verify the rotations of the coordinates θ, ϕ, ξ_+, and ξ_- in (23.33), and thereby verify the rotation operators in (23.36a)–(23.36c).

3. Starting from the integral representation for the spherical Bessel function $j_l(t)$ given in (23.62), show, by repeatedly integrating by parts so as to eliminate the $(1-\mu^2)^l$ factor, that
$$j_l(t) = t^l \left(-\frac{1}{t}\frac{d}{dt}\right)^l \frac{\sin t}{t} \tag{23.82}$$
[this is (21.137)], and from this give explicit forms for j_l for $l = 0, \ldots 3$.

4. The expansion
$$F(x,t) \equiv \frac{1}{\sqrt{1-2xt+t^2}} = \sum_{l=0}^{\infty} t^l P_l(x), \quad |t| < 1, \tag{23.83}$$
is usually referred to as the generating function for Legendre's polynomials. From it all the properties of these polynomials may be derived. In particular,

   (a) by differentiating $F(x,t)$ with respect to t, derive the recurrence relation for the Legendre polynomials,
$$(2l+1)xP_l(x) = (l+1)P_{l+1}(x) + lP_{l-1}(x), \quad l = 1,2,3,\ldots, \tag{23.84}$$

(b) and by differentiating with respect to x, derive the differential equation satisfied by P_l:

$$(1-x^2)P_l''(x) - 2xP_l'(x) + l(l+1)P_l(x) = 0. \tag{23.85}$$

5. By multiplying the differential equation for $P_l(\cos\gamma)$, (23.85), by $P_{l'}$, show that

$$\int_{-1}^{1} d(\cos\gamma) P_l(\cos\gamma) P_{l'}(\cos\gamma) = 0 \quad \text{if} \quad l' \neq l. \tag{23.86}$$

Then by integrating

$$\int_{=1}^{1} d(\cos\gamma) \frac{1}{r^2 + r'^2 - 2rr'\cos\gamma}, \tag{23.87}$$

and using the generating function (23.83) show that

$$\int_{-1}^{1} d(\cos\gamma) P_l(\cos\gamma) P_{l'}(\cos\gamma) = \frac{2}{2l+1}\delta_{ll'}, \tag{23.88}$$

which is the orthonormality relation (22.47).

6. As with the spherical Bessel function of the first kind, the spherical Neumann function may be defined in terms of cylinder functions of half-integer order by

$$n_l(x) = \sqrt{\frac{\pi}{2x}} N_{l+1/2}(x). \tag{23.89}$$

Using the result of Problem 18.10, compute n_0 and n_1 in terms of sines and cosines.

7. Using the results derived in Problem 21.13, derive the orthonormalization condition for spherical Bessel functions on a finite interval, $0 \leq \rho \leq a$,

$$\int_0^a d\rho\, \rho^2 j_l\left(\gamma_{ln}\frac{\rho}{a}\right) j_l\left(\gamma_{lm}\frac{\rho}{a}\right) = \delta_{nm}\frac{1}{2}a^3 \left[j_{l+1}(\gamma_{ln})\right]^2, \tag{23.90}$$

where γ_{ln} is the nth zero of j_l.

8. Recognize that the completeness properties of $j_0(k\rho)$ on an infinite and on a finite interval [recall (21.93)] are equivalent to previously established results, in terms of sine functions and imaginary exponentials.

9. Prove that

$$\begin{aligned} &kJ_m(k\rho)J_m(k\rho') - kJ_{m+1}(k\rho)J_{m+1}(k\rho') \\ &= \frac{d}{dk}\left[\frac{k}{\rho+\rho'}\left(J_{m+1}(k\rho)J_m(k\rho') + J_m(k\rho)J_{m+1}(k\rho')\right)\right]. \end{aligned} \tag{23.91}$$

Use this result to show that if the $J_m(k\rho)$, $k : 0 \to \infty$ are complete, so also are the $J_{m+1}(k\rho)$, i.e., the completeness of all the $J_{l+1/2}(k\rho)$, $l = 1, 2, \ldots$, follows from that of the $J_{1/2}(k\rho)$. [You might also want to think about a convergence factor or something equivalent.]

24

Multipoles

In terms of the above discussion of spherical harmonics, we now make a general analysis of the potential, due to a given charge distribution, $\rho(\mathbf{r}')$, outside of that distribution. (See Fig. 24.1.) The potential is given by (1.5),

$$\phi(\mathbf{r}) = \int (d\mathbf{r}') \frac{\rho(\mathbf{r}')}{|\mathbf{r} - \mathbf{r}'|}, \tag{24.1}$$

where, for convenience, we will choose the origin to lie within the charge distribution. If $\mathbf{r}$ is large compared to the characteristic dimensions of the charge distribution, we may expand Coulomb's potential as follows:

$$\begin{aligned} \frac{1}{|\mathbf{r} - \mathbf{r}'|} &= \frac{1}{r} - \mathbf{r}' \cdot \boldsymbol{\nabla} \frac{1}{r} + \frac{1}{2}(\mathbf{r}' \cdot \boldsymbol{\nabla})^2 \frac{1}{r} + \dots \\ &= \frac{1}{r} + \frac{\mathbf{r} \cdot \mathbf{r}'}{r^3} + \frac{1}{2}\frac{1}{r^5}\mathbf{r} \cdot (3\mathbf{r}'\mathbf{r}' - \mathbf{1}r'^2) \cdot \mathbf{r} + \dots, \end{aligned} \tag{24.2}$$

so that the potential, in its leading behavior for large distances, has the form

$$\phi(\mathbf{r}) = \frac{e}{r} + \frac{\mathbf{r} \cdot \mathbf{d}}{r^3} + \frac{1}{2}\frac{1}{r^5}\mathbf{r} \cdot \mathbf{q} \cdot \mathbf{r} + \dots. \tag{24.3}$$

Occurring here are the first three moments of the charge distribution,

$$e = \int (d\mathbf{r}')\, \rho(\mathbf{r}'), \tag{24.4a}$$

$$\mathbf{d} = \int (d\mathbf{r}')\, \mathbf{r}' \rho(\mathbf{r}'), \tag{24.4b}$$

$$\mathbf{q} = \int (d\mathbf{r}') (3\mathbf{r}'\mathbf{r}' - \mathbf{1}r'^2) \rho(\mathbf{r}'), \tag{24.4c}$$

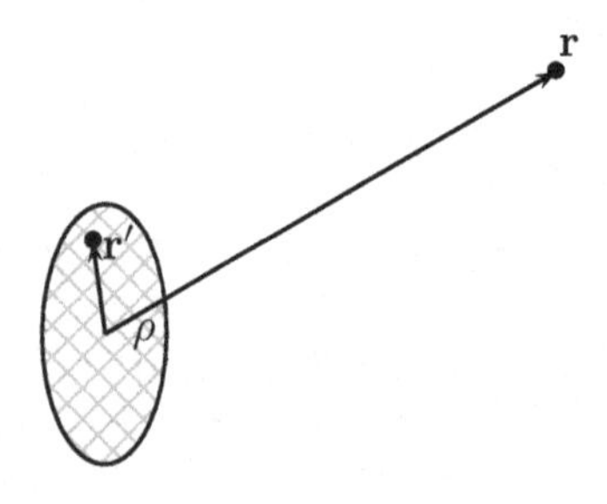

FIGURE 24.1
Geometry of field point and source point for a bounded charge distribution.

DOI: 10.1201/9781003057369-24

which are the total charge, the dipole moment vector, and the quadrupole moment dyadic, respectively.

Using this potential, we can now calculate the interaction energy of the charge distribution with an additional point charge e_1 located at a point $\mathbf{r}$ lying far outside the charge distribution:

$$U = e_1\phi(\mathbf{r}) = \frac{ee_1}{r} + \mathbf{d}\cdot\frac{e_1\mathbf{r}}{r^3} + \frac{1}{2}e_1\frac{1}{r^5}\mathbf{r}\cdot\mathbf{q}\cdot\mathbf{r} + \dots. \tag{24.5}$$

We may alternatively interpret (24.5) as the interaction energy of the various moments of the charge distribution with the field produced by e_1 at the origin, that is

$$U = e\phi - \mathbf{d}\cdot\mathbf{E} + \frac{1}{6}\boldsymbol{\nabla}\cdot\mathbf{q}\cdot\mathbf{E} + \dots \tag{24.6}$$

where

$$\phi = \frac{e_1}{r}, \tag{24.7a}$$

$$\mathbf{E} = \frac{e_1(-\mathbf{r})}{r^3}. \tag{24.7b}$$

(We have seen this form of the dipole interaction energy before, in Section 4.2.) Note that the trace of $\mathbf{q}$ is zero, $\sum_i q_{ii} = 0$. This is a starting point for considering the interaction of one charge distribution with another charge distribution. For example, if one had a dipole $\mathbf{d}_1$ rather than a charge, e_1, interacting with a charge distribution which had only a dipole moment, $\mathbf{d}_2$, the interaction energy deduced from (24.6) would be

$$U = -\mathbf{d}_2\cdot\left[-\boldsymbol{\nabla}\frac{\mathbf{r}\cdot\mathbf{d}_1}{r^3}\right] = -\frac{3\mathbf{r}\cdot\mathbf{d}_1\,\mathbf{r}\cdot\mathbf{d}_2 - \mathbf{d}_1\cdot\mathbf{d}_2 r^2}{r^5}, \tag{24.8}$$

which is the dipole-dipole interaction.

Although this approach could obviously be continued indefinitely, it rapidly becomes unwieldy for higher multipoles. A systematic approach can be based on the use of spherical harmonics. Outside a charge distribution ($r > r'$), the potential (24.1) can be expanded in spherical harmonics according to (23.23); that is, the expansion

$$\frac{1}{|\mathbf{r}-\mathbf{r}'|} = \sum_{lm}\frac{r'^l}{r^{l+1}}\sqrt{\frac{4\pi}{2l+1}}Y_{lm}(\theta,\phi)\sqrt{\frac{4\pi}{2l+1}}Y^*_{lm}(\theta',\phi'), \tag{24.9}$$

implies

$$\phi(\mathbf{r}) = \sum_{lm}\frac{1}{r^{l+1}}\sqrt{\frac{4\pi}{2l+1}}Y_{lm}(\theta,\phi)\rho_{lm}. \tag{24.10}$$

Here, the multipole moments, ρ_{lm}, are defined by

$$\rho_{lm} = \int(d\mathbf{r}')\,r'^l\sqrt{\frac{4\pi}{2l+1}}Y^*_{lm}(\theta',\phi')\rho(\mathbf{r}'). \tag{24.11}$$

The connection with the previous definition, (24.4a) and (24.4b), is, for example, given by [see (22.30b)]

$$l = 0: \quad \rho_{00} = e, \tag{24.12a}$$

$$l = 1: \quad \rho_{11} = -\frac{1}{\sqrt{2}}(d_x - id_y), \tag{24.12b}$$

$$\rho_{10} = d_z, \tag{24.12c}$$

$$\rho_{1-1} = \frac{1}{\sqrt{2}}(d_x + id_y). \tag{24.12d}$$

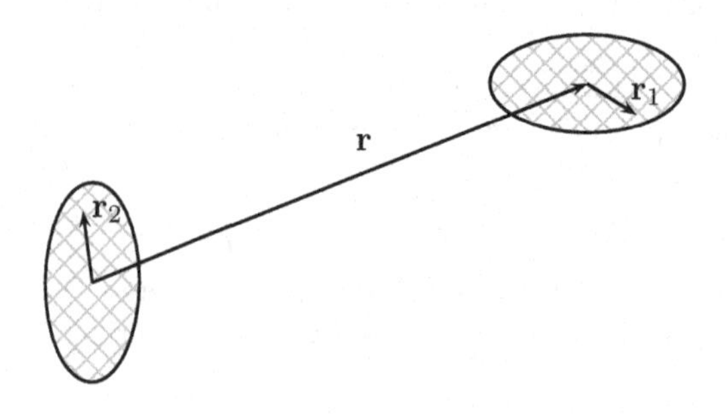

FIGURE 24.2
Two interacting charge distributions.

Now we return to the consideration of the energy of interaction of a charge distribution, $\rho(\mathbf{r})$, with an external potential, $\phi(\mathbf{r})$:

$$U = \int (d\mathbf{r})\, \rho(\mathbf{r})\phi(\mathbf{r}). \tag{24.13}$$

Since the potential is produced by sources outside of the charge distribution, it can be expanded in terms of spherical harmonics,

$$\phi(\mathbf{r}) = \sum_{lm} r^l Y_{lm}(\theta, \phi)\sqrt{\frac{4\pi}{2l+1}}\phi_{lm}, \tag{24.14}$$

ϕ_{lm} being the expansion coefficients. Inserting this multipole expansion for the potential back into (24.13) and using the definition (24.11) for the multipole moments, we obtain the simple expression for the energy of interaction

$$U = \sum_{lm} \rho^*_{lm}\phi_{lm}, \tag{24.15}$$

generalizing (24.6).

Rather than expressing the interaction energy in the unsymmetrical form (24.15), let us formulate the energy in terms of the interaction of the charge multipole moments of each distribution; that is, we seek a generalization of the dipole-dipole interaction, (24.8). If we let $\mathbf{r}_1$ and $\mathbf{r}_2$ be measured from points within ρ_1 and ρ_2, respectively, while $\mathbf{r}$ measures the distance between these two origins, as illustrated in Fig. 24.2, the interaction energy can be written as

$$E = \int (d\mathbf{r}_1)(d\mathbf{r}_2)\frac{\rho_1(\mathbf{r}_1)\rho_2(\mathbf{r}_2)}{|\mathbf{r} + \mathbf{r}_1 - \mathbf{r}_2|}. \tag{24.16}$$

Since the two charge distributions are non-overlapping, we can expand the denominator occurring here in a double Taylor series:

$$\frac{1}{|\mathbf{r} + \mathbf{r}_1 - \mathbf{r}_2|} = \sum_{l_1 l_2} \frac{(\mathbf{r}_1 \cdot \boldsymbol{\nabla})^{l_1}}{l_1!}\frac{(-\mathbf{r}_2 \cdot \boldsymbol{\nabla})^{l_2}}{l_2!}\frac{1}{r}. \tag{24.17}$$

We already know that, for $r > r'$,

$$\frac{1}{|\mathbf{r} - \mathbf{r}'|} = \sum_{lm} \frac{r'^l}{r^{l+1}}\frac{4\pi}{2l+1}Y_{lm}(\theta, \phi)Y^*_{lm}(\theta', \phi') = \sum_l \frac{(-\mathbf{r}' \cdot \boldsymbol{\nabla})^l}{l!}\frac{1}{r}, \tag{24.18a}$$

or, if we take the complex conjugate and equate powers of $\mathbf{r}'$,

$$\frac{(-\mathbf{r}'\cdot\boldsymbol{\nabla})^l}{l!}\frac{1}{r}=\frac{4\pi}{2l+1}\sum_m r'^l Y_{lm}(\theta',\phi')r^{-l-1}Y^*_{lm}(\theta,\phi). \tag{24.18b}$$

Further, recall the generating function for the spherical harmonics, (22.22),

$$\frac{(\mathbf{r}'\cdot\mathbf{a})^l}{l!}=r'^l\sum_m\sqrt{\frac{4\pi}{2l+1}}Y_{lm}(\theta',\phi')\psi_{lm}, \tag{24.19}$$

which is valid for $\mathbf{a}^2=0$. Thus it is permissible to replace $\mathbf{a}$ by a gradient,

$$\mathbf{a}\to\boldsymbol{\nabla}, \tag{24.20}$$

as long as the derivatives act on $1/r$,

$$\mathbf{a}^2\to\nabla^2\frac{1}{r}=0,\quad r>0. \tag{24.21}$$

In this way, a comparison of (24.18b) and (24.19) gives the identity

$$\sqrt{\frac{4\pi}{2l+1}}r^{-l-1}Y^*_{lm}(\theta,\phi)=(-1)^l\psi_{lm}\frac{1}{r}, \tag{24.22}$$

where ψ_{lm} is now regarded as a differential operator, constructed according to (22.21) from $\boldsymbol{\nabla}$. (We might recall from Section 23.2 that Y^*_{lm} and ψ_{lm} transform the same way under rotations.) Using (24.19) twice with the above replacement, we obtain

$$\begin{aligned}\frac{(\mathbf{r}_1\cdot\boldsymbol{\nabla})^{l_1}}{l_1!}\frac{(-\mathbf{r}_2\cdot\boldsymbol{\nabla})^{l_2}}{l_2!}\frac{1}{r}&=(-1)^{l_2}r_1^{l_1}r_2^{l_2}\\&\times\sum_{m_1m_2}\sqrt{\frac{4\pi}{2l_1+1}}Y_{l_1m_1}(\theta_1,\phi_1)\sqrt{\frac{4\pi}{2l_2+1}}Y_{l_2m_2}(\theta_2,\phi_2)\psi_{l_1m_1}\psi_{l_2m_2}\frac{1}{r}.\end{aligned} \tag{24.23}$$

According to the definition of ψ_{lm}, (22.21), the product of two of these functions is

$$\psi_{l_1m_1}\psi_{l_2m_2}=C_{l_1l_2m_1m_2}\psi_{l_1+l_2,m_1+m_2}, \tag{24.24a}$$

where

$$C_{l_1l_2m_1m_2}=\left[\frac{(l_1+l_2+m_1+m_2)!\,(l_1+l_2-m_1-m_2)!}{(l_1+m_1)!\,(l_1-m_1)!\,(l_2+m_2)!\,(l_2-m_2)!}\right]^{1/2}. \tag{24.24b}$$

Then we evaluate the derivative structure in (24.23) by means of (24.22):

$$\psi_{l_1+l_2,m_1+m_2}\frac{1}{r}=(-1)^{l_1+l_2}\sqrt{\frac{4\pi}{2(l_1+l_2)+1}}r^{-l_1-l_2-1}Y^*_{l_1+l_2,m_1+m_2}(\theta,\phi). \tag{24.25}$$

Combining (24.16), (24.17), (24.23), (24.24a), and (24.25), taking the complex conjugate, and identifying ρ_{lm}, (24.11), we find for the energy of interaction

$$U=\sum_{l_1l_2m_1m_2}(-1)^{l_1}\left[\frac{4\pi}{2(l_1+l_2)+1}\right]^{1/2}\frac{C_{l_1l_2m_1m_2}}{r^{l_1+l_2+1}}(\rho_1)_{l_1m_1}Y_{l_1+l_2,m_1+m_2}(\theta,\phi)(\rho_2)_{l_2m_2}. \tag{24.26}$$

If we were to set $l_1=l_2=1$ we would rederive the dipole-dipole interaction, (24.8). However, this is a completely general result for the interaction between two arbitrary non-overlapping charge distributions, of a remarkably simple and compact form.

24.1 Problems for Chapter 24

1. Derive the explicit connection between the quadrupole moment $\mathbf{q}$ defined by (24.4c) and ρ_{2m} defined by (24.11).
2. Check that, indeed, (24.15) generalizes (24.6) by showing

$$U = \sum_{lm} \rho_{lm}^* \phi_{lm} = e\phi - \mathbf{d} \cdot \mathbf{E} + \frac{1}{6} \boldsymbol{\nabla} \cdot \mathbf{q} \cdot \mathbf{E} + \ldots. \tag{24.27}$$

3. Derive (24.8) from (24.26).
4. A spherical shell of radius a centered at the origin has a charge density lying entirely on the surface, having the form:

$$\rho(\mathbf{r}) = a\delta(r - a)\cos\theta, \tag{24.28}$$

in polar coordinates, where $z = r\cos\theta$. Calculate the total charge, the electric dipole moment, and the electric quadrupole moment of this charge distribution. Calculate all components of ρ_{lm}, given by (24.11).

5. Compute the thermal averaged interaction energy between two dipoles, of dipole moment d_1 and d_2, separated by a distance r, in the high temperature limit. Express the result in terms of the average electric polarizabilities, defined by

$$\langle \mathbf{d} \rangle_T = \alpha \mathbf{E}. \tag{24.29}$$

[Hints: Recall that the polarizability of a permanent dipole at high temperature is given by (5.67). Compute the thermal average value of dipole interaction energy in the same high-temperature limit, using rotational invariance, *e.g.*, $\langle \mathbf{d}_1 \mathbf{d}_1 \rangle = \frac{1}{3} \mathbf{1} d_1^2$. Supply a factor of $1/2$ to account for the fact that the two dipoles polarize each other. The answer is

$$\langle U \rangle_T \sim -\frac{3\alpha_1\alpha_2}{r^6} kT, \quad T \to \infty. \tag{24.30}$$

This result holds whether the dipoles are permanent or induced. This is the high-temperature, or classical, limit of the van der Waals [25], or Casimir-Polder [26] interaction between molecules.]

6. Develop directly the expression for the energy of interaction, analogous to (24.8), of two electric quadrupoles. Show that your result is consistent with the general expression (24.26).
7. Recall that the solid harmonic is related to the spherical harmonic by

$$\mathcal{Y}_{lm}(\mathbf{r}) = r^l Y_{lm}(\theta, \phi), \tag{24.31}$$

since it is a homogeneous polynomial of degree l. Similarly, we may define a "solid" Legendre polynomial by

$$\mathcal{P}_l(\mathbf{r} \cdot \mathbf{r}') = r^l P_l(\cos\gamma) r'^l, \tag{24.32}$$

which is homogeneous of degree l in $\mathbf{r}$ and $\mathbf{r}'$. The addition theorem reads

$$\frac{2l+1}{4\pi} \mathcal{P}_l(\mathbf{r} \cdot \mathbf{r}') = \sum_m \mathcal{Y}_{lm}(\mathbf{r}) \mathcal{Y}_{lm}^*(\mathbf{r}'), \tag{24.33}$$

where (check this)

$$P_l(\mathbf{r}\cdot\mathbf{r}') = \frac{(2l)!}{(l!)^2 2^l}(\mathbf{r}\cdot\mathbf{r}')^l + O(r^2 r'^2), \tag{24.34}$$

where the latter notation means the remaining terms have increasing powers of $r^2 r'^2$. Now replace $\mathbf{r}$ by $\boldsymbol{\nabla}$ (with an operand $1/r$ understood) and arrive at

$$\frac{(2l)!}{l!\,2^l}\psi_{lm}(\boldsymbol{\nabla}) = \sqrt{\frac{4\pi}{2l+1}}Y^*_{lm}(\boldsymbol{\nabla}). \tag{24.35}$$

Check this for $m = 0$.

8. Consider

$$\frac{(\mathbf{a}\cdot\boldsymbol{\nabla})^l}{l!}\frac{1}{r}, \quad \mathbf{a}^2 = 0, \tag{24.36}$$

and conclude that

$$Y_{lm}(\boldsymbol{\nabla})\frac{1}{r} = (-1)^l\frac{(2l)!}{l!\,2^l}\frac{1}{r^{2l+1}}Y_{lm}(\mathbf{r}), \tag{24.37}$$

from which the result (24.22) follows. Compare with the result of Problems 22.10 and 22.13.

25

Conducting Sphere and Dielectric Ball

25.1 Interior of Conducting Spherical Shell

The spherical harmonics are useful in solving problems possessing spherical symmetry. In this chapter, we will solve a few such problems. First, we wish to find Green's function *inside* a hollow conducting sphere of radius a, which is grounded, that is, the potential is zero on its surface. As usual, the Green's function equation is

$$-\nabla^2 G(\mathbf{r}, \mathbf{r}') = 4\pi\delta(\mathbf{r} - \mathbf{r}'). \tag{25.1}$$

The solution must be expressible in terms of spherical harmonics as

$$r < r': \quad G = \sum_{lm} r^l Y_{lm}(\theta, \phi) A_{lm}, \tag{25.2a}$$

$$r' < r \leq a: \quad G = \sum_{lm} \left(\frac{1}{r^{l+1}} - \frac{r^l}{a^{2l+1}} \right) Y_{lm}(\theta, \phi) B_{lm}, \tag{25.2b}$$

where we have imposed the boundary conditions that

$$G = \text{ finite at } r = 0, \tag{25.3a}$$

and

$$G = 0 \text{ at } r = a. \tag{25.3b}$$

To determine the expansion coefficients, A_{lm} and B_{lm}, we use the equations for the continuity of G, (23.17a),

$$r'^l A_{lm} = \left(\frac{1}{r'^{l+1}} - \frac{r'^l}{a^{2l+1}} \right) B_{lm}, \tag{25.4a}$$

and for the discontinuity of $\frac{\partial}{\partial r} G$, (23.17b), [this uses (23.22)]

$$\sum_{lm} \left[\left(\frac{l+1}{r'^l} + \frac{l r'^{l+1}}{a^{2l+1}} \right) B_{lm} + l r'^{l+1} A_{lm} \right] Y_{lm}(\theta, \phi) = 4\pi \sum_{lm} Y_{lm}(\theta, \phi) Y^*_{lm}(\theta', \phi'), \tag{25.4b}$$

at $r = r'$. Solving (25.4a) by introducing C_{lm} defined by

$$A_{lm} = \left(\frac{1}{r'^{l+1}} - \frac{r'^l}{a^{2l+1}} \right) C_{lm}, \tag{25.5a}$$

$$B_{lm} = r'^l C_{lm}, \tag{25.5b}$$

we find, from (25.4b),

$$C_{lm} = \frac{4\pi}{2l+1} Y^*_{lm}(\theta', \phi'). \tag{25.5c}$$

DOI: 10.1201/9781003057369-25

Therefore, Green's function is

$$G = \sum_l \left(\frac{r_<^l}{r_>^{l+1}} - \frac{r^l r'^l}{a^{2l+1}} \right) P_l(\cos\gamma), \tag{25.6}$$

where we have used the addition theorem (23.24), γ being the angle between $\mathbf{r}$ and $\mathbf{r}'$, (23.8). Noticing that

$$\frac{r^l r'^l}{a^{2l+1}} = \frac{a}{r'} \frac{r^l}{(a^2/r')^{l+1}}, \quad \frac{a^2}{r'} > a \geq r, \tag{25.7}$$

we can perform the summation in (25.6) by use of (23.10):

$$G = \frac{1}{|\mathbf{r} - \mathbf{r}'|} - \frac{a}{r'} \frac{1}{|\mathbf{r} - \overline{\mathbf{r}}'|} \tag{25.8}$$

where $\overline{\mathbf{r}}'$ locates the so-called image point,

$$\overline{\mathbf{r}}' = \left(\frac{a^2}{r'}, \theta', \phi' \right), \tag{25.9}$$

which, of course, lies outside the sphere. Thus we have achieved for the sphere the analog of the image solution given for the conducting plane in (17.32).

What is the induced charge density on the inside surface of the sphere? This charge density is proportional to the radial electric field, according to (13.62),

$$4\pi\sigma = -E_r = \left. \frac{\partial}{\partial r} G \right|_{r=a}, \tag{25.10}$$

since the normal is inward, and so in the negative radial direction. Differentiating (25.6) with respect to $r = r_>$, we obtain

$$4\pi\sigma = -\sum_l (2l+1) \frac{1}{a^2} \left(\frac{r'}{a} \right)^l P_l(\cos\gamma). \tag{25.11a}$$

Alternatively, we could use the image charge form of Green's function, (25.8), to derive

$$4\pi\sigma = -\frac{1}{a^2} \frac{1 - (r'/a)^2}{[1 - 2(r'/a)\cos\gamma + (r'/a)^2]^{3/2}}, \tag{25.11b}$$

which indicates that as $r' \to a$, the only significant charge buildup is near $\gamma = 0$. The total charge can be computed from (25.11a) by use of the orthonormality condition, (22.47), which, for $l' = 0$, implies

$$\int_0^\pi \sin\gamma \, d\gamma \, P_l(\cos\gamma) = 2\delta_{l0}. \tag{25.12}$$

Therefore, only the $l = 0$ term in (25.11a) contributes to the total charge:

$$Q = \int dS \, \sigma = \int a^2 \sin\gamma \, d\gamma \, d\phi \left[-\frac{1}{4\pi a^2} \right] = -1, \tag{25.13}$$

which is the expected result. Of course, it is possible to use (25.11b) to compute the total charge, but that is more elaborate. (See Problems 25.1 and 25.2.)

25.1.1 Bessel Function Representation

Another approach to the interior of the sphere exploits the completeness of the spherical Bessel functions over the finite range between $r = 0$ and $r = a$, as inferred from (21.114),

$$\frac{1}{r^2}\delta(r-r') = \sum_{n=1}^{\infty} \frac{2}{a^3} \frac{r^{-1/2} J_{l+1/2}(\gamma_{ln} r/a) r'^{-1/2} J_{l+1/2}(\gamma_{ln} r'/a)}{[a^{-1/2} J'_{l+1/2}(\gamma_{ln})]^2}$$
$$= \sum_{n=1}^{\infty} \frac{2}{a^3} \frac{j_l(\gamma_{ln} r/a) j_l(\gamma_{ln} r'/a)}{[j'_l(\gamma_{ln})]^2}, \tag{25.14}$$

where now the γ_{ln} are the roots of j_l:

$$j_l(\gamma_{ln}) = 0. \tag{25.15}$$

Exhibited in (25.14) are the radial functions that are orthonormal in the sense of the integral

$$\int_0^a dr\, r^2 R_{ln}(r) R_{ln'}(r) = \delta_{nn'}, \tag{25.16}$$

namely

$$R_{ln}(r) = \left(\frac{2}{a^3}\right)^{1/2} \frac{j_l(\gamma_{ln} r/a)}{j'_l(\gamma_{ln})}. \tag{25.17}$$

This orthonormality property can be verified easily for $l = 0$, $j_0(t) = t^{-1}\sin t$, where (25.15) reads

$$\sin\gamma_{0n} = 0, \quad \gamma_{0n} \neq 0, \tag{25.18a}$$

or

$$\gamma_{0n} = n\pi, \quad n = 1, 2, \dots. \tag{25.18b}$$

Then we have

$$R_{0n}(r) = \left(\frac{2}{a^3}\right)^{1/2} (-1)^n \frac{a}{r} \sin\frac{n\pi}{a} r \tag{25.19}$$

and the integral in (25.16) becomes

$$\int_0^a dr\, r^2 \frac{2}{a^3} (-1)^{n-n'} \frac{a^2}{r^2} \sin\frac{n\pi r}{a} \sin\frac{n'\pi r}{a} = 2(-1)^{n-n'} \int_0^1 dt \sin n\pi t \sin n'\pi t = \delta_{nn'}. \tag{25.20}$$

The orthonormal radial functions are combined with the orthonormal spherical harmonics:

$$\phi_{lmn}(\mathbf{r}) = Y_{lm}(\theta,\phi) R_{ln}(r), \tag{25.21}$$

to produce the three-dimensional set of complete,

$$\delta(\mathbf{r}-\mathbf{r}') = \sum_{lmn} \phi_{lmn}(\mathbf{r}) \phi^*_{lmn}(\mathbf{r}'), \tag{25.22a}$$

orthonormal,

$$\int (d\mathbf{r})\, \phi^*_{lmn}(\mathbf{r}) \phi_{l'm'n'}(\mathbf{r}) = \delta_{ll'}\delta_{mm'}\delta_{nn'}, \tag{25.22b}$$

functions that obey the differential equation

$$(\nabla^2 + \gamma_{ln}^2/a^2)\phi_{lmn}(\mathbf{r}) = 0. \tag{25.23}$$

Consequently, the solution of the differential equation for Green's function, $G(\mathbf{r},\mathbf{r}')$, is given by

$$G(\mathbf{r},\mathbf{r}') = 4\pi \sum_{lmn} \frac{\phi_{lmn}(\mathbf{r})\phi^*_{lmn}(\mathbf{r}')}{\gamma^2_{lm}/a^2} = \sum_{ln} (2l+1)P_l(\cos\gamma)\frac{R_{ln}(r)R_{ln}(r')}{\gamma^2_{ln}/a^2}. \tag{25.24}$$

Comparison with (25.6) then supplies the relation

$$\sum_{n=1}^{\infty} \frac{R_{ln}(r)R_{ln}(r')}{\gamma^2_{ln}/a^2} = \frac{1}{2l+1} r^l_< \left(\frac{1}{r^{l+1}_>} - \frac{r^l_>}{a^{2l+1}}\right), \tag{25.25}$$

presenting equivalent forms for the solution of the radial equation

$$-\left(\frac{d^2}{dr^2} + \frac{2}{r}\frac{d}{dr} - \frac{l(l+1)}{r^2}\right) g_l(r,r') = \frac{1}{r^2}\delta(r-r'), \tag{25.26a}$$

with

$$g_l(a,r) = 0. \tag{25.26b}$$

Approximations for the roots of the spherical Bessel functions are obtained from (21.110) by writing m as $l+\frac{1}{2}$,

$$\gamma_{ln} \approx (l+2n)\frac{\pi}{2} - \frac{1}{\pi}\frac{l(l+1)}{l+2n} - \frac{1}{3\pi^3}\frac{l(l+1)}{(l+2n)^3}[7l(l+1)-6], \tag{25.27}$$

as illustrated by

$$\gamma_{1n} \approx \left(n+\frac{1}{2}\right)\pi - \frac{1}{\pi}\frac{1}{n+1/2} - \frac{2}{3\pi^3}\frac{1}{(n+1/2)^3}. \tag{25.28}$$

The approximation to γ_{11} produced in this way is 4.4938, which is in excess by 0.009% of the correct number,

$$\gamma_{11} = 4.4934. \tag{25.29}$$

Incidentally, as in the discussion of (21.114), the leading term in (25.27) is used to convert the summation of (25.25) into an integral, in the limit $a \to \infty$, thereby recovering what is displayed in (23.76a).

Some useful statements about the γ_{ln} are obtained from (25.25) through a general process that is most simply illustrated by setting $r' = r$ and integrating (r^2dr) from 0 to a. The normalization of the R_{ln} then produces

$$\sum_{n=1}^{\infty} \frac{a^2}{\gamma^2_{ln}} = \frac{1}{2l+1}\int_0^a dr\, r^2\left(\frac{1}{r} - \frac{r^{2l}}{a^{2l+1}}\right) = \frac{a^2}{2l+1}\left(\frac{1}{2} - \frac{1}{2l+3}\right), \tag{25.30a}$$

or

$$\sum_{n=1}^{\infty} \frac{1}{\gamma^2_{ln}} = \frac{1}{2}\frac{1}{2l+3}. \tag{25.30b}$$

For $l=0$, $\gamma_{0n} = n\pi$, this yields a familiar summation,

$$\sum_{n=1}^{\infty} \frac{1}{(n\pi)^2} = \frac{1}{6}, \tag{25.31}$$

which is usually derived from the pole expansion of $\cot t$,

$$\cot t = \sum_{n=-\infty}^{\infty} \frac{1}{t-n\pi} = \frac{1}{t} - 2t\sum_{n=1}^{\infty}\frac{1}{(n\pi)^2 - t^2}; \tag{25.32}$$

the comparison of the t terms in the small t expansion of the two sides yields (25.31).

The second in the unlimited sequence of such statements considers the square of (25.25), integrated over both r and r'. Here, the orthonormality of the R_{ln} yields

$$\sum_{n=1}^{\infty} \frac{a^4}{\gamma_{ln}^4} = 2\frac{1}{(2l+1)^2} \int_0^a dr\, r^2 \int_0^r dr' r'^2 r'^{2l} \left(\frac{1}{r^{l+1}} - \frac{r^l}{a^{2l+1}}\right)^2$$
$$= 2\frac{1}{(2l+1)^2}\frac{1}{2l+3} \int_0^a dr\, r^3 \left(1 - \left(\frac{r}{a}\right)^{2l+1}\right)^2, \tag{25.33}$$

where the symmetry in r and r' permits one to simply double the contribution of the integrals for $r' < r$. The outcome of this integration is given by

$$\sum_{n=1}^{\infty} \frac{1}{\gamma_{ln}^4} = \frac{1}{2}\frac{1}{(2l+3)^2}\frac{1}{2l+5}, \tag{25.34}$$

which is illustrated, for $l = 0$, by

$$\sum_{n=1}^{\infty} \frac{1}{(n\pi)^4} = \frac{1}{90}, \tag{25.35}$$

as also follows from the t^3 terms of (25.32).

One naturally asks for a generalization of the expansion (25.32) to $l \neq 0$. It is given by the pole expansion of $j_l'(t)/j_l(t)$ which, for $l = 0$, is $\cot t - 1/t$. First, we need to notice that $j_l(t)$, as a numerical multiple of $t^{-1/2}J_{l+1/2}(t)$, behaves as t^l for small t. That gives the pole expansion in terms of the roots of j_l, γ_{ln}, as

$$\frac{j_l'(t)}{j_l(t)} = \frac{l}{t} - 2t\sum_{n=1}^{\infty} \frac{1}{(\gamma_{ln})^2 - t^2}. \tag{25.36}$$

The required power series for $j_l(t)$ can be derived from (21.132) and (23.63),

$$j_l(t) = t^l \left(-\frac{1}{t}\frac{d}{dt}\right)^l \frac{\sin t}{t}$$
$$= t^l \left(\frac{1}{t}\frac{d}{dt}\right)^l \sum_{k=0}^{\infty} (-1)^k \frac{t^{2l+2k}}{(2l+1+2k)!}$$
$$= (2t)^l \sum_{k=0}^{\infty} (-1)^k \frac{(l+k)!}{k!\,(2l+1+2k)!} t^{2k}, \tag{25.37}$$

or, writing out the first few terms,

$$j_l(t) = (2t)^l \frac{l!}{(2l+1)!}\left[1 - \frac{1}{2}\frac{1}{2l+3}t^2 + \frac{1}{8}\frac{1}{(2l+3)(2l+5)}t^4 + \dots\right]. \tag{25.38}$$

Then the expansion produced by differentiating the logarithm of $j_l(t)$ is

$$\frac{j_l'(t)}{j_l(t)} = \frac{l}{t} - \frac{t}{2l+3} - \frac{t^3}{(2l+3)^2(2l+5)} + \dots, \tag{25.39}$$

from which we regain (25.30b) and (25.34) by examining the t and t^3 terms in the power series expansion of (25.36). We leave it to the reader to carry out the next step and derive the sum of the inverse sixth powers of the γ_{ln} [Problem 25.3].

One use of these summations,

$$\sigma_l^{(r)} = \sum_{n=1}^{\infty} \frac{1}{\gamma_{ln}^{2r}}, \quad r = 1, 2, \ldots \tag{25.40}$$

comes from the evident inequality

$$\sigma_l^{(r)} > \frac{1}{\gamma_{l1}^{2r}}, \tag{25.41}$$

or

$$\gamma_{l1} > \frac{1}{(\sigma_l^{(r)})^{1/2r}}; \tag{25.42}$$

these are lower bounds to the smallest of the γ's for a given l.

As an example, consider (25.31), (25.35), and the next member of the sequence,

$$\sum_{n=1}^{\infty} \frac{1}{(n\pi)^6} = \frac{1}{945}, \tag{25.43}$$

which yield, successively,

$$\pi > 2.4495;\ 3.0801;\ 3.1326;\ \ldots, \tag{25.44}$$

rapidly converging from below to the limiting value. One can also improve the last member of such a sequence by an approximate extrapolation to the limit that employs it and one or more of the preceding numbers. (See Problem 25.4.) We defer the discussion of another application of such sums, one that focuses on the second smallest γ value. (See Chapter 27.)

25.2 Exterior of Conducting Sphere

Now let us use the surface charge density in setting up the conditions to be satisfied for a unit charge at the point $\mathbf{r}'$ exterior to a conducting sphere S of radius a, where the origin of coordinates is taken to be the center of the sphere:

$$\begin{aligned} G(\mathbf{r}, \mathbf{r}') &= \frac{1}{|\mathbf{r} - \mathbf{r}'|} + \int_S dS_1 \frac{\sigma(\mathbf{r}_1)}{|\mathbf{r} - \mathbf{r}_1|}, \\ G(\mathbf{r}, \mathbf{r}') &= 0, \quad \mathbf{r} \text{ on } S. \end{aligned} \tag{25.45}$$

Implicit in this is the restriction to the external region, $r, r' > a$.

The individual potentials that appear here are

$$\frac{1}{|\mathbf{r} - \mathbf{r}'|} = \sum_{l=0}^{\infty} \frac{r_<^l}{r_>^{l+1}} \frac{4\pi}{2l+1} \sum_{m=-l}^{l} Y_{lm}(\theta, \phi) Y_{lm}^*(\theta', \phi') \tag{25.46a}$$

and $(r > a)$

$$\int_S dS_1 \frac{\sigma(\mathbf{r}_1)}{|\mathbf{r} - \mathbf{r}_1|} = \sum_{l=0}^{\infty} \frac{a^l}{r^{l+1}} \frac{4\pi}{2l+1} \sum_m Y_{lm}(\theta, \phi) \sigma_{lm}, \tag{25.46b}$$

where

$$\sigma_{lm} = \int_S dS_1 Y_{lm}^*(\theta_1, \phi_1) \sigma(\mathbf{r}_1). \tag{25.47}$$

The requirement that G, the sum of these components, vanishes for all θ and ϕ at $r = a$ leads immediately to

$$0 = \frac{a^l}{r'^{l+1}} Y^*_{lm}(\theta', \phi') + \frac{1}{a}\sigma_{lm}, \tag{25.48}$$

and we have learned that

$$G(\mathbf{r}, \mathbf{r}') = \sum_{l=0}^{\infty} \left(r^l_< - \frac{a^{2l+1}}{r^{l+1}_<} \right) \frac{1}{r^{l+1}_>} P_l(\cos\gamma), \tag{25.49a}$$

which is the analog of (25.6).

We move toward an equivalent form by writing this as

$$G(\mathbf{r}, \mathbf{r}') = \frac{1}{|\mathbf{r} - \mathbf{r}'|} - \frac{a}{r'} \sum_{l=0}^{\infty} \frac{\bar{r}'^l}{r^{l+1}} P_l(\cos\gamma), \tag{25.49b}$$

where

$$\bar{r}' = \frac{a^2}{r'} < a. \tag{25.49c}$$

Then, with the introduction of the vector $\bar{\mathbf{r}}'$, with magnitude $\bar{r}'$ and the direction of $\mathbf{r}'$, presents Green's function in the image form

$$G(\mathbf{r}, \mathbf{r}') = \frac{1}{|\mathbf{r} - \mathbf{r}'|} - \frac{a}{r'} \frac{1}{|\mathbf{r} - \bar{\mathbf{r}}'|}, \tag{25.50}$$

the same as (25.8).

With the information contained in (25.48), that

$$\sigma_{lm} = -\frac{a^{l+1}}{r'^{l+1}} Y^*_{lm}(\theta', \phi'), \tag{25.51}$$

we can construct the surface charge density. According to the definition (25.47) and the completeness of the spherical harmonics, it is $[dS_1 = a^2 d\Omega]$

$$\sigma(\mathbf{r}) = \sum_{lm} Y_{lm}(\theta, \phi) \frac{1}{a^2} \sigma_{lm} = -\frac{1}{4\pi a^2} \sum_{l=0}^{\infty} (2l+1) \left(\frac{a}{r'}\right)^{l+1} P_l(\cos\gamma). \tag{25.52}$$

One should notice, however, that an aspect of the surface charge distribution, the total induced charge, is given directly by σ_{00}:

$$Q_{\text{ind}} = \int dS\, \sigma(\mathbf{r}) = \sqrt{4\pi}\sigma_{00} = -\frac{a}{r'}; \tag{25.53}$$

it is just the image charge displayed in (25.50).

An alternative approach to the charge induced on the conducting surface is provided by the normal component of the electric field,

$$\sigma(\mathbf{r}) = \frac{1}{4\pi} \left(-\frac{\partial}{\partial r} G(\mathbf{r}, \mathbf{r}') \right) \bigg|_{r=a}. \tag{25.54}$$

The introduction of the Green's function form (25.49a) immediately reproduces (25.52). Now let's use the image version, (25.50), which we write as

$$G(\mathbf{r}, \mathbf{r}') = \frac{1}{[r'^2 - 2rr'\cos\gamma + r^2]^{1/2}} - \frac{1}{[(rr'/a)^2 - 2rr'\cos\gamma + a^2]^{1/2}} \tag{25.55}$$

thereby making quite transparent that G vanishes at $r = a$. This gives [*cf.*(25.11b)]

$$\sigma(\mathbf{r}) = -\frac{1}{4\pi a^2}\frac{a}{r'}\frac{1-(a/r')^2}{[1-2(a/r')\cos\gamma+(a/r')^2]^{3/2}}, \tag{25.56}$$

and we note particularly the ratios of the surface charge density to its average value, $(-a/r')/4\pi a^2$, at the point nearest the unit charge:

$$\gamma = 0: \quad \frac{1+\frac{a}{r'}}{\left(1-\frac{a}{r'}\right)^2} > 1, \tag{25.57}$$

and at the point farthest from the unit charge:

$$\gamma = \pi: \quad \frac{1-\frac{a}{r'}}{\left(1+\frac{a}{r'}\right)^2} < 1. \tag{25.58}$$

Of course, we also have

$$\frac{1}{Q_{\text{ind}}}\int dS\,\sigma(\mathbf{r}) = \frac{1}{2}\int_{-1}^{1} d(\cos\gamma)\frac{1-(a/r')^2}{[1-2(a/r')\cos\gamma+(a/r')^2]^{3/2}} = 1. \tag{25.59}$$

The equivalence of the two surface density forms, (25.52) and (25.56), as presented by $(t = a/r')$

$$t < 1: \quad \frac{1-t^2}{[1-2\mu t+t^2]^{3/2}} = \sum_{l=0}^{\infty}(2l+1)t^l P_l(\mu), \tag{25.60}$$

another generating function, which is especially interesting in the limit $t \to 1$. For $\mu < 1$, the limiting value of the left-hand side is zero [as illustrated by (25.58)], whereas with $\mu = 1$ [see (25.57)] the limit is infinite. And it is the content of (25.59) that, independently of the value of t, half of the μ-integral of the left-hand side, from -1 to 1, is equal to one. In short, we have learned that

$$\sum_{l=0}^{\infty}(2l+1)P_l(\mu) = 2\delta(1-\mu), \tag{25.61}$$

where unity in the argument of the delta function is to be understood as $1-0$. This can be recognized as the completeness relation (23.22), rewritten with the aid of the addition theorem (23.24),

$$\sum_{l=0}^{\infty}\frac{2l+1}{4\pi}P_l(\cos\gamma) = \frac{1}{\sin\theta}\delta(\theta-\theta')\delta(\phi-\phi'), \tag{25.62a}$$

where the right-hand side is expressed in terms of γ as

$$\frac{1}{\sin\gamma}\delta(\gamma)\frac{1}{2\pi} = \frac{1}{2\pi}\delta(1-\cos\gamma). \tag{25.62b}$$

As has been done so often before, we use G to determine the energy of interaction between a charge e, stationed at $\mathbf{r}_0$, and the conducting sphere:

$$\begin{aligned} E_{\text{int}} &= \frac{1}{2}e^2[G(\mathbf{r},\mathbf{r}') - G_0(\mathbf{r},\mathbf{r}')]\Big|_{\mathbf{r},\mathbf{r}'\to\mathbf{r}_0} \\ &= -\frac{1}{2}e^2\frac{a}{r_0}\frac{1}{r_0-(a^2/r_0)} \\ &= -\frac{1}{2}e^2\frac{a}{r_0^2-a^2}. \end{aligned} \tag{25.63}$$

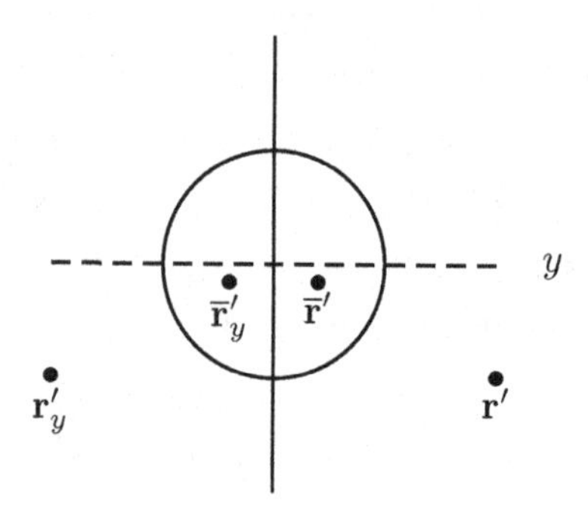

FIGURE 25.1
Sphere bisected by plane. Shown are the locations of the physical charge at $\mathbf{r}'$ and the image charges at $\bar{\mathbf{r}}'$, $\mathbf{r}'_y$, and $\bar{\mathbf{r}}'_y$.

This is just the Coulomb energy between the charge e and the image charge $-(a/r_0)e$, multiplied by the characteristic factor of 1/2. [Recall Section 19.5.] The magnitude of the force of attraction between charge and conductor,

$$\begin{aligned} F = \frac{\partial}{\partial r_0} E_{\text{int}} &= e^2 \frac{a r_0}{(r_0^2 - a^2)^2} \\ &= \frac{|e||-(a/r_0)e|}{[r_0 - (a^2/r_0)]^2}, \end{aligned} \tag{25.64}$$

is that between charge e and its image charge.

25.3 Conducting Plate and Hemispherical Boss

Let the conducting sphere be bisected by a plane conducting sheet of unlimited extent, thereby producing the situation of a hemispherical boss raised over a plane surface. We choose the coordinate system, with its origin at the center of the sphere, such that the conducting sheet lies in the x-z plane. The range of spherical coordinates for this exterior problem is: $r > a$, $\pi > \theta > 0$, $\pi > \phi > 0$. Now, G is required to vanish at $r = a$, the surface of the boss, and at $\phi = 0, \pi$, the surface of the sheet. As for the latter requirement, we already know that it implies the substitution [(19.54a), with $z, z' \to \phi, \phi'$; $a \to \pi$; $n \to m$]

$$\delta(\phi - \phi') \to \frac{2}{\pi} \sum_{m=1}^{\infty} \sin m\phi \sin m\phi' = \sum_{m=-\infty}^{\infty} \frac{1}{2\pi} \left[e^{im(\phi-\phi')} - e^{im(\phi+\phi')} \right], \tag{25.65}$$

which exhibits the equivalent image charge structure at reversed values of ϕ', or of y'. Accordingly, if images in the plane $y = 0$ are indicated by the subscript y, Green's function for this external situation is (see Fig. 25.1)

$$G(\mathbf{r}, \mathbf{r}') = \frac{1}{|\mathbf{r} - \mathbf{r}'|} - \frac{1}{|\mathbf{r} - \mathbf{r}'_y|} - \frac{a}{r'} \left(\frac{1}{|\mathbf{r} - \bar{\mathbf{r}}'|} - \frac{1}{|\mathbf{r} - \bar{\mathbf{r}}'_y|} \right). \tag{25.66}$$

As an application, let charge e be placed on the positive y-axis, at distance $r_0 > a$ from the origin. The image charges are also located on the y-axis, their charges and y-coordinates being: $-e$, $-r_0$; $-e(a/r_0)$, a^2/r_0; $e(a/r_0)$, $-a^2/r_0$. The magnitude of the attractive force between charge e and the conductors is, then,

$$F = e^2 \left[\frac{1}{(2r_0)^2} + \frac{a/r_0}{[r_0 - (a^2/r_0)]^2} - \frac{a/r_0}{[r_0 + (a^2/r_0)]^2} \right]. \tag{25.67}$$

One notes that as r_0 approaches a the first and third terms of F tend to cancel, leaving just the force associated with the sphere, whereas, in the limit of large r_0/a, it is the first term, associated with the conducting plane, that dominates,

$$a/r_0 \ll 1: \quad F \approx e^2 \left[\frac{1}{(2r_0)^2} + 4\frac{a^3}{r_0^5} \right]. \tag{25.68}$$

It should also be observed that, while the sum of the second and third terms in (25.67) is positive so that the force always exceeds that produced by the conducting plane alone, a similar remark about the first and third terms, and the force produced by the conducting sphere alone, requires that r_0 be sufficiently large. Indeed, the contrary circumstance,

$$\frac{a/r_0}{[r_0 + (a^2/r_0)]^2} > \frac{1}{(2r_0)^2}, \tag{25.69a}$$

or $(x = r_0/a)$

$$\frac{4x^3}{(x^2+1)^2} > 1 \tag{25.69b}$$

occurs in the range

$$1 < x < 3.383; \tag{25.70}$$

here the effect of the conducting plate is to reduce the attractive force that the conducting sphere alone produces. The largest reduction occurs for $r_0/a = 2.297$ where it is 8.7%.

25.4 Dielectric Ball

To move beyond the consideration of conductors as examples of the use of spherical harmonics in solving Green's function problems with spherical symmetry, we consider a dielectric ball of radius a, with a unit point charge *outside*, as shown in Fig. 25.2. In this case, the Green's function equation is

$$r > a: \quad -\nabla^2 G(\mathbf{r}, \mathbf{r}') = 4\pi\delta(\mathbf{r} - \mathbf{r}'), \tag{25.71a}$$

$$r < a: \quad -\nabla^2 G(\mathbf{r}, \mathbf{r}') = 0, \tag{25.71b}$$

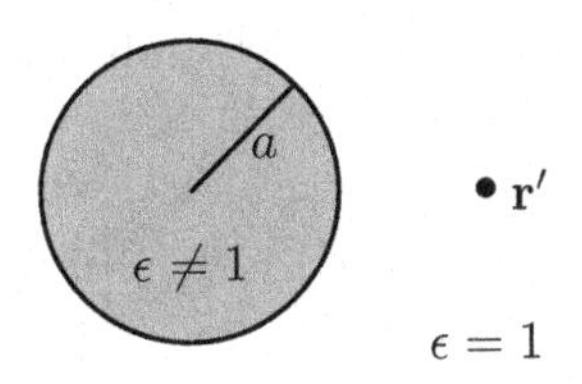

FIGURE 25.2
Geometry of dielectric ball with source point outside.

where we will take ϵ to be a constant in the interior of the ball, $r < a$. The boundary conditions at $r = a$ are, from (13.54) and (13.55b),

$$G \text{ is continuous,} \tag{25.72a}$$

and

$$\left[-\frac{\partial}{\partial r}G\right]_{r=a+0} = \left[-\epsilon\frac{\partial}{\partial r}G\right]_{r=a-0}. \tag{25.72b}$$

The conditions on G at $r = r'$ are as given in (23.17a) and (23.17b),

$$G \text{ is continuous,} \tag{25.73a}$$

and

$$\left[-r^2\frac{\partial}{\partial r}G\right]_{r=r'-0}^{r=r'+0} = 4\pi\frac{1}{\sin\theta}\delta(\theta-\theta')\delta(\phi-\phi'). \tag{25.73b}$$

As is familiar by now, the solution in the three regions has the form

$$r < a: \quad G = \sum_{lm} r^l Y_{lm}(\theta,\phi)A_{lm}, \tag{25.74a}$$

$$r > r': \quad G = \sum_{lm} r^{-l-1} Y_{lm}(\theta,\phi)D_{lm}, \tag{25.74b}$$

$$a < r < r': \quad G = \sum_{lm} (r^l B_{lm} + r^{-l-1}C_{lm})Y_{lm}(\theta,\phi). \tag{25.74c}$$

It is very easy to find the expansion coefficients by use of (25.72a)–(25.73b):

$$A_{lm} = \frac{4\pi Y_{lm}^*(\theta',\phi')}{l(\epsilon+1)+1}r'^{-l-1}, \tag{25.75a}$$

$$B_{lm} = \frac{4\pi Y_{lm}^*(\theta',\phi')}{2l+1}r'^{-l-1}, \tag{25.75b}$$

$$C_{lm} = -\frac{(\epsilon-1)l}{l(\epsilon+1)+1}\frac{4\pi Y_{lm}^*(\theta',\phi')}{2l+1}\frac{a^{2l+1}}{r'^{l+1}}, \tag{25.75c}$$

$$D_{lm} = C_{lm} + \frac{4\pi Y_{lm}^*(\theta',\phi')}{2l+1}r'^l. \tag{25.75d}$$

Green's function, outside the ball, is therefore found to be

$$r, r' > a: \quad G(\mathbf{r},\mathbf{r}') = \frac{1}{|\mathbf{r}-\mathbf{r}'|} - \sum_{l=1}^{\infty}\frac{(\epsilon-1)l}{l(\epsilon+1)+1}\frac{a^{2l+1}}{r^{l+1}r'^{l+1}}P_l(\cos\gamma). \tag{25.76}$$

We now ask what is the leading behavior of this potential when the separation between the point charge and the ball is large compared to the radius of the ball, $r' \gg a$. Since the lth term in the sum behaves as $(a/r')^{l+1}$, only small values of l contribute. The leading contribution arises from $l = 1$,

$$r' \gg a: \quad G(\mathbf{r},\mathbf{r}') \sim \frac{1}{|\mathbf{r}-\mathbf{r}'|} - \frac{\epsilon-1}{\epsilon+2}\frac{a^3}{r^2r'^2}\cos\gamma. \tag{25.77}$$

Since γ is the angle between $\mathbf{r}$ and $\mathbf{r}'$,

$$\cos\gamma = \frac{\mathbf{r}\cdot\mathbf{r}'}{rr'}, \tag{25.78}$$

this asymptotic form of Green's function can be rewritten as

$$G(\mathbf{r},\mathbf{r}') = \frac{1}{|\mathbf{r}-\mathbf{r}'|} + \frac{\mathbf{r}}{r^3}\cdot\mathbf{d}, \tag{25.79}$$

the two terms of which have simple physical interpretations. The first term is due to the point charge while the second is the potential arising from the induced electric dipole moment of the ball [*cf.*(24.3)]. The latter is identified from (25.77) to be

$$\mathbf{d} = \frac{\epsilon-1}{\epsilon+2}a^3\left(-\frac{\mathbf{r}'}{r'^3}\right), \tag{25.80}$$

where $-\mathbf{r}'/r'^3$ is interpreted as the electric field, $\mathbf{E}_0$, at the center of the ball (in the absence of the dielectric) produced by the unit point charge. Since this electric field is essentially constant over the ball, we recognize that the electric dipole moment induced in a dielectric ball of radius a by a constant electric field $\mathbf{E}_0$ is

$$\mathbf{d} = \frac{\epsilon-1}{\epsilon+2}a^3\mathbf{E}_0. \tag{25.81}$$

The proportionality constant between the electric field and the induced dipole moment is called the electric polarizability,

$$\alpha = \frac{\epsilon-1}{\epsilon+2}a^3, \tag{25.82}$$

proportional to the volume of the ball.

Finally, we write the expression for Green's function inside the ball:

$$\left.\begin{matrix} r<a \\ r'>a \end{matrix}\right\}: \quad G(\mathbf{r},\mathbf{r}') = \sum_{l=0}^{\infty}\frac{r^l}{r'^{l+1}}\frac{2l+1}{l(\epsilon+1)+1}P_l(\cos\gamma). \tag{25.83}$$

Again, in the situation in which the point charge is located far from the ball, $r' \gg a$, low values of l predominate:

$$G \sim \frac{1}{r'} + \frac{3}{\epsilon+2}\frac{\mathbf{r}\cdot\mathbf{r}'}{r'^3} = \frac{1}{r'} - \frac{3}{\epsilon+2}\mathbf{r}\cdot\mathbf{E}_0 = \frac{1}{r'} - \mathbf{r}\cdot\mathbf{E}, \tag{25.84}$$

where we identify the electric field in the dielectric as the negative gradient of G so that the field $\mathbf{E}$ in the dielectric,

$$\mathbf{E} = \frac{3}{\epsilon+2}\mathbf{E}_0, \tag{25.85}$$

is less than the applied field $\mathbf{E}_0$ if $\epsilon > 1$. This is equivalent to (5.86), (5.92)—See Problem 25.7.

25.4.1 Interior of Ball

We illustrate a slightly different method of computing Green's function for the circumstance when the point charge is inside the dielectric ball, $r' < a$. We recall that if all space were filled with a medium of dielectric constant ϵ, the Coulomb potential would be reduced by the factor ϵ. Then, if we define the reduced Green's function by

$$G(\mathbf{r},\mathbf{r}') = \sum_{l=0}^{\infty} g_l(r,r')P_l(\cos\gamma), \tag{25.86}$$

we recognize that surface charge density on the sphere induced by that Coulomb potential at $r_> = a$ is proportional to $(r_< = r')^l$, and so we have the forms for the reduced Green's functions in the two regions:

$$r' < a,\, r < a: \quad g_l(r,r') = \frac{1}{\epsilon}\frac{r_<^l}{r_>^{l+1}} + \rho_l \frac{1}{\epsilon}\frac{r^l r'^l}{a^{2l+1}},$$
$$r' < a,\, r > a: \quad g_l(r,r') = \tau_l \frac{1}{\epsilon}\frac{r'^l}{r^{l+1}}. \tag{25.87}$$

Here ρ_l and τ_l are numerical coefficients to be determined. Now we impose the boundary conditions at $r = a$: the continuity of the potential that is Green's function, or

$$1 + \rho_l = \tau_l, \tag{25.88}$$

and (in the absence of a surface distribution of free charge) the continuity of the radial component of $\mathbf{D}$, or

$$\epsilon(l + 1 - l\rho_l) = (l+1)\tau_l. \tag{25.89}$$

Combining these equations, we obtain

$$\rho_l = \frac{(\epsilon-1)(l+1)}{(\epsilon+1)l+1}, \qquad \tau_l = \epsilon\frac{2l+1}{(\epsilon+1)l+1}, \tag{25.90}$$

and so the reduced Green's function is

$$r' < a,\, r < a: \quad g_l(r,r') = \frac{1}{\epsilon} r_<^l \left(\frac{1}{r_>^{l+1}} + \frac{(\epsilon-1)(l+1)}{(\epsilon+1)l+1}\frac{r_>^l}{a^{2l+1}} \right),$$
$$r' < a,\, r > a: \quad g_l(r,r') = \frac{2l+1}{(\epsilon+1)l+1}\frac{r'^l}{r^{l+1}}. \tag{25.91}$$

For comparison, let us write down the reduced Green's function for the previously considered situation in which the point charge is exterior to the ball:

$$r' > a,\, r > a: \quad g_l(r,r') = \left(r_<^l - \frac{(\epsilon-1)l}{(\epsilon+1)l+1}\frac{a^{2l+1}}{r_<^{l+1}} \right)\frac{1}{r_>^{l+1}},$$
$$r' > a,\, r < a: \quad g_l(r,r') = \frac{2l+1}{(\epsilon+1)l+1}\frac{r^l}{r'^{l+1}}. \tag{25.92}$$

One recognizes here the symmetry in r and r' of $g_l(r,r')$, in particular, that between the two situations $r < a$, $r' > a$, and $r > a$, $r' < a$; it is the radial aspect of the symmetry of $G(\mathbf{r},\mathbf{r}')$ in $\mathbf{r}$ and $\mathbf{r}'$. As an illustration of that symmetry consider $r = 0$, $r' > a$, where all g_l vanish except

$$g_0(r = 0, r' > a) = \frac{1}{r'}. \tag{25.93}$$

Thus, according to (25.86), the potential, at the center of the dielectric ball, that is produced by a unit point charge at distance $r' > a$ is just $1/r'$, independently of the value of the dielectric constant. We can understand that through the symmetry of Green's function, which asserts that what we have just considered is also the potential produced at $r' > a$ by a unit charge at the center of the ball. But Gauss' theorem, in the form

$$\oint d\mathbf{S}\cdot\mathbf{D} = 4\pi \int (d\mathbf{r})\,\rho, \tag{25.94}$$

assures us that the radially directed electric field at distance r', in the vacuum surrounding a spherically symmetrical arrangement containing a unit charge, is indeed just that derived from the potential $1/r'$.

25.5 Problems for Chapter 25

1. Derive (25.11b).
2. Use (25.11b) to compute the total charge on the sphere.
3. By carrying out the expansion in (25.39) to order t^5, find an expression for the sum $\sum_n \gamma_{ln}^{-6}$. As an example, establish (25.43).
4. Verify the lower bounds for π shown in (25.44) and carry out the improvement referred to at the end of Section 25.1.1.
5. Construct Green's function, from its differential equation, for the region *exterior* to a conducting sphere. [That is, use the method given in Section 25.1.] Give the image interpretation of the result. What is the physical significance of the two leading terms when one point is very far from the sphere?
6. Calculate Green's function for a dielectric ball when the point charge is *inside* the ball, $r' < a$, using the direct discontinuity method given in Section 25.4.
7. Recall our discussion of the Clausius-Mossotti equation in Chapter 5. Show that $\mathbf{E}_{\text{driving}}$ discussed there may be identified with $\mathbf{E}_0$, and derive the formula for the induced dipole moment, (25.81), from the consideration in Chapter 5.
8. Consider the limit in which $\epsilon \to \infty$ so that the dielectric ball discussed in Section 25.4 may be regarded as a conductor. Give the form of the Green's function in that situation for $r' > a$ by taking the $\epsilon \to \infty$ limit of (25.76) and (25.83). Show that indeed the interior of the ball is an equipotential region, but not one of zero potential. Also, show that in this case there is zero charge on the conductor. Show that by adding a suitable charge distribution to the ball one can recover the situation considered in Section 25.2.
9. Consider a dielectric cavity, that is, a spherical region with $\epsilon = 1$ embedded in an otherwise uniform dielectric medium. As in the case of plane surfaces, this is not a new situation. Green's function can be obtained, first, by replacing ϵ by $\frac{1}{\epsilon}$, and then increasing the scale of all dielectric constants by the factor of ϵ, which decreases G by the same factor. Obtain the reduced Green's function g_l in the interior of the spherical boundary, $r < a$, for r' either inside or outside the cavity. In the latter case, by letting $r' \gg a$, show that the field within the cavity is uniform,
$$\mathbf{E} = \frac{3\epsilon}{2\epsilon + 1}\mathbf{E}(0), \tag{25.95}$$
where $\mathbf{E}(0)$ is the field that the unit charge would produce at the origin if there were no cavity in the dielectric medium. Also, discuss the limit $\epsilon \to \infty$.
10. What is the statement of completeness for the functions $\Theta_{lm}(\theta)$, as inferred from that of the $Y_{lm}(\theta, \phi)$? Consider a conducting sphere that is bisected by an infinitely thin conducting partition, as discussed in Section 25.3. Find Green's function for the *interior* of the conducting hemisphere.

26

Dielectrics and Conductors

26.1 Variational Principle

In Chapter 13, we investigated the stationary properties of the electrostatic energy when only dielectrics are present, that is, we had a stationary principle,

$$\delta E = 0, \tag{26.1}$$

where [see (13.10)]

$$E = \int (d\mathbf{r}) \left[\rho\phi + \frac{\epsilon}{4\pi} \left(\mathbf{E} \cdot \boldsymbol{\nabla}\phi + \frac{1}{2}\mathbf{E}^2 \right) \right]. \tag{26.2}$$

We now wish to generalize this situation to include conductors as well. The new feature here is the existence of surface charges on various conductors implying an additional contribution to the energy:

$$E = \int (d\mathbf{r}) \left[\rho\phi + \frac{\epsilon}{4\pi} \left(\mathbf{E} \cdot \boldsymbol{\nabla}\phi + \frac{1}{2}\mathbf{E}^2 \right) \right] + \sum_{i=1}^{n} \int dS_i \, \sigma\phi, \tag{26.3a}$$

where the volume integral extends over all space exterior to the conductors and the surface integral is over all of the conductors, σ being the surface charge density. (See Fig. 26.1.) This energy functional is to be supplemented by the condition that the total charge on each conductor,

$$Q_i = \int dS_i \, \sigma, \quad i = 1, 2, \ldots, n, \tag{26.3b}$$

is fixed. The electrostatic problem is completely specified by the location of the conductors and dielectrics, the free volume charge density, ρ, and the charge on each conductor, Q_i.

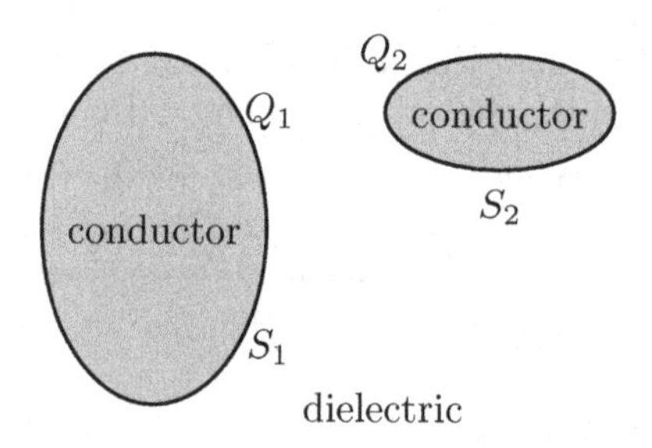

FIGURE 26.1
Conducting surfaces embedded in a dielectric medium.

DOI: 10.1201/9781003057369-26

Note, in particular, that the surface charge density, σ, is to be determined dynamically. The change of the energy under variations of ϕ, $\mathbf{E}$, and σ is

$$\delta E = \int (d\mathbf{r}) \left[\rho\delta\phi + \frac{\epsilon}{4\pi}(\delta\mathbf{E}\cdot\boldsymbol{\nabla}\phi + \mathbf{E}\cdot\boldsymbol{\nabla}\delta\phi + \mathbf{E}\cdot\delta\mathbf{E})\right] + \sum_i \int dS_i[\delta\sigma\phi + \sigma\delta\phi], \quad (26.4a)$$

which is subject to the condition that Q_i be constant, that is

$$\int dS_i\,\delta\sigma = 0. \quad (26.4b)$$

We rewrite the $\boldsymbol{\nabla}\delta\phi$ term by means of an integration by parts, which makes use of the identity

$$\frac{\mathbf{D}}{4\pi}\cdot\boldsymbol{\nabla}\delta\phi = \boldsymbol{\nabla}\cdot\left(\frac{\mathbf{D}}{4\pi}\delta\phi\right) - \delta\phi\frac{\boldsymbol{\nabla}\cdot\mathbf{D}}{4\pi}. \quad (26.5)$$

The implied surface integral here cannot be discarded since now there are contributions arising from the surfaces of the conductors. If we let $\mathbf{n}_i$ be the outward normal on the ith conductor, this surface term is

$$\int (d\mathbf{r})\,\boldsymbol{\nabla}\cdot\left(\frac{\mathbf{D}}{4\pi}\delta\phi\right) = -\sum_i \int dS_i\frac{\mathbf{n}_i\cdot\mathbf{D}}{4\pi}\delta\phi. \quad (26.6)$$

The variation in the energy, (26.4a), now reads

$$\begin{aligned}\delta E = &\int (d\mathbf{r}) \left[\rho\delta\phi + \frac{\epsilon}{4\pi}\delta\mathbf{E}\cdot\boldsymbol{\nabla}\phi - \delta\phi\frac{\boldsymbol{\nabla}\cdot\mathbf{D}}{4\pi} + \frac{1}{4\pi}\mathbf{D}\cdot\delta\mathbf{E}\right] \\ &+\sum_i \int dS_i\left[-\frac{\mathbf{n}_i\cdot\mathbf{D}}{4\pi}\delta\phi + \delta\sigma\phi + \sigma\delta\phi\right]. \end{aligned} \quad (26.7)$$

The requirement that the energy be stationary under independent variations in ϕ and $\mathbf{E}$ then implies, in the interior of the dielectric,

$$\delta\phi: \quad \boldsymbol{\nabla}\cdot\mathbf{D} = 4\pi\rho, \quad (26.8a)$$

$$\delta\mathbf{E}: \quad \mathbf{E} = -\boldsymbol{\nabla}\phi, \quad (26.8b)$$

while just outside the surface of the conductors,

$$\delta\phi: \quad \mathbf{n}\cdot\mathbf{D} = 4\pi\sigma. \quad (26.8c)$$

Finally, the variation in the surface charge density requires

$$\delta E = \sum_i \int dS_i\,\delta\sigma\phi = 0 \quad (26.9)$$

which is subject to the restriction (26.4b), implying that each conductor is an equipotential surface,

$$\phi = \text{constant on } S_i = \phi_i. \quad (26.10)$$

Thus, the stationary action principle, based on the energy functional (26.3a), yields all the physical laws governing electrostatics in the presence of conductors and dielectrics.

26.2 Restricted Forms of the Variational Principle

As in Chapter 13, there are two restricted forms of the variational principle we may discuss. In the first, we take the electric field as being defined by

$$\mathbf{E} = -\boldsymbol{\nabla}\phi, \tag{26.11}$$

so that the energy functional becomes

$$E[\phi,\sigma] = \int (d\mathbf{r}) \left[\rho\phi - \frac{\epsilon}{8\pi}(\boldsymbol{\nabla}\phi)^2\right] + \sum_i \int dS_i\, \sigma\phi. \tag{26.12}$$

The independent variables are ϕ and σ, the latter of which is subject to the condition (26.3b). For the second form, $\mathbf{D}$ is regarded as an independent variable, subject to the condition

$$\boldsymbol{\nabla}\cdot\mathbf{D} = 4\pi\rho, \quad \text{inside dielectric,} \tag{26.13a}$$

while σ is determined by

$$\mathbf{n}\cdot\mathbf{D} = 4\pi\sigma, \quad \text{on surface } S_i. \tag{26.13b}$$

To rewrite the energy as a functional of $\mathbf{D}$ only, we integrate by parts on the $\mathbf{D}\cdot\boldsymbol{\nabla}\phi$ term in (26.3a) and use (26.13a) and (26.13b) to obtain

$$E[\mathbf{D}] = \int (d\mathbf{r}) \frac{1}{8\pi}\frac{\mathbf{D}^2}{\epsilon}, \tag{26.14a}$$

while the subsidiary condition (26.3b) becomes

$$Q_i = \int dS_i \frac{\mathbf{n}\cdot\mathbf{D}}{4\pi}. \tag{26.14b}$$

In (26.14a), we identify the integrand as the energy density of the field.

Let us now verify that the second restricted form of the variational principle correctly describes the electrostatic situation under consideration. For a finite change in $\mathbf{D}$, $\mathbf{D}\to\mathbf{D}+\Delta\mathbf{D}$, the change in the energy functional (26.14a) is

$$\Delta E = \int (d\mathbf{r}) \frac{1}{4\pi}\frac{\mathbf{D}}{\epsilon}\cdot\Delta\mathbf{D} + \int (d\mathbf{r}) \frac{1}{8\pi}\frac{(\Delta\mathbf{D})^2}{\epsilon}, \tag{26.15}$$

while the constraints read

$$\boldsymbol{\nabla}\cdot\Delta\mathbf{D} = 0, \tag{26.16a}$$

$$\int dS_i \frac{\mathbf{n}\cdot\Delta\mathbf{D}}{4\pi} = 0. \tag{26.16b}$$

The stationary condition requires that the integral linear in $\Delta\mathbf{D}$ in (26.15) vanishes. To incorporate the constraint (26.16a), we add to (26.15) the volume integral

$$0 = \int (d\mathbf{r}) \frac{\phi(\mathbf{r})}{4\pi}\boldsymbol{\nabla}\cdot\Delta\mathbf{D} = -\sum_i \int dS_i \frac{\mathbf{n}\cdot\Delta\mathbf{D}}{4\pi}\phi - \int (d\mathbf{r}) \frac{\boldsymbol{\nabla}\phi}{4\pi}\cdot\Delta\mathbf{D} \tag{26.17}$$

where $\phi(\mathbf{r})$ is an arbitrary function. (This is a Lagrange multiplier.) Likewise, to incorporate the constraint (26.16b), we add to (26.15) a sum of surface integrals,

$$0 = \sum_i \phi_i \int dS_i \frac{\mathbf{n}\cdot\Delta\mathbf{D}}{4\pi} \tag{26.18}$$

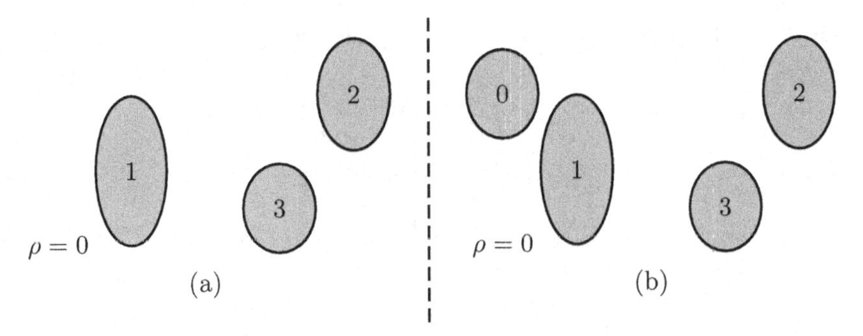

FIGURE 26.2
Introduction of an uncharged conductor into a region where there is no free charge density.

where ϕ_i is an arbitrary constant. In the resulting form of ΔE, the variations $\Delta\mathbf{D}$ can be regarded as independent so that the stationary principle implies, in the volume,

$$\frac{\mathbf{D}}{\epsilon} = \mathbf{E} = -\boldsymbol{\nabla}\phi, \tag{26.19a}$$

while, on the surfaces,

$$\phi = \phi_i. \tag{26.19b}$$

In this way, we recover the full set of equations for electrostatics. Moreover, (26.15) also tells us that, in going from the correct field configuration to any other, $\Delta E > 0$, that is, the physical field minimizes the energy functional (26.14a). This is a statement of Thomson's Theorem[1]: The charges on the surfaces of conductors always readjust themselves in such a way that each conductor becomes an equipotential surface and the total energy of the system is a minimum.

26.3 Introduction of Additional Conductor

We now consider a region of space with dielectric constant $\epsilon(\mathbf{r})$ bounded by an array of conductors into which we introduce an uncharged conductor at a location where there is no free charge density. We are interested in the change of energy in going from the initial configuration (a) to the final configuration (b). (In the following, the subscript 0 refers to the introduced conductor.) See Fig. 26.2. The energy for (a) is

$$E_a = \int_V (d\mathbf{r})\, \frac{\mathbf{D}_a^2}{8\pi\epsilon}, \tag{26.20}$$

where V is the volume exterior to the conductors and the charge on the ith conductor is

$$\int dS_i \frac{\mathbf{n}\cdot\mathbf{D}_a}{4\pi} = Q_i. \tag{26.21}$$

For (b) the energy is

$$E_b = \int_{V-V_0} (d\mathbf{r})\, \frac{\mathbf{D}_b^2}{8\pi\epsilon}, \tag{26.22}$$

[1]William Thomson (Lord Kelvin), 1848.

where now the volume occupied by conductor 0 (V_0) is also excluded, and the charges on the conductors are

$$\int dS_i \frac{\mathbf{n}\cdot\mathbf{D}_b}{4\pi} = Q_i, \tag{26.23a}$$

$$\int dS_0 \frac{\mathbf{n}\cdot\mathbf{D}_b}{4\pi} = 0. \tag{26.23b}$$

The energy, for case (a), satisfies the following inequality,

$$E_a = \left[\int_{V-V_0} + \int_{V_0}\right] (d\mathbf{r})\, \frac{\mathbf{D}_a^2}{8\pi\epsilon} > \int_{V-V_0} (d\mathbf{r})\, \frac{\mathbf{D}_a^2}{8\pi\epsilon}. \tag{26.24}$$

Although $\mathbf{D}_a$ is not the correct field for (b), it is an allowable trial function to use in the energy functional, (26.14a), because it satisfies all the necessary conditions:

$$\boldsymbol{\nabla}\cdot\mathbf{D}_a = 4\pi\rho, \tag{26.25a}$$

$$\int dS_i \frac{\mathbf{n}\cdot\mathbf{D}_a}{4\pi} = Q_i, \tag{26.25b}$$

$$\int dS_0 \frac{\mathbf{n}\cdot\mathbf{D}_a}{4\pi} = \int_{V_0} (d\mathbf{r})\, \frac{\boldsymbol{\nabla}\cdot\mathbf{D}_a}{4\pi} = 0, \tag{26.25c}$$

since, by hypothesis, the region V_0 originally had no charge ($\rho = 0$). According to Thomson's Theorem, the correct field yields a minimum value of the energy functional,

$$E_b = \int_{V-V_0} (d\mathbf{r})\, \frac{\mathbf{D}_b^2}{8\pi\epsilon} < \int_{V-V_0} (d\mathbf{r})\, \frac{\mathbf{D}_a^2}{8\pi\epsilon}, \tag{26.26}$$

implying, upon comparison with (26.24),

$$E_a > E_b, \tag{26.27}$$

which states that the introduction of an uncharged conductor into a charge-free region lowers the energy of the system.

26.4 Alternate Variational Principle

In the first restricted version of the variational principle, (26.12), the charges on the conductors, (26.3b), are specified. For some purposes, it is more convenient to regard the potentials, ϕ_i, on the surfaces of the conductors as specified, rather than the charges, Q_i. For simplicity we will assume that there is no volume charge density, $\rho = 0$. In order to obtain a new form of the energy functional, we note that the stationary property of (26.12) under the replacement

$$\phi \to \lambda\phi, \tag{26.28}$$

for λ infinitesimally different from unity, implies

$$0 = \left.\frac{\partial E}{\partial \lambda}\right|_{\lambda=1} = 2\int (d\mathbf{r})\, \left(-\frac{\epsilon}{8\pi}\right) (\boldsymbol{\nabla}\phi)^2 + \sum_i \int dS_i\, \sigma\phi. \tag{26.29}$$

Consequently, the energy functional is

$$E = \frac{1}{2}\sum_i \int dS_i \, \sigma\phi, \tag{26.30}$$

which becomes, for the actual field values on the surfaces, $\phi = \phi_i$,

$$E = \frac{1}{2}\sum_i Q_i\phi_i. \tag{26.31}$$

Therefore, we obtain another energy functional by combining (26.31) and (26.12) (this is a Legendre transformation),

$$E = \sum_i Q_i\phi_i - \sum_i \int dS_i \, \sigma\phi + \int (d\mathbf{r}) \, \frac{\epsilon}{8\pi}(\boldsymbol{\nabla}\phi)^2, \tag{26.32}$$

or, using (26.3b),

$$E[\phi, \sigma] = \sum_i \int dS_i \sigma(\phi_i - \phi) + \int (d\mathbf{r}) \, \frac{\epsilon}{8\pi}(\boldsymbol{\nabla}\phi)^2. \tag{26.33}$$

Here we regard ϕ and σ to be the variables while ϕ_i is specified [note that here we impose no subsidiary restriction on σ]. Under variations in ϕ and σ the energy changes by

$$\delta E = \sum_i \int dS_i[\delta\sigma(\phi_i - \phi) - \sigma\delta\phi] + \int (d\mathbf{r}) \, \frac{\epsilon}{4\pi}(\boldsymbol{\nabla}\phi) \cdot (\boldsymbol{\nabla}\delta\phi), \tag{26.34}$$

which becomes

$$\delta E = \sum_i \int dS_i \left[\delta\sigma(\phi_i - \phi) - \sigma\delta\phi + \frac{\mathbf{n}\cdot\mathbf{D}}{4\pi}\delta\phi\right] + \int (d\mathbf{r}) \, \frac{\boldsymbol{\nabla}\cdot\mathbf{D}}{4\pi}\delta\phi, \tag{26.35}$$

by identifying $\mathbf{D}$ and integrating by parts. The stationary principle, $\delta E = 0$, implies, from the volume part of (26.35),

$$\delta\phi : \quad \boldsymbol{\nabla}\cdot\mathbf{D} = 0, \tag{26.36a}$$

and from the surface part,

$$\delta\phi : \quad \mathbf{n}\cdot\mathbf{D} = 4\pi\sigma, \tag{26.36b}$$

$$\delta\sigma : \quad \phi = \phi_i. \tag{26.36c}$$

These are the correct equations of electrostatics when there is no volume charge density.

26.5 Green's Function

In our study of electrostatics, we have found Green's functions to be of great use. We will here introduce Green's function in the presence of conductors that are grounded, that is, $\phi_i = 0$; the corresponding differential equation is

$$-\boldsymbol{\nabla}\cdot[\epsilon(\mathbf{r})\boldsymbol{\nabla}G(\mathbf{r}, \mathbf{r}')] = 4\pi\delta(\mathbf{r} - \mathbf{r}'), \tag{26.37a}$$

with the boundary condition

$$G(\mathbf{r},\mathbf{r}') = 0 \quad \text{for } \mathbf{r} \text{ on } S_i. \tag{26.37b}$$

We will show that this Green's function can be used to solve the electrostatics problem in which the potentials on the conductors are specified. We wish to consider a situation for which the free charge density is zero,

$$\boldsymbol{\nabla}\cdot\mathbf{D} = -\boldsymbol{\nabla}\cdot(\epsilon\boldsymbol{\nabla}\phi) = 4\pi\rho = 0, \tag{26.38a}$$

while the potential, ϕ_i on each conducting surface, S_i, is constant,

$$\phi = \phi_i \text{ on } S_i. \tag{26.38b}$$

If we multiply (26.37a) by $\phi(\mathbf{r})$ and (26.38a) by $G(\mathbf{r},\mathbf{r}')$ and subtract, we obtain

$$-\phi(\mathbf{r})\boldsymbol{\nabla}\cdot[\epsilon(\mathbf{r})\boldsymbol{\nabla}G(\mathbf{r},\mathbf{r}')] + G(\mathbf{r},\mathbf{r}')\boldsymbol{\nabla}\cdot(\epsilon(\mathbf{r})\boldsymbol{\nabla}\phi(\mathbf{r})) = 4\pi\delta(\mathbf{r}-\mathbf{r}')\phi(\mathbf{r}). \tag{26.39a}$$

Since the left-hand side of (26.39a) is a divergence,

$$\boldsymbol{\nabla}\cdot[G(\mathbf{r},\mathbf{r}')\epsilon(\mathbf{r})\boldsymbol{\nabla}\phi(\mathbf{r}) - \phi(\mathbf{r})\epsilon(\mathbf{r})\boldsymbol{\nabla}G(\mathbf{r},\mathbf{r}')] = 4\pi\delta(\mathbf{r}-\mathbf{r}')\phi(\mathbf{r}), \tag{26.39b}$$

when we integrate over the entire volume, V, exterior to all of the conductors, we obtain an integral over a surface S which is made up of all the surfaces of the individual conductors, S_i,

$$\begin{aligned} 4\pi\phi(\mathbf{r}') &= \int_S d\mathbf{S}\cdot[G(\mathbf{r},\mathbf{r}')\epsilon(\mathbf{r})\boldsymbol{\nabla}\phi(\mathbf{r}) - \phi(\mathbf{r})\epsilon(\mathbf{r})\boldsymbol{\nabla}G(\mathbf{r},\mathbf{r}')] \\ &= -\sum_i \int dS_i[G(\mathbf{r},\mathbf{r}')\epsilon(\mathbf{r})\mathbf{n}_i\cdot\boldsymbol{\nabla}\phi(\mathbf{r}) - \phi(\mathbf{r})\epsilon(\mathbf{r})\mathbf{n}_i\cdot\boldsymbol{\nabla}G(\mathbf{r},\mathbf{r}')]. \end{aligned} \tag{26.40}$$

The negative sign occurs since $d\mathbf{S}$ is directed out of the volume V, and so into the conductors, while $\mathbf{n}_i$ is the outward normal for the ith conductor. Deleted here is the surface at infinity for which

$$\begin{aligned} dS &\sim R^2, \\ G &\lesssim \frac{1}{R}, \quad |\boldsymbol{\nabla}G| \lesssim \frac{1}{R^2}, \\ \phi &\lesssim \frac{1}{R}, \quad |\boldsymbol{\nabla}\phi| \lesssim \frac{1}{R^2}, \end{aligned} \tag{26.41}$$

so that the corresponding contribution goes to zero as the volume gets arbitrarily large. Now imposing the boundary conditions, (26.37b) and (26.38b), we obtain the desired expression for the potential,

$$\phi(\mathbf{r}) = \sum_i \phi_i \int dS_i' \frac{\epsilon(\mathbf{r}')}{4\pi}\mathbf{n}_i'\cdot\boldsymbol{\nabla}'G(\mathbf{r},\mathbf{r}'), \tag{26.42}$$

where we have interchanged the roles of $\mathbf{r}$ and $\mathbf{r}'$ and used the symmetry property of G, (14.7). (See Problem 26.2.) Therefore, if we know G (the potential due to a point charge with zero potential on the conductors), we can calculate the potential, $\phi(\mathbf{r})$, due to arbitrarily specified potentials on the conductors. [This result, and its generalization to the situation when the free charge is not zero, was anticipated in Problem 14.3.]

26.6 Capacitance

Once we know the potential, we can compute the surface charge density on the ith conductor by using

$$\begin{aligned}\sigma_i &= \frac{1}{4\pi}\mathbf{n}_i \cdot (-\epsilon \boldsymbol{\nabla}\phi) \\ &= -\sum_j \phi_j \int dS'_j \frac{\epsilon(\mathbf{r})}{4\pi}\frac{\epsilon(\mathbf{r}')}{4\pi}(\mathbf{n}_i \cdot \boldsymbol{\nabla})(\mathbf{n}'_j \cdot \boldsymbol{\nabla}')G(\mathbf{r},\mathbf{r}'). \end{aligned} \tag{26.43}$$

The total charge on S_i is therefore

$$Q_i = \int dS_i\, \sigma_i = -\sum_j \phi_j \int dS_i\, dS'_j \frac{\epsilon(\mathbf{r})}{4\pi}\frac{\epsilon(\mathbf{r}')}{4\pi}(\mathbf{n}_i \cdot \boldsymbol{\nabla})(\mathbf{n}'_j \cdot \boldsymbol{\nabla}')G(\mathbf{r},\mathbf{r}'). \tag{26.44}$$

Occurring here are the coefficients of capacitance, C_{ij}, defined by

$$C_{ij} = -\int dS_i dS'_j \frac{\epsilon(\mathbf{r})}{4\pi}\frac{\epsilon(\mathbf{r}')}{4\pi}(\mathbf{n}_i \cdot \boldsymbol{\nabla})(\mathbf{n}'_j \cdot \boldsymbol{\nabla}')G(\mathbf{r},\mathbf{r}'), \tag{26.45}$$

which are purely geometrical quantities that are symmetric in i and j,

$$C_{ij} = C_{ji}. \tag{26.46}$$

The total charge on the ith conductor is thus simply written as

$$Q_i = \sum_j C_{ij}\phi_j. \tag{26.47}$$

The energy of a system of charged conductors can be expressed in terms of the coefficients of capacitance by means of (26.31),

$$E = \frac{1}{2}\sum_i Q_i\phi_i = \frac{1}{2}\sum_{ij}\phi_i C_{ij}\phi_j. \tag{26.48}$$

There is a consistency check between this expression and the variational principle which employs (26.33). Suppose we vary the potential on conductor i by an amount $\delta\phi_i$. Such a change induces variations in σ and ϕ but the resulting change in the energy from these induced variations is of second order due to the stationary principle. So the first-order variation in the energy arises only from the explicit variation of ϕ_i,:

$$\delta E = \int dS_i\, \sigma\delta\phi_i = \delta\phi_i Q_i, \tag{26.49a}$$

or,

$$\frac{\partial E}{\partial \phi_i} = Q_i. \tag{26.49b}$$

This result is in agreement with that obtained from (26.48),

$$\frac{\partial E}{\partial \phi_i} = \sum_j C_{ij}\phi_j = Q_i, \tag{26.50}$$

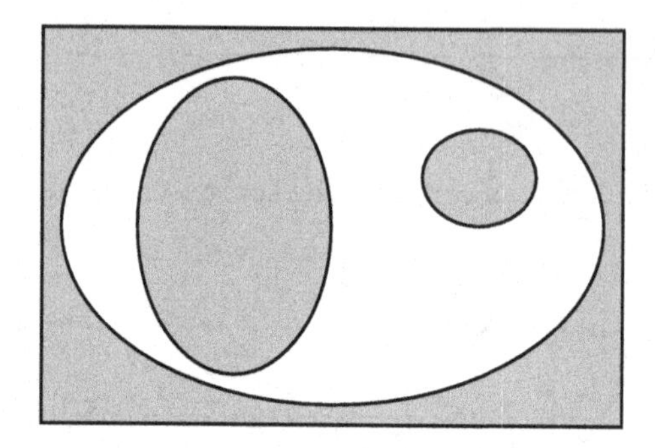

FIGURE 26.3
Finite dielectric bounded by conductors. The shaded regions represent conductors.

where we make use of (26.47).

Suppose the system consists of a finite region bounded by conducting surfaces, as in Fig. 26.3, that is, there is no surface at infinity. The total charge induced on such a system of grounded conductors by a unit point charge at any interior point, $\mathbf{r}'$, is

$$-\frac{1}{4\pi}\sum_i \int dS_i \epsilon(\mathbf{r})\mathbf{n}_i \cdot \boldsymbol{\nabla} G(\mathbf{r},\mathbf{r}') = -\frac{1}{4\pi}\int (d\mathbf{r})\,(-\boldsymbol{\nabla})\cdot[\epsilon(\mathbf{r})\boldsymbol{\nabla} G(\mathbf{r},\mathbf{r}')] = -1, \tag{26.51}$$

where we have used the first line of (26.43), with ϕ replaced by G, as well as the differential equation satisfied by Green's function, (26.37a). This implies that the coefficients of capacitance, C_{ij}, (26.45), satisfy

$$\begin{aligned}\sum_i C_{ij} &= -\int dS_j' \frac{\epsilon(\mathbf{r}')}{4\pi}(\mathbf{n}_j' \cdot \boldsymbol{\nabla}')\sum_i \int dS_i \frac{\epsilon(\mathbf{r})}{4\pi}(\mathbf{n}_i \cdot \boldsymbol{\nabla})G(\mathbf{r},\mathbf{r}')\\ &= -\int dS_j' \frac{\epsilon(\mathbf{r}')}{4\pi}(\mathbf{n}_j' \cdot \boldsymbol{\nabla}')(1) = 0,\end{aligned} \tag{26.52}$$

that is, the sum of all the coefficients of capacitance referring to a given conductor vanishes,

$$\sum_i C_{ij} = \sum_j C_{ij} = 0. \tag{26.53}$$

Consequently, the total charge on the conductors is zero when there is no volume charge present:

$$\sum_i Q_i = \sum_{ij} C_{ij}\phi_j = 0. \tag{26.54}$$

Furthermore, for this system, only relative values of the potential are significant. If we were to add a common constant to all potentials, all charges would remain the same:

$$Q_i = \sum_j C_{ij}(\phi_j + \text{constant}) = \sum_j C_{ij}\phi_j. \tag{26.55}$$

As a simple example, consider a closed system bounded by only two conductors. In this case, in order to satisfy (26.46) and (26.53), we must have

$$C_{11} = -C_{21} = -C_{12} = C_{22} \equiv C, \tag{26.56}$$

where C is called the capacitance of the system. The charges on the two conductors are

$$Q_1 = -Q_2 = C(\phi_1 - \phi_2) = CV \tag{26.57}$$

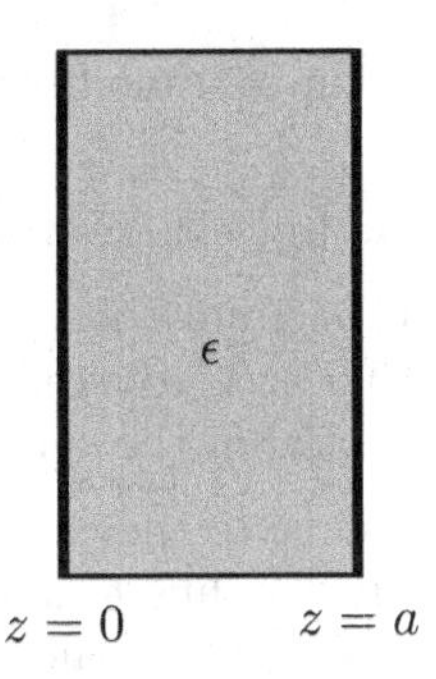

FIGURE 26.4
Parallel plate capacitor. Here, the shaded area represents the dielectric medium between the plates.

where V is the potential difference between the two conductors, while the energy is

$$E = \frac{1}{2}\sum_{ij} \phi_i C_{ij} \phi_j = \frac{1}{2} C V^2. \tag{26.58}$$

As an application of these ideas, consider a capacitor constructed from two parallel conducting plates of area A. The separation of the plates, a, is assumed to be small compared to the transverse extent of the plates, $a \ll \sqrt{A}$, the approximate Green's function therefore being that of two infinite plates as shown in Fig. 26.4 [*cf.* (19.11)]. The material between the plates is characterized by a dielectric constant, ϵ . The above discussion applies to this situation so that the system has a capacitance C,

$$C = C_{11} = -\int dS dS' \frac{\epsilon^2}{(4\pi)^2} \left(\frac{\partial}{\partial z}\right)\left(\frac{\partial}{\partial z'}\right) G \bigg|_{z,z'=0}. \tag{26.59}$$

The first surface integral here was previously evaluated in (19.15a) and (19.17a):

$$\int dS \left(-\frac{1}{4\pi}\right) \frac{\partial}{\partial z} G(\mathbf{r}, \mathbf{r}') \bigg|_{z=0} = -\frac{1}{\epsilon}\left(1 - \frac{z'}{a}\right), \tag{26.60}$$

where we have now included the presence of ϵ in (26.37a). The remaining surface integral is trivial,

$$C = \frac{\epsilon}{4\pi a} \int dS' = \frac{\epsilon A}{4\pi a}, \tag{26.61}$$

yielding the well-known result for a parallel plate capacitor, derived in Section 13.4.

26.7 Problems for Chapter 26

1. Use the stress tensor to evaluate, in terms of the surface charge density, the force per unit area on the surface of a conductor placed in an electric field. Check the result by considering the energy of a parallel plate capacitor with prescribed charges. Reverse the last argument to find the capacitance of a spherical capacitor.
2. Prove the symmetry of Green's function for a system of conductors embedded in a dielectric medium.

3. How is the discussion of Section 26.4 modified if the volume charge density. ρ, is not zero?
4. How is (26.42) modified if the free charge density ρ is not zero?
5. Use (26.45) to calculate the capacitance of a perfectly conducting sphere, using the exterior Green's function (25.50). Verify the result by an elementary argument, based on computing the potential of the sphere relative to the value at infinity.
6. Use (26.45) and (26.56) to calculate the capacitance of a circular cylindrical capacitor of inner radius a, outer radius b, and overall length L, $L \gg a, b$, filled with material of dielectric constant ϵ. Calculate $G(\mathbf{r}, \mathbf{r}')$ following the differential equation (discontinuity) approach discussed in Section 21.4, beginning with (21.66).
7. (a) Consider first two parallel infinite conducting plates separated by a distance a. If one plate occupies the plane $z = 0$, the second $z = a$, the reduced Green's function between the plates is given by (19.55). Determine the capacitance of the system from (26.45). After you do the integrals, you are left with a formally divergent sum. You may recognize that it may be defined in terms of the derivative with respect to z at $z = a$ of the sum

$$\sum_{n=1}^{\infty} \frac{\sin(n\pi z/a)}{n\pi/2} = 1 - \frac{z}{a}. \tag{26.62}$$

Alternatively, you may define it by the Riemann zeta function (19.26), with $\zeta(0) = -1/2$. Either way, you should obtain the known result.

(b) Now consider a conducting cube of side a, with one corner at the origin. Find the Dirichlet Green's function in the interior of the cube in terms of the eigenfunctions

$$\psi_{lmn}(x, y, z) = \left(\frac{2}{a}\right)^{3/2} \sin\frac{l\pi x}{a} \sin\frac{m\pi y}{a} \sin\frac{n\pi z}{a}. \tag{26.63}$$

Again, compute the capacitance between two opposite faces of the cube, in terms of an infinite sum over l, m, and n. Reduce the triple sum to a double sum using

$$\sum_{n=1}^{\infty} \frac{(-1)^n}{n^2 + y^2} = \frac{\pi}{2y} \frac{1}{\sinh \pi y} - \frac{1}{2y^2}. \tag{26.64}$$

The resulting sum is very convergent and is accurately given by its first term. Compare the result with that found for a parallel plate capacitor, in the unphysical case where $A = a^2$.

8. The inner surface of a sphere of radius a is kept at a potential $\varphi_a(\theta, \phi)$. Use the result (26.40) to show that the interior potential, when no internal charges are present, is

$$\varphi(r, \theta, \phi) = \frac{a^3}{4\pi}\left(1 - \left(\frac{r}{a}\right)^2\right) \int_0^{2\pi} d\phi' \int_0^{\pi} \sin\theta' d\theta' \frac{\varphi_a(\theta', \phi')}{[a^2 + r^2 - 2ar\cos\gamma]^{3/2}}, \tag{26.65}$$

where γ is the angle between the direction specified by θ, ϕ, and that specified by θ', ϕ',

$$\cos\gamma = \cos\theta\cos\theta' + \sin\theta\sin\theta'\cos(\phi - \phi'). \tag{26.66}$$

9. Prove that the electrostatic potential at any point in charge-free space is equal to the average of the potential on any spherical surface surrounding that point. How is this consistent with the result of the previous problem? (Such a potential, satisfying Laplace's equation, is called harmonic.)

27

Modes and Variations

The interior of a conducting sphere, and the interiors of conducting cylinders, truncated at both ends by conducting sheets, are examples of conducting cavities—finite regions within which electric fields are contained. Green's function for the volume V bounded by the surface S is characterized by

$$
\begin{aligned}
-\boldsymbol{\nabla}\cdot[\epsilon(\mathbf{r})\boldsymbol{\nabla}G(\mathbf{r},\mathbf{r}')] &= 4\pi\delta(\mathbf{r}-\mathbf{r}'),\\
\mathbf{r}\text{ on }S: \quad G(\mathbf{r},\mathbf{r}') &= 0,
\end{aligned}
\tag{27.1}
$$

as is appropriate to the presence of a dielectric medium in V. We have already provided examples (for $\epsilon = 1$) of the construction of G [for example, see (21.14)], as

$$G(\mathbf{r},\mathbf{r}') = 4\pi\sum_\alpha \frac{\phi_\alpha(\mathbf{r})\phi_\alpha^*(\mathbf{r}')}{\gamma_\alpha^2}, \tag{27.2}$$

where the $\phi_\alpha(\mathbf{r})$ obey

$$-\boldsymbol{\nabla}\cdot[\epsilon(\mathbf{r})\boldsymbol{\nabla}\phi_\alpha(\mathbf{r})] = \gamma_\alpha^2\phi_\alpha(\mathbf{r}), \tag{27.3}$$

and constitute a complete:

$$\sum_\alpha \phi_\alpha(\mathbf{r})\phi_\alpha^*(\mathbf{r}') = \delta(\mathbf{r}-\mathbf{r}'), \tag{27.4a}$$

and orthonormal:

$$\int_V (d\mathbf{r})\,\phi_\alpha^*(\mathbf{r})\phi_\beta(\mathbf{r}) = \delta_{\alpha\beta}, \tag{27.4b}$$

set of functions.

The functions $\phi_\alpha(\mathbf{r})$ describe the various possible field configurations, or modes, such that the charge density is proportional to the potential that it produces,

$$4\pi\rho_\alpha(\mathbf{r}) = \gamma_\alpha^2\phi_\alpha(\mathbf{r}). \tag{27.5}$$

The coefficients of proportionality,

$$\lambda_\alpha = \gamma_\alpha^2, \tag{27.6}$$

are called eigenvalues, and constitute a fundamental aspect of the modes. We shall now describe an approximation method for the estimation of the lowest eigenvalues and the associated modes.

The starting point is the stationary energy expression for a given charge density in the interior of a volume V, subject to the boundary condition

$$\mathbf{r}\text{ on }S: \quad \phi(\mathbf{r}) = 0, \tag{27.7a}$$

namely, [(13.14)]

$$E[\phi] = \int_V (d\mathbf{r})\left[\rho(\mathbf{r})\phi(\mathbf{r}) - \frac{\epsilon(\mathbf{r})}{8\pi}(\boldsymbol{\nabla}\phi(\mathbf{r}))^2\right]. \tag{27.7b}$$

DOI: 10.1201/9781003057369-27

We review the effect of an infinitesimal variation,

$$\delta E[\phi] = \int_V (d\mathbf{r})\,\delta\phi(\mathbf{r})\left[\rho(\mathbf{r}) + \frac{1}{4\pi}\boldsymbol{\nabla}\cdot(\epsilon(\mathbf{r})\boldsymbol{\nabla}\phi(\mathbf{r}))\right] - \oint_S d\mathbf{S}\cdot\delta\phi(\mathbf{r})\frac{\epsilon(\mathbf{r})}{4\pi}\boldsymbol{\nabla}\phi(\mathbf{r}), \tag{27.8}$$

noting that $\delta E[\phi] = 0$ when $\phi(\mathbf{r})$ obeys the appropriate differential equation, and when $\delta\phi(\mathbf{r}) = 0$ on S, corresponding to the imposition of the boundary condition (27.7a). And, for finite variations we have, corresponding to the minus sign of the quadratic term in (27.7b),

$$E[\phi] \le \frac{1}{2}\int (d\mathbf{r})(d\mathbf{r}')\rho(\mathbf{r})G(\mathbf{r},\mathbf{r}')\rho(\mathbf{r}'), \tag{27.9}$$

the upper bound being the energy (14.10). It is also occasionally useful to add a surface integral to (27.7b) [corresponding to the surface integral in (26.12)],

$$\oint_S dS\,\phi(\mathbf{r})\frac{\epsilon(\mathbf{r})}{4\pi}\mathbf{n}\cdot\boldsymbol{\nabla}\phi(\mathbf{r}), \tag{27.10}$$

where $\mathbf{n}$ is the outwardly directed normal ($d\mathbf{S} = \mathbf{n}dS$). The additional terms thereby produced in $\delta E[\phi]$ are

$$\oint_S dS\left[\delta\phi\frac{\epsilon}{4\pi}\mathbf{n}\cdot\boldsymbol{\nabla}\phi + \phi\frac{\epsilon}{4\pi}\delta(\mathbf{n}\cdot\boldsymbol{\nabla}\phi)\right], \tag{27.11}$$

the first of which cancels the surface integral term of (27.8). Accordingly, this $E[\phi]$ is stationary for infinitesimal variations about a solution of the differential equation for ϕ that obeys the boundary condition, without resort to surface restrictions on the variations. One does, however, lose the general inequality of (27.9), for finite variations.

For our immediate purposes, however, we want to retain that inequality, into which we introduce two modifications. First, let us supply ϕ with a scale factor, κ,

$$\phi(\mathbf{r}) \to \kappa\phi(\mathbf{r}), \tag{27.12}$$

which is to be chosen to maximize $E[\phi]$. Now,

$$E[\kappa\phi] = \kappa\int(d\mathbf{r})\rho\phi - \kappa^2\int(d\mathbf{r})\frac{\epsilon}{8\pi}(\boldsymbol{\nabla}\phi)^2 \tag{27.13}$$

vanishes at $\kappa = 0$, and initially rises linearly with increasing κ until the negative quadratic term dominates, thus producing a maximum at the value of κ given by

$$\frac{\partial}{\partial\kappa}E[\kappa\phi] = \int(d\mathbf{r})\rho\phi - 2\kappa\int(d\mathbf{r})\,\frac{\epsilon}{8\pi}(\boldsymbol{\nabla}\phi)^2 = 0. \tag{27.14}$$

Thus, (27.9) yields the scale independent inequality (divided by 2π)

$$\frac{[\int(d\mathbf{r})\,\rho\phi]^2}{\int(d\mathbf{r})\,\epsilon(\boldsymbol{\nabla}\phi)^2} \le \int(d\mathbf{r})(d\mathbf{r}')\rho(\mathbf{r})\frac{1}{4\pi}G(\mathbf{r},\mathbf{r}')\rho(\mathbf{r}'). \tag{27.15}$$

Then we subtract a common term from both sides

$$\frac{[\int(d\mathbf{r})\,\rho\phi]^2}{\int(d\mathbf{r})\,\epsilon(\boldsymbol{\nabla}\phi)^2} - \frac{1}{\lambda}\int(d\mathbf{r})\,\rho^2 \le \int(d\mathbf{r})\,(d\mathbf{r}')\rho(\mathbf{r})\left[\frac{1}{4\pi}G(\mathbf{r},\mathbf{r}') - \frac{1}{\lambda}\delta(\mathbf{r}-\mathbf{r}')\right]\rho(\mathbf{r}'), \tag{27.16}$$

where λ remains to be chosen. Its role becomes more apparent on recognizing that

$$\frac{1}{4\pi}G(\mathbf{r},\mathbf{r}') - \frac{1}{\lambda}\delta(\mathbf{r}-\mathbf{r}') = \sum_\alpha \phi_\alpha(\mathbf{r})\phi_\alpha^*(\mathbf{r}')\left[\frac{1}{\gamma_\alpha^2} - \frac{1}{\lambda}\right]. \tag{27.17}$$

Of all the $\lambda_\alpha = \gamma_\alpha^2$ there is a smallest one, λ_1. Let $\lambda = \lambda_1$. Then the summation on the right-hand side of (27.17) begins with the mode (or modes) with the next smallest value of λ_α, λ_2, where

$$\lambda_\alpha \geq \lambda_2: \quad \frac{1}{\lambda_\alpha} - \frac{1}{\lambda_1} < 0. \tag{27.18}$$

In consequence, the right-hand side of (27.16) is

$$\sum_{\lambda_\alpha > \lambda_1} \left| \int (d\mathbf{r}) \rho \phi_\alpha \right|^2 \left(\frac{1}{\lambda_\alpha} - \frac{1}{\lambda_1} \right) \leq 0, \tag{27.19}$$

the equality sign holding only if ρ is orthogonal to all the ϕ_α with $\lambda_\alpha > \lambda_1$. The inequality (27.16) now states that

$$\lambda_1 \leq \int (d\mathbf{r}) \epsilon (\boldsymbol{\nabla}\phi)^2 \frac{\int (d\mathbf{r})\, \rho^2}{[\int (d\mathbf{r})\, \rho\phi]^2}, \tag{27.20}$$

which supplies an upper limit to λ_1, one that involves two *arbitrary* functions [ϕ, of course, must obey the boundary condition]. For a given ϕ, we can choose ρ to minimize the right-hand side of (27.20). There is a general inequality, analogous to that for a pair of vectors, called the Cauchy-Schwarz-Bunyakovskii[1] inequality,

$$\left[\int (d\mathbf{r}) \rho\phi \right]^2 \leq \int (d\mathbf{r}) \rho^2 \int (d\mathbf{r}) \phi^2, \tag{27.21}$$

the equality sign applying only when ρ is a constant multiple of ϕ—just the situation described in (27.5). This gives

$$\lambda_1 \leq \frac{\int (d\mathbf{r}) \epsilon (\boldsymbol{\nabla}\phi)^2}{\int (d\mathbf{r}) \phi^2}, \tag{27.22}$$

where ϕ is to be chosen so as to minimize the right-hand side. That will be accomplished when ϕ, which is a constant multiple of ρ, is orthogonal to all the ϕ_α, $\lambda_\alpha > \lambda_1$, and therefore is a mode function with eigenvalue λ_1.

Note that if ρ were chosen to be a constant multiple of $-\boldsymbol{\nabla} \cdot (\epsilon \boldsymbol{\nabla}\phi)$, rather than of ϕ, we should generally get a larger value in (27.20), which is the inequality

$$\frac{\int (d\mathbf{r}) [\boldsymbol{\nabla} \cdot (\epsilon \boldsymbol{\nabla}\phi)]^2}{\int (d\mathbf{r})\, \epsilon (\boldsymbol{\nabla}\phi)^2} \geq \frac{\int (d\mathbf{r}) \epsilon (\boldsymbol{\nabla}\phi)^2}{\int (d\mathbf{r})\, \phi^2}; \tag{27.23}$$

the equality sign appears only when ϕ obeys the differential equation of (27.3) and is therefore a mode function. (For proof of this, and the general Cauchy-Schwarz-Bunyakovskii inequality, see Problem 27.1.) Perhaps we should also inject here the more general observation that if

$$\lambda[\phi] = \frac{\int (d\mathbf{r}) \epsilon (\boldsymbol{\nabla}\phi)^2}{\int (d\mathbf{r}) \phi^2} \tag{27.24}$$

is required to be stationary, [$\delta\lambda = 0$] so that

$$\delta \left[\int (d\mathbf{r})\, \epsilon (\boldsymbol{\nabla}\phi)^2 - \lambda \int (d\mathbf{r})\, \phi^2 \right] = 2 \int (d\mathbf{r})\, \delta\phi [-\boldsymbol{\nabla} \cdot (\epsilon \boldsymbol{\nabla}\phi) - \lambda\phi] = 0, \tag{27.25}$$

one infers that ϕ is a mode function having the stationary value of λ as its eigenvalue. It is only the lowest eigenvalue, λ_1, for which the stationary value is a minimum. [This is Rayleigh's principle.[2]]

---

[1] Hermann Amandus Schwarz, 1843–1921; Viktor Yakovievich Bunyakovskii, 1804–1889.
[2] Lord Rayleigh, John William Strutt, 1842–1919.

As a prelude to another application of the inequality (27.16), let us, for a given ϕ, choose ρ according to

$$4\pi\rho = -\boldsymbol{\nabla}\cdot(\epsilon\boldsymbol{\nabla}\phi) \equiv D\phi. \tag{27.26}$$

Then we have satisfied the differential equation connecting ϕ with ρ, and the equality sign in (27.16) applies. Indeed, one can verify directly that both sides of this relation now coincide with

$$\frac{1}{(4\pi)^2}\left[\int (d\mathbf{r})\epsilon(\boldsymbol{\nabla}\phi)^2 - \frac{1}{\lambda}\int (d\mathbf{r})(D\phi)^2\right]. \tag{27.27}$$

In the next step we add some infinitesimal $\delta\rho$ to the ρ of (27.26). That would require an infinitesimal change of ϕ to maintain (27.26). However, the stationary property tells us that ignoring the need for such a $\delta\phi$ only introduces an error of second order. In short, (27.16), with the equality sign, continues to hold on introducing the infinitesimal change of ρ, or

$$\begin{aligned}\frac{1}{4\pi}\int (d\mathbf{r})\,\delta\rho\,\phi - \frac{1}{\lambda}\int (d\mathbf{r})\,\delta\rho\,\rho \\ = \int (d\mathbf{r})(d\mathbf{r}')\,\delta\rho(\mathbf{r})\left[\frac{1}{4\pi}G(\mathbf{r},\mathbf{r}') - \frac{1}{\lambda}\delta(\mathbf{r}-\mathbf{r}')\right]\rho(\mathbf{r}'),\end{aligned} \tag{27.28}$$

which also incorporates the relation

$$\frac{1}{2}\int (d\mathbf{r})\,\rho\phi = \int (d\mathbf{r})\,\frac{\epsilon}{8\pi}(\boldsymbol{\nabla}\phi)^2. \tag{27.29}$$

[The use of the latter is unnecessary if one returns to the $E[\phi]$ of (27.7b), as already noted in Chapter 14.]

Now we choose $\delta\rho$ to be an infinitesimal multiple of

$$4\pi(D\phi - \lambda_1\phi). \tag{27.30}$$

Omitting that infinitesimal factor, we then get

$$\frac{1}{4\pi}\int (d\mathbf{r})\,\delta\rho\,\phi = \int (d\mathbf{r})\,\epsilon(\boldsymbol{\nabla}\phi)^2 - \lambda_1\int (d\mathbf{r})\,\phi^2 \tag{27.31}$$

and

$$\begin{aligned}\int (d\mathbf{r})\,\delta\rho\,\rho = \int (d\mathbf{r})(D\phi)^2 - \lambda_1\int (d\mathbf{r})\,\epsilon(\boldsymbol{\nabla}\phi)^2 \\ = \Delta^2\int (d\mathbf{r})\,\phi^2 + \frac{\int (d\mathbf{r})\,\epsilon(\boldsymbol{\nabla}\phi)^2}{\int (d\mathbf{r})\,\phi^2}\left[\int (d\mathbf{r})\,\epsilon(\boldsymbol{\nabla}\phi)^2 - \lambda_1\int (d\mathbf{r})\,\phi^2\right],\end{aligned} \tag{27.32a}$$

where

$$\Delta^2 = \frac{\int (d\mathbf{r})(D\phi)^2}{\int (d\mathbf{r})\,\phi^2} - \left[\frac{\int (d\mathbf{r})\,\epsilon(\boldsymbol{\nabla}\phi)^2}{\int (d\mathbf{r})\,\phi^2}\right]^2 \geq 0, \tag{27.32b}$$

according to the inequality (27.23). So within a factor of $\int (d\mathbf{r})\,\phi^2$, the left-hand side of (27.28), multiplied by λ, is thus given by

$$(\lambda[\phi] - \lambda_1)(\lambda - \lambda[\phi]) - \Delta^2, \tag{27.33}$$

which employs the notation (27.24).

We encounter, on the right-hand side of (27.28), when we use the eigenfunction expansions (27.2) and (27.4a), integrals of the type (the infinitesimal factor in $\delta\rho$ is again omitted)

$$\frac{1}{4\pi}\int (d\mathbf{r})\,\delta\rho\,\phi_\alpha = \int (d\mathbf{r})(D\phi - \lambda_1\phi)\phi_\alpha = \int (d\mathbf{r})\,\phi(D-\lambda_1)\phi_\alpha$$
$$= (\lambda_\alpha - \lambda_1)\int (d\mathbf{r})\,\phi\phi_\alpha \tag{27.34a}$$

and

$$4\pi\int (d\mathbf{r})\,\phi_\alpha^*\rho = \int (d\mathbf{r})\,\phi_\alpha^* D\phi = \int (d\mathbf{r})\,(D\phi_\alpha)^*\phi$$
$$= \lambda_\alpha\int (d\mathbf{r})\,\phi_\alpha^*\phi, \tag{27.34b}$$

in which the partial integrations employed to shift $D = -\boldsymbol{\nabla}\cdot\epsilon\boldsymbol{\nabla}$ from one function in an integral to the other are validated by the boundary condition that ϕ and the ϕ_α obey. The result for the right-hand side of (27.28), also multiplied by λ, is

$$\sum_{\lambda_\alpha \geq \lambda_2}\left|\int (d\mathbf{r})\,\phi_\alpha^*\phi\right|^2(\lambda_\alpha - \lambda_1)(\lambda - \lambda_\alpha), \tag{27.35}$$

where the presence of the factor $\lambda_\alpha - \lambda_1$ removes any contribution associated with the lowest eigenvalue. Then, if we choose λ to be a lower limit to λ_2, $\underline{\lambda}_2 \leq \lambda_2$, and note that

$$\lambda_\alpha \geq \lambda_2, \quad \underline{\lambda}_2 - \lambda_\alpha \leq 0, \tag{27.36}$$

we are assured that (27.35) is a negative quantity (≤ 0). That statement, in conjunction with (27.33) for $\lambda = \underline{\lambda}_2$ produces the inequality

$$(\lambda[\phi] - \lambda_1)(\underline{\lambda}_2 - \lambda[\phi]) \leq \Delta^2. \tag{27.37}$$

The useful aspect of this inequality emerges when $\lambda[\phi]$, an upper limit to λ_1, and $\underline{\lambda}_2$, a lower limit to λ_2, are both sufficiently close to their respective referents that

$$\underline{\lambda}_2 - \lambda[\phi] > 0. \tag{27.38}$$

Then the inequality supplies a positive upper bound for $\lambda[\phi] - \lambda_1$, which is a lower bound to λ_1. Indeed, λ_1 is now bracketed as

$$\lambda[\phi] \geq \lambda_1 \geq \lambda[\phi] - \frac{\Delta^2}{\underline{\lambda}_2 - \lambda[\phi]}. \tag{27.39}$$

We have used the stationary property of the energy as a guide to these results. But, once seen, they can be given somewhat shorter derivations, as suggested by the structures of (27.19) and (27.35). Instead of Green's function, we employ the completeness of the mode functions $\phi_\alpha(\mathbf{r})$, writing the arbitrary $\phi(\mathbf{r})$ as the linear combination

$$\phi(\mathbf{r}) = \sum_\alpha C_\alpha\phi_\alpha(\mathbf{r}), \tag{27.40}$$

which we also subject to term by term differentiation,

$$(D-\lambda)\phi(\mathbf{r}) = \sum_\alpha C_\alpha(\lambda_\alpha - \lambda)\phi_\alpha(\mathbf{r}). \tag{27.41}$$

The complex conjugates of the expansions of these real functions are required as well. First, let $\lambda = \lambda_1$ and then use the orthonormality of the ϕ_α to evaluate

$$\int (d\mathbf{r})\, \phi(D - \lambda_1)\phi = \sum_{\lambda_\alpha > \lambda_1} |C_\alpha|^2(\lambda_\alpha - \lambda_1) \geq 0; \tag{27.42}$$

this is the inequality

$$\lambda[\phi] \geq \lambda_1. \tag{27.43}$$

Then consider the product of the expansions, (27.41), referring to $\lambda = \lambda_1$ and $\lambda = \underline{\lambda}_2$, and get

$$\int (d\mathbf{r})(D - \lambda_1)\phi(D - \underline{\lambda}_2)\phi = \sum_{\lambda_\alpha \geq \lambda_2} |C_\alpha|^2(\lambda_\alpha - \lambda_1)(\lambda_\alpha - \underline{\lambda}_2) \geq 0, \tag{27.44a}$$

which is

$$\int (d\mathbf{r})\, (D\phi)^2 - (\lambda_1 + \underline{\lambda}_2) \int (d\mathbf{r})\, \epsilon(\boldsymbol{\nabla}\phi)^2 + \lambda_1\underline{\lambda}_2 \int (d\mathbf{r})\, \phi^2 \geq 0, \tag{27.44b}$$

or

$$\Delta^2 + (\lambda[\phi])^2 - (\lambda_1 + \underline{\lambda}_2)\lambda[\phi] + \lambda_1\underline{\lambda}_2 = (\lambda[\phi] - \lambda_1)(\lambda[\phi] - \underline{\lambda}_2) + \Delta^2 \geq 0; \tag{27.44c}$$

this is the inequality (27.37).

The usefulness of the lower bound to λ_1 depends on the availability of a reasonably good lower bound to λ_2, one that lies significantly above λ_1. We have already alluded to a possibility Section 25.1, the knowledge, preferably for some subset of modes, of summations of the type

$$r = 1, 2, \cdots: \quad \sum_\alpha \frac{1}{\lambda_\alpha^r} > \frac{1}{\lambda_1^r} + \frac{1}{\lambda_2^r} > \frac{1}{\overline{\lambda}_1^r} + \frac{1}{\lambda_2^r}. \tag{27.45}$$

The last statement above recognizes that the inequality is maintained if λ_1 is replaced by a larger value, an upper limit, $\overline{\lambda}_1$. And, if the final relation were solved for λ_2 as an equation, what would emerge is necessarily a smaller number than the true λ_2; it is a lower limit, $\underline{\lambda}_2$. We shall illustrate this procedure shortly.

27.1 A Comparison Method

We have remarked on the stationary property of $\lambda[\phi]$, (27.24). In this discussion the boundary condition was respected by the infinitesimal variation $\delta\phi$. But we can remove that restriction by proceeding as in (27.7b), (27.10), which replaces $\lambda[\phi]$ with

$$\lambda[\phi] = \frac{\int_V (d\mathbf{r})\, \epsilon(\boldsymbol{\nabla}\phi)^2 - 2\oint_S dS\, \epsilon\phi\mathbf{n} \cdot \boldsymbol{\nabla}\phi}{\int_V (d\mathbf{r})\, \phi^2}. \tag{27.46}$$

Now we want to examine what happens to λ_α, the eigenvalue associated with the (unnormalized) mode function ϕ_α, when the boundary surface S is altered infinitesimally by a displacement $\delta n(\mathbf{r})$ along the normal to the original surface. According to the stationary property in which the boundary condition need not be maintained, it suffices to extrapolate the original mode function into the new region, or terminate it within the initial region. Thus we have

$$\delta \int_V (d\mathbf{r})\, \epsilon(\boldsymbol{\nabla}\phi_\alpha)^2 = \oint_S dS\, \delta n\, \epsilon \left(\frac{\partial\phi_\alpha}{\partial n}\right)^2, \tag{27.47}$$

where

$$\frac{\partial}{\partial n}\phi_\alpha = \mathbf{n}\cdot\boldsymbol{\nabla}\phi_\alpha, \tag{27.48}$$

the contribution of gradients tangential to the surface being of second order (because the mode functions vanish on the undisplaced surface), and

$$\delta\oint_S dS\,\epsilon\phi_\alpha\mathbf{n}\cdot\boldsymbol{\nabla}\phi_\alpha = \oint_S dS\,\delta n\,\epsilon\left(\frac{\partial\phi_\alpha}{\partial n}\right)^2, \tag{27.49}$$

whereas

$$\delta\int_V (d\mathbf{r})\,\phi_\alpha^2 = 0, \tag{27.50}$$

inasmuch as the change here is of second order. That gives

$$\delta\lambda_\alpha = -\frac{\oint_S dS\,\delta n\,\epsilon(\partial\phi_\alpha/\partial n)^2}{\int_V(d\mathbf{r})\,\phi_\alpha^2} \tag{27.51}$$

as the net variation: Any outward displacement generally lowers the value of λ_α; any inward displacement generally increases it.

One application of (27.51) refers to situations where λ_α has a known dependence on geometry. Then the normalization of the mode function ϕ_α—the requirement that the denominator in (27.51) be unity—can be achieved in terms of the surface behavior of the mode function. This is illustrated by a sphere, of radius a, where [(25.21),which refers to $\epsilon = 1$]

$$\phi_{lmn}(\mathbf{r}) = Y_{lm}(\theta,\phi)R_{ln}(r) \tag{27.52a}$$

and

$$R_{ln}(r) = C_{ln}j_l(\gamma_{ln}r/a), \quad j_l(\gamma_{ln}) = 0; \tag{27.52b}$$

here C_{ln} is the normalization constant left unspecified by the radial differential equation. In this example we know that $\lambda_{ln} = \gamma_{ln}^2/a^2$ varies inversely with a^2. Hence, on choosing $\delta n = \delta a$, (27.51) reads

$$-2\frac{\delta a}{a}\gamma_{ln}^2 = -a^2\delta a[C_{ln}\gamma_{ln}j_l'(\gamma_{ln})]^2, \tag{27.53a}$$

or

$$C_{ln} = \left(\frac{2}{a^3}\right)^{1/2}\frac{1}{j_l'(\gamma_{ln})}, \tag{27.53b}$$

in agreement with (25.17).

Before pressing on let's use (27.51) to fill a small gap in Section 21.3, where we promised a simpler verification of the normalization constant in $\phi_0(\mu)$, the mode function of lowest eigenvalue for an equilateral triangle. That function is ($\epsilon = 1$)

$$\phi_0(\mu) = C\sum_{a=1}^{3}\sin\frac{2\pi}{h}(\mu_a + r) = C\left[\sin\frac{2\pi}{h}(y+r)\right.$$
$$\left. + \sin\frac{2\pi}{h}\left(-\frac{3^{1/2}}{2}x - \frac{1}{2}y + r\right) + \sin\frac{2\pi}{h}\left(\frac{3^{1/2}}{2}x - \frac{1}{2}y + r\right)\right], \tag{27.54}$$

according to (21.54) and (21.26c). The dimensions of the triangle are specified alternatively by the height h, the radius of the inscribed circle $r = \frac{1}{3}h$, or the base $a = (2/3^{1/2})h$. Thus, the lowest eigenvalue is given equivalently as [(21.53)]

$$\lambda_1 = \left(\frac{2\pi}{h}\right)^2 = \frac{4}{3}\left(\frac{2\pi}{a}\right)^2 = \left(\frac{2\pi}{3r}\right)^2. \tag{27.55}$$

In the two-dimensional version of (27.51) the surface integral is replaced by a line integral

$$\delta\lambda_1 = -\oint_s ds\,\delta n \left(\frac{\partial\phi_1}{\partial n}\right)^2 = -3\int_{-a/2}^{a/2} dx\,\delta n \left(\frac{\partial\phi_1}{\partial y}\right)^2\bigg|_{y=-r}, \tag{27.56}$$

where the latter form exploits the threefold symmetry of the triangular sides. It is clear from the geometry that the uniform displacement δn needed to maintain the triangular shape is

$$\delta n = \delta r = \frac{1}{3}\delta h. \tag{27.57}$$

Now, the derivative appearing in (27.56) is

$$\begin{aligned}\frac{\partial\phi_1}{\partial y}\bigg|_{y=-r} &= C\frac{2\pi}{h}\left[1 - \frac{1}{2}\cos\frac{2\pi}{h}\left(-\frac{3^{1/2}}{2}x + \frac{h}{2}\right) - \frac{1}{2}\cos\frac{2\pi}{h}\left(\frac{3^{1/2}}{2}x + \frac{h}{2}\right)\right]\\ &= C\frac{2\pi}{h}\left[1 + \cos\frac{2\pi}{a}x\right],\end{aligned} \tag{27.58}$$

which immediately yields

$$\int_{-a/2}^{a/2} dx \left(\frac{\partial\phi_1}{\partial y}\right)^2\bigg|_{y=-r} = \lambda_1\frac{3}{2}aC^2. \tag{27.59}$$

Then (27.56) gives

$$-2\frac{\delta h}{h}\lambda_1 = -\delta h\,\lambda_1\frac{3}{2}aC^2, \tag{27.60}$$

and

$$C = \frac{2}{(3ha)^{1/2}}, \tag{27.61}$$

as stated in (21.54).

We turn from these quantitative uses of (27.51) to its qualitative side. An outward displacement of the boundary of a cavity decreases the eigenvalue of any mode. If the given region can be transformed in this way into another one, with known properties, and a desired mode identified among those of the modified region, the latter eigenvalue provides a lower bound to the eigenvalue of interest. As a first example, consider a sphere of radius [to avoid confusion] R, and the enclosing cube of side $2R$. The mode functions of the sphere that have $l = 0$ are spherically symmetrical [$Y_{00} = (4\pi)^{-1/2}$], and we look for their counterparts in the modes of the cube that are symmetrical in the three orthogonal directions defined by the cube. Accordingly, we should have [(21.15)]

$$\lambda_{0n} > 3\left(\frac{n\pi}{2R}\right)^2, \quad n = 1, 2, \ldots, \tag{27.62a}$$

or

$$\gamma_{0n} > \frac{3^{1/2}}{2}\frac{n\pi}{R} = 0.866\frac{n\pi}{R}, \tag{27.62b}$$

which is true, for [(25.18b)] $\gamma_{0n} = n\pi/R$.

With cylindrical conductors, attention focuses on the two-dimensional cross section. Let's compare a circle of radius R with an enclosing equilateral triangle, for which R is the radius of the inscribed circle, and with an enclosing square, of side equal to $2R$. (See Figure 27.1.) The modes of the circle with $m = 0$ are axially symmetrical; we look for the first few

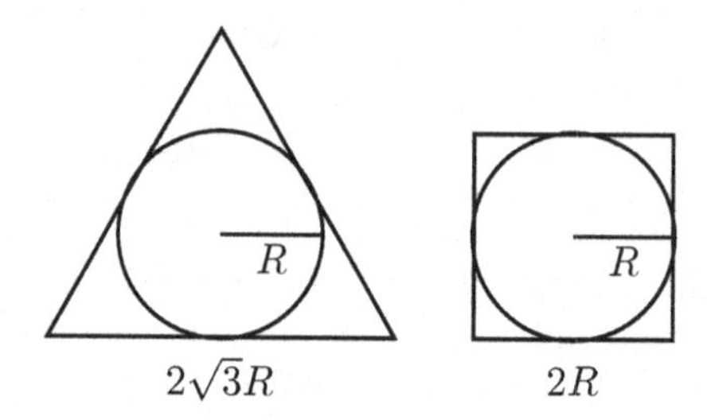

FIGURE 27.1
A circle enclosed by an equilateral triangle, and by a square.

counterparts among the real mode functions of the equilateral triangle, specifically those with $l_1 = l_2$, say, where [(21.52)]

$$\gamma_l^2 = \frac{2}{3}\left(\frac{\pi}{3R}\right)^2 (l_1^2 + l_1^2 + 4l_1^2) = \left(\frac{2\pi}{3R} l_1\right)^2 . \tag{27.63}$$

Thus, we expect that

$$\gamma_{01} > \frac{2\pi}{3R} = \frac{2.094}{R}, \quad \gamma_{02} > \frac{4\pi}{3R} = \frac{4.189}{R}, \tag{27.64}$$

and indeed [(21.111b), (21.112b)] $\gamma_{01}R = 2.405$, $\gamma_{02}R = 5.520$. The analogous consideration for the somewhat more closely fitting square gives

$$\gamma_{0n}^2 > 2\left(\frac{n\pi}{2R}\right)^2 , \tag{27.65}$$

and

$$\gamma_{01}R > 2^{-1/2}\pi = 2.221, \quad \gamma_{02}R > 2^{1/2}\pi = 4.443. \tag{27.66}$$

A lower limit to γ_{11} is provided by the first example of an unsymmetrical mode of the square, [$l = 1$, $m = 2$, with $a = b$ in (21.8)]

$$\gamma_{11}^2 > \left(\frac{\pi}{2R}\right)^2 + \left(\frac{\pi}{R}\right)^2 = \frac{5}{4}\left(\frac{\pi}{R}\right)^2 , \tag{27.67}$$

and

$$\gamma_{11}R > \frac{5^{1/2}}{2}\pi = 3.512, \tag{27.68}$$

to be compared with [(21.113b)] $\gamma_{11}R = 3.832$.

27.2 Iteration

A (real) mode function, say one with the smallest eigenvalue, λ_1, obeys the boundary condition, and the differential equation

$$-\nabla^2\phi_1(\mathbf{r}) = \lambda_1\phi_1(\mathbf{r}), \tag{27.69}$$

where, for simplicity, we restrict this discussion to $\epsilon = 1$. Suppose one has picked some initial approximation, $\phi_1^{(0)}(\mathbf{r})$, which satisfies the boundary condition but not the differential equation. Is there a procedure for systematically improving that initial choice? We begin our affirmative response by defining $\phi_1^{(1)}(\mathbf{r})$ as the solution of the differential equation

$$-\nabla^2\phi_1^{(1)}(\mathbf{r}) = \phi_1^{(0)}(\mathbf{r}) \tag{27.70}$$

that satisfies the boundary condition. It is the first step in a process of iteration wherein an approximation is introduced on the right-hand side of (27.69) and the equation solved (with a change of scale) to produce an improved approximation. The general statement of the iteration process is

$$-\nabla^2\phi_1^{(n+1)}(\mathbf{r}) = \phi_1^{(n)}(\mathbf{r}). \tag{27.71}$$

We use the nth iterate in the stationary expression (27.24) to define

$$\lambda_1^{(n)} \equiv \lambda[\phi_1^{(n)}] = \frac{[2n-1]}{[2n]} \geq \lambda_1. \tag{27.72}$$

This has involved the rearrangement

$$\int (d\mathbf{r})\,(\boldsymbol{\nabla}\phi_1^{(n)})^2 = \int (d\mathbf{r})\,(-)\nabla^2\phi_1^{(n)}\phi_1^{(n)} = \int (d\mathbf{r})\,\phi_1^{(n-1)}\phi_1^{(n)}, \tag{27.73}$$

as justified by the boundary condition, and introduces the useful notation

$$[m+n] = \int (d\mathbf{r})\phi_1^{(m)}\phi_1^{(n)}. \tag{27.74}$$

The stated dependence on only the sum of the indices follows from the relations

$$\begin{aligned}\int (d\mathbf{r})\phi_1^{(m)}\phi_1^{(n)} &= \int (d\mathbf{r})\phi_1^{(m)}(-)\nabla^2\phi_1^{(n+1)} = \int (d\mathbf{r})(-)\nabla^2\phi_1^{(m)}\phi_1^{(n+1)} \\ &= \int (d\mathbf{r})\phi_1^{(m-1)}\phi_1^{(n+1)}.\end{aligned} \tag{27.75}$$

Now we demonstrate that the iteration process does converge to the correct eigenvalue and mode function, provided that the initial choice $\phi_1^{(0)}$ is not orthogonal to the desired mode function. We begin by applying the inequality (27.23) with $\phi = \phi_1^{(n+1)}$

$$\frac{[2n]}{[2n+1]} \geq \frac{[2n+1]}{[2n+2]}. \tag{27.76}$$

Then we use the version of the Cauchy-Schwarz-Bunyakovskii inequality, analogous to (27.21), that refers to vector fields,

$$\left[\int (d\mathbf{r})\boldsymbol{\nabla}\phi\cdot\boldsymbol{\nabla}\psi\right]^2 \leq \int (d\mathbf{r})(\boldsymbol{\nabla}\phi)^2\int (d\mathbf{r})(\boldsymbol{\nabla}\psi)^2. \tag{27.77}$$

The choice $\phi = \phi_1^{(n)}$, $\psi = \phi_1^{(n+1)}$ yields

$$[2n]^2 \leq [2n-1][2n+1] \tag{27.78a}$$

or

$$\frac{[2n-1]}{[2n]} \geq \frac{[2n]}{[2n+1]}. \tag{27.78b}$$

The combination of the two inequalities, (27.76) and (27.78b), is expressed by

$$\lambda_1^{(n)} \geq \lambda_1^{(n+1/2)} \geq \lambda_1^{(n+1)} \geq \lambda_1, \tag{27.79}$$

where

$$\lambda_1^{(n+1/2)} = \frac{[2n]}{[2n+1]} \tag{27.80}$$

is indeed obtained from (27.72) by the formal substitution $n \to n+\frac{1}{2}$. Thus, the sequence of approximations to λ_1 is monotonically decreasing, but cannot be less than λ_1—it approaches a limit, $\mu \geq \lambda_1$. Preparatory to a discussion of that limit, let us remark that one solution of the iteration equation given in (27.71) can be displayed with the aid of Green's function:

$$\phi_1^{(n+1)}(\mathbf{r}) = \int (d\mathbf{r}') \frac{1}{4\pi} G(\mathbf{r},\mathbf{r}') \phi_1^{(n)}(\mathbf{r}'). \tag{27.81}$$

On applying the inequality (27.21), we then learn that

$$\left(\phi_1^{(n+1)}(\mathbf{r})\right)^2 \leq \int (d\mathbf{r}') \left[\frac{1}{4\pi} G(\mathbf{r},\mathbf{r}')\right]^2 [2n]. \tag{27.82}$$

The only question about the integral involving the square of Green's function is its existence for a small region that includes the point $\mathbf{r}' = \mathbf{r}$. In such a region, however, Green's function is dominated by Coulomb's potential, and the integral of $|\mathbf{r}' - \mathbf{r}|^{-2}$ does exist. Accordingly, (27.82) presents us with a bound of the form

$$|\phi_1^{(n+1)}(\mathbf{r})| \leq C(\mathbf{r})[2n]^{1/2}. \tag{27.83}$$

Now consider the following series, for positive β,

$$F_1(\mathbf{r}) = \sum_{n=0}^{\infty} \beta^n \phi_1^{(n+1)}(\mathbf{r}). \tag{27.84}$$

We can say that

$$|F_1(\mathbf{r})| \leq \sum_{n=0}^{\infty} \beta^n |\phi_1^{(n+1)}(\mathbf{r})| \leq C(\mathbf{r}) \sum_{n=0}^{\infty} \beta^n [2n]^{1/2}, \tag{27.85}$$

and so the function defined by the infinite summation will exist if the ratio of successive terms in this upper bound approaches a limit less than unity,

$$\beta \lim_{n\to\infty} \left(\frac{[2n+2]}{[2n]}\right)^{1/2} = \beta \lim_{n\to\infty} \frac{1}{\left[\lambda_1^{(n+1/2)} \lambda_1^{(n+1)}\right]^{1/2}} < 1, \tag{27.86}$$

which is to say, provided

$$\beta < \mu. \tag{27.87}$$

The function $F_1(\mathbf{r})$ obeys the differential equation

$$-\nabla^2 F_1(\mathbf{r}) = \sum_{n=0}^{\infty} \beta^n \phi_1^{(n)}(\mathbf{r}) = \beta F_1(\mathbf{r}) + \phi_1^{(0)}(\mathbf{r}). \tag{27.88}$$

Let us multiply it by the mode function $\phi_1(\mathbf{r})$, and integrate, to get

$$\begin{aligned} \int (d\mathbf{r}) \phi_1(\mathbf{r}) \phi_1^{(0)}(\mathbf{r}) &= \int (d\mathbf{r})\, \phi_1 (-\nabla^2 - \beta) F_1 = \int (d\mathbf{r})\, F_1 (-\nabla^2 - \beta) \phi_1 \\ &= (\lambda_1 - \beta) \int (d\mathbf{r})\, F_1 \phi_1. \end{aligned} \tag{27.89}$$

Here is where we prove that $\mu = \lambda_1$, provided the initial choice $\phi_1^{(0)}$ is not orthogonal to the actual mode function ϕ_1—the left-hand side of (27.89) does not vanish. Assume the converse, that $\mu > \lambda_1$. Then, according to (27.87), the function $F_1(\mathbf{r})$ exists for $\beta = \lambda_1$. But with that value of β the right-hand side of (27.89) does vanish. The contradiction shows that $\mu = \lambda_1$. And we know that $\lambda[\phi] = \lambda_1$ can only be achieved with $\phi = \phi_1$. Perhaps the following afterthought will be helpful. One can verify, with the aid of the completeness of the mode functions, that the solution of the differential equation (27.88) for F_1 is

$$F_1(\mathbf{r}) = \sum_\alpha \frac{\phi_\alpha(\mathbf{r})}{\lambda_\alpha - \beta} \int (d\mathbf{r}')\phi_\alpha(\mathbf{r}')\phi_1^{(0)}(\mathbf{r}'). \tag{27.90}$$

Here, clearly displayed, is the first singularity that appears with increasing β, at $\beta = \lambda_1$, unless this term is missing because $\phi_1^{(0)}$ is orthogonal to ϕ_1. In the latter circumstance the first singularity—and the number to which the sequence of approximants converges—appears at λ_2 unless....

How rapidly does the approximation sequence approach λ_1? To answer that let's look at the inequality (27.44b), with $\phi = \phi_1^{(n+1)}$:

$$[2n] - (\lambda_1 + \underline{\lambda}_2)[2n+1] + \lambda_1\underline{\lambda}_2[2n+2] > 0, \tag{27.91a}$$

or

$$\lambda_1^{(n+1/2)}\lambda_1^{(n+1)} - (\lambda_1 + \underline{\lambda}_2)\lambda_1^{(n+1)} + \lambda_1\underline{\lambda}_2 \geq 0, \tag{27.91b}$$

which we present as

$$\frac{\lambda_1^{(n+1/2)} - \lambda_1}{\lambda_1^{(n+1)} - \lambda_1} \geq \frac{\underline{\lambda}_2}{\lambda_1^{(n+1)}}. \tag{27.92}$$

Another such inequality is produced by replacing one ϕ-function in (27.44a) by $D\phi$. The effect in (27.91a) is to change $2n$ into $2n-1$, and that yields

$$\frac{\lambda_1^{(n)} - \lambda_1}{\lambda_1^{(n+1/2)} - \lambda_1} \geq \frac{\underline{\lambda}_2}{\lambda_1^{(n+1/2)}}. \tag{27.93}$$

The product of these two inequalities then states that

$$\frac{\lambda_1^{(n)} - \lambda_1}{\lambda_1^{(n+1)} - \lambda_1} \geq \frac{\underline{\lambda}_2^2}{\lambda_1^{(n+1/2)}\lambda_1^{(n+1)}}. \tag{27.94}$$

The last result shows that the error of the $(n+1)$th iteration is smaller than that of the nth iteration by a number that (for $\underline{\lambda}_2 = \lambda_2$) approaches $(\lambda_1/\lambda_2)^2 = (\gamma_1/\gamma_2)^4$ as the iteration proceeds. Thus, the larger the ratio λ_2/λ_1, the more rapid is the convergence. We must point out here that the second eigenvalue referred to in this convergence criterion can well exceed the second smallest eigenvalue of the cavity. This occurs in the presence of spatial symmetry properties that permit the decomposition of the modes into different symmetry classes. If the initial function $\phi_1^{(0)}$ possesses the particular symmetry that is characteristic of ϕ_1 so also will the successive approximations $\phi_1^{(n)}$; every member of this sequence is automatically orthogonal to modes of other symmetry classes, and the relevant second eigenvalue is that of a mode having the same symmetry as ϕ_1. Furthermore, in consequence of this orthonormality, the method under development applies to the lowest eigenvalue of each symmetry class, independently.

These remarks are illustrated by the circular cylinder. A real mode function has the angular dependence $\cos m\phi$ or $\sin m\phi$, and modes associated with different m values are orthogonal; each value of m is independent. Accordingly, in discussing the lowest eigenvalue of the $m = 0$ class, λ_{01}, the relevant second eigenvalue is not the actual next larger one, λ_{11}, such that $(\lambda_{11}/\lambda_{01})^2 = (3.832/2.405)^4 = 6.445$, but rather it is λ_{02}, with $(\lambda_{02}/\lambda_{01})^2 = (5.520/2.405)^4 = 27.75$. [Recall the eigenvalues given in (21.111b), (21.112b), and (21.113b).]

As we have already noted, the inequality (27.37) provides a lower limit to λ_1 if a reasonably accurate $\underline{\lambda}_2$ is available. The evaluation of Δ^2 [(27.32b)] for $\phi = \phi_1^{(n)}$ is

$$\begin{aligned}\Delta^2 &= \frac{[2n-2]}{[2n]} - \left(\frac{[2n-1]}{[2n]}\right)^2\\ &= \left(\lambda_1^{(n-1/2)} - \lambda_1^{(n)}\right)\lambda_1^{(n)},\end{aligned} \tag{27.95}$$

and we learn from (27.39) how λ_1 is bounded at the nth iteration stage:

$$\lambda_1^{(n)} > \lambda_1 > \lambda_1^{(n)} - \frac{\lambda_1^{(n-1/2)} - \lambda_1^{(n)}}{\underline{\lambda}_2/\lambda_1^{(n)} - 1}; \tag{27.96}$$

the lower limit also follows from (27.91b), with $n \to n-1$. As always, the corresponding statement at the $(n+1/2)$th stage is produced by formal substitution.

We now want to recognize how the iteration process can be used to optimize the bounds within which λ_1 is confined. Suppose that iteration has advanced so far that (27.94) can be presented as

$$\frac{\lambda_1^{(n)} - \lambda_1}{\lambda_1^{(n+1)} - \lambda_1} \approx \left(\frac{\lambda_2}{\lambda_1}\right)^2, \tag{27.97}$$

which employs the optimum choice, $\underline{\lambda}_2 = \lambda_2$, for that is what the process selects (we return to this point shortly). This asymptotic ratio of iteration errors shows that the approach to the limit is as

$$\lambda_1^{(n)} \approx \lambda_1 + a\left(\frac{\lambda_1}{\lambda_2}\right)^{2n}, \quad a > 0. \tag{27.98}$$

Now we use

$$\lambda_1^{(n-1/2)} - \lambda_1^{(n)} \approx a\left(\frac{\lambda_1}{\lambda_2}\right)^{2n}\left(\frac{\lambda_2}{\lambda_1} - 1\right) \tag{27.99}$$

to learn the asymptotic behavior of $\underline{\lambda}_1^{(n)}$, the lower limit given in (27.96),

$$\begin{aligned}\underline{\lambda}_1^{(n)} &\approx \lambda_1 + a\left(\frac{\lambda_1}{\lambda_2}\right)^{2n} - a\left(\frac{\lambda_1}{\lambda_2}\right)^{2n}\frac{\lambda_2 - \lambda_1}{\underline{\lambda}_2 - \lambda_1}\\ &= \lambda_1 - \frac{\lambda_2 - \underline{\lambda}_2}{\underline{\lambda}_2 - \lambda_1}a\left(\frac{\lambda_1}{\lambda_2}\right)^{2n}.\end{aligned} \tag{27.100}$$

Thus, if one can merely pick $\underline{\lambda}_2$ such that

$$\underline{\lambda}_2 > \frac{1}{2}(\lambda_1 + \lambda_2), \tag{27.101}$$

the error of the lower limit at any (sufficiently advanced) stage is less than that of the upper limit. And one can improve matters, for the asymptotic ratio

$$\frac{\lambda_1^{(n-1/2)} - \lambda_1^{(n)}}{\lambda_1^{(n)} - \lambda_1^{(n+1/2)}} \approx \frac{\lambda_2}{\lambda_1} \tag{27.102}$$

provides an internally generated estimate of λ_2 that can be used for $\underline{\lambda}_2$. The closer the latter is to λ_2, the more rapidly will the successive lower limits converge, according to (27.100). [As we shall see in a moment, with $\underline{\lambda}_2 = \lambda_2$, the error of $\underline{\lambda}_1^{(n)}$ is dominated by a multiple of $(\lambda_1/\lambda_3)^{2n}$.] And the latter equation also shows that too large a choice for $\underline{\lambda}_2$ will betray itself—then the "lower limits" will converge from above!

Before illustrating all this in a reasonably favorable situation, we shall supply additional assurances concerning the convergence of the iteration process. We begin by representing $\phi_1^{(0)}$ in terms of the complete set of ϕ_α,

$$\phi_1^{(0)}(\mathbf{r}) = \sum_\alpha C_\alpha \phi_\alpha(\mathbf{r}). \tag{27.103}$$

Then the first iteration, the solution of (27.70), is

$$\phi_1^{(1)}(\mathbf{r}) = \sum_\alpha \frac{C_\alpha}{\lambda_\alpha} \phi_\alpha(\mathbf{r}), \tag{27.104}$$

and, in general,

$$\phi_1^{(n)}(\mathbf{r}) = \sum_\alpha \frac{C_\alpha}{(\lambda_\alpha)^n} \phi_\alpha(\mathbf{r}). \tag{27.105}$$

Notice that the evaluation of the integral defining $[m+n]$, (27.74), confirms its dependence on the sum of the indices,

$$[m+n] = \sum_\alpha \frac{C_\alpha^2}{(\lambda_\alpha)^{m+n}}. \tag{27.106}$$

Now we have

$$\lambda_1^{(n)} = \frac{[2n-1]}{[2n]} = \frac{\sum_\alpha \lambda_\alpha (C_\alpha/\lambda_\alpha^n)^2}{\sum_\alpha (C_\alpha/\lambda_\alpha^n)^2} \tag{27.107a}$$

or

$$\lambda_1^{(n)} = \lambda_1 + \frac{\sum_\alpha (\lambda_\alpha - \lambda_1)(C_\alpha/\lambda_\alpha^n)^2}{\sum_\alpha (C_\alpha/\lambda_\alpha^n)^2}. \tag{27.107b}$$

As the iteration proceeds, the summations in the latter version will be dominated increasingly by the leading terms:

$$\lambda_1^{(n)} \approx \lambda_1 + \frac{(\lambda_2 - \lambda_1)(C_2/\lambda_2^n)^2}{(C_1/\lambda_1^n)^2} = \lambda_1 + a\left(\frac{\lambda_1}{\lambda_2}\right)^{2n}, \tag{27.108a}$$

where

$$a = (\lambda_2 - \lambda_1)\left(\frac{C_2}{C_1}\right)^2, \tag{27.108b}$$

which is indeed of the form (27.98), with λ_2 entering automatically as the optimum choice of $\underline{\lambda}_2$.

It is also worth noting what would happen to $\underline{\lambda}_1^{(n)}$ if one managed to choose $\underline{\lambda}_2 = \lambda_2$:

$$\underline{\lambda}_1^{(n)} = \lambda_1^{(n)} - \frac{\lambda_1^{(n-1/2)} - \lambda_1^{(n)}}{(\lambda_2/\lambda_1^{(n)}) - 1} = \frac{\lambda_2 - \lambda_1^{(n-1/2)}}{\lambda_2 - \lambda_1^{(n)}} \lambda_1^{(n)}. \tag{27.109}$$

Now we need to carry (27.108a) a step farther,

$$\begin{aligned}\lambda_1^{(n)} &\approx \lambda_1 + \frac{(\lambda_2 - \lambda_1)(C_2/\lambda_2^n)^2 + (\lambda_3 - \lambda_1)(C_3/\lambda_3^n)^2}{(C_1/\lambda_1^n)^2 + (C_2/\lambda_2^n)^2} \\ &\approx \lambda_1 + a\left(\frac{\lambda_1}{\lambda_2}\right)^{2n} - \frac{a^2}{\lambda_2 - \lambda_1}\left(\frac{\lambda_1}{\lambda_2}\right)^{4n} + b\left(\frac{\lambda_1}{\lambda_3}\right)^{2n},\end{aligned} \tag{27.110}$$

in which

$$b = (\lambda_3 - \lambda_1)(C_3/C_1)^2. \tag{27.111}$$

We find that

$$\underline{\lambda}_1^{(n)} \approx \lambda_1 - \frac{\lambda_3 - \lambda_2}{\lambda_2 - \lambda_1} b \left(\frac{\lambda_1}{\lambda_3}\right)^{2n}, \tag{27.112}$$

and thus the correct localization of λ_2 is signaled by a marked increase (to the extent that $\lambda_3 > \lambda_2$) in the convergence rate of the lower limit.

27.3 Example

Consider the parallel plate geometry, $0 \leq z \leq a$, for which the lowest mode function is $\sin\frac{\pi z}{a}$, with corresponding eigenvalue $\gamma_1^2 = \pi^2/a^2$. First consider the variational bound (27.22). We must choose a trial function which vanishes at $z = 0,\, a$. The simplest example is

$$\phi_1^{(0)}(z) = z(a-z), \tag{27.113}$$

so

$$\begin{aligned}\lambda_1^{(0)} &= \frac{\int_0^a dz\,\left[\frac{\partial}{\partial z}(z(a-z))\right]^2}{\int_0^a dz\, z^2(a-z)^2} = \frac{1}{a^2}\frac{\int_0^1 dt(1-4t+4t^2)}{\int_0^1 dt(t^2-2t^3+t^4)}\\ &= \frac{1}{a^2}\frac{\frac{1}{3}}{\frac{1}{30}} = \frac{10}{a^2};\end{aligned} \tag{27.114}$$

indeed, $\sqrt{10} = 3.162$ is only 0.7% bigger than π. To get the next iterated bound, we have to solve (27.70), or

$$-\frac{d^2}{dz^2}\phi_1^{(1)} = z(a-z); \tag{27.115}$$

the solution which vanishes at $z = 0,\, a$ is

$$\phi_1^{(1)} = \frac{1}{12}z^4 - \frac{a}{6}z^3 + \frac{a^3}{12}z. \tag{27.116}$$

So, the second iterate is

$$\lambda_1^{(1)} = \frac{[1]}{[2]}, \tag{27.117a}$$

where

$$[1] = \int_0^a dz\,\phi_1^{(0)}\phi_1^{(1)} = a^7\int_0^1 dt\,t(1-t)\left(\frac{t^4}{12} - \frac{t^3}{6} + \frac{t}{12}\right) = \frac{17}{5040}a^7, \tag{27.117b}$$

and

$$[2] = \int_0^a dz\,(\phi_1^{(1)})^2 = a^9\int_0^1 dt\left(\frac{t^4}{12} - \frac{t^3}{6} + \frac{t}{12}\right)^2 = \frac{31}{90720}a^9, \tag{27.117c}$$

which gives

$$\lambda_1^{(1)} = \frac{306}{31}\frac{1}{a^2}, \quad \gamma_1^{(1)} = \frac{3.14181}{a}, \tag{27.118}$$

only 0.007% too high. As for the lower bound, let us simply take $\underline{\lambda}_2 = \lambda_2 = (2\pi/a)^2$, and then set $n = 1$ on the right-hand side of (27.96):

$$\underline{\lambda}_1^{(1)} = \lambda_1^{(1)} - \frac{\lambda_1^{(1/2)} - \lambda_1^{(1)}}{(2\pi)^2/a^2\lambda_1^{(1)} - 1}, \tag{27.119a}$$

where

$$\lambda_1^{(1/2)} = \frac{[0]}{[1]} = \frac{168}{17}\frac{1}{a^2}, \tag{27.119b}$$

so

$$\underline{\lambda}_1^{(1)} = \frac{9.8672}{a^2}, \quad \underline{\gamma}_1^{(1)} = \frac{3.14121}{a}, \tag{27.120}$$

only 0.01% below the correct value.

27.4 Problems for Chapter 27

1. Prove the Cauchy-Schwarz-Bunyakovskii inequality as follows: For two complex function ϕ and ψ, and a complex number λ, note that

$$\int (d\mathbf{r})|\phi(\mathbf{r}) - \lambda\psi(\mathbf{r})|^2 \geq 0. \tag{27.121}$$

Differentiate this with respect to λ, and determine the value of λ which minimizes the integral. Show then that

$$\left|\int (d\mathbf{r})\phi^*(\mathbf{r})\psi(\mathbf{r})\right|^2 \leq \int (d\mathbf{r})|\phi(\mathbf{r})|^2 \int (d\mathbf{r})|\psi(\mathbf{r})|^2, \tag{27.122}$$

which is the desired inequality. When does equality hold? From this, derive (27.22) and (27.23).

2. Derive the approximate expression for the lower bound for the lowest eigenvalue, (27.112). [Hint: show that if $b = 0$, $\underline{\lambda}_1$ given by (27.109) would be $\lambda_1 + O((\lambda_1/\lambda_2)^{4n-1})$, and then by taking the derivative with respect to b, show that (27.112) holds up to a multiplicative factor of $1 + O\left((\lambda_1/\lambda_2)^{2n}\right)$.]

3. Consider a circle of radius R, for which the lowest eigenvalue corresponds to $\gamma_{01}R = 2.40483$. Using the trial function $\phi_1^{(0)} = R - \rho$, where ρ is the radial coordinate, compute the corresponding upper bound $\gamma_1^{(0)}$. Do better by solving the differential equation (27.70) for $\phi_1^{(1)}$, and thereby find the improved upper bound $\gamma_1^{(1)}$. If you assume the lower bound to the second eigenvalue to be given by $\underline{\gamma}_2 = 5$, compute the lower bound to the first eigenvalue, or $\underline{\gamma}_1^{(1)}$. How is that estimate changed if the exact second eigenvalue $\gamma_{02} = 5.520$ is used instead?

4. Carry out similar estimates for the lowest eigenvalue of a spherical cavity.

5. We have two alternative forms for the reduced Green's function for the interior of a conducting circular cylinder, (21.65) and (21.90). Set $\rho = \rho'$ and integrate to obtain

$$\sum_{n=1}^{\infty} \frac{1}{k^2 + \gamma_{mn}^2} = \int_o^{\infty} d\rho\, \rho\, I_m(k\rho) \left[K_m(k\rho) - I_m(k\rho)\frac{K_m(ka)}{I_m(ka)}\right]. \tag{27.123}$$

Do the integral on ρ, using the differential equation satisfied by the modified Bessel functions, to obtain its value

$$\frac{a}{2k}\left[\frac{I'_m(ka)}{I_m(ka)} - \frac{m}{ka}\right], \tag{27.124}$$

and thus obtain the relation

$$\frac{I'_m(t)}{I_m(t)} = \frac{m}{t} + 2t\sum_{n=1}^{\infty}\frac{1}{t^2+(\gamma_{mn}a)^2}. \tag{27.125}$$

This is quite understandable, since it exhibits the pole expansion of the logarithmic derivative of $I_m(t)$, and the imaginary roots of $I_m(t)$ at $\pm\gamma_{mn}a$. On the other hand, we have the series expansion (20.58), which yields

$$\frac{I'_m(t)}{I_m(t)} = \frac{m}{t} + \frac{1}{2}\frac{t}{m+1} - \frac{1}{8}\frac{t^3}{(m+1)^2(m+2)} + \dots \tag{27.126}$$

By expanding (27.125) in powers of t, obtain the sum rules:

$$\sum_{n=1}^{\infty}\frac{1}{\gamma_{mn}^2 a^2} = \frac{1}{4}\frac{1}{m+1}, \tag{27.127a}$$

$$\sum_{n=1}^{\infty}\frac{1}{\gamma_{mn}^4 a^4} = \frac{1}{16}\frac{1}{(m+1)^2(m+2)}, \tag{27.127b}$$

These sum rules are useful for setting bounds on eigenvalues. They coincide with those given in (25.30b) and (25.34) when we set $m = l + 1/2$, as we would expect.

6. A circular cylinder of radius a is bounded by inscribed and circumscribed square cylinders (that is, cylinders of square cross section that are wholly inside and wholly outside the circular cylinder, and touch on four lines). Show that indeed the lowest value of γ is intermediate between those of the two squares, in accordance with the theorem of Section 27.1.

Part III

Magnetostatics

28

Magnetostatics

28.1 Variational Principle

We now return to the general action principle of electrodynamics, (10.9), before the specialization to electrostatics. Recall at the beginning of Chapter 13 we isolated the terms referring to the electromagnetic field, and omitted time derivative terms. In this way we obtained a separation of static electric and magnetic energies, given by (13.2a) and (13.2b). The former was the basis for our ensuing discussion of electrostatics. The second describes magnetostatics, which is the subject of our investigation here. Notice that this separation is possible only because of the condition

$$\frac{\partial}{\partial t}\mathbf{A} = \mathbf{0}; \tag{28.1}$$

otherwise, electric and magnetic effects are interrelated. Analogously to our incorporation of dielectrics in electrostatics (see Chapter 13), we here pass to a macroscopic description of fields in permeable media. The energy, (13.2b), for these circumstances, becomes

$$E[\mathbf{A},\mathbf{B}] = -\int (d\mathbf{r})\left[\frac{1}{c}\mathbf{J}\cdot\mathbf{A} + \frac{1}{4\pi\mu}\left(-\mathbf{B}\cdot\boldsymbol{\nabla}\times\mathbf{A} + \frac{1}{2}\mathbf{B}^2\right)\right], \tag{28.2}$$

where μ is the permeability of the medium. We now have to check that the stationary principle applied to this form of the energy yields the correct equations of magnetostatics. We are to regard $\mathbf{A}$ and $\mathbf{B}$ as the independent variables, so the variation of the energy is

$$\delta E = -\int (d\mathbf{r})\left[\frac{1}{c}\mathbf{J}\cdot\delta\mathbf{A} - \frac{1}{4\pi}\frac{\mathbf{B}}{\mu}\cdot\boldsymbol{\nabla}\times\delta\mathbf{A} + \frac{1}{4\pi\mu}\delta\mathbf{B}\cdot(\mathbf{B} - \boldsymbol{\nabla}\times\mathbf{A})\right]. \tag{28.3}$$

From the coefficient of $\delta\mathbf{B}$, we obtain

$$\mathbf{B} = \boldsymbol{\nabla}\times\mathbf{A}, \tag{28.4a}$$

which is equivalent to

$$\boldsymbol{\nabla}\cdot\mathbf{B} = 0. \tag{28.4b}$$

By making use of the identity

$$\frac{\mathbf{B}}{\mu}\cdot\boldsymbol{\nabla}\times\delta\mathbf{A} = \boldsymbol{\nabla}\cdot\left(\delta\mathbf{A}\times\frac{\mathbf{B}}{\mu}\right) + \delta\mathbf{A}\cdot\left(\boldsymbol{\nabla}\times\frac{\mathbf{B}}{\mu}\right), \tag{28.5}$$

and discarding the implied surface integral, we find from the vanishing of the coefficient of $\delta\mathbf{A}$,

$$\boldsymbol{\nabla}\times\mathbf{H} = \frac{4\pi}{c}\mathbf{J}, \tag{28.6}$$

DOI: 10.1201/9781003057369-28

a consequence of which is that only steady currents occur here:

$$\boldsymbol{\nabla}\cdot\mathbf{J}=0. \tag{28.7}$$

As appropriate to macroscopic media, we have introduced the magnetic field,

$$\mathbf{H}=\frac{\mathbf{B}}{\mu}. \tag{28.8}$$

Thus we have recovered Maxwell's equations in the static limit, (28.4b) and (28.6).

As in electrostatics, there is a restricted version of the stationary principle for the energy. If we take

$$\mathbf{B}=\boldsymbol{\nabla}\times\mathbf{A} \tag{28.9}$$

as the definition of $\mathbf{B}$, the expression for the energy becomes

$$E[\mathbf{A}]=-\int(d\mathbf{r})\left[\frac{1}{c}\mathbf{J}\cdot\mathbf{A}-\frac{1}{8\pi\mu}(\boldsymbol{\nabla}\times\mathbf{A})^2\right]. \tag{28.10}$$

Regarding this as stationary under variations in $\mathbf{A}$, we derive the equation satisfied by the vector potential,

$$\boldsymbol{\nabla}\times\left(\frac{1}{\mu}\boldsymbol{\nabla}\times\mathbf{A}\right)=\frac{4\pi}{c}\mathbf{J}, \tag{28.11}$$

which coincides with (28.6).

Proceeding in a manner parallel to the corresponding discussion in electrostatics (see Section 13.2), we consider a change in the permeability, $\delta\mu(\mathbf{r})$. Because of the stationary property, the only first-order variation in the energy arises from the explicit appearance of μ in (28.10):

$$\delta E=-\int(d\mathbf{r})\,\frac{\delta\mu}{\mu^2}\frac{1}{8\pi}(\boldsymbol{\nabla}\times\mathbf{A})^2=-\int(d\mathbf{r})\,\frac{\delta\mu}{8\pi}\mathbf{H}^2, \tag{28.12}$$

which is the analog of (13.36a). In particular, by considering $\delta\mu$ to arise from a displacement of the material, we infer the force on the (inhomogeneous) permeable medium to be [cf. (13.40)]

$$\mathbf{F}=-\int(d\mathbf{r})\,\frac{\mathbf{H}^2}{8\pi}\boldsymbol{\nabla}\mu; \tag{28.13}$$

a diamagnetic material, with $\mu<1$, is repelled from a region of stronger magnetic field.

28.2 Boundary Conditions

The simplest example of an inhomogeneous magnetic material occurs when μ is discontinuous across an interface. First we consider the boundary conditions that $\mathbf{B}$ and $\mathbf{A}$ must satisfy across the interface, as illustrated in Fig. 28.1. The fact that $\mathbf{B}$ is the curl of $\mathbf{A}$ implies that the tangential component of $\mathbf{A}$, $\mathbf{A}_t$ must be continuous across the boundary, in order that $\mathbf{B}_t$ be finite:

$$\mathbf{n}_1\times\mathbf{A}_1+\mathbf{n}_2\times\mathbf{A}_2=\mathbf{0}, \tag{28.14a}$$

or,

$$\mathbf{n}_1\times(\mathbf{A}_1-\mathbf{A}_2)=\mathbf{0}, \tag{28.14b}$$

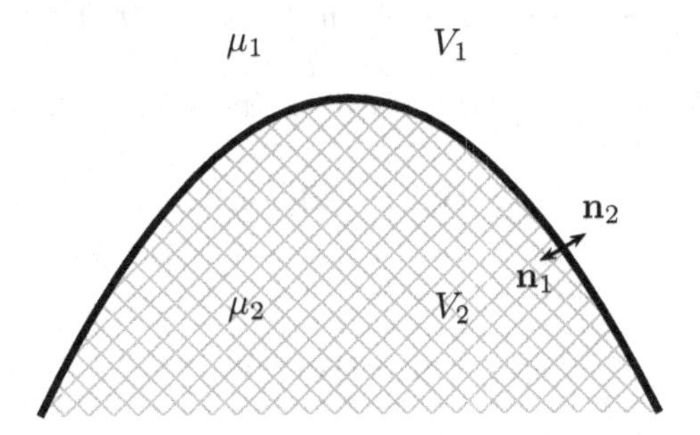

FIGURE 28.1
Discontinuity in permeability across an interface.

where $\mathbf{n}_1$ ($\mathbf{n}_2$) is the outward normal to V_1 (V_2) so that $\mathbf{n}_2 = -\mathbf{n}_1$. The relation, (28.14a), is true for all points on the surface. Thus, when we take the divergence of this expression, in which only tangential components of $\boldsymbol{\nabla}$ occur, we find

$$\mathbf{n}_1 \cdot \mathbf{B}_1 + \mathbf{n}_2 \cdot \mathbf{B}_2 = 0, \tag{28.15a}$$

or

$$\mathbf{n}_1 \cdot (\mathbf{B}_1 - \mathbf{B}_2) = 0, \tag{28.15b}$$

that is, the normal component of $\mathbf{B}$ is continuous across the boundary. [We may regard this as a surface version of $\boldsymbol{\nabla} \cdot \mathbf{B} = 0$.]

If we include the possibility that there is a surface current, $\mathbf{K}$, on the boundary between V_1 and V_2, we must amend the energy expression, (28.2), to read

$$E = -\int (d\mathbf{r}) \left[\frac{1}{c}\mathbf{J} \cdot \mathbf{A} + \frac{1}{4\pi\mu}\left(-\mathbf{B} \cdot \boldsymbol{\nabla}\times\mathbf{A} + \frac{1}{2}\mathbf{B}^2\right)\right] - \int dS \frac{1}{c}\mathbf{K} \cdot \mathbf{A}. \tag{28.16}$$

In our previous discussion of the variation in the energy, we discarded the surface integral [see (28.5)]; this is no longer permissible because of the presence of the boundary. Consequently, there is a new contribution to the variation of the energy arising from the occurrence of the interface,

$$\begin{aligned}\delta E &= \int_{V_1} (d\mathbf{r})\, \boldsymbol{\nabla} \cdot \left(\delta\mathbf{A}_1 \times \frac{\mathbf{H}_1}{4\pi}\right) + \int_{V_2} (d\mathbf{r})\, \boldsymbol{\nabla} \cdot \left(\delta\mathbf{A}_2 \times \frac{\mathbf{H}_2}{4\pi}\right) - \int dS \frac{1}{c}\mathbf{K} \cdot \delta\mathbf{A} \\ &= \int dS \left[\mathbf{n}_1 \cdot \frac{\delta\mathbf{A}_1 \times \mathbf{H}_1}{4\pi} + \mathbf{n}_2 \cdot \frac{\delta\mathbf{A}_2 \times \mathbf{H}_2}{4\pi} - \frac{1}{c}\mathbf{K} \cdot \delta\mathbf{A}\right] \\ &= -\int dS\, \delta\mathbf{A} \cdot \left[\frac{\mathbf{n}_1 \times \mathbf{H}_1}{4\pi} + \frac{\mathbf{n}_2 \times \mathbf{H}_2}{4\pi} + \frac{1}{c}\mathbf{K}\right].\end{aligned} \tag{28.17}$$

Here we have used the fact that $\mathbf{A}_t$ must be continuous,

$$\delta\mathbf{A}_{1t} = \delta\mathbf{A}_{2t} = \delta\mathbf{A}_t. \tag{28.18}$$

We then conclude from the stationary principle on the surface that

$$\mathbf{n}_1 \times \mathbf{H}_1 + \mathbf{n}_2 \times \mathbf{H}_2 + \frac{4\pi}{c}\mathbf{K} = \mathbf{0}, \tag{28.19a}$$

or

$$\mathbf{n}_1 \times (\mathbf{H}_1 - \mathbf{H}_2) + \frac{4\pi}{c}\mathbf{K} = \mathbf{0}. \tag{28.19b}$$

When no surface current is present, $\mathbf{K} = \mathbf{0}$, this states that the tangential component of $\mathbf{H}$ is continuous. If, in addition, we have $\mu_2 \gg \mu_1$ [idealized as $\mu_2 \to \infty$; we might call this a perfect magnetic conductor (see Chapter 52)], the magnetic field in medium 2 goes to zero,

$$\mathbf{H}_2 = \frac{1}{\mu_2}\mathbf{B}_2 \to \mathbf{0}, \tag{28.20}$$

so that $\mathbf{H}_1$ is normal to the surface,

$$\mathbf{n}_1 \times \mathbf{H}_1 = \mathbf{0}. \tag{28.21}$$

This is the same condition satisfied by the electric field at the surface of a conductor.

28.3 Vector Potential

The fundamental equation of magnetostatics is (28.11),

$$\boldsymbol{\nabla} \times \left(\frac{1}{\mu} \boldsymbol{\nabla} \times \mathbf{A} \right) = \frac{4\pi}{c}\mathbf{J}, \tag{28.22}$$

which reduces for vacuum ($\mu = 1$) to

$$\boldsymbol{\nabla}(\boldsymbol{\nabla} \cdot \mathbf{A}) - \nabla^2 \mathbf{A} = \frac{4\pi}{c}\mathbf{J}. \tag{28.23}$$

This equation can be simplified by using the fact that it is invariant under a gauge transformation [see (10.53)],

$$\mathbf{A} \to \mathbf{A} + \boldsymbol{\nabla}\lambda. \tag{28.24}$$

Because of this gauge freedom, we can usually choose some particular gauge to simplify the problem at hand. In the present situation, the convenient choice of gauge is one for which

$$\boldsymbol{\nabla} \cdot \mathbf{A} = 0, \tag{28.25}$$

called the radiation, Coulomb, or transverse gauge. To show that it is always possible to choose this gauge, suppose we start with a vector potential, $\mathbf{A}_0$, which does not satisfy this condition,

$$\boldsymbol{\nabla} \cdot \mathbf{A}_0 \neq 0. \tag{28.26}$$

It is possible to make a gauge transformation,

$$\mathbf{A}_0 \to \mathbf{A} = \mathbf{A}_0 + \boldsymbol{\nabla}\lambda, \tag{28.27}$$

such that (28.25) is satisfied, that is

$$\boldsymbol{\nabla} \cdot (\mathbf{A}_0 + \boldsymbol{\nabla}\lambda) = 0, \tag{28.28}$$

for then λ is a solution to Poisson's equation,

$$-\nabla^2 \lambda = \boldsymbol{\nabla} \cdot \mathbf{A}_0. \tag{28.29}$$

In the radiation gauge, (28.23) becomes

$$-\nabla^2 \mathbf{A} = \frac{4\pi}{c}\mathbf{J}, \tag{28.30}$$

the solution of which is

$$\mathbf{A}(\mathbf{r}) = \frac{1}{c}\int (d\mathbf{r}')\frac{\mathbf{J}(\mathbf{r}')}{|\mathbf{r}-\mathbf{r}'|}, \tag{28.31}$$

in precise analogy to the solution of the electrostatic problem,

$$-\nabla^2\phi = 4\pi\rho. \tag{28.32}$$

As a consistency check, we verify explicitly that (28.31) satisfies the radiation gauge condition (28.25),

$$\begin{aligned}\boldsymbol{\nabla}\cdot\mathbf{A}(\mathbf{r}) &= \frac{1}{c}\int (d\mathbf{r}')\left(\boldsymbol{\nabla}\frac{1}{|\mathbf{r}-\mathbf{r}'|}\right)\cdot\mathbf{J}(\mathbf{r}') = -\frac{1}{c}\int (d\mathbf{r}')\left(\boldsymbol{\nabla}'\frac{1}{|\mathbf{r}-\mathbf{r}'|}\right)\cdot\mathbf{J}(\mathbf{r}')\\ &= \frac{1}{c}\int (d\mathbf{r}')\frac{1}{|\mathbf{r}-\mathbf{r}'|}\boldsymbol{\nabla}'\cdot\mathbf{J}(\mathbf{r}') = 0\end{aligned} \tag{28.33}$$

where we have used (28.7), and the fact that the current distribution is localized.

Once we have the vector potential, we can compute the magnetic induction $\mathbf{B}$:

$$\mathbf{B} = \boldsymbol{\nabla}\times\mathbf{A} = \frac{1}{c}\boldsymbol{\nabla}\times\int (d\mathbf{r}')\frac{\mathbf{J}(\mathbf{r}')}{|\mathbf{r}-\mathbf{r}'|} = \frac{1}{c}\int (d\mathbf{r}')\mathbf{J}(\mathbf{r}')\times\frac{\mathbf{r}-\mathbf{r}'}{|\mathbf{r}-\mathbf{r}'|^3}. \tag{28.34}$$

For a point charge moving with velocity $\mathbf{v}$, the electric current is

$$\mathbf{J}(\mathbf{r}) = e\mathbf{v}\delta(\mathbf{r}-\mathbf{R}), \tag{28.35}$$

where $\mathbf{R}$ is the position of the particle, which produces the magnetic field

$$\mathbf{B} = \frac{\mathbf{v}}{c}\times\frac{e(\mathbf{r}-\mathbf{R})}{|\mathbf{r}-\mathbf{R}|^3}. \tag{28.36}$$

This has the form

$$\mathbf{B} = \frac{\mathbf{v}}{c}\times\mathbf{E}, \tag{28.37}$$

which was our starting point for introducing magnetic fields in Chapter 1. [Of course, we have now transcended the domain of magnetostatics since $\partial\mathbf{A}/\partial t \neq \mathbf{0}$. However, since $(1/c)(\partial\mathbf{A}/\partial t)$ is of order v^2/c^2, (28.36) is the correct magnetic field to first order in v/c. For the general case, see Chapter 34.]

In the following three brief chapters, we will develop some applications of magnetostatics. Many situations in magnetostatics can be attacked by evident variations on the techniques developed for electrostatics.

28.4 Problems for Chapter 28

1. Revisit Problem 7.2e, which asked you to compute the magnetic field corresponding to the potentials

$$\mathbf{A}(\mathbf{r}) = \frac{\mathbf{a}\times\mathbf{r}}{r^3}, \quad \phi = 0, \quad r > 0. \tag{28.38}$$

   Show that $\mathbf{B}$ can be written in the form

$$\mathbf{B} = \boldsymbol{\nabla}\phi + \mathbf{a}\nabla^2\psi, \tag{28.39}$$

and identify the functions ϕ and ψ. What is $\nabla^2\psi$? From the equation of magnetostatics

$$\boldsymbol{\nabla}\times\mathbf{B} = \frac{4\pi}{c}\mathbf{J}, \tag{28.40}$$

determine the current density belonging to this magnetic field. Use this current to work out the magnetic dipole moment from

$$\boldsymbol{\mu} = \frac{1}{2c}\int (d\mathbf{r})\,\mathbf{r}\times\mathbf{J}. \tag{28.41}$$

(Hint: integrate by parts.)

2. In 1935 brothers Fritz (1900-1954) and Heinz (1907–1970) London proposed that a superconductor is characterized by

$$\mathbf{j} + \frac{c}{4\pi\lambda_L^2}\mathbf{A} = \boldsymbol{\nabla}\chi.$$

Show that this leads to ($\mu = 1$)

$$\nabla^2\mathbf{B} = \frac{1}{\lambda_L^2}\mathbf{B},$$

which implies the Meissner[1] effect, that a uniform magnetic field cannot exist inside a superconductor. λ_L is called the London penetration depth, and has the typical value of 10^{-5} cm.

[1] Walther Meissner (1882–1974); the name of Robert Ochsenfeld (1901–1993) should be added.

29

Macroscopic Current Distributions

The simplest example of a macroscopic current is that which flows in a long straight wire. We will take the wire to lie along the z axis, carrying a current I that flows in the $+z$ direction. We will let the direction of current flow be denoted, generally, by $\mathbf{n}$. See Fig. 29.1. We wish to find the magnetic induction $\mathbf{B}$ produced by this current. Since $\mathbf{B}$ is independent of z, without loss of generality we evaluate it at $z = 0$. For a wire with negligible cross section, any volume integral involving the current density becomes a line integral

$$\int (d\mathbf{r})\, \mathbf{J}(\mathbf{r}') \cdots = \int dz'\, dS'\, \mathbf{n} J(\mathbf{r}') \cdots = \int dz'\, \mathbf{n} I \ldots . \tag{29.1}$$

The expression for the magnetic induction, (28.34), a distance ρ from the wire, is then reduced to

$$\mathbf{B} = \frac{I}{c} \int dz' \left(\boldsymbol{\nabla} \frac{1}{\sqrt{\rho^2 + z'^2}} \right) \times \mathbf{n}. \tag{29.2}$$

For a long wire of length $2L$, $L \gg \rho$, the integral occurring here is

$$\int_{-L}^{L} dz' \frac{1}{\sqrt{\rho^2 + z'^2}} = 2 \int_0^L \frac{dz'}{\sqrt{z'^2 + \rho^2}} \approx 2 \left(\ln \frac{L}{\rho} + \text{constant} \right), \tag{29.3}$$

the gradient of which is

$$\boldsymbol{\nabla} \left(2 \ln \frac{L}{\rho} + \text{constant} \right) = -2 \frac{\boldsymbol{\nabla} \rho}{\rho} = -\frac{2}{\rho} \hat{\boldsymbol{\rho}}, \tag{29.4}$$

where $\hat{\boldsymbol{\rho}}$ is a unit vector in the radial direction. The magnetic field produced by this wire is therefore

$$\mathbf{B} = \frac{2I}{c\rho} (\mathbf{n} \times \hat{\boldsymbol{\rho}}), \tag{29.5}$$

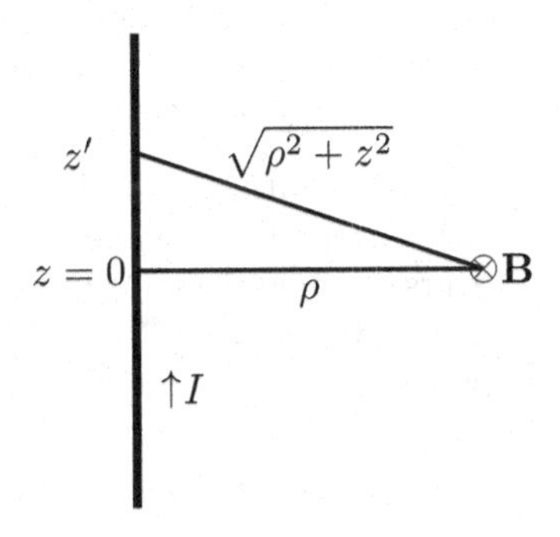

FIGURE 29.1
Magnetic field produced by a current-carrying wire.

DOI: 10.1201/9781003057369-29

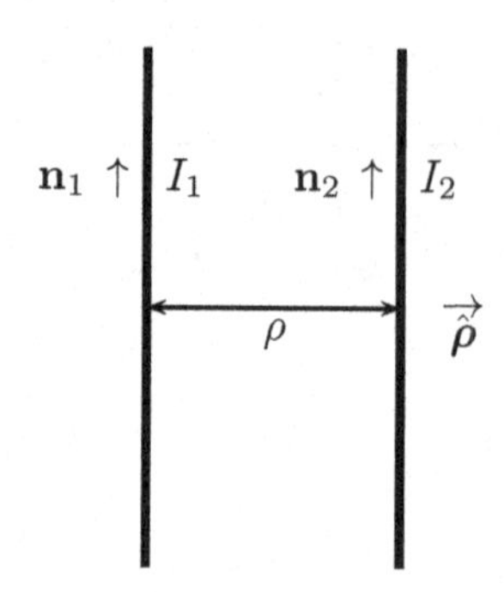

FIGURE 29.2
Force between parallel, current-carrying wires.

which is concentric with the wire, and in the sense given by the right-hand rule, that is, if the thumb of the right hand points in the direction of current flow $\mathbf{n}$, the fingers curl in the sense of $\mathbf{B}$.

The force exerted on a current distribution by a magnetic field is [see (8.6)]

$$\mathbf{F} = \frac{1}{c}\int (d\mathbf{r})\, \mathbf{J}\times\mathbf{B}, \tag{29.6}$$

which, when specialized to current flowing in a long straight wire, becomes

$$\mathbf{F} = \int dz\, \frac{I}{c}\mathbf{n}\times\mathbf{B}. \tag{29.7}$$

One possibility is that the magnetic field is produced by a second, parallel, current-carrying wire, as shown in Fig. 29.2. Of course, the total force acting on I_2 due to the magnetic field produced by I_1 is unbounded. The quantity of interest is the force per unit length on I_2,

$$\frac{\text{force}}{\text{length}} = \frac{I_2}{c}\mathbf{n}_2\times\mathbf{B} = \frac{I_2}{c}\frac{2I_1}{c\rho}\mathbf{n}_2\times(\mathbf{n}_1\times\hat{\boldsymbol{\rho}}). \tag{29.8}$$

For parallel flowing currents, $\mathbf{n}_1 = \mathbf{n}_2$, so

$$\frac{\text{force}}{\text{length}} = -\frac{2I_1I_2}{c^2}\frac{\hat{\boldsymbol{\rho}}}{\rho}, \tag{29.9}$$

that is, the force is attractive. If the currents flow in opposite senses, the force is repulsive.

We can also obtain the above result by recalling that the force is the negative gradient of the energy,

$$\mathbf{F} = -\boldsymbol{\nabla} E, \tag{29.10}$$

where the energy is given by (28.10), with $\mu = 1$. Equation (28.10) is quite analogous to the electrostatic energy, (13.14), except for the overall sign, which implies that the sense of attraction or repulsion is reversed when we go from static charge distributions to steady current flows. By integrating by parts on the $(\boldsymbol{\nabla}\times\mathbf{A})$ term and then using the differential equation (28.11), we may rewrite the magnetostatic energy as

$$E = -\frac{1}{2c}\int (d\mathbf{r})\, \mathbf{J}\cdot\mathbf{A}. \tag{29.11}$$

[Notice that (29.11) is gauge invariant, since under a gauge transformation,

$$\mathbf{A} \to \mathbf{A} + \boldsymbol{\nabla}\lambda, \tag{29.12a}$$

the energy does not change,

$$\delta E = -\frac{1}{2c}\int (d\mathbf{r})\,\mathbf{J}\cdot\boldsymbol{\nabla}\lambda = 0, \tag{29.12b}$$

since we can integrate by parts and use (28.7).] Introducing the explicit form for $\mathbf{A}$, (28.31), we can write the energy in terms of the current density alone,

$$E = -\frac{1}{2}\int (d\mathbf{r})\,(d\mathbf{r}')\frac{\frac{1}{c}\mathbf{J}(\mathbf{r})\cdot\frac{1}{c}\mathbf{J}(\mathbf{r}')}{|\mathbf{r}-\mathbf{r}'|}, \tag{29.13}$$

which is analogous to the electrostatic result, (14.10), or (1.1), except that its sign is opposite. For the case of two current distributions,

$$\mathbf{J}(\mathbf{r}) = \mathbf{J}_1(\mathbf{r}) + \mathbf{J}_2(\mathbf{r}), \tag{29.14}$$

the energy expression contains self energies as well as the mutual interaction energy. We are here interested only in the latter, which is

$$E = -\int (d\mathbf{r})\,(d\mathbf{r}')\frac{\frac{1}{c}\mathbf{J}_1(\mathbf{r})\cdot\frac{1}{c}\mathbf{J}_2(\mathbf{r}')}{|\mathbf{r}-\mathbf{r}'|}, \tag{29.15}$$

as it is the sole term that contributes to the force, (29.10). For straight wires, this becomes

$$E = -\int dz\,dz'\,\frac{I_1 I_2}{c^2}\frac{\mathbf{n}_1\cdot\mathbf{n}_2}{|\mathbf{r}-\mathbf{r}'|}. \tag{29.16}$$

For parallel wires with currents flowing in the same sense,

$$\mathbf{n}_1\cdot\mathbf{n}_2 = 1, \quad |\mathbf{r}-\mathbf{r}'| = \sqrt{\rho^2 + (z-z')^2}, \tag{29.17}$$

the integration over z' is evaluated as in (29.3) so that

$$E = -\frac{I_1 I_2}{c^2}2\left(\ln\frac{L}{\rho} + \text{constant}\right)\int dz = -\frac{I_1 I_2}{c^2}2\left(\ln\frac{L}{\rho} + \text{constant}\right)2L, \tag{29.18}$$

where we have used the restriction $L/\rho \gg 1$. The force can now be calculated from (29.10), or, since E depends only on ρ,

$$\mathbf{F} = -\left(\frac{\partial}{\partial\rho}E\right)\hat{\boldsymbol{\rho}}. \tag{29.19}$$

The force per unit length is therefore

$$\frac{F}{2L} = -\frac{2I_1 I_2}{c^2}\frac{\hat{\boldsymbol{\rho}}}{\rho}, \tag{29.20}$$

which is our previous result, (29.9).

29.1 Magnetic Energy. Coefficients of Inductance

We have mentioned repeatedly the difference in sign between magnetostatic energy and electrostatic energy. This might cause some confusion. In particular, we might recall the form of the magnetic energy density given in (8.13),

$$U_m = \frac{1}{8\pi}\mathbf{B}\cdot\mathbf{H}. \tag{29.21}$$

If the volume integral of this is evaluated in terms of prescribed currents, using (28.34) and (28.6), we obtain a current-current interaction of the form (29.13), *but with the opposite sign.* What is going on?

Equation (29.21) is correct, but not relevant to the situation of *statics*, where the current has no time dependence. The difference is attributable to the energy required to build up the magnetic field, which is a dynamical consideration. If we are interested in the force between steady currents, the energy (29.15) is the relevant quantity to derive the force through (29.10). [The reader is also referred to the independent derivation of the magnetostatic energy presented, in the context of radiation, in Section 36.3.]

For a system of current-carrying wires, with currents $I_i\mathbf{n}_i$, the coefficients of inductance L_{ij} are defined in terms of the energy by

$$E = -\frac{1}{2}\sum_{ij} L_{ij} I_i I_j, \tag{29.22}$$

analogously to the way the coefficients of capacitance were defined in (26.48). We can express L_{ij} in terms of a geometrical integral, for wires of negligible cross section, by introducing local coordinates along the wires, ξ_i so that

$$\mathbf{J}(\mathbf{r}) = \sum_i I_i \mathbf{n}_i(\xi_i)\delta(\mathbf{r}_{\perp i}), \tag{29.23}$$

where $\mathbf{r}_{\perp i}$ is perpendicular to $\mathbf{n}_i(\xi_i)$. The coefficients of inductance are then given by

$$L_{ij} = \frac{1}{c^2}\int d\xi_i d\xi_j \frac{\mathbf{n}_i(\xi_i)\cdot\mathbf{n}_j(\xi_j)}{|\mathbf{r}_i - \mathbf{r}_j|}, \tag{29.24}$$

a formula discovered by Franz Neumann in 1845.

For closed filamentary current loops, we can also write the energy (29.11) as

$$E = -\frac{1}{2c}\sum_i I_i \oint d\boldsymbol{\xi}_i\cdot\mathbf{A} = -\frac{1}{2c}\sum_i I_i\Phi_i \tag{29.25}$$

in terms of the magnetic flux penetrating the ith loop,

$$\Phi_i = \int_{S_i} d\mathbf{S}\cdot\mathbf{B}. \tag{29.26}$$

Thus that flux is given by

$$\Phi_i = c\sum_j L_{ij} I_j, \tag{29.27}$$

analogous to (26.47), which is the basis of elementary circuit analysis.[1]

Examples of calculation of the inductance for simple systems are given in the problems.

[1] For example, Faraday's law, (1.66), gives the emf in the ith circuit as $\mathcal{E}_i = \oint d\boldsymbol{\xi}_i\cdot\mathbf{E} = -\frac{1}{c}\int d\mathbf{S}_i\cdot\dot{\mathbf{B}} = -\sum_j L_{ij}\dot{I}_j$. However, such circuit analysis goes beyond statics, dealing as it does with alternating currents. In this case, as noted above, the energy associated with an inductor L is positive, $E_L = \frac{1}{2}LI^2$, and thus an LC oscillator, for example, can be understood in terms of the energy interchange between that in the inductor, and that in the capacitor, $E_C = \frac{1}{2}Q^2/C$.

29.2 Problems for Chapter 29

1. Do the integral in (29.3) exactly, and derive the approximate result given there.
2. Derive (29.11) from (28.10) by scaling $\mathbf{A}$,

$$\mathbf{A} \to \lambda\mathbf{A}, \quad \lambda = \text{ constant}, \tag{29.28}$$

and then using the stationary property of $E[\mathbf{A}]$.

3. As an alternative to the derivation given in this chapter for the field of a long straight wire, proceed as follows. Start from the differential equation satisfied by the vector potential in the radiation gauge, (28.30), and solve the equation in cylindrical coordinates, where the z axis coincides with the direction of the wire. Show that the cylindrically symmetric solution has the form

$$\mathbf{A} = \hat{\mathbf{z}}(a + b\ln\rho), \tag{29.29}$$

where a and b are constants, and ρ is the distance from the wire. To determine the constant b, integrate the differential equation satisfied by $\mathbf{A}$ over the volume of a circular cylinder concentric with the wire, of radius r and height h. Then, from $\mathbf{A}$ compute the magnetic field in magnitude and direction everywhere outside the wire. Verify this result by recasting the magnetostatic equation (28.40) into an integral equation, Ampère's law, around a closed loop, and solving for $\mathbf{B}$ using symmetry considerations.

4. Calculate the mutual inductance $M = 2L_{12}$ of a pair of parallel wires of length l separated by a distance a. Do the integral exactly, and examine the limit $l \gg a$, where

$$M \approx \frac{4l}{c^2}\left(\ln\frac{2l}{a} - 1\right). \tag{29.30}$$

5. Calculate the self-inductance of a straight wire of length l and radius b if $l \gg b$. Do so using an argument originally due to Maxwell: Consider the self-inductance of the wire as arising from the mutual inductance of each pair of filaments that make up the wire so that

$$L = \int \frac{(d\mathbf{r}_{\perp 1})(d\mathbf{r}_{\perp 2})}{(\pi b^2)^2}\frac{2l}{c^2}\left(\ln\frac{2l}{\rho_{12}} - 1\right), \tag{29.31}$$

where $(d\mathbf{r}_{\perp 1,2})$ are the cross-sectional area elements of the two filaments, and ρ_{12} is the distance between the two filaments. The integral over $\ln\rho$ may be carried out by noticing that $-2\ln\rho$ is the potential of a unit point charge at the origin in two dimensions (two-dimensional Green's function) so that

$$\phi_2 = -2\int (d\mathbf{r}_{\perp 1})\ln\rho_{12} \tag{29.32}$$

is the potential due to a uniform charge distribution. But the latter must satisfy the two-dimensional Poisson's equation [*cf.* (18.16)], so for $\rho < b$,

$$\phi = A + B\ln\rho - \pi\rho^2,$$

By doing the integral at the center of the wire, determine the constants, and thereby find the self-inductance

$$L = \frac{2l}{c^2}\left(\ln\frac{2l}{b} - \frac{3}{4}\right). \tag{29.33}$$

6. Calculate the self-inductance of a two-wire transmission line which consists of parallel wires of great length l and of radius b, separated by a distance a, traversed by equal currents in opposite directions. The self-inductance L is given by the energy

$$E = -\frac{1}{2}LI^2, \tag{29.34}$$

where I is the common current.

7. Show that the self-inductance of a circular loop of wire of radius r and cross-sectional radius b is ($b \ll r$)

$$L = \frac{4\pi r}{c^2}\left(\ln\frac{8r}{b} - \frac{7}{4}\right). \tag{29.35}$$

30

Magnetic Multipoles

30.1 Magnetic Dipole Moment

We now direct our attention to the magnetic field produced by a confined current distribution. If we wish to evaluate the vector potential far outside the current distribution, $|\mathbf{r}| \gg$ extent of the region of current flow (where the origin of the coordinate system is located in the current distribution), we may use the first two terms in the expansion (24.2),

$$\frac{1}{|\mathbf{r}-\mathbf{r}'|} = \frac{1}{r} + \frac{\mathbf{r}\cdot\mathbf{r}'}{r^3} + \dots, \tag{30.1}$$

in the expression for $\mathbf{A}$, (28.31). The resulting expansion for the vector potential is then

$$\mathbf{A}(\mathbf{r}) = \frac{1}{r}\int (d\mathbf{r}')\frac{1}{c}\mathbf{J}(\mathbf{r}') + \frac{\mathbf{r}}{r^3}\cdot\int (d\mathbf{r}')\,\mathbf{r}'\left(\frac{1}{c}\mathbf{J}(\mathbf{r}')\right) + \dots, \tag{30.2}$$

which is analogous to the expansion for ϕ, (24.3). From current conservation for steady currents,

$$\boldsymbol{\nabla}\cdot\mathbf{J}(\mathbf{r}) = 0, \tag{30.3}$$

the first term of (30.2) vanishes for a confined current distribution:

$$\mathbf{0} = \int (d\mathbf{r})\,\mathbf{r}\boldsymbol{\nabla}\cdot\mathbf{J} = \int (d\mathbf{r})\,[\boldsymbol{\nabla}\cdot(\mathbf{J}\mathbf{r}) - \mathbf{J}] = -\int (d\mathbf{r})\,\mathbf{J}, \tag{30.4}$$

so that there is no $1/r$ term in the expansion of $\mathbf{A}(\mathbf{r})$. Physically, there is no Coulomb-like potential for magnetism, when magnetic charge is not present. The leading term in the vector potential expansion is therefore

$$\mathbf{A}(\mathbf{r}) = \frac{1}{r^3}\frac{1}{c}\int (d\mathbf{r}')\,\mathbf{r}\cdot\mathbf{r}'\mathbf{J}(\mathbf{r}') + \dots. \tag{30.5}$$

To evaluate this integral, we again use (30.3) and consider the integral

$$0 = \int (d\mathbf{r})\,x_i x_j \sum_{k=1}^{3} \nabla_k J_k = -\int (d\mathbf{r})\,(x_j J_i + x_i J_j), \tag{30.6}$$

or, in a dyadic notation,

$$\int (d\mathbf{r})\,[\mathbf{r}\mathbf{J} + \mathbf{J}\mathbf{r}] = \mathbf{0}. \tag{30.7}$$

Using this fact, we make the following rearrangement:

$$\begin{aligned}\int (d\mathbf{r}')\,\mathbf{r}\cdot\mathbf{r}'\mathbf{J} &= \frac{1}{2}\int (d\mathbf{r}')\,\mathbf{r}\cdot[(\mathbf{r}'\mathbf{J} + \mathbf{J}\mathbf{r}') + (\mathbf{r}'\mathbf{J} - \mathbf{J}\mathbf{r}')]\\ &= \frac{1}{2}\int (d\mathbf{r}')\,\mathbf{r}\cdot(\mathbf{r}'\mathbf{J} - \mathbf{J}\mathbf{r}') = \frac{1}{2}\int (d\mathbf{r}')(\mathbf{r}'\times\mathbf{J})\times\mathbf{r}.\end{aligned} \tag{30.8}$$

DOI: 10.1201/9781003057369-30

The leading term of the vector potential now becomes

$$\mathbf{A}(\mathbf{r}) = \frac{\boldsymbol{\mu}\times\mathbf{r}}{r^3}, \tag{30.9}$$

where $\boldsymbol{\mu}$ is the magnetic dipole moment, defined by

$$\boldsymbol{\mu} = \frac{1}{2c}\int (d\mathbf{r})\,\mathbf{r}\times\mathbf{J}(\mathbf{r}). \tag{30.10}$$

[For a point charge, the current density is given by (28.35), so the magnetic dipole moment is

$$\boldsymbol{\mu} = \frac{e}{2c}\mathbf{R}\times\mathbf{v}, \tag{30.11}$$

which obviously generalizes to a system of point charges, in agreement with (7.24).]

The leading contribution to the magnetic field can now be calculated from this vector potential,

$$\mathbf{B} = \boldsymbol{\nabla}\times\left(\boldsymbol{\mu}\times\frac{\mathbf{r}}{r^3}\right) = \boldsymbol{\mu}\boldsymbol{\nabla}\cdot\frac{\mathbf{r}}{r^3} - (\boldsymbol{\mu}\cdot\boldsymbol{\nabla})\frac{\mathbf{r}}{r^3} = 4\pi\boldsymbol{\mu}\delta(\mathbf{r}) - \boldsymbol{\nabla}\left(\frac{\boldsymbol{\mu}\cdot\mathbf{r}}{r^3}\right), \tag{30.12}$$

since

$$-\boldsymbol{\nabla}\frac{1}{r} = \frac{\mathbf{r}}{r^3}, \tag{30.13}$$

and consequently

$$\mathbf{0} = \boldsymbol{\mu}\times\left(\boldsymbol{\nabla}\times\frac{\mathbf{r}}{r^3}\right) = \boldsymbol{\nabla}\left(\frac{\boldsymbol{\mu}\cdot\mathbf{r}}{r^3}\right) - (\boldsymbol{\mu}\cdot\boldsymbol{\nabla})\frac{\mathbf{r}}{r^3}. \tag{30.14}$$

The delta function term in $\mathbf{B}$ is necessary in order to satisfy $\boldsymbol{\nabla}\cdot\mathbf{B} = 0$:

$$\boldsymbol{\nabla}\cdot\mathbf{B} = 4\pi(\boldsymbol{\mu}\cdot\boldsymbol{\nabla})\delta(\mathbf{r}) + \nabla^2(\boldsymbol{\mu}\cdot\boldsymbol{\nabla})\frac{1}{r} = 0. \tag{30.15}$$

For $r > 0$, this magnetic field, (30.12), has the same form as that of the electric field produced by an electric dipole moment, which is contained in (24.8), that is,

$$\mathbf{B}(\mathbf{r}) = \frac{3\mathbf{r}\boldsymbol{\mu}\cdot\mathbf{r} - r^2\boldsymbol{\mu}}{r^5}. \tag{30.16}$$

In general, this is only the leading contribution, since there are higher multipoles. We will not, however, explore these further here.

30.2 Rotating Charged Spherical Shell

An example for which the dipole expression is exact beyond a certain distance results if a charge e is distributed uniformly over a spherical shell of radius a that is rotating with angular velocity $\boldsymbol{\omega}$. If we choose the origin to be at the center of the sphere, the velocity of a point $\mathbf{r}'$ on the surface is

$$\mathbf{v} = \boldsymbol{\omega}\times\mathbf{r}'. \tag{30.17}$$

The current density is

$$\mathbf{J} = \rho\mathbf{v}, \tag{30.18}$$

where here the charge density is entirely concentrated on the surface,

$$\int (d\mathbf{r}')\rho \cdots = \oint dS'\,\sigma \ldots, \tag{30.19}$$

where the surface charge density is constant,

$$\sigma = \frac{e}{4\pi a^2}. \tag{30.20}$$

The expression for the vector potential, (28.31), becomes

$$\mathbf{A}(\mathbf{r}) = \frac{1}{c}\oint dS' \frac{e}{4\pi a^2}\frac{\boldsymbol{\omega}\times\mathbf{r}'}{|\mathbf{r}-\mathbf{r}'|}, \tag{30.21}$$

where $\mathbf{r}$ may be either inside or outside the sphere.

We first calculate $\mathbf{A}$ inside the sphere, that is, for $|\mathbf{r}| < a = |\mathbf{r}'|$. Recalling the expansion, (23.10),

$$\frac{1}{|\mathbf{r}-\mathbf{r}'|} = \sum_{l=0}^{\infty} \frac{r^l}{a^{l+1}} P_l(\cos\gamma), \tag{30.22}$$

and noting that $\mathbf{r}'$ occurring in the numerator of (30.21) is related to Y_{1m}, we see that the surface integral selects only $l = 1$:

$$\oint dS'\, Y^*_{lm} Y_{1m'} \sim \delta_{l1}\delta_{mm'}. \tag{30.23}$$

Therefore, only the $l = 1$ term in (30.22) will contribute to the integral (30.21),

$$\frac{1}{|\mathbf{r}-\mathbf{r}'|} \overset{l=1}{\rightarrow} \frac{r}{a^2}\frac{\mathbf{r}\cdot\mathbf{r}'}{rr'} = \frac{1}{a^3}\mathbf{r}\cdot\mathbf{r}', \tag{30.24}$$

yielding for the vector potential,

$$\mathbf{A}(\mathbf{r}) = \frac{e}{4\pi a^2 c}\oint dS'\,(\boldsymbol{\omega}\times\mathbf{r}')\frac{\mathbf{r}'\cdot\mathbf{r}}{a^3}. \tag{30.25}$$

Using spherical symmetry, we easily evaluate the integral over the dyadic to be

$$\oint dS'\,\mathbf{r}'\mathbf{r}' = \oint dS'\,\frac{1}{3}\mathbf{1}r'^2 = \frac{1}{3}a^2 4\pi a^2\mathbf{1}. \tag{30.26}$$

Therefore, the vector potential inside the sphere is

$$\mathbf{A} = \frac{e}{3ac}(\boldsymbol{\omega}\times\mathbf{r}) \equiv \frac{1}{2}\mathbf{B}\times\mathbf{r}, \tag{30.27}$$

where, using the result of (7.20), we identify the magnetic field $\mathbf{B}$ as

$$\mathbf{B} = \frac{2e}{3ac}\boldsymbol{\omega}, \tag{30.28}$$

which is uniform inside the sphere.

We now calculate $\mathbf{A}$ outside the sphere, where $|\mathbf{r}| > a = |\mathbf{r}'|$, and the appropriate expansion is

$$\frac{1}{|\mathbf{r}-\mathbf{r}'|} = \sum_{l=0}^{\infty} \frac{a^l}{r^{l+1}} P_l(\cos\gamma) \overset{l=1}{\rightarrow} \frac{a}{r^2}\frac{\mathbf{r}\cdot\mathbf{r}'}{rr'} = \frac{1}{r^3}\mathbf{r}\cdot\mathbf{r}', \tag{30.29}$$

since, as before, only $l = 1$ contributes to the surface integral. The calculation proceeds as above except for a factor of a^3/r^3 with the result

$$\mathbf{A} = \frac{ea^2}{3c}\frac{\boldsymbol{\omega}\times\mathbf{r}}{r^3} = \frac{\boldsymbol{\mu}\times\mathbf{r}}{r^3}, \tag{30.30}$$

which, upon comparison with (30.9), allows us to identify the magnetic dipole moment as

$$\boldsymbol{\mu} = \frac{ea^2}{3c}\boldsymbol{\omega}. \tag{30.31}$$

The magnetic field $\mathbf{B}$ is then given by (30.16).

Notice that $\mathbf{B}$ is discontinuous across the spherical shell because there is a surface current density. The values of $\mathbf{B}$ just outside and just inside the surface ($\mathbf{n}$ = outward normal = $\mathbf{r}/a$) are

$$r = a + 0: \quad \mathbf{B}_+ = \frac{3(\boldsymbol{\mu}\cdot\mathbf{n})\mathbf{n} - \boldsymbol{\mu}}{a^3} = \frac{ea^2}{3c}\frac{3(\boldsymbol{\omega}\cdot\mathbf{n})\mathbf{n} - \boldsymbol{\omega}}{a^3}, \tag{30.32a}$$

$$r = a - 0: \quad \mathbf{B}_- = \frac{2e}{3ca}\boldsymbol{\omega}, \tag{30.32b}$$

so the discontinuity in $\mathbf{B}$ is

$$\mathbf{B}_+ - \mathbf{B}_- = \frac{e}{ca}[(\boldsymbol{\omega}\cdot\mathbf{n})\mathbf{n} - \boldsymbol{\omega}] = \frac{e}{ca}\mathbf{n}\times(\mathbf{n}\times\boldsymbol{\omega}). \tag{30.33}$$

Now recall that in vacuum ($\mathbf{B} = \mathbf{H}$), the normal component of $\mathbf{B}$ is continuous while the tangential component is discontinuous if there is a surface current. Written in the notation above, these boundary conditions [(28.15b) and (28.19a)] read

$$\mathbf{n}\cdot(\mathbf{B}_+ - \mathbf{B}_-) = 0, \tag{30.34a}$$

$$\mathbf{n}\times(\mathbf{B}_+ - \mathbf{B}_-) = \frac{4\pi}{c}\mathbf{K}. \tag{30.34b}$$

Obviously, (30.34a) is satisfied. From (30.34b) and (30.33), we calculate the surface current density,

$$\mathbf{K} = \frac{c}{4\pi}\mathbf{n}\times\left[\frac{e}{ca}\mathbf{n}\times(\mathbf{n}\times\boldsymbol{\omega})\right] = \frac{e}{4\pi a}\boldsymbol{\omega}\times\mathbf{n}, \tag{30.35a}$$

in agreement with the direct result ($\mathbf{r}' = a\mathbf{n}$)

$$\mathbf{K} = \sigma\mathbf{v} = \frac{e}{4\pi a^2}(\boldsymbol{\omega}\times\mathbf{n})a = \frac{e}{4\pi a}\boldsymbol{\omega}\times\mathbf{n}. \tag{30.35b}$$

30.3 Problems for Chapter 30

1. Derive (30.31) directly from (30.10).
2. Show that a perfectly conducting sphere of radius a placed in a constant magnetic field $\mathbf{B}$ acquires a magnetic moment,

$$\boldsymbol{\mu} = -\frac{1}{2}a^3\mathbf{B}. \tag{30.36a}$$

This says the magnetic polarizability of a perfectly conducting sphere is

$$\beta = -\frac{1}{2}a^3, \tag{30.36b}$$

which is $-\frac{1}{2}$ the electric polarizability of the same sphere—See (25.82) with $\epsilon \to \infty$.

31

Magnetic Scalar Potential

We now return to the macroscopic situation with a steady current flowing in a permeable medium characterized by a constant μ so that the vector potential is

$$\mathbf{A}(\mathbf{r}) = \frac{\mu}{c}\int(d\mathbf{r}')\frac{\mathbf{J}(\mathbf{r}')}{|\mathbf{r}-\mathbf{r}'|}, \tag{31.1}$$

in the Coulomb gauge. In particular, consider a current I flowing in a closed loop, as shown in Fig. 31.1 so that the volume integral is to be replaced by a line integral,

$$\int(d\mathbf{r}')\mathbf{J}\cdots = \int d\mathbf{r}'I\ldots, \tag{31.2}$$

where $d\mathbf{r}'$ is a directed line element tangential to the wire, in the direction of the current flow. Now the vector potential, (31.1), becomes

$$\mathbf{A}(\mathbf{r}) = \frac{\mu}{c}I\oint\frac{d\mathbf{r}'}{|\mathbf{r}-\mathbf{r}'|}, \tag{31.3a}$$

which implies for the magnetic induction,

$$\mathbf{B} = \frac{\mu}{c}I\boldsymbol{\nabla}\times\oint\frac{d\mathbf{r}'}{|\mathbf{r}-\mathbf{r}'|}, \tag{31.3b}$$

or, for the magnetic field,

$$\mathbf{H} = \frac{I}{c}\oint\boldsymbol{\nabla}\frac{1}{|\mathbf{r}-\mathbf{r}'|}\times d\mathbf{r}' = \frac{I}{c}\oint d\mathbf{r}'\times\boldsymbol{\nabla}'\frac{1}{|\mathbf{r}-\mathbf{r}'|}. \tag{31.3c}$$

It is convenient to rewrite $\mathbf{H}$ in terms of a surface integral instead of a line integral. We make use of Stokes' theorem (1.78) for a vector field, $\mathbf{V}$, which reads

$$\oint_C d\mathbf{r}'\cdot\mathbf{V} = \int_S d\mathbf{S}'\cdot(\boldsymbol{\nabla}'\times\mathbf{V}), \tag{31.4}$$

where S is any surface which has the contour C as its boundary, where the orientation of the surface is given by the right-hand rule. If we replace

$$\mathbf{V}\to\mathbf{V}\times\mathbf{a} \tag{31.5}$$

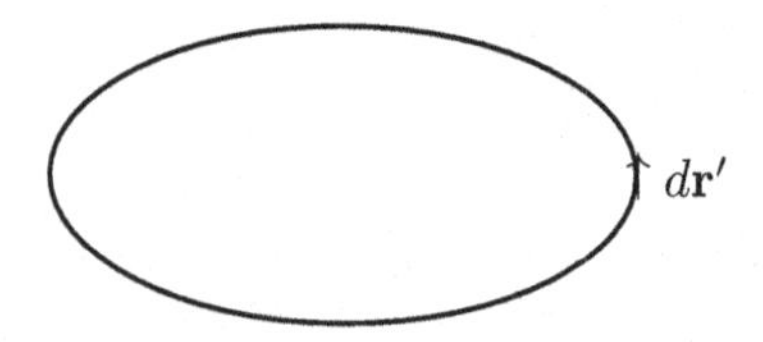

FIGURE 31.1
Closed current loop.

DOI: 10.1201/9781003057369-31

where $\mathbf{a}$ is an arbitrary constant vector, Stokes' theorem becomes

$$\oint_C d\mathbf{r}'\times\mathbf{V}\cdot\mathbf{a} = \int_S d\mathbf{S}'\cdot\boldsymbol{\nabla}'\times(\mathbf{V}\times\mathbf{a}) = \int_S d\mathbf{S}'\cdot[(\mathbf{a}\cdot\boldsymbol{\nabla}')\mathbf{V} - \mathbf{a}\boldsymbol{\nabla}'\cdot\mathbf{V}]. \tag{31.6}$$

The identity

$$\mathbf{a}\times(\boldsymbol{\nabla}'\times\mathbf{V}) = \boldsymbol{\nabla}'(\mathbf{a}\cdot\mathbf{V}) - (\mathbf{a}\cdot\boldsymbol{\nabla}')\mathbf{V}, \tag{31.7}$$

allows us to rewrite (31.6) as

$$\oint_C d\mathbf{r}'\times\mathbf{V}\cdot\mathbf{a} = \int_S d\mathbf{S}'\cdot[\boldsymbol{\nabla}'(\mathbf{V}\cdot\mathbf{a}) - \mathbf{a}\times(\boldsymbol{\nabla}'\times\mathbf{V}) - \mathbf{a}\boldsymbol{\nabla}'\cdot\mathbf{V}]. \tag{31.8}$$

Furthermore, if everywhere on the surface S, the vector field satisfies

$$\boldsymbol{\nabla}'\cdot\mathbf{V} = 0, \quad \text{and} \quad \boldsymbol{\nabla}'\times\mathbf{V} = \mathbf{0}, \tag{31.9}$$

(31.8) reduces to

$$\oint_C d\mathbf{r}'\times\mathbf{V} = \int_S (d\mathbf{S}'\cdot\boldsymbol{\nabla}')\mathbf{V}. \tag{31.10}$$

We will apply this result to rewrite (31.3c) for which

$$\mathbf{V} = \boldsymbol{\nabla}'\frac{1}{|\mathbf{r}-\mathbf{r}'|} = -\boldsymbol{\nabla}\frac{1}{|\mathbf{r}-\mathbf{r}'|}, \tag{31.11}$$

which satisfies the conditions (31.9) as long as $\mathbf{r}\neq\mathbf{r}'$, that is, at points outside the wire. We therefore find

$$\mathbf{H} = -\boldsymbol{\nabla}\phi_m, \tag{31.12}$$

where the magnetic scalar potential, ϕ_m, is

$$\phi_m(\mathbf{r}) = \frac{I}{c}\int_{S'} d\mathbf{S}'\cdot\boldsymbol{\nabla}'\frac{1}{|\mathbf{r}-\mathbf{r}'|}, \tag{31.13}$$

where S' is a surface bounded by the current loop. According to (1.16), the surface integral,

$$-\int_{S'} d\mathbf{S}'\cdot\boldsymbol{\nabla}'\frac{1}{|\mathbf{r}-\mathbf{r}'|} = -\int_{S'} d\mathbf{S}'\cdot\frac{\mathbf{r}-\mathbf{r}'}{|\mathbf{r}-\mathbf{r}'|^3} = \Omega, \tag{31.14}$$

is the solid angle subtended by the current loop at the observation point so that

$$\phi_m = -\frac{I}{c}\Omega. \tag{31.15}$$

Therefore, a scalar potential for $\mathbf{H}$ exists for points not on the wire, consistent with the Maxwell equation, (28.6),

$$\boldsymbol{\nabla}\times\mathbf{H} = \mathbf{0}. \tag{31.16}$$

However, this scalar potential is not single-valued. We consider the integral of $\mathbf{H}$ around a closed path, C, which does not touch the wire,

$$\oint_C d\mathbf{r}\cdot\mathbf{H} = -\oint_C d\mathbf{r}\cdot\boldsymbol{\nabla}\phi_m = -\phi_m\big|_{\circlearrowleft}. \tag{31.17}$$

The naive anticipation is that (31.17) would be zero. An alternative calculation of this quantity can be made using Stokes' theorem:

$$\oint_C d\mathbf{r}\cdot\mathbf{H} = \int_S d\mathbf{S}\cdot(\boldsymbol{\nabla}\times\mathbf{H}) = \int_S d\mathbf{S}\cdot\frac{4\pi}{c}\mathbf{J} = \begin{cases} \pm\frac{4\pi}{c}I, & \text{YES,} \\ 0, & \text{NO,} \end{cases} \tag{31.18}$$

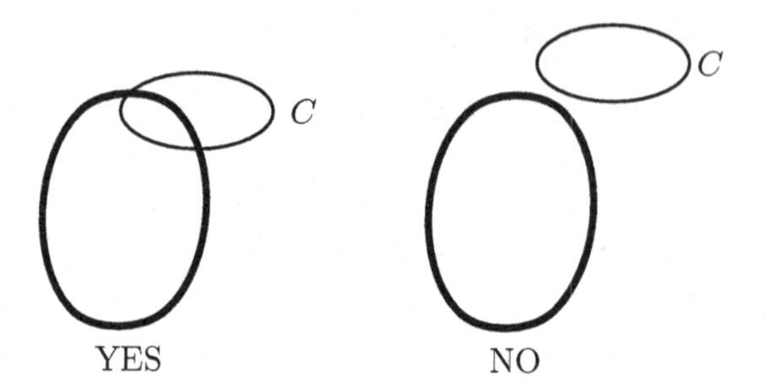

FIGURE 31.2
Topology of curve C relative to current loop, given by the closed dark curve. In the first panel it is understood that the curve C threads through the current loop.

where YES means the wire passes once through the surface S, bounded by the path C, (the $\pm$ sign refers to the relative orientations of $d\mathbf{S}$ and $\mathbf{J}$) while NO means the wire does not pass through the surface. Some examples of this are supplied by the illustrations in Fig. 31.2. Therefore, contrary to our naive expectation,

$$\phi_m \oint = \begin{cases} \mp\frac{4\pi}{c} I, & \text{YES}, \\ 0, & \text{NO}, \end{cases} \tag{31.19a}$$

or

$$\Omega \oint = \begin{cases} \pm 4\pi, & \text{YES}, \\ 0, & \text{NO}, \end{cases} \tag{31.19b}$$

which means that ϕ_m is a multivalued function, as required by the fact that $\boldsymbol{\nabla}\times\mathbf{H}$ is not zero everywhere. The discontinuity found in (31.19b) corresponds to the fact that when one crosses the surface S' defined by the current loop and used in the evaluation of the solid angle in Ω, there is a change of 4π.

Very far away from the current loop, the solid angle subtended by it is, if we assume that the points of S' are localized in the vicinity of the loop

$$\Omega = -\frac{\mathbf{r}\cdot\mathbf{S}'}{r^3}, \quad \mathbf{S}' = \int d\mathbf{S}', \tag{31.20}$$

so the corresponding magnetic field is

$$\mathbf{H} = -\boldsymbol{\nabla}\left(\frac{\boldsymbol{\mu}\cdot\mathbf{r}}{r^3}\right), \tag{31.21}$$

which upon comparison with (30.12) identifies the magnetic moment of the current loop to be

$$\boldsymbol{\mu} = \frac{I}{c}\mathbf{S}. \tag{31.22a}$$

We obtain the same result if we use the definition of the magnetic moment, (30.10),

$$\boldsymbol{\mu} = \frac{1}{2c}\int (d\mathbf{r}')\,\mathbf{r}'\times\mathbf{J}(\mathbf{r}') = \frac{I}{c}\frac{1}{2}\oint \mathbf{r}'\times d\mathbf{r}' = \frac{I}{c}\mathbf{S}, \tag{31.22b}$$

where we have used (31.8) to evaluate the line integral, which is also obvious geometrically.

31.1 Problems for Chapter 31

1. Use (31.15) to calculate the magnetic field produced by a circular current loop along the symmetry axis perpendicular to the loop.

2. Using either (31.15) or (28.31), compute the magnetic field produced by a circular current loop of radius a at an arbitrary point. Express the answer in terms of complete elliptic integrals. In cylindrical coordinates, the results are

$$H_\rho = -\frac{I}{c}\frac{2z}{\rho}\frac{1}{\sqrt{(a+\rho)^2+z^2}}\left[K(k) - \frac{a^2+\rho^2+z^2}{(a-\rho)^2+z^2}E(k)\right], \qquad (31.23a)$$

$$H_z = \frac{I}{c}\frac{2}{\sqrt{(a+\rho)^2+z^2}}\left[K(k) + \frac{a^2-\rho^2-z^2}{(a-\rho)^2+z^2}E(k)\right], \qquad (31.23b)$$

where the complete elliptic integrals of the first and second kind are, respectively,

$$K(k) = \int_0^{\pi/2} d\psi \frac{1}{\sqrt{1-k^2\sin^2\psi}}, \qquad (31.24a)$$

$$E(k) = \int_0^{\pi/2} d\psi \sqrt{1-k^2\sin^2\psi}, \qquad (31.24b)$$

and

$$k^2 = \frac{4a\rho}{(a+\rho)^2+z^2}. \qquad (31.25)$$

3. By taking the large distance limit of the result found in Problem 2, determine the magnetic dipole moment of a circular current loop. Compare the result with (31.22b).

4. Revisit the problem of the magnetic field produced by a rotating spherical shell of charge discussed in Chapter 30. Show that inside the sphere, the magnetic field is given by a magnetic scalar potential, $\mathbf{B} = -\boldsymbol{\nabla}\phi_m$, where ϕ_m must have the form $\phi_m = c + dz$, $z = a\cos\theta$, where a is the radius of the sphere, centered at the origin, and θ is the polar angle with respect to the z axis, the axis of rotation. By use of the boundary condition (30.34a), relate the constant d to μ. Insert this result into the second boundary condition (30.34b) and thereby determine both d and $\boldsymbol{\mu}$, and then give the magnetic field both inside and outside the rotating shell.

5. Consider a static magnetic field $\mathbf{B}$ in a region with permeability $\mu = 1$ and no electric field present. The volume V is bounded by a closed superconducting wall S. Let the only electric currents present be on or outside the surface S. The only property of a superconductor needed is that, within the superconducting material, the magnetic field is zero. Show that $\mathbf{B}$ may be derived from a magnetic scalar potential ϕ_m at all points inside the volume V, and then use the boundary condition at the surface S to show that the normal derivative of ϕ_m vanishes at the surface. Then show that $\mathbf{B}$ vanishes at all interior points in V.

6. Derive the result of Problem 30.2 by introducing a magnetic scalar potential in the region external to the sphere, and solving for that potential using the perfect conductor boundary condition $\hat{\mathbf{n}}\cdot\mathbf{B} = 0$ at the surface of the sphere, where $\hat{\mathbf{n}}$ is the normal to the spherical surface. Use a spherical harmonic expansion, and determine the nonzero constant by requiring at large distances the applied field $\mathbf{B}_0$ be recovered. Read off the induced dipole moment from (31.21).

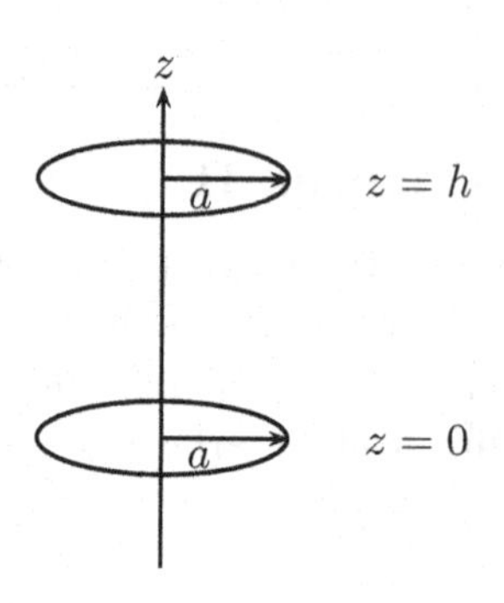

FIGURE 31.3
Two parallel current loops sharing a common axis. The radius of each loop is a, and they are separated by a distance h.

7. Consider two circular current loops of radius a perpendicular to a common axis, the z axis, as shown in Fig. 31.3. Let one loop be centered at the origin, the other at $z = h$. Let a current I be flowing in the first loop in a counterclockwise sense, and the same current I be flowing in the second loop in a clockwise sense. Recalling the magnetic scalar potential of a single loop is given by (31.15), where Ω is the solid angle subtended by the current loop, give an expression for ϕ_m on the z axis, for $z \gg a$, ignoring terms of order a^4/z^4. If further $z \gg h$, give an expression for ϕ_m through order h/z. Compute the magnetic field on the axis when $z \gg a, h$. What is the magnetic dipole moment of this configuration? The field produced by this arrangement is asymptotically a magnetic quadrupole. In analogy with an electric quadrupole, the magnetic scalar potential for a magnetic quadrupole field has the form

$$\phi_m(\mathbf{r}) = \frac{1}{2}\frac{\mathbf{r}\cdot\mathbf{Q}\cdot\mathbf{r}}{r^5}, \tag{31.26}$$

where $\mathbf{Q}$ is the magnetic quadrupole dyadic, which is symmetric, $Q_{ij} = Q_{ji}$. For such a magnetic scalar potential, work out the corresponding magnetic field. Now infer the value of Q_{zz} in this case considered here. How is it related to the magnetic dipole moment of a single loop? Is that connection plausible?

32

Steady Currents and Dissipation

32.1 Variational Principles for Current

Suppose we have a medium with an isotropic conductivity $\sigma(\mathbf{r}) \geq 0$, which is defined in terms of the linear relation between the electric field and the electric current:

$$\mathbf{J} = \sigma \mathbf{E}. \tag{32.1}$$

In statics, the electric field is derived from a scalar potential,

$$\mathbf{E} = -\boldsymbol{\nabla}\phi, \tag{32.2}$$

and in addition the current density is divergenceless,

$$\boldsymbol{\nabla} \cdot \mathbf{J} = 0, \tag{32.3}$$

which is the static version of the local conservation of charge (1.42). This implies the potential obeys the equation

$$\boldsymbol{\nabla} \cdot \sigma(-\boldsymbol{\nabla}\phi) = 0. \tag{32.4}$$

The rate at which the field does work on the current, or the power of Joule[1] heating, is given by

$$P = \int (d\mathbf{r})\, \mathbf{J} \cdot \mathbf{E} = \int (d\mathbf{r})\, \sigma E^2 = \int (d\mathbf{r}) \frac{J^2}{\sigma}. \tag{32.5}$$

The following construction of the power,

$$P[\mathbf{J}, \phi] = \int (d\mathbf{r}) \left[2\mathbf{J} \cdot (-\boldsymbol{\nabla}\phi) - \frac{J^2}{\sigma} \right], \tag{32.6}$$

is stationary with respect to small variations in the current density and the potential, leading to the following equations holding in the volume

$$\delta\mathbf{J}: \quad \mathbf{J} = \sigma(-\boldsymbol{\nabla}\phi) = \sigma\mathbf{E}, \tag{32.7a}$$

$$\delta\phi: \quad \boldsymbol{\nabla} \cdot \mathbf{J} = 0. \tag{32.7b}$$

Omitted here was a surface contribution

$$\delta P = -2 \int (d\mathbf{r}) \boldsymbol{\nabla} \cdot (\mathbf{J}\, \delta\phi) = -2 \oint_S dS \mathbf{n} \cdot \mathbf{J}\, \delta\phi, \tag{32.8}$$

where S is the surface surrounding the region of interest, and $\mathbf{n}$ is the outward normal. Let us suppose that we insert, as part of that boundary, conducting surfaces S_i, electrodes, on which the potential is specified:

$$\phi = \phi_i = \text{constant on } S_i, \tag{32.9}$$

[1] James Prescott Joule, 1818–1889.

DOI: 10.1201/9781003057369-32

that is, $\delta\phi$ is zero on S_i. Let us denote the rest of the boundary as S':

$$S = S' + \sum_i S_i, \tag{32.10}$$

so that (32.8) is

$$\delta P = -2 \int_{S'} dS\, \mathbf{n} \cdot \mathbf{J}\, \delta\phi, \tag{32.11}$$

where $\delta\phi$ is arbitrary. This will be zero, $\delta P = 0$, if

$$\mathbf{n} \cdot \mathbf{J} = 0 \text{ on } S', \tag{32.12}$$

which is to say, no current flows out of or into the region of interest, except through the electrodes.

With prescribed potentials ϕ_i on the electrode surfaces S_i, $P[\mathbf{J}, \phi]$ is stationary for variations about $\mathbf{J} = \sigma\mathbf{E}$, $\boldsymbol{\nabla} \cdot \mathbf{J} = 0$, $\mathbf{n} \cdot \mathbf{J}$ on S'. Suppose we now accept (32.1) as given. Then the two terms in (32.6) combine to give

$$P[\phi] = \int (d\mathbf{r})\, \sigma(-\boldsymbol{\nabla}\phi)^2, \tag{32.13}$$

which when varied with respect to ϕ yields the current conditions (32.3) and (32.12). Because $P[\phi]$ is stationary for infinitesimal variations, for finite variations about the true potential ϕ_0,

$$P[\phi_0 + \delta\phi] = P + \int (d\mathbf{r})\, \sigma(-\boldsymbol{\nabla}\delta\phi)^2 > P; \tag{32.14}$$

the correct potential minimizes the power loss, with given potentials ϕ_i on the electrodes. On the other hand, if we accept (32.3) and (32.12), we have

$$P[\mathbf{J}] = -2 \sum_i \phi_i \int_{S_i} dS\, \mathbf{n} \cdot \mathbf{J} - \int (d\mathbf{r}) \frac{J^2}{\sigma}, \tag{32.15}$$

which is stationary for infinitesimal variations in $\mathbf{J}$ (Problem 32.1), while for finite variations,

$$P[\mathbf{J}_0 + \delta\mathbf{J}] = P - \int (d\mathbf{r}) \frac{(\delta\mathbf{J})^2}{\sigma} < P, \tag{32.16}$$

so the correct current makes P a maximum, for given ϕ_i.

From the latter form (32.15), by rescaling the current $\mathbf{J} \to \lambda\mathbf{J}$ and applying the stationary principle at $\lambda = 1$ we learn that

$$0 = -2 \sum_i \phi_i \int_{S_i} dS\, \mathbf{n} \cdot \mathbf{J} - 2 \int (d\mathbf{r}) \frac{J^2}{\sigma}, \tag{32.17}$$

or

$$P = \int (d\mathbf{r}) \frac{J^2}{\sigma} = \sum_i \phi_i I_i, \tag{32.18}$$

where

$$I_i = -\int_{S_i} dS\, \mathbf{n} \cdot \mathbf{J} \tag{32.19}$$

is the current input at the ith electrode. Of course,

$$\sum_i I_i = -\int_S dS\, \mathbf{n} \cdot \mathbf{J} = -\int (d\mathbf{r}) \boldsymbol{\nabla} \cdot \mathbf{J} = 0\,. \tag{32.20}$$

And therefore only potential differences matter in (32.18), i.e.,

$$\sum_i (\phi_i + \text{constant}) I_i = \sum_i \phi_i I_i. \tag{32.21}$$

Another stationary principle begins with

$$\begin{aligned} P &= 2\sum_i \phi_i I_i - \int (d\mathbf{r}) \left[2\mathbf{J} \cdot (-\boldsymbol{\nabla}\phi) - \frac{J^2}{\sigma} \right] \\ &= -2\int_S dS\, \mathbf{n} \cdot \mathbf{J}\phi + \int (d\mathbf{r}) \left[2\boldsymbol{\nabla} \cdot (\mathbf{J}\phi) - 2\boldsymbol{\nabla} \cdot \mathbf{J}\phi + \frac{J^2}{\phi} \right], \end{aligned} \tag{32.22}$$

which leads us to

$$P[\mathbf{J}, \phi] = \int (d\mathbf{r}) \left[\frac{J^2}{\sigma} - 2\phi \boldsymbol{\nabla} \cdot \mathbf{J} \right]. \tag{32.23a}$$

Here

$$I_i = -\int_{S_i} dS\, \mathbf{n} \cdot \mathbf{J} \text{ is specified, and } \quad \mathbf{n} \cdot \mathbf{J} = 0 \text{ on } S'. \tag{32.23b}$$

If (32.23a) undergoes infinitesimal variations, we recover the appropriate equations

$$\delta\phi: \quad \boldsymbol{\nabla} \cdot \mathbf{J} = 0, \quad \delta\mathbf{J}: \quad \mathbf{J} = \sigma(-\boldsymbol{\nabla}\phi). \tag{32.24}$$

The latter follows from the volume variation. What is left is a surface integral:

$$\delta P = -2\int (d\mathbf{r}) \boldsymbol{\nabla} \cdot (\phi\, \delta\mathbf{J}) = -2\int dS\, \phi\, \mathbf{n} \cdot \delta\mathbf{J}, \tag{32.25a}$$

subject to

$$\int_{S_i} dS\, \mathbf{n} \cdot \delta\mathbf{J} = 0, \quad \mathbf{n} \cdot \delta\mathbf{J} = 0 \text{ on } S'. \tag{32.25b}$$

Therefore,

$$\delta P = -2\sum_i \int_{S_i} dS\, \mathbf{n} \cdot \delta\mathbf{J}\, \phi = 0, \tag{32.26}$$

subject to the first constraint in (32.25b), from which we conclude that on S_i, $\phi = \phi_i$, a constant.

Next, impose

$$\mathbf{J} = \sigma(-\boldsymbol{\nabla}\phi) \tag{32.27}$$

in (32.23a). Then we have

$$\begin{aligned} P[\phi] &= \int (d\mathbf{r}) \left[\sigma(-\boldsymbol{\nabla}\phi)^2 - 2\phi \boldsymbol{\nabla} \cdot (\sigma(-\boldsymbol{\nabla}\phi)) \right] = -2\int_S dS\, \phi\, \mathbf{n} \cdot \sigma(-\boldsymbol{\nabla}\phi) - \int (d\mathbf{r})\, \sigma(-\boldsymbol{\nabla}\phi)^2 \\ &= -2\sum_i \int_{S_i} dS\, \phi\, \mathbf{n} \cdot \sigma(-\boldsymbol{\nabla}\phi) - \int (d\mathbf{r})\, \sigma(-\boldsymbol{\nabla}\phi)^2. \end{aligned} \tag{32.28a}$$

Here we have used on S'

$$0 = \mathbf{n} \cdot \mathbf{J} = \mathbf{n} \cdot \sigma(-\boldsymbol{\nabla}\phi). \tag{32.28b}$$

Now, under a finite variation around the true potential ϕ_0,

$$P[\phi_0 + \delta\phi] = P - 2\sum_i \int_{S_i} dS\, \delta\phi\, \mathbf{n} \cdot \sigma(-\boldsymbol{\nabla}\delta\phi) - \int (d\mathbf{r})\, \sigma(-\boldsymbol{\nabla}\delta\phi)^2. \tag{32.29}$$

Now if we only demand that $\delta\phi$ be constant on S_i, that is, ϕ is constant but not necessarily the correct potential, the first integral here vanishes:

$$-\int_{S_i} dS\,\delta\phi\,\mathbf{n}\cdot\sigma(-\boldsymbol{\nabla}\delta\phi) = -\delta\phi\int_{S_i} dS\,\mathbf{n}\cdot\sigma(-\boldsymbol{\nabla}\delta\phi) = \delta\phi\,\delta I_i = 0, \tag{32.30}$$

so $P[\phi] < P$; for a given I_i the correct ϕ makes P maximum. On the other hand, when we set $\boldsymbol{\nabla}\cdot\mathbf{J} = 0$ in (32.23a), then

$$P[\mathbf{J}] = \int(d\mathbf{r})\frac{J^2}{\sigma}, \tag{32.31}$$

and under a finite variation

$$P[\mathbf{J}+\delta\mathbf{J}] = P + \int(d\mathbf{r})\frac{(\delta\mathbf{J})^2}{\sigma} > P, \tag{32.32}$$

so the correct $\mathbf{J}$ makes P a minimum, again subject to given I_i.

32.2 Green's Functions

Suppose we now introduce a Green's function for the potential so that the steady current condition

$$\boldsymbol{\nabla}\cdot\mathbf{J} = \boldsymbol{\nabla}\cdot\sigma(-\boldsymbol{\nabla}\phi) = 0, \tag{32.33}$$

corresponds to the following Green's function equation

$$-\boldsymbol{\nabla}\cdot[\sigma(\mathbf{r})\boldsymbol{\nabla}G(\mathbf{r},\mathbf{r}')] = \delta(\mathbf{r}-\mathbf{r}'). \tag{32.34}$$

The boundary conditions on the Green's function are

$$\mathbf{n}\cdot\boldsymbol{\nabla}G = 0 \text{ on } S', \quad G = 0 \text{ on } S_i, \tag{32.35}$$

Dirichlet boundary conditions on S_i and Neumann ones on S'. Multiply the Green's function equation (32.34) by $\phi(\mathbf{r})$ and the potential equation (32.33) by $G(\mathbf{r},\mathbf{r}')$ and subtract:

$$-\phi(\mathbf{r})\boldsymbol{\nabla}\cdot[\sigma(\mathbf{r})\boldsymbol{\nabla}G(\mathbf{r},\mathbf{r}')] + G(\mathbf{r},\mathbf{r}')\boldsymbol{\nabla}\cdot[\sigma(\mathbf{r})\boldsymbol{\nabla}\phi(\mathbf{r})] = \delta(\mathbf{r}-\mathbf{r}')\phi(\mathbf{r}'). \tag{32.36}$$

The left-hand side of this equation is a total divergence:

$$\boldsymbol{\nabla}\cdot[G(\mathbf{r},\mathbf{r}')\sigma(\mathbf{r})\boldsymbol{\nabla}\phi(\mathbf{r}) - \phi(\mathbf{r})\sigma(\mathbf{r})\boldsymbol{\nabla}G(\mathbf{r},\mathbf{r}')], \tag{32.37}$$

so when we integrate over the volume we obtain

$$\begin{aligned}\phi(\mathbf{r}') &= \int_{S'+\sum_i S_i} dS\,\mathbf{n}\cdot[G(\mathbf{r},\mathbf{r}')\sigma(\mathbf{r})\boldsymbol{\nabla}\phi(\mathbf{r}) - \phi(\mathbf{r})\sigma(\mathbf{r})\boldsymbol{\nabla}G(\mathbf{r},\mathbf{r}')]\\ &= -\sum_i \phi_i\int_{S_i} dS\,\sigma(\mathbf{r})\mathbf{n}\cdot\boldsymbol{\nabla}G(\mathbf{r},\mathbf{r}'),\end{aligned} \tag{32.38}$$

because on S'

$$\mathbf{n}\cdot\boldsymbol{\nabla}\phi = 0, \quad \mathbf{n}\cdot\boldsymbol{\nabla}G(\mathbf{r},\mathbf{r}') = 0, \tag{32.39}$$

while on S_i

$$\phi = \phi_i, \quad G = 0. \tag{32.40}$$

Changing variables, and using the reciprocity relation [Problem 32.2]

$$G(\mathbf{r},\mathbf{r}') = G(\mathbf{r}',\mathbf{r}), \tag{32.41}$$

we obtain

$$\phi(\mathbf{r}) = \sum_j \phi_j \int_{S_j} dS'\,\sigma(\mathbf{r}')(-\mathbf{n}'\cdot\boldsymbol{\nabla}')G(\mathbf{r},\mathbf{r}'). \tag{32.42}$$

Now compute the current input by the ith electrode,

$$I_i = -\int_{S_i} dS\,\mathbf{n}\cdot\mathbf{J} = -\int_{S_i} dS\,\sigma\,\mathbf{n}\cdot(-\boldsymbol{\nabla}\phi). \tag{32.43}$$

Inserting (32.42) into this we obtain a linear relation between the current at the ith electrode and the potential on the jth:

$$I_i = \sum_j G_{ij}\phi_j \tag{32.44}$$

where the "coefficients of conductance" are given by

$$G_{ij} = -\int_{S_i} dS \int_{S_j} dS'\,\sigma(\mathbf{r})\mathbf{n}\cdot\boldsymbol{\nabla}\sigma(\mathbf{r}')\mathbf{n}'\cdot\boldsymbol{\nabla}'G(\mathbf{r},\mathbf{r}') = G_{ji}. \tag{32.45}$$

Then we can write the power as

$$P = \sum_i \phi_i I_i = \sum_{ij} \phi_i G_{ij}\phi_j. \tag{32.46}$$

From this we see

$$\frac{\partial P}{\partial \phi_i} = 2I_i, \tag{32.47}$$

which is consistent with (32.11). We already know from (32.20) that $\sum_i I_i = 0$, for arbitrary ϕ_i. Therefore,

$$\sum_i G_{ij} = 0. \tag{32.48}$$

A direct proof of this is as follows:

$$\begin{aligned}\sum_i G_{ij} &= \int_{S_j} dS'\,\sigma(\mathbf{r}')\mathbf{n}'\cdot\boldsymbol{\nabla}' \int_S dS\,\mathbf{n}\cdot\sigma(\mathbf{r})(-\boldsymbol{\nabla})G(\mathbf{r},\mathbf{r}')\\ &= \int_{S_j} dS'\,\sigma(\mathbf{r}')\mathbf{n}'\cdot\boldsymbol{\nabla}' \int (d\mathbf{r})\boldsymbol{\nabla}\cdot\left[\sigma(\mathbf{r})(-\boldsymbol{\nabla})G(\mathbf{r},\mathbf{r}')\right]\\ &= 0,\end{aligned} \tag{32.49}$$

since the last volume integral is 1.

Consider the case of two electrodes. Then

$$G_{11} + G_{21} = 0, \quad G_{12} + G_{22} = 0. \tag{32.50}$$

Thus there is a single conductance, which we can take to be

$$G = -G_{12} = -G_{21} = G_{11} = G_{22}. \tag{32.51}$$

The current in the two electrodes is

$$I_1 = -I_2 = G_{11}\phi_1 + G_{12}\phi_2 = G(\phi_1 - \phi_2) = GV = I. \tag{32.52}$$

The resistance is defined as the inverse of the conductance, $G = 1/R$. Then the power is

$$P = \sum_i I_i \phi_i = IV = GV^2 = RI^2. \tag{32.53}$$

As an example, consider a cylindrical conductor, of arbitrary cross section, with σ varying arbitrarily across the section. The Green's function equation (32.34) is

$$-\frac{\partial}{\partial z}\sigma\frac{\partial}{\partial z}G - \boldsymbol{\nabla}_\perp \cdot (\sigma \boldsymbol{\nabla}_\perp G) = \delta(z-z')\delta(\mathbf{r}_\perp - \mathbf{r}'_\perp). \tag{32.54}$$

Now when we integrate this over the cross section of the cylinder we encounter

$$\int (d\mathbf{r}_\perp)\boldsymbol{\nabla}_\perp \cdot (\sigma\boldsymbol{\nabla}_\perp G) = \oint_C ds\, \sigma\, \mathbf{n}\cdot\boldsymbol{\nabla}_\perp G = 0, \tag{32.55}$$

where the line integral is extended over the circumference C of the cylinder, on which (32.39) applies, and $\mathbf{n}$ is perpendicular to the walls of the cylinder. We are left with, under the assumption that σ does not depend on z,

$$-\frac{\partial^2}{\partial z^2}\int (d\mathbf{r}_\perp)\,\sigma G = \delta(z-z'). \tag{32.56}$$

The solution of this equation is (Problem 32.3)

$$\int (d\mathbf{r}_\perp)\,\sigma G(\mathbf{r},\mathbf{r}') = \frac{z_<(L-z_>)}{L}, \tag{32.57}$$

where L is the length of the cylinder. We have imposed the boundary condition (32.40) $G=0$ at $z=0, L$. Then the conductance is obtained from (32.45)

$$G = \frac{1}{R} = -\int_{z,z'=0}(d\mathbf{r}'_\perp)\,\sigma(\mathbf{r}'_\perp)\frac{\partial}{\partial z}\frac{\partial}{\partial z'}\left[-\frac{zz'}{L}\right] = \frac{1}{L}\int (d\mathbf{r}_\perp)\,\sigma(\mathbf{r}_\perp), \tag{32.58}$$

which is reminiscent of the corresponding formula for the capacitance (26.61), generalized to a spatially varying permittivity,

$$C = \frac{1}{4\pi a}\int (d\mathbf{r}_\perp)\,\epsilon, \tag{32.59}$$

a being the separation between the parallel plates of a capacitor, filled with a dielectric of permittivity ϵ.[2]

32.3 Problems for Chapter 32

1. Use Lagrange multipliers to show that $P[\mathbf{J}]$, (32.15), is stationary for infinitesimal variations.

[2]A few comments about units are in order. We have not included the usual 4π factor in the definition of the Green's function, because, in Gaussian units, the electric susceptibilty, in which the conductivity resides, differs from $(\epsilon - 1) = 4\pi\chi_e$ by precisely that factor. Corresponding, the definition of the coefficients of capacitance introduced in Chapter 26 differ from those for the coefficients of conductance by factors of 4π, and the conductance formula (32.58) also does not have such a factor. Finally, because we are contemplating steady, not alternating, currents here, factors of $1/2$, which arise from temporal averaging, do not appear in formulas such as (32.53).

2. Prove the reciprocity relation (32.41) directly from the differential equation (32.34) and the boundary conditions (32.35).
3. Directly solve (32.56) for (32.57).

33

Magnetic Charge II

In the previous chapters, we have considered the magnetic fields produced by steady currents with some attention to the attendant vector potential. As we have indicated at various points, an alternative source of a static magnetic field would be static magnetic charge, if such exists. We would here like to consider a few consequences for the vector potential corresponding to such a magnetic field.

Let a magnetic charge, g, be located at the origin so that the magnetic field satisfies

$$\boldsymbol{\nabla}\cdot\mathbf{B} = 4\pi g\delta(\mathbf{r}), \tag{33.1}$$

which has the solution

$$\mathbf{B} = g\frac{\mathbf{r}}{r^3} = -\boldsymbol{\nabla}\frac{g}{r}. \tag{33.2}$$

Away from the origin, $\mathbf{B}$ is divergenceless,

$$\boldsymbol{\nabla}\cdot\mathbf{B} = 0, \tag{33.3a}$$

so we would once again expect $\mathbf{B}$ to be the curl of a vector potential,

$$\mathbf{B} = \boldsymbol{\nabla}\times\mathbf{A}. \tag{33.3b}$$

However, this cannot be true everywhere since

$$\oint d\mathbf{S}\cdot\boldsymbol{\nabla}\times\mathbf{A} = \int (d\mathbf{r})\boldsymbol{\nabla}\cdot\boldsymbol{\nabla}\times\mathbf{A} = 0, \tag{33.4}$$

while (33.1) implies for a closed surface surrounding the magnetic charge

$$\oint d\mathbf{S}\cdot\mathbf{B} = \int (d\mathbf{r})\boldsymbol{\nabla}\cdot\mathbf{B} = 4\pi g. \tag{33.5}$$

We now want to find a vector potential that satisfies (33.3b) almost everywhere. The simplest possibility is that this equation fails to hold on a line, which we may take to be the $+z$ axis. We apply Stokes' theorem in the form

$$\oint_C \mathbf{A}\cdot d\mathbf{r} = -\int_S \mathbf{B}\cdot d\mathbf{S}, \tag{33.6}$$

where C is a circle of constant polar angle θ on a sphere of radius r about the origin and S is the lower portion of the spherical surface bounded by C. (See Fig. 33.1.) Equation (33.6) holds since (33.3b) is true everywhere on S. [The minus sign appears because we use the outward normal to the surface S.] The surface integral follows trivially from (33.2),

$$\int_S \mathbf{B}\cdot d\mathbf{S} = \frac{g}{r^2}2\pi r^2(1+\cos\theta). \tag{33.7}$$

An obvious solution of (33.6) is then

$$\mathbf{A} = A_\phi\hat{\boldsymbol{\phi}}, \tag{33.8a}$$

DOI: 10.1201/9781003057369-33

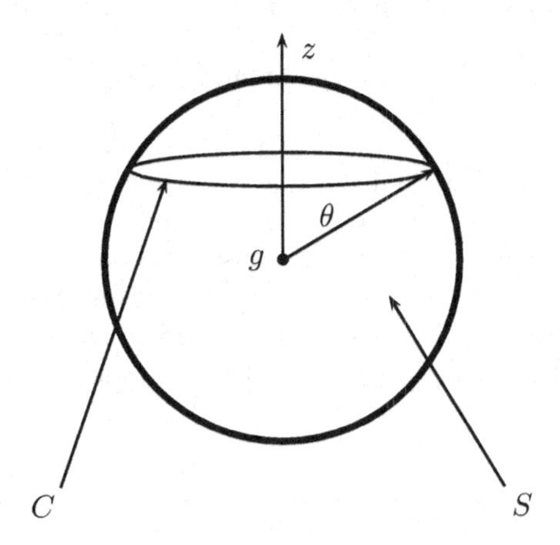

FIGURE 33.1
Sphere surrounding a point magnetic charge.

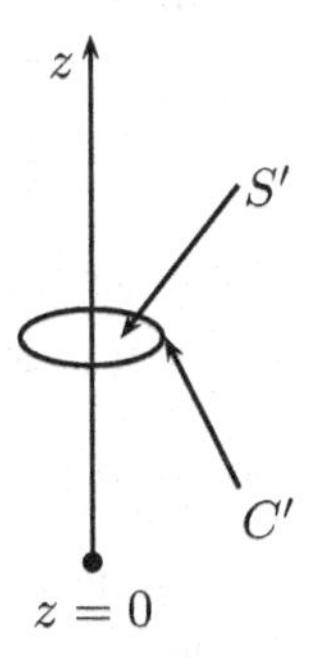

FIGURE 33.2
Circle surrounding string.

where

$$A_\phi = -\frac{g}{r}\frac{1+\cos\theta}{\sin\theta}. \tag{33.8b}$$

The structure of the singularity on the z axis is now isolated by taking the limit $\theta \to 0$,

$$\oint_{C'} d\mathbf{r} \cdot \mathbf{A} = \int_{S'} d\mathbf{S}' \cdot \boldsymbol{\nabla}\times\mathbf{A} = -4\pi g, \tag{33.9}$$

where C' is an infinitesimal circle about the z axis and S' is the enclosed area, as shown in Fig. 33.2. Since (33.9) shows that $(\boldsymbol{\nabla}\times\mathbf{A})_z$ has the singularity $-4\pi g\delta(x)\delta(y)$ on the $+z$ axis, we conclude that the magnetic field can be expressed everywhere by

$$\mathbf{B} = \boldsymbol{\nabla}\times\mathbf{A} + 4\pi g\delta(x)\delta(y)\eta(z)\hat{\mathbf{z}}, \tag{33.10}$$

where $\eta(z)$ is the (Heaviside) step function,

$$\eta(z) = \begin{cases} 1, z > 0, \\ 0, z < 0. \end{cases} \tag{33.11}$$

This result can be confirmed by noting that $\mathbf{B}$ has the correct divergence,

$$\boldsymbol{\nabla}\cdot\mathbf{B} = 0 + 4\pi g\delta(x)\delta(y)\delta(z), \tag{33.12}$$

or, alternatively, that (33.5) is consistent with (33.4). The vector potential (33.8) is an example of a class of potentials that yield the correct magnetic field except for a one-dimensional set of points, a curve. On this curve, called a string, $\mathbf{A}$ is singular, whereas the magnetic field is regular, being the curl of $\mathbf{A}$ plus a compensating singularity on the string. (In quantum field theory, it can be shown that reorientation of the string is a kind of singular gauge transformation and is only unphysical provided Dirac's quantization condition is satisfied, $eg = n\hbar c/2$, where n is an integer. Recall Problem 3.10. A review of this can be found in Ref. [27].)

33.1 Problems for Chapter 33

1. More generally, a string can point in a fixed direction $\hat{\mathbf{n}}$. Then

$$\mathbf{A} = -\frac{g}{r}\frac{\hat{\mathbf{n}}\times\mathbf{r}}{r - \hat{\mathbf{n}}\cdot\mathbf{r}}. \tag{33.13}$$

   Show that $\mathbf{B} = \boldsymbol{\nabla}\times\mathbf{A}$ almost everywhere, and give the generalization of (33.10).

2. Consider the vector potential given in spherical coordinates by

$$\mathbf{A}(r,\theta,\phi) = -\cos\theta\boldsymbol{\nabla}\phi. \tag{33.14}$$

   Using both geometrical and analytical arguments, determine the corresponding magnetic field. Compute the line integral

$$\oint_C d\mathbf{r}\cdot\mathbf{A}, \tag{33.15}$$

   where C is a circle about the z axis as shown in Fig. 33.3. What is the limit of this line integral as $\theta \to 0$, $\theta \to \pi$?

3. There have been many, so far unsuccessful, attempts to detect magnetic monopoles. The most unambiguous method of detection involves the generation

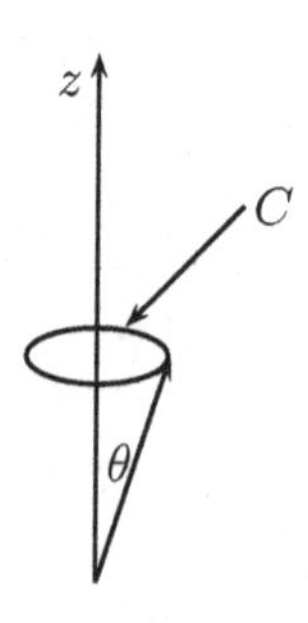

FIGURE 33.3
Contour for line integral in Problem 2.

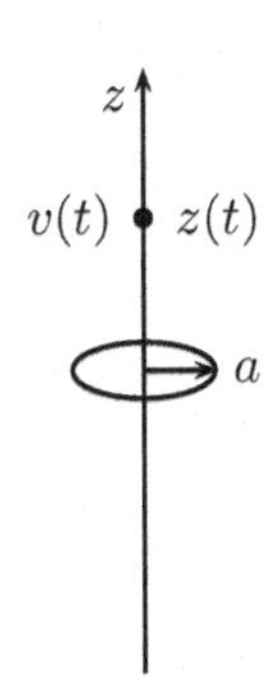

FIGURE 33.4
Schematic of a monopole detector

of a persistent current in a superconducting loop. A schematic of such a detector is given in Fig. 33.4. Imagine a particle carrying magnetic charge g travels along the z axis, and passes through a circular conducting loop of radius a in the x-y plane. At time t the particle is at height $z(t)$ above the loop, traveling with non-relativistic speed $v(t)$ along the $-z$ direction. Take the origin to lie at the center of the loop, and suppose the monopole crosses the plane of the loop at $t = 0$, $z(0) = 0$. Suppose in the distant past the monopole was infinitely far above the loop, $z(-\infty) = \infty$, and in the far future it will be infinitely far below the loop, $z(\infty) = -\infty$.

(a) Derive a formula giving the emf $\mathcal{E} = \oint_{\text{loop}} \mathbf{E} \cdot d\mathbf{l}$ induced in the loop at time t in terms of the magnetic flux subtended by the loop, $\Phi = \int_S d\mathbf{S} \cdot \mathbf{B}$ (where S is the surface bounded by the loop), the magnetic charge g, and its position and velocity.

(b) If the loop has negligible resistance (in practice it's a superconductor) but has inductance L, the current induced in the loop at time t may be determined by the formula

$$L\frac{d}{dt}I(t) = \mathcal{E}(t). \tag{33.16}$$

Compute the magnetic flux passing through the loop at time t, and derive the following formula for the current in terms is $z(t)$:

$$I(t) = \frac{2\pi g}{Lc}\left(1 - \frac{z(t)}{\sqrt{a^2 + z(t)^2}}\right). \tag{33.17}$$

In particular, derive $I(-\infty)$, $I(0)$, and $I(\infty)$. Derive the latter result directly from the formula found in part 3a without computing the magnetic flux.

(c) How is the above result changed if the particle contains electric charge as well as magnetic charge?

(d) A significant background is the small magnetic dipole moments embedded in the particle, even for one composed of nonmagnetic materials. Suppose instead of a monopole, the particle contains a magnetic dipole moment $\boldsymbol{\mu}$. Compute the current signal in that case from the field produced by a magnetic dipole,

$$\mathbf{B} = \frac{3(\boldsymbol{\mu} \cdot \mathbf{r})\mathbf{r} - r^2\boldsymbol{\mu}}{r^5}. \tag{33.18}$$

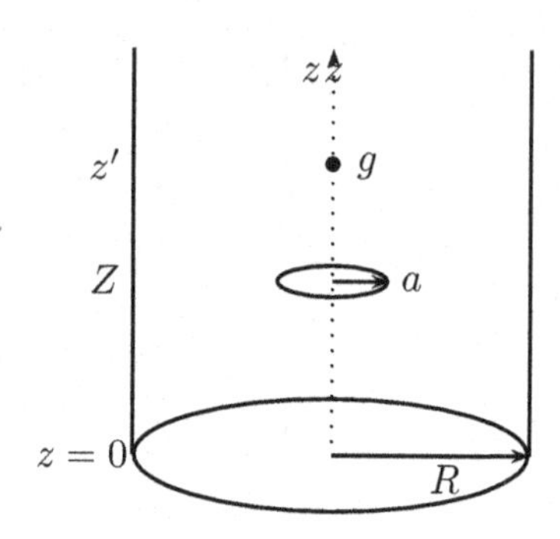

FIGURE 33.5
Diagram of monopole detector, showing superconducting loop enclosed in superconducting can.

4. More realistically, the detector used in the University of Oklahoma experiment to search for magnetic monopoles [28] enclosed the superconducting loop in a superconducting can, to keep out interference, as sketched in Fig. 33.5. The figure shows a perfectly conducting right circular cylinder of radius R of infinite length, with axis along the z axis, and with a perfectly conducting end cap at $z = 0$. Use cylindrical coordinates ρ, θ, and z. What are the boundary conditions for the magnetic field $\mathbf{B}$ on the boundaries? Suppose a magnetic pole of strength g is placed on the z axis at $z = z' > 0$. This could either be a magnetic monopole or one pole of a very long electromagnet ("pseudopole"). Consider a circular conducting loop of radius $a < R$ centered on the axis of the cylinder and perpendicular to that axis, with center at $z = Z$. Inside the cylinder and outside the loop show that $\mathbf{B}$ is derivable from a magnetic scalar potential, $\mathbf{B} = -\boldsymbol{\nabla}\phi_m$, provided that the time variation is negligible. Write the equation that ϕ satisfies in cylindrical coordinates. Show that the solution of this equation can be written the following separated form:

$$\phi_m(\rho, \theta, z) = 16\pi g \int_0^\infty \frac{dk}{2\pi} \cos kz \cos kz' \sum_{m=-\infty}^{\infty} \frac{1}{2\pi} e^{im(\theta-\theta')} G(\rho, \rho'), \tag{33.19}$$

where we've generalized the position of the monopole to be (z', ρ', θ'). Here, G is a one-dimensional Green's function satisfying

$$\left(\frac{1}{\rho}\frac{\partial}{\partial \rho}\rho\frac{\partial}{\partial \rho} - \frac{m^2}{\rho^2} - k^2\right) G(\rho, \rho') = -\frac{1}{\rho}\delta(\rho - \rho'). \tag{33.20}$$

Thus obtain the following formula for the magnetic flux subtended by the loop:

$$\Phi = \int_{\text{loop}} d\mathbf{S} \cdot \mathbf{B} = 4\pi g \Bigg\{ \eta(z - z') - \frac{2}{\pi}\frac{a}{R} \int_0^\infty dx \sin\frac{xZ}{R} \cos\frac{xz'}{R} \left[K_1(xa/R) - I_1(xa/R)\frac{K_1(x)}{I_1(x)} \right] \Bigg\}. \tag{33.21}$$

Now, suppose that the pole is *slowly* moved from $z' = +\infty$ to $z' = z_0$, $Z > z_0 > 0$. Show from Maxwell's equation (2.8) that the emf induced in the loop is

$$\mathcal{E} = \oint d\mathbf{l} \cdot \mathbf{E} = -\frac{1}{c}\frac{d\Phi}{dt} + \frac{4\pi}{c} g\delta(t), \tag{33.22}$$

where $t = 0$ is the time at which the pole passes through the plane of the loop. The net change in emf gives rise to a persistent current I in the superconducting loop,

$$LI = \int_{-\infty}^{\infty} dt\, \mathcal{E}, \tag{33.23}$$

where L is the inductance of the loop. Now suppose that the final position of the pole is near the bottom cap of the can, $z_0 \ll R$. Show that if the walls of the cylinder are very far from the loop, $Z/R \gg 1$, $a/R \ll 1$, the cylinder becomes irrelevant, and the result of Problem 3 is recovered,

$$\int_{-\infty}^{\infty} dt\, \mathcal{E} = \frac{4\pi g}{c}. \tag{33.24}$$

Show that if the condition a/R being small is relaxed, we obtain instead

$$\int_{-\infty}^{\infty} dt\, \mathcal{E} = \frac{4\pi g}{c}\left(1 - \frac{a^2}{R^2}\right), \tag{33.25}$$

so the signal is maximized by making the loop as small as possible, relative to the radius of the cylinder.

Part IV

Electromagnetic Radiation

34

Retarded Green's Function

34.1 Potentials and Gauges

In the previous chapters, we have primarily confined ourselves to the discussion of electrostatics and magnetostatics. We will now study in general how time-dependent electromagnetic fields are produced by arbitrary charges and currents. In vacuum, we recall that Maxwell's equations are [see (1.61)]

$$\boldsymbol{\nabla}\times\mathbf{B}=\frac{1}{c}\frac{\partial}{\partial t}\mathbf{E}+\frac{4\pi}{c}\mathbf{j}, \tag{34.1a}$$

$$\boldsymbol{\nabla}\cdot\mathbf{E}=4\pi\rho, \tag{34.1b}$$

$$-\boldsymbol{\nabla}\times\mathbf{E}=\frac{1}{c}\frac{\partial}{\partial t}\mathbf{B}, \tag{34.1c}$$

$$\boldsymbol{\nabla}\cdot\mathbf{B}=0, \tag{34.1d}$$

where ρ is the charge, and $\mathbf{j}$ is the current density, and we have assumed that no magnetic charge is present. Notice that the local conservation law

$$\boldsymbol{\nabla}\cdot\mathbf{j}+\frac{\partial}{\partial t}\rho=0, \tag{34.2}$$

is not an independent statement, but is derivable from (34.1a) and (34.1b).

To solve Maxwell's equations, we first recognize that the last two equations, (34.1c) and (34.1d), make no reference to charge or current, and they can be identically satisfied by introducing potentials through the definitions

$$\mathbf{B}=\boldsymbol{\nabla}\times\mathbf{A}, \tag{34.3a}$$

$$\mathbf{E}=-\frac{1}{c}\frac{\partial}{\partial t}\mathbf{A}-\boldsymbol{\nabla}\phi. \tag{34.3b}$$

As we have observed previously, in Section 10.5, the potentials $\mathbf{A}$ and ϕ are not uniquely defined. Since the magnetic field is the curl of $\mathbf{A}$, it is unchanged when a gradient is added to $\mathbf{A}$,

$$\mathbf{A}\to\mathbf{A}+\boldsymbol{\nabla}\lambda, \tag{34.4a}$$

where λ is an arbitrary function. In order that this new choice of vector potential not alter the electric field, (34.3b), it is necessary to simultaneously replace the scalar potential by

$$\phi\to\phi-\frac{1}{c}\frac{\partial}{\partial t}\lambda. \tag{34.4b}$$

This new set of potentials, (34.4a) and (34.4b), is as acceptable as the original one since only the fields $\mathbf{B}$ and $\mathbf{E}$ are physically measurable quantities.[1] This arbitrariness in the choice of

[1]The reader might object that in the Aharonov-Bohm effect (Yakir Aharonov, 1932–, and David Bohm, 1917–1992) the line integral $\oint d\mathbf{l}\cdot\mathbf{A}$ is measured. However, this line integral is equal to the magnetic flux through the closed loop. This effect was first predicted ten years before Aharonov and Bohm, in 1949, by Werner Ehrenberg and Raymond Siday.

DOI: 10.1201/9781003057369-34

potentials is called the gauge freedom of the theory, while the corresponding transformations are called gauge transformations. In the following, we will exploit this freedom in the process of solving the differential equations for the potentials.

Upon substituting the constructions of **B** and **E** in terms of potentials, (34.3a) and (34.3b), into the first set of Maxwell's equations, we find, from (34.1a),

$$\boldsymbol{\nabla}\times(\boldsymbol{\nabla}\times\mathbf{A}) = \frac{1}{c}\frac{\partial}{\partial t}\left(-\boldsymbol{\nabla}\phi - \frac{1}{c}\frac{\partial}{\partial t}\mathbf{A}\right) + \frac{4\pi}{c}\mathbf{j}, \tag{34.5}$$

or

$$-\left(\nabla^2 - \frac{1}{c^2}\frac{\partial^2}{\partial t^2}\right)\mathbf{A} = -\boldsymbol{\nabla}\left(\boldsymbol{\nabla}\cdot\mathbf{A} + \frac{1}{c}\frac{\partial}{\partial t}\phi\right) + \frac{4\pi}{c}\mathbf{j}, \tag{34.6a}$$

and, from (34.1b),

$$-\nabla^2\phi - \frac{1}{c}\frac{\partial}{\partial t}(\boldsymbol{\nabla}\cdot\mathbf{A}) = 4\pi\rho. \tag{34.6b}$$

This is a pair of coupled second-order differential equations for **A** and ϕ, which may be simplified by utilizing the gauge freedom in defining the potentials. The two most convenient and common choices of gauge are discussed below.

1. The radiation gauge (or Coulomb gauge) is defined by the condition

$$\boldsymbol{\nabla}\cdot\mathbf{A} = 0. \tag{34.7}$$

That we can always make this choice was shown in Section 28.3. In this gauge, (34.6a) and (34.6b) reduce to

$$-\nabla^2\phi = 4\pi\rho, \tag{34.8a}$$

$$-\Box^2\mathbf{A} = \frac{4\pi}{c}\mathbf{j} - \frac{1}{c}\frac{\partial}{\partial t}(\boldsymbol{\nabla}\phi), \tag{34.8b}$$

where

$$\Box^2 = \nabla^2 - \frac{1}{c^2}\frac{\partial^2}{\partial t^2} \tag{34.9}$$

is the d'Alembertian. (Jean d'Alembert's *Traité de Dynamique* was published in 1758.) The equation for ϕ, (34.8a), is just the same as that in electrostatics (hence the origin of the term "Coulomb gauge") so that ϕ is, in principle, known. The structure on the right-hand side of (34.8b) is proportional to an effective current, the second term of which is present in order that it be divergenceless:

$$\boldsymbol{\nabla}\cdot\left[\mathbf{j} - \boldsymbol{\nabla}\left(\frac{1}{4\pi}\frac{\partial}{\partial t}\phi\right)\right] = \boldsymbol{\nabla}\cdot\mathbf{j} - \frac{1}{4\pi}\frac{\partial}{\partial t}(\nabla^2\phi) = \boldsymbol{\nabla}\cdot\mathbf{j} + \frac{\partial}{\partial t}\rho = 0, \tag{34.10}$$

where the last equality follows from charge conservation, (34.2). This relation also entails the consistency of the choice of the radiation gauge in that if we set $\boldsymbol{\nabla}\cdot\mathbf{A}$ equal to zero at one time, it remains zero for all time, since

$$-\Box^2(\boldsymbol{\nabla}\cdot\mathbf{A}) = \frac{4\pi}{c}\boldsymbol{\nabla}\cdot\left[\mathbf{j} - \boldsymbol{\nabla}\left(\frac{1}{4\pi}\frac{\partial}{\partial t}\phi\right)\right] = 0. \tag{34.11}$$

2. The Lorenz gauge[2] condition is a relation between vector and scalar potentials,

$$\boldsymbol{\nabla}\cdot\mathbf{A} + \frac{1}{c}\frac{\partial}{\partial t}\phi = 0. \tag{34.12}$$

[2]This gauge condition, due to Ludvig Lorenz (1867), is often mistakenly ascribed to Henrik Lorentz, including by the present authors in the first edition of this book. The confusion is understandable due to the Lorentz-invariance of the condition.

In this gauge, the equations for **A** and ϕ have the symmetrical form,

$$-\Box^2 \mathbf{A} = \frac{4\pi}{c}\mathbf{j}, \tag{34.13a}$$

$$-\Box^2 \phi = 4\pi\rho. \tag{34.13b}$$

The consistency of this gauge choice again follows from the fact that charge is conserved,

$$-\Box^2 \left(\boldsymbol{\nabla}\cdot\mathbf{A} + \frac{1}{c}\frac{\partial}{\partial t}\phi\right) = \frac{4\pi}{c}\left(\boldsymbol{\nabla}\cdot\mathbf{j} + \frac{\partial\rho}{\partial t}\right) = 0. \tag{34.14}$$

34.2 Green's Function in the Lorenz Gauge

In the following, we will solve the differential equations, (34.13a) and (34.13b), for the potentials in the Lorenz gauge. Since the potentials are linearly related to their sources, they may be expressed in terms of a Green's function,

$$\phi(\mathbf{r},t) = \int (d\mathbf{r}')\, dt'\, G(\mathbf{r}-\mathbf{r}', t-t')\rho(\mathbf{r}',t'), \tag{34.15a}$$

$$\mathbf{A}(\mathbf{r},t) = \int (d\mathbf{r}')\, dt'\, G(\mathbf{r}-\mathbf{r}', t-t')\frac{1}{c}\mathbf{j}(\mathbf{r}',t'). \tag{34.15b}$$

This Green's function, $G(\mathbf{r}-\mathbf{r}', t-t')$, is a function only of relative positions and times because of translational invariance in unbounded space. Since ϕ satisfies (34.13b), this Green's function obeys the differential equation

$$-\Box^2 G(\mathbf{r}-\mathbf{r}', t-t') = 4\pi\delta(\mathbf{r}-\mathbf{r}')\delta(t-t'), \tag{34.16}$$

which is a four-dimensional generalization of the three-dimensional Green's function equation we studied in electrostatics,

$$-\nabla^2 G(\mathbf{r}-\mathbf{r}') = 4\pi\delta(\mathbf{r}-\mathbf{r}'). \tag{34.17}$$

To solve (34.16), we will analyze its time dependence by making use of the exponential representations (Fourier transforms in time)

$$\delta(t-t') = \int_{-\infty}^{\infty} \frac{d\omega}{2\pi} e^{-i\omega(t-t')}, \tag{34.18a}$$

$$G(\mathbf{r}-\mathbf{r}', t-t') = \int_{-\infty}^{\infty} \frac{d\omega}{2\pi} e^{-i\omega(t-t')} G_\omega(\mathbf{r}-\mathbf{r}'), \tag{34.18b}$$

where G_ω satisfies the three-dimensional differential equation, the inhomogeneous Helmholtz equation

$$-\left(\nabla^2 + \frac{\omega^2}{c^2}\right) G_\omega(\mathbf{r}-\mathbf{r}') = 4\pi\delta(\mathbf{r}-\mathbf{r}'). \tag{34.19}$$

In the static limit, $\omega \to 0$, (34.19) reduces to (34.17), the solution of which is Coulomb's potential, (15.3):

$$G_{\omega=0}(\mathbf{r}-\mathbf{r}') = \frac{1}{|\mathbf{r}-\mathbf{r}'|}. \tag{34.20}$$

Since G_ω depends only on $\mathbf{r}-\mathbf{r}'$, we may set $\mathbf{r}'=\mathbf{0}$, without loss of generality in the following discussion. Also, since we are now looking for a spherically symmetrical solution for G_ω, it is natural to use a spherical coordinate system in which the Laplacian here reduces to

$$\nabla^2 \to \frac{1}{r^2}\frac{d}{dr}\left(r^2\frac{d}{dr}\right). \tag{34.21}$$

Therefore, for $r>0$, we wish to solve the homogeneous equation

$$\left[\frac{1}{r^2}\frac{d}{dr}\left(r^2\frac{d}{dr}\right)+\frac{\omega^2}{c^2}\right]G_\omega(\mathbf{r})=0, \tag{34.22}$$

subject to the boundary condition that there is a point charge at the origin. The consequence of this requirement is most conveniently extracted by integrating (34.19) over a sphere S of vanishing radius r_0 about the origin,

$$4\pi=-\int(d\mathbf{r})\boldsymbol{\nabla}\cdot(\boldsymbol{\nabla}G_\omega)=-\oint_S dS\,\nabla_r G_\omega=-4\pi r^2\frac{d}{dr}G_\omega\bigg|_{r_0\to 0} \tag{34.23a}$$

or

$$-r^2\frac{d}{dr}G_\omega(\mathbf{r})\bigg|_{r_0\to 0}=1. \tag{34.23b}$$

[We have noted that the ω^2/c^2 term in the differential equation does not contribute to the integral since

$$\frac{\omega^2}{c^2}G_\omega\sim\frac{1}{r},\quad \text{as}\quad r\to 0, \tag{34.24}$$

which has vanishing volume integral as $r_0\to 0$.] To solve (34.22), we introduce g_ω, defined by

$$G_\omega=\frac{1}{r}g_\omega, \tag{34.25}$$

which satisfies the differential equation

$$\left(\frac{d^2}{dr^2}+\frac{\omega^2}{c^2}\right)g_\omega(r)=0,\quad \text{for}\quad r>0, \tag{34.26}$$

where we have used

$$r^2\frac{d}{dr}G_\omega=r\frac{d}{dr}g_\omega-g_\omega, \tag{34.27a}$$

$$\frac{1}{r^2}\frac{d}{dr}\left(r^2\frac{d}{dr}G_\omega\right)=\frac{1}{r}\frac{d^2}{dr^2}g_\omega. \tag{34.27b}$$

The independent solutions of (34.26) have the form

$$g_\omega\sim e^{\pm i\omega r/c}, \tag{34.28a}$$

and the corresponding forms for G_ω are

$$G_\omega(r)=\frac{C}{r}e^{\pm i\omega r/c},\quad \text{for}\quad r>0. \tag{34.28b}$$

For either choice of + or − sign, the constant C is determined by the boundary condition (34.23b) to be

$$C=1. \tag{34.29}$$

Therefore, we have two fundamental solutions to (34.19),

$$G_\omega(\mathbf{r}-\mathbf{r}') = \frac{1}{|\mathbf{r}-\mathbf{r}'|}e^{\pm i\omega|\mathbf{r}-\mathbf{r}'|/c}, \tag{34.30}$$

while from (34.18b), we now obtain the space-time form of the Green's functions,

$$\begin{aligned} G(\mathbf{r}-\mathbf{r}', t-t') &= \int_{-\infty}^{\infty} \frac{d\omega}{2\pi}\frac{1}{|\mathbf{r}-\mathbf{r}'|}e^{i\omega[\pm|\mathbf{r}-\mathbf{r}'|/c-(t-t')]} \\ &= \frac{1}{|\mathbf{r}-\mathbf{r}'|}\delta\left(\pm\frac{1}{c}|\mathbf{r}-\mathbf{r}'|-(t-t')\right). \end{aligned} \tag{34.31}$$

What is implied by the use of the + or − sign in (34.31)? The choice of the + sign leads to the retarded Green's function,

$$G_{\text{ret}}(\mathbf{r}-\mathbf{r}', t-t') = \frac{1}{|\mathbf{r}-\mathbf{r}'|}\delta\left(\frac{1}{c}|\mathbf{r}-\mathbf{r}'|-(t-t')\right), \tag{34.32}$$

which is nonvanishing when

$$t = t' + \frac{1}{c}|\mathbf{r}-\mathbf{r}'|. \tag{34.33}$$

This means that the signal propagates with the speed of light c from the source (at time t') to the observer (at time t); the effect occurs later than the cause. If we pick the − sign, we obtain the advanced Green's function

$$G_{\text{adv}}(\mathbf{r}-\mathbf{r}', t-t') = \frac{1}{|\mathbf{r}-\mathbf{r}'|}\delta\left(\frac{1}{c}|\mathbf{r}-\mathbf{r}'|+(t-t')\right), \tag{34.34}$$

which is nonzero when

$$t = t' - \frac{1}{c}|\mathbf{r}-\mathbf{r}'|, \tag{34.35}$$

the signal arriving at the observer before it is emitted by the source. Since the latter is not in accordance with elementary ideas of causality, we adopt the retarded Green's function as the solution which satisfies the correct time boundary condition. (Actually, both retarded and advanced Green's functions are useful in physics.[3] We can now obtain explicit expressions for the potentials by substituting (34.32) into (34.15a) and (34.15b),

$$\phi(\mathbf{r}, t) = \int (d\mathbf{r}')\, dt'\, \frac{\delta\left(\frac{1}{c}|\mathbf{r}-\mathbf{r}'|-(t-t')\right)}{|\mathbf{r}-\mathbf{r}'|}\rho(\mathbf{r}', t'), \tag{34.36a}$$

$$\mathbf{A}(\mathbf{r}, t) = \int (d\mathbf{r}')\, dt'\, \frac{\delta\left(\frac{1}{c}|\mathbf{r}-\mathbf{r}'|-(t-t')\right)}{|\mathbf{r}-\mathbf{r}'|}\frac{1}{c}\mathbf{j}(\mathbf{r}', t'). \tag{34.36b}$$

Integrating over t', we obtain the so-called retarded or Liénard-Wiechert[4] potentials (published in 1898 and 1900, respectively),

$$\phi(\mathbf{r}, t) = \int (d\mathbf{r}')\frac{1}{|\mathbf{r}-\mathbf{r}'|}\rho\left(\mathbf{r}', t - \frac{1}{c}|\mathbf{r}-\mathbf{r}'|\right), \tag{34.37a}$$

$$\mathbf{A}(\mathbf{r}, t) = \int (d\mathbf{r}')\frac{1}{|\mathbf{r}-\mathbf{r}'|}\frac{1}{c}\mathbf{j}\left(\mathbf{r}', t - \frac{1}{c}|\mathbf{r}-\mathbf{r}'|\right). \tag{34.37b}$$

These results are elementary generalizations of the potentials for electrostatics and magnetostatics, but now reflecting the finite propagation speed of light.

---

[3]In fact, this observation led Richard Feynman (1918–1988) to the discovery of his celebrated propagator.
[4]Alfred-Marie Liénard (1869–1958) and Emil Johann Wiechert (1861–1928).

34.3 Problems for Chapter 34

1. A particle with charge e moves along the z axis with constant speed v. Its coordinates are

$$x(t) = 0, \quad y(t) = 0, \quad z(t) = vt. \tag{34.38}$$

Construct the potentials, ϕ and $\mathbf{A}$, in the Lorenz gauge by solving the differential equations (34.13a) and (34.13b), noting that the only variables are x, y, and $z - vt$. The result is

$$\phi = \frac{e}{\sqrt{(z-vt)^2 + \left(1 - \frac{v^2}{c^2}\right)(x^2+y^2)}}, \qquad \mathbf{A} = \frac{\mathbf{v}}{c}\phi. \tag{34.39}$$

(A particle in uniform motion is not very different from a particle at rest.)

2. What are the electric and magnetic fields implied by Problem 1?

3. Obtain the results of the preceding two problems by performing a Lorentz transformation of the Coulomb potential from a frame in which a charge e is at rest, to one in which it is moving with velocity $\mathbf{v}$. Use the transformation of potentials, (11.78), and of the fields, (11.80), (11.81).

4. From Maxwell's equations, without introducing potentials, show that the electric and magnetic fields satisfy the inhomogeneous wave equations

$$-\Box^2\mathbf{E} = 4\pi\left(-\boldsymbol{\nabla}\rho - \frac{1}{c^2}\frac{\partial}{\partial t}\mathbf{j}\right), \tag{34.40a}$$

$$-\Box^2\mathbf{B} = \frac{4\pi}{c}\boldsymbol{\nabla}\times\mathbf{j}. \tag{34.40b}$$

Extend this result to magnetic charges and currents.

5. The charge and current densities of a point charge are given by

$$\begin{Bmatrix}\rho\\ \mathbf{j}\end{Bmatrix}(\mathbf{r},t) = e\begin{Bmatrix}1\\ \mathbf{v}(t)\end{Bmatrix}\delta(\mathbf{r}-\mathbf{r}(t)). \tag{34.41}$$

From the expressions (34.36a) and (34.36b) for the retarded potentials, derive

$$\phi(\mathbf{r},t) = \frac{e}{|\mathbf{r}-\mathbf{r}(t')| - [\mathbf{r}-\mathbf{r}(t')]\cdot\frac{\mathbf{v}(t')}{c}}, \tag{34.42a}$$

$$\mathbf{A}(\mathbf{r},t) = \frac{\mathbf{v}(t')}{c}\phi(\mathbf{r},t), \tag{34.42b}$$

where the retarded time satisfies

$$t' = t - \frac{|\mathbf{r}-\mathbf{r}(t')|}{c}.$$

These are often called the Liénard-Wiechert potentials.

6. From Problem 5 compute $\mathbf{E}$ and $\mathbf{B}$. Express the answers as the sum of a "velocity" part (involving $\mathbf{v}$ only, and asymptotic to $1/r^2$, and an "acceleration" part (proportional to $\dot{\mathbf{v}}$ and asymptotic to $1/r$). Only the latter is significant for radiation.

7. Show that in the four-vector notation of Chapter 12, the construction (34.15) can be written more compactly as

$$A^\mu(x) = \int (dx')G(x-x')\frac{1}{c}j^\mu(x'), \tag{34.43}$$

where x and x' represent space-time points, $x^\mu = (ct, x^i)$. In this same notation, the Lorenz condition is simply

$$\partial_\mu A^\mu = 0, \tag{34.44}$$

so that when this is satisfied, the four-vector potential satisfies

$$-\partial^2 A^\mu = \frac{4\pi}{c}j^\mu. \tag{34.45}$$

The scalar Green's function satisfies

$$-\partial^2 G(x-x') = 4\pi\delta(x-x'), \quad \partial^2 = \partial_\lambda\partial^\lambda = \Box^2. \tag{34.46}$$

8. Show that the four-potential produced by a charged particle with four-velocity v^μ is

$$A^\mu(x) = \int ds'\eta(x^0 - x^{0\prime}(s'))2\delta[(x-x'(s'))^2]ev^\mu(s') = -\frac{ev^\mu(s')}{(x-x'(s'))v(s')}. \tag{34.47}$$

Here η is the unit step function. Make explicit what is left implicit in this result. Write this result in $3+1$ dimensional notation and compare with the Liénard-Wiechert potentials (34.42).

9. Solve (34.16) for G by writing the space-time Green's function as a Fourier transform

$$G(\mathbf{r}-\mathbf{r}', t-t') = \int \frac{(d\mathbf{k})}{(2\pi)^3}\frac{d\omega}{2\pi}e^{i\mathbf{k}\cdot(\mathbf{r}-\mathbf{r}')}e^{-i\omega(t-t')}g(\mathbf{k},\omega). \tag{34.48}$$

From the differential equation (34.16) determine $g(\mathbf{k},\omega)$. The Fourier transform g possesses singularities when $\omega = \pm|\mathbf{k}|c$, so it must be defined there. Define g at the singularities so that

$$\int_{-\infty}^{\infty}\frac{d\omega}{2\pi}e^{-i\omega(t-t')}g(\mathbf{k},\omega), \tag{34.49}$$

and hence G, vanish when $t < t'$ so that this gives the retarded Green's function. Use Cauchy's theorem to evaluate this last integral for $t > t'$. You should obtain

$$\frac{4\pi c}{k}\sin kc(t-t'), \quad t > t'. \tag{34.50}$$

Finally, carry out the integration of $\mathbf{k}$ in (34.48), first doing the integral over the angle θ between $\mathbf{k}$ and $\mathbf{R} = \mathbf{r}-\mathbf{r}'$. [Take $(d\mathbf{k}) = k^2dk\,2\pi\,d(\cos\theta)$.] You should obtain (34.32) for the retarded Green's function.

10. A point charge in n Euclidean dimensions corresponds to Green's function $G^{(n)}$, which satisfies the differential equation (note here we have removed the ubiquitous 4π from the differential equation)

$$-\left(\sum_{k=1}^{n}\frac{\partial^2}{\partial x_k^2}\right)G^{(n)}(x_1,\dots,x_n) = \delta(x_1)\cdots\delta(x_n). \tag{34.51}$$

The solution to this equation can be written as the Fourier transform

$$G^{(n)}(x_1,\dots,x_n) = \int \prod_{m=1}^{n} \frac{dk_m}{2\pi} \exp\left(i\sum_{m=1}^{n} k_m x_m\right) \frac{1}{\sum_{m=1}^{n} k_m^2}. \tag{34.52}$$

Evaluate this, by making use of the exponential representation

$$\frac{1}{\sum k_m^2} = \int_0^\infty ds\, e^{-s\sum k_m^2}. \tag{34.53}$$

Express the answer as a function of $R^2 = \sum_{m=1}^{n} x_m^2$. Verify, as special cases,

$$G^{(3)} = \frac{1}{4\pi R}, \quad G^{(4)} = \frac{1}{4\pi^2 R^2}. \tag{34.54}$$

11. If we integrate the above Green's function over one coordinate, we obtain Green's function in one lower dimension,

$$\int_{-\infty}^{\infty} dx_n\, G^{(n)}(x_1,\dots,x_n) = G^{(n-1)}(x_1,\dots,x_{n-1}). \tag{34.55}$$

Check this explicitly for $n = 4$, and then in general, from the explicit answer in the previous problem.

12. Solve for the retarded Green's function for the two-dimensional wave equation, which satisfies

$$\left(-\nabla^2 + \frac{1}{c^2}\frac{\partial^2}{\partial t^2}\right) G(\mathbf{r}-\mathbf{r}', t-t') = 4\pi\delta(\mathbf{r}-\mathbf{r}')\delta(t-t'), \quad \nabla^2 = \frac{\partial^2}{\partial x^2} + \frac{\partial^2}{\partial y^2}, \tag{34.56}$$

in the following three ways.

(a) The retarded Green's function vanishes unless the causality condition is satisfied,

$$c(t-t') > |\mathbf{r}-\mathbf{r}'|. \tag{34.57}$$

When this condition is satisfied, the Green's function must be a function of $c^2(t-t')^2 - (\mathbf{r}-\mathbf{r}')^2$ by virtue of Lorentz invariance. Thus, the solution must be of the form

$$G(\mathbf{r}-\mathbf{r}', t-t') = f\big(c^2(t-t')^2 - (\mathbf{r}-\mathbf{r}')^2\big)\, \eta\left(t-t' - \frac{|\mathbf{r}-\mathbf{r}'|}{c}\right), \tag{34.58a}$$

By dimensional considerations, show that

$$G(\mathbf{r}-\mathbf{r}', t-t') = A\frac{1}{\sqrt{c^2(t-t')^2 - (\mathbf{r}-\mathbf{r}')^2}}\eta\left(t-t' - \frac{|\mathbf{r}-\mathbf{r}'|}{c}\right), \tag{34.58b}$$

and from the differential equation, determine the constant A. [Hint: you may integrate the differential equation for Green's function over a small circle about the origin.]

(b) Recall the form (34.32) for the 3 + 1-dimensional retarded Green's function. The two-dimensional Green's function can be obtained from this by integrating over z, according to (34.55). [Why does the fact the latter is expressed in Euclidean space not matter?] Thereby, obtain the same result as in Problem 12a.

(c) Introduce the time Fourier transform,

$$G_\omega(\mathbf{r}-\mathbf{r}') = \int_{-\infty}^{\infty} dt\, e^{i\omega(t-t')} G(\mathbf{r}-\mathbf{r}', t-t'). \tag{34.59}$$

Show that, by rotational invariance, G_ω satisfies ($k = |\omega|/c$)

$$\left(\frac{1}{r}\frac{d}{dr} r \frac{d}{dr} + k^2\right) G_\omega(\mathbf{r}-\mathbf{r}') = 0 \quad \text{for} \quad \mathbf{r} \neq \mathbf{r}', \tag{34.60a}$$

and

$$r\frac{d}{dr} G_\omega(\mathbf{r}-\mathbf{r}')\bigg|_{R=|\mathbf{r}-\mathbf{r}'|=-0}^{R=|\mathbf{r}-\mathbf{r}'|=+0} = -2. \tag{34.60b}$$

The solution to this differential equation, with appropriate boundary conditions at infinity, is a Bessel function of the first kind, the so-called Hankel function,

$$H_0^{(1)}(k|\mathbf{r}-\mathbf{r}'|) = J_0(k|\mathbf{r}-\mathbf{r}'|) + iN_0(k|\mathbf{r}-\mathbf{r}'|), \tag{34.61}$$

This is because $H_0^{(1)}$ represents an outgoing cylindrical wave,

$$H_0^{(1)}(x) \sim \sqrt{\frac{2}{\pi x}} e^{i(x-\pi/4)}, \quad x \to +\infty. \tag{34.62}$$

Determine the numerical coefficient B required to obtain Green's function:

$$G_\omega(\mathbf{r}-\mathbf{r}') = BH_0^{(1)}(k|\mathbf{r}-\mathbf{r}'|), \tag{34.63}$$

using the behavior for small argument,

$$J_0(x) \sim 1, \quad N_0(x) \sim \frac{2}{\pi} \ln x, \quad x \to +0. \tag{34.64}$$

Now undo the Fourier transform, using ($b, a > 0$)

$$\int_0^\infty dx\, J_0(ax) \sin bx = \frac{1}{\sqrt{b^2 - a^2}} \eta(b-a), \tag{34.65a}$$

$$\int_0^\infty dx\, N_0(ax) \cos bx = -\frac{1}{\sqrt{b^2 - a^2}} \eta(b-a), \tag{34.65b}$$

and thereby find $G(\mathbf{r}-\mathbf{r}', t-t')$.

(d) Comment on the form of Green's function. Does the signal travel faster than the speed of light? Do all portions travel at speed c? Is there a wake?

13. The 4-dimensional Euclidean Green's function in Problem 10 satisfies

$$-\sum_{k=1}^{4} \left(\frac{\partial}{\partial x_k}\right)^2 G = \delta(x_1)\cdots\delta(x_4), \tag{34.66}$$

and is explicitly

$$G = \frac{1}{4\pi^2} \frac{1}{\sum_{k=1}^4 x_k^4}. \tag{34.67}$$

By making the complex replacement

$$x_4 \to ict \equiv \lim_{\epsilon \to +0} e^{i(\pi/2-\epsilon)} ct, \tag{34.68}$$

show that $D_+ = iG$ satisfies the differential equation

$$-\left(\nabla^2 - \frac{1}{c^2}\frac{\partial^2}{\partial t^2}\right) D_+(\mathbf{r}, t) = \delta(\mathbf{r})\delta(ct), \tag{34.69}$$

where $\mathbf{r} = (x_1, x_2, x_3)$. Starting from the solution in Problem 10, show that

(a)

$$D_+ = \frac{i}{4\pi^2}\frac{1}{r^2 - (ct)^2 + i\epsilon}\bigg|_{\epsilon\to+0}, \tag{34.70a}$$

and

(b)

$$\operatorname{Re} D_+ = \frac{1}{4\pi}\delta\left(r^2 - (ct)^2\right) = \frac{1}{2}\left(\frac{\delta(r - ct)}{4\pi r} + \frac{\delta(r + ct)}{4\pi r}\right), \tag{34.70b}$$

where the two terms here are the retarded and advanced Green's functions, respectively, apart from an overall factor of $1/4\pi c$.[5]

(c) From the Fourier representation

$$G = \int \frac{(d\mathbf{k})dk_4}{(2\pi)^4}\frac{e^{i(\mathbf{k}\cdot\mathbf{r} + k_4 x_4)}}{\mathbf{k}^2 + k_4^2}, \tag{34.71}$$

make the replacements (complex rotations)

$$x_4 \to e^{i(\pi/2 - \epsilon)}ct, \quad k_4 \to e^{-i(\pi/2-\epsilon)}\left(-\frac{\omega}{c}\right) \tag{34.72}$$

to obtain

$$D_+ = \int \frac{(d\mathbf{k})(d\omega/c)}{(2\pi)^4}\frac{e^{i(\mathbf{k}\cdot\mathbf{r} - \omega t)}}{k^2 - \frac{\omega^2}{c^2} - i\epsilon}\bigg|_{\epsilon\to+0}. \tag{34.73}$$

(d) Show that

$$\int_{-\infty}^{\infty} \frac{d\omega}{c}\frac{1}{2\pi}e^{-i\omega t}\frac{1}{\mathbf{k}^2 - \frac{\omega^2}{c^2} - i\epsilon} = i\frac{e^{-i|\mathbf{k}|c|t|}}{2|\mathbf{k}|}, \tag{34.74}$$

and consequently

$$D_+(\mathbf{r}, t) = i\int \frac{(d\mathbf{k})}{(2\pi)^3}\frac{1}{2|\mathbf{k}|}e^{i(\mathbf{k}\cdot\mathbf{r} - c|\mathbf{k}||t|)}. \tag{34.75}$$

(e) Show that D_+ given in part (d) satisfies

$$\frac{1}{c}\frac{\partial}{\partial t}D_+(\mathbf{r}, t)\Big|_{t=-0}^{t=+0} = \delta(\mathbf{r}) \tag{34.76a}$$

and

$$\left(\nabla^2 - \frac{1}{c^2}\frac{\partial^2}{\partial t^2}\right) D_+(\mathbf{r}, t) = 0 \tag{34.76b}$$

for $t \neq t'$. Consequently, D_+ obeys the correct differential equation.

[5] D_+ is usually called the Feynman propagator.

(f) By integrating over $\mathbf{k}$, obtain the alternative representation

$$D_+(\mathbf{r}, t) = \frac{1}{c}\int_{-\infty}^{\infty} \frac{d\omega}{2\pi} e^{-i\omega t} \frac{e^{i|\omega| r/c}}{4\pi r}. \tag{34.77}$$

Note that the Fourier transform of D_+ is a function of $|\omega|$, while the retarded and advanced Green's functions are respectively functions of ω and $-\omega$. The replacement $|\omega| \to -|\omega|$ leads to a function called D_-. The factor $\exp(i|\omega|r/c)$ characterizes an outgoing wave.

(g) From (f), directly derive the result of part (b).

35

Radiation—Field Point of View

35.1 Asymptotic Potentials and Fields

As we have seen, the distinction between static electric and magnetic fields and those produced by time-varying charges and currents is that, in the latter case, we must take into account the finite propagation speed of light. The fact that the time of emission is different from the time of detection is the basis for the existence of electromagnetic radiation, as we will now see. Sufficiently near the source, retardation effects can be neglected. That is, if ρ and $\mathbf{j}$ do not change appreciably over a time scale of $|\mathbf{r}-\mathbf{r}'|/c$, the time of emission, t', can be effectively replaced by the time of detection, t, in (34.37a) and (34.37b), which is to say that the potentials are not very different from those occurring in statics. On the contrary, far away from the source, retardation effects become important. Choosing the origin of coordinates to lie inside the charge distribution, having characteristic length a, we may use the expansion, for $r \gg a$,

$$|\mathbf{r}-\mathbf{r}'| = \sqrt{r^2 - 2\mathbf{r}\cdot\mathbf{r}' + r'^2} = r - \mathbf{n}\cdot\mathbf{r}' + O\left(\frac{1}{r}\right), \tag{35.1}$$

to derive the asymptotic form of the potentials in a Lorenz gauge [*cf.* (34.37a) and (34.37b)],

$$\phi(\mathbf{r},t) \approx \frac{1}{r}\int (d\mathbf{r}')\rho\left(\mathbf{r}', t - \frac{r}{c} + \frac{1}{c}\mathbf{n}\cdot\mathbf{r}'\right), \tag{35.2a}$$

$$\mathbf{A}(\mathbf{r},t) \approx \frac{1}{r}\int (d\mathbf{r}')\frac{1}{c}\mathbf{j}\left(\mathbf{r}', t - \frac{r}{c} + \frac{1}{c}\mathbf{n}\cdot\mathbf{r}'\right), \tag{35.2b}$$

where

$$\mathbf{n} = \frac{\mathbf{r}}{r} \tag{35.3}$$

is the unit vector in the direction toward the observation point. In the above equations, the $\mathbf{n}\cdot\mathbf{r}'$ term in the expansion of $1/|\mathbf{r}-\mathbf{r}'|$ has been deleted since it gives rise to a $1/r^2$ term in the potential, while it has been retained in the expression for the time of emission, t',

$$t' \approx t - \frac{r}{c} + \frac{1}{c}\mathbf{n}\cdot\mathbf{r}' \equiv t_r. \tag{35.4}$$

The last term in t_r reflects the finite amount of time it takes radiation to propagate across the source, which can be significant if the source distribution changes rapidly, or, more precisely, when a typical frequency of oscillation of the source distribution is of order c/a.

The fields at large distances can now be calculated by substituting (35.2a) and (35.2b) into (34.3a) and (34.3b), and by using the evaluation

$$\begin{aligned}\boldsymbol{\nabla}\left[\frac{1}{r}f\left(t - \frac{r}{c} + \frac{1}{c}\mathbf{n}\cdot\mathbf{r}'\right)\right] &= -\frac{\mathbf{n}}{r^2}f(t_r) - \left[\frac{\mathbf{n}}{c} + \frac{1}{c}\frac{1}{r}\mathbf{n}\times(\mathbf{n}\times\mathbf{r}')\right]\frac{\partial}{\partial t}\left[\frac{1}{r}f(t_r)\right] \\ &= -\frac{\mathbf{n}}{c}\frac{\partial}{\partial t}\left[\frac{1}{r}f(t_r)\right] + O\left(\frac{1}{r^2}\right).\end{aligned} \tag{35.5}$$

DOI: 10.1201/9781003057369-35

We see that because of the appearance of r in the time dependences, the fields behave as $1/r$ rather than the behavior $1/r^2$ characteristic of statics, and in particular, for $r \gg a$, the field strengths are

$$\mathbf{B}(\mathbf{r},t) \approx -\frac{1}{c}\mathbf{n}\times\frac{1}{r}\int(d\mathbf{r}')\frac{1}{c}\frac{\partial}{\partial t}\mathbf{j}(\mathbf{r}',t_r), \tag{35.6a}$$

$$\mathbf{E}(\mathbf{r},t) \approx \frac{\mathbf{n}}{c}\frac{1}{r}\int(d\mathbf{r}')\frac{\partial}{\partial t}\rho(\mathbf{r}',t_r) - \frac{1}{c}\frac{1}{r}\int(d\mathbf{r}')\frac{1}{c}\frac{\partial}{\partial t}\mathbf{j}(\mathbf{r}',t_r). \tag{35.6b}$$

These two terms in (35.6b) can be further combined by using the local charge conservation condition (34.2),

$$\frac{\partial}{\partial t}\rho(\mathbf{r}',t_r) = -\boldsymbol{\nabla}'\cdot\mathbf{j}(\mathbf{r}',t_r) + \frac{\mathbf{n}}{c}\cdot\frac{\partial}{\partial t}\mathbf{j}(\mathbf{r}',t_r), \tag{35.7}$$

where we have used the fact that the divergence operator in (34.2) acts only on the spatial arguments of $\mathbf{j}$, while $\boldsymbol{\nabla}'$ in (35.7) also differentiates the $\mathbf{r}'$ dependence of t_r, where from (35.4) $\boldsymbol{\nabla}'t_r = \mathbf{n}/c$. The first integral in (35.6b) may then be simplified through the use of

$$\int(d\mathbf{r}')\boldsymbol{\nabla}'\cdot\mathbf{j} = 0, \tag{35.8}$$

since the charge distribution is bounded, and the remaining terms involving the time derivative of $\mathbf{j}$ can be combined by means of the identity

$$\mathbf{n}(\mathbf{n}\cdot\mathbf{V}) - \mathbf{V} = \mathbf{n}\times(\mathbf{n}\times\mathbf{V}) \tag{35.9}$$

to read

$$\mathbf{E}(\mathbf{r},t) = \mathbf{n}\times\left[\frac{\mathbf{n}}{c}\times\frac{1}{r}\int(d\mathbf{r}')\frac{1}{c}\frac{\partial}{\partial t}\mathbf{j}(\mathbf{r}',t_r)\right] = -\mathbf{n}\times\mathbf{B}(\mathbf{r},t). \tag{35.10}$$

We observe that, far from the source distribution, $\mathbf{E}$ and $\mathbf{B}$ are perpendicular to each other,

$$\mathbf{E} = -\mathbf{n}\times\mathbf{B}, \quad \mathbf{B} = \mathbf{n}\times\mathbf{E}, \tag{35.11a}$$

are perpendicular to the direction of propagation $\mathbf{n}$, and have equal magnitude,

$$E^2 = B^2. \tag{35.11b}$$

These are the same characteristics seen in Section 3.4, where we considered the propagation of electromagnetic waves along a single direction, in terms of the flow of energy and momentum. (See Fig. 35.1.)

35.2 Angular Distribution of Radiated Power

Next we ask at what rate does this time-varying charge and current distribution radiate energy. The amount of energy flowing across a unit area per unit time is expressed by Poynting's vector, (3.4b),

$$\mathbf{S} = \frac{c}{4\pi}\mathbf{E}\times\mathbf{B}, \tag{35.12}$$

which points in the direction of propagation $\mathbf{n}$. Substituting the asymptotic expressions for the fields, (35.6a) and (35.10), into (35.12), we may write the rate of energy radiated per unit area in terms of the current distribution:

$$\mathbf{n}\cdot\mathbf{S} = \frac{c}{4\pi}(\mathbf{n}\times\mathbf{E})\cdot\mathbf{B} = \frac{c}{4\pi}\mathbf{B}^2 = \frac{1}{4\pi r^2}\frac{1}{c^3}\left[\mathbf{n}\times\int(d\mathbf{r}')\frac{\partial}{\partial t}\mathbf{j}(\mathbf{r}',t_r)\right]^2. \tag{35.13}$$

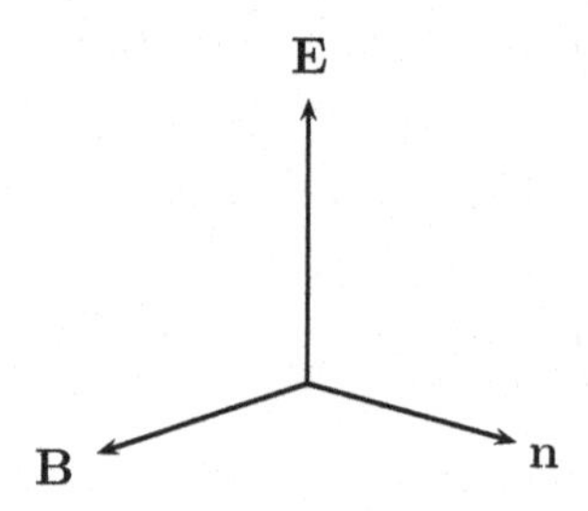

FIGURE 35.1
Electric and magnetic fields for a wave propagating in the direction $\mathbf{n}$.

Rather than the energy crossing an element of area dS, we would instead like the energy radiated into a solid angle $d\Omega$,

$$d\Omega = \frac{dS}{r^2}, \tag{35.14}$$

since the latter measure is independent of how far away the observer is from the source. Therefore, the amount of energy radiated per unit time (the power) per unit solid angle in the direction $\mathbf{n}$ is

$$\frac{dP}{d\Omega} = \frac{1}{4\pi c^3}\left[\mathbf{n}\times\int (d\mathbf{r}')\frac{\partial}{\partial t}\mathbf{j}(\mathbf{r}', t_r)\right]^2, \tag{35.15}$$

while the total power radiated is obtained by integrating this over all solid angles,

$$P = \int d\Omega\left(\frac{dP}{d\Omega}\right). \tag{35.16}$$

This finite energy flow at large distances is a consequence of the $1/r$ behavior of the fields, which, in turn, arises from the time variation of the current density.

35.3 Radiation by an Accelerated Charged Particle

Let us first apply the above general result, (35.15), to the simple example of a particle, with charge e, moving with a velocity $\mathbf{v}$, small compared with the speed of light, $v/c \ll 1$. If $\mathbf{R}(t')$ is the position of the charged particle, the corresponding current density is

$$\mathbf{j}(\mathbf{r}', t') = e\mathbf{v}(t')\delta(\mathbf{r}' - \mathbf{R}(t')). \tag{35.17}$$

For this situation, the time of emission, (35.4), of the radiation may be approximated by

$$t_r = t - \frac{r}{c} + \frac{1}{c}\mathbf{n}\cdot\mathbf{R}(t') \approx t - \frac{r}{c} \equiv t_e, \quad \text{for } \frac{v}{c} \ll 1, \tag{35.18}$$

since $|\mathbf{R}(t')|$ is bounded by v times a characteristic time. Therefore, the integral in (35.15) is immediately evaluated to be

$$\int (d\mathbf{r}')\frac{\partial}{\partial t}\mathbf{j}(\mathbf{r}', t_r) \approx \frac{d}{dt}\int (d\mathbf{r}')\,\mathbf{j}(\mathbf{r}', t_e) = e\frac{d\mathbf{v}}{dt_e}(t_e), \tag{35.19}$$

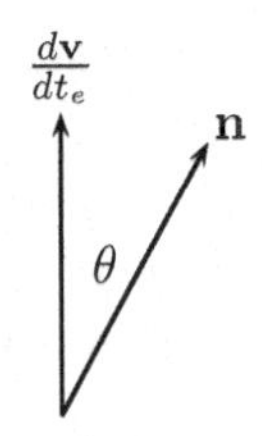

FIGURE 35.2
Orientation of acceleration at emission time to direction of observation.

implying that radiation is produced whenever a charged particle is accelerated. (However, see Chapter 40.) The angular distribution of the radiated power is then given by, from (35.15) and (35.19),

$$\frac{dP}{d\Omega} \approx \frac{1}{4\pi c^3}\left(\mathbf{n}\times e\frac{d\mathbf{v}}{dt}\right)^2 = \frac{e^2}{4\pi c^3}\left[\left(\frac{d\mathbf{v}}{dt}\right)^2 - \left(\mathbf{n}\cdot\frac{d\mathbf{v}}{dt}\right)^2\right]$$
$$= \frac{e^2}{4\pi c^3}\left(\frac{d\mathbf{v}(t_e)}{dt_e}\right)^2 \sin^2\theta, \tag{35.20}$$

where θ is the angle between the direction of observation $\mathbf{n}$ and the direction of the acceleration at the emission time t_e. (See Fig. 35.2.) Evidently there is no radiation emitted along the direction of the acceleration. By employing the angular integral

$$\int \frac{d\Omega}{4\pi}\sin^2\theta = \frac{2}{3}, \tag{35.21}$$

we obtain the total radiated power:

$$P = \frac{2e^2}{3c^3}\left(\frac{d\mathbf{v}}{dt}\right)^2, \quad \text{for } \frac{v}{c} \ll 1, \tag{35.22}$$

which is called the Larmor[1] formula.

35.4 Dipole Radiation

Next, we generalize the above discussion to a system consisting of many charged particles. For a small system in which particles are all moving with low velocities, the time of emission, t_e, does not vary significantly over the current distribution. Consequently, the integral in (35.15) becomes

$$\int (d\mathbf{r}')\frac{\partial}{\partial t}\mathbf{j}(\mathbf{r}', t_r) \approx \frac{d}{dt}\int (d\mathbf{r}')\,\mathbf{j}(\mathbf{r}', t_e), \tag{35.23}$$

which is evaluated by means of an identity, derived from current conservation:

$$\mathbf{0} = \int (d\mathbf{r})\,\mathbf{r}\left[\boldsymbol{\nabla}\cdot\mathbf{j} + \frac{\partial}{\partial t}\rho\right] = \int (d\mathbf{r})\left[\boldsymbol{\nabla}\cdot(\mathbf{j}\,\mathbf{r}) - \mathbf{j} + \mathbf{r}\frac{\partial}{\partial t}\rho\right]$$
$$= -\int (d\mathbf{r})\,\mathbf{j}(\mathbf{r}, t) + \frac{d}{dt}\int (d\mathbf{r})\,\mathbf{r}\,\rho(\mathbf{r}, t). \tag{35.24}$$

[1] Joseph Larmor (1857–1942).

Recalling the definition of the electric dipole moment,

$$\mathbf{d}(t) = \int (d\mathbf{r})\,\mathbf{r}\,\rho(\mathbf{r},t), \tag{35.25}$$

we recognize (35.23) as the second time derivative of $\mathbf{d}(t)$. Therefore, the angular distribution is given by

$$\frac{dP}{d\Omega} \approx \frac{1}{4\pi c^3}(\mathbf{n}\times\ddot{\mathbf{d}})^2, \quad \text{for } \frac{v}{c} \ll 1, \tag{35.26}$$

while the total power is

$$P \approx \frac{2}{3c^3}(\ddot{\mathbf{d}})^2, \tag{35.27}$$

where we have used a dot to denote time differentiation. Radiation described by these formulæ is called electric dipole radiation. For a single charged particle

$$\mathbf{d} = e\mathbf{r}, \quad \ddot{\mathbf{d}} = e\frac{d\mathbf{v}}{dt}, \tag{35.28}$$

and (35.20) and (35.22) are recovered, as expected. However, for a system of n charged particles, the electric dipole moment is

$$\mathbf{d} = \sum_{a=1}^{n} e_a \mathbf{r}_a, \tag{35.29}$$

so the power radiated is not additive, but exhibits interference effects:

$$\frac{dP}{d\Omega} = \frac{1}{4\pi c^3}\left(\mathbf{n}\times\sum_{a=1}^{n} e_a\ddot{\mathbf{r}}_a\right)^2. \tag{35.30}$$

Let us now make a better approximation by keeping the v/c correction arising from the $\frac{1}{c}\mathbf{n}\cdot\mathbf{r}'$ term in the retarded time. The integral over the current in (35.15) now becomes

$$\begin{aligned}\int (d\mathbf{r}')\,\mathbf{j}(\mathbf{r}',t_e+\frac{1}{c}\mathbf{n}\cdot\mathbf{r}') &\approx \dot{\mathbf{d}}(t_e) + \int (d\mathbf{r}')\frac{1}{c}\mathbf{n}\cdot\mathbf{r}'\frac{\partial}{\partial t_e}\mathbf{j}(\mathbf{r}',t_e) + \dots \\ &= \dot{\mathbf{d}}(t_e) + \frac{1}{c}\frac{d}{dt}\int (d\mathbf{r}')\,\mathbf{n}\cdot\left[\frac{1}{2}(\mathbf{r}'\,\mathbf{j}+\mathbf{j}\,\mathbf{r}') + \frac{1}{2}(\mathbf{r}'\,\mathbf{j}-\mathbf{j}\,\mathbf{r}')\right] \\ &= \dot{\mathbf{d}}(t_e) - \frac{d}{dt}\int (d\mathbf{r}')\,\mathbf{n}\times\left[\frac{1}{2c}\mathbf{r}'\times\mathbf{j}(\mathbf{r}',t_e)\right] \\ &= \dot{\mathbf{d}}(t_e) - \mathbf{n}\times\dot{\boldsymbol{\mu}}(t_e),\end{aligned} \tag{35.31}$$

where $\boldsymbol{\mu}$ is the magnetic dipole moment, (30.10). [Here we have neglected the contribution due to

$$\frac{1}{2}(\mathbf{r}'\,\mathbf{j}+\mathbf{j}\,\mathbf{r}') \tag{35.32}$$

since this is an electric quadrupole moment effect,

$$\int (d\mathbf{r}')[x_i' j_j + x_j' j_i] = -\int (d\mathbf{r}')\,x_i' x_j'\,\boldsymbol{\nabla}'\cdot\mathbf{j} = \frac{d}{dt}\int (d\mathbf{r}')\,x_i' x_j'\rho \to \frac{1}{3}\dot{\mathsf{q}}_{ij}, \tag{35.33}$$

from (24.4c), since the unit dyadic evidently does not contribute to the radiated power because $\mathbf{n}\times\mathbf{n} = \mathbf{0}$. See Problem 9 for a discussion of electric quadrupole radiation.] The angular distribution of the radiated power is therefore more accurately given by

$$\frac{dP}{d\Omega} = \frac{1}{4\pi c^3}[\mathbf{n}\times(\ddot{\mathbf{d}} - \mathbf{n}\times\ddot{\boldsymbol{\mu}})]^2 = \frac{1}{4\pi c^3}[(\mathbf{n}\times\ddot{\mathbf{d}})^2 + (\mathbf{n}\times\ddot{\boldsymbol{\mu}})^2 + 2\mathbf{n}\cdot(\ddot{\mathbf{d}}\times\ddot{\boldsymbol{\mu}})], \tag{35.34}$$

where the last term represents interference between $\mathbf{d}$ and $\boldsymbol{\mu}$, which, since it depends linearly on $\mathbf{n}$, does not contribute to the total radiated power:

$$P = \frac{2}{3c^3}[(\ddot{\mathbf{d}})^2 + (\ddot{\boldsymbol{\mu}})^2]. \tag{35.35}$$

The behavior of the fields at large distances can be obtained by substituting (35.31) into (35.6a) and (35.10),

$$\mathbf{B} \sim -\frac{1}{c^2}\frac{1}{r}\mathbf{n}\times(\ddot{\mathbf{d}} - \mathbf{n}\times\ddot{\boldsymbol{\mu}}), \tag{35.36a}$$

$$\mathbf{E} \sim -\mathbf{n}\times\mathbf{B} \sim \frac{1}{c^2}\frac{1}{r}\mathbf{n}\times(\ddot{\boldsymbol{\mu}} + \mathbf{n}\times\ddot{\mathbf{d}}). \tag{35.36b}$$

Note that these results are invariant under the replacements $\mathbf{E} \to \mathbf{B}$ and $\mathbf{B} \to -\mathbf{E}$ together with $\mathbf{d} \to \boldsymbol{\mu}$ and $\boldsymbol{\mu} \to -\mathbf{d}$, which is a manifestation of the dual symmetry discussed in Chapter 2. We here have an indication of the connection between the directions of the electric field for electric and magnetic dipole radiation.

35.5 Potentials in Radiation Gauge

To this point, we have discussed radiation by use of the Lorenz gauge. However, we do have the freedom to choose an arbitrary gauge without affecting the physical results. As an illustration, let us here consider the radiation gauge, which exhibits a certain physical simplicity. In this gauge, the potentials satisfy the differential equations (34.8a) and (34.8b), that is

$$-\nabla^2\phi = 4\pi\rho, \tag{35.37a}$$

$$-\Box^2\mathbf{A} = \frac{4\pi}{c}\left(\mathbf{j} - \frac{1}{4\pi}\boldsymbol{\nabla}\frac{\partial}{\partial t}\phi\right), \tag{35.37b}$$

while the vector potential is subject to the gauge condition (34.7),

$$\boldsymbol{\nabla}\cdot\mathbf{A} = 0. \tag{35.38}$$

The electric and magnetic fields are obtained from these potentials by the relations (34.3a) and (34.3b). In order to make contact with what has gone before, we consider the fields at large distances, which are those of interest for radiation. The solution of (35.37a) is the Coulomb potential,

$$\phi(\mathbf{r},t) = \int (d\mathbf{r}')\frac{\rho(\mathbf{r}',t)}{|\mathbf{r}-\mathbf{r}'|}, \tag{35.39}$$

which, asymptotically, behaves as

$$\phi \sim \frac{e}{r}, \tag{35.40}$$

with e the total charge. Since the gradient of this is inversely proportional to the square of the distance,

$$-\boldsymbol{\nabla}\phi \sim \frac{e}{r^3}\mathbf{r}, \tag{35.41}$$

we can neglect the scalar potential in computing the radiation fields, which decrease only as $1/r$. Therefore, the vector potential alone determines the radiation fields:

$$\mathbf{E} \sim -\frac{1}{c}\frac{\partial \mathbf{A}}{\partial t}, \tag{35.42a}$$

$$\mathbf{B} = \boldsymbol{\nabla}\times\mathbf{A}. \tag{35.42b}$$

We also note that the gauge condition (35.38) enforces the transversality of these fields. That is, as a consequence of $\boldsymbol{\nabla}\cdot\mathbf{A} = 0$, we recover the scalar Maxwell equations outside the sources,

$$\boldsymbol{\nabla}\cdot\mathbf{B} = 0 \quad \text{and} \quad \boldsymbol{\nabla}\cdot\mathbf{E} = 0, \tag{35.43}$$

which, by virtue of (35.5), supplies the relations

$$\mathbf{n}\cdot\mathbf{B} = 0 \quad \text{and} \quad \mathbf{n}\cdot\mathbf{E} = 0, \tag{35.44}$$

while $\mathbf{E} \perp \mathbf{B}$ follows immediately from (35.42a) and (35.42b).

To solve (35.37b) for the vector potential, we first write the solution to (35.37a) symbolically as

$$\phi = \frac{1}{-\nabla^2}4\pi\rho, \tag{35.45}$$

the time derivative of which is

$$\frac{\partial}{\partial t}\phi = \frac{1}{-\nabla^2}4\pi\frac{\partial}{\partial t}\rho = \frac{1}{\nabla^2}4\pi\boldsymbol{\nabla}\cdot\mathbf{j}. \tag{35.46}$$

Consequently, we may rewrite (35.37b) as

$$-\Box^2\mathbf{A} = \frac{4\pi}{c}\left(\mathbf{1} - \frac{\boldsymbol{\nabla}\boldsymbol{\nabla}}{\nabla^2}\right)\cdot\mathbf{j}, \tag{35.47}$$

which makes the radiation gauge condition (35.38) transparent. The solution to (35.47) may be obtained from that of (34.13a) by applying the operator

$$\mathbf{1} - \frac{\boldsymbol{\nabla}\boldsymbol{\nabla}}{\nabla^2} \tag{35.48}$$

to (34.37b):

$$\mathbf{A}(\mathbf{r},t) = \left(\mathbf{1} - \frac{\boldsymbol{\nabla}\boldsymbol{\nabla}}{\nabla^2}\right)\cdot\int(d\mathbf{r}')\frac{\frac{1}{c}\mathbf{j}(\mathbf{r}',t-\frac{1}{c}|\mathbf{r}-\mathbf{r}'|)}{|\mathbf{r}-\mathbf{r}'|}. \tag{35.49}$$

At large distances, by making use of (35.1) and (35.5), we have effectively the replacement

$$\boldsymbol{\nabla} \to -\frac{\mathbf{n}}{c}\frac{\partial}{\partial t} \tag{35.50}$$

so that the operator $\mathbf{1} - \boldsymbol{\nabla}\boldsymbol{\nabla}/\nabla^2$ can be replaced by

$$\mathbf{1} - \frac{\boldsymbol{\nabla}\boldsymbol{\nabla}}{\nabla^2} \to \mathbf{1} - \frac{\left(-\frac{\mathbf{n}}{c}\frac{\partial}{\partial t}\right)\left(-\frac{\mathbf{n}}{c}\frac{\partial}{\partial t}\right)}{\left(-\frac{\mathbf{n}}{c}\frac{\partial}{\partial t}\right)^2} = \mathbf{1} - \mathbf{n}\mathbf{n}. \tag{35.51}$$

Notice that this symbolic notation is convenient when $1/\nabla^2$ can be computed simply. By making use of (35.9) and (35.1), we obtain the asymptotic form of the vector potential, in the radiation gauge, to be

$$\mathbf{A}(\mathbf{r},t) \sim (\mathbf{1}-\mathbf{nn})\cdot\frac{1}{cr}\int(d\mathbf{r}')\,\mathbf{j}(\mathbf{r}',t_r) = -\mathbf{n}\times\left[\mathbf{n}\times\frac{1}{cr}\int(d\mathbf{r}')\,\mathbf{j}(\mathbf{r}',t_r)\right]. \tag{35.52}$$

The resulting electric and magnetic fields are precisely the same as those found in the Lorenz gauge, (35.6a) and (35.10).

35.6 Problems for Chapter 35

1. A particle, of charge e and mass m, moves with speed v, $v/c \ll 1$, in a uniform magnetic field $\mathbf{B}$. Suppose the motion is confined to the plane perpendicular to $\mathbf{B}$. Calculate the power radiated P in terms of B and v, and show that

$$P = -\frac{dE}{dt} = \gamma E, \tag{35.53}$$

where E is the energy of the particle, and find γ. Since then

$$E(t) = E(0)e^{-\gamma t}, \tag{35.54}$$

$1/\gamma$ is the mean lifetime of the motion. For an electron, find $1/\gamma$ in seconds for a magnetic field of 10^4 gauss. (The answer is given in Appendix A.)

2. A nonrelativistic particle of charge e and mass m moves in a Hooke's law potential (a linear oscillator) with natural frequency ω_0. Again find P, the power radiated. Recall that for such motion, the time-averaged kinetic and potential energy satisfy

$$\overline{T} = \overline{V} = \frac{1}{2}E. \tag{35.55}$$

Show then that the power radiated, averaged over one cycle is

$$P = -\frac{dE}{dt} = \gamma E, \tag{35.56}$$

and find γ. Compute $1/\gamma$ in seconds when ω_0 is 10^{15} sec^{-1} (a characteristic atomic frequency, corresponding to visible light).

3. An electron of charge e and mass m moves in a circular orbit under Coulomb forces produced by a proton. The average potential energy is related to the total energy by

$$E = \frac{1}{2}\overline{V}. \tag{35.57}$$

Suppose, as it radiates, the electron continues to move on a circle, and calculate the power radiated, and thereby $-dE/dt$, as a function of E (the relation is no longer linear). Integrate this result, and find how long it takes for the energy to change from E_2 to E_1. In a finite time the electron reaches the center, so calculate how long it takes the electron to hit the proton if it starts from an initial radius of $r_{\text{initial}} = 10^{-8}$cm. (This instability was one of the reasons for the discovery of quantum mechanics.)

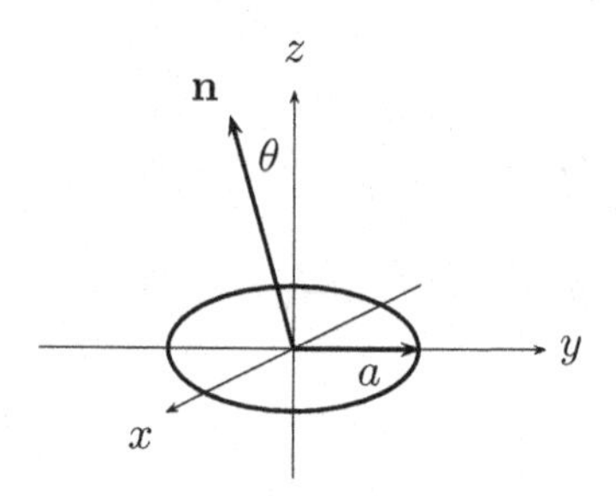

FIGURE 35.3
A current loop perpendicular to the z axis. The normal direction $\mathbf{n}$ lies in the x-z plane.

4. Consider a nonrelativistic particle of charge e moving in a circular orbit of radius R and with angular velocity ω_0. Calculate the dipole moment $\mathbf{d}(t)$ of this charge distribution. Then compute the instantaneous power radiated at an angle ϕ above the plane of the orbit using the nonrelativistic Larmor formula (35.26). What is this result when averaged over one cycle? Integrate the result over all angles ϕ and show that that you obtain coincidence with that calculated from (35.27). Calculate the magnetic dipole moment in this situation and show that there is no magnetic dipole radiation.

5. A circular loop of wire of negligible cross section and radius a lies in the x-y plane as shown in Figure 35.3. The loop carries a sinusoidally varying current,
$$I(t) = I_0 \sin \omega_0 t. \tag{35.58}$$
In cylindrical coordinates write down the charge and current density. Calculate the electric and magnetic dipole moments relative to the center of the loop. In the dipole approximation work out the angular distribution of radiation from (35.34). Work out the angular distribution of radiated power using the general formula (35.15). Express your answer in terms of Bessel functions, particularly J_0 and J_1. Rederive the dipole result by considering a small system, for which $\omega_0 a \ll 1$, and using the leading behavior of the Bessel functions for small argument,
$$J_0(x) \approx 1 - \frac{x^2}{2}, \quad J_1(x) \approx \frac{x}{2}, \quad x \to 0. \tag{35.59}$$

6. Charge e is distributed uniformly over the surface of a sphere of radius a, which is rotating about an axis with constant angular velocity Ω. Compute the power radiated, either by applying a general method or by considering electric and magnetic dipole radiation.

7. In the text, we derived the radiation fields from the scalar and vector potentials. Alternatively, we could solve the wave equations for the electric and magnetic fields, (34.40) by using the retarded Green's function (34.32). Show that very far from the source, assumed to be bounded, where $1/|\mathbf{r} - \mathbf{r}'| \approx 1/r$, and the time of emission is $t' \approx t_r \equiv t - \frac{r}{c} - \frac{\mathbf{n} \cdot \mathbf{r}'}{c}$, we obtain the results (35.6).

8. Derive an alternative form for the angular distribution of radiated power, (35.15),
$$\frac{dP}{d\Omega} = \frac{1}{4\pi c}\left(\left[\int (d\mathbf{r}') \frac{1}{c}\frac{\partial}{\partial t}\mathbf{j}(\mathbf{r}', t_r)\right]^2 - \left[\int (d\mathbf{r}') \frac{\partial}{\partial t}\rho(\mathbf{r}', t_r)\right]^2\right). \tag{35.60}$$

9. Carry out the multipole expansion initiated in Section 35.4 and obtain the fields and the power radiated in electric quadrupole radiation. If only an electric quadrupole exists, the fields at a given frequency ω are then

$$\mathbf{B} \sim -\frac{1}{3c^3 r}\mathbf{n}\times\dddot{\mathbf{q}}\cdot\mathbf{n}, \quad \mathbf{E} \sim -\mathbf{n}\times\mathbf{B}. \tag{35.61}$$

The power radiated into a given solid angle is

$$\frac{dP}{d\Omega} = \frac{1}{144\pi c^5}\left[\mathbf{n}\times\dddot{\mathbf{q}}\cdot\mathbf{n}\right]^2 \tag{35.62}$$

and the total power radiated is

$$P = \frac{1}{180c^5}\sum_{ij}\dddot{q}_{ij}^{\,2}. \tag{35.63}$$

Here the quadrupole moment is defined by (24.4c):

$$\mathbf{q} = \int (d\mathbf{r}')(3\mathbf{r}'\mathbf{r}' - \mathbf{1}r^2)\rho(\mathbf{r}'). \tag{35.64}$$

10. More generally, a system will possess electric and magnetic dipole moments as well as an electric quadrupole, and there is interference between the contributions. Obtain a generalization of (35.34) when $\mathbf{d}$, $\boldsymbol{\mu}$, and $\mathbf{q}$ are all present. The interference disappears from the total power radiated:

$$P = \frac{2}{3c^3}(\ddot{\mathbf{d}})^2 + \frac{2}{3c^3}(\ddot{\boldsymbol{\mu}})^2 + \frac{1}{180c^5}(\dddot{\mathbf{q}})^2, \tag{35.65}$$

where the last term is understood in the sense of (35.63). [Hints:

$$\int \frac{d\Omega}{4\pi} n_i n_j = \frac{1}{3}\delta_{ij}. \tag{35.66a}$$

What is

$$\int \frac{d\Omega}{4\pi} n_i n_j n_k n_k ?. \tag{35.66b}$$

Also recall

$$\epsilon_{ijk}\epsilon_{ilm} = \delta_{jl}\delta_{km} - \delta_{jm}\delta_{kl}.] \tag{35.67}$$

11. Let v^μ be the four-vector velocity of a physical system, $v^\mu = \gamma(c, \mathbf{v})$, where $\gamma = (1 - v^2/c^2)^{-1/2}$. Use the invariance of $k_\mu v^\mu$ in relating ω', a frequency observed when the system is at rest, to quantities measured when the system is in motion along the z axis with velocity v. [Recall (12.88).] Show that the invariant

$$I = \frac{dp^\mu}{d\tau}\frac{dp_\mu}{d\tau} - \left(mc\frac{k_\mu dp^\mu/d\tau}{k_\mu p^\mu}\right)^2 \tag{35.68a}$$

is written as

$$I = \left(\frac{E}{mc^2}\right)^2\left[(\dot{\mathbf{p}})^2 - \left(\frac{\dot{E}}{c}\right)^2 - \left(\frac{mc^2}{E}\right)^2\frac{(\mathbf{n}\cdot\dot{\mathbf{p}} - \dot{E}/c)^2}{(1 - \mathbf{n}\cdot\mathbf{p}c/E)^2}\right], \tag{35.68b}$$

where m is the rest mass, and $\mathbf{n}$ is the direction of $\mathbf{k}$.

12. Verify that the energy radiated per unit proper time into a unit solid angle, by a system that is momentarily at rest, is given, in any inertial coordinate frame, by the invariant expression

$$-\frac{d^2p^\mu}{d\tau\, d\Omega'}v_\mu, \tag{35.69a}$$

where v^μ is is the velocity four-vector of the system, and $d\Omega'$ is the solid angle element in the rest frame. Then use the relation between the energy and the momentum of the radiation moving in direction $\mathbf{n}$ to write the above radiation quantity, for a system moving with velocity $\mathbf{v}$, as

$$\frac{d^2E}{dt\, d\Omega'}\frac{1-\mathbf{n}\cdot\mathbf{v}/c}{1-v^2/c^2}. \tag{35.69b}$$

13. The power radiated in the direction $\mathbf{n}$, per unit solid angle $d\Omega'$, by an accelerated charge that is momentarily at rest is given by (35.20). Then combine this with (35.69b), (12.94), and (35.68b) to produce the power radiated into a solid angle $d\Omega$:

$$\frac{dP}{d\Omega} = \frac{e^2}{4\pi m^2c^3}\left(\frac{mc^2}{E}\right)^2\left[\frac{\dot{\mathbf{p}}^2-(\dot{E}/c)^2}{(1-\mathbf{n}\cdot\mathbf{p}c/E))^3} - \left(\frac{mc^2}{E}\right)^2\frac{(\mathbf{n}\cdot\dot{\mathbf{p}}-\dot{E}/c)^2}{(1-\mathbf{n}\cdot\mathbf{p}c/E)^5}\right]. \tag{35.70a}$$

Substitute $\mathbf{v}=\mathbf{p}c^2/E$ to recast this as

$$\frac{dP}{d\Omega} = \frac{e^2}{4\pi c^3}\left[\frac{\dot{\mathbf{v}}^2}{(1-\mathbf{n}\cdot\mathbf{v}/c)^3} + \frac{2}{c}\frac{\mathbf{n}\cdot\dot{\mathbf{v}}\,\mathbf{v}\cdot\dot{\mathbf{v}}}{(1-\mathbf{n}\cdot\mathbf{v}/c)^4} - \left(1-\frac{v^2}{c^2}\right)\frac{(\mathbf{n}\cdot\dot{\mathbf{v}})^2}{(1-\mathbf{n}\cdot\mathbf{v}/c)^5}\right]. \tag{35.70b}$$

14. For a more elementary derivation of the above formula, start from (35.20). Use the emission time expression,

$$t' = t - \frac{1}{c}|\mathbf{r}-\mathbf{r}(t')|, \quad \frac{dt}{dt'} = 1 - \frac{\mathbf{v}(t')\cdot\mathbf{n}(t')}{c}, \tag{35.71}$$

to derive

$$\mathbf{v}(t') = \mathbf{v}(t)[1-\mathbf{n}(t')\cdot\mathbf{v}(t')/c]. \tag{35.72}$$

Then verify that (35.70b) follows, written in terms ot the emission time t'.

15. Use the result (35.70b) to show, for the case of linear acceleration, that is, where $\dot{\mathbf{v}}$ is in the same direction as $\mathbf{v}$, which makes an angle θ with respect to the direction of observation, that ($\beta = v/c$)

$$-\frac{d^2E}{dt\, d\Omega} = \frac{e^2}{4\pi c^3}\left(\frac{d\mathbf{v}}{dt}\right)^2\frac{\sin^2\theta}{(1-\beta\cos\theta)^5}. \tag{35.73}$$

Integrate this over all solid angles to arrive at the energy loss rate for this circumstance.

36

Radiation—Source Point of View

36.1 Conservation of Energy

Having examined the radiation fields, we turn our attention to an examination of the source of the radiated energy. Energy and momentum are transferred from the charges to the electromagnetic field; the rate at which the current does work on the field is

$$-\mathbf{j}\cdot\mathbf{E} = \frac{\partial}{\partial t}\left(\frac{\mathbf{E}^2+\mathbf{B}^2}{8\pi}\right) + \boldsymbol{\nabla}\cdot\left(\frac{c}{4\pi}\mathbf{E}\times\mathbf{B}\right), \tag{36.1}$$

which is the local statement of energy conservation, (3.5). When (36.1) is integrated over a large volume enclosing the charge and current distributions, the conservation of total energy follows:

$$\int (d\mathbf{r})\,(-\mathbf{j}\cdot\mathbf{E}) = \frac{d}{dt}\int (d\mathbf{r})\,\frac{\mathbf{E}^2+\mathbf{B}^2}{8\pi} + \oint d\mathbf{S}\cdot\frac{c}{4\pi}\mathbf{E}\times\mathbf{B} = \frac{d}{dt}E + P, \tag{36.2}$$

or stated in words, the rate at which the charged particles transfer energy to the electromagnetic field is equal to the sum of the rate of increase of the total electromagnetic energy, E, in the volume, and the rate of flow of energy, P, out of the surface bounding the volume. Equation (36.2) gives us an alternative way of calculating the radiated power, P, by computing the rate at which energy is transferred to the fields,

$$\int (d\mathbf{r})\,(-\mathbf{j}\cdot\mathbf{E}), \tag{36.3}$$

and discarding total time derivative terms, which are not associated with radiation. From this point of view, we need to know the electric field inside the current distribution, in contrast to the previous discussion, in which we computed the radiated power by evaluating the fields far from the source.

36.2 Dipole Radiation

To illustrate this method, we again consider dipole radiation produced by a small charge distribution, for which the nonrelativistic approximation is valid, that is, all particle speeds are small compared to the speed of light, $v/c \ll 1$. In this limit, we will require an expression for the field $\mathbf{E}$ accurate to order $1/c^3$, which means, from the definition (34.3b), that the scalar potential ϕ must be expanded in powers of $1/c$ up to order $1/c^3$, while the vector potential $\mathbf{A}$ need only be expanded up to order $1/c^2$. In the Lorenz gauge, the appropriate

DOI: 10.1201/9781003057369-36

expansion of the scalar potential (34.37a) is

$$\begin{aligned}\phi(\mathbf{r},t) &\approx \int (d\mathbf{r}')\frac{\rho(\mathbf{r}',t)}{|\mathbf{r}-\mathbf{r}'|} - \int (d\mathbf{r}')\frac{\frac{1}{c}|\mathbf{r}-\mathbf{r}'|\frac{\partial}{\partial t}\rho}{|\mathbf{r}-\mathbf{r}'|} \\ &\quad + \int (d\mathbf{r}')\frac{\frac{1}{2}\left(\frac{1}{c}|\mathbf{r}-\mathbf{r}'|\right)^2\frac{\partial^2}{\partial t^2}\rho}{|\mathbf{r}-\mathbf{r}'|} - \int (d\mathbf{r}')\frac{\frac{1}{6}\left(\frac{1}{c}|\mathbf{r}-\mathbf{r}'|\right)^3\frac{\partial^3}{\partial t^3}\rho}{|\mathbf{r}-\mathbf{r}'|} + \dots \\ &= \int (d\mathbf{r}')\frac{\rho(\mathbf{r}',t)}{|\mathbf{r}-\mathbf{r}'|} + \frac{1}{2c^2}\int (d\mathbf{r}')|\mathbf{r}-\mathbf{r}'|\frac{\partial^2}{\partial t^2}\rho(\mathbf{r}',t) \\ &\quad - \frac{1}{6c^3}\int (d\mathbf{r}')(\mathbf{r}-\mathbf{r}')^2\frac{\partial^3}{\partial t^3}\rho(\mathbf{r}',t) + \dots,\end{aligned} \tag{36.4}$$

where we have omitted the second term on the right-hand side of the first line because the total charge e is conserved,

$$\int (d\mathbf{r}')\frac{\partial}{\partial t}\rho = \frac{d}{dt}e = 0. \tag{36.5}$$

Similarly expanding the vector potential (34.37b), we find

$$\mathbf{A}(\mathbf{r},t) \approx \frac{1}{c}\int (d\mathbf{r}')\frac{\mathbf{j}(\mathbf{r}',t)}{|\mathbf{r}-\mathbf{r}'|} - \frac{1}{c^2}\int (d\mathbf{r}')\frac{\partial}{\partial t}\mathbf{j}(\mathbf{r}',t) + \dots = \frac{1}{c}\int (d\mathbf{r}')\frac{\mathbf{j}(\mathbf{r}',t)}{|\mathbf{r}-\mathbf{r}'|} - \frac{1}{c^2}\ddot{\mathbf{d}} + \dots, \tag{36.6}$$

where we have used (35.24) and (35.25) for the electric dipole moment $\mathbf{d}$. The contribution of the $1/c^3$ term in ϕ to $\mathbf{E}$ can be simplified as follows:

$$-\boldsymbol{\nabla}\left[-\frac{1}{6c^3}\int (d\mathbf{r}')(\mathbf{r}-\mathbf{r}')^2\frac{\partial^3}{\partial t^3}\rho(\mathbf{r}',t)\right] = \left(\frac{d}{dt}\right)^3\frac{1}{3c^3}\int (d\mathbf{r}')(\mathbf{r}-\mathbf{r}')\rho(\mathbf{r}',t) = -\frac{1}{3c^3}\dddot{\mathbf{d}}. \tag{36.7}$$

The expression for the energy transfer, (36.3), becomes, in this approximation,

$$\begin{aligned}-\int (d\mathbf{r})\,\mathbf{j}\cdot\mathbf{E} &= -\int (d\mathbf{r})\,\mathbf{j}\cdot\left(-\boldsymbol{\nabla}\phi - \frac{1}{c}\frac{\partial}{\partial t}\mathbf{A}\right) = \int (d\mathbf{r})\left(\frac{\partial\rho}{\partial t}\phi + \frac{1}{c}\mathbf{j}\cdot\frac{\partial}{\partial t}\mathbf{A}\right) \\ &= \frac{d}{dt}\left\{\frac{1}{2}\int (d\mathbf{r})(d\mathbf{r}')\frac{\rho(\mathbf{r},t)\rho(\mathbf{r}',t)}{|\mathbf{r}-\mathbf{r}'|} + \frac{1}{4c^2}\int (d\mathbf{r})(d\mathbf{r}')\frac{\partial}{\partial t}\rho(\mathbf{r},t)\frac{\partial}{\partial t}\rho(\mathbf{r}',t)|\mathbf{r}-\mathbf{r}'|\right\} \\ &\quad + \frac{1}{3c^3}\dot{\mathbf{d}}\cdot\dddot{\mathbf{d}} + \frac{d}{dt}\left[\frac{1}{2c^2}\int (d\mathbf{r})(d\mathbf{r}')\frac{\mathbf{j}(\mathbf{r},t)\cdot\mathbf{j}(\mathbf{r}',t)}{|\mathbf{r}-\mathbf{r}'|}\right] - \frac{1}{c^3}\dot{\mathbf{d}}\cdot\dddot{\mathbf{d}} \end{aligned} \tag{36.8a}$$

$$\to -\frac{2}{3c^3}\dot{\mathbf{d}}\cdot\dddot{\mathbf{d}} \tag{36.8b}$$

$$= \frac{2}{3c^3}(\ddot{\mathbf{d}})^2 + \frac{d}{dt}\left(-\frac{2}{3c^3}\dot{\mathbf{d}}\cdot\ddot{\mathbf{d}}\right) \to \frac{2}{3c^3}(\ddot{\mathbf{d}})^2 = P, \tag{36.8c}$$

where we have again used (35.24) and have set aside total time derivative terms, which do not contribute to the radiation. The power radiated, (36.8c), is the same as that found in the preceding chapter, given by (35.27).

36.3 Hamiltonian

As a byproduct of this source approach we may identify the total electromagnetic energy of the system, to order $1/c^2$, by comparing (36.2) with the total time derivative terms in

(36.8a),

$$E(t) = \frac{1}{2}\int (d\mathbf{r})(d\mathbf{r}')\frac{\rho(\mathbf{r},t)\rho(\mathbf{r}',t)}{|\mathbf{r}-\mathbf{r}'|} + \frac{1}{2c^2}\int (d\mathbf{r})(d\mathbf{r}')\frac{\mathbf{j}(\mathbf{r},t)\cdot\mathbf{j}(\mathbf{r}',t)}{|\mathbf{r}-\mathbf{r}'|} + \frac{1}{4c^2}\int (d\mathbf{r})(d\mathbf{r}')\left(\frac{\partial}{\partial t}\rho(\mathbf{r},t)\frac{\partial}{\partial t}\rho(\mathbf{r}',t)\right)|\mathbf{r}-\mathbf{r}'| + \dots, \tag{36.9}$$

where the first two terms have the form of the electrostatic and magnetostatic field energies.

However, the sign of the current-current interaction is opposite to that of (29.13). The resolution of this apparent discrepancy requires the third term in (36.9), which we rewrite by means of the local charge conservation condition, (34.2):

$$\begin{aligned}
&\frac{1}{4c^2}\int (d\mathbf{r})(d\mathbf{r}')\boldsymbol{\nabla}\cdot\mathbf{j}(\mathbf{r},t)\boldsymbol{\nabla}'\cdot\mathbf{j}(\mathbf{r}',t)|\mathbf{r}-\mathbf{r}'| \\
&= -\frac{1}{4c^2}\int (d\mathbf{r})(d\mathbf{r}')\left(\boldsymbol{\nabla}'\cdot\mathbf{j}(\mathbf{r}',t)\,\mathbf{j}(\mathbf{r},t)\cdot\frac{\mathbf{r}-\mathbf{r}'}{|\mathbf{r}-\mathbf{r}'|}\right) \\
&= \frac{1}{4c^2}\int (d\mathbf{r})(d\mathbf{r}')\,\mathbf{j}(\mathbf{r}',t)\cdot\boldsymbol{\nabla}'\left[\mathbf{j}(\mathbf{r},t)\cdot\frac{\mathbf{r}-\mathbf{r}'}{|\mathbf{r}-\mathbf{r}'|}\right] \\
&= \frac{1}{4c^2}\int (d\mathbf{r})(d\mathbf{r}')\left\{-\frac{\mathbf{j}(\mathbf{r},t)\cdot\mathbf{j}(\mathbf{r}',t)}{|\mathbf{r}-\mathbf{r}'|} + \frac{[(\mathbf{r}-\mathbf{r}')\cdot\mathbf{j}(\mathbf{r},t)][(\mathbf{r}-\mathbf{r}')\cdot\mathbf{j}(\mathbf{r}',t)]}{|\mathbf{r}-\mathbf{r}'|^3}\right\},
\end{aligned} \tag{36.10}$$

which, when combined with the first two terms in (36.9) yields

$$E(t) = \frac{1}{2}\int (d\mathbf{r})(d\mathbf{r}')\frac{\rho(\mathbf{r},t)\rho(\mathbf{r}',t)}{|\mathbf{r}-\mathbf{r}'|} + \frac{1}{4c^2}\int (d\mathbf{r})(d\mathbf{r}')\left\{\frac{\mathbf{j}(\mathbf{r},t)\cdot\mathbf{j}(\mathbf{r}',t)}{|\mathbf{r}-\mathbf{r}'|} + \frac{[(\mathbf{r}-\mathbf{r}')\cdot\mathbf{j}(\mathbf{r},t)][(\mathbf{r}-\mathbf{r}')\cdot\mathbf{j}(\mathbf{r}',t)]}{|\mathbf{r}-\mathbf{r}'|^3}\right\}. \tag{36.11}$$

For point charges, the charge and current densities are

$$\rho(\mathbf{r},t) = \sum_a e_a\delta(\mathbf{r}-\mathbf{r}_a(t)), \tag{36.12a}$$

$$\mathbf{j}(\mathbf{r},t) = \sum_a e_a\mathbf{v}_a\delta(\mathbf{r}-\mathbf{r}_a(t)), \tag{36.12b}$$

so the terms in (36.11) referring to the mutual interaction of the particles are

$$E_{\text{field}} = \frac{1}{2}\sum_{a\neq b}\frac{e_ae_b}{r_{ab}} + \frac{1}{4c^2}\sum_{a\neq b}e_ae_b\left[\frac{\mathbf{v}_a\cdot\mathbf{v}_b}{r_{ab}} + \frac{(\mathbf{r}_{ab}\cdot\mathbf{v}_a)(\mathbf{r}_{ab}\cdot\mathbf{v}_b)}{r_{ab}^3}\right], \tag{36.13}$$

where

$$\mathbf{r}_{ab} \equiv \mathbf{r}_a - \mathbf{r}_b. \tag{36.14}$$

To this we must add the particle kinetic energy, from (11.11),

$$E_{\text{particle}} = \sum_a m_ac^2\left(\left(1-\frac{v_a^2}{c^2}\right)^{-1/2} - 1\right) \approx \sum_a\left(\frac{1}{2}m_av_a^2 + \frac{3}{8}m_a\frac{v_a^4}{c^2} + \dots\right), \quad \frac{v_a^2}{c^2} \ll 1, \tag{36.15}$$

in order to obtain the total energy of the system:

$$E_{\text{total}} = E_{\text{field}} + E_{\text{particle}}. \tag{36.16}$$

We now wish to describe this mechanical system in Hamiltonian language. We recall [from (9.15)] that the Hamiltonian is related to the Lagrangian by

$$H = \sum_a \mathbf{p}_a \cdot \mathbf{v}_a - L(\{\mathbf{r}_b\}, \{\mathbf{v}_b\}), \tag{36.17}$$

where the canonical momentum $\mathbf{p}_a$ is defined by

$$\mathbf{p}_a = \frac{\partial L(\{\mathbf{r}_b\}, \{\mathbf{v}_b\})}{\partial \mathbf{v}_a}. \tag{36.18}$$

To determine the Lagrangian, we substitute (36.18) into (36.17),

$$H = \sum_a \frac{\partial L(\{\mathbf{r}_b\}, \{\mathbf{v}_b\})}{\partial \mathbf{v}_a} \cdot \mathbf{v}_a - L(\{\mathbf{r}_b\}, \{\mathbf{v}_b\}), \tag{36.19}$$

where H is given in terms of $\mathbf{v}$ by (36.16). Equating terms on either side of (36.19) that are independent of $\mathbf{v}$, quadratic in $\mathbf{v}$, and quartic in $\mathbf{v}$, we find for the Lagrangian

$$L = \sum_a \left(\frac{1}{2}m_a v_a^2 + \frac{1}{8}m_a \frac{v_a^4}{c^2}\right) + \frac{1}{4c^2}\sum_{\substack{a,b\\a\neq b}} e_a e_b \left(\frac{\mathbf{v}_a \cdot \mathbf{v}_b}{r_{ab}} + \frac{(\mathbf{r}_{ab}\cdot\mathbf{v}_a)(\mathbf{r}_{ab}\cdot\mathbf{v}_b)}{r_{ab}^3}\right) - \frac{1}{2}\sum_{\substack{a,b\\a\neq b}} \frac{e_a e_b}{r_{ab}}. \tag{36.20}$$

Note that the particle term in (36.20) agrees with the expansion of (11.20). The canonical momenta are given by (36.18),

$$\mathbf{p}_a = m_a \mathbf{v}_a + \frac{1}{2}m_a \frac{v_a^2 \mathbf{v}_a}{c^2} + \frac{1}{2c^2}\sum_{b\neq a} e_a e_b \left[\frac{\mathbf{v}_b}{r_{ab}} + \frac{\mathbf{r}_{ab}(\mathbf{r}_{ab}\cdot\mathbf{v}_b)}{r_{ab}^3}\right], \tag{36.21}$$

in terms of which the Hamiltonian of the system is

$$H = \frac{1}{2}\sum_{\substack{a,b\\a\neq b}} \frac{e_a e_b}{r_{ab}} + \sum_a \left(\frac{p_a^2}{2m_a} - \frac{1}{8}\frac{p_a^4}{m_a^3 c^2}\right) - \frac{1}{4c^2}\sum_{\substack{a,b\\a\neq b}} \frac{e_a e_b}{m_a m_b}\left[\frac{\mathbf{p}_a\cdot\mathbf{p}_b}{r_{ab}} + \frac{(\mathbf{r}_{ab}\cdot\mathbf{p}_a)(\mathbf{r}_{ab}\cdot\mathbf{p}_b)}{r_{ab}^3}\right], \tag{36.22}$$

where we have consistently kept terms up to order $1/c^2$. This form of the Hamiltonian is appropriate for small systems and has application in both atomic and nuclear physics. It is called the Darwin[1] Hamiltonian when it is applied classically, and the Breit[2] Hamiltonian when it is applied quantum mechanically (with an accompanying re-expression in terms of Dirac matrices).

From the general discussion given in Chapter 10, the canonical momentum can be expressed in terms of the vector potential

$$\mathbf{p}_a = \frac{m_a \mathbf{v}_a}{\sqrt{1 - v_a^2/c^2}} + \frac{e_a}{c}\mathbf{A}(\mathbf{r}_a, t), \tag{36.23}$$

where we have used (11.11) in generalizing (10.12b). Upon comparison with (36.21), we obtain an explicit form of the vector potential, $\mathbf{A}(\mathbf{r}_a, t)$, which, when generalized to an arbitrary position, reads

$$\begin{aligned}\mathbf{A}(\mathbf{r}, t) &= \sum_b \frac{e_b}{2c}\left[\frac{\mathbf{v}_b}{|\mathbf{r}-\mathbf{r}_b|} + \frac{[(\mathbf{r}-\mathbf{r}_b)\cdot\mathbf{v}_b](\mathbf{r}-\mathbf{r}_b)}{|\mathbf{r}-\mathbf{r}_b|^3}\right]\\ &= \sum_b \frac{e_b}{c}\left[\frac{\mathbf{v}_b}{|\mathbf{r}-\mathbf{r}_b|} - \frac{1}{2}\boldsymbol{\nabla}(\mathbf{v}_b\cdot\boldsymbol{\nabla}|\mathbf{r}-\mathbf{r}_b|)\right].\end{aligned} \tag{36.24}$$

[1] Charles Galton Darwin, 1867–1962.

[2] Gregory Alfredovitch Breit-Schneider, 1899–1981.

We notice that this vector potential satisfies the radiation gauge condition (34.7):

$$\boldsymbol{\nabla}\cdot\mathbf{A} = \sum_b \frac{e_b}{c}\left[\mathbf{v}_b\cdot\boldsymbol{\nabla}\frac{1}{|\mathbf{r}-\mathbf{r}_b|} - \frac{1}{2}\mathbf{v}_b\cdot\boldsymbol{\nabla}(\nabla^2|\mathbf{r}-\mathbf{r}_b|)\right] = 0, \tag{36.25}$$

since

$$\nabla^2|\mathbf{r}-\mathbf{r}_b| = \frac{2}{|\mathbf{r}-\mathbf{r}_b|}, \tag{36.26}$$

and, as required, it also satisfies the differential equation (34.8b) to order $1/c^2$:

$$\begin{aligned} -\Box^2\mathbf{A} \approx -\nabla^2\mathbf{A} &= \sum_b \frac{e_b}{c}\left[4\pi\mathbf{v}_b\delta(\mathbf{r}-\mathbf{r}_b) + \frac{1}{2}\boldsymbol{\nabla}\left(\mathbf{v}_b\cdot\boldsymbol{\nabla}\frac{2}{|\mathbf{r}-\mathbf{r}_b|}\right)\right] \\ &\approx \frac{4\pi}{c}\mathbf{j}(\mathbf{r},t) - \boldsymbol{\nabla}\frac{1}{c}\frac{\partial}{\partial t}\phi(\mathbf{r},t). \end{aligned} \tag{36.27}$$

Furthermore, if we rewrite the last term of (36.22) in terms of $\mathbf{A}$, (36.24), we obtain

$$-\frac{1}{2c}\sum_a e_a\mathbf{v}_a\cdot\mathbf{A}(\mathbf{r}_a,t) = -\frac{1}{2c}\int(d\mathbf{r})\,\mathbf{j}(\mathbf{r},t)\cdot\mathbf{A}(\mathbf{r},t), \tag{36.28}$$

which has the form of the magnetostatic field energy, (29.11), which is therefore correctly given by (36.9). As we commented in Chapter 29, because of the negative sign in (36.28), "like" currents attract each other.

36.4 Problems for Chapter 36

1. Verify the Lagrangian (36.20) and the Hamiltonian (36.22).
2. Fill in the steps in the derivation of (36.27).
3. Obtain the result (35.65) using the source viewpoint as discussed in Section 36.2, that is, by expanding in powers of $1/c$.
4. The total energy transferred from the current to the electromagnetic field is

$$E = -\int_{-\infty}^{\infty} dt\int(d\mathbf{r})\mathbf{E}(\mathbf{r},t)\cdot\mathbf{j}(\mathbf{r},t). \tag{36.29}$$

   Write this in terms of Fourier transforms. Insert the solution to the wave equations satisfied by $\mathbf{E}(\mathbf{r};\omega)$ (34.40) to obtain

$$E = 4\pi\int\frac{d\omega}{2\pi}\frac{(d\mathbf{k})}{(2\pi)^3}\left(\frac{-i\omega}{c}\right)\frac{[|\mathbf{j}(\mathbf{k},\omega)|^2 - c^2|\rho(\mathbf{k},\omega)|^2]}{k^2-(\omega+i\epsilon)^2/c^2}. \tag{36.30}$$

37

Models of Antennas

We first discuss a simplified model of an antenna, and then a more realistic model of a center-fed antenna. As we will see, the radiation patterns are quite different in the two cases.

37.1 Simplified Model

We have been discussing a small system, in which the time delay effects are not great. For an example of the opposite situation, let us consider an oversimplified model of an antenna. In this model we have a wire of length l, and of negligible cross section, carrying a current density flowing in the z direction:

$$J_z = I\delta(x)\delta(y)\sin\omega t, \quad -\frac{l}{2} \leq z \leq \frac{l}{2}, \tag{37.1}$$

which has the property

$$\boldsymbol{\nabla}\cdot\mathbf{J} = \frac{\partial}{\partial z}J_z = \begin{cases} 0 & \text{for} \quad -\frac{l}{2} < z < \frac{l}{2}, \\ \neq 0 & \text{for} \quad z = \pm\frac{l}{2}. \end{cases} \tag{37.2}$$

From the local charge conservation condition (34.2),

$$0 = \frac{\partial}{\partial t}\rho + \frac{\partial}{\partial z}J_z, \tag{37.3}$$

we see that (37.2) implies that there is an oscillating charge density at both ends of the antenna. In any realistic model, J_z will depend on z. We lack this dependence since our model assumes that the antenna is fed at every point along its length. Even though this model is oversimplified, it possesses many of the significant characteristics of a real antenna. To compute the power radiated, (35.15), we evaluate the integral

$$\begin{aligned} &\int (d\mathbf{r}')\frac{1}{c}\frac{\partial}{\partial t}J_z\left(\mathbf{r}', t - \frac{r}{c} + \frac{1}{c}\mathbf{n}\cdot\mathbf{r}'\right) \\ &= \frac{\omega}{c}I\int_{-l/2}^{l/2} dz' \cos\omega\left(t - \frac{r}{c} + \frac{1}{c}z'\cos\theta\right) = \frac{\omega}{c}I\int_{-l/2}^{l/2} dz' \cos\omega\left(t - \frac{r}{c}\right)\cos\left(\frac{\omega z'}{c}\cos\theta\right) \\ &= \frac{2\frac{\omega}{c}I\cos\omega\left(t - \frac{r}{c}\right)\sin\left(\frac{\omega l}{2c}\cos\theta\right)}{\frac{\omega}{c}\cos\theta}, \end{aligned} \tag{37.4}$$

where, as indicated in Fig. 37.1, θ denotes the angle between the direction of observation and the antenna. The angular distribution of the radiated power, at the observation time t, is then

$$\frac{dP(t)}{d\Omega} = \frac{1}{\pi c}\sin^2\theta\frac{I^2\cos^2\omega\left(t - \frac{r}{c}\right)\sin^2\left(\frac{\omega l}{2c}\cos\theta\right)}{\cos^2\theta}, \tag{37.5}$$

DOI: 10.1201/9781003057369-37

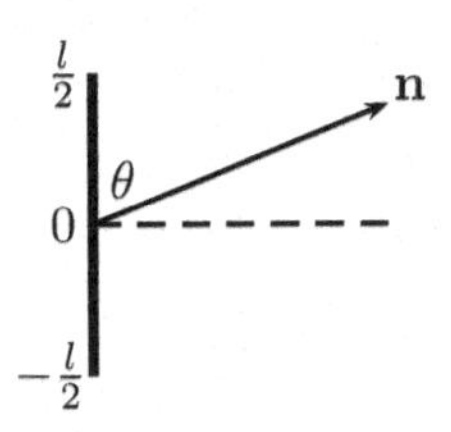

FIGURE 37.1
Geometry of linear antenna.

which, when averaged over one cycle of oscillation, becomes

$$\frac{dP}{d\Omega} = \frac{I^2}{2\pi c} \frac{\sin^2\theta \sin^2\left(\frac{\omega l}{2c}\cos\theta\right)}{\cos^2\theta}. \tag{37.6}$$

To rewrite (37.6) in terms of more convenient parameters, we recognize that, far from the antenna, the fields oscillate periodically both in space and in time:

$$E, H \sim \left\{\begin{matrix}\cos\\ \sin\end{matrix}\right\}\left(\omega t - \frac{\omega}{c}r\right) = \left\{\begin{matrix}\cos\\ \sin\end{matrix}\right\}\left(2\pi\nu t - \frac{2\pi\nu}{c}r\right), \tag{37.7}$$

so we identify the relation between the frequency ν (Hertz) and wavelength λ to be

$$\lambda\nu = c, \quad \omega = 2\pi\nu, \tag{37.8a}$$

or

$$\frac{\omega}{c} = \frac{2\pi}{\lambda}. \tag{37.8b}$$

Therefore, the radiation of the system is characterized by two parameters: the length of the antenna, l, and the wavelength of the radiation, λ. The combination that appears in the expression for the power, (37.6), is

$$\frac{\omega}{c}\frac{l}{2} = \frac{\pi l}{\lambda}, \tag{37.9}$$

in terms of which the angular distribution is

$$\frac{dP}{d\Omega} = \frac{I^2}{2\pi c} \frac{\sin^2\theta \sin^2\left(\frac{\pi l}{\lambda}\cos\theta\right)}{\cos^2\theta}. \tag{37.10}$$

In particular, in the direction perpendicular to the antenna, $\theta = \frac{\pi}{2}$, the radiated power is proportional to the square of the length of the antenna,

$$\left.\frac{dP}{d\Omega}\right|_{\theta=\pi/2} = \frac{I^2}{2\pi c}\left(\frac{\pi l}{\lambda}\right)^2. \tag{37.11}$$

To appreciate the characteristic features of this radiation, we will consider the application of this general formula, (37.10), to three special circumstances:

1. $\lambda \gg l$.

   For a short antenna, $l \ll \lambda$, the approximation

$$\frac{\sin^2\left(\frac{\pi l}{\lambda}\cos\theta\right)}{\cos^2\theta} \approx \left(\frac{\pi l}{\lambda}\right)^2 \tag{37.12}$$

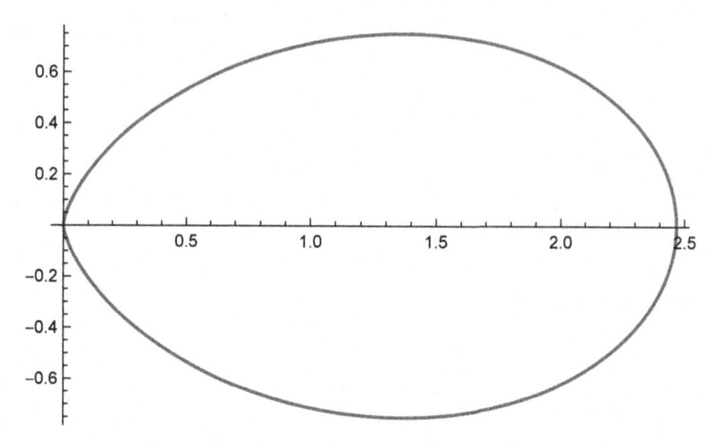

FIGURE 37.2
Radiation pattern produced by a short antenna, $l < \lambda$. In this and the following figures axial symmetry about the z-axis (the vertical axis–the line of the antenna) is to be understood. This particular polar plot is for $l/\lambda = 0.5$. In this, and subsequent figures, the factor $I^2/(2\pi c)$ has been removed.

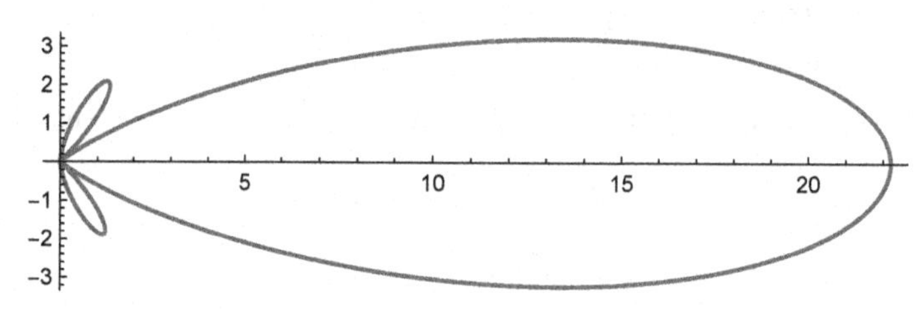

FIGURE 37.3
Radiation from an antenna of intermediate length, $\lambda < l < 2\lambda$. This diagram is meant to be understood schematically only. Actually, the side lobes are far smaller than indicated, so to make them visible, they have been multiplied by a factor of 10. This polar plot is for $l/\lambda = 1.5$.

holds true for all angles. The resulting radiation pattern may be alternatively derived [Problem 37.1] from the dipole radiation formula, (35.26), which is appropriate to a small system:

$$\frac{dP}{d\Omega} \approx \frac{I^2}{2\pi c}\left(\frac{\pi l}{\lambda}\right)^2 \sin^2\theta. \tag{37.13}$$

2. $\lambda > l$.

   When $\frac{\pi l}{\lambda} < \pi$, the argument of the factor $\sin^2\left(\frac{\pi l}{\lambda}\cos\theta\right)$ goes from 0 to something less than π when the angle θ varies from $\frac{\pi}{2}$ to 0. Therefore, the only angles at which the power radiated vanishes are 0 and π, so the radiation pattern has a single lobe. (See Fig. 37.2.)

3. $\lambda < l < 2\lambda$.

   When $\pi < \frac{\pi l}{\lambda} < 2\pi$ there is an additional zero in the radiated power at the angle $\theta = \cos^{-1}\left(\frac{\lambda}{l}\right)$. Consequently the radiation pattern exhibits both a main lobe and two side lobes. See Fig. 37.3. Evidently, as l/λ increases, more and more side lobes appear.

The total power radiated by this antenna may be obtained by integrating (37.10) over all angles:

$$P = \int_0^\pi 2\pi \sin\theta \, d\theta \frac{I^2}{2\pi c} \frac{\sin^2\theta \sin^2\left(\frac{\pi l}{\lambda}\cos\theta\right)}{\cos^2\theta} = \frac{I^2}{c}\int_{-\pi/2}^{\pi/2} \cos\chi \, d\chi \cos^2\chi \frac{\sin^2\left(\frac{\pi l}{\lambda}\sin\chi\right)}{\sin^2\chi}$$

$$= \frac{2I^2}{c}\left(\frac{\pi l}{\lambda}\right)\int_0^{\pi l/\lambda} dz \frac{\sin^2 z}{z^2}\left[1 - \frac{z^2}{(\pi l/\lambda)^2}\right], \tag{37.14}$$

where we have made the successive changes of variables,

$$\chi = \frac{\pi}{2} - \theta, \quad z = \frac{\pi l}{\lambda}\sin\chi. \tag{37.15}$$

If $\frac{\pi l}{\lambda} \gg 1$, the second term in the square brackets in (37.14) is negligible compared with the first for the significant values of z, so we find

$$P \approx \frac{2I^2}{c}\frac{\pi l}{\lambda}\int_0^\infty dz \frac{\sin^2 z}{z^2} = \frac{\pi^2 I^2}{c}\frac{l}{\lambda}, \tag{37.16}$$

where we have used the integral

$$\int_0^\infty dz \frac{\sin^2 z}{z^2} = \int_0^\infty dz \frac{\sin 2z}{z} = \int_0^\infty dt \frac{\sin t}{t} = \frac{\pi}{2}. \tag{37.17}$$

Here we observe that the total radiated power, (37.16), increases linearly with l, while the power radiated in the direction perpendicular to the antenna, (37.11), is proportional to l^2; that is, as the length of the antenna increases, a larger and larger fraction of the radiated power is concentrated near $\theta = \pi/2$.

To see how much energy is radiated into a very small angular range near $\chi = 0$ (or $\theta = \pi/2$), we consider the power radiated into the main lobe by a long antenna

$$0 < \chi < \frac{\lambda}{l} \ll 1. \tag{37.18}$$

Following the same procedure used to obtain the total power radiated, (37.14), we find for the total power radiated into the main lobe

$$P_{\text{main lobe}} \approx \frac{2I^2}{c}\int_0^{\lambda/l} d\chi \frac{\sin^2\left(\frac{\pi l}{\lambda}\chi\right)}{\chi^2} = \frac{2I^2}{c}\frac{\pi l}{\lambda}\int_0^\pi dz \frac{\sin^2 z}{z^2}. \tag{37.19}$$

The fraction of the energy radiated into the main lobe is obtained by taking the ratio of (37.19) to (37.16):

$$\frac{2}{\pi}\int_0^\pi dz \frac{\sin^2 z}{z^2} = \frac{2}{\pi}\int_0^{2\pi} dt \frac{\sin t}{t} = \frac{2}{\pi}\int_0^\infty dt \frac{\sin t}{t} - \frac{2}{\pi}\int_{2\pi}^\infty dt \frac{\sin t}{t}$$

$$= 1 - \frac{1}{\pi^2} + \frac{1}{2\pi^4} - \cdots = 0.9028, \tag{37.20}$$

where the infinite series is derived by integrating by parts repeatedly. Over 90% of the power is radiated into the main lobe, which has angular width λ/l, implying that the radiation from the antenna is highly directional, a feature characteristic of large systems. In contrast, small systems, for which the dipole approximation is valid, typically have angular distributions proportional to $\sin^2\theta$.

37.2 Center-Fed Antenna

The previous model is greatly oversimplified, of course, and the results (see Problem 37.2) greatly de-emphasize the importance of radiation into the side lobes. Therefore, we turn to the consideration of what might seem to be a somewhat more realistic model, that of a *center-fed, linear antenna*, described by

$$J_z = I \sin\left(\frac{kl}{2} - k|z|\right) \delta(x)\delta(y) \sin \omega t, \quad |z| < l/2, \tag{37.21}$$

where now we have introduced the wavenumber $k = \omega/c = 2\pi/\lambda$. Now, when we repeat the steps in (37.4), we find

$$\begin{aligned}
&\int (d\mathbf{r}') \frac{1}{c}\frac{\partial}{\partial t} J_z(\mathbf{r}', t - \frac{r}{c} + \frac{1}{c}\mathbf{n}\cdot\mathbf{r}') \\
&= kI \int_{-l/2}^{l/2} dz' \sin\left(\frac{kl}{2} - k|z'|\right) \cos\omega\left(t - \frac{r}{c} + \frac{z'}{c}\cos\theta\right) \\
&= 2kI \int_0^{l/2} dz' \sin\left(\frac{kl}{2} - kz'\right) \cos\left(kz'\cos\theta\right) \cos\omega\left(t - \frac{r}{c}\right) \\
&= kI \int_0^{l/2} dz' \left[\sin k\left(z'(\cos\theta - 1) + \frac{l}{2}\right) - \sin k\left(z'(\cos\theta + 1) - \frac{l}{2}\right)\right] \cos\omega\left(t - \frac{r}{c}\right) \\
&= \frac{2I\cos\omega\left(t - \frac{r}{c}\right)}{\sin^2\theta}\left[\cos\left(\frac{kl}{2}\cos\theta\right) - \cos\frac{kl}{2}\right].
\end{aligned} \tag{37.22}$$

The power radiated into a given solid angle is then given by (35.15), or

$$\frac{dP}{d\Omega} = \frac{1}{4\pi c}\sin^2\theta \frac{4I^2\cos^2\omega\left(t - \frac{r}{c}\right)}{\sin^4\theta}\left[\cos\left(\frac{kl}{2}\cos\theta\right) - \cos\frac{kl}{2}\right]^2, \tag{37.23}$$

which, when averaged over one cycle, gives

$$\frac{dP}{d\Omega} = \frac{I^2}{2\pi c}\frac{1}{\sin^2\theta}\left[\cos\left(\frac{kl}{2}\cos\theta\right) - \cos\frac{kl}{2}\right]^2. \tag{37.24}$$

Notice for a short antenna, $\lambda \gg l$, we recover the dipole formula:

$$\frac{dP}{d\Omega} = \frac{(I/2)^2}{2\pi c}\left(\frac{\pi l}{\lambda}\right)^4 \sin^2\theta, \tag{37.25}$$

which may alternatively be derived from the dipole radiation formula, (35.26). See Problem 37.1.

The angular distribution of the power radiated, $dP/d\Omega$, vanishes whenever

$$\cos\theta = \pm\left(1 - \frac{4\pi n}{kl}\right), \quad n = 1, 2, \ldots. \tag{37.26}$$

Thus side lobes appear in the radiation pattern whenever kl passes through a multiple of 2π. Plots of $dP/d\Omega$ are given in Figs. 37.4 and 37.5. It will be noted that now as $l/\lambda \to \infty$, substantial energy is radiated into the extreme side lobes, increasingly near the z axis, the antenna direction. See Problem 37.5. This is very similar to the radiation pattern produced by impulsive scattering—see Chapter 40. Therefore, such an antenna is most useful in the half- or full-wave regime.

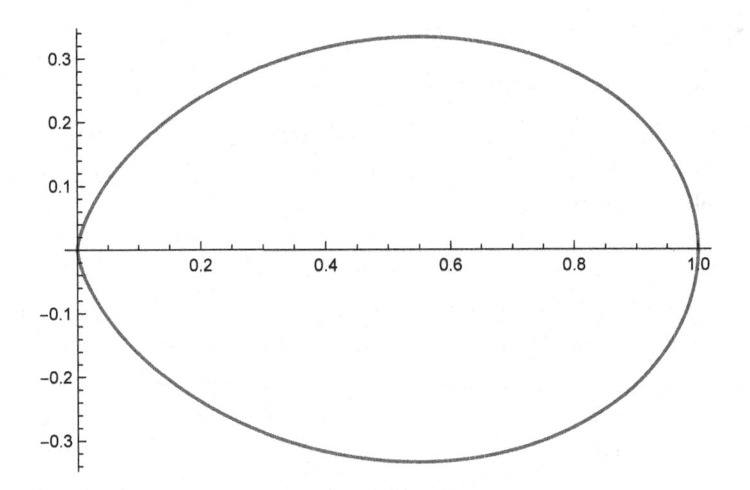

FIGURE 37.4
Radiation pattern produced by center-fed antenna for $kl = \pi$. This is called a half-wave antenna because $l = \lambda/2$, or $l/\lambda = 1/2$.

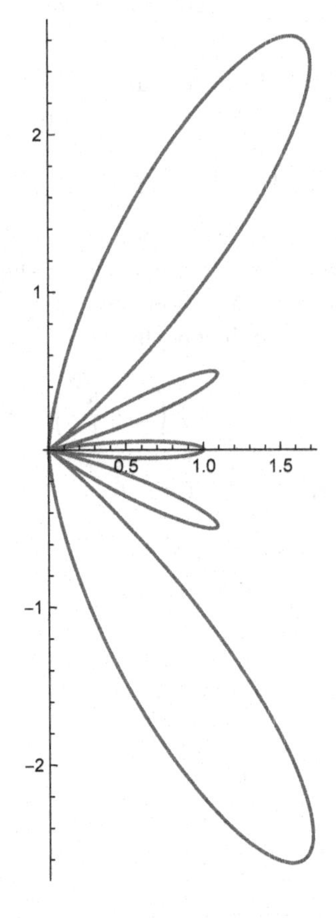

FIGURE 37.5
Radiation pattern produced by center-fed antenna for $kl = 5\pi$, or $l/\lambda = 5/2$.

37.3 Problems for Chapter 37

1. Derive (37.13) and (37.25) from the dipole radiation formula, (35.26).
2. Using Mathematica, Maple, Mathcad, or any other computer program of your choice, plot accurately the radiation pattern produced by the simplified antenna for $l = \lambda/2$, $l = 3\lambda/2$, $l = 5\lambda$, and $l = 13\lambda/2$. Calculate the fraction of power in the main lobe and check how closely the limit (37.20) is approached.
3. Verify the result (37.20) by computing a sufficient number of terms in the series, and by direct numerical integration. Give an error estimate for your answer.
4. Obtain formulas for the angular distribution of radiated power for a half-wave center fed antenna, $kl = \pi$, and for a full-wave antenna. Plot the latter, and compare with Figure 37.4.
5. Consider (37.24) for $kl \gg 1$. Compute the total power radiated in this limit, and compare with the power radiated in the two extreme side lobes, corresponding to $n = [l/\lambda]$ and $n = 1$ in (37.26), where $[x]$ denotes the largest integer less than or equal to x.
6. Compare the power radiated at $\theta = \pi/2$ for the center-fed antenna to that for the simple model described by (37.11). What happens in particular for $kl = n\pi$, where n is an integer?
7. A straight wire of negligible thickness and infinite length carries a current that varies in time and with distance z along the wire, according to the relation
$$I(z,t) = I_0 \cos(\kappa z - \omega t).$$
It is placed in an infinite, uniform medium. Prove that this current does not radiate if the propagation constant κ exceeds k, the intrinsic wavenumber of the external medium. Verify that radiation does occur if $\kappa < k$, and that the time-averaged power radiated per unit length of the wire is given by
$$\frac{k\pi}{2c} I_0^2 \left(1 - \frac{\kappa^2}{k^2}\right).$$
Note that this agrees with (37.16) if $\kappa = 0$.

38

Spectral Distribution of Radiation

38.1 Spectral and Angular Distribution

In the previous chapters, we discussed the angular distribution of the radiation produced by a time varying charge and current distribution. Here we will turn our attention to the spectral characteristics of this radiation, that is, its dependence on frequency, or wavelength. To investigate this dependence, we return to our starting point, the potentials in the Lorenz gauge, given by (34.36a) and (34.36b),

$$\phi(\mathbf{r},t) = \int (d\mathbf{r}')\, dt' \frac{\delta\left(\frac{1}{c}|\mathbf{r}-\mathbf{r}'| - (t-t')\right)}{|\mathbf{r}-\mathbf{r}'|}\rho(\mathbf{r}',t'), \tag{38.1a}$$

$$\mathbf{A}(\mathbf{r},t) = \int (d\mathbf{r}')\, dt' \frac{\delta\left(\frac{1}{c}|\mathbf{r}-\mathbf{r}'| - (t-t')\right)}{|\mathbf{r}-\mathbf{r}'|}\frac{1}{c}\mathbf{j}(\mathbf{r}',t'). \tag{38.1b}$$

In deriving these results, we had used the spectral representation [*cf.* (34.31)]

$$\delta\left(\frac{1}{c}|\mathbf{r}-\mathbf{r}'| - (t-t')\right) = \int_{-\infty}^{\infty} \frac{d\omega}{2\pi} e^{i\omega[|\mathbf{r}-\mathbf{r}'|/c-(t-t')]}. \tag{38.2}$$

If we now reinsert (38.2) into (38.1a) and (38.1b), and carry out the t' integration by introducing the temporal Fourier transform

$$\int_{-\infty}^{\infty} dt'\, e^{i\omega t'} f(t') = f(\omega), \tag{38.3a}$$

$$\int_{-\infty}^{\infty} \frac{d\omega}{2\pi} e^{-i\omega t'} f(\omega) = f(t'), \tag{38.3b}$$

we obtain the Fourier transformed versions of (38.1a) and (38.1b):

$$\phi(\mathbf{r},\omega) = \int (d\mathbf{r}') \frac{e^{i\omega|\mathbf{r}-\mathbf{r}'|/c}}{|\mathbf{r}-\mathbf{r}'|}\rho(\mathbf{r}',\omega), \tag{38.4a}$$

$$\mathbf{A}(\mathbf{r},\omega) = \int (d\mathbf{r}') \frac{e^{i\omega|\mathbf{r}-\mathbf{r}'|/c}}{|\mathbf{r}-\mathbf{r}'|}\frac{1}{c}\mathbf{j}(\mathbf{r}',\omega). \tag{38.4b}$$

[Of course, these expressions follow immediately from (34.30).] We observe that if $f(t')$ is a real function of t', its Fourier transform, $f(\omega)$, satisfies the condition

$$f(\omega)^* = f(-\omega); \tag{38.5}$$

consequently,

$$f(\omega)^* f(\omega) = |f(\omega)|^2 = f(-\omega)f(\omega) \tag{38.6}$$

is a real positive number, symmetric under the interchange $\omega \to -\omega$. This implies that the algebraic sign of ω is not significant, since only its magnitude enters into physical quantities.

DOI: 10.1201/9781003057369-38

Let us focus our attention on the radiation fields, far from the sources. Following the procedure given in Section 35.1, in particular, using the expansion (35.1) for $|\mathbf{r}-\mathbf{r}'|$ in (38.4a) and (38.4b), we obtain the asymptotic expression for the potentials, in terms of spatial Fourier transforms,

$$\phi(\mathbf{r},\omega)\sim\frac{e^{i\omega r/c}}{r}\int(d\mathbf{r}')e^{-i\omega\mathbf{n}\cdot\mathbf{r}'/c}\rho(\mathbf{r}',\omega), \tag{38.7a}$$

$$\mathbf{A}(\mathbf{r},\omega)\sim\frac{e^{i\omega r/c}}{r}\int(d\mathbf{r}')e^{-i\omega\mathbf{n}\cdot\mathbf{r}'/c}\frac{1}{c}\mathbf{j}(\mathbf{r}',\omega), \tag{38.7b}$$

where $\mathbf{n}=\mathbf{r}/r$. Evidently, the effectiveness of radiation with a given wavelength and direction of propagation depends upon the Fourier analysis of the time and spatial dependences of the charges and currents. In the exponential, the term $\omega\mathbf{n}\cdot\mathbf{r}'/c$, which is of the order of the ratio of the size of the system to the wavelength of the radiation, is significant for all but small systems. We call

$$\mathbf{k}=\frac{\omega}{c}\mathbf{n} \tag{38.8}$$

the propagation vector, in terms of which the potentials are written as Fourier transforms in space and time,

$$\begin{Bmatrix}\phi\\ \mathbf{A}\end{Bmatrix}(\mathbf{r},\omega)\sim\frac{e^{i\omega r/c}}{r}\begin{Bmatrix}\rho\\ \frac{1}{c}\mathbf{j}\end{Bmatrix}(\mathbf{k},\omega), \tag{38.9}$$

where, for example,

$$\rho(\mathbf{k},\omega)=\int(d\mathbf{r}')\,e^{-i\mathbf{k}\cdot\mathbf{r}'}\rho(\mathbf{r}',\omega). \tag{38.10}$$

The corresponding field strengths can now be computed from the time Fourier transforms of (34.3a) and (34.3b):

$$\mathbf{E}(\mathbf{r},\omega)=i\frac{\omega}{c}\mathbf{A}(\mathbf{r},\omega)-\boldsymbol{\nabla}\phi(\mathbf{r},\omega), \tag{38.11a}$$

$$\mathbf{B}(\mathbf{r},\omega)=\boldsymbol{\nabla}\times\mathbf{A}(\mathbf{r},\omega), \tag{38.11b}$$

where we have used the effective replacement

$$\frac{\partial}{\partial t}\to-i\omega, \tag{38.12a}$$

because

$$\int_{-\infty}^{\infty}dt\,e^{i\omega t}\frac{\partial}{\partial t}F(\mathbf{r},t)=-i\omega F(\mathbf{r},\omega), \tag{38.12b}$$

provided that $F(\mathbf{r},t)\to0$ as $t\to\pm\infty$. Physically, the time boundary conditions state that in the infinite past and the infinite future, nothing is happening: what is significant for our observation takes place in a finite time interval only. The measure of significant variation in r is the wavelength, $\lambda=2\pi c/\omega$, as indicated by the effective replacement for the gradients in (38.11a) and (38.11b),

$$\boldsymbol{\nabla}\to i\mathbf{k} \tag{38.13}$$

[recall (35.5)]. Consequently, the asymptotic forms of the electric and magnetic fields are

$$\mathbf{E}(\mathbf{r},\omega)\sim i\frac{\omega}{c}\frac{e^{i\omega r/c}}{r}\frac{1}{c}\mathbf{j}(\mathbf{k},\omega)-i\frac{\omega}{c}\frac{e^{i\omega r/c}}{r}\mathbf{n}\rho(\mathbf{k},\omega), \tag{38.14a}$$

$$\mathbf{B}(\mathbf{r},\omega)\sim i\frac{\omega}{c}\frac{e^{i\omega r/c}}{r}\mathbf{n}\times\frac{1}{c}\mathbf{j}(\mathbf{k},\omega). \tag{38.14b}$$

By using the Fourier transformed version of the local charge conservation condition, (34.2),

$$\omega\rho(\mathbf{k},\omega) = \mathbf{k}\cdot\mathbf{j}(\mathbf{k},\omega), \tag{38.15}$$

we may rewrite the second term of (38.14a) as

$$-\frac{i}{c}\frac{e^{i\omega r/c}}{r}\mathbf{n}\,\mathbf{k}\cdot\mathbf{j}(\mathbf{k},\omega) = -\frac{i\omega}{c}\frac{e^{i\omega r/c}}{r}\mathbf{n}\,\mathbf{n}\cdot\frac{1}{c}\mathbf{j}(\mathbf{k},\omega). \tag{38.16}$$

The electric field now becomes

$$\begin{aligned}\mathbf{E}(\mathbf{r},\omega) &\sim \frac{e^{i\omega r/c}}{r}i\frac{\omega}{c}(\mathbf{1}-\mathbf{n}\,\mathbf{n})\cdot\frac{1}{c}\mathbf{j}(\mathbf{k},\omega) = -\frac{e^{i\omega r/c}}{r}i\frac{\omega}{c}\mathbf{n}\times\left[\mathbf{n}\times\frac{1}{c}\mathbf{j}(\mathbf{k},\omega)\right]\\ &\sim -\mathbf{n}\times\mathbf{B}(\mathbf{r},\omega),\end{aligned} \tag{38.17}$$

which reconfirms (35.11a).

Before proceeding, we remark that the relation between the two terms in (38.14a) can also be obtained by using the Lorenz gauge condition; this is not surprising since the consistency of the Lorenz gauge depends upon current conservation [see (34.14)]. The Fourier transform of (34.12) reads

$$\boldsymbol{\nabla}\cdot\mathbf{A}(\mathbf{r},\omega) - i\frac{\omega}{c}\phi(\mathbf{r},\omega) = 0, \tag{38.18}$$

which becomes, upon using the asymptotic replacement (38.13),

$$ik[\mathbf{n}\cdot\mathbf{A}(\mathbf{r},\omega) - \phi(\mathbf{r},\omega)] = 0, \tag{38.19a}$$

or

$$\phi(\mathbf{r},\omega) \sim \mathbf{n}\cdot\mathbf{A}(\mathbf{r},\omega), \tag{38.19b}$$

so from (38.11a) the reduction (38.17) follows.

The instantaneous flux of energy, at a particular time t, is given by Poynting's vector

$$\mathbf{S}(\mathbf{r},t) = \frac{c}{4\pi}\mathbf{E}(\mathbf{r},t)\times\mathbf{B}(\mathbf{r},t), \tag{38.20}$$

so the total radiated energy crossing a unit area of surface normal to $\mathbf{S}$ is

$$\begin{aligned}\int_{-\infty}^{\infty} dt\,\mathbf{S}(\mathbf{r},t) &= \frac{c}{4\pi}\int_{-\infty}^{\infty} dt\int_{-\infty}^{\infty}\frac{d\omega}{2\pi}\mathbf{E}(\mathbf{r},\omega)^* e^{i\omega t}\times\mathbf{B}(\mathbf{r},t) = \frac{c}{4\pi}\int_{-\infty}^{\infty}\frac{d\omega}{2\pi}\mathbf{E}(\mathbf{r},\omega)^*\times\mathbf{B}(\mathbf{r},\omega)\\ &= \mathbf{n}\frac{c}{4\pi}\int_{-\infty}^{\infty}\frac{d\omega}{2\pi}|\mathbf{B}(\mathbf{r},\omega)|^2,\end{aligned} \tag{38.21}$$

where we use (38.17), and the fact that $\mathbf{n}\cdot\mathbf{B} = 0$. The energy flows in the direction of $\mathbf{n}$ and the energy radiated per unit area perpendicular to this direction is

$$\int_{-\infty}^{\infty} dt\,\mathbf{n}\cdot\mathbf{S}(\mathbf{r},t) = \frac{c}{4\pi^2}\int_0^{\infty} d\omega\,|\mathbf{B}(\mathbf{r},\omega)|^2, \tag{38.22}$$

where we have used the symmetry property (38.6). As before, it is more useful to consider the total energy radiated into the solid angle $d\Omega$ [see (35.14)],

$$\int_{-\infty}^{\infty} dt\,(\mathbf{n}\cdot\mathbf{S})r^2 d\Omega = d\Omega\,r^2\frac{c}{4\pi^2}\int_0^{\infty} d\omega\,|\mathbf{B}(\mathbf{r},\omega)|^2 \equiv d\Omega\frac{dE}{d\Omega} \equiv d\Omega\int_0^{\infty} d\omega\frac{d^2E}{d\omega d\Omega}, \tag{38.23}$$

where [*cf.* Problem 38.1]

$$\frac{d^2E}{d\omega d\Omega} = \frac{\omega^2}{4\pi^2c^3}\left|\mathbf{n}\times\mathbf{j}(\mathbf{k},\omega)\right|^2 = \frac{\omega^2}{4\pi^2c}\left\{\left|\frac{1}{c}\mathbf{j}(\mathbf{k},\omega)\right|^2 - |\rho(\mathbf{k},\omega)|^2\right\} \tag{38.24}$$

is the general expression for the spectral distribution, the energy radiated per unit frequency per unit solid angle in the direction of observation $\mathbf{n}$. This equation is the analog of (35.15) for $dP/d\Omega$.

38.2 Spectral Distribution for Dipole Radiation

As an application of the general result, (38.24), we consider a small system with typical length a much smaller than the reduced wavelength of the radiation, that is, $\lambdabar \equiv \lambda/2\pi \gg a$. Then, the exponential factor in the Fourier transforms in (38.7a) and (38.7b) can be approximated by unity, since

$$\frac{\omega}{c}\mathbf{n}\cdot\mathbf{r}' = \frac{1}{\lambdabar}\mathbf{n}\cdot\mathbf{r}' \ll 1, \tag{38.25}$$

whence the Fourier transform of the current becomes that of the time derivative of the electric dipole moment, according to (35.24) and (35.25):

$$\mathbf{j}(\mathbf{k},\omega) = \int dt\, e^{i\omega t}\frac{d}{dt}\mathbf{d}(t) = -i\omega\mathbf{d}(\omega). \tag{38.26}$$

Then, the spectral distribution becomes

$$\frac{d^2E}{d\omega d\Omega} \approx \frac{\omega^2}{4\pi^2c^3}\left|\mathbf{n}\times\dot{\mathbf{d}}(\omega)\right|^2 = \frac{\omega^2}{4\pi^2c^3}\sin^2\theta\left|\dot{\mathbf{d}}(\omega)\right|^2, \tag{38.27}$$

where θ is the angle between the observation direction and the direction of $\mathbf{d}(\omega)$. For the small system discussed here, the only reference to the direction of observation, $\mathbf{n}$, occurs as a multiplicative factor, implying the $\sin\theta$ behavior exhibited above, characteristic of dipole radiation. For larger systems, $\mathbf{n}$ also enters in the exponential so that the angular distribution could be completely different. The total energy radiated per unit frequency range can be obtained by integrating (38.27) over all angles,

$$\frac{dE}{d\omega} = \int d\Omega\frac{d^2E}{d\omega d\Omega} = \frac{2}{3}\frac{\omega^2}{\pi c^3}\left|\dot{\mathbf{d}}(\omega)\right|^2, \tag{38.28}$$

where we have used (35.21).

Suppose we further specialize to a single point charge in nonrelativistic motion, corresponding to the current density

$$\mathbf{j}(\mathbf{r},t) = e\mathbf{v}(t)\delta(\mathbf{r}-\mathbf{r}(t)), \tag{38.29}$$

in which case, of course,

$$\dot{\mathbf{d}}(t) = e\mathbf{v}(t), \tag{38.30a}$$

of which the Fourier transform is

$$\dot{\mathbf{d}}(\omega) = e\mathbf{v}(\omega). \tag{38.30b}$$

The energy radiated per unit frequency interval, $dE/d\omega$, is

$$\frac{dE}{d\omega} = \frac{2}{3}\frac{e^2}{\pi c^3}\omega^2|\mathbf{v}(\omega)|^2 = \frac{2}{3}\frac{e^2}{\pi c^3}|\dot{\mathbf{v}}(\omega)|^2, \tag{38.31}$$

where $\dot{\mathbf{v}}(\omega)$ is the Fourier transform of $\dot{\mathbf{v}}(t)$:

$$\dot{\mathbf{v}}(\omega) = \int_{-\infty}^{\infty} dt\, e^{i\omega t}\dot{\mathbf{v}}(t) = -i\omega\mathbf{v}(\omega). \tag{38.32}$$

The total energy radiated,

$$E_{\text{rad}} = \int_0^\infty \frac{dE}{d\omega}d\omega = \frac{2}{3}\frac{e^2}{\pi c^3}\int_0^\infty d\omega\,\omega^2|\mathbf{v}(\omega)|^2, \tag{38.33}$$

can also be obtained from the Larmor formula, (35.22):

$$E_{\rm rad} = \int_{-\infty}^{\infty} dt\, P(t) = \frac{2}{3}\frac{e^2}{c^3}\int_{-\infty}^{\infty} dt\, |\dot{\mathbf{v}}(t)|^2 = \frac{2}{3}\frac{e^2}{\pi c^3}\int_0^{\infty} d\omega\, \omega^2 |\mathbf{v}(\omega)|^2, \tag{38.34}$$

as expected. In the above, we have used (38.32), and the theorem

$$\int_{-\infty}^{\infty} dt\, |f(t)|^2 = \int_{-\infty}^{\infty} \frac{d\omega}{2\pi}|f(\omega)|^2 = \int_0^{\infty} \frac{d\omega}{\pi}|f(\omega)|^2. \tag{38.35}$$

(This is the Parseval[1]-Plancherel[2] identity.) Thus we see that we can calculate the total energy radiated by a small system through the use either of the spectral distribution, (38.28), or of the Larmor formula for the power, (35.27). The equivalence of these two descriptions is demonstrated generally in Problem 38.1, where the spectral distribution $d^2E/d\omega d\Omega$, (38.24), is derived directly from the angular distribution of radiated power, $dP(t)/d\Omega$, (35.15).

38.3 Damped Harmonic Motion

As a further simple application of the spectral distribution, consider a model in which a charged particle undergoes damped motion in a Hooke's law potential,

$$\ddot{\mathbf{r}} = -\omega_0^2 \mathbf{r} - \gamma\dot{\mathbf{r}}, \tag{38.36}$$

where, due to the radiation produced by the accelerating charged particle, there is a damping force represented by $-\gamma\dot{\mathbf{r}}$, which we will assume to be small:

$$\frac{\gamma}{\omega_0} \ll 1. \tag{38.37}$$

This model is often taken as an oversimplified description of a bound electron inside an atom. (Recall Section 5.2.) Given the initial conditions that at $t = 0$, the particle has a displacement $\mathbf{a}$ from the force center and has zero velocity, the solution to (38.36) when (38.37) holds is approximately

$$\mathbf{r}(t) \approx \mathbf{a}\cos\omega_0 t\, e^{-\gamma t/2}, \quad \text{for} \quad t > 0, \tag{38.38}$$

which exhibits the fact that many oscillations are necessary before significant damping occurs. The velocity of the particle,

$$\mathbf{v}(t) \approx -\mathbf{a}\,\omega_0 \sin\omega_0 t\, e^{-\gamma t/2}, \quad \text{for} \quad t > 0, \tag{38.39}$$

has the Fourier transform

$$\begin{aligned}\mathbf{v}(\omega) &= \int_0^{\infty} dt\,(-\mathbf{a}\,\omega_0)e^{i\omega t}\sin\omega_0 t\, e^{-\gamma t/2} = \frac{i}{2}\mathbf{a}\,\omega_0\int_0^{\infty} dt\left[-e^{i(\omega-\omega_0+i\gamma/2)t} + e^{i(\omega+\omega_0+i\gamma/2)t}\right] \\ &= \frac{\omega_0\mathbf{a}}{2}\left(\frac{1}{\omega-\omega_0+i\gamma/2} - \frac{1}{\omega+\omega_0+i\gamma/2}\right).\end{aligned} \tag{38.40}$$

Without loss of generality, we may assume $\omega > 0$ here, in which case the two terms are very different. Only the first denominator can be small, implying that radiation is predominantly

[1]Marc-Antoine Parseval, 1755-1836.
[2]Michel Plancherel, 1885-1967.

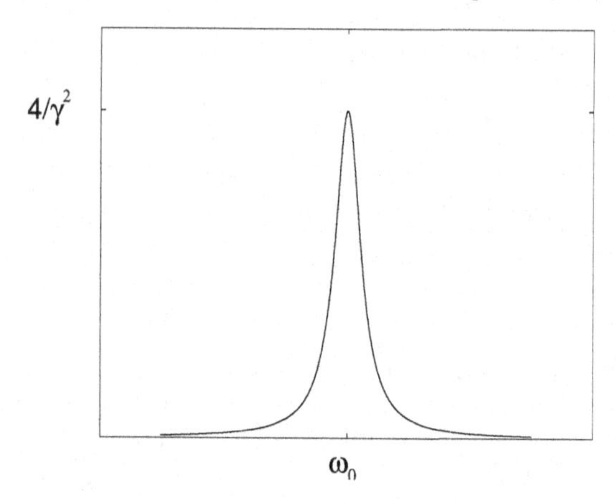

FIGURE 38.1
Lorentzian line shape for the energy radiated per unit frequency range.

emitted with frequencies $\omega \approx \omega_0$. Therefore, we approximate the square of the magnitude of (38.40) by

$$|\mathbf{v}(\omega)|^2 \approx \frac{\omega_0^2 a^2}{4} \frac{1}{(\omega - \omega_0)^2 + \gamma^2/4}, \tag{38.41}$$

implying for the energy radiated per unit frequency range, (38.31),

$$\frac{dE}{d\omega} \approx \frac{2}{3} \frac{e^2}{\pi c^3} \omega_0^2 \frac{\omega_0^2 a^2}{4} \frac{1}{(\omega - \omega_0)^2 + \gamma^2/4}, \quad \text{for} \quad \omega \sim \omega_0. \tag{38.42}$$

The behavior of $[(\omega - \omega_0)^2 + \gamma^2/4]^{-1}$ is plotted in Fig. 38.1, which exhibits what is called the Lorentzian line shape. Since when $|\omega - \omega_0| = \gamma/2$, the intensity is half that at ω_0, where the intensity is maximum, γ is the width of the spectrum at half-maximum intensity. In the limit of negligible damping, $\gamma \to 0$, only one frequency, ω_0, is emitted, with infinite intensity. In general, a range of frequencies is emitted, with γ being a measure of the sharpness of the spectrum. That is, if the system decays slowly (γ small), the emission line is very narrow, while if it decays rapidly (γ large), it is very broad.

To demonstrate that no significant amount of energy is radiated outside of this sharp peak, we evaluate the total energy radiated in the peak by integrating (38.42) over all frequencies:

$$\begin{aligned} E_{\text{rad}} &= \int_0^\infty d\omega \, \frac{dE}{d\omega} \approx \frac{2}{3} \frac{e^2}{\pi c^3} \omega_0^2 \left(\frac{\omega_0^2 a^2}{4} \right) \int_0^\infty \frac{d\omega}{(\omega - \omega_0)^2 + \gamma^2/4} \\ &\approx \frac{2}{3} \frac{e^2}{\pi c^3} \omega_0^2 \frac{\omega_0^2 a^2}{4} \frac{2\pi}{\gamma}, \end{aligned} \tag{38.43}$$

where we have used the integral

$$\int_0^\infty \frac{d\omega}{(\omega - \omega_0)^2 + \gamma^2/4} = \int_{-2\omega_0/\gamma}^\infty \frac{(\gamma/2) dx}{(\gamma^2/4)(1 + x^2)} \approx \frac{2}{\gamma} \int_{-\infty}^\infty \frac{dx}{1 + x^2} = \frac{2\pi}{\gamma}. \tag{38.44}$$

We wish to compare this radiated energy to the original energy of the oscillator. Since the latter is given by

$$E_{\text{initial}} = \frac{1}{2} m \omega_0^2 a^2, \tag{38.45}$$

the radiated energy, (38.43), can be rewritten as

$$E_{\text{rad}} = E_{\text{initial}} \left(\frac{2}{3} \frac{e^2}{mc^3} \frac{\omega_0^2}{\gamma} \right). \tag{38.46}$$

Assuming that the Lorentzian peak adequately accounts for the energy radiated and that there are no other forms of energy dissipation, we learn, from the conservation of energy,

$$\gamma = \frac{2}{3} \frac{e^2}{mc^3} \omega_0^2. \tag{38.47}$$

Since this is the same result found in Problem 35.2 where we calculated the power radiated by the oscillator and identified γ from

$$P = -\frac{dE}{dt} = \gamma E, \quad E = \text{energy of oscillator}, \tag{38.48}$$

we conclude that (38.42) is an adequate representation of the energy spectrum. Besides thus demonstrating that most of the radiated energy is contained within the peak, we have also established self-consistency in that the damping of the oscillator is shown to arise from the reaction of the radiation back on the radiating system.

38.4 Problems for Chapter 38

1. Derive the expression (38.24) for the spectral distribution directly from the power spectrum $\frac{dP}{d\Omega}(t)$ given in Problem 35.8.
2. Suppose γ is not small so that (38.36) must be solved exactly. Calculate $dE/d\omega$, E_{rad} without approximation and compare the latter to (38.43).
3. Repeat the analysis in Section 38.1 in the Coulomb gauge. In particular show that the asymptotic vector potential is

$$\mathbf{A}(\mathbf{r}, \omega) \sim \frac{e^{i\omega r/c}}{r} \frac{1}{c} \left[\mathbf{j}(\mathbf{k}, \omega) - \mathbf{n} c \rho(\mathbf{k}, \omega) \right]. \tag{38.49a}$$

   Using current conservation show that

$$\mathbf{A}(\mathbf{r}, \omega) \sim \frac{e^{i\omega r/c}}{r} \frac{1}{c} \mathbf{n} \times (\mathbf{j} \times \mathbf{n}). \tag{38.49b}$$

   From this derive the formulas for **B** and **E**, (38.14b) and (38.17).
4. Derive a formula for the collisional broadening of spectral lines by computing the spectrum of a weakly damped oscillator that vibrates freely for only a finite time T. Assume that the probability of an oscillation time in excess of a particular value decreases exponentially,

$$\exp\left(-\frac{1}{2} \gamma_c T \right), \tag{38.50}$$

   and perform a statistical average.

5. Consider a charged particle undergoing acceleration for a finite time, where the velocity is (a is a constant)

$$\mathbf{v}(t) = \begin{cases} \hat{\mathbf{x}} a \cos \frac{\pi t}{\tau}, & -\frac{\tau}{2} < t < \frac{\tau}{2}, \\ 0, & \text{otherwise.} \end{cases} \tag{38.51}$$

Compute the Fourier transform of this as in (38.32) and find the spectral distribution of radiated energy from (38.31). At what frequencies is the spectral distribution zero? Is it infinite anywhere? Obtain an expression for the total energy radiated, in terms of the acceleration at $t = -\tau/2 + \epsilon$, $\epsilon \to +0$. Make the result look as similar as possible to Larmor's formula.

39

Power Spectrum and Čerenkov Radiation

39.1 Macroscopic Power Spectrum

In the previous chapter, we derived a general expression for the spectral distribution, (38.24), which accounts for the radiation from all times, $t \to -\infty$ to $t \to +\infty$. Here we wish to obtain a generalization applicable to a limited epoch. In so doing, we must note that time and frequency are complementary; that is, a time interval of many periods is required in order to identify a corresponding frequency. We first rewrite (38.24) in the space-time form:

$$\frac{d^2E}{d\omega d\Omega} = \frac{\omega^2}{4\pi^2 c^3}\left[\mathbf{n}\times\int dt\, e^{i\omega t}\mathbf{j}(\mathbf{k},t)\right]^* \cdot \left[\mathbf{n}\times\int dt'\, e^{i\omega t'}\mathbf{j}(\mathbf{k},t')\right], \tag{39.1}$$

and focus our attention on the part involving time integrations:

$$\int dt\, dt'\, e^{-i\omega(t-t')}\mathbf{j}(\mathbf{k},t)^*\,\mathbf{j}(\mathbf{k},t') = \int dT\, d\tau\, e^{-i\omega\tau}\mathbf{j}(\mathbf{k},T+\tau/2)^*\,\mathbf{j}(\mathbf{k},T-\tau/2), \tag{39.2}$$

where we have introduced the average time and the time difference,

$$T = \frac{1}{2}(t+t'), \quad \tau = t-t', \quad dt\, dt' = dT\, d\tau. \tag{39.3}$$

From the exponential structure of (39.2), we infer that the important range of τ that contributes to the integral is of order $1/\omega$, thus setting the time scale for the emission of radiation. This microscopic time scale may be much smaller than macroscopic time intervals; for example, for visible light, $\tau \sim 10^{-15}$ sec. The time T is then interpreted as the average (macroscopic) time of emission, which can be specified only to within a time of order τ. Substituting (39.2) into (39.1), we write

$$\frac{d^2E}{d\omega d\Omega} = \int dT\, \frac{d^2P(T)}{d\omega d\Omega}, \tag{39.4}$$

from which we infer the power spectrum at time T,

$$\frac{d^2P(T)}{d\omega d\Omega} = \frac{\omega^2}{4\pi^2 c}\int_{-\infty}^{\infty} d\tau\, e^{-i\omega\tau}\left[\mathbf{n}\times\frac{1}{c}\mathbf{j}(\mathbf{k},T+\tau/2)^*\right]\cdot\left[\mathbf{n}\times\frac{1}{c}\mathbf{j}(\mathbf{k},T-\tau/2)\right], \tag{39.5a}$$

or, alternatively, using the second form of (38.24),

$$\begin{aligned}\frac{d^2P(T)}{d\omega d\Omega} = \frac{\omega^2}{4\pi^2 c}\int_{-\infty}^{\infty} d\tau\, e^{-i\omega\tau}\bigg\{&\frac{1}{c}\mathbf{j}(\mathbf{k},T+\tau/2)^*\cdot\frac{1}{c}\mathbf{j}(\mathbf{k},T-\tau/2)\\ &-\rho(\mathbf{k},T+\tau/2)^*\rho(\mathbf{k},T-\tau/2)\bigg\}.\end{aligned} \tag{39.5b}$$

DOI: 10.1201/9781003057369-39

39.2 Čerenkov Radiation

As an application of (39.5b), we consider the radiation produced by a charged particle moving with constant velocity $\mathbf{v}$, for which the charge and current densities are

$$\rho(\mathbf{r},t) = e\delta(\mathbf{r}-\mathbf{v}t), \quad \mathbf{j}(\mathbf{r},t) = e\mathbf{v}\delta(\mathbf{r}-\mathbf{v}t). \tag{39.6}$$

The Fourier transforms in (39.5b) are trivially evaluated:

$$\int (d\mathbf{r}) e^{\pm i\omega\mathbf{n}\cdot\mathbf{r}/c} \left\{ \begin{matrix} \rho(\mathbf{r}, T\pm\tau/2) \\ \frac{1}{c}\mathbf{j}(\mathbf{r}, T\pm\tau/2) \end{matrix} \right\} = e^{\pm i\omega\mathbf{n}\cdot\mathbf{v}(T\pm\tau/2)/c} \left\{ \begin{matrix} e \\ \frac{e}{c}\mathbf{v} \end{matrix} \right\}. \tag{39.7}$$

When these are substituted into (39.5b), the T dependence disappears, as expected, and we obtain

$$\begin{aligned} \frac{d^2P}{d\omega d\Omega} &= \frac{\omega^2}{4\pi^2 c}\int_{-\infty}^{\infty} d\tau\, e^{-i\omega\tau} e^2\left(\frac{v^2}{c^2}-1\right) e^{i\omega\mathbf{n}\cdot\mathbf{v}\tau/c} \\ &= \frac{\omega^2}{4\pi^2 c} e^2\left(\frac{v^2}{c^2}-1\right)\int_{-\infty}^{\infty} d\tau\, e^{-i\omega\tau(1-\mathbf{n}\cdot\mathbf{v}/c)} \\ &= \frac{\omega^2 e^2}{4\pi^2 c}\left(\frac{v^2}{c^2}-1\right) 2\pi\delta\left(\omega(1-\frac{v}{c}\cos\theta)\right), \end{aligned} \tag{39.8}$$

where θ is the angle between the direction of observation, $\mathbf{n}$, and the velocity of the particle, $\mathbf{v}$. The δ function implies that there is no radiation, since

$$\frac{v}{c}\cos\theta < 1. \tag{39.9}$$

This is the familiar result that a charged particle moving with a constant velocity in vacuum does not radiate.

However, if it were possible that

$$\frac{v}{c} > 1, \tag{39.10}$$

the argument of the δ function could vanish, and radiation would be emitted by the charged particle. Is there any way of effectively satisfying (39.10)? In a medium, light can move with a speed, c', less than c, and correspondingly the speed of a particle can be greater than c'. Now does the particle radiate? Recall that the macroscopic Maxwell's equations, (4.56), for a medium with dielectric constant ϵ and magnetic permeability μ, can be put into vacuum form (1.61b), by the redefinitions (recall Section 8.1b)

$$\mathbf{E}' = \sqrt{\epsilon}\mathbf{E}, \quad \mathbf{H}' = \sqrt{\mu}\mathbf{H}, \quad c' = \frac{c}{\sqrt{\epsilon\mu}}, \quad \rho' = \frac{1}{\sqrt{\epsilon}}\rho, \quad \mathbf{J}' = \frac{1}{\sqrt{\epsilon}}\mathbf{J}. \tag{39.11}$$

Therefore, the power radiated when a charged particle is moving with constant velocity in a nonmagnetic medium ($\mu = 1$) of index of refraction $n = \sqrt{\epsilon}$ can be obtained immediately from (39.8) by the substitutions $e \to e/n$ and $c \to c/n$:

$$\frac{d^2P}{d\omega d\Omega} = \frac{\omega^2}{4\pi^2(c/n)}\left(\frac{e}{n}\right)^2\left(\frac{v^2}{(c/n)^2}-1\right) 2\pi\delta\left(\omega\left(1-\frac{v}{(c/n)}\cos\theta\right)\right). \tag{39.12}$$

Thus, indeed there is radiation if the condition

$$\frac{nv}{c}\cos\theta = 1, \tag{39.13a}$$

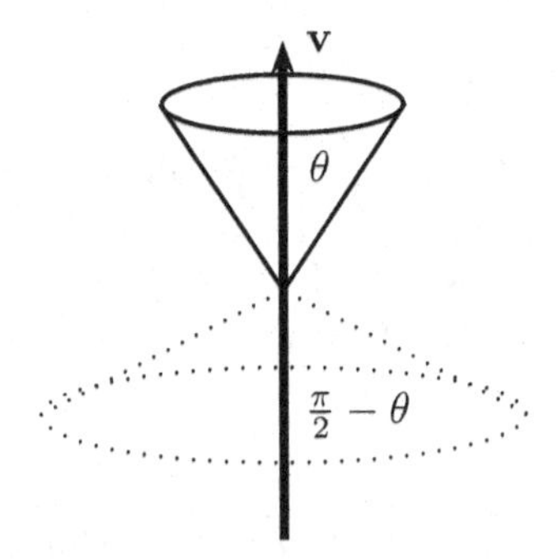

FIGURE 39.1
Cone of Čerenkov radiation produced at a definite wavelength by a particle traveling with a velocity **v**, v greater than the speed of light in the medium. Shown by the dotted curves are the wave front or shock wave orthogonal to the radiation direction, as discussed in Problems 5 and 6.

or

$$\cos\theta = \frac{c}{nv} < 1, \tag{39.13b}$$

is satisfied. Here we see that, for a charged particle moving with a constant velocity inside a medium characterized by an index of refraction $n > 1$, electromagnetic radiation can be emitted if the criterion

$$v > \frac{c}{n} \tag{39.14}$$

is satisfied. Such a medium can be easily found for fast particles. The radiation is emitted on a cone described by (39.13b) (see Fig. 39.1), and because of its unique characteristics, is especially suited for determining the velocities of relativistic charged particles. This phenomenon is called Čerenkov radiation. We emphasize that the condition (39.13b) can only be satisfied when $n > 1$. Because the index of refraction depends on frequency, that is, media are dispersive, this means that the condition (39.14) can only hold for a finite range of frequencies (typically, in the optical region). Moreover, this dispersion implies that different frequencies are emitted at different angles. Čerenkov radiation is commonly seen in water-moderated nuclear reactors as blue light surrounding the core. The effect is used for particle detection and identification, for example in the Super-Kamiokande neutrino observatory in Hida, Japan.

Historically, this radiation was first observed by Marie Curie in 1910, studied deliberately by Lucien Mallet in the 1920s, but only definitively explored by Pavel Alekseyevich Čerenkov from 1934 through 1938.[1] The theoretical explanation was given by Il'ja Mikhailovich Frank and Igor Evgen'evich Tamm in 1937. For a complete account of the history and application of the Čerenkov effect through the 1950s see Ref. [29]. For a more recent review, see Ref. [30].

The frequency spectrum of the radiated power can be obtained by integrating (39.12) over all angles,

$$\begin{aligned}\frac{dP}{d\omega} &= \int d\Omega \frac{d^2P}{d\omega d\Omega} = \omega^2 \frac{e^2}{nc}\frac{n^2v^2}{c^2}\left(1 - \frac{c^2}{n^2v^2}\right)\int_{-1}^{1} d(\cos\theta)\,\delta\Big(\omega\left(1 - \frac{nv}{c}\cos\theta\right)\Big) \\ &= \omega\frac{e^2 v}{c^2}\left(1 - \frac{c^2}{n^2(\omega)v^2}\right), \quad \text{if} \quad n(\omega) > \frac{c}{v},\end{aligned} \tag{39.15}$$

[1]This work was carried out under the direction of Sergei Ivanovich Vavilov (1891-1951), so the effect should be more properly called Vavilov-Čerenkov radiation.

and the total radiated power is

$$-\frac{dE}{dt} = P = \int_0^\infty d\omega \, \frac{dP}{d\omega} = \int d\omega \, \omega \frac{e^2 v}{c^2} \left(1 - \frac{c^2}{n^2(\omega) v^2}\right), \tag{39.16a}$$

where E is the energy of the particle. In practice, it is more convenient to consider the energy lost per unit distance traveled by the particle, since this is what can be directly measured by a Čerenkov counter:

$$-\frac{dE}{dz} = \int d\omega \, \omega \frac{e^2}{c^2} \left(1 - \frac{c^2}{n^2(\omega) v^2}\right). \tag{39.16b}$$

In (39.16a) and (39.16b) it is understood that the ω integration extends only over the range where $n(\omega) > c/v$. Finally we note that detection is a quantum process, involving photons. The energy of a photon of frequency $\nu = \omega/2\pi$ is $h\nu = \hbar\omega$, where $h = 2\pi\hbar$ is Planck's constant. Therefore, the number of photons emitted per unit length is

$$\frac{dN}{dz} = \alpha \int \frac{d\omega}{c} \left(1 - \frac{c^2}{n^2(\omega) v^2}\right), \tag{39.16c}$$

where the fine structure constant α is defined by

$$\alpha = \frac{e^2}{\hbar c} \approx \frac{1}{137}. \tag{39.17}$$

To obtain an order of magnitude estimate, we suppose that over a certain range of ω, $\Delta\omega$, the $1 - c^2/n^2 v^2$ factor is of order 1:

$$\frac{dN}{dz} \sim \alpha \frac{\Delta\omega}{c} \sim \frac{\alpha}{\lambda\!\!\!^{-}} = \frac{1}{137\lambda\!\!\!^{-}}, \tag{39.18}$$

where we have noted that, typically, the range of wavelengths is of the same order of magnitude as the wavelengths themselves. Equation (39.18) implies that, roughly speaking, in a distance of 137 wavelengths, one photon is emitted. In the visible spectrum, where $\lambda \sim 10^{-5}$ cm, about 10^3 photons/cm are emitted. (A more accurate estimate is 10^2 photons/cm.)

39.3 Problems for Chapter 39

1. Čerenkov light of a given wavelength is emitted on a cone of half-angle θ_c. Show that for small θ_c,

$$\theta_c \approx \sqrt{2\left(1 - \frac{1}{n\beta}\right)}, \tag{39.19}$$

where βc is the velocity of the particle moving in a medium with index of refraction n.

2. Show that the number of photons produced per unit path length of a particle with charge Ze and per unit energy interval of the photons, dE, is

$$\frac{d^2N}{dE\,dz} = \frac{\alpha Z^2}{\hbar c} \sin^2\theta_c = \frac{\alpha^2 Z^2}{r_0 m_e c^2} \left(1 - \frac{1}{\beta^2 n^2(E)}\right), \tag{39.20a}$$

where the *classical electron radius* $r_0 = e^2/m_e c^2$. For $Z = 1$, show that

$$\frac{d^2N}{dE\,dz} \approx 370 \sin^2\theta_c(E)\,\mathrm{eV}^{-1}\mathrm{cm}^{-1}. \tag{39.20b}$$

Equivalently, show that

$$\frac{d^2N}{dz\,d\lambda} = \frac{2\pi\alpha Z^2}{\lambda^2}\left(1 - \frac{1}{\beta^2 n^2(\lambda)}\right). \tag{39.20c}$$

3. Čerenkov detectors are often used in high-energy experiments to distinguish particles. Modern "ring-imaging" Čerenkov detectors can, for example, distinguish electrons from pions with momenta below 5 Gev/c. Discuss why such a detector cannot discriminate between particles of different mass if the momenta are too high, and make an estimate of the discrimination limit for e/π, π/K, K/p, assuming $\sin^2\theta_c$ can be measured to 0.1% accuracy.

4. An electric charge e moves faster than the speed of light in a medium characterized by $\epsilon = 1$, $\mu > 1$. Find the energy radiated per unit length. Repeat for a magnetic charge g and a medium with $\epsilon > 1$, $\mu = 1$.

5. The Lorenz gauge potentials for a particle of charge e moving uniformly in a dispersionless dielectric medium with permittivity ϵ are determined by

$$\mathbf{A} = e\frac{\mathbf{v}}{c}\phi, \quad -\left(\nabla^2 - \frac{\epsilon}{c^2}\frac{\partial^2}{\partial t^2}\right)\phi = 4\pi\frac{e}{\epsilon}\delta(\mathbf{r} - \mathbf{v}t). \tag{39.21}$$

Verify this. (Warning: The Lorenz gauge condition differs somewhat from its vacuum form.) Let $\mathbf{v}$ point along the z axis. Write the frequency transform of the differential equation, solve it, and perform the inversion that gives $\phi(\mathbf{r}, t)$. Consider the circumstances of Čerenkov radiation and show that ϕ, $\mathbf{A}$, and, therefore, the fields are zero if

$$vt - z < |\mathbf{r}_\perp|\sqrt{\beta^2\epsilon - 1}, \quad \beta = \frac{v}{c}. \tag{39.22}$$

Recognize in this the conical surface of the trailing "shock wave" and demonstrate that the radiation moving perpendicularly to that surface has the expected behavior.

6. This Problem sketches yet another approach to deriving Čerenkov radiation.

   (a) Starting from the Coulomb potential of a charge at the origin at rest, perform a Lorentz transformation to a frame in which the particle is moving with velocity $\mathbf{v}$ in the z direction, according to (11.43), and, in the Lorenz gauge, $A^\mu = (\phi, \mathbf{A})$ transform the same way.

   (b) Now using the substitution

   $$c \to \frac{c}{n}, \quad e \to \frac{e}{n}, \tag{39.23}$$

   obtain the potential for a particle moving through a medium with index of refraction n.

   (c) If $n > 1$ show there is, for a sufficiently large v, an angle θ between $\mathbf{r} - \mathbf{v}t$ and $\mathbf{v}$ at which the potentials become singular. Express $\sin\theta$ as a function of v/c and n.

(d) Intepret the singularity lines as the trailing wave front (shock wave), as in Problem 5.

7. (a) Show that the power emitted at macroscopic time T at the frequency ω can be expressed as

$$\frac{dP(T)}{d\omega} = \frac{\omega^2}{\pi c}\int_{-\infty}^{\infty} d\tau\, e^{-i\omega\tau}\int (d\mathbf{r})(d\mathbf{r}')\frac{\sin\frac{\omega}{c}|\mathbf{r}-\mathbf{r}'|}{\frac{\omega}{c}|\mathbf{r}-\mathbf{r}'|}$$
$$\times\left[\frac{1}{c}\mathbf{j}\left(\mathbf{r},T+\frac{\tau}{2}\right)\cdot\frac{1}{c}\mathbf{j}\left(\mathbf{r}',T-\frac{\tau}{2}\right)-\rho\left(\mathbf{r},T+\frac{\tau}{2}\right)\rho\left(\mathbf{r}',T-\frac{\tau}{2}\right)\right]. \tag{39.24}$$

(b) If c/ω is large compared to a typical length characterizing the source, we can expand the above sine function. Thus derive the dipole radiation formula (35.27). [As in (36.8a)–(36.8c), you can omit total time derivatives.]

(c) For a system with $\ddot{\mathbf{d}} = \mathbf{0}$, derive the next leading approximations for the radiated power:

$$\text{Magnetic dipole:}\quad P=\frac{2}{3}\frac{1}{c^3}(\ddot{\boldsymbol{\mu}})^2 \quad \text{[see (35.35)]}, \tag{39.25a}$$
$$\text{Electric quadrupole:}\quad P=\frac{1}{180}\frac{1}{c^5}\sum_{ij}(\dddot{\mathsf{q}}_{ij})^2 \quad \text{[see (35.63])}. \tag{39.25b}$$

8. Consider Maxwell's equations with both electric ($\rho_e, \mathbf{j}_e$) and magnetic ($\rho_m, \mathbf{j}_m$) charges. Derive second-order differential equations for $\mathbf{E}$ and $\mathbf{B}$. Show that

$$\mathbf{E} = -\boldsymbol{\nabla}\phi_e - \frac{1}{c}\frac{\partial}{\partial t}\mathbf{A}_e - \boldsymbol{\nabla}\times\mathbf{A}_m, \tag{39.26a}$$
$$\mathbf{B} = -\boldsymbol{\nabla}\phi_m - \frac{1}{c}\frac{\partial}{\partial t}\mathbf{A}_m + \boldsymbol{\nabla}\times\mathbf{A}_e, \tag{39.26b}$$

and exhibit the differential equations for these potentials in the Lorenz gauge, and in the radiation gauge.

9. Solve for the above potentials in some gauge, and find the asymptotic radiation field. Now what is the relationship between $\mathbf{E}$ and $\mathbf{B}$? Construct the spectral-angular distribution of the radiated power. How does it change when electric ($\mathcal{E}$) and magnetic ($\mathcal{M}$) quantities are redefined according to

$$\mathcal{E} \to \mathcal{E}\cos\phi + \mathcal{M}\sin\phi, \quad \mathcal{M} \to \mathcal{M}\cos\phi - \mathcal{E}\sin\phi? \tag{39.27}$$

10. A nonrelativistic particle of mass m and charge e moves in a circle of radius R with angular velocity ω_0. Assume the system to be small, so the dipole approximation is valid. Further assume that the radiation per cycle is small so that ω_0 and R may be regarded as constant. Using the Larmor formula (35.22), obtain an expression for the energy radiated in one cycle. Use the formula for the energy radiated per unit frequency, (38.31), to obtain an expression for the power spectrum $dP/d\omega$ in terms of R and ω_0. Describe the shape of the frequency spectrum. By integrating over ω, reproduce the result found from the Larmor formula. [Hint: You will encounter

$$[\delta(\omega-\omega_0)]^2 = \delta(\omega-\omega_0)\int\frac{dt}{2\pi}e^{i(\omega-\omega_0)t} = \delta(\omega-\omega_0)\frac{T}{2\pi}, \tag{39.28}$$

where T is the total time of existence of the system.] Now, start with the expression for the power spectrum radiated into a given solid angle, (39.5a), and by inserting the appropriate current density and making the dipole approximation, and, in addition, averaging over one cycle so that, for example, $\overline{\cos\omega_0 T} = 0$, find a formula for $d^2P/d\omega d\Omega$ in terms of ω_0, R, and θ, the angle between the perpendicular to the orbit and the direction of observation. Verify the result found above for the power spectrum by integrating over θ.

11. Consider a particle of charge e undergoing one-dimensional simple harmonic motion along the z axis:

$$z = a\cos\omega_0 t, \qquad v_z = -a\omega\sin\omega_0 t. \tag{39.29}$$

From the corresponding current density, obtain the power radiated at a given macroscopic time T at frequency ω into an element of solid angle $d\Omega$ as given by (39.5a), where $\mathbf{n}$ is the direction of observation, which makes an angle θ with respect to the z axis. Compute $dP/d\Omega$ by using the generating function for the Bessel functions, (18.27). Average over one cycle using

$$\langle e^{i(m-m')\omega_0 t}\rangle_{\text{one cycle}} = \delta_{mm'}, \tag{39.30}$$

and then by carrying out the integral over τ, find

$$\frac{dP(\omega)}{d\Omega} = \sum_{m=1}^{\infty}\delta(\omega - m\omega_0)\frac{dP_m}{d\Omega}, \tag{39.31}$$

and give the formula for $dP_m/d\Omega$ in terms of J_m. [Hint: Use the recursion formula (18.69b).] What can you say about the polarization of the radiation?

40

Constant Acceleration and Impulsive Scattering

In this chapter, we will discuss radiation by a uniformly accelerated particle, and by one which undergoes a sudden change in velocity, an impulse. Although it might, at first glance, appear that these two extreme situations are rather unrelated, we will see, particularly in the Problems, that the considerations for radiation are closely connected.

40.1 Radiation by a Uniformly Accelerated Particle

Next, let us examine the characteristics of the radiation emitted by a uniformly accelerated charged particle. For simplicity we will assume that the particle is nonrelativistic (which remains valid only for a finite length of time), but the conclusions we will draw are often thought to be independent of that simplification. (See Problem 40.3.) The equations of motion of such a particle are

$$\mathbf{r}(t) = \frac{1}{2}\mathbf{a}t^2, \tag{40.1a}$$

$$\mathbf{v}(t) = \mathbf{a}t, \tag{40.1b}$$

in terms of which we construct the current density

$$\mathbf{j}(\mathbf{r}, t) = e\mathbf{v}(t)\delta(\mathbf{r} - \mathbf{r}(t)). \tag{40.2}$$

The power spectrum, at time T, (39.5a), becomes

$$\begin{aligned}\frac{d^2P(T)}{d\omega d\Omega} &= \frac{\omega^2}{4\pi^2c^3}\int_{-\infty}^{\infty} d\tau\, e^{-i\omega\tau}(\mathbf{n}\times\mathbf{a})^2 e^2[T^2 - (\tau^2/4)] \\ &\quad\times \exp\left[i\frac{\omega}{c}\mathbf{n}\cdot\frac{\mathbf{a}}{2}\left([T + (\tau/2)]^2 - [T - (\tau/2)]^2\right)\right] \\ &= \frac{\omega^2e^2}{4\pi^2c^3}(\mathbf{n}\times\mathbf{a})^2\int_{-\infty}^{\infty} d\tau\, \exp\left[-i\omega\tau\left(1 - \frac{1}{c}\mathbf{n}\cdot\mathbf{v}(T)\right)\right][T^2 - (\tau/2)^2] \\ &= \frac{\omega^2e^2}{2\pi c^3}(\mathbf{n}\times\mathbf{a})^2\left\{T^2\delta\left(\omega(1 - \mathbf{n}\cdot\mathbf{v}/c)\right) + \frac{1}{4}\delta''\left(\omega(1 - \mathbf{n}\cdot\mathbf{v}/c)\right)\right\},\end{aligned} \tag{40.3}$$

where we have identified the second derivative of the δ-function according to

$$\int_{-\infty}^{\infty} d\tau\, \tau^2 e^{-i\tau x} = -2\pi\delta''(x). \tag{40.4}$$

For $\omega > 0$, the argument of the δ-functions in (40.3) never vanishes,

$$\frac{d^2P(T)}{d\omega d\Omega} = 0, \quad \text{for} \quad \omega > 0. \tag{40.5}$$

DOI: 10.1201/9781003057369-40

The power, (40.3), is nonzero only for $\omega = 0$, which corresponds to static fields. Such fields do not correspond to radiation, and a uniformly accelerated charged particle does not radiate. On the other hand, the total radiated power can be computed from the Larmor formula, (35.22):

$$P = \frac{2}{3}\frac{e^2}{c^3}(\dot{\mathbf{v}})^2 = \frac{2}{3}\frac{e^2}{c^3}a^2 \neq 0, \tag{40.6}$$

which seems to indicate that a uniformly accelerated charge does radiate. How can we reconcile the above two seemingly contradictory results, (40.5) and (40.6)? Actually there is no logical contradiction, because a constant power radiated would appear to correspond to a power spectrum which consists only of zero frequency, which, to reiterate, does not represent radiation. The Larmor formula thus is not applicable to this situation. (But, see later.)

It is instructive to consider this radiation process from the source point of view, discussed in Section 36.2. Remember that, there, we obtained the following expression for the rate at which charges in a small, nonrelativistic system do work on the electromagnetic field [*cf.* (36.8b)],

$$-\int (d\mathbf{r})\mathbf{E}\cdot\mathbf{j} = \frac{d}{dt}E - \frac{2}{3c^3}\dot{\mathbf{d}}\cdot\dddot{\mathbf{d}}, \tag{40.7}$$

where we identified E as the electromagnetic field energy while the remaining term is the power radiated. To obtain the Larmor formula, we neglected a further total time derivative. If we now use the above expression for the power radiated,

$$P = -\frac{2}{3c^3}\dot{\mathbf{d}}\cdot\dddot{\mathbf{d}} = -\frac{2}{3}\frac{e^2}{c^3}\mathbf{v}\cdot\ddot{\mathbf{v}}, \tag{40.8}$$

since the dipole moment of a point charge is

$$\mathbf{d} = e\mathbf{r}, \tag{40.9}$$

we immediately see that there is no radiation produced by a uniformly accelerated charged particle.

Having persuaded you that there is no radiation produced by uniform acceleration, we now enlarge the picture and recognize that radiation is associated with the whole history of a process, not just a particular period of time. What we have assumed so far is a situation, in which, for all time, the particle undergoes uniform acceleration, which, if nothing else, violates our nonrelativistic treatment since eventually the velocity of the particle will become comparable with the speed of light. The essential point to recognize is that uniform acceleration for all time is an idealization of the realistic situation in which uniform acceleration is only experienced for a finite time interval.

As an example, we consider the acceleration of a charged particle by a linear accelerator. Suppose originally the particle moves with a small constant speed $\mathbf{v}(0)$. At a particular time $t = 0$, it enters a region where a uniform electric field is applied, causing the particle to undergo a constant acceleration $\mathbf{a}$. After a period of time T, the particle is ejected from the accelerator with a velocity

$$\mathbf{v}(T) = \mathbf{v}(0) + \mathbf{a}T, \tag{40.10}$$

which is still assumed to be small compared to the speed of light. This acceleration process is represented in Fig. 40.1. The spectral distribution of the energy radiated by this accelerated particle is given by (38.31), which can be rewritten in the form

$$\frac{dE}{d\omega} = \frac{2}{3}\frac{e^2}{c^3}\frac{1}{\pi}\frac{1}{\omega^2}|\ddot{\mathbf{v}}(\omega)|^2, \tag{40.11}$$

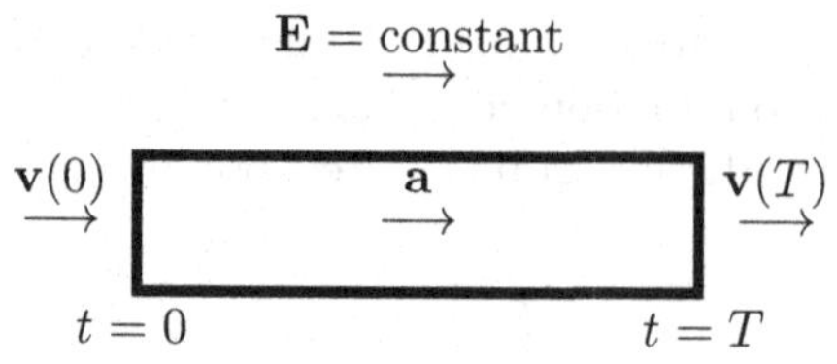

$\mathbf{E} = \text{constant}$ $\longrightarrow$

$\mathbf{v}(0)$ $\longrightarrow$ $\mathbf{a}$ $\longrightarrow$ $\mathbf{v}(T)$ $\longrightarrow$

$t = 0$ $t = T$

FIGURE 40.1
Simple model of a linear accelerator.

since, from (38.32),

$$\dot{\mathbf{v}}(\omega) = -i\omega\mathbf{v}(\omega), \tag{40.12a}$$

and

$$\ddot{\mathbf{v}}(\omega) = -i\omega\dot{\mathbf{v}}(\omega). \tag{40.12b}$$

The quantity $\mathbf{v}(\omega)$ can be calculated by noting that since the acceleration is discontinuous at $t = 0$ and $t = T$, the derivative of $\mathbf{a}$ has an impulse at these two times:

$$\begin{aligned}\ddot{\mathbf{v}}(\omega) &= \int_{-\infty}^{\infty} dt\, e^{i\omega t}\ddot{\mathbf{v}}(t) = \int_{-\infty}^{\infty} dt\, e^{i\omega t}[\mathbf{a}\delta(t) - \mathbf{a}\delta(t-T)] = \mathbf{a}\left(1 - e^{i\omega T}\right)\\ &= -2i\mathbf{a}\, e^{i\omega T/2}\sin\left(\frac{1}{2}\omega T\right).\end{aligned} \tag{40.13}$$

The square of the magnitude of (40.13) is

$$|\ddot{\mathbf{v}}(\omega)|^2 = 4a^2\sin^2\frac{1}{2}\omega T, \tag{40.14}$$

so that the spectral distribution of radiated energy is

$$\frac{dE}{d\omega} = \frac{2}{3}\frac{e^2}{c^3}\frac{1}{\pi}\frac{4a^2}{\omega^2}\sin^2\frac{1}{2}\omega T, \tag{40.15}$$

which is nonzero in general. Of course, there is no contradiction between this result and (40.5), since we are no longer talking about uniform acceleration for all time. We also observe that the contributions from the changes of acceleration at $t = 0$ and $t = T$ interfere with each other. This is incompatible with locality in time, that is, we cannot specify at which time the radiation is emitted.

To see what are the important frequencies being radiated, we plot the frequency distribution, (40.15), as a function of ω in Fig. 40.2. The significant range of frequencies is set by the time of acceleration, T,

$$\omega \sim 1/T, \tag{40.16}$$

that is, as T becomes longer, $dE/d\omega$ is more and more concentrated at low values of ω. In the limit when $T \to \infty$, we recover the situation of uniform acceleration discussed above. Moreover, we see again that the emission spectrum is not time analyzable, since the relation (40.16) implies that it takes a time of the order of the whole process to determine the frequency. The total energy radiated is obtained from (40.15) by integrating over all frequencies,

$$\begin{aligned}E_{\text{rad}} &= \int_0^{\infty} d\omega\,\frac{dE}{d\omega} = \frac{2}{3}\frac{e^2}{c^3}\frac{4a^2}{\pi}\int_0^{\infty}\frac{d\omega}{\omega^2}\sin^2\frac{\omega T}{2} = \frac{2}{3}\frac{e^2}{c^3}a^2T\left(\frac{2}{\pi}\int_0^{\infty}\frac{dx}{x^2}\sin^2 x\right)\\ &= \frac{2}{3}\frac{e^2}{c^3}a^2T,\end{aligned} \tag{40.17}$$

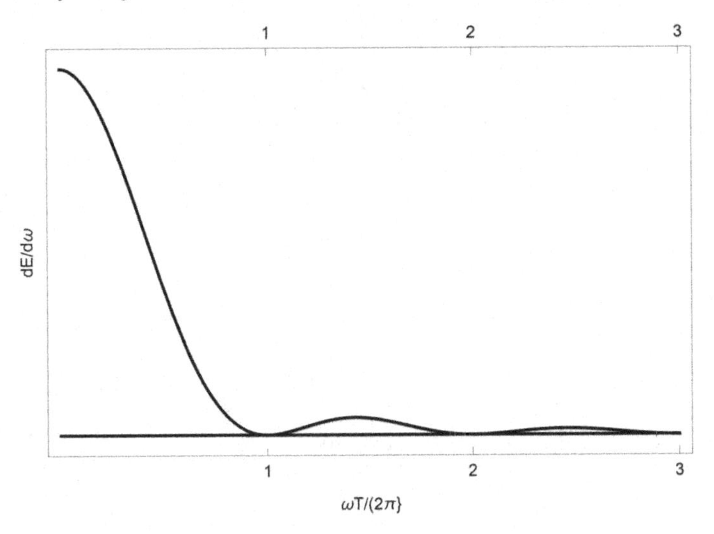

FIGURE 40.2
Spectral distribution of energy radiated by accelerator in Fig. 40.1, shown as a function of $\omega T/2\pi$.

where we have used the integral (37.17). From (40.17), we observe that the total energy radiated per unit time is

$$\frac{E_{\rm rad}}{T} = \frac{2}{3}\frac{e^2}{c^3}a^2, \tag{40.18}$$

which is precisely that obtained from the Larmor formula, (40.6), which refers only to uniform acceleration.

Finally let us calculate, from the source point of view, the radiated power using (40.8),

$$P = -\frac{2}{3}\frac{e^2}{c^3}\mathbf{v}\cdot\ddot{\mathbf{v}} = -\frac{2}{3}\frac{e^2}{c^3}\mathbf{a}\cdot[\mathbf{v}(0)\delta(t) - \mathbf{v}(T)\delta(t-T)], \tag{40.19}$$

which incorrectly attributes the radiation entirely to the beginning and the end of the acceleration process. However, the total radiated energy,

$$E_{\rm rad} = \int_{-\infty}^{\infty} dt\, P = \frac{2}{3}\frac{e^2}{c^3}\mathbf{a}\cdot[\mathbf{v}(T) - \mathbf{v}(0)] = \frac{2}{3}\frac{e^2}{c^3}a^2 T, \tag{40.20}$$

is the same as (40.17).

Thus we have seen that the question of whether there is radiation produced by a uniformly accelerated charge is only properly answered by taking into account the beginning and ending of the acceleration. Facetiously, we may say that a uniformly accelerated charge radiates because it is not uniformly accelerated.

Radiation by a uniformly accelerated charge has a long, controversial, and continuing history. Besides the work of Pauli [31] we should refer to Rohrlich [32, 33] and Boulware [34]. For a recent perspective, see Landulfo et al. [35], which describes some of the earlier literature. As we have seen, one controversy concerns the interpretation of the Larmor formula,

$$P_L = \frac{2}{3c^3}\ddot{\mathbf{d}}^2, \tag{40.21a}$$

versus that of the Abraham-Lorentz formula

$$P_{AL} = -\frac{2}{3c^3}\dot{\mathbf{d}}\cdot\dddot{\mathbf{d}}. \tag{40.21b}$$

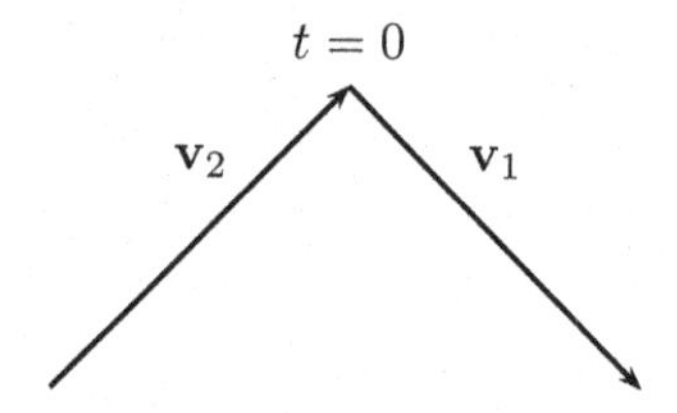

FIGURE 40.3
Impulsive scattering.

The difference between these two powers, seen in (36.8c), is the time derivative of what is called the Schott[1] energy [36]. For example, Refs. [37, 38].

For discussion of the relativistic situation, see the Problems at the end of this chapter.

40.2 Radiation by Impulsive Scattering

As an idealization of a scattering process, consider one in which a charged particle abruptly changes its velocity from a constant value $\mathbf{v}_2$ to another constant value $\mathbf{v}_1$. (See Fig. 40.3.) We can think of this as an idealized limit of bremsstrahlung.[2] We will calculate the radiation produced by this accelerated particle, bearing in mind that this description can be realistic only for radiation of sufficiently low frequencies since long wavelengths cannot probe the detailed character of the particle motion. The charge and current densities before and after the deflection act, which is assumed to take place at $t = 0$, are

$$\rho = e\delta(\mathbf{r} - \mathbf{v}_2 t), \quad \mathbf{j} = e\mathbf{v}_2\delta(\mathbf{r} - \mathbf{v}_2 t), \quad \text{for} \quad t < 0, \tag{40.22a}$$

and

$$\rho = e\delta(\mathbf{r} - \mathbf{v}_1 t), \quad \mathbf{j} = e\mathbf{v}_1\delta(\mathbf{r} - \mathbf{v}_1 t), \quad \text{for} \quad t > 0. \tag{40.22b}$$

The spectral distribution for the radiated energy can be computed by substituting (40.22a) and (40.22b) into (38.24), where we encounter the integral

$$\begin{aligned}\mathbf{j}(\mathbf{k},\omega) &= \int (d\mathbf{r})\, dt\, e^{i\omega t} e^{-i\omega \mathbf{n}\cdot\mathbf{r}/c}\mathbf{j}(\mathbf{r},t)\\ &= e\mathbf{v}_2\int_{-\infty}^{0} dt\, e^{i\omega t(1-\mathbf{n}\cdot\mathbf{v}_2/c)} + e\mathbf{v}_1\int_0^{\infty} dt\, e^{i\omega t(1-\mathbf{n}\cdot\mathbf{v}_1/c)}\\ &= i\frac{e}{\omega}\left(\frac{\mathbf{v}_1}{1-\mathbf{n}\cdot\mathbf{v}_1/c} - \frac{\mathbf{v}_2}{1-\mathbf{n}\cdot\mathbf{v}_2/c}\right).\end{aligned} \tag{40.23}$$

Here we have used the effective evaluations

$$\int_0^{\infty} dt\, e^{i\lambda t} = \frac{i}{\lambda}, \quad \int_{-\infty}^{0} dt\, e^{i\lambda t} = -\frac{i}{\lambda}, \tag{40.24}$$

since physical quantities cannot depend on what transpired at infinitely remote times. As a consistency check of (40.24), we can calculate the corresponding integral on ρ directly,

$$\rho(\mathbf{k},\omega) = i\frac{e}{\omega}\left(\frac{1}{1-\mathbf{n}\cdot\mathbf{v}_1/c} - \frac{1}{1-\mathbf{n}\cdot\mathbf{v}_2/c}\right), \tag{40.25}$$

[1] George Adolphus Schott, 1868–1937.

[2] "Braking radiation," due to deflection of charged particles by the electric fields of nuclei, for example.

or by means of the charge conservation condition, (38.15),

$$\rho(\mathbf{k},\omega) = \frac{\mathbf{k}}{\omega}\cdot\mathbf{j}(\mathbf{k},\omega). \tag{40.26}$$

Indeed, if we take the scalar product of (40.23) with $\mathbf{n}/c$, we have (40.25):

$$i\frac{e}{\omega}\left(\frac{\frac{1}{c}\mathbf{n}\cdot\mathbf{v}_1}{1-\mathbf{n}\cdot\mathbf{v}_1/c} - \frac{\frac{1}{c}\mathbf{n}\cdot\mathbf{v}_2}{1-\mathbf{n}\cdot\mathbf{v}_2/c}\right) = i\frac{e}{\omega}\left(\frac{1}{1-\mathbf{n}\cdot\mathbf{v}_1/c} - \frac{1}{1-\mathbf{n}\cdot\mathbf{v}_2/c}\right). \tag{40.27}$$

We now immediately obtain the spectral distribution from (38.24),

$$\frac{d^2E}{d\omega d\Omega} = \frac{e^2}{4\pi^2c^3}\left|\mathbf{n}\times\left(\frac{\mathbf{v}_1}{1-\mathbf{n}\cdot\mathbf{v}_1/c} - \frac{\mathbf{v}_2}{1-\mathbf{n}\cdot\mathbf{v}_2/c}\right)\right|^2. \tag{40.28}$$

Since (40.28) is independent of the frequency, the implied total radiated energy is unbounded. This unphysical result is due to the idealization that the scattering occurs instantaneously, that is, our assumption that the particle changes its velocity abruptly. Realistically, this change occurs over some finite period of time, T so that our result, (40.28) holds only when $\omega \ll 1/T$. In the nonrelativistic limit, where $|\mathbf{v}_{1,2}/c| \ll 1$, (40.28) reduces to

$$\frac{d^2E}{d\omega d\Omega} \approx \frac{e^2}{4\pi^2c^3}|\mathbf{n}\times(\mathbf{v}_1-\mathbf{v}_2)|^2 = \frac{e^2}{4\pi^2c^3}(\mathbf{v}_1-\mathbf{v}_2)^2\sin^2\theta, \tag{40.29}$$

where θ is the angle between $\mathbf{v}_1-\mathbf{v}_2$ and $\mathbf{n}$. Integrating this over all angles, we obtain the energy radiated at the frequency ω

$$\frac{dE}{d\omega} = \frac{2}{3}\frac{e^2}{\pi c^3}(\mathbf{v}_1-\mathbf{v}_2)^2, \tag{40.30}$$

which can also be easily obtained from (38.31), since the Fourier transform of the derivative here is simply

$$\dot{\mathbf{v}}(\omega) = \int_{-\infty}^{\infty} dt\, e^{i\omega t}\dot{\mathbf{v}}(t) = \int_{-\infty}^{\infty} dt\, e^{i\omega t}(\mathbf{v}_1-\mathbf{v}_2)\delta(t) = \mathbf{v}_1-\mathbf{v}_2. \tag{40.31}$$

Because either denominator in (40.28) has the structure

$$1-\frac{1}{c}\mathbf{n}\cdot\mathbf{v}_i = 1-\frac{v_i}{c}\cos\theta_i, \tag{40.32}$$

where $\mathbf{v}_i$ is either $\mathbf{v}_1$ or $\mathbf{v}_2$, and θ_i is the angle between $\mathbf{n}$ and $\mathbf{v}_i$, we see that the radiation in the ultrarelativistic limit ($|\mathbf{v}_i| \sim c$) is preferentially emitted near the direction of the velocity of the particle (the forward direction), either before or after the scattering act. On the other hand, this behavior is softened by the numerator factor $|\mathbf{n}\times\mathbf{v}_i| = v_i\sin\theta_i$, which forbids radiation in the exactly forward direction, $\theta_i = 0$. In Figs. 40.4 and 40.5 we plot the spectral distribution (40.28) in and out of the scattering plane for relativistic particles.

In either region of significant radiation, characterized by $\theta_i \ll 1$, only one term in (40.28) makes a major contribution (in the following we drop the subscript i):

$$\begin{aligned}\left(\frac{\mathbf{n}\times\mathbf{v}/c}{1-\frac{v}{c}\cos\theta}\right)^2 &= \left(\frac{v}{c}\right)^2\sin^2\theta\frac{1}{\left(1-\frac{v}{c}\cos\theta\right)^2} \approx \frac{\theta^2}{\left(1-\frac{v}{c}+\frac{\theta^2}{2}\right)^2} \approx \frac{4\theta^2}{\left(1-\frac{v^2}{c^2}+\theta^2\right)^2}\\ &\approx \begin{cases} 4/\theta^2, & \text{if}\quad \sqrt{1-(v/c)^2} \ll \theta \ll 1,\\ 0, & \quad\text{if}\quad \theta = 0.\end{cases}\end{aligned} \tag{40.33}$$

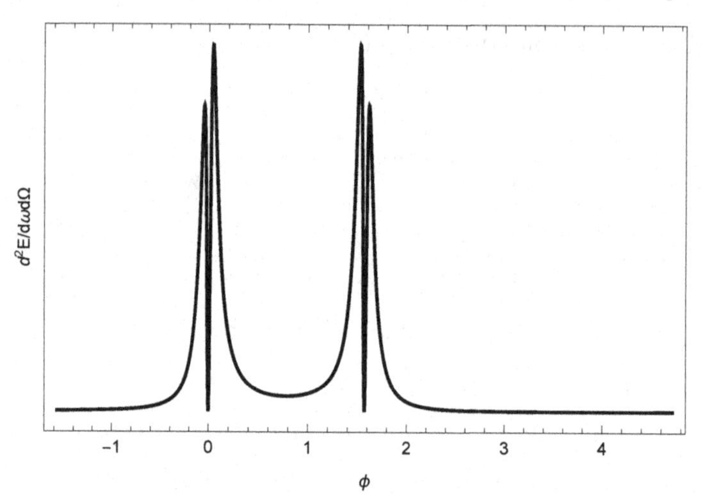

FIGURE 40.4
The spectral distribution of radiated energy in the plane of scattering, for a scattering angle of $\pi/2$, as a function of the angle ϕ in the plane, where the initial velocity is in the $\phi = 0$ direction. In this and the following graph, the velocity of the particle both before and after the scattering is 0.99c. For definition of the coordinate system used, see Problem 40.8.

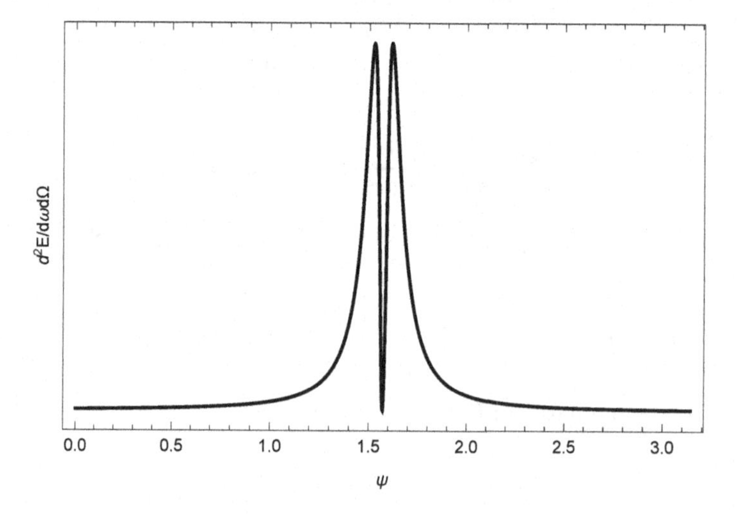

FIGURE 40.5
The spectral distribution of radiated energy perpendicular to the plane of scattering, for a scattering angle of $\pi/2$, as a function of the angle perpendicular to the plane, for the azimuthal angle being that of the direction of the particle after scattering.

Therefore, the maximum intensity occurs at

$$\theta = \sqrt{1 - (v/c)^2} = \frac{mc^2}{E} \tag{40.34}$$

(which we have implicitly assumed to be small compared to the scattering angle), and most of the radiation is emitted in a small angular range near this angle. As the particle moves faster, the peaking of the radiation is more pronounced. This is a fundamental difference between radiation produced by relativistic ($v \sim c$) and nonrelativistic ($v \ll c$) particles. We will see this again as we now turn to a discussion of synchrotron radiation.

40.3 Problems for Chapter 40

1. By considering the Liénard-Wiechert fields found in Problems 34.5 and 34.6, demonstrate, in the nonrelativistic approximation, that no radiation zone is formed by a uniformly accelerated particle. Do so by showing that the acceleration part of **B** is static.
2. Consider a particle that undergoes nonuniform motion for only a finite period of time, $t_1 < t < t_2$. Otherwise, we have

$$\mathbf{r}(t) = \begin{cases} \mathbf{R}_1 + \mathbf{v}_1(t - t_1), & t \le t_1, \\ \mathbf{R}_2 + \mathbf{v}_2(t - t_2), & t \ge t_2. \end{cases} \tag{40.35}$$

   (a) Write $\mathbf{j}(\mathbf{k}, \omega)$ and $\rho(\mathbf{k}, \omega)$ as integrals over the time interval $t_1 \le t \le t_2$.
   (b) Show that charge conservation, (38.15), is satisfied.
   (c) Derive the general formula for $dE/d\Omega$ as an integral over the time interval $t_1 \le t \le t_2$,

$$\frac{dE}{d\Omega} = \frac{e^2}{4\pi c^3} \int_{t_1}^{t_2} dt \frac{1}{1 - \mathbf{n} \cdot \mathbf{v}(t)/c} \left(\frac{d}{dt} \frac{\mathbf{n} \times \mathbf{v}(t)}{1 - \mathbf{n} \cdot \mathbf{v}(t)/c} \right)^2 . \tag{40.36}$$

   This can be viewed as a generalization of (35.20).
   (d) Integrate over Ω in the above general formula for $dE/d\Omega$ and thereby derive the general formula for the total energy radiated,

$$E = \frac{2e^2}{3c} \int_{t_1}^{t_2} \frac{dt}{(1 - \beta^2)^3} \left[(1 - \beta^2) \dot{\boldsymbol{\beta}}^2 + (\boldsymbol{\beta} \cdot \dot{\boldsymbol{\beta}})^2 \right], \tag{40.37}$$

   where $\boldsymbol{\beta} = \mathbf{v}(t)/c$. This can be viewed as a generalization of (35.22).
3. Apply the results of the previous Problem to treat the radiation by a relativistic, uniformly accelerated[3] particle, undergoing so-called hyperbolic motion, for a finite time interval,

$$z = a \cosh \frac{c\tau}{a}, \quad ct = a \sinh \frac{c\tau}{a}, \quad -\tau_0 \le \tau \le \tau_0. \tag{40.38}$$

   (a) By considering short times (or $\tau \approx 0$), make connection with the nonrelativistic formulas and identify the parameter a.
   (b) Show that

$$\left.\frac{d^2 E}{d\omega\, d\Omega}\right|_{\mathbf{n} \cdot \hat{\mathbf{z}} = 0} = \frac{e^2}{\pi^2 c} \left[\omega \frac{a}{c} K_1 \left(\omega \frac{a}{c} \right) \right]^2 \tag{40.39}$$

[3] According to Pauli [31] "In relativistic kinematics one will naturally describe as 'uniformly accelerated' a motion for which the acceleration in a system, moving with the medium or particle, is always of the same magnitude. The system is a different one at each instant; for one and the same Galilean system, the acceleration of such a motion is not constant in time."

in the $\tau_0 \to \infty$ limit. Use the second integral representation in (20.14b) and note that

$$K_1(x) = -K_0'(x) \tag{40.40}$$

[*cf.* (20.62b)].

(c) For $\theta \neq 0, \pi$ ($\mathbf{n} \cdot \hat{\mathbf{z}} = \cos\theta$) and $\tau_0 \to \infty$, show that

$$\frac{dE}{d\Omega} = \frac{3e^2}{32a}\frac{1}{\sin^3\theta}. \tag{40.41}$$

(d) Show that the total energy radiated is

$$E = \frac{4}{3}\frac{e^2}{a}\frac{\beta_m}{\sqrt{1-\beta_m^2}}, \tag{40.42}$$

where β_m is the maximum value of v/c ($= |\mathbf{v}(\pm\tau_0)|/c$). In terms of the nonrelativistic acceleration, the Larmor formula is recovered:

$$\frac{E}{2T} = \frac{2}{3}\frac{e^2}{c^3}A^2 \tag{40.43}$$

where $A = c^2/a$ is the nonrelativistic acceleration, and the total time of the acceleration is $2T = 2\frac{a}{c}\sinh\frac{c\tau_0}{a}$.

(e) Comment on the following quotation from Pauli [31]:

> Hyperbolic motion thus constitutes a special case, for which there is no formation of a wave zone nor any corresponding radiation. (Radiation, on the other hand, does occur when two uniform, rectilinear, motions are connected by a "portion" of hyperbolic motion.)

4. Reconsider Problem 1 for hyperbolic motion as described by (40.38). Discuss the behavior of the "acceleration" electric and magnetic fields, and whether a signal of radiation can be discerned there.

5. Consider a particle undergoing a constant acceleration g for a finite time, $-t_0 \leq t \leq t_0$, during which time its velocity changes from $-v_0$ ($= -gt_0$) to $+v_0$. [Here, the "constant acceleration" means $d^2\mathbf{r}/dt^2 =$ constant, so the hyperbolic motion described in Problem 3 is not being considered. Hence, we require that $|g|t_0 < c$, but not that the motion be nonrelativistic.]

   (a) Let g be equal to the acceleration at $t = 0$ for Problem 3 and find a relation between g and a.

   (b) By means of the result (40.37), calculate the total energy radiated. Express it as a function of β_m ($= v_0/c$).

   (c) Show that the nonrelativistic limit of this result agrees with the corresponding limit of Problem 3d.

6. This problem concerns the radiation of a massless scalar field by a uniformly radiated scalar source. This problem was first worked out by Ren and Weinberg [39]. In that case, instead of a four-vector source, the electric current density, we have a scalar density,

$$\rho(\mathbf{r}, t) = \frac{e}{\gamma(t)}\delta(\mathbf{r} - \mathbf{r}(t)), \quad \gamma(t) = \frac{1}{\sqrt{1 - v^2(t)}}. \tag{40.44}$$

(Explain the appearance of the $1/\gamma$ factor.) Apart from this modification, one can proceed in close analogy with the electromagnetic calculation. The asymptotic field at large distance from the source should be, in comparison with (35.2a),

$$\phi(\mathbf{r},t) \sim \frac{1}{r}\int d(\mathbf{r}')\rho\left(\mathbf{r}', t - \frac{r}{c} + \frac{1}{c}\mathbf{n}\cdot\mathbf{r}'\right). \tag{40.45}$$

(a) The major subtlety here is the ambiguity in the energy-momentum tensor. Following the procedure outlined in Problem 12.21, show that the energy-momentum or stress tensor for a massless scalar field is

$$T^{\mu\nu} = \frac{1}{4\pi}\left[\partial^\mu\phi\partial^\nu\phi - \frac{1}{2}\partial^\lambda\phi\partial_\lambda\phi g^{\mu\nu} - \xi(\partial^\mu\partial^\nu - g^{\mu\nu}\partial^2)\phi^2\right]. \tag{40.46}$$

The first two terms might be dubbed the "canonical" stress tensor, while the last, "conformal," term arises because there is an ambiguity in the general coordinate transformations from which the stress tensor is derived. A totally arbitrary identically conserved term can be added to the stress tensor without affecting global quantities such as the total energy or momentum. This, apparently, was first observed by Callen, Coleman, and Jackiw [40]. The arbitrary parameter ξ can be freely chosen: The value $\xi = 0$ yields the canonical stress tensor, while $\xi = \frac{1}{6}$ gives the conformal stress tensor, which has some theoretical advantages, in softening divergences in quantum field theory. (The trace $T^\mu_\mu = 0$ in that case.) But, outside of gravity, the choice of ξ should be without effect.

(b) Now use the radiation-zone field (40.45) to compute the energy flux vector

$$S_i = T^{0i} = \frac{1}{4\pi}\left[\partial^0\phi(\mathbf{r},t)\partial^i\phi(\mathbf{r},t) - \xi\partial^0\partial^i\phi^2(\mathbf{r},t)\right], \tag{40.47}$$

to find the time-integrated flux in the direction $\mathbf{n}$

$$\int_{-\infty}^{\infty} dt\,\mathbf{S}\cdot\mathbf{n} = \frac{1}{4\pi c r^2}\int\frac{d\omega}{2\pi}\omega^2|\rho(\mathbf{k},\omega)|^2, \quad \mathbf{k} = \frac{\omega}{c}\mathbf{n}. \tag{40.48}$$

Evidently, the conformal correction does not contribute to the radiated flux.

(c) Now show that the angular distribution of radiated energy is

$$\begin{aligned}\frac{dE}{d\Omega} &= \frac{1}{4\pi^2 c}\int_0^\infty d\omega\,\omega^2|\rho(\mathbf{k},\omega)|^2\\ &= \frac{e^2}{4\pi c}\int_{t_1}^{t_2}\frac{dt}{1-\mathbf{n}\cdot\boldsymbol{\beta}(t)}\left[\frac{d}{dt}\left(\frac{1}{\gamma(t)}\frac{1}{1-\mathbf{n}\cdot\boldsymbol{\beta}(t)}\right)\right]^2,\end{aligned} \tag{40.49}$$

slightly different from (40.36) because of the relativistic dilation factor in (40.44).

(d) Now integrate over all solid angles (directions of $\mathbf{n}$) to obtain precisely one-half the formula (40.37). Therefore, the energy emitted by scalar radiation from a particle undergoing hyperbolic motion for a finite time $2T = 2\frac{a}{c}\sinh\frac{c\tau_0}{a}$ is exactly one-half that given by the Larmor formula (40.43).

7. Consider a charged particle accelerated from rest by a constant electric field $\mathbf{E}$, and show, from the result (40.37), that the rate of radiation, dE/dt, does not change with increasing energy. This is very different from the power radiated in a circular accelerator, as we shall see in Chapter 41, and suggests why the successor to the LHC will most likely be a linear collider.

8. Consider the formula for the spectral distribution for impulsive radiation (40.28). Introduce spherical polar coordinates to describe the velocities $\mathbf{v}_1$, $\mathbf{v}_2$ and the direction of observation $\mathbf{n}$, for example,

$$\mathbf{v}_1 = v_1\hat{\mathbf{x}}, \tag{40.50a}$$
$$\mathbf{v}_2 = v_2(\hat{\mathbf{x}}\cos\theta + \hat{\mathbf{y}}), \quad \text{while} \tag{40.50b}$$
$$\mathbf{n} = \hat{\mathbf{x}}\sin\psi\cos\phi + \hat{\mathbf{y}}\sin\psi\sin\phi + \hat{\mathbf{z}}\cos\psi. \tag{40.50c}$$

Using this parameterization, and the computer program of your choice, verify the distributions seen in Figs. 40.4 and 40.5. Study the distribution as a function of the velocities of the particle before and after the impulse, and the direction of observation. How closely does the approximation (40.33) describe the exact expression?

9. Consider a particle undergoing an instantaneous reversal in direction, changing from velocity $\mathbf{v}$ to velocity $-\mathbf{v}$ in negligible time.

   (a) Derive the following formula for the number of photons with energy $\hbar\omega$ emitted into a frequency interval $d\omega$ and into an element of solid angle $d\Omega$, making an angle θ with respect to the direction specified by $\mathbf{v}$, with the fine structure constant being $\alpha = e^2/\hbar c$:

$$\frac{d^2N}{d\Omega\, d\omega} = \frac{\alpha}{4\pi^2}\frac{1}{\omega}\left\{2\frac{1+\beta^2}{1-(\beta\cos\theta)^2}\right.$$
$$\left. -(1-\beta^2)\left[\frac{1}{(1-\beta\cos\theta)^2}+\frac{1}{(1+\beta\cos\theta)^2}\right]\right\}. \tag{40.51}$$

   (b) What is the result of integrating this over all angles, for any $\beta < 1$? Do you recover the known photon spectrum for $\beta \ll 1$? What does it become for $\beta \approx 1$? Can you understand this result by looking at the approximate form of (40.51) for $\beta \approx 1$, $\theta \ll 1$, $\pi - \theta \ll 1$?

   (c) Now, suppose the charged particle stops on impact. Find the analog of (40.51). Again, integrate it over all angles and look at the limits $\beta \ll 1$ and $\beta \approx 1$. Are the last two results what you expected? Explain.

41

Synchrotron Radiation I

A particle, of mass μ[1] and charge e, moving in a circular orbit, undergoes centripetal acceleration and therefore radiates electromagnetic energy. The physical circumstances under which this process occurs arise in synchrotrons, in which charged particles are guided in a circle by external magnetic fields. In this and the following two chapters, we will explore the characteristics of this synchrotron radiation. We begin by considering some kinematics.

41.1 Motion of a Charged Particle in a Homogeneous Magnetic Field

If E is the energy of the charged particle and $\mathbf{p}$ is its momentum, then the (relativistic) equations of motion of the particle in a magnetic field $\mathbf{B}$ are

$$\frac{d\mathbf{p}}{dt} = \frac{e}{c}\mathbf{v}\times\mathbf{B}, \tag{41.1a}$$

$$\frac{dE}{dt} = 0, \tag{41.1b}$$

where the momentum is related to the velocity, $\mathbf{v}$, by

$$\mathbf{p} = \frac{\mu\mathbf{v}}{\sqrt{1-\beta^2}} = \frac{E}{c^2}\mathbf{v}, \quad \beta \equiv \frac{v}{c}, \tag{41.2a}$$

and the energy is

$$E = \frac{\mu c^2}{\sqrt{1-\beta^2}}. \tag{41.2b}$$

Since the magnetic force is always perpendicular to the direction of motion, no work is done on the particle, and consequently the energy of the particle is conserved, as stated by (41.1b). This fact, together with (41.2a), enables us to rewrite (41.1a) in the form

$$\frac{E}{c^2}\frac{d\mathbf{v}}{dt} = \frac{e}{c}\mathbf{v}\times\mathbf{B}, \tag{41.3a}$$

or

$$\frac{d\mathbf{v}}{dt} = \left(-\frac{ec}{E}\mathbf{B}\right)\times\mathbf{v}. \tag{41.3b}$$

This implies a constant deflection of the velocity vector, that is, $\mathbf{v}$ precesses with angular velocity $\boldsymbol{\omega}_0$,

$$\boldsymbol{\omega}_0 = -\frac{ec}{E}\mathbf{B}, \tag{41.4a}$$

[1]We use μ not m here for the rest mass of the particle, so as to avoid confusion with the harmonic number m.

DOI: 10.1201/9781003057369-41

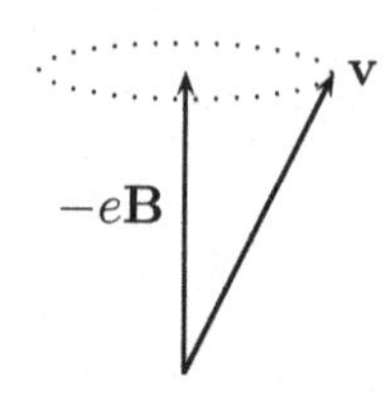

FIGURE 41.1
Precession of velocity vector around a constant magnetic field.

about the direction of $-e$ times the magnetic field. (See Fig. 41.1.) The angular speed of this precession is the Larmor frequency,

$$\omega_0 = \frac{|e|c}{E}B, \tag{41.4b}$$

which reduces, in the nonrelativistic limit, to the cyclotron frequency,

$$\omega_0 \approx \frac{|e|B}{\mu c}, \quad \frac{v}{c} \ll 1. \tag{41.5}$$

The component of the velocity parallel to the magnetic field, according to (41.3b), is constant. In practice, we constrain the motion such that $\mathbf{v}$ is perpendicular to $\mathbf{B}$. Then the particle moves with angular speed ω_0 in a circle of radius R,

$$R = \frac{v}{\omega_0} = \frac{\beta E}{|e|B}. \tag{41.6}$$

The relation between the momentum of the particle and the radius of the circular orbit,

$$p = \frac{E}{c^2}v = \frac{|e|}{c}BR, \tag{41.7}$$

supplies us with a practical method of measuring the momentum of a relativistic particle. When the information thus obtained is coupled with that derived from a Čerenkov counter, which measures the speed of the particle according to (39.13a), we can determine the mass of the particle:

$$\mu = \frac{\sqrt{1-\beta^2}}{v}p. \tag{41.8}$$

41.2 Spectrum of Synchrotron Radiation

We now proceed to calculate the radiation emitted by a charged particle moving in a circle. For a point particle, the charge and current densities are given by

$$\rho(\mathbf{r}, t) = e\delta(\mathbf{r} - \mathbf{r}(t)), \tag{41.9a}$$
$$\mathbf{j}(\mathbf{r}, t) = e\mathbf{v}(t)\delta(\mathbf{r} - \mathbf{r}(t)), \tag{41.9b}$$

where $\mathbf{r}(t)$ is the position vector of the particle at time t, and $\mathbf{v}(t)$ is its velocity. Substituting (41.9a) and (41.9b) into (39.5b), we obtain the spectrum of the power, emitted at time T,

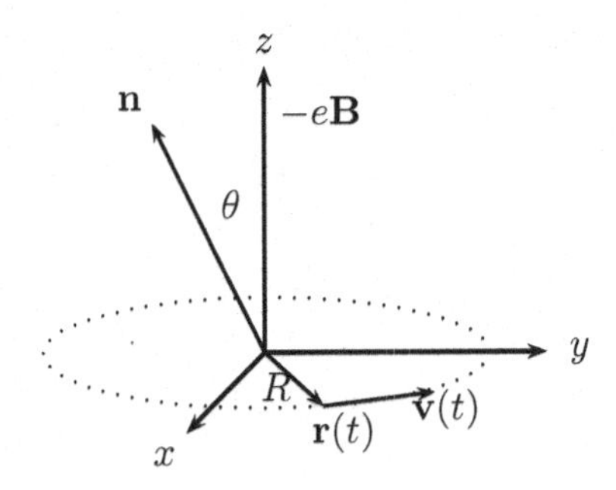

FIGURE 41.2
Diagram of charged particle moving in a circle in a magnetic field. The coordinate system used is shown.

into the element of solid angle in the direction $\mathbf{n}$,

$$\frac{d^2P(T)}{d\omega d\Omega} = \frac{\omega^2 e^2}{4\pi^2 c}\int_{-\infty}^{\infty} d\tau\, e^{-i\omega\tau}\left[\frac{1}{c^2}\mathbf{v}(T+\tau/2)\cdot\mathbf{v}(T-\tau/2) - 1\right]$$
$$\times \exp\left\{i\frac{\omega}{c}\mathbf{n}\cdot[\mathbf{r}(T+\tau/2) - \mathbf{r}(T-\tau/2)]\right\}. \tag{41.10}$$

This description in terms of an average macroscopic time can be important, for then it is possible to consider the effect of a slow alteration in the parameters describing the motion.

It is convenient to choose the coordinate system such that the particle is moving in a circle of radius R about the origin in the xy plane, and the magnetic field $\mathbf{B}$ is directed along the z direction. Also, without loss of generality, we choose the observation direction, $\mathbf{n}$, to lie in the xz plane, making an angle θ with the $+z$ axis, as shown in Fig. 41.2. Then we have, for a convenient choice of initial conditions,

$$\mathbf{n} = (\sin\theta, 0, \cos\theta), \tag{41.11a}$$
$$x(t) = R\cos\omega_0 t, \tag{41.11b}$$
$$y(t) = R\sin\omega_0 t, \tag{41.11c}$$
$$z(t) = 0, \tag{41.11d}$$

where ω_0 and R are given by (41.4b) and (41.6), respectively, The corresponding velocity is given by

$$v_x(t) = -v\sin\omega_0 t, \tag{41.12a}$$
$$v_y(t) = v\cos\omega_0 t, \tag{41.12b}$$
$$v_z(t) = 0, \tag{41.12c}$$

where the speed v satisfies

$$v = R\omega_0. \tag{41.13}$$

Using these explicit representations for $\mathbf{r}(t)$ and $\mathbf{v}(t)$, we may simplify the integrand of (41.10) by means of the following:

$$\frac{1}{c^2}\mathbf{v}(T+\tau/2)\cdot\mathbf{v}(T-\tau/2) = \beta^2\cos\omega_0\tau \tag{41.14}$$

(a fact which is apparent geometrically), and

$$\mathbf{n}\cdot\mathbf{r}(T\pm\tau/2)=\sin\theta\, x(T\pm\tau/2)=R\sin\theta\cos\omega_0(T\pm\tau/2), \tag{41.15a}$$

$$e^{\pm i(\omega/c)\mathbf{n}\cdot\mathbf{r}(T\pm\tau/2)}=e^{\pm i(\omega/c)R\sin\theta\cos\omega_0(T\pm\tau/2)}$$
$$=\sum_{m=-\infty}^{\infty}(\pm i)^m e^{\pm im\omega_0(T\pm\tau/2)}J_m\left(\frac{\omega}{c}R\sin\theta\right). \tag{41.15b}$$

In the last equation, we have used the generating function for the Bessel functions of integer order, (18.27),

$$e^{iz\cos\phi}=\sum_{m=-\infty}^{\infty}i^m e^{im\phi}J_m(z). \tag{41.16}$$

With these evaluations, we obtain the power spectrum, (41.10), in the form

$$\frac{d^2P(T)}{d\omega d\Omega}=\frac{\omega^2e^2}{4\pi^2c}\int_{-\infty}^{\infty}d\tau\, e^{-i\omega\tau}[\beta^2\cos\omega_0\tau-1]$$
$$\times\sum_{m=-\infty}^{\infty}\sum_{m'=-\infty}^{\infty}i^m e^{im\omega_0(T+\tau/2)}J_m\left(\frac{\omega}{c}R\sin\theta\right)$$
$$\times(-i)^{m'}e^{-im'\omega_0(T-\tau/2)}J_{m'}\left(\frac{\omega}{c}R\sin\theta\right). \tag{41.17}$$

Now we recall that the emission time T represents an average over many periods of the motion, since many oscillations are required to identify a frequency. Here, it is sufficient to consider the average of (41.17) over one period, by using the relation

$$\langle e^{i(m-m')\omega_0T}\rangle_{\text{one period}}=\delta_{mm'}. \tag{41.18}$$

After this averaging procedure, (41.17) becomes

$$\frac{d^2P}{d\omega d\Omega}=\frac{\omega^2e^2}{4\pi^2c}\int_{-\infty}^{\infty}d\tau\, e^{-i\omega\tau}[\beta^2\cos\omega_0\tau-1]\sum_{m=-\infty}^{\infty}e^{im\omega_0\tau}\left[J_m\left(\frac{\omega}{c}R\sin\theta\right)\right]^2. \tag{41.19}$$

Incidentally, we remark that this result, (41.19), can also be obtained by combining the two exponentials directly as follows:

$$\langle e^{i(\omega/c)\mathbf{n}\cdot[\mathbf{r}(T+\tau/2)-\mathbf{r}(T-\tau/2)]}\rangle_{\text{one period}}=\langle e^{i(\omega/c)R\sin\theta(-2\sin\omega_0T\sin\frac{1}{2}\omega_0\tau)}\rangle_{\text{one period}}$$
$$=J_0\left(\frac{2\omega R}{c}\sin\theta\sin\frac{\omega_0\tau}{2}\right)=\sum_{m=-\infty}^{\infty}\left[J_m\left(\frac{\omega R}{c}\sin\theta\right)\right]^2e^{im\omega_0\tau}. \tag{41.20}$$

Here we have used the integral representation of J_0, (18.4),

$$J_0(x)=\int_0^{2\pi}\frac{d\phi}{2\pi}e^{ix\cos\phi}, \tag{41.21}$$

and the addition theorem for the Bessel functions, (18.65),

$$J_0\left(k\sqrt{\rho^2+\rho'^2-2\rho\rho'\cos(\phi-\phi')}\right)=\sum_{m=-\infty}^{\infty}J_m(k\rho)e^{im\phi}J_m(k\rho')e^{-im\phi'}, \tag{41.22}$$

with $\rho=\rho'=R$, $k=(\omega/c)\sin\theta$, and $\phi-\phi'=\omega_0\tau$.

Exploiting the exponential representation for $\cos\omega_0\tau$,

$$\cos\omega_0\tau = \frac{1}{2}\left[e^{i\omega_0\tau} + e^{-i\omega_0\tau}\right], \tag{41.23}$$

we can easily perform the τ integration in (41.19) by means of the integral

$$\int_{-\infty}^{\infty} d\tau\, e^{-i\omega\tau} e^{im\omega_0\tau} = 2\pi\delta(\omega - m\omega_0). \tag{41.24}$$

Consequently, the spectrum is discrete in that only harmonics of the Larmor frequency, (41.4b), are radiated. That is, the emitted frequencies are multiples of ω_0, $\omega = m\omega_0$, where the integer m is positive since $\omega > 0$. Thus the power spectrum can be written as a sum of the contributions of each harmonic,

$$\frac{d^2P(T)}{d\omega d\Omega} = \sum_{m=1}^{\infty} \delta(\omega - m\omega_0)\frac{dP_m(T)}{d\Omega}, \tag{41.25}$$

where the power radiated into the mth harmonic is

$$\frac{dP_m}{d\Omega} = \frac{m^2\omega_0^2 e^2}{2\pi c}\left[\frac{\beta^2}{2}\left(J_{m+1}^2 + J_{m-1}^2\right) - J_m^2\right], \tag{41.26}$$

the argument of the Bessel functions being

$$\frac{\omega R}{c}\sin\theta = m\beta\sin\theta. \tag{41.27}$$

In (41.25) the T dependence could arise from any slow variation in the parameters ω_0 and β. We may further simplify (41.26) by using the following recurrence relations for Bessel's functions, found in Problem 18.4,

$$J_{m-1}(z) - J_{m+1}(z) = 2J_m'(z), \tag{41.28a}$$

$$J_{m-1}(z) + J_{m+1}(z) = \frac{2m}{z}J_m(z), \tag{41.28b}$$

which imply that

$$\frac{1}{2}\left[(J_{m+1})^2 + (J_{m-1})^2\right] - \frac{1}{\beta^2}(J_m)^2 = (J_m')^2 + \left(\frac{J_m}{\beta\tan\theta}\right)^2. \tag{41.29}$$

Therefore, the angular distribution for the power radiated into the mth harmonic is

$$\frac{dP_m}{d\Omega} = \frac{\omega_0}{2\pi}\frac{e^2}{R}\beta^3 m^2\left\{\left[J_m'(m\beta\sin\theta)\right]^2 + \left[\frac{J_m(m\beta\sin\theta)}{\beta\tan\theta}\right]^2\right\}. \tag{41.30}$$

41.3 Total Power Emitted into the mth Harmonic

To obtain the total power radiated into the mth harmonic, we could integrate (41.30) over all angles. However, we find it simpler to return to the general expression (41.10) and perform the integration over the angles first, since there all the angular dependence is in the exponential factor. If we define $\mathbf{s}$ to be

$$\mathbf{s} = \mathbf{r}(T + \tau/2) - \mathbf{r}(T - \tau/2), \tag{41.31a}$$

$$|\mathbf{s}| = \sqrt{R^2 + R^2 - 2R^2 \cos\omega_0\tau} = 2R|\sin\frac{1}{2}\omega_0\tau|, \tag{41.31b}$$

the angular integration we encounter in this way is

$$\begin{aligned}\int d\Omega\, e^{i(\omega/c)\mathbf{n}\cdot\mathbf{s}} &= 2\pi\int_0^\pi \sin\chi d\chi\, e^{i(\omega/c)|\mathbf{s}|\cos\chi} = 4\pi\frac{\sin\omega|\mathbf{s}|/c}{\omega|\mathbf{s}|/c}\\ &= \frac{2\pi c}{\omega R}\frac{\sin\left(\frac{2\omega}{c}R\sin\frac{\omega_0\tau}{2}\right)}{\sin\frac{\omega_0\tau}{2}}.\end{aligned} \tag{41.32}$$

The resulting frequency distribution of the radiated power is, from (41.10),

$$\begin{aligned}\frac{dP}{d\omega} &= \frac{\omega}{2\pi}\frac{e^2}{R}\int_{-\infty}^{\infty} d\tau\, e^{-i\omega\tau}(\beta^2\cos\omega_0\tau - 1)\frac{\sin\left(\frac{2\omega}{c}R\sin\frac{\omega_0\tau}{2}\right)}{\sin\frac{\omega_0\tau}{2}}\\ &\equiv \frac{\omega}{2\pi}\frac{e^2}{R}\int_{-\infty}^{\infty} d\tau\, e^{-i\omega\tau} f(\omega_0\tau).\end{aligned} \tag{41.33}$$

The challenge now is to carry out the τ integration. We observe that $f(\omega_0\tau)$ is a periodic function,

$$f(\omega_0\tau) = f(\omega_0\tau + 2n\pi), \quad n = \text{integer}, \tag{41.34}$$

so that it may be represented by a Fourier series,

$$f(\omega_0\tau) = \sum_{m=-\infty}^{\infty} e^{im\omega_0\tau} f_m, \tag{41.35a}$$

where the Fourier coefficient f_m is given by

$$f_m = \int_{-\pi}^{\pi}\frac{d\phi}{2\pi}e^{-im\phi}f(\phi). \tag{41.35b}$$

Inserting the representation (41.35a) into (41.33), and carrying out the then trivial τ integration, (41.24), we obtain the spectrum of the radiated power:

$$\frac{dP}{d\omega} = \sum_{m=1}^{\infty}\delta(\omega - m\omega_0)P_m, \tag{41.36}$$

where the total power radiated in the mth harmonic is expressed as a Fourier coefficient,

$$\begin{aligned}P_m &= \frac{e^2}{R}m\omega_0\int_{-\pi}^{\pi}\frac{d\phi}{2\pi}e^{-im\phi}(\beta^2\cos\phi - 1)\frac{\sin\left(2m\beta\sin\frac{\phi}{2}\right)}{\sin\frac{\phi}{2}}\\ &= \frac{e^2}{R}m\omega_0\int_0^\pi\frac{d\phi}{\pi}\cos m\phi\left(\beta^2 - 1 - 2\beta^2\sin^2\frac{\phi}{2}\right)\frac{\sin\left(2m\beta\sin\frac{\phi}{2}\right)}{\sin\frac{\phi}{2}}.\end{aligned} \tag{41.37}$$

Not surprisingly, this can be rewritten in terms of Bessel functions. Starting from the integral representation (18.29), equivalent to (41.16),

$$i^m J_m(z) = \int_{-\pi}^{\pi}\frac{d\phi}{2\pi}e^{-im\phi}e^{iz\cos\phi}, \tag{41.38a}$$

we make the substitution $\phi \to \phi - \pi/2$, leading to

$$J_m(z) = \int_0^\pi \frac{d\phi}{\pi} \cos(z \sin\phi - m\phi). \tag{41.38b}$$

Breaking this integral at $\pi/2$ and substituting $\phi \to \pi - \phi$ for the range $\pi/2$ to π, we finally find

$$J_m(z) = \int_0^{\pi/2} \frac{d\phi}{\pi} [\cos(z \sin\phi - m\phi) + \cos(z\sin\phi + m\phi - m\pi)]. \tag{41.39}$$

This supplies the following integral representation for the Bessel functions of even order:

$$\begin{aligned} J_{2m}(z) &= \int_0^{\pi/2} \frac{d\phi}{\pi} [\cos(z\sin\phi - 2m\phi) + \cos(z\sin\phi + 2m\phi)] \\ &= \int_0^\pi \frac{d\phi}{\pi} \cos\left(z \sin\frac{\phi}{2}\right) \cos m\phi, \end{aligned} \tag{41.40}$$

where $\phi \to \frac{1}{2}\phi$ in the second line. The integral in (41.37) can now be expressed in terms of the derivative and the integral of J_{2m}, which are represented by

$$J'_{2m}(z) = -\int_0^\pi \frac{d\phi}{\pi} \sin\frac{\phi}{2} \cos m\phi \sin\left(z\sin\frac{\phi}{2}\right), \tag{41.41a}$$

and

$$\int_0^x dz\, J_{2m}(z) = \int_0^\pi \frac{d\phi}{\pi} \cos m\phi \frac{\sin\left(x \sin\frac{\phi}{2}\right)}{\sin\frac{\phi}{2}}. \tag{41.41b}$$

Therefore, the total power radiated into the mth harmonic is

$$P_m = \frac{e^2}{R} m\omega_0 \left[2\beta^2 J'_{2m}(2m\beta) - (1-\beta^2)\int_0^{2m\beta} dz\, J_{2m}(z)\right]. \tag{41.42}$$

41.4 Total Radiated Power

To find the total radiated power, we could sum (41.42) over all m. However, as before, it is much simpler to return to an earlier stage, this time to (41.33), and first perform the frequency integration, before doing the τ integration. An intermediate form for the radiated power is

$$\begin{aligned} P &= \int_0^\infty d\omega \frac{dP}{d\omega} = \frac{1}{2}\int_{-\infty}^\infty d\omega \frac{dP}{d\omega} \\ &= \frac{e^2}{R}\frac{1}{2\pi}\int_{-\infty}^\infty d\tau \frac{\beta^2 \cos\omega_0\tau - 1}{\sin\frac{1}{2}\omega_0\tau} \frac{1}{2}\int_{-\infty}^\infty d\omega\, \omega\, e^{-i\omega\tau} \frac{1}{2i}\left[e^{i\frac{2\omega}{c}R\sin\frac{\omega_0\tau}{2}} - e^{-i\frac{2\omega}{c}R\sin\frac{\omega_0\tau}{2}}\right] \\ &= \frac{e^2}{R}\frac{\omega_0}{4}\int_{-\infty}^\infty d\phi \frac{\beta^2\cos\phi - 1}{\sin\frac{\phi}{2}}\left[\delta'\left(\phi - 2\beta\sin\frac{\phi}{2}\right) - \delta'\left(\phi + 2\beta\sin\frac{\phi}{2}\right)\right], \end{aligned} \tag{41.43}$$

where we have made use of the representation

$$\int_{-\infty}^{\infty} d\omega\, \omega\, e^{-i\omega\lambda} = i\frac{d}{d\lambda}\int_{-\infty}^{\infty} d\omega\, e^{-i\omega\lambda} = 2\pi i \delta'(\lambda). \tag{41.44}$$

Equation (41.43) can be evaluated in a straightforward manner. However, it is simpler to combine the delta functions there by means of

$$\frac{1}{2\beta\sin\frac{\phi}{2}}\left[\delta'\left(\phi + 2\beta\sin\frac{\phi}{2}\right) - \delta'\left(\phi - 2\beta\sin\frac{\phi}{2}\right)\right] = \int_{-1}^{1} d\lambda\,\delta''(y), \tag{41.45}$$

where

$$y \equiv \phi + 2\lambda\beta\sin\frac{\phi}{2}, \tag{41.46}$$

so that we may rewrite the total power as

$$P = -\frac{e^2}{2R}\omega_0\beta\int_{-1}^{1} d\lambda\int_{-\infty}^{\infty} dy \frac{\beta^2\cos\phi - 1}{1+\lambda\beta\cos\frac{\phi}{2}}\delta''(y). \tag{41.47}$$

Here the y integration can be performed by integrating by parts twice and noting that the support of the delta function occurs at $\phi = 0$:

$$\begin{aligned}\frac{d^2}{dy^2}\left(\frac{\beta^2\cos\phi - 1}{1+\lambda\beta\cos\frac{\phi}{2}}\right)\bigg|_{y=0} &= \frac{1}{1+\lambda\beta\cos\frac{\phi}{2}}\frac{d}{d\phi}\left(\frac{1}{1+\lambda\beta\cos\frac{\phi}{2}}\frac{d}{d\phi}\frac{\beta^2\cos\phi - 1}{1+\lambda\beta\cos\frac{\phi}{2}}\right)\bigg|_{\phi=0} \\ &= \frac{1}{(1+\lambda\beta)^2}\frac{d^2}{d\phi^2}\left(\frac{\beta^2\cos\phi - 1}{1+\lambda\beta\cos\frac{\phi}{2}}\right)\bigg|_{\phi=0} = \frac{2}{(1+\lambda\beta)^3}\left(-\frac{\beta^2}{2} + \frac{\lambda\beta}{8}\frac{\beta^2 - 1}{1+\lambda\beta}\right). \end{aligned} \tag{41.48}$$

The remaining λ integral can now be easily evaluated, yielding (see Problem 41.1)

$$\begin{aligned} P &= -\frac{e^2}{R}\omega_0\beta\int_{-1}^{1} d\lambda \frac{1}{(1+\lambda\beta)^3}\left(-\frac{\beta^2}{2} + \frac{\lambda\beta}{8}\frac{\beta^2-1}{1+\lambda\beta}\right) \\ &= \frac{2}{3}\frac{e^2}{R}\omega_0\frac{\beta^3}{(1-\beta^2)^2} = \frac{2}{3}\frac{e^2}{R}\omega_0\beta^3\left(\frac{E}{\mu c^2}\right)^4, \end{aligned} \tag{41.49}$$

which is the exact result for the total radiated power.

In the nonrelativistic limit, $\beta \ll 1$, (41.49) reduces to the Larmor formula, (35.22),

$$P_{\text{N.R.}} = \frac{2}{3}\frac{e^2}{R}\omega_0\beta^3 = \frac{2}{3}\frac{e^2}{c^3}\left(\frac{d\mathbf{v}}{dt}\right)^2, \tag{41.50}$$

because, for circular motion, the centripetal acceleration is

$$\left|\frac{d\mathbf{v}}{dt}\right| = \frac{v^2}{R} = v\omega_0. \tag{41.51}$$

For a relativistic particle, the power radiated is larger by the factor $(E/mc^2)^4$, signifying that high energy charged particles moving in circular orbits emit substantial synchrotron radiation. In one period, $T = 2\pi/\omega_0$, the total energy radiated is

$$\Delta E = PT = \frac{4\pi}{3}\frac{e^2}{R}\beta^3\left(\frac{E}{\mu c^2}\right)^4. \tag{41.52}$$

In practical units, if we express energy in electron-volts (eV), and the radius of the accelerator in meters, then for $\beta \approx 1$, the energy loss per cycle by an electron is

$$\Delta E(\text{keV}) = 88.4 \frac{E^4(\text{GeV})}{R(\text{m})}, \tag{41.53}$$

where

$$1 \text{ GeV} = 10^6 \text{ keV} = 10^9 \text{ eV}. \tag{41.54}$$

Inserting typical numbers for an electron synchrotron, $R = 10$ meters and $E = 10$ GeV, we find for the energy loss per cycle

$$\Delta E = 88.4 \text{ MeV}, \tag{41.55}$$

which is quite substantial. For this reason electron synchrotrons are impractical for energies greater than ~ 10 GeV. On the other hand, the radiation from any existing or projected proton synchrotron is quite negligible, being smaller, for the same energy, than electron synchrotron radiation by the factor $(m_e/m_p)^4 = 8.8 \times 10^{-14}$. For the LHC, where the energy per beam is $E = 6.8$ TeV and $R = 4.3$ km, the energy loss per cycle is only about 3.9 keV.

41.5 Problems for Chapter 41

1. Carry out the λ integration by which (41.49) is obtained from (41.48).
2. Provide an alternative derivation of the total power emitted in synchrotron radiation, (41.49), by proceeding as follows: Starting from (41.10) let

$$\mathbf{n} \cdot (\mathbf{r} - \mathbf{r}') = -\lambda |\mathbf{r} - \mathbf{r}'| \tag{41.56a}$$

where

$$\mathbf{r} = \mathbf{r}(T + \tau/2), \quad \mathbf{r}' = \mathbf{r}(T - \tau/2), \tag{41.56b}$$

thereby obtaining the frequency distribution

$$\frac{dP}{d\omega} = \frac{\omega^2 e^2}{2\pi c} \int_{-\infty}^{\infty} d\tau \, e^{-i\omega\tau} \left(\frac{\mathbf{v}(T + \tau/2) \cdot \mathbf{v}(T - \tau/2)}{c^2} - 1 \right) \int_{-1}^{1} d\lambda \, e^{-i(\omega/c)\lambda |\mathbf{r} - \mathbf{r}'|}, \tag{41.57}$$

whence, by integration over ω, the total power is obtained,

$$P = -\frac{e^2}{2c} \int_{-\infty}^{\infty} d\tau \int_{-1}^{1} d\lambda \left(\frac{\mathbf{v} \cdot \mathbf{v}'}{c^2} - 1 \right) \delta''(\tau + \lambda |\mathbf{r} - \mathbf{r}'|/c). \tag{41.58}$$

Specialize this general result to circular motion, and let $\omega_0 \tau = \phi$, to obtain (41.47) and hence (41.49). (In this way, the second derivative of the δ function emerges automatically.)
3. Verify the numerical result for the energy loss per cycle, (41.53). (1 esu unit of potential = 300 volts—see Appendix A.)
4. What strength of magnetic field (in Tesla) is required to keep a relativistic electron of energy E (in GeV) moving in a circle of radius R (in meters)?

5. An electron is raised to energy E (in GeV), while moving in a circle of radius R (meters). Then the power input is cut off. Assuming that the radius is maintained constant, how long (in milliseconds) will it take for the energy to drop to half its maximum value?

6. Derive the total power radiated, (41.49), from the result (40.37).

7. It is easy to see where the relativistic factor $(E/\mu c^2)^4$ comes from in the formula for the total energy emitted in synchrotron radiation, (41.52). Argue that because the power $P = dE/dt$ is a relativistic invariant (why is this?) the generalization of the Larmor formula must be

$$P = \frac{2}{3}\frac{e^2}{c^3}\left(\frac{d^2x^\lambda}{d\tau^2}\frac{d^2x_\lambda}{d\tau^2}\right) = \frac{2}{3}\frac{e^2}{\mu^2c^3}\left(\frac{dp^\lambda}{d\tau}\frac{dp_\lambda}{d\tau}\right), \tag{41.59}$$

where the space-time position of the particle is described by the four-vector

$$x^\lambda = (ct, \mathbf{x}), \quad x_\lambda = (-ct, \mathbf{x}), \quad \lambda \in \{0, 1, 2, 3\}, \tag{41.60}$$

$p^\lambda = (E/c, \mathbf{p})$ is the four-vector momentum of the particle, and the proper-time interval is

$$d\tau = \sqrt{dt^2 - d\mathbf{x}^2/c^2}. \tag{41.61}$$

Show that if the power is written in terms of the velocity $\mathbf{v} = d\mathbf{x}/dt$, and when $|\mathbf{v}| = v = \beta c$ is constant,

$$P = \frac{2}{3}\frac{e^2}{c^3}\frac{1}{(1-\beta)^2}\left(\frac{d\mathbf{v}}{dt}\right)^2 = \frac{2}{3}\frac{e^2}{R}\omega_0\beta^3\left(\frac{E}{\mu c^2}\right)^4, \tag{41.62}$$

where the last applies to a circular orbit of radius R, where

$$\frac{d\mathbf{v}}{dt} = \boldsymbol{\omega}_0 \times \mathbf{v}. \tag{41.63}$$

8. Recall our consideration of the work done by a current source on the electromagnetic field in Chapter 36 leading to (36.8b). This implies there must be a reaction force back on the charged particle:

$$\mathbf{F}_{\text{rr}} \cdot \mathbf{v} = \frac{2}{3}\frac{e^2}{c^3}\ddot{\mathbf{v}} \cdot \mathbf{v}, \quad \mathbf{F}_{\text{rr}} = \frac{2}{3}\frac{e^2}{c^3}\ddot{\mathbf{v}}. \tag{41.64}$$

This force is called radiation reaction. Thus, the nonrelativistic equation of motion is

$$\mu\frac{d\mathbf{v}}{dt} = e\left(\mathbf{E} + \frac{\mathbf{v}}{c}\times\mathbf{B}\right) + \frac{2}{3}\frac{e^2}{c^3}\frac{d^2\mathbf{v}}{dt^2}. \tag{41.65}$$

By multiplying by $\mathbf{v}$ obtain the energy equation

$$\frac{d}{dt}\left[\frac{1}{2}\mu v^2 - \frac{2}{3}\frac{e^2}{c^3}\frac{d}{dt}\frac{v^2}{2}\right] = e\mathbf{v} \cdot \mathbf{E} - \frac{2}{3}\frac{e^2}{c^3}\left(\frac{d\mathbf{v}}{dt}\right)^2. \tag{41.66}$$

The last term here is, of course, the Larmor formula. Show that the relativistic generalization of these energy and momentum equations is

$$\frac{d}{d\tau}p^\mu = \frac{e}{c}F^{\mu\nu}\frac{dx_\nu}{d\tau} - \frac{2}{3}\frac{e^2}{c^5}\left(\frac{d^2x_\nu}{d\tau^2}\frac{d^2x^\nu}{d\tau^2}\right)\frac{dx^\mu}{d\tau}, \tag{41.67}$$

where

$$p^\mu = \mu\frac{dx^\mu}{d\tau} - \frac{2}{3}\frac{e^2}{c^3}\frac{d^2x^\mu}{d\tau^2}. \tag{41.68}$$

Take the time component of (41.67) and obtain for the energy loss by a charged particle, or the power radiated,

$$P = -\left(\frac{dE}{dt}\right)_{\rm rad} = \frac{2}{3}\frac{e^2}{\mu^2c^3}\left(\frac{E}{\mu c^2}\right)^2\left[\left(\frac{d\mathbf{p}}{dt}\right)^2 - \left(\frac{E/c}{dt}\right)^2\right]. \tag{41.69}$$

Show that this relativistic Larmor formula leads, once again, to the total energy emitted in synchrotron radiation, (41.52).

42

Synchrotron Radiation II—Polarization

The polarization state of an electromagnetic wave is determined by the direction of its electric field of which there are two independent possibilities, each normal to the direction of propagation, $\mathbf{n}$. The results obtained in the previous chapter referred to the sum of powers radiated into both polarization states. We now investigate the polarization characteristics of synchrotron radiation. One way in which this information is practically useful lies in studying astrophysical objects; for example, the Crab Nebula is inferred to emit synchrotron radiation, because of the unique polarization characteristics of the latter. In order to calculate the power radiated into each polarization state, we return to the basic formula (39.5a). However, we recall that the latter equation was obtained from (38.23) in which energy flow was expressed in terms of $|\mathbf{B}|$. It is desirable to shift the emphasis from $|\mathbf{B}|$ to $|\mathbf{E}|$, since it is the electric field that is measured when polarization is determined. We do this by noting that since $\mathbf{E}$ and $\mathbf{B}$ are related by (38.17), (39.5a) can be equivalently written as

$$\frac{d^2P(T)}{d\omega d\Omega} = \frac{\omega^2}{4\pi^2 c}\int_{-\infty}^{\infty} d\tau\, e^{-i\omega\tau}\left[-\mathbf{n}\times\left(\mathbf{n}\times\frac{1}{c}\mathbf{j}(\mathbf{k},T+\tau/2)^*\right)\right] \cdot\left[-\mathbf{n}\times\left(\mathbf{n}\times\frac{1}{c}\mathbf{j}(\mathbf{k},T-\tau/2)\right)\right], \tag{42.1}$$

where the terms in the square brackets are proportional to the electric field strength.

To isolate the effect of each polarization state, we look at the two components of $\mathbf{E}$, in the plane perpendicular to $\mathbf{n}$, separately. We choose the coordinate system (which is naturally set by the synchrotron) as before, with $\mathbf{n}$ given by (41.11a). We define two polarization orientations as follows: For "parallel polarization" the electric field is in the direction of $\mathbf{e}_{\parallel}$, which lies in the orbital plane and is perpendicular to $\mathbf{n}$, that is, $\mathbf{e}_{\parallel}$ points in the $+y$ direction,

$$\mathbf{e}_{\parallel} = (0,1,0). \tag{42.2a}$$

The "perpendicular polarization" vector, $\mathbf{e}_{\perp}$, is perpendicular to both $\mathbf{e}_{\parallel}$ and $\mathbf{n}$, with $\mathbf{e}_{\parallel}$, $\mathbf{e}_{\perp}$, and $\mathbf{n}$ forming a right-handed system of unit vectors as shown in Fig. 42.1:

$$\mathbf{e}_{\perp} = (-\cos\theta, 0, \sin\theta). \tag{42.2b}$$

[Terminological note: we might also refer to these polarization states as follows: $\perp$ polarization could also be termed a transverse electric (TE) or an H mode, because the electric field is entirely in the plane defined by $\mathbf{n}$ and $\hat{\mathbf{z}}$, while the magnetic find has a component perpendicular to that plane; and $\parallel$ polarization is transverse magnetic (TM) or E polarization, because the behavior of the field is reversed.[1]]

It is sufficient to compute the radiation produced in a single polarization state since the result of the previous section then supplies the power radiated in the other state. It is simpler to consider the $\mathbf{e}_{\parallel}$ polarization. The partial intensity with this polarization arises

[1] $\parallel$ polarization is also denoted by p, and $\perp$ by s for *senkrecht*.

DOI: 10.1201/9781003057369-42

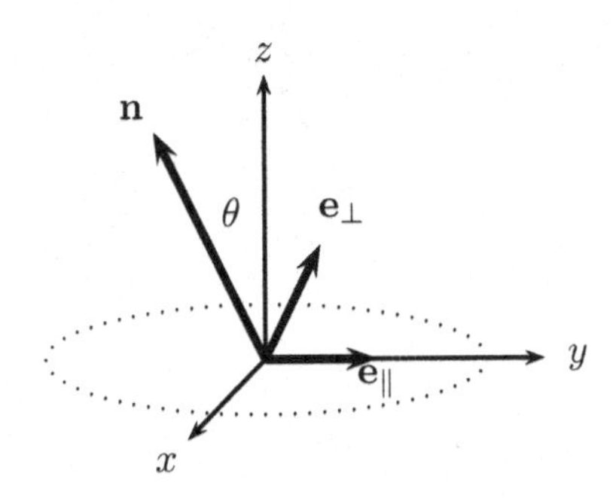

FIGURE 42.1
Polarization vectors for synchrotron radiation. Both $\mathbf{e}_{\|}$ and $\mathbf{e}_{\perp}$ are perpendicular to the direction of observation $\mathbf{n}$.

from the product of y components of the vectors in square brackets in (42.1). Since $\mathbf{n}$ lies in the xz plane, this component is

$$[-\mathbf{n}\times(\mathbf{n}\times\mathbf{j})]_y = [\mathbf{j} - \mathbf{n}(\mathbf{n}\cdot\mathbf{j})]_y = j_y, \tag{42.3}$$

and the resulting contribution to the power spectrum radiated in the parallel polarization state is

$$\left(\frac{d^2P(T)}{d\omega d\Omega}\right)_{\|} = \frac{\omega^2}{4\pi^2 c}\int_{-\infty}^{\infty} d\tau\, e^{-i\omega\tau}\int(d\mathbf{r})(d\mathbf{r}')\, e^{i(\omega/c)\mathbf{n}\cdot(\mathbf{r}-\mathbf{r}')}$$
$$\times\frac{1}{c}j_y(\mathbf{r}, T+\tau/2)\frac{1}{c}j_y(\mathbf{r}', T-\tau/2). \tag{42.4}$$

For synchrotron radiation, the current density is described by (41.9b) so that (42.4) reduces to

$$\left(\frac{d^2P(T)}{d\omega d\Omega}\right)_{\|} = \frac{\omega^2 e^2}{4\pi^2 c}\int_{-\infty}^{\infty} d\tau\, e^{-i\omega\tau} e^{i(\omega/c)\mathbf{n}\cdot[\mathbf{r}(T+\tau/2)-\mathbf{r}(T-\tau/2)]}$$
$$\times\frac{1}{c^2}v_y(T+\tau/2)v_y(T-\tau/2). \tag{42.5}$$

We now can evaluate (42.5) by following closely the procedure given in the previous chapter. Therefore, instead of the factor

$$\frac{1}{c^2}\mathbf{v}(T+\tau/2)\cdot\mathbf{v}(T-\tau/2) - 1 = \beta^2\cos\omega_0\tau - 1 \tag{42.6}$$

appearing in (41.10), we now encounter, because the position vector and velocity are given by (41.11b)–(41.11d) and (41.12a)–(41.12c), respectively,

$$\frac{1}{c^2}v_y(T+\tau/2)v_y(T-\tau/2) = \beta^2\cos\omega_0(T+\tau/2)\cos\omega_0(T-\tau/2)$$
$$= \frac{\beta^2}{2}(\cos 2\omega_0 T + \cos\omega_0\tau). \tag{42.7}$$

The time average of (42.5) involves not only (41.20) but also the new evaluation [most easily done starting with (41.15b)],

$$\langle\cos 2\omega_0 T\, e^{i(\omega/c)\mathbf{n}\cdot[\mathbf{r}(T+\tau/2)-\mathbf{r}(T-\tau/2)]}\rangle_{\text{one cycle}}$$
$$= -\sum_{m=-\infty}^{\infty} e^{im\omega_0\tau} J_{m+1}\left(\frac{\omega R}{c}\sin\theta\right) J_{m-1}\left(\frac{\omega R}{c}\sin\theta\right). \tag{42.8}$$

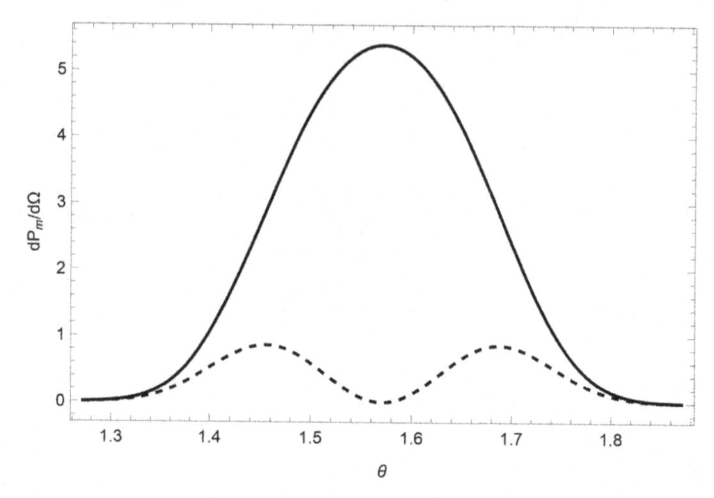

FIGURE 42.2
The angular power distributions for the two polarizations given in (42.12). The factor $\frac{\omega_0}{2\pi}\frac{e^2}{R}$ has been pulled out of what is plotted. The solid curve is for the $\|$ polarization, while the dashed curve is for $\perp$ polarization. The curves shown are for $\beta = 0.99$ and for $m = 356$, which is the characteristic value of m, m_c, found in the next chapter. It will be seen that the radiation is predominantly polarized in the plane of the orbit.

Therefore, instead of the combination

$$(\beta^2 \cos\omega_0\tau - 1) \sum_{m=-\infty}^{\infty} e^{im\omega_0\tau} \left[J_m \left(\frac{\omega}{c} R \sin\theta \right) \right]^2, \tag{42.9}$$

appearing in (41.19), we have here, for parallel polarization, the term

$$\frac{1}{2}\beta^2 \sum_{m=-\infty}^{\infty} e^{im\omega_0\tau} [-J_{m+1}J_{m-1} + \cos\omega_0\tau\, J_m^2]. \tag{42.10}$$

The τ integration now leads to the power radiated into the mth harmonic being proportional to

$$\frac{\beta^2}{2}\left(\frac{J_{m+1}^2 + J_{m-1}^2}{2} - J_{m+1}J_{m-1} \right) = \frac{1}{4}\beta^2(-J_{m+1} + J_{m-1})^2 = \beta^2 (J_m')^2, \tag{42.11}$$

where we have used the recurrence relation (41.28a). Thus, the two terms in (41.30) represent, respectively, the radiation in the states characterized by $\mathbf{e}_\|$ and $\mathbf{e}_\perp$:

$$\left(\frac{dP_m}{d\Omega} \right)_\| = \frac{\omega_0}{2\pi}\frac{e^2}{R}\beta^3 m^2 [J_m'(m\beta\sin\theta)]^2, \tag{42.12a}$$

$$\left(\frac{dP_m}{d\Omega} \right)_\perp = \frac{\omega_0}{2\pi}\frac{e^2}{R}\beta^3 m^2 \left[\frac{J_m(m\beta\sin\theta)}{\beta\tan\theta} \right]^2. \tag{42.12b}$$

We plot these distributions in Fig. 42.2.

42.1 Spin Polarization

Of course, the emission of radiation has a reaction on the charged particle. Particularly interesting is the effect on the spin polarization of the electron. Sokolov and Ternov [41] showed that the spins tend to become antialigned with the magnetic field, reaching a maximum polarization of $\frac{8}{5\sqrt{3}} = 92.4\%$. Schwinger's take on this appeared a decade later [42]. Bell and Leinaas [43] showed that this polarization is related to the Unruh effect. These interesting phenomena are quantum mechanical, and so lie outside the scope of this book.

42.2 Problems for Chapter 42

1. Fill in the details leading to (42.8).
2. Compute the power radiated into the mth harmonic for a given polarization, $(P_m)_\parallel$ and $(P_m)_\perp$, starting from (42.1). In this case, we must take the direction of observation to lie in an arbitrary direction,

$$\mathbf{n} = (x, y, z) = (\sin\theta\cos\phi, \sin\theta\sin\phi, \cos\theta), \tag{42.13}$$

and therefore, that $\parallel$ polarization means that only the $\hat{\boldsymbol{\phi}}$ component of $\mathbf{j}$ enters so that

$$[-\mathbf{n}\times(\mathbf{n}\times\mathbf{j})]\cdot\hat{\boldsymbol{\phi}} = \mathbf{j}\cdot\hat{\boldsymbol{\phi}}, \quad \hat{\boldsymbol{\phi}} = (-\sin\phi, \cos\phi, 0). \tag{42.14}$$

One form of the answer is

$$(P_m)_\parallel = \frac{e^2}{R} m\omega_0\beta \left[2\beta J'_{2m}(2m\beta) + \beta\int_0^{2m\beta} dz\, J_{2m}(z) - 2m\int_0^{2m\beta}\frac{dz}{z}J_{2m}(z)\right], \tag{42.15a}$$

$$(P_m)_\perp = \frac{e^2}{R} m\omega_0 \left[2m\beta\int_0^{2m\beta}\frac{dz}{z}J_{2m}(z) - \int_0^{2m\beta} dz\, J_{2m}(z)\right]. \tag{42.15b}$$

[Hint: In integrating over ϕ it is convenient to perform a coordinate rotation about the x axis. That is, in (42.13) replace $y \to z$, $z \to -y$, $x \to x$. In this way the exponent becomes independent of ϕ, and the integrals are greatly simplified.]

3. Compute the total radiated power for a given polarization, $P_\parallel$ and $P_\perp$. The result is

$$P_\parallel = \frac{6+\beta^2}{8}P, \quad P_\perp = \frac{2-\beta^2}{8}P, \tag{42.16}$$

so nonrelativistically, and ultrarelativistically, respectively, the ratio of power in the two polarizations is

$$\frac{P_\parallel}{P_\perp} = 3 \quad (\beta \ll 1), \qquad \frac{P_\parallel}{P_\perp} = 7 \quad (\beta \approx 1). \tag{42.17}$$

[Hint: Do the ϕ integral, as described in Problem 2, then the frequency integral, followed by the τ integral.]

43

Synchrotron Radiation III—High Energies

In the previous chapters, we developed general formulas that describe the total power radiated by a synchrotron, and the spectral and angular distributions of that radiation. The radiation into the different polarization modes was also worked out. Here we give the asymptotic behavior for high energies, and provide a qualitative understanding of the phenomena. We conclude this discussion with some historical notes.

43.1 Range of Important Harmonics

For most practical applications, we are interested in the frequency spectrum of the power radiated by high energy electrons. In the limit $\beta \to 1$ the power radiated into the mth harmonic, (41.42), is approximately

$$P_m \approx \omega_0 \frac{e^2}{R} 2m J'_{2m}(2m), \tag{43.1}$$

which becomes for large harmonic number, $m \gg 1$,

$$P_m \sim \frac{3^{1/6}}{\pi} \Gamma(2/3) \omega_0 \frac{e^2}{R} m^{1/3}, \tag{43.2}$$

where we have used the asymptotic form of $J'_{2m}(2m)$,

$$J'_{2m}(2m) \sim \frac{3^{1/6}}{2\pi} \Gamma(2/3) m^{-2/3}, \quad m \gg 1, \tag{43.3}$$

which is already good to within 15% for $m = 1$. We will derive (43.3) in the following section. The total power radiated is thus roughly estimated by summing (43.2) over all harmonics:

$$P = \sum_{m=1}^{\infty} P_m \sim \sum_{m=1}^{m_c} m^{1/3}, \tag{43.4}$$

where we have noted that the continuing increase in the power radiated into higher and higher harmonics must break down for sufficiently large m, since the total power radiated, (41.49), is finite. Consequently, we have cut off the sum at m_c, the critical harmonic number. In terms of this cutoff, the total power radiated is

$$P \sim m_c^{4/3}, \tag{43.5}$$

which, on the other hand, is, by (41.49), proportional to

$$P \sim \left(\frac{E}{\mu c^2}\right)^4. \tag{43.6}$$

DOI: 10.1201/9781003057369-43

Therefore, we conclude that the order of magnitude of m_c is

$$m_c \sim \left(\frac{E}{\mu c^2}\right)^3. \tag{43.7a}$$

Roughly speaking, then, the maximum frequency of radiation is of the order

$$\omega_c \sim \omega_0 \left(\frac{E}{\mu c^2}\right)^3, \tag{43.7b}$$

which implies that the shortest (reduced) wavelength ($\lambda\!\!\!^{-} = \lambda/(2\pi)$) radiated is about

$$\lambda\!\!\!^{-}_c \sim \left(\frac{mc^2}{E}\right)^3 R, \tag{43.7c}$$

where we have used the relations

$$\omega = \frac{c}{\lambda\!\!\!^{-}}, \quad \text{and} \quad \omega_0 \approx \frac{c}{R}. \tag{43.8}$$

For an electron, the shortest wavelength radiated is approximately

$$\lambda\!\!\!^{-}_c(\text{Å}) \sim \frac{R(\text{m})}{E^3(\text{GeV})}. \tag{43.9}$$

where one Angstrom, Å $= 10^{-8}$ cm, is a characteristic x-ray wavelength. Thus, the synchrotron radiation produced by a high energy electron is characterized by very large harmonic numbers; and consequently one gets visible, ultraviolet, x-ray, and even γ radiation from a typical accelerator.

43.2 Asymptotic Form for $J'_{2m}(2m)$.

We now furnish a derivation of the asymptotic formula for $J'_{2m}(2m)$, (43.3). Starting from the integral representation for the Bessel function [see (41.38b)]

$$J_{2m}(z) = \int_0^\pi \frac{d\phi}{\pi} \cos(z\sin\phi - 2m\phi), \tag{43.10}$$

we obtain

$$J'_{2m}(z) = -\int_0^\pi \frac{d\phi}{\pi} \sin\phi \sin(z\sin\phi - 2m\phi), \tag{43.11a}$$

which, when we set $z = 2m$, becomes

$$J'_{2m}(2m) = -\int_0^\pi \frac{d\phi}{\pi} \sin\phi \sin[2m(\sin\phi - \phi)]. \tag{43.11b}$$

If $2m$ is large, the integrand of (43.11b) oscillates rapidly as ϕ varies, leading to destructive interference except near $\phi = 0$. (This is the basis of stationary phase or saddle point evaluations, which we will discuss in the next Section.) The main contribution comes from that value of ϕ satisfying the stationary condition

$$\frac{d}{d\phi}(\sin\phi - \phi) = 0, \tag{43.12a}$$

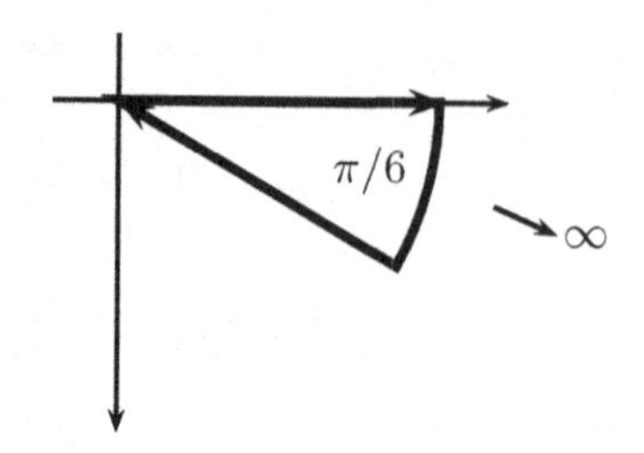

FIGURE 43.1
Change of contour used in evaluating (43.14).

which implies, as expected,

$$\cos\phi = 1, \quad \text{or} \quad \phi = 0. \tag{43.12b}$$

Therefore, making use of the approximations, for $\phi \ll 1$,

$$\sin\phi - \phi \approx -\frac{\phi^3}{6}, \tag{43.13a}$$

$$\sin\phi \approx \phi, \tag{43.13b}$$

and noting the range of integration can be extended to infinity with negligible error, we obtain

$$\begin{aligned} J'_{2m}(2m) &\sim -\operatorname{Im}\int_0^\infty \frac{d\phi}{\pi}\phi\, e^{-im\phi^3/3} = -\operatorname{Im}\left(\frac{3}{m}\right)^{2/3}\frac{e^{-i\pi/3}}{\pi}\int_0^\infty dt\left(\frac{1}{3}t^{-2/3}\right)t^{1/3}e^{-t} \\ &= -\operatorname{Im}\left(\frac{3}{m}\right)^{2/3}\frac{\Gamma(2/3)}{3\pi}e^{-i\pi/3} = \frac{3^{1/6}}{2\pi}\frac{\Gamma(2/3)}{m^{2/3}}, \quad \text{for} \quad m \gg 1, \end{aligned} \tag{43.14}$$

which is just (43.3). In the above evaluation, we have used Cauchy's theorem to perform a change of contour, as shown in Fig. 43.1, and have used the definition of the gamma function (20.18):

$$\Gamma(z) = \int_0^\infty dt\, t^{z-1}e^{-t}. \tag{43.15}$$

Notice that (43.14) is valid for m either integer or half-integer.

43.3 Spectral Distribution

We now want to improve upon the qualitative discussion of the high energy power spectrum given in Section 43.1. As we have previously remarked, the approximation (43.2) breaks down for sufficiently large m. This can be traced to the fact that, in (41.42), β is not exactly equal to 1. Consequently, both terms there contribute but we will concentrate on the first one as it contains all the essential characteristics of the radiation. (See Problem 43.2.) Thus we seek an asymptotic expression for $J'_{2m}(2m\beta)$, starting from the integral representation (43.11a):

$$J'_{2m}(2m\beta) = -\int_0^\pi \frac{d\phi}{\pi}\sin\phi\sin 2m(\beta\sin\phi - \phi). \tag{43.16}$$

As before, in the limit when $\beta \to 1$, $m \gg 1$, the main contribution comes from the region near $\phi = 0$. Therefore, we may expand the integrand in (43.16) as follows:

$$\begin{aligned}\sin\phi\sin 2m(\beta\sin\phi-\phi) &\approx \phi\sin 2m\left(\beta\left[\phi-\frac{\phi^3}{3!}\right]-\phi\right)=\phi\sin\left(2m\left[-\phi(1-\beta)-\frac{1}{6}\beta\phi^3\right]\right)\\ &\approx -\phi\sin\left[m\left((1-\beta^2)\phi+\frac{1}{3}\phi^3\right)\right]\\ &= -\sqrt{1-\beta^2}x\sin\left[m(1-\beta^2)^{3/2}\left(x+\frac{1}{3}x^3\right)\right],\end{aligned}\tag{43.17}$$

where we have introduced the change of scale

$$\phi = \sqrt{1-\beta^2}x. \tag{43.18}$$

As a result, in this limit, (43.16) can be approximated by

$$\begin{aligned}J'_{2m}(2m\beta) &\sim (1-\beta^2)\int_0^\infty \frac{dx}{\pi}x\sin\left(m(1-\beta^2)^{3/2}\left(x+\frac{1}{3}x^3\right)\right)\\ &= \frac{(1-\beta^2)}{\pi}\,\mathrm{Im}\int_0^\infty dx\,x\,e^{im(1-\beta^2)^{3/2}(x+x^3/3)}.\end{aligned}\tag{43.19}$$

For m fixed and β approaching unity in such a way that $m(1-\beta^2)^{3/2} \ll 1$, the significant contribution to (43.19) comes from the region where x is large, and (43.19) reduces to (43.14):

$$J'_{2m}(2m\beta) \sim (1-\beta^2)\int_0^\infty \frac{dx}{\pi}x\sin\left(\frac{m}{3}(1-\beta^2)^{3/2}x^3\right) = \int_0^\infty \frac{d\phi}{\pi}\phi\sin\left(\frac{m}{3}\phi^3\right), \tag{43.20}$$

where all reference to the speed of the particle has disappeared. However, for sufficiently large m, the parameter $m(1-\beta^2)^{3/2}$ becomes large, and the integrand undergoes rapid oscillations in x except near the stationary points, which satisfy

$$\frac{d}{dx}\left(x+\frac{1}{3}x^3\right) = 1 + x^2 = 0; \tag{43.21a}$$

that is, the stationary phase points are located at

$$x = \pm i. \tag{43.21b}$$

By extending the region of integration from $-\infty$ to $+\infty$, we evaluate (43.19) asymptotically by following the standard procedure of the saddle point method (or the method of steepest descents). We deform the contour of integration so that it passes through the stationary point $x = i$, because then the dominant contribution comes from the vicinity of that point. (See Fig. 43.2.) We further deform the contour so the steepest-descent path, as shown in the figure so that away from the saddle point, the integrand is real and decreases exponentially fast. See Problem 43.3. In the neighborhood of $x = i$, we let

$$x = i + \xi, \tag{43.22}$$

where ξ is real, to take advantage of the saddle point character. For small ξ

$$x + \frac{1}{3}x^3 = (i+\xi) + \frac{1}{3}(i+\xi)^3 \approx i\left(\frac{2}{3}+\xi^2\right), \tag{43.23}$$

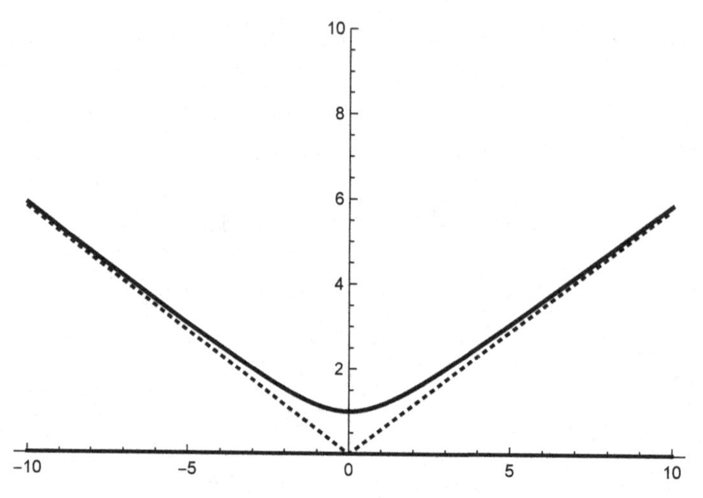

FIGURE 43.2
Stationary phase contour for evaluation of (43.19). The original contour of integration, along the real axis, is distorted into the steepest-descents path in the complex plane, passing through the stationary point $x = i$ and going to ∞ asymptotically with phase $\pi/6$, $5\pi/6$. For details see Problem 43.3.

so that the exponential factor in (43.19) becomes

$$e^{-\frac{2}{3}m(1-\beta^2)^{3/2}}e^{-m(1-\beta^2)^{3/2}\xi^2}, \tag{43.24}$$

which falls off exponentially on both sides of $x = i$. The resulting Gaussian integral in (43.19) leads to the following asymptotic form:

$$J'_{2m}(2m\beta) \sim \frac{1}{2}\frac{(1-\beta^2)^{1/4}}{\sqrt{\pi m}}e^{-\frac{2}{3}m(1-\beta^2)^{3/2}}, \quad m(1-\beta^2)^{3/2} \gg 1. \tag{43.25}$$

Thus, for very large harmonic numbers, the power spectrum decreases exponentially in contrast to the behavior for smaller values of m where it increases like $m^{1/3}$. The transition between these two regimes occurs near the critical harmonic number, m_c , for which

$$m_c(1-\beta^2)^{3/2} \equiv 1, \tag{43.26}$$

or

$$m_c = (1-\beta^2)^{-3/2} = \left(\frac{E}{\mu c^2}\right)^3, \tag{43.27}$$

supporting our previous estimate, (43.7a). The bulk of the radiation is emitted with harmonic numbers near m_c. The qualitative shape of the spectrum is shown in Fig. 43.3.

43.4 Angular Distribution

We now go back one more stage and examine the angular distribution in the high energy limit. For the radiation in the plane of the orbit ($\theta = \pi/2$), (41.30) becomes

$$\frac{dP_m}{d\Omega} = \frac{\omega_0}{2\pi}\frac{e^2}{R}\beta^3 m^2[J'_m(m\beta)]^2. \tag{43.28}$$

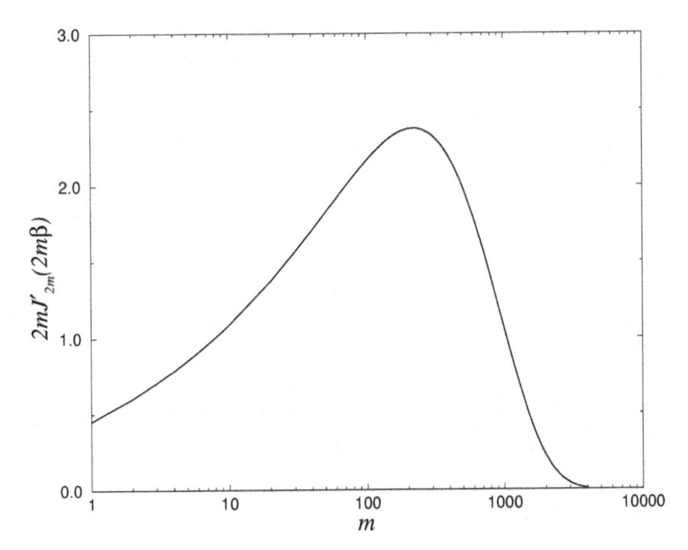

FIGURE 43.3
Sketch of power emitted into mth harmonic as a function of m. What is actually plotted is $2mJ'_{2m}(2m\beta)$ for $\beta = 0.99$. In this case $m_c = 356$.

If we further let $\beta \to 1$ and $m_c \gg m \gg 1$, we find from (43.3) and (43.2),

$$\frac{dP_m}{2\pi\, d\theta} \sim \frac{\omega_0}{2\pi}\frac{e^2}{R}m^2\left[\frac{3^{1/6}}{2\pi}\Gamma(2/3)(2/m)^{2/3}\right]^2 \sim \omega_0\frac{e^2}{R}m^{2/3} \sim \frac{P_m}{m^{-1/3}}. \tag{43.29}$$

Here we compare P_m, which increases as $m^{1/3}$, with $dP_m/d\theta$, which behaves as $m^{2/3}$, from which it is evident that the radiation, for large harmonic numbers, is confined to a small angular range around $\theta = \pi/2$ of width

$$\Delta\theta \sim m^{-1/3}. \tag{43.30}$$

Since most of the radiation is emitted with harmonic numbers in the neighborhood of

$$m_c \sim \left(\frac{E}{\mu c^2}\right)^3, \tag{43.31}$$

the radiation is concentrated in an angular range about the plane of the electron orbit of the order

$$\Delta\theta \sim \frac{\mu c^2}{E} = \sqrt{1-\beta^2}. \tag{43.32}$$

A plot of the distribution in θ is given in Fig. 43.4.

This radiation, concentrated around $\theta = \pi/2$, is predominantly polarized in the plane of the orbit, according to (42.12a) and (42.12b). More precisely, it was shown in Problem 42.3 that in the ultrarelativistic limit the ratio of the power radiated with parallel polarization to that with perpendicular polarization is

$$\frac{P_\parallel}{P_\perp} = 7, \quad \text{for} \quad \beta \approx 1, \tag{43.33}$$

while in the nonrelativistic limit,

$$\frac{P_\parallel}{P_\perp} = 3, \quad \text{for} \quad \beta \ll 1. \tag{43.34}$$

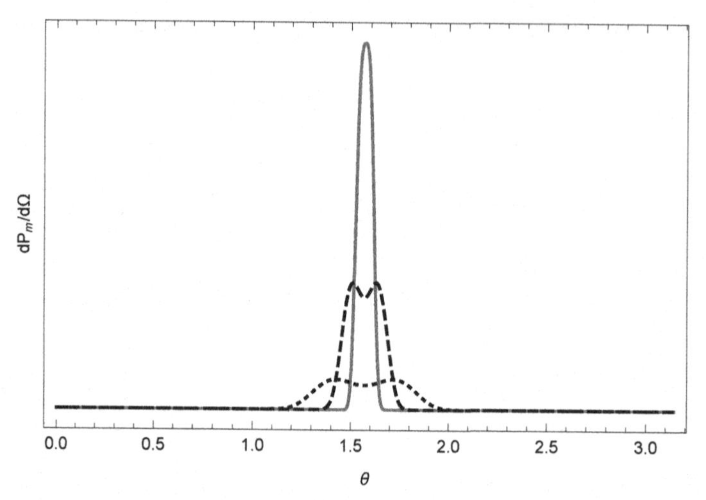

FIGURE 43.4
Angular distribution of synchrotron radiation for $\beta = 0.999$, from (41.30). The different curves are for $m = 100$ (dotted), $m = 1000$ (dashed), and $m = 10,000$ (solid). Note that the peak becomes higher and narrower as m increases, and the small dip in the center disappears.

That is, for any value of β, the radiation is strongly polarized, the degree of polarization increasing with the speed of the charged particle. This characteristic distinguishes synchrotron radiation from thermal radiation, and was, for example, the clue to understanding the origin of the nonthermal radiation emitted by the Crab Nebula.

43.5 Qualitative Description

One might ask whether there is a more direct way of seeing that, for synchrotron radiation produced by a high energy charged particle, the characteristic frequency emitted is so much larger than the Larmor frequency, as stated by (43.7b), and that the angular distribution is so strongly peaked in the forward direction, as suggested by (43.32). We here wish to emphasize the simple physics behind these striking features. For a point charge, the current density, $\mathbf{j}(\mathbf{r}, t)$, is given by (41.9b) so that the vector potential, in the Lorenz gauge, can be computed from (34.37b) to be (see Problem 34.5)

$$\mathbf{A}(\mathbf{r}, t) = \frac{\frac{e}{c}\mathbf{v}(t')}{|\mathbf{r} - \mathbf{r}(t')| \left[1 - \frac{\mathbf{n}}{c} \cdot \mathbf{v}(t')\right]}, \tag{43.35}$$

where $\mathbf{r}(t')$ is the position vector of the particle at the emission time t', while t is the detection time, which are related by (34.33),

$$t = t' + \frac{1}{c}|\mathbf{r} - \mathbf{r}(t')|, \tag{43.36}$$

and $\mathbf{n}$ is the direction of observation

$$\mathbf{n} = \frac{\mathbf{r} - \mathbf{r}(t')}{|\mathbf{r} - \mathbf{r}(t')|}. \tag{43.37}$$

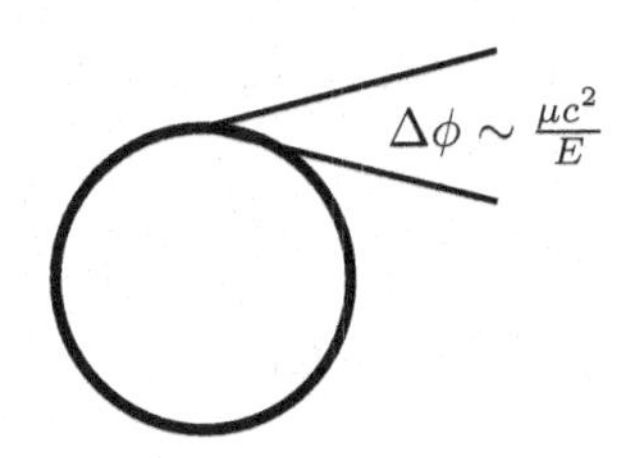

FIGURE 43.5
Directionality of synchrotron radiation. The radiation is predominantly radiated in the cone indicated.

If we let ϕ be the angle between $\mathbf{n}$ and $\mathbf{v}$, the factor

$$\frac{1}{1-\frac{\mathbf{n}\cdot\mathbf{v}}{c}}=\frac{1}{1-\beta\cos\phi}, \tag{43.38}$$

for $\beta\to 1$, is dominated by the small angle region, so we may approximate it by

$$\frac{1}{1-\beta\,(1-\phi^2/2+\dots)}\approx\frac{1}{\frac{1}{2}(1-\beta^2+\phi^2)}. \tag{43.39}$$

As we saw in greater detail for impulsive scattering [see (40.34)], the radiation is therefore concentrated in a narrow angular range near the forward direction

$$\phi\sim\sqrt{1-\beta^2}=\frac{\mu c^2}{E}, \tag{43.40}$$

which is just the behavior found in the preceding section.

Next, we seek a qualitative understanding of the characteristic frequencies radiated by a high energy charged particle moving in a circle with frequency ω_0. Because of the strong directionality of the emitted radiation as expressed by (43.40), the radiation detected at a particular point only arises from a small portion of the orbit, or equivalently, is only emitted during a time interval small compared with the period of revolution, $2\pi/\omega_0$. (See Fig. 43.5.) This effective emission time interval is of order $(2\pi/\omega_0)\phi$, so the important frequencies radiated are $\sim$ (emission time)$^{-1}$ $\sim$ $(\omega_0/2\pi)(1/\phi)$. That is, the smaller the time interval involved in the emission, the higher the frequency emitted. Therefore, a typical frequency emitted is

$$\omega_e\sim\frac{\omega_0}{\sqrt{1-\beta^2}}=\frac{E}{\mu c^2}\omega_0. \tag{43.41}$$

However, this is not what the observer sees since detection time intervals are not equal to emission time intervals. To see the connection between these intervals, the Doppler effect [Christian Doppler (1803–1853)], we recall that these two times are related by (43.36), which implies for the respective time intervals

$$dt'=dt-\frac{1}{c}\mathbf{n}\cdot[-\mathbf{v}(t')dt'] \tag{43.42}$$

or

$$dt=\left(1-\frac{1}{c}\mathbf{n}\cdot\mathbf{v}\right)dt'\sim(1-\beta^2)dt',\quad \beta\approx 1,\quad \phi\approx\sqrt{1-\beta^2}. \tag{43.43}$$

This is the origin of the denominator factor appearing in (43.35). From (43.43), we see that the time interval of detection is shorter than that of emission by a factor $1-\beta^2$, implying that the detection frequency, ω_d, is higher than the emission frequency by a factor $(1-\beta^2)^{-1}$:

$$\omega_d \sim \frac{1}{(1-\beta^2)}\omega_e \sim \frac{1}{(1-\beta^2)^{3/2}}\omega_0 = \omega_0 \left(\frac{\mathcal{E}}{\mu c^2}\right)^3, \tag{43.44}$$

which is the characteristic frequency found in (43.27).

43.6 Historical Note

The theory of radiation from electrons in betatrons was worked out by Schwinger in 1945, but only an abstract was published at the time [44]. A full publication had to wait until 1949 [8]. An earlier unpublished manuscript by Schwinger was transcribed by M. Furman [9]. Independently, D. Ivanenko, I. Pomeranchuk, A. A. Sokolov worked out the spectral and angular properties of the radiation [45, 46] For the history of the development of this subject and many further details, with a particularly Russian perspective, see A. A. Sokolov and I. M. Ternov [47]. For more recent treatises, see [48, 49].

Experimentally, the phenomena were first observed by Floyd Haber in 1947, working with Irving Langmuir and Herbert Pollock at the General Electric Laboratory in Schenectady, who finally realized what was seen was "Ivanenko and Pomeranchuk radiation."

43.7 Problems for Chapter 43

1. Evaluate the power radiated into the mth harmonic, (41.42), for $m=1$, $\beta\to 1$, and compare with the corresponding value obtained from the asymptotic formula (43.3).
2. Improve on the discussion of the high energy limit in Section 43.3 by considering the contribution from the integral in (41.42).
3. This problem provides details of the method of steepest descents described in Section 43.4. The integral evaluated in (43.19) can be written as

$$\frac{1}{2}\int_{-\infty}^{\infty} dx\, x e^{i\zeta\phi(x)}, \quad \phi(x) = x + \frac{1}{3}x^3, \quad \zeta = m(1-\beta^2)^{3/2}. \tag{43.45}$$

   We find the path of steepest descents (PSD) passing through the saddle point at $x=i$ by writing $x=i+\xi$, where the real and imaginary parts of ξ are given by $\xi=\xi_1+i\xi_2$. We require $i\phi$ to be real on the PSD, so show that that means

$$\xi_2 = \sqrt{1+\xi_1^2/3} - 1, \tag{43.46}$$

   and that along the PSD $i\phi \sim -\frac{8}{9\sqrt{3}}\xi_1^3$ for large ξ_1, so the integrand becomes exponentially small. Further, you should note that the distortion of the integrand into the complex plane is necessitated by the requirement that the integral in (43.19) be convergent.
4. For an ultrarelativistic electron, the radiation is concentrated near the orbital plane, $\theta=\pi/2$, so the quantity $\epsilon = 1-\beta^2\sin^2\theta$ becomes very small. Show that

asymptotically the following relation holds for the angular distribution for the power radiated into the mth harmonic:

$$\frac{dP_m}{d\Omega} \sim \omega_0 \frac{e^2}{R}\frac{m^2}{6\pi^3}\left[\epsilon^2 K_{2/3}^2\left(\frac{m}{3}\epsilon^{3/2}\right) + \epsilon\cos^2\theta K_{1/3}^2\left(\frac{m}{3}\epsilon^{3/2}\right)\right],$$

in the limit $m \to \infty$, $\epsilon \to 0$. Here we encounter the integrals

$$K_{1/3}(\zeta) = \pi\sqrt{\frac{3}{z}}\mathrm{Ai}(z) = \sqrt{\frac{3}{z}}\int_0^\infty dt\,\cos\left(zt + \frac{t^3}{3}\right),$$
$$K_{2/3}(\zeta) = -\pi\frac{\sqrt{3}}{z}\mathrm{Ai}'(z) = \frac{\sqrt{3}}{z}\int_0^\infty dt\,t\,\sin\left(zt + \frac{t^3}{3}\right), \qquad (43.47a)$$

where $z = (3\zeta/2)^{2/3}$, and Ai is the Airy[1] function.

5. Show that the asymptotic behavior of the Bessel functions is more uniformly captured by

$$J_{2m}(2m\beta) \sim \frac{1}{\sqrt{3}\pi}\left(\frac{3}{2m}\right)^{1/3}\xi^{1/3}K_{1/3}(\xi), \quad m \gg 1, \qquad (43.48a)$$
$$J'_{2m}(2m\beta) \sim \frac{1}{\sqrt{3}\pi}\left(\frac{3}{2m}\right)^{2/3}\xi^{1/3}K_{2/3}(\xi), \quad m \gg 1, \qquad (43.48b)$$

where $\xi = (2m/3)(\mu c^2/E)^3$. Hence the asymptotic formula for the power radiated in the mth harmonic is

$$P_m \sim \omega_0\frac{e^2}{R}\left(\frac{2m}{3}\right)^{1/3}\frac{\sqrt{3}}{2\pi}\left[2K_{2/3}(\xi) - \int_\xi^\infty dx\,K_{1/3}(\xi)\right]. \qquad (43.49)$$

Show that the limits (43.3) and (43.25) follow from this.

6. Compute the total power from (43.49) by replacing the sum on m by an integral on ξ (why is this justified?). Use the integral

$$\int_0^\infty dt, t^{\mu-1}K_\nu(t) = 2^{\mu-2}\Gamma\left(\frac{\mu-\nu}{2}\right)\Gamma\left(\frac{\mu+\nu}{2}\right), \quad \Re(\mu\pm\nu) > 0, \qquad (43.50)$$

and obtain (41.49) except for the factor of β^3, which is unity in the ultrarelativistic limit.

7. The asymptotic formula for the power radiated by a synchrotron into the mth harmonic is given above, in (43.49), where the modified Bessel function combination given there can be alternatively written as

$$2K_{2/3}(\xi) - \int_\xi^\infty d\eta\,K_{1/3}(\eta) = \int_\xi^\infty d\eta\,K_{5/3}(\eta). \qquad (43.51)$$

Show that when we are considering parallel polarization, this combination is replaced by

$$\frac{1}{2}\left[3K_{2/3}(\xi) - \int_\xi^\infty d\eta\,K_{1/3}(\eta)\right] = \frac{1}{4}\int_\xi^\infty d\eta\,\left[3K_{5/3}(\eta) + K_{1/3}(\eta)\right], \qquad (43.52)$$

[1] George Airy, 1801–1892.

while the remainder describes perpendicular polarization,

$$\frac{1}{4}\left[3K_{2/3}(\xi) - \int_\xi^\infty d\eta\, K_{1/3}(\eta)\right] = \frac{1}{3}\int_\xi^\infty \frac{d\eta}{\eta} K_{5/3}(\eta)$$
$$= \frac{1}{4}\int_\xi^\infty d\eta\left[K_{5/3}(\eta) - K_{1/3}(\eta)\right]. \quad (43.53)$$

Show that the 7 : 1 ratio seen in (42.17) emerges on performing the frequency integral.

8. Synchrotron radiation is closely confined to the orbit, because the Bessel functions of high order are very small if the argument is appreciably less than the order. Make this quantitative by using (43.48). and introducing $\psi = \pi/2 - \theta$ between the point of observation and the plane of the orbit, which we suppose to be small. Then write the power radiated into a unit angular range about the angle ψ, in the mth harmonic, as

$$\frac{dP_m}{d\Omega}(\psi) = \frac{3}{\pi^2}\frac{\omega_0}{2\pi}\frac{e^2}{R}\left(\frac{m}{3}\right)^{2/3}\left[\xi^{4/3}K_{2/3}^2(\xi) + \psi^2\left(\frac{m}{3}\right)^{2/3}\xi^{2/3}K_{1/3}^2(\xi)\right], \quad (43.54)$$

for $m \gg 1$ and $\psi \ll 1$, where

$$\psi = \frac{m}{3}(1 - \beta^2\sin^2\theta)^{3/2} \approx \frac{m}{3}\left[\psi^2 + \left(\frac{\mu c^2}{E}\right)^2\right]^{3/2}. \quad (43.55)$$

Using this result, reproduce the values shown in Fig. 43.4, and the qualitative features discussed in Section 43.4.

44

Propagation in a Dielectric Medium

In this chapter, we will discuss the reflection and refraction of light by a dielectric interface. The situation of an imperfect conductor will be treated in the following chapter.

44.1 Equations for the Normal Modes

We now turn from the mechanisms by which electromagnetic radiation (or light) is produced to how it propagates in a material medium, described at a macroscopic level. Quite quickly, we shall specialize to the case of a dielectric, with inhomogeneity in a single direction only. We begin by restating Maxwell's equations for a macroscopic medium,

$$\boldsymbol{\nabla}\times\mathbf{H}=\frac{1}{c}\dot{\mathbf{D}}+\frac{4\pi}{c}\mathbf{J}, \tag{44.1a}$$

$$-\boldsymbol{\nabla}\times\mathbf{E}=\frac{1}{c}\dot{\mathbf{B}}, \tag{44.1b}$$

$$\boldsymbol{\nabla}\cdot\mathbf{D}=4\pi\rho, \tag{44.1c}$$

$$\boldsymbol{\nabla}\cdot\mathbf{B}=0. \tag{44.1d}$$

Before proceeding, we note that here the scalar equations, (44.1c), (44.1d), are not independent of the first set, (44.1a), (44.1b), since taking the divergence of the latter, and making use of the current conservation condition, (34.2), we find

$$0=\frac{\partial}{\partial t}(\boldsymbol{\nabla}\cdot\mathbf{D}-4\pi\rho), \tag{44.2a}$$

$$0=\frac{\partial}{\partial t}(\boldsymbol{\nabla}\cdot\mathbf{B}). \tag{44.2b}$$

Thus, excluding statics, the divergence equations, (44.1c), (44.1d), are subsumed within the curl equations, (44.1a), (44.1b).

To concentrate our attention on light of a definite frequency ω, we introduce the Fourier transform

$$\int_{-\infty}^{\infty} dt\, e^{i\omega t}F(\mathbf{r},t)=F(\mathbf{r},\omega), \tag{44.3}$$

which amounts to the replacement in (44.1a) and (44.1b) of

$$\frac{\partial}{\partial t}\to -i\omega, \tag{44.4}$$

yielding for the vector equations

$$\boldsymbol{\nabla}\times\mathbf{H}=-\frac{i\omega}{c}\mathbf{D}+\frac{4\pi}{c}\mathbf{J}, \tag{44.5a}$$

$$-\boldsymbol{\nabla}\times\mathbf{E}=-\frac{i\omega}{c}\mathbf{B}. \tag{44.5b}$$

DOI: 10.1201/9781003057369-44

Specifically, we assume a linear, isotropic, dispersive medium, for which

$$\mathbf{B} = \mu\mathbf{H}, \quad \mathbf{D} = \epsilon\mathbf{E}, \tag{44.6}$$

where $\mu(\mathbf{r},\omega)$ and $\epsilon(\mathbf{r},\omega)$ are complex functions. We will further assume

$$\mu(\mathbf{r},\omega) \approx 1, \tag{44.7}$$

since a ferromagnet cannot follow the rapid oscillations of the electromagnetic field in a light wave. It will be sufficient for our purposes to suppose that spatial variation of ϵ occurs in the z direction only,

$$\epsilon(\mathbf{r},\omega) = \epsilon(z,\omega). \tag{44.8}$$

Since the material is then translationally invariant in x and y, we introduce corresponding spatial Fourier transforms:

$$\int dx\, dy\, e^{-i(k_x x + k_y y)} F(\mathbf{r},\omega) \equiv \int (d\mathbf{r}_\perp) e^{-i\mathbf{k}_\perp \cdot \mathbf{r}_\perp} F(\mathbf{r},\omega) \equiv F(z, \mathbf{k}_\perp, \omega). \tag{44.9}$$

Letting $\mathbf{n}$ be the unit vector pointing in the z direction, we replace the gradient operator by

$$\boldsymbol{\nabla} = \boldsymbol{\nabla}_\perp + \mathbf{n}\frac{\partial}{\partial z} \to i\mathbf{k}_\perp + \mathbf{n}\frac{\partial}{\partial z}. \tag{44.10}$$

It is natural to project Maxwell's equations, (44.5), onto spaces parallel to, and perpendicular to, $\mathbf{n}$. The z components of (44.5) become, using a highly redundant notation,

$$-i\mathbf{k}_\perp \cdot (\mathbf{n}\times\mathbf{B}_\perp) = -\frac{i\omega\epsilon}{c} E_z + \frac{4\pi}{c} J_z, \tag{44.11a}$$

$$i\mathbf{k}_\perp \cdot (\mathbf{n}\times\mathbf{E}_\perp) = -\frac{i\omega}{c} B_z, \tag{44.11b}$$

while the $\perp$ (x, y) components appear as

$$-\frac{\partial}{\partial z}\mathbf{B}_\perp + i\mathbf{k}_\perp B_z = -\frac{i\omega\epsilon}{c}\mathbf{n}\times\mathbf{E}_\perp + \frac{4\pi}{c}\mathbf{n}\times\mathbf{J}_\perp, \tag{44.12a}$$

$$\frac{\partial}{\partial z}\mathbf{E}_\perp - i\mathbf{k}_\perp E_z = -\frac{i\omega}{c}\mathbf{n}\times\mathbf{B}_\perp. \tag{44.12b}$$

For $\omega \neq 0$, (44.11) determines E_z and B_z in terms of $\mathbf{E}_\perp$ and $\mathbf{B}_\perp$:

$$B_z = -\frac{c}{\omega}\mathbf{k}_\perp \cdot (\mathbf{n}\times\mathbf{E}_\perp), \tag{44.13a}$$

$$E_z = \frac{c}{\omega\epsilon}\mathbf{k}_\perp \cdot (\mathbf{n}\times\mathbf{B}_\perp) - i\frac{4\pi}{\omega\epsilon} J_z. \tag{44.13b}$$

Inserting these expressions into the $\perp$ set, (44.12), we have the following equations for the $\perp$ components:

$$\frac{\partial}{\partial z}\mathbf{B}_\perp - \frac{i\omega\epsilon}{c}\mathbf{n}\times\mathbf{E}_\perp + i\mathbf{k}_\perp\frac{c}{\omega}\mathbf{k}_\perp \cdot (\mathbf{n}\times\mathbf{E}_\perp) = -\frac{4\pi}{c}\mathbf{n}\times\mathbf{J}_\perp, \tag{44.14a}$$

$$\frac{\partial}{\partial z}\mathbf{E}_\perp + \frac{i\omega}{c}\mathbf{n}\times\mathbf{B}_\perp - i\mathbf{k}_\perp\frac{c}{\omega\epsilon}\mathbf{k}_\perp \cdot (\mathbf{n}\times\mathbf{B}_\perp) = \mathbf{k}_\perp\frac{4\pi}{\omega\epsilon} J_z, \tag{44.14b}$$

which mix $\mathbf{B}_\perp = (B_x, B_y)$ with $\mathbf{n}\times\mathbf{E}_\perp = (-E_y, E_x)$, and $\mathbf{E}_\perp = (E_x, E_y)$ with $\mathbf{n}\times\mathbf{B}_\perp = (-B_y, B_x)$. Next, as an alternative form of the same equations, we take the cross product of the above equations with $\mathbf{n}$,

$$\frac{\partial}{\partial z}\mathbf{n}\times\mathbf{B}_\perp + \frac{i\omega\epsilon}{c}\mathbf{E}_\perp + i\mathbf{n}\times\mathbf{k}_\perp\frac{c}{\omega}\mathbf{k}_\perp\cdot(\mathbf{n}\times\mathbf{E}_\perp) = \frac{4\pi}{c}\mathbf{J}_\perp, \tag{44.15a}$$

$$\frac{\partial}{\partial z}\mathbf{n}\times\mathbf{E}_\perp - \frac{i\omega}{c}\mathbf{B}_\perp - i\mathbf{n}\times\mathbf{k}_\perp\frac{c}{\omega\epsilon}\mathbf{k}_\perp\cdot(\mathbf{n}\times\mathbf{B}_\perp) = \mathbf{n}\times\mathbf{k}_\perp\frac{4\pi}{\omega\epsilon}J_z, \tag{44.15b}$$

where we have noted that for any vector $\mathbf{V}_\perp$ in the xy plane, $-\mathbf{n}\times(\mathbf{n}\times\mathbf{V}_\perp) = \mathbf{V}_\perp$. Now we project these vector equations on $\mathbf{k}_\perp$. From (44.14a) and (44.15b), we find a system of equations relating $\mathbf{k}_\perp\cdot\mathbf{B}_\perp$ and $\mathbf{k}_\perp\cdot(\mathbf{n}\times\mathbf{E}_\perp)$:

$$\frac{\partial}{\partial z}\mathbf{k}_\perp\cdot\mathbf{B}_\perp - \frac{i\omega\epsilon}{c}\left(1-\frac{k_\perp^2c^2}{\omega^2\epsilon}\right)\mathbf{k}_\perp\cdot(\mathbf{n}\times\mathbf{E}_\perp) = -\frac{4\pi}{c}\mathbf{k}_\perp\cdot(\mathbf{n}\times\mathbf{J}_\perp), \tag{44.16a}$$

$$\frac{\partial}{\partial z}\mathbf{k}_\perp\cdot(\mathbf{n}\times\mathbf{E}_\perp) - \frac{i\omega}{c}\mathbf{k}_\perp\cdot\mathbf{B}_\perp = 0, \tag{44.16b}$$

and, from (44.14b) and (44.15a), a system relating $\mathbf{k}_\perp\cdot\mathbf{E}_\perp$ and $\mathbf{k}_\perp\cdot(\mathbf{n}\times\mathbf{B}_\perp)$:

$$\frac{\partial}{\partial z}\mathbf{k}_\perp\cdot\mathbf{E}_\perp + \frac{i\omega}{c}\left(1-\frac{k_\perp^2c^2}{\omega^2\epsilon}\right)\mathbf{k}_\perp\cdot(\mathbf{n}\times\mathbf{B}_\perp) = k_\perp^2\frac{4\pi}{\omega\epsilon}J_z, \tag{44.17a}$$

$$\frac{\partial}{\partial z}\mathbf{k}_\perp\cdot(\mathbf{n}\times\mathbf{B}_\perp) + \frac{i\omega\epsilon}{c}\mathbf{k}_\perp\cdot\mathbf{E}_\perp = \frac{4\pi}{c}\mathbf{k}_\perp\cdot\mathbf{J}_\perp. \tag{44.17b}$$

The vectors $\mathbf{n}$ and $\mathbf{k}_\perp$ define a plane, called the plane of incidence so that the system (44.16a) and (44.16b) relates the component of $\mathbf{B}_\perp$ in this plane to the component $\mathbf{E}_\perp$ perpendicular to this plane, and vice versa for system (44.17a), (44.17b). That is, if we take $\mathbf{k}_\perp$ in the x direction, we have equations governing B_x and E_y, and B_y and E_x, respectively. These perpendicular components belong together physically [see Section 8.2 and also the relativistic transformations in (11.80) and (11.81)]. By combining (44.16a) with (44.16b), and (44.17a) with (44.17b), we convert these systems of first-order differential equations into second-order ones:

$$\left[\frac{\partial^2}{\partial z^2}+\frac{\omega^2\epsilon}{c^2}-k_\perp^2\right]\mathbf{k}_\perp\cdot(\mathbf{n}\times\mathbf{E}_\perp) = -\frac{i\omega}{c}\frac{4\pi}{c}\mathbf{k}_\perp\cdot(\mathbf{n}\times\mathbf{J}_\perp), \tag{44.18a}$$

$$\frac{\partial}{\partial z}\left[\frac{1}{\epsilon}\frac{\partial}{\partial z}\mathbf{k}_\perp\cdot(\mathbf{n}\times\mathbf{B}_\perp)\right] + \left(\frac{\omega^2}{c^2}-\frac{k_\perp^2}{\epsilon}\right)\mathbf{k}_\perp\cdot(\mathbf{n}\times\mathbf{B}_\perp)$$
$$= -\frac{i\omega}{c}k_\perp^2\frac{4\pi}{\omega\epsilon}J_z + \frac{\partial}{\partial z}\left(\frac{4\pi}{\epsilon c}\mathbf{k}_\perp\cdot\mathbf{J}_\perp\right), \tag{44.18b}$$

where we must remember that ϵ is a function of z. From these components, perpendicular to the plane of incidence, those in the plane of incidence can be obtained from (44.16b) and (44.17b), and the longitudinal components (the z-components) from (44.13).

When ϵ is a constant function of z, the differential operators in these last two equations are the same, namely

$$\frac{\partial^2}{\partial z^2}+\frac{\omega^2\epsilon}{c^2}-k_\perp^2. \tag{44.19}$$

We further assume that ϵ is real. Then, depending on the sign of $\omega^2\epsilon/c^2 - k_\perp^2$, the solutions to the corresponding homogeneous differential equations are different:

$$\frac{\omega^2\epsilon}{c^2}-k_\perp^2<0:\quad \exp\left(\pm\sqrt{k_\perp^2-\frac{\omega^2\epsilon}{c^2}}z\right), \tag{44.20a}$$

$$\frac{\omega^2\epsilon}{c^2}-k_\perp^2>0:\quad \exp\left(\pm i\sqrt{\frac{\omega^2\epsilon}{c^2}-k_\perp^2}z\right). \tag{44.20b}$$

The first possibility is essentially that discussed in electrostatics, in that there is no wave propagation, and so we will not consider it further here. (See Chapter 16.) For the second possibility, we do have propagating waves. In terms of

$$k_z = \sqrt{\frac{\omega^2\epsilon}{c^2} - k_\perp^2}, \tag{44.21}$$

we construct a plane wave by combining the imaginary exponential structure in (44.20b) with the transverse spatial dependence $e^{i\mathbf{k}_\perp \cdot \mathbf{r}_\perp}$:

$$e^{i\mathbf{k}_\perp \cdot \mathbf{r}_\perp} e^{ik_z z} = e^{i\mathbf{k}\cdot\mathbf{r}}, \tag{44.22}$$

which represents a plane wave moving in the direction $\mathbf{k}$, since the phase is constant on a plane perpendicular to $\mathbf{k}$. The wavelength, λ, is defined as the distance over which the phase advances by 2π, so

$$|\mathbf{k}|\lambda = 2\pi, \quad \text{or} \quad |\mathbf{k}| = \frac{2\pi}{\lambda} \equiv \frac{1}{\lambda\!\!\!^{-}}; \tag{44.23}$$

$|\mathbf{k}|$ is called the wavenumber, $\mathbf{k}$ the propagation vector, and $\lambda\!\!\!^{-}$ the reduced wavelength. Including the time factor $e^{-i\omega t}$, we find the dependence of a plane wave on space and time to be

$$e^{i(\mathbf{k}\cdot\mathbf{r}-\omega t)}. \tag{44.24}$$

The surface of constant phase advances with time so that

$$\mathbf{k}\cdot\frac{d\mathbf{r}}{dt} = \omega; \tag{44.25}$$

the phase speed thus is

$$v = \frac{\omega}{|\mathbf{k}|} = \omega\lambda\!\!\!^{-} = \nu\lambda. \tag{44.26}$$

From (44.21), we see that the square of the wavenumber is

$$|\mathbf{k}|^2 = k_\perp^2 + k_z^2 = \frac{\omega^2\epsilon}{c^2}, \tag{44.27}$$

so that the phase speed is

$$v = \frac{c}{\sqrt{\epsilon}} = \frac{c}{n}, \tag{44.28}$$

$$n = \sqrt{\epsilon}, \tag{44.29}$$

as we saw in Section 8.2.

44.2 Reflection and Refraction: ⊥ Polarization

We now specialize the above discussion to a situation in which the inhomogeneity in the dielectric constant is due to a plane interface between two dielectric substances:

$$z > 0: \quad \epsilon(z) = \epsilon_2, \tag{44.30a}$$

$$z < 0: \quad \epsilon(z) = \epsilon_1. \tag{44.30b}$$

Since the relation between field and source is linear, the field produced by a prescribed current can be expressed in terms of an appropriate Green's function. Green's function corresponding to (44.18a) satisfies

$$-\left(\frac{\partial^2}{\partial z^2}+\frac{\omega^2\epsilon}{c^2}-k_\perp^2\right)g(z,z')=\delta(z-z'), \tag{44.31}$$

which, as we noted above, refers to the component of $\mathbf{E}_\perp$ perpendicular to the plane of incidence, defined by $\mathbf{k}_\perp$ and $\mathbf{n}$, and so we will call this the $\perp$ polarization. (Other notations for this polarization is transverse electric (TE), H, or s.) We imagine that the source lies in medium 2, that is $z'>0$. The boundary conditions at the interface are that $\mathbf{k}_\perp\cdot(\mathbf{n}\times\mathbf{E}_\perp)$ and $\mathbf{k}_\perp\cdot\mathbf{B}_\perp$ be continuous, which impose the conditions that g and, by (44.16b), $\partial g/\partial z$ be continuous at $z=0$. For $z\to\pm\infty$ we must have outgoing waves, since the source is localized at z'. The solutions to (44.31) in the three regions have the forms

$$z>z': \quad g=Ae^{ik_{z2}z}, \tag{44.32a}$$
$$z'>z>0: \quad g=Be^{ik_{z2}z}+Ce^{-ik_{z2}z}, \tag{44.32b}$$
$$z<0: \quad g=De^{-ik_{z1}z}, \tag{44.32c}$$

where the subscripts 1 and 2 refer to the value of k_z, as defined by (44.21), in medium 1 and 2, respectively. Note that these equations are very similar to those considered in electrostatics; in fact, when in both dielectrics

$$\frac{\omega^2\epsilon}{c^2}-k_\perp^2<0, \tag{44.33}$$

we make the replacement

$$\sqrt{\frac{\omega^2\epsilon}{c^2}-k_\perp^2}\to i\sqrt{k_\perp^2-\frac{\omega^2\epsilon}{c^2}} \tag{44.34}$$

in (44.32a)–(44.32c), and thereby recover the same system in (16.8a)–(16.8c), except that, here, the wavenumbers are different in the two media.

We now determine the coefficients A through D by satisfying the boundary conditions stated above as well as the discontinuity at $z=z'$. At $z=0$, continuity of g yields the relation

$$B+C=D, \tag{44.35a}$$

while continuity of $\partial g/\partial z$ requires

$$B-C=-\frac{k_{z1}}{k_{z2}}D. \tag{44.35b}$$

The combination of these supplies

$$B=\frac{1}{2}\left(1-\frac{k_{z1}}{k_{z2}}\right)D, \tag{44.36a}$$
$$C=\frac{1}{2}\left(1+\frac{k_{z1}}{k_{z2}}\right)D. \tag{44.36b}$$

(Note that if $\epsilon_1=\epsilon_2$, these coefficients are simply $B=0$ and $C=D$, which expresses the obvious fact that the entire wave is transmitted, with no reflection, since there is no interface.) The presence of the source term in (44.31) imposes the conditions that, at $z=z'$,

$$g \text{ is continuous,} \quad \text{and} \quad \left[-\frac{\partial}{\partial z}g\right]_{z=z'-0}^{z=z'+0}=1. \tag{44.37}$$

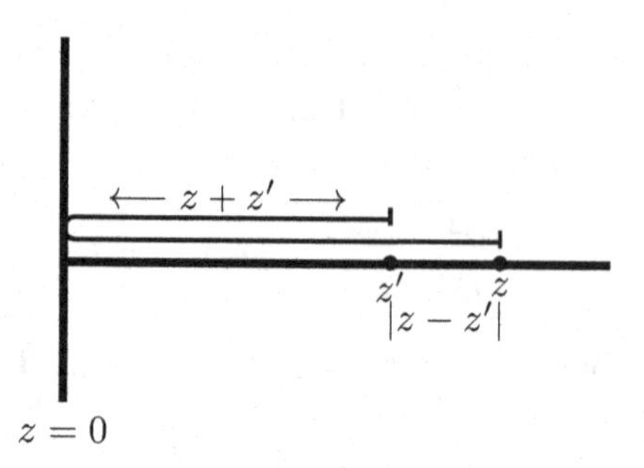

FIGURE 44.1
Interpretation of direct and reflected waves.

Consequently, there are two additional equations for the constants,

$$Ae^{ik_{z2}z'} = Be^{ik_{z2}z'} + Ce^{-ik_{z2}z'}, \tag{44.38a}$$

$$-Ae^{ik_{z2}z'} + Be^{ik_{z2}z'} - Ce^{-ik_{z2}z'} = \frac{1}{ik_{z2}}. \tag{44.38b}$$

Equations (44.36) and (44.38) can be solved to yield, successively,

$$C = \frac{i}{2k_2}e^{ik_2z'}, \tag{44.39a}$$

$$B = \frac{k_2 - k_1}{k_2 + k_1}\frac{i}{2k_2}e^{ik_2z'}, \tag{44.39b}$$

$$D = \frac{2k_2}{k_2 + k_1}\frac{i}{2k_2}e^{ik_2z'}, \tag{44.39c}$$

$$A = \frac{k_2 - k_1}{k_2 + k_1}\frac{i}{2k_2}e^{ik_2z'} + \frac{i}{2k_2}e^{-ik_2z'}, \tag{44.39d}$$

where we have simplified the notation by dropping all the z subscripts on the k's. Combining these results, we may now write down the Green's function in the two regions,

$$z > 0: \quad g(z, z') = \frac{i}{2k_2}e^{ik_2|z-z'|} + r\frac{i}{2k_2}e^{ik_2(z+z')}, \tag{44.40a}$$

$$z < 0: \quad g(z, z') = t\frac{i}{2k_2}e^{-ik_1z}e^{ik_2z'}, \tag{44.40b}$$

in terms of the reflection and transmission coefficients,

$$r = \frac{k_2 - k_1}{k_2 + k_1}, \tag{44.41a}$$

$$t = \frac{2k_2}{k_2 + k_1}. \tag{44.41b}$$

The two terms in (44.40a) have a simple physical interpretation. The first refers to the direct wave from the source, since $|z - z'|$ is (the z-projection of) the distance from the source, while the second represents the wave reflected from the interface, since $z + z'$ is (the z-projection of) the distance from the source to the interface and back to the observation point z. (See Fig. 44.1.) The following algebraic identities relating t and r,

$$t = 1 + r, \tag{44.42a}$$

$$k_1t = k_2(1 - r), \tag{44.42b}$$

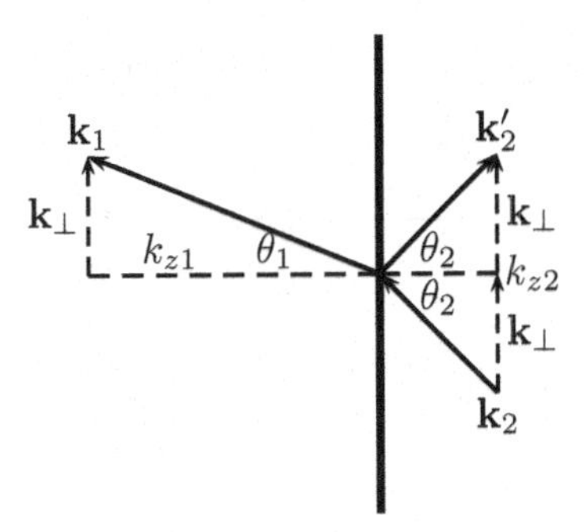

FIGURE 44.2
Incident, reflected, and transmitted propagation vectors.

restate the continuity of g and g' at $z = 0$, which in turn are consequences of the continuity of the tangential components of **E** and **B** across the interface. Moreover, on combining (44.42a) and (44.42b) to give

$$k_1 t^2 = k_2(1 - r^2), \tag{44.43}$$

we have a statement of conservation of energy, as we will show in Section 44.5.

We are here primarily interested in how a plane interface reflects and transmits light, irrespective of how the radiation is produced. Accordingly, for regions to the left of the source, we regroup the factors of Green's function suggestively:

$$0 < z < z': \quad g(z,z') = \left(e^{-ik_2 z} + re^{ik_2 z}\right)\left(\frac{i}{2k_2}e^{ik_2 z'}\right), \tag{44.44a}$$

$$z < 0: \quad g(z,z') = te^{-ik_1 z}\left(\frac{i}{2k_2}e^{ik_2 z'}\right). \tag{44.44b}$$

The common factors in parentheses refer to how the wave is produced, which is not of interest to us at the moment. The other remaining factors describe the reflection and transmission of a plane wave: unit amplitude incident on the interface, amplitude r reflected, amplitude t transmitted.

We recall that Green's function, $g(z, z')$, describes only the z-dependence of the wave propagation; the x and y dependence is contained in the factor $e^{i\mathbf{k}_\perp \cdot \mathbf{r}_\perp}$ [recall (44.22)]. Since the transverse momenta are unaltered, while the longitudinal momenta (k_z) change from k_{z2} to k_{z1} as the wave crosses the interface, we have the geometrical picture shown in Fig. 44.2 for the incident, reflected, and transmitted propagation vectors. Under reflection k_z merely changes sign, so the angle of incidence equals the angle of reflection. The relation between the wavenumber and the index of refraction is, from (44.27),

$$|\mathbf{k}| = \frac{\omega n}{c}, \tag{44.45}$$

with $n = \sqrt{\epsilon}$. The continuity of $\mathbf{k}_\perp$ at the interface,

$$|\mathbf{k}_2| \sin\theta_2 = |\mathbf{k}_1| \sin\theta_1, \tag{44.46}$$

then implies Snell's law of refraction (discovered in 1621),[1]

$$n_1 \sin\theta_1 = n_2 \sin\theta_2, \tag{44.47}$$

relating the angle of incidence θ_2 to the angle of refraction θ_1.

[1]Willebrord Snellius, 1580–1626. First discovered by Ibn Sahl in 984.

We may rewrite the reflection and transmission coefficients given in (44.41a) and (44.41b) in terms of the angles θ_1 and θ_2 by using

$$\frac{k_z}{|\mathbf{k}_\perp|} = \cot\theta, \tag{44.48}$$

to obtain

$$r = \frac{\sin(\theta_1 - \theta_2)}{\sin(\theta_1 + \theta_2)}, \tag{44.49a}$$

and

$$t = \frac{2\sin\theta_1\cos\theta_2}{\sin(\theta_1 + \theta_2)}, \tag{44.49b}$$

but the wavenumber form is usually preferable. This is particularly evident for normal incidence, where $\theta_1 = \theta_2 = 0$ and (44.49a)–(44.49b) is indeterminate. On the other hand the wavenumber form, (44.41a)–(44.41b), is well defined, since when $k_\perp = 0$, $k_z = \frac{\omega}{c}n$, which yields

$$r = -\frac{n_1 - n_2}{n_1 + n_2}, \tag{44.50a}$$

$$t = \frac{2n_2}{n_1 + n_2}. \tag{44.50b}$$

Also from the wavenumber form it is easy to see that if $n_1/n_2 \to \infty$, the coefficients become

$$r = -1, \quad t = 0; \tag{44.51}$$

this corresponds to the situation of total reflection by a perfect conductor. We verify this last statement by recalling the plane wave part of the Green's function between the source and the interface, (44.44a):

$$0 < z < z' : \quad g \propto e^{-ik_{2z}z} + re^{ik_{2z}z}. \tag{44.52}$$

When (44.51) is satisfied, g vanishes at $z = 0$,

$$g \propto (1 + r) = 0, \tag{44.53}$$

which expresses the vanishing of the tangential component of $\mathbf{E}$ on the surface, characteristic of a perfect conductor.

44.3 Reflection and Refraction: $\|$ Polarization

We now turn to a consideration of the second polarization state, in which $\mathbf{E}_\perp$ lies in the plane of incidence and $\mathbf{B}_\perp$ is perpendicular to that plane. We will call this the $\|$ polarization (or transverse magnetic (TM), E, p), in contrast to the $\perp$ case discussed above.

Green's function in this situation, which relates $\mathbf{k}_\perp \cdot (\mathbf{n}\times\mathbf{B}_\perp)$ to its source, satisfies, according to (44.18b),

$$-\frac{\partial}{\partial z}\left(\frac{1}{\epsilon}\frac{\partial}{\partial z}g\right) - \frac{k_z^2}{\epsilon}g = \delta(z - z'), \tag{44.54}$$

where, again,

$$k_z^2 = \frac{\omega^2}{c^2}\epsilon - k_\perp^2. \tag{44.55}$$

With the plane interface between the two dielectric media located at $z = 0$, the form of the solution in the three regions, when $z' > 0$, is

$$z > z': \quad g = Ae^{ik_2 z}, \tag{44.56a}$$
$$0 < z < z': \quad g = Be^{ik_2 z} + Ce^{-ik_2 z}, \tag{44.56b}$$
$$z < 0: \quad g = De^{-ik_1 z}, \tag{44.56c}$$

where, as before, the z subscript is suppressed. Here, the boundary conditions at $z = 0$ are the continuity of g (since $\mathbf{B}_\perp$ is continuous) and the continuity of $\frac{1}{\epsilon}g'$ [since $\mathbf{E}_\perp$, which is related to g by (44.17b), is continuous], which furnish the relations

$$B + C = D, \tag{44.57a}$$
$$B - C = -\frac{\epsilon_2}{\epsilon_1}\frac{k_1}{k_2}D, \tag{44.57b}$$

implying

$$C = \frac{1}{2}\left(1 + \frac{\epsilon_2}{\epsilon_1}\frac{k_1}{k_2}\right)D, \tag{44.58a}$$
$$B = \frac{1}{2}\left(1 - \frac{\epsilon_2}{\epsilon_1}\frac{k_1}{k_2}\right)D. \tag{44.58b}$$

At $z = z'$, g is continuous, while from (44.54)

$$-\frac{1}{\epsilon_2}\frac{\partial}{\partial z}g\Bigg|_{z=z'-0}^{z=z'+0} = 1, \tag{44.59}$$

so

$$Be^{ik_2 z'} + Ce^{-ik_2 z'} = Ae^{ik_2 z'}, \tag{44.60a}$$

and

$$ik_2\left(-Ae^{ik_2 z'} + Be^{ik_2 z'} - Ce^{-ik_2 z'}\right) = \epsilon_2. \tag{44.60b}$$

The solution to this system of equations is

$$C = \frac{i\epsilon_2}{2k_2}e^{ik_2 z'}, \tag{44.61a}$$
$$B = \frac{1 - \frac{\epsilon_2}{\epsilon_1}\frac{k_1}{k_2}}{1 + \frac{\epsilon_2}{\epsilon_1}\frac{k_1}{k_2}}\frac{i\epsilon_2}{2k_2}e^{ik_2 z'}, \tag{44.61b}$$
$$D = \frac{2}{1 + \frac{\epsilon_2}{\epsilon_1}\frac{k_1}{k_2}}\frac{i\epsilon_2}{2k_2}e^{ik_2 z'}, \tag{44.61c}$$
$$A = \frac{1 - \frac{\epsilon_2}{\epsilon_1}\frac{k_1}{k_2}}{1 + \frac{\epsilon_2}{\epsilon_1}\frac{k_1}{k_2}}\frac{i\epsilon_2}{2k_2}e^{ik_2 z'} + \frac{i\epsilon_2}{2k_2}e^{-ik_2 z'}. \tag{44.61d}$$

These are as in the other polarization, as given by (44.39a)–(44.39d), except that in the amplitudes multiplying the exponentials, where we had k_z , we now have k_z/ϵ. Green's function is thus

$$z > 0: \quad g = \frac{i\epsilon_2}{2k_2}e^{ik_2|z-z'|} + r\frac{i\epsilon_2}{2k_2}e^{ik_2(z+z')}, \tag{44.62a}$$
$$z < 0: \quad g = t\frac{i\epsilon_2}{2k_2}e^{-ik_1 z}e^{ik_2 z'}. \tag{44.62b}$$

Green's functions for the two polarizations, $\perp$ and $\|$, are given by (44.40a)–(44.40b) and (44.62a)–(44.62b), respectively, with the corresponding reflection and transmission coefficients given by

$$\begin{aligned} &\perp \text{ mode} && \| \text{ mode} \\ r &= -\frac{k_1 - k_2}{k_1 + k_2}, \qquad & r &= -\frac{k_1/\epsilon_1 - k_2/\epsilon_2}{k_1/\epsilon_1 + k_2/\epsilon_2}, \qquad (44.63a) \\ t &= \frac{2k_2}{k_1 + k_2}, \qquad & t &= \frac{2k_2/\epsilon_2}{k_1/\epsilon_1 + k_2/\epsilon_2}. \qquad (44.63b) \end{aligned}$$

In both cases the algebraic relation (44.42a) holds true,

$$t = 1 + r, \tag{44.64}$$

expressing the continuity of $\mathbf{E}_\perp$ and of $\mathbf{B}_\perp$ in the $\perp$ and $\|$ polarizations, respectively, but the statement of energy conservation is different in the two situations [recall (44.43)]:

$$\begin{aligned} &\perp \text{ mode} && \| \text{ mode} \\ k_1 t^2 &= k_2(1 - r^2), \qquad & \frac{k_1}{\epsilon_1} t^2 &= \frac{k_2}{\epsilon_2}(1 - r^2), \qquad (44.65) \end{aligned}$$

where, again, the second relation is obtained from the first by the substitution $k_z \to k_z/\epsilon$.

It is particularly interesting to ask when the reflection coefficient can vanish. For the $\perp$ polarization, $r = 0$ only when $k_{z1} = k_{z2}$, or $\epsilon_1 = \epsilon_2$, that is, in the absence of an interface. However, for the $\|$ polarization, another solution to $r = 0$ exists. The vanishing of the reflection coefficient requires

$$\frac{k_{z1}}{\epsilon_1} = \frac{k_{z2}}{\epsilon_2}, \tag{44.66}$$

or, since

$$\frac{k_z^2}{\epsilon^2} = \frac{\omega^2}{c^2}\frac{1}{\epsilon} - \frac{k_\perp^2}{\epsilon^2}, \tag{44.67}$$

the condition becomes

$$\left(\frac{\epsilon_2 - \epsilon_1}{\epsilon_1 \epsilon_2}\right)\left(\frac{\omega^2}{c^2} - \frac{\epsilon_2 + \epsilon_1}{\epsilon_1 \epsilon_2} k_\perp^2\right) = 0. \tag{44.68}$$

The new possibility for r vanishing occurs when

$$\frac{\omega}{c} = \sqrt{\frac{\epsilon_1 + \epsilon_2}{\epsilon_1 \epsilon_2}}|\mathbf{k}_\perp|. \tag{44.69}$$

Geometrically, since

$$|\mathbf{k}_\perp| = \frac{\omega}{c} n \sin\theta, \quad n = \sqrt{\epsilon}, \tag{44.70}$$

(44.69) is satisfied when the angle of incidence, θ_2 equals θ_B, where

$$\sin\theta_B = \sqrt{\frac{\epsilon_1}{\epsilon_1 + \epsilon_2}}, \tag{44.71a}$$

or,

$$\tan\theta_B = \frac{n_1}{n_2}. \tag{44.71b}$$

At the incident angle θ_B, the $\|$ mode is completely transmitted so that the reflected wave is completely polarized perpendicular to the plane of incidence. The phenomenon was discovered by Sir David Brewster about the year 1800, and in consequence θ_B is called Brewster's angle.

To express r and t again in terms of the angle of incidence θ_2 and the angle of refraction θ_1, we use the geometrical relations

$$k_z = |\mathbf{k}_\perp| \cot\theta, \quad n\sin\theta = \frac{c}{\omega}|\mathbf{k}_\perp|, \tag{44.72}$$

to write

$$\frac{k_z}{\epsilon} = \frac{(\omega/c)^2}{|\mathbf{k}_\perp|}\frac{1}{2}\sin 2\theta. \tag{44.73}$$

Consequently, for the || mode, the reflection and transmission coefficients can be given as (see Problem 44.1)

$$r = -\frac{\tan(\theta_1 - \theta_2)}{\tan(\theta_1 + \theta_2)}, \tag{44.74a}$$

$$t = \frac{2\sin\theta_2\cos\theta_2}{\sin(\theta_1+\theta_2)\cos(\theta_1-\theta_2)}. \tag{44.74b}$$

Here we observe, from (44.74a), that Brewster's angle occurs when the sum of the angle of incidence and the angle of refraction forms a right angle,

$$\theta_1 + \theta_2 = \pi/2. \tag{44.75}$$

For normal incidence ($\theta_1 = \theta_2 = 0$), as before it is convenient to return to (44.63) and employ

$$\frac{k_z}{\epsilon} = \frac{\omega}{nc}. \tag{44.76}$$

For the two polarizations we have [recall (44.50a) and (44.50b)]

$$\begin{array}{cc} \perp \text{ mode} & \| \text{ mode} \\ r = -\dfrac{n_1 - n_2}{n_1 + n_2}, & r = \dfrac{n_1 - n_2}{n_1 + n_2}, \qquad (44.77\text{a}) \\ t = \dfrac{2n_2}{n_1 + n_2}, & t = \dfrac{2n_1}{n_1 + n_2}. \qquad (44.77\text{b}) \end{array}$$

But for normal incidence there can be no physical difference between the two polarizations, since the plane of incidence is not defined. To reconcile the two forms in (44.77a), we recognize that the reflection and transmission coefficients for the || mode refer to **B**, while those for the $\perp$ mode refer to **E** [see (44.18a) and (44.18b)]. Under reflection, **E** reverses its sense relative to **B**, since the direction of propagation is reversed. (See Fig. 44.3.) So the two forms for the reflection coefficient in (44.77) are physically equivalent. To reach the same conclusion for the transmission coefficients, we recall from Section 8.2 that the magnitudes of the electric and magnetic fields in the plane wave are connected by

$$\epsilon|\mathbf{E}|^2 = |\mathbf{B}|^2, \tag{44.78}$$

which relates transmission coefficients referring to electric and magnetic fields by

$$t_E = \frac{\sqrt{\epsilon_2}}{\sqrt{\epsilon_1}} t_B = \frac{n_2}{n_1} t_B, \tag{44.79}$$

which explains the ratio of the two forms for t in (44.77).

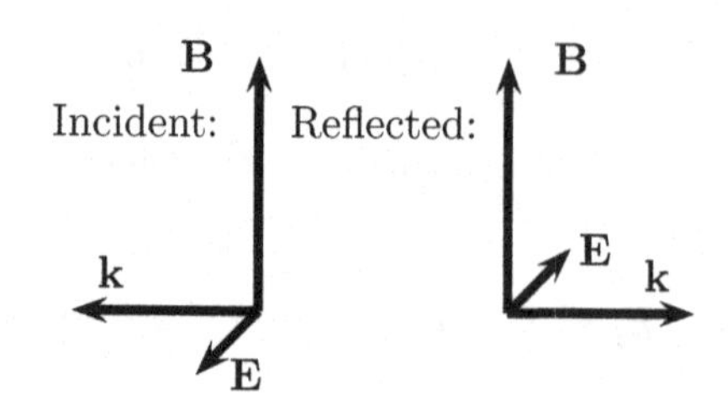

FIGURE 44.3
Relation between field orientations for incident and reflected waves.

44.4 Total Internal Reflection

We now return to Snell's law, (44.47),

$$n_1 \sin\theta_1 = n_2 \sin\theta_2. \tag{44.80}$$

Suppose we consider a plane wave going from a region of greater dielectric constant to a region with a smaller dielectric constant, that is

$$n_1 < n_2 \Rightarrow \sin\theta_1 > \sin\theta_2. \tag{44.81}$$

For this arrangement, there are incident angles satisfying

$$n_2 \sin\theta_2 > n_1, \tag{44.82}$$

for which Snell's law can be satisfied only if

$$\sin\theta_1 > 1. \tag{44.83}$$

What does this mean physically? It is helpful to consider the angle of transition into this new regime, where

$$n_2 \sin\theta_2 = n_1, \tag{44.84a}$$

or

$$\sin\theta_1 = 1. \tag{44.84b}$$

At exactly this angle, the transmitted wave travels along the interface since

$$|\mathbf{k}_1| = |\mathbf{k}_\perp|, \quad k_{z1} = 0. \tag{44.85}$$

Now as θ_2 is increased, $|\mathbf{k}_\perp|^2$ becomes greater than $\frac{\omega^2}{c^2}\epsilon_1$ so that

$$k_{z1} = \sqrt{\frac{\omega^2}{c^2}\epsilon_1 - k_\perp^2} \tag{44.86}$$

becomes imaginary,

$$k_{z1}^2 < 0 \quad \text{or} \quad k_{z1} = i\kappa, \tag{44.87}$$

where κ is real. The propagation of the wave in medium 1 is now described by

$$e^{ik_{z1}|z|} = e^{-\kappa|z|}, \tag{44.88}$$

a decreasing exponential: the field penetrates only a short distance into region 1, and no energy is transmitted into that region, it being all reflected. This last fact follows from the form of the reflection coefficients,

$$\perp: \quad r = \frac{k_{z2} - i\kappa_1}{k_{z2} + i\kappa_1}, \tag{44.89a}$$

$$\parallel: \quad r = \frac{k_{z2}/\epsilon_2 - i\kappa_1/\epsilon_1}{k_{z2}/\epsilon_2 + i\kappa_1/\epsilon_1}, \tag{44.89b}$$

both of which have unit magnitude,

$$|r|^2 = 1. \tag{44.90}$$

In the next section, we will prove that this implies that all the energy is reflected. We have here the situation of total (internal) reflection.

44.5 Energy Conservation

We now turn to a general consideration of energy flow across the plane interface. Recall Poynting's vector, (8.3), which has as the component normal to the interface

$$S_z = \frac{c}{4\pi}\mathbf{n}\cdot(\mathbf{E}\times\mathbf{H}), \tag{44.91}$$

where here

$$\mathbf{H} = \mathbf{B}. \tag{44.92}$$

From this the total energy per unit area flowing in the $+z$ direction is computed to be

$$\int dt\, S_z(\mathbf{r},t) = \frac{c}{4\pi}\int_{-\infty}^{\infty}\frac{d\omega}{2\pi}\mathbf{n}\cdot\mathbf{E}(\mathbf{r},\omega)^*\times\mathbf{B}(\mathbf{r},\omega) = \frac{c}{4\pi}2\,\mathrm{Re}\int_0^{\infty}\frac{d\omega}{2\pi}\mathbf{n}\cdot\mathbf{E}(\mathbf{r},\omega)^*\times\mathbf{B}(\mathbf{r},\omega). \tag{44.93}$$

We next integrate over the xy plane by introducing the Fourier transform in the $\mathbf{k}_\perp$ space:

$$\int dx\, dy\, dt\, S_z = \frac{c}{2\pi}\,\mathrm{Re}\int_0^{\infty}\frac{d\omega}{2\pi}\int\frac{(d\mathbf{k}_\perp)}{(2\pi)^2}\mathbf{n}\cdot\mathbf{E}(z,\mathbf{k}_\perp,\omega)^*\times\mathbf{B}(z,\mathbf{k}_\perp,\omega). \tag{44.94}$$

The integrand in (44.94) states that the flow of energy in the $+z$ direction per unit frequency interval and per unit $\mathbf{k}_\perp$ volume is

$$\frac{c}{2\pi}\,\mathrm{Re}\left[\mathbf{n}\cdot\mathbf{E}(z,\mathbf{k}_\perp,\omega)^*\times\mathbf{B}(z,\mathbf{k}_\perp,\omega)\right]. \tag{44.95}$$

We adopt a coordinate system such that $\mathbf{k}_\perp$ is in the x-direction so that the y-axis is perpendicular to the plane of incidence. Then for $\perp$ polarization, the flow of energy in the $-z$ direction is

$$\frac{c}{2\pi}\,\mathrm{Re}[-\mathbf{n}\cdot\mathbf{E}^*\times\mathbf{B}] = \frac{c}{2\pi}\,\mathrm{Re}\,E_y^* B_x. \tag{44.96}$$

From the relation between B_x and E_y, (44.16b),

$$B_x = i\frac{c}{\omega}\frac{\partial}{\partial z}E_y, \tag{44.97}$$

we see that (44.96) is

$$\frac{c^2}{2\pi\omega}\,\mathrm{Re}\;E_y^* i\frac{\partial}{\partial z}E_y. \tag{44.98}$$

Now we use (44.44b) for the transmitted field ($z < 0$):

$$E_y = te^{-ik_1 z}, \tag{44.99a}$$

$$i\frac{\partial}{\partial z}E_y = k_1 E_y, \tag{44.99b}$$

so the energy flowing in medium 1 away from the interface is proportional to

$$\text{Re } E_y^* i\frac{\partial}{\partial z}E_y \propto \text{Re}(k_1|t|^2). \tag{44.100}$$

For the case of total reflection, k_1 is purely imaginary, so no energy is transmitted into region 1 as stated in the preceding section.

In region 2, $z > 0$, (44.44a) expresses the incident and reflected fields as

$$E_y = e^{-ik_2 z} + re^{ik_2 z}, \tag{44.101a}$$

so that

$$i\frac{\partial}{\partial z}E_y = k_2\left(e^{-ik_2 z} - re^{ik_2 z}\right). \tag{44.101b}$$

Just outside the interface, at $z = 0+$, where

$$E_y = 1 + r, \tag{44.102a}$$

$$i\frac{\partial}{\partial z}E_y = k_2(1 - r), \tag{44.102b}$$

the flow of energy toward the interface is given by

$$\frac{c^2}{2\pi\omega}\text{ Re } E_y^* i\frac{\partial}{\partial z}E_y \propto \text{Re } k_2(1 - |r|^2 + r^* - r). \tag{44.103}$$

Since we are considering an incident plane wave, with real k_z, this becomes

$$k_2(1 - |r|^2). \tag{44.104}$$

The quantities (44.100) (at $z = 0-$) and (44.104) (at $z = 0+$) therefore must be identical for two reasons:

1. The tangential components of $\mathbf{E}$, $\mathbf{B}$ must be continuous, since there are no surface currents. [For k_1 real, we derived this equality earlier in (44.43).]
2. Energy is conserved, since there is no source of energy on the surface.

Thus, if k_1 is real, the statement of energy conservation is

$$k_1|t|^2 = k_2(1 - |r|^2), \tag{44.105}$$

while if k_1 is imaginary, there is no flow of energy in region 1 and

$$0 = k_2(1 - |r|^2) \tag{44.106a}$$

or

$$|r|^2 = 1, \tag{44.106b}$$

which is the situation of total reflection considered in Section 44.4.

For $\|$ polarization, the flow of energy is given in terms of

$$-\mathbf{n}\cdot\mathbf{E}^*\times\mathbf{B} = -E_x^* B_y, \tag{44.107}$$

where, using (44.17b) in the absence of currents,

$$E_x = \frac{c}{i\omega\epsilon}\frac{\partial}{\partial z}B_y. \tag{44.108}$$

Thus the flow of energy is proportional to

$$\text{Re}[-B_y^* E_x] \propto \text{Re}\ B_y^* \frac{1}{\epsilon} i \frac{\partial}{\partial z} B_y, \tag{44.109}$$

with the same constant of proportionality as before. Since now the transmission and reflection coefficients, and the Green's function, refer to B_y, we may obtain the result by simply replacing

$$k_z \to \frac{k_z}{\epsilon} \tag{44.110}$$

in the previous form. And so, again, the $\|$ form of (44.65) expresses the conservation of energy when k_1 is real; generally, for this polarization, the conservation of energy for a wave reflected and transmitted by an interface between plane dielectrics is stated by

$$\text{Re}\left(\frac{k_1}{\epsilon_1}|t|^2\right) = \frac{k_2}{\epsilon_2}(1 - |r|^2). \tag{44.111}$$

44.6 Problems for Chapter 44

1. Verify (44.74a) and (44.74b).
2. Two parallel dielectric media are backed by a perfect electric conductor as shown in Fig. 44.4. (A perfect electric conductor is characterized by the fact that the tangential component of **E** vanishes on the surface.) A source, to the left of the first interface, initiates an incident plane wave described by

$$e^{ik_2(z+a/2)} \tag{44.112}$$

with $\perp$ polarization. (The phase is so chosen as to make the exponential equal 1 at the first interface.) The reflected wave is represented in terms of

$$r_a e^{-ik_2(z+a/2)}. \tag{44.113}$$

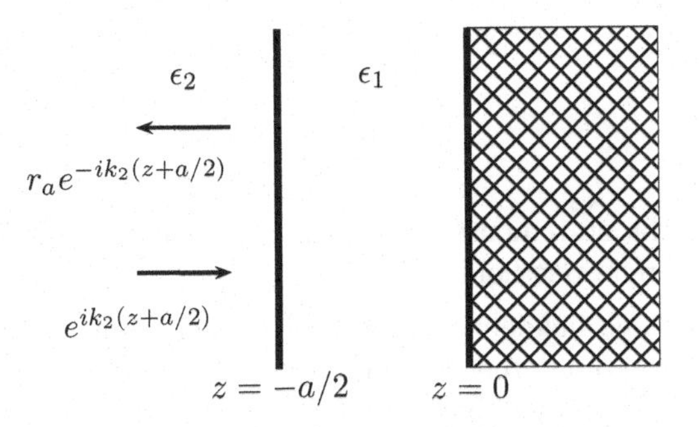

FIGURE 44.4
Parallel dielectrics backed by conductor, marked by crosshatching.

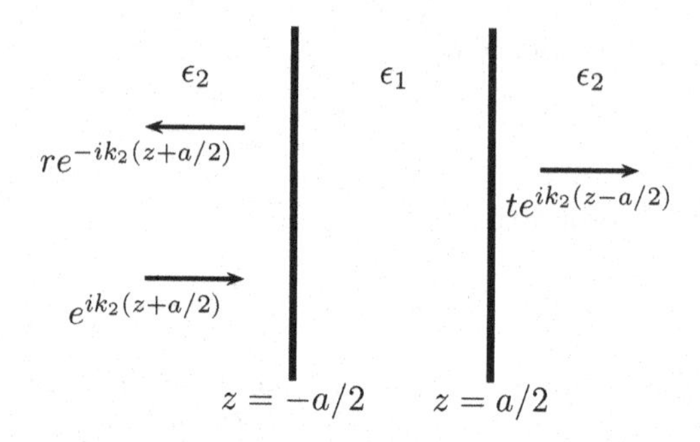

FIGURE 44.5
Three parallel dielectrics.

Find the reflection coefficient r_a. What is the form of the statement of energy conservation here?

3. Consider the same situation given in Problem 2 except that the perfect electric conductor is replaced by a perfect magnetic conductor, for which the tangential component of **B** vanishes on the surface. (In the coordinate system given in Section 44.5, this means that $\partial E_y/\partial z = 0$ on the magnetic conductor for $\perp$ polarization.) Calculate the reflection coefficient for this situation, r_b. What is the form of energy conservation here?

4. Consider two parallel dielectric interfaces as shown in Fig. 44.5. A plane wave, incident form the left, given by

$$e^{ik_2(z+a/2)}, \tag{44.114a}$$

gives rise to a reflected wave,

$$re^{-ik_2(z+a/2)}, \tag{44.114b}$$

and a transmitted wave,

$$te^{ik_2(z-a/2)}. \tag{44.114c}$$

Calculate the reflection and transmission coefficients, r and t. Again, what form does energy conservation take here?

5. Compare the results of Problem 4 with those of Problems 2 and 3, and deduce the relations

$$r = \frac{1}{2}(r_a + r_b) \tag{44.115a}$$

$$t = \frac{1}{2}(r_b - r_a). \tag{44.115b}$$

These relate the statements of energy conservation in the different situations. Why do these relations hold?

6. Show that a wave incident normally on a plane boundary between two media is transmitted without reflection if the ratio of dielectric constant to magnetic permeability is the same for the two media.

45

Reflection by an Imperfect Conductor

We here briefly consider the reflection and transmission of electromagnetic radiation by a conductor. A conductor is characterized by Ohm's law, (5.10),

$$\mathbf{J}_{\text{cond}} = \sigma \mathbf{E}, \tag{45.1}$$

where $\mathbf{J}_{\text{cond}}$ is the conduction current and σ is the conductivity. The effect of this current can be incorporated into the previous discussion by noting that Maxwell's equation for the curl of $\mathbf{H}$ now becomes

$$\boldsymbol{\nabla}\times\mathbf{H} = \frac{1}{c}\frac{\partial}{\partial t}\mathbf{D} + \frac{4\pi}{c}\mathbf{J}_{\text{cond}} + \frac{4\pi}{c}\mathbf{J}, \tag{45.2}$$

where $\mathbf{J}$ is a source, external to the conductor, for $\mathbf{E}$ and $\mathbf{H}$. For fields of a definite frequency, that is, whose time dependence is given by $e^{-i\omega t}$, this equation becomes

$$\boldsymbol{\nabla}\times\mathbf{H} = -\frac{i\omega}{c}\left(\epsilon + i\frac{4\pi}{\omega}\sigma\right)\mathbf{E} + \frac{4\pi}{c}\mathbf{J}. \tag{45.3}$$

This shows that the conductor can be described by an effective dielectric constant,

$$\epsilon + i\frac{4\pi}{\omega}\sigma, \tag{45.4}$$

which is complex, expressing the dissipative nature of the conductor. (This was implicit in the considerations in Section 5.2.) According to (44.21), the component of the propagation vector normal to the surface of a planar conductor is then also complex:

$$k_z = \left[\frac{\omega^2}{c^2}\left(\epsilon + i\frac{4\pi}{\omega}\sigma\right) - k_\perp^2\right]^{1/2}. \tag{45.5}$$

The simplest situation is that of a good conductor, where the conduction current dominates the displacement current, that is

$$\frac{4\pi\sigma}{\omega} \gg \epsilon. \tag{45.6}$$

In the following we will assume that σ is real, which can be approximately valid only for sufficiently low frequencies, as (5.11) states. Since there is wave propagation outside the conductor, $k_\perp^2 \leq \omega^2/c^2$, and (45.5) is, approximately,

$$k_z \approx \sqrt{\frac{\omega^2}{c^2} i\frac{4\pi}{\omega}\sigma} = \frac{\omega}{c}\sqrt{\frac{4\pi\sigma}{\omega}}\frac{1+i}{\sqrt{2}}, \tag{45.7}$$

or

$$k_z \approx \frac{1+i}{\delta}, \tag{45.8a}$$

where

$$\delta = \frac{c}{\sqrt{2\pi\omega\sigma}}. \tag{45.8b}$$

DOI: 10.1201/9781003057369-45

The z dependence of the propagation of a plane wave through the conductor is then given by

$$e^{ik_z|z|} = e^{-|z|/\delta}e^{i|z|/\delta}. \tag{45.9}$$

The first factor on the right side of (45.9) represents the exponential attenuation of the wave; the amplitude is reduced by a factor of e^{-1} in the distance δ. This characteristic distance δ is called the skin depth. As indicated by (45.1), the current in a conductor is confined to a region within a distance $\sim \delta$ of the surface, a distance which becomes smaller as the frequency increases. For a good conductor characterized by (45.6), the ratio of the skin depth to the reduced wavelength of the impinging radiation, $\lambda\!\!\!^{-} \approx c/\omega$, is very small:

$$\frac{\delta}{\lambda\!\!\!^{-}} \approx \sqrt{\omega/2\pi\sigma} \ll 1. \tag{45.10}$$

Thus, only radiation of long wavelengths (low frequencies) penetrates the conductor appreciably.

The discussion of transmission and reflection given in the previous chapter remains essentially unchanged except for the replacement of k_z by (45.8a) inside the conductor. Therefore, for $\perp$ polarization, the reflection coefficient for a plane wave impinging from a dielectric (labeled 2) onto a flat conductor (labeled 1) is, from (44.41a) (again the z subscript is suppressed),

$$r = \frac{k_2 - k_1}{k_2 + k_1} = \frac{k_2 - \frac{1+i}{\delta}}{k_2 + \frac{1+i}{\delta}} \approx -[1 - k_2\delta(1-i)], \tag{45.11}$$

since

$$\frac{1}{\delta} \gg k_2. \tag{45.12}$$

For a perfect conductor, $\sigma \to \infty$, $\delta \to 0$, and

$$r = -1 \tag{45.13}$$

as we saw earlier in (44.51). The fractional amount of power reflected by a good conductor is

$$|r|^2 \approx 1 - 2k_2\delta, \tag{45.14}$$

so that $2k_2\delta$ is the relative power entering the conductor. For a perfect conductor there is total reflection, with no energy absorbed by the conductor.

Another way of deriving this result, (45.14), is to return to the equation for $\mathbf{E}_\perp$, (44.14b),

$$\frac{\partial}{\partial z}\mathbf{E}_\perp + i\frac{\omega}{c}\mathbf{n}\times\mathbf{B}_\perp - \frac{ic}{\omega\epsilon}\mathbf{k}_\perp[\mathbf{k}_\perp\cdot(\mathbf{n}\times\mathbf{B}_\perp)] = (\text{external currents}), \tag{45.15}$$

in which the effects of conduction currents are incorporated in the replacement

$$\epsilon \to \epsilon + i\frac{4\pi\sigma}{\omega}, \tag{45.16}$$

which has a magnitude much greater than that of $\epsilon \sim 1$ inside a good conductor. Under such circumstances, the third term in (45.15) is negligible, implying, inside the conductor,

$$\frac{\partial}{\partial z}\mathbf{E}_\perp + i\frac{\omega}{c}\mathbf{n}\times\mathbf{B}_\perp \approx \mathbf{0}. \tag{45.17}$$

The derivative here is proportional to $\mathbf{E}_\perp$, since the electric field in the conductor attenuates exponentially according to (45.9),

$$\frac{\partial}{\partial z}\mathbf{E}_\perp = \frac{1-i}{\delta}\mathbf{E}_\perp, \tag{45.18}$$

which implies that inside a good conductor,

$$\mathbf{E}_\perp \approx \frac{1-i}{2}\frac{\omega\delta}{c}\mathbf{n}\times\mathbf{B}_\perp, \tag{45.19}$$

relating the tangential components of the electric and magnetic fields. This relation is valid for an arbitrarily shaped surface as long as the radius of curvature of the surface is much larger than the skin depth δ. For a perfect conductor, this supplies the boundary condition that the tangential component of the electric field vanishes on the surface,

$$\mathbf{E}_\perp = \mathbf{0}. \tag{45.20}$$

The energy flow in the $-z$ direction just inside the conductor is now seen to be proportional to [recall (44.95)]

$$-\operatorname{Re}\mathbf{n}\cdot\mathbf{E}^*\times\mathbf{B} = \operatorname{Re}\frac{1+i}{2}\frac{\omega\delta}{c}(\mathbf{n}\times\mathbf{B}_\perp^*)\cdot(\mathbf{n}\times\mathbf{B}_\perp) = \frac{\omega\delta}{2c}|\mathbf{B}_\perp|^2. \tag{45.21}$$

To find the fractional amount of energy absorbed by a good conductor, we approximately evaluate $\mathbf{B}_\perp$ as though $\delta = 0$ (a perfect conductor), and compare (45.21) at $z = 0-$ to the incident flux at $z = 0+$. The latter is constructed in terms of the fields (normalized so that the amplitude of the incident magnetic field is unity), for the two independent polarization states,

$$\begin{aligned}\perp:\quad & B_x = e^{-ik_2z} + e^{ik_2z},\\ & E_y = \frac{\omega}{ck_2}\left(e^{-ik_2z} - e^{ik_2z}\right),\end{aligned} \tag{45.22a}$$

$$\begin{aligned}\|:\quad & B_y = e^{-ik_2z} + e^{ik_2z},\\ & E_x = -\frac{ck_2}{\omega\epsilon_2}\left(e^{-ik_2z} - e^{ik_2z}\right).\end{aligned} \tag{45.22b}$$

Here we have used (44.97) and (44.108) to relate B_x to E_y, and E_x to B_y, respectively. The incident energy flux arises from the terms proportional to e^{-ik_2z}:

$$\begin{aligned}\text{Incident flux} &\propto \operatorname{Re}\left\{\frac{\omega}{ck_2}|B_{x\text{inc}}|^2, \frac{ck_2}{\omega\epsilon_2}|B_{y\text{inc}}|^2\right\}\\ &= \frac{1}{4}\left\{\frac{\omega}{ck_2}|B_x(z=0)|^2, \frac{ck_2}{\omega\epsilon_2}|B_y(z=0)|^2\right\}.\end{aligned} \tag{45.23}$$

The tangential magnetic field is continuous since, for an imperfect conductor, there is no localized surface current. Consequently, to determine the fractional amount of power absorbed, we take the ratio of (45.21) to (45.23):

$$\perp:\quad \frac{\omega\delta/2c}{\omega/4ck_2} = 2k_2\delta, \tag{45.24a}$$

$$\|:\quad \frac{\omega\delta/2c}{ck_2/4\omega\epsilon_2} = 2\left(\frac{\omega}{c}\right)^2\frac{\epsilon_2}{k_2}\delta. \tag{45.24b}$$

Of course, (45.24a) agrees with (45.14). (See also Problem 45.1.)

45.1 Problems for Chapter 45

1. Derive the analog of (45.14) for $\|$ polarization.

2. Show that the rate of power loss per unit area by an electromagnetic wave impinging on a good conductor is

$$\frac{dP_{\rm loss}}{dS} = \frac{1}{4\sigma\delta}|\mathbf{K}|^2,$$

where $\mathbf{K}$ is the effective surface current,

$$\mathbf{K} = \int_0^\infty d|z|\,\mathbf{J}.$$

46

Cylindrical Coordinates

A waveguide, in which conducting walls force a wave to propagate in a single direction (say along the z axis), motivates a consideration of a $2+1$ dimensional break-up of Green's function for a wave propagating in vacuum with a definite frequency ω. This is analogous to our earlier discussion in electrostatics. (See Chapters 15–21.) Furthermore, by deriving alternative expressions for Green's function, we are able to obtain additional mathematical properties of Bessel's functions.

46.1 $2+1$ Dimensional Decomposition of Green's Function

The retarded Green's function satisfying (34.19), the inhomogeneous Helmholtz equation,

$$-\left(\nabla^2+\left(\frac{\omega}{c}\right)^2\right)G_\omega(\mathbf{r},\mathbf{r}')=4\pi\delta(\mathbf{r}-\mathbf{r}'), \tag{46.1}$$

is given by (34.30), that is

$$G_\omega(\mathbf{r},\mathbf{r}')=\frac{e^{i\omega|\mathbf{r}-\mathbf{r}'|/c}}{|\mathbf{r}-\mathbf{r}'|}. \tag{46.2}$$

In deriving this expression we treated all directions on the same footing; if instead we single out the z axis, by introducing a Fourier transform in the $\perp$ (transverse) space,

$$G_\omega(\mathbf{r},\mathbf{r}')=4\pi\int\frac{(d\mathbf{k}_\perp)}{(2\pi)^2}e^{i\mathbf{k}_\perp\cdot(\mathbf{r}-\mathbf{r}')_\perp}g(z,z'), \tag{46.3}$$

(46.1) is satisfied providing the reduced Green's function $g(z,z')$ obeys

$$\left[-\frac{\partial}{\partial z^2}+k_\perp^2-\left(\frac{\omega}{c}\right)^2\right]g(z,z')=\delta(z-z'). \tag{46.4}$$

We have seen this differential equation before; it is (44.31) with $\epsilon=1$, and its solution is given by (44.40a) with $r=0$:

$$g(z,z')=\frac{i}{2k_z}e^{ik_z|z-z'|}, \tag{46.5}$$

where, for $\omega>0$,

$$k_z=\begin{cases}\sqrt{\frac{\omega^2}{c^2}-k_\perp^2}, & \text{if } \frac{\omega}{c}>|\mathbf{k}_\perp|,\\ i\sqrt{k_\perp^2-\frac{\omega^2}{c^2}}, & \text{if } \frac{\omega}{c}<|\mathbf{k}_\perp|.\end{cases} \tag{46.6}$$

In statics ($\omega=0$) the second possibility in (46.6) occurs. [See (15.22).] Without loss of generality, we set $\mathbf{r}'=\mathbf{0}$, and define $k=\omega/c>0$. Then, comparing (46.2) with (46.3) and

DOI: 10.1201/9781003057369-46

(46.5), we obtain the identity

$$\frac{e^{ikr}}{r} = 4\pi \int \frac{(d\mathbf{k}_\perp)}{(2\pi)^2} e^{i\mathbf{k}_\perp \cdot \mathbf{r}_\perp} \frac{ie^{i\sqrt{k^2-k_\perp^2}|z|}}{2\sqrt{k^2-k_\perp^2}}, \tag{46.7}$$

where the square root is defined by (46.6). It is convenient to introduce a cylindrical coordinate system, with

$$\mathbf{r} = (\rho, 0, z), \quad r = \sqrt{z^2+\rho^2}, \quad \mathbf{k}_\perp = (\lambda, \phi), \tag{46.8}$$

from which follow

$$e^{i\mathbf{k}_\perp \cdot \mathbf{r}_\perp} = e^{i\lambda\rho\cos\phi}, \tag{46.9a}$$

$$(d\mathbf{k}_\perp) = \lambda\, d\lambda\, d\phi. \tag{46.9b}$$

Making use of the integral representation for the Bessel function J_0, (18.4),

$$\int_0^{2\pi} \frac{d\phi}{2\pi} e^{i\lambda\rho\cos\phi} = J_0(\lambda\rho), \tag{46.10}$$

we find, from (46.7), the identity

$$\frac{e^{ik\sqrt{\rho^2+z^2}}}{\sqrt{\rho^2+z^2}} = i\int_0^\infty \lambda\, d\lambda\, J_0(\lambda\rho) \frac{e^{i\sqrt{k^2-\lambda^2}|z|}}{\sqrt{k^2-\lambda^2}}. \tag{46.11}$$

In the static limit ($k = 0$, $\sqrt{k^2-\lambda^2} \to i\lambda$), we recover the previously derived result (18.6),

$$\frac{1}{\sqrt{\rho^2+z^2}} = \int_0^\infty d\lambda\, J_0(\lambda\rho) e^{-\lambda|z|}. \tag{46.12}$$

46.2 Three-Dimensional Fourier Representation

An alternative representation for Green's function can be obtained by taking the three-dimensional Fourier transform of (46.1), which immediately leads to the formal solution

$$G_\omega(\mathbf{r}, \mathbf{r}') = 4\pi \int \frac{(d\mathbf{k})}{(2\pi)^3} \frac{e^{i\mathbf{k}\cdot(\mathbf{r}-\mathbf{r}')}}{\mathbf{k}^2 - \omega^2/c^2}. \tag{46.13}$$

The above expression is not well defined since the integrand has a singularity wherever $|\mathbf{k}| = \omega/c$, corresponding to real wave propagation. To assign meaning to this expression, we must specify the boundary conditions. Recall that the retarded and advanced Green's functions have the explicit forms given in (34.30),

$$G_{\text{ret}} = \frac{1}{|\mathbf{r}-\mathbf{r}'|} e^{i\omega|\mathbf{r}-\mathbf{r}'|/c}, \tag{46.14a}$$

$$G_{\text{adv}} = \frac{1}{|\mathbf{r}-\mathbf{r}'|} e^{-i\omega|\mathbf{r}-\mathbf{r}'|/c}. \tag{46.14b}$$

Mathematically, both signs of i are possible; a consideration of physical conditions is necessary to select the appropriate solution. If we let the wavenumber be complex, which is the situation occurring in a conductor [for example, see (45.8a)]:

$$\frac{\omega}{c} \to \frac{\omega_1}{c} + \frac{i\omega_2}{c}, \quad \omega_1 > 0, \tag{46.15}$$

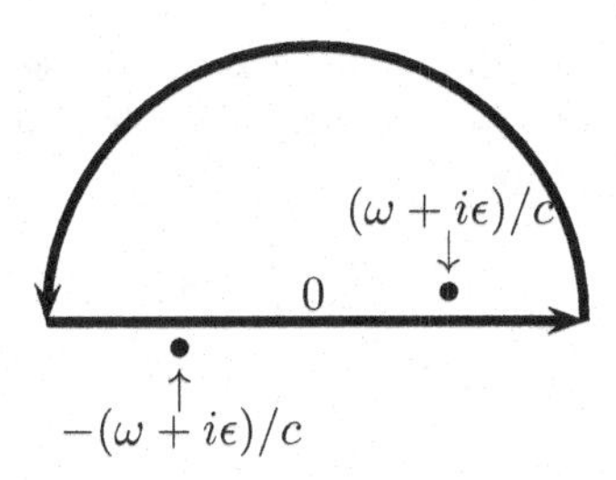

FIGURE 46.1
Contour for evaluating (46.18).

the retarded and advanced Green's functions become, respectively,

$$G_{\text{ret}} \to \frac{1}{R} e^{i\omega_1 R/c} e^{-\omega_2 R/c}, \tag{46.16a}$$

$$G_{\text{adv}} \to \frac{1}{R} e^{-i\omega_1 R/c} e^{\omega_2 R/c}, \tag{46.16b}$$

where $R = |\mathbf{r} - \mathbf{r}'|$. As $R \to \infty$, G_{adv} increases without bound, while the physically correct solution, G_{ret}, goes to zero. Thus the device for defining G_{ret} in terms of the representation (46.13), which builds in the boundedness requirement, is to replace ω there by $\omega \to \omega + i\epsilon$, $\epsilon \to +0$:

$$G_{\text{ret}}(\mathbf{r}, \mathbf{r}') = 4\pi \int \frac{(d\mathbf{k})}{(2\pi)^3} \frac{e^{i\mathbf{k}\cdot(\mathbf{r}-\mathbf{r}')}}{\mathbf{k}^2 - \left(\frac{\omega+i\epsilon}{c}\right)^2}\Bigg|_{\epsilon\to+0}. \tag{46.17}$$

We can quickly check the validity of this result by explicitly evaluating the integral using spherical coordinates [recall the spherical average of a plane wave seen, for example, in (41.32)]:

$$\begin{aligned} G_{\text{ret}}(\mathbf{r}, \mathbf{r}') &= 4\pi \int_0^\infty \frac{4\pi k^2 dk}{(2\pi)^3} \frac{\sin kR}{kR} \frac{1}{k^2 - \left(\frac{\omega+i\epsilon}{c}\right)^2} = \frac{2}{\pi}\frac{1}{R} \int_0^\infty dk\, k \frac{\sin kR}{k^2 - \left(\frac{\omega+i\epsilon}{c}\right)^2} \\ &= -\frac{2}{\pi}\frac{1}{R}\frac{d}{dR} \int_0^\infty dk \frac{\cos kR}{k^2 - \left(\frac{\omega+i\epsilon}{c}\right)^2} = -\frac{1}{\pi}\frac{1}{R}\frac{d}{dR} \int_{-\infty}^\infty dk \frac{e^{ikR}}{k^2 - \left(\frac{\omega+i\epsilon}{c}\right)^2}, \end{aligned} \tag{46.18}$$

since the third integrand is even in k. Equation (46.18) can be evaluated by expressing it in terms of a contour integral in the complex k plane, as shown in Fig. 46.1. For both $\omega > 0$ and $\omega < 0$, we close the contour in the upper half plane, since the infinite semicircle then makes no contribution (because $R > 0$). Then by the residue theorem, we find, because only the pole at $k = \omega/c + i\epsilon$ is enclosed,

$$G_{\text{ret}}(\mathbf{r}, \mathbf{r}') = -\frac{1}{\pi}\frac{1}{R}\frac{d}{dR} 2\pi i \frac{e^{i\omega R/c}}{2\omega/c} = \frac{1}{R} e^{i\omega R/c}, \tag{46.19}$$

which is the retarded Green's function given in (46.14a).

46.3 Hankel Functions

We have thus established the equivalence between two representations for G_ω, namely (46.2) and (46.17),

$$\frac{1}{r} e^{i\omega r/c} = 4\pi \int \frac{(d\mathbf{k})}{(2\pi)^3} \frac{e^{i\mathbf{k}\cdot\mathbf{r}}}{k^2 - \left(\frac{\omega+i\epsilon}{c}\right)^2}\Bigg|_{\epsilon\to+0}. \tag{46.20}$$

Note that if we now perform the integral on k_z, and use the integral representation for J_0, (46.10), we rederive the identity (46.11). (See Problem 46.1.) If, instead, we rewrite (46.20) in the form of a $1+2$ dimensional decomposition,

$$\frac{1}{r}e^{i\omega r/c} = 4\pi\int_{-\infty}^{\infty}\frac{dk_z}{2\pi}\int\frac{(d\mathbf{k}_\perp)}{(2\pi)^2}\frac{e^{i\mathbf{k}_\perp\cdot\mathbf{r}_\perp}e^{ik_z z}}{k_\perp^2+k_z^2-\left(\frac{\omega+i\epsilon}{c}\right)^2}, \tag{46.21}$$

and first integrate over the angle associated with $\mathbf{k}_\perp$, we find

$$\frac{1}{r}e^{i\omega r/c} = 4\pi\int_{-\infty}^{\infty}\frac{dk_z}{2\pi}e^{ik_z z}\int_0^{\infty}\frac{\lambda\,d\lambda}{2\pi}\frac{J_0(\lambda\rho)}{\lambda^2-\left(\left(\frac{\omega+i\epsilon}{c}\right)^2-k_z^2\right)}, \tag{46.22}$$

which, in the static limit, $\omega=0$, we have already seen in (20.25).

If we substitute the representation (46.22) into the Green's function equation, (46.1), and use the Laplacian in cylindrical coordinates, (18.16), we immediately see that the λ integral obeys Bessel's equation

$$\left(\frac{d^2}{d\rho^2}+\frac{1}{\rho}\frac{d}{d\rho}+\kappa^2\right)\int_0^{\infty}d\lambda\,\lambda\frac{J_0(\lambda\rho)}{\lambda^2-\kappa^2}=0, \tag{46.23}$$

for $\rho\neq 0$. We may also obtain this result explicitly from the differential equation satisfied by J_0, (18.17a), since

$$\begin{aligned}&\left(\frac{d^2}{d\rho^2}+\frac{1}{\rho}\frac{d}{d\rho}+\kappa^2\right)\int_0^{\infty}d\lambda\,\lambda\frac{J_0(\lambda\rho)}{\lambda^2-\kappa^2}\\ &=\int_0^{\infty}d\lambda\,\lambda\frac{1}{\lambda^2-\kappa^2}(-\lambda^2+\kappa^2)J_0(\lambda\rho)=0,\quad \text{for}\quad \rho\neq 0,\end{aligned} \tag{46.24}$$

where we have used (18.48),

$$\int_0^{\infty}d\lambda\,\lambda\,J_0(\lambda\rho)=\frac{1}{\rho}\delta(\rho). \tag{46.25}$$

With the exception of the origin, here is another solution of Bessel's equation, which is called the zeroth order Hankel function of the first kind, defined by

$$\int_0^{\infty}\lambda\,d\lambda\frac{J_0(\lambda\rho)}{\lambda^2-\left[\left(\frac{\omega+i\epsilon}{c}\right)^2-k_z^2\right]}=\frac{\pi i}{2}H_0^{(1)}\left(\sqrt{\frac{\omega^2}{c^2}-k_z^2}\,\rho\right). \tag{46.26}$$

When $\omega/c<|k_z|$, the square root in (46.26) becomes imaginary, $i\sqrt{k_z^2-\omega^2/c^2}$, and in this case, the Hankel function becomes the modified Bessel function, (20.3a),

$$\frac{\pi i}{2}H_0^{(1)}\left(\sqrt{\frac{\omega^2}{c^2}-k_z^2}\,\rho\right)\to K_0\left(\sqrt{k_z^2-\frac{\omega^2}{c^2}}\,\rho\right). \tag{46.27}$$

In terms of the Hankel function defined by (46.26), the equality (46.22) reads

$$\begin{aligned}\frac{1}{r}e^{i\omega r/c}&=\frac{i}{2}\int_{-\infty}^{\infty}dk_z e^{ik_z z}H_0^{(1)}\left(\sqrt{\frac{\omega^2}{c^2}-k_z^2}\,\rho\right)\\ &=i\int_0^{\infty}dk_z\cos(k_z z)H_0^{(1)}\left(\sqrt{\frac{\omega^2}{c^2}-k_z^2}\,\rho\right).\end{aligned} \tag{46.28}$$

The representation (46.11) has its ρ dependence expressed in terms of $J_0(\lambda\rho)$, which is a slowly damped oscillatory function, while the associated z dependence is either oscillatory or exponentially damped, depending on whether the square root is real or imaginary. On the other hand, the form (46.28) represents the function in terms of oscillatory functions of z, while the corresponding ρ dependence is either oscillatory or exponentially decreasing. Let us extract the Hankel function in (46.28) by taking the Fourier transform in z:

$$\int_{-\infty}^{\infty} dz \frac{e^{ik\sqrt{\rho^2+z^2}}}{\sqrt{\rho^2+z^2}} e^{-ik_z z} = i\pi H_0^{(1)}\left(\sqrt{k^2-k_z^2}\,\rho\right), \tag{46.29}$$

where $k = \omega/c$. Then by setting $k_z = 0$, we obtain another integral representation for the Hankel function,

$$i\pi H_0^{(1)}(k\rho) = \int_{-\infty}^{\infty} dz \frac{e^{ik\sqrt{\rho^2+z^2}}}{\sqrt{\rho^2+z^2}}. \tag{46.30}$$

We will now use this representation to find the asymptotic form of the Hankel function for $k\rho \gg 1$. As we will justify *a posteriori*, the main contribution of the integral comes from small values of z/ρ so that we may expand the square root in the exponential as

$$\sqrt{\rho^2+z^2} \approx \rho + \frac{z^2}{2\rho}, \tag{46.31}$$

and approximate (46.30) by

$$i\pi H_0^{(1)}(k\rho) \sim \frac{1}{\rho} e^{ik\rho} \int_{-\infty}^{\infty} dz\, e^{-kz^2/2i\rho} = i\sqrt{\frac{2\pi}{k\rho}} e^{i(k\rho-\pi/4)}, \quad k\rho \gg 1. \tag{46.32}$$

Here the Gaussian integral is evaluated as

$$\int_{-\infty}^{\infty} dz\, e^{-kz^2/2i\rho} = \sqrt{\pi\frac{2i\rho}{k}}. \tag{46.33}$$

The consistency of our approximation hinges on the fact that the dominant contribution to the integral, (46.33), comes from small values of z/ρ, for which

$$\frac{z}{\rho} \sim \sqrt{\frac{1}{k\rho}} \ll 1. \tag{46.34}$$

As an application of (46.32), we may obtain the asymptotic behavior of J_0 by taking the imaginary part of (46.26) for $k_z = 0$ and using the representation given in Problem 46.2a:

$$J_0(k\rho) = \text{Re}\, H_0^{(1)}(k\rho) \sim \sqrt{\frac{2}{\pi}\frac{1}{k\rho}} \cos(k\rho - \pi/4). \tag{46.35}$$

This is a result, derived in an alternative way in Problem 46.2b, which we first saw in (18.37).

As with the previous Bessel functions, we can ask whether $H_0^{(1)}(kP)$, with P given by (18.2), possesses an addition theorem. We might expect an affirmative answer in view of (46.27) and (20.56), the addition theorem for K_0. Since $H_0^{(1)}$ satisfies the same differential equation as J_0, we can follow the development in Section 20.1, leading to the addition theorem for J_0, (20.47). In particular, we have the Taylor series expansion [*cf.* (20.42)]

$$H_0^{(1)}(kP) = e^{-\mathbf{r}'_\perp \cdot \boldsymbol{\nabla}_\perp} H_0^{(1)}(k\rho), \tag{46.36}$$

where, similarly to the K_0 discussion, $\rho' < \rho$ because $H_0^{(1)}(kP)$ has a singularity at $P = 0$. The introduction of the polar angle Φ, (20.44), remains unchanged as does the expansion of $\exp(-\mathbf{r}'_\perp \cdot \boldsymbol{\nabla}_\perp)$, (20.45). Consequently, we have

$$H_0^{(1)}(kP) = \sum_{m=-\infty}^{\infty} i^m e^{-im\phi'} J_m(k\rho') e^{im\Phi} H_0^{(1)}(k\rho). \tag{46.37}$$

This becomes the addition theorem for $H_0^{(1)}$ (recall $\rho' < \rho$)

$$H_0^{(1)}(kP) = \sum_{m=-\infty}^{\infty} e^{im\phi} H_m^{(1)}(k\rho) e^{-im\phi'} J_m(k\rho'), \tag{46.38}$$

with the introduction of the mth order Hankel functions of the first kind,

$$e^{im\phi} H_m^{(1)}(k\rho) = i^m e^{im\Phi} H_0^{(1)}(k\rho) = \left[\mp\frac{1}{k}\left(\frac{\partial}{\partial x} \pm i\frac{\partial}{\partial y}\right)\right]^{\pm m} H_0^{(1)}(k\rho), \tag{46.39}$$

for $m \geq 0$ ($m \leq 0$), respectively. Notice that since the whole construction is identical to the J_m construction, $H_m^{(1)}$ satisfies the same differential equation, (18.32a), and the same recurrence relations, (16.71) and (16.72), as does J_m.

46.4 Problems for Chapter 46

1. Derive (46.11) from (46.20), as suggested.
2. (a) Using the representation

$$\lim_{\epsilon\to 0} \operatorname{Im} \frac{1}{x - i\epsilon} = \lim_{\epsilon\to 0} \frac{\epsilon}{x^2 + \epsilon^2} = \pi\delta(x), \tag{46.40}$$

show that $\operatorname{Re} H_0^{(1)}(t) = J_0(t)$.

(b) Use the result of Problem 46.2a to derive

$$\begin{aligned} J_0(t) &= \frac{2}{\pi}\int_0^\infty dx \frac{\sin\sqrt{t^2 + x^2}}{\sqrt{t^2 + x^2}} = \frac{2}{\pi}\int_0^\infty d\theta \, \sin(t\cosh\theta) \\ &= \frac{2}{\pi}\int_1^\infty dy \frac{\sin ty}{\sqrt{y^2 - 1}}, \end{aligned} \tag{46.41}$$

and hence derive the asymptotic form

$$J_0(t) \sim \sqrt{\frac{2}{\pi t}} \cos(t - \pi/4), \quad (t \gg 1). \tag{46.42}$$

Reconcile the third form of $J_0(t)$ above with the fact that $J_0(0) = 1$.

3. (a) Recalling Problem 20.2, show that

$$\operatorname{Re} H_m^{(1)}(k\rho) = J_m(k\rho). \tag{46.43a}$$

(b) Show that

$$\operatorname{Im} H_m^{(1)}(k\rho) = N_m(k\rho). \tag{46.43b}$$

4. In general, Hankel functions of the first and second kind are defined by

$$H_\nu^{(1)}(x) = J_\nu(x) + iN_\nu(x) \tag{46.44a}$$
$$H_\nu^{(2)}(x) = J_\nu(x) - iN_\nu(x). \tag{46.44b}$$

Show from Problem 18.10 that these have the leading asymptotic behaviors as $x \to \infty$:

$$H_\nu^{(1,2)}(x) \sim \left(\frac{2}{\pi x}\right)^{1/2} e^{\pm i(x-\nu\pi/2-\pi/4)}. \tag{46.45}$$

Thus, $H^{(1)}$, $H^{(2)}$ correspond to outgoing or incoming cylindrical (or spherical) waves, respectively.

5. Consider unbounded two-dimensional space. Find Green's function for Helmholtz' equation

$$(\nabla^2 + k^2)G_k(\mathbf{r}-\mathbf{r}') = -4\pi\delta(\mathbf{r}-\mathbf{r}') \tag{46.46}$$

in three ways. (Why is G_k translationally invariant?)

(a) By directly solving the equation in polar coordinates, noticing that there is no preferred direction, and selecting the solution of the homogeneous equation that is singular, $\sim -2\ln|\mathbf{r}-\mathbf{r}'|$, as $|\mathbf{r}-\mathbf{r}'| \to 0$, and represents, at large distances, an outgoing wave. [Answer, which was already worked out in Problem 34.12c: $i\pi H_0^{(1)}(k|\mathbf{r}-\mathbf{r}'|)$.]

(b) Expand Green's function in terms of eigenfunctions, in polar coordinates, ρ, ϕ. [Prove that the normalized eigenfunctions are

$$\sqrt{\frac{k}{2\pi}} e^{im\phi} J_m(k\rho), \tag{46.47}$$

where k is a continuous variable.]

(c) Write, in polar coordinates,

$$G_k(\mathbf{r}-\mathbf{r}') = \sum_{m=-\infty}^{\infty} \frac{1}{2\pi} e^{im(\phi-\phi')} p_m(\rho,\rho'), \tag{46.48}$$

and solve the resulting differential equation for p_m, by using the discontinuity method. Note that the boundary condition for large ρ is to be an outgoing wave, so the appropriate behavior there is given by $H_m^{(1)}(k\rho)$.

6. Show that the relation between the modified Bessel function K_m, and the Hankel function $H_m^{(1)}$ is

$$K_m(z) = \frac{i\pi}{2} e^{i\pi m/2} H_m^{(1)}(iz), \quad -\pi < \arg z \le \frac{\pi}{2}. \tag{46.49}$$

Use (46.39) and (20.57). This result is already given in (20.73b).

47

Waveguides

The theory of microwave cavities and their application to radar technology was largely developed at the MIT Radiation Laboratory during World War II. A key player in that development was Julian Schwinger. Lectures he gave contemporaneously on the subject are included in part in Ref. [1]. which may be consulted for much more detail on the subject. In this chapter, we consider some of the elementary aspects of electromagnetic waveguides, which we will see is closely tied to the electrostatic mode theory discussed in Chapter 27. Throughout this chapter, we assume the waveguide is filled with a homogeneous dielectric with permittivity ϵ; the system is supposed to be translationally invariant in the z direction.

47.1 E and H Modes

In this chapter, we want to introduce the theory of waveguides, especially relevant for microwave radiation. We start from the analysis in Chapter 44, where now we write (44.13) and (44.14a), (44.14b) as

$$B_z = \frac{ic}{\omega}\boldsymbol{\nabla}\cdot\mathbf{n}\times\mathbf{E}, \tag{47.1a}$$

$$E_z = -\frac{ic}{\omega\epsilon}\boldsymbol{\nabla}\cdot\mathbf{n}\times\mathbf{B} - i\frac{4\pi}{\omega\epsilon}J_z, \tag{47.1b}$$

$$\frac{\partial}{\partial z}\mathbf{E}_\perp = -\frac{i\omega}{c}\left(\mathbf{1}_\perp + \frac{c^2}{\omega^2\epsilon}\boldsymbol{\nabla}_\perp\boldsymbol{\nabla}_\perp\right)\cdot\mathbf{n}\times\mathbf{B} - i\frac{4\pi}{\omega\epsilon}\boldsymbol{\nabla}_\perp J_z, \tag{47.1c}$$

$$\frac{\partial}{\partial z}\mathbf{B}_\perp = \frac{i\omega\epsilon}{c}\left(\mathbf{1}_\perp + \frac{c^2}{\omega^2\epsilon}\boldsymbol{\nabla}_\perp\boldsymbol{\nabla}_\perp\right)\cdot\mathbf{n}\times\mathbf{E} - \frac{4\pi}{c}\mathbf{n}\times\mathbf{J}, \tag{47.1d}$$

where we have used the gradient operator $\boldsymbol{\nabla}_\perp$ in place of $i\mathbf{k}_\perp$, that is, we are undoing the transverse Fourier transform. Now in the transform space the two independent transverse vectors are $\mathbf{k}_\perp$ and $\mathbf{n}\times\mathbf{k}_\perp$, so correspondingly, the transverse fields $\mathbf{E}_\perp$ and $\mathbf{B}_\perp$ can be written in terms of two scalar functions each,[1]

$$\mathbf{E}_\perp = -\boldsymbol{\nabla}_\perp V' - \boldsymbol{\nabla}\times\mathbf{n}V'', \tag{47.2a}$$

$$\mathbf{B}_\perp = \boldsymbol{\nabla}\times\mathbf{n}I' - \boldsymbol{\nabla}_\perp I'', \tag{47.2b}$$

and then, since

$$\mathbf{n}\times(\boldsymbol{\nabla}\times\mathbf{n}) = \boldsymbol{\nabla} - \mathbf{n}\,\mathbf{n}\cdot\boldsymbol{\nabla} = \boldsymbol{\nabla}_\perp, \tag{47.3}$$

[1]The resemblance of the V and I symbols to voltage and current is not coincidental: See Ref. [1] and Problem 47.21.

DOI: 10.1201/9781003057369-47

the transversely rotated fields are

$$\mathbf{n}\times\mathbf{E}_\perp = \boldsymbol{\nabla}\times\mathbf{n}V' - \boldsymbol{\nabla}_\perp V'', \tag{47.4a}$$

$$\mathbf{n}\times\mathbf{B}_\perp = \boldsymbol{\nabla}_\perp I' + \boldsymbol{\nabla}\times\mathbf{n}I''. \tag{47.4b}$$

When we insert these constructions into (47.1c), we obtain, for the $\boldsymbol{\nabla}_\perp$ and the $\boldsymbol{\nabla}\times\mathbf{n}$ components,

$$\frac{\partial}{\partial z}V' = \frac{i\omega}{c}\left(1+\frac{c^2}{\omega^2\epsilon}\boldsymbol{\nabla}_\perp^2\right)I' + i\frac{4\pi}{\omega\epsilon}J_z, \tag{47.5a}$$

$$\frac{\partial}{\partial z}V'' = \frac{i\omega}{c}I''. \tag{47.5b}$$

We proceed similarly for the $\partial\mathbf{B}_\perp/\partial z$ equation, (47.1d), where we introduce similar components for $\mathbf{J}$:

$$\mathbf{n}\times\mathbf{J} = \boldsymbol{\nabla}\times\mathbf{n}J' - \boldsymbol{\nabla}_\perp J'', \tag{47.6}$$

which are determined by

$$\boldsymbol{\nabla}_\perp\cdot\mathbf{n}\times\mathbf{J} = -\boldsymbol{\nabla}_\perp^2 J'', \tag{47.7a}$$

$$-\boldsymbol{\nabla}\cdot\mathbf{J}_\perp = \boldsymbol{\nabla}_\perp^2 J'. \tag{47.7b}$$

The resulting equations are

$$\frac{\partial}{\partial z}I' = i\frac{\omega\epsilon}{c}V' - \frac{4\pi}{c}J', \tag{47.8a}$$

$$\frac{\partial}{\partial z}I'' = i\frac{\omega\epsilon}{c}\left(1+\frac{c^2}{\omega^2\epsilon}\nabla_\perp^2\right)V'' - \frac{4\pi}{c}J''. \tag{47.8b}$$

We thus have four coupled differential equations for the four functions I', I'', V', V''. Once those functions are found, the longitudinal fields can be found from (47.1a) and (47.1b):

$$E_z = -i\frac{c}{\omega\epsilon}\nabla_\perp^2 I' - i\frac{4\pi}{\omega\epsilon}J_z, \tag{47.9a}$$

$$B_z = -i\frac{c}{\omega}\nabla_\perp^2 V''. \tag{47.9b}$$

Because there is no coupling between the sets I', V' and I'', V'' we have two sets of independent modes, characterized by the values of E_z and B_z, respectively.

1. If $I'', V'' = 0$, and hence $B_z = 0$, $E_z \neq 0$, these are called the E modes (or "transverse magnetic," TM, modes).
2. If $I', V' = 0$, and hence $E_z = 0$ (outside the current distribution), $B_z \neq 0$, these are called the H modes (or "transverse electric," TE, modes).

(These correspond to what were referred to as $\parallel$ (p) and $\perp$ (s) polarization, respectively, in Chapter 44.) For these two modes, the electric and magnetic fields are given simply in terms of I', V', and I'', V'', respectively, outside the current distribution, as

$$\text{E modes:}\quad \mathbf{E}_\perp = -\boldsymbol{\nabla}_\perp V', \quad E_z = -i\frac{c}{\omega\epsilon}\nabla_\perp^2 I',$$

$$\mathbf{n}\times\mathbf{B} = \boldsymbol{\nabla}_\perp I', \quad B_z = 0, \tag{47.10a}$$

$$\text{H modes:}\quad \mathbf{B}_\perp = -\boldsymbol{\nabla}_\perp I'', \quad E_z = 0,$$

$$\mathbf{n}\times\mathbf{E} = -\boldsymbol{\nabla}_\perp V'', \quad B_z = -i\frac{c}{\omega}\nabla_\perp^2 V''. \tag{47.10b}$$

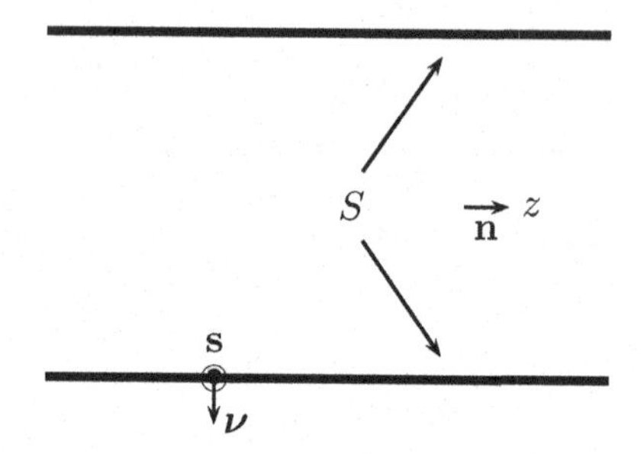

FIGURE 47.1
Cylindrical waveguide with arbitrarily shaped surface S. Shown also is the outward normal to the surface, $\boldsymbol{\nu}$, the circumferential tangent vector $\mathbf{s}$, and the waveguide direction $\mathbf{n}$, which form a right-handed triad of unit vectors, satisfying $\mathbf{s} \times \boldsymbol{\nu} = \mathbf{n}$.

47.2 Boundary Conditions

When we discussed reflection and refraction in Chapter 44, the transverse character of the plane waves was $\exp(i\mathbf{k}_\perp \cdot \mathbf{r}_\perp)$, which meant that we could make the replacement $\nabla_\perp^2 \to -k_\perp^2$. Now, however, we have in mind a 2+1 dimensional situation, translationally invariant in the z direction, but with (for simplicity) a perfectly conducting boundary surface S in the transverse directions, as sketched in Figure 47.1. The boundary condition on the surface S is that the tangential component of $\mathbf{E}$ must vanish, or $E_z = 0$ and the circumferential component of $\mathbf{E}_\perp$ must vanish. Thus, for the E modes this means that on the surface,

$$\text{E modes:} \qquad \nabla_\perp^2 I' = 0, \quad V' = \text{constant on } S. \tag{47.11}$$

For the H modes, E_z is already zero, so only the tangential component of $\mathbf{E}_\perp$ is required to vanish, or

$$\text{H modes:} \qquad \boldsymbol{\nu} \cdot \boldsymbol{\nabla}_\perp V'' = 0 \text{ on } S, \tag{47.12}$$

where $\boldsymbol{\nu}$ is the outward normal to the cylindrical boundary.

The various modes are characterized by functions satisfying

$$(\nabla_\perp^2 + \gamma^2)\phi = 0, \tag{47.13}$$

where the eigenvalue $-\gamma^2$ of the transverse Laplacian operator is the generalization of $-k_\perp^2$ in the case of unbounded transverse space. The E modes are characterized by a set of functions $\{\phi_a\}$ which satisfy the condition $\gamma_a^2 \phi_a = 0$, or

$$\phi_a = 0 \quad \text{on } S, \quad \gamma_a^2 \neq 0 \tag{47.14a}$$

$$\phi_a = \text{constant on } S, \quad \gamma_a^2 = 0. \tag{47.14b}$$

The H modes are characterized by a set of functions $\{\psi_a\}$ satisfying

$$\frac{\partial}{\partial \nu}\psi_a = 0 \quad \text{on } S. \tag{47.15}$$

Thus, we have Dirichlet and Neumann boundary conditions, respectively, for the E and H modes. For the E modes we write the eigenfunction expansion

$$V'(\mathbf{r}_\perp, z) = \sum_a \phi_a(\mathbf{r}_\perp) V'_a(z), \tag{47.16a}$$

$$I'(\mathbf{r}_\perp, z) = \sum_a \phi_a(\mathbf{r}_\perp) I'_a(z), \tag{47.16b}$$

while the H modes have the expansion

$$V''(\mathbf{r}_\perp, z) = \sum_a \psi_a(\mathbf{r}_\perp) V''_a(z), \tag{47.17a}$$

$$I''(\mathbf{r}_\perp, z) = \sum_a \psi_a(\mathbf{r}_\perp) I''_a(z). \tag{47.17b}$$

If $\gamma_a^2 = 0$, then the E modes have both $E_z = 0$ and $B_z = 0$, and, since both electric and magnetic fields are transverse in this case, we call it merely a T mode (or a TEM mode). Since such a mode is both an E and an H mode, we must have

$$\mathbf{E} = \boldsymbol{\nabla}_\perp \phi = \boldsymbol{\nabla} \times \mathbf{n} \psi, \tag{47.18}$$

which indeed implies the required Laplacian equation for ϕ,

$$\nabla_\perp^2 \phi = 0. \tag{47.19}$$

The boundary conditions for the T modes are that ϕ is constant on S, or

$$\frac{\partial \phi}{\partial s} = -\frac{\partial \psi}{\partial \nu} = 0, \tag{47.20}$$

where $d\mathbf{s}$ is a tangent vector to the surface. If we write the differential relation between ϕ and ψ in Cartesian coordinates,

$$\frac{\partial}{\partial x}\phi = \frac{\partial}{\partial y}\psi, \quad \frac{\partial}{\partial y}\phi = -\frac{\partial}{\partial x}\psi, \tag{47.21}$$

we see that these are just the Riemann conditions satisfied by the real and the imaginary parts of an analytic function; that is, because

$$\frac{\partial}{\partial x}(\phi + i\psi) = \frac{1}{i}\frac{\partial}{\partial y}(\phi + i\psi), \tag{47.22}$$

we see that

$$\phi + i\psi = F(x + iy), \tag{47.23}$$

ϕ and ψ are the real and imaginary part of an analytic function.

How general is the existence of T modes? Because they are characterized by a harmonic function (that is, a solution of Laplace's equation), if the region enclosed by S is simply connected, ϕ = constant, and hence there is no electric field. The proof of this assertion follows from the divergence theorem:

$$\int_\Sigma dS\, (\boldsymbol{\nabla}_\perp \phi)^2 = \int_\Sigma dS\, \boldsymbol{\nabla}_\perp \cdot (\phi \boldsymbol{\nabla}_\perp \phi) = \oint_C ds \left(\phi \frac{\partial}{\partial \nu}\phi\right), \tag{47.24}$$

where Σ is the cross-sectional area of the waveguide, and C is the circumference. Because ϕ is constant ($= \phi_c$) on the boundary, we conclude

$$\int_\Sigma dS\, (\boldsymbol{\nabla}_\perp \phi)^2 = \phi_c \oint_C ds \frac{\partial}{\partial \nu}\phi = \phi_c \int_\Sigma dS\, \boldsymbol{\nabla}_\perp^2 \phi = 0, \tag{47.25}$$

which implies that $\boldsymbol{\nabla}_\perp \phi = 0$, $\phi = \phi_c$ everywhere in the waveguide.

FIGURE 47.2
Cross section of waveguide consisting of inner and outer conductors. The waveguide is the shaded region between the two conducting surfaces.

However, the above conclusion does not hold if the bounding surface is disconnected, as shown in Figure 47.2. In this case, ϕ can take on different constant values on the inner and outer conductors. The most familiar example of this geometry, which permits T modes, is that of a coaxial cable.

Let us consider two simple examples of T modes.

1. For parallel plates, lying in the x-z plane, the analytic function is merely a linear one,

$$\phi + i\psi = F(x + iy) = -i(x + iy)V, \quad \text{or} \quad \phi = Vy,\ \psi = -Vx. \tag{47.26}$$

Note that ϕ is constant on the surface and ($\hat{\boldsymbol{\nu}} = -\hat{\mathbf{y}}$)

$$\frac{\partial\psi}{\partial\nu} = -\frac{\partial}{\partial y}Vx = 0, \tag{47.27}$$

as should be the case.

2. For a coaxial cable lying along the z axis, we take

$$\phi + i\psi = C\ln(x + iy) = C\ln\rho e^{i\theta}, \tag{47.28}$$

where θ is the azimuthal angle. Thus

$$\phi = C\ln\rho, \quad \psi = C\theta. \tag{47.29}$$

Again, the boundary condition $\partial\psi/\partial\nu = 0$ is satisfied:

$$\frac{\partial\psi}{\partial\nu} = \frac{\partial\psi}{\partial\rho} = 0. \tag{47.30}$$

47.3 Modes

Let us summarize where we are. There are three types of modes: E, H (with the eigenvalue $\gamma \neq 0$) and T (with $\gamma = 0$). The modes are characterized by the equations

$$(\boldsymbol{\nabla}_\perp^2 + \gamma^2)\begin{Bmatrix}\phi \\ \psi\end{Bmatrix} = 0. \tag{47.31}$$

For each mode, labeled by a, the z-dependence is given by the functions I'_a, V'_a (E) or I''_a, V''_a (H), which satisfy the equations [(47.5a), (47.8a), and (47.5b), (47.8b)]

$$\text{E:} \quad \frac{\partial V'_a}{\partial z} = \frac{i\omega}{c}\left(1 - \frac{\gamma_a^2 c^2}{\omega^2 \epsilon}\right) I'_a, \quad \frac{\partial I'_a}{\partial z} = \frac{i\omega\epsilon}{c} V'_a, \tag{47.32a}$$

$$\text{H:} \quad \frac{\partial V''_a}{\partial z} = \frac{i\omega}{c} I''_a, \quad \frac{\partial I''_a}{\partial z} = \frac{i\omega\epsilon}{c}\left(1 - \frac{\gamma_a^2 c^2}{\omega^2 \epsilon}\right) V''_a. \tag{47.32b}$$

Note that if $\gamma_a = 0$, these two sets of equations are identical, that is, the E and H modes reduce to the common T mode, where

$$\text{T:} \quad \frac{\partial V_a}{\partial z} = \frac{i\omega}{c} I_a, \quad \frac{\partial I_a}{\partial z} = \frac{i\omega\epsilon}{c} V_a, \tag{47.33}$$

which implies the second-order equation

$$\left[\frac{d^2}{dz^2} + \left(\frac{\omega\sqrt{\epsilon}}{c}\right)^2\right] \begin{Bmatrix} V_a \\ I_a \end{Bmatrix} = 0, \tag{47.34}$$

the solutions of which are $e^{\pm ikz}$ with $k = \omega\sqrt{\epsilon}/c$, reflecting the expected speed of light in the medium, $c/\sqrt{\epsilon}$.

Let us henceforth confine ourselves to a situation where there are no T modes. Consider a rectangular cavity, with sides a (along the x axis) and b (along the y axis). For the E modes, where ϕ must vanish on the boundaries, we have

$$\phi_{mn} \propto \sin\frac{m\pi x}{a} \sin\frac{n\pi y}{b}, \quad m, n = 1, 2, 3, \ldots. \tag{47.35}$$

Correspondingly, the eigenvalues are

$$\gamma_{mn}^2 = \left(\frac{m\pi}{a}\right)^2 + \left(\frac{n\pi}{b}\right)^2 \geq \gamma_{11}^2, \tag{47.36}$$

a relation noted in Chapter 21. In contrast, for the H modes, the normal derivative must vanish, so

$$\psi_{mn} \propto \cos\frac{m\pi x}{a} \cos\frac{n\pi y}{b}, \quad m, n = 0, 1, 2, \ldots (m, n \text{ not both } = 0). \tag{47.37}$$

In this case the eigenvalue has a smaller bound,

$$\gamma_{mn}^2 = \left(\frac{m\pi}{a}\right)^2 + \left(\frac{n\pi}{b}\right)^2 \geq \gamma_{10}^2, \text{ when } a > b. \tag{47.38}$$

The z dependence, for either mode, is characterized by the same differential equation,

$$\left(\frac{d^2}{dz^2} + k^2 - \gamma^2\right) \begin{Bmatrix} V' \\ I'' \end{Bmatrix} = 0, \tag{47.39}$$

for which the solutions are

$$e^{\pm ik_z z}, \quad k_z = \sqrt{k^2 - \gamma^2}, \quad k = \frac{\omega}{c}\sqrt{\epsilon} > \gamma. \tag{47.40}$$

We see here the appearance of a cutoff wavenumber; if $k^2 < \gamma^2$ there is no propagation. To have any propagation at all, we must have

$$k > \frac{\pi}{a}, \tag{47.41}$$

or, in terms of the wavelength $\lambda = 2\pi/k$,

$$\lambda < 2a, \tag{47.42}$$

where a is the larger of the two sides of the rectangular cavity.

Let us look at the longest wavelength mode, H_{10}. The corresponding mode function is $\psi_{10} \propto \cos(\pi x/a)$. Of course, $E_z = 0$, and because

$$\mathbf{E}_\perp = \boldsymbol{\nabla}_\perp \times \mathbf{n}\psi, \tag{47.43}$$

$E_x = 0$ as well, leaving only

$$E_y \propto \sin\frac{\pi x}{a}. \tag{47.44}$$

On the other hand,

$$B_z \propto \nabla_\perp^2 \psi \propto \cos\frac{\pi x}{a}, \tag{47.45a}$$

while

$$B_y \propto \frac{\partial}{\partial y}\psi = 0, \quad B_x \propto \frac{\partial}{\partial x}\psi \propto \sin\frac{\pi x}{a}. \tag{47.45b}$$

Of course, the above relation between E and B follows directly from Maxwell's equation

$$\boldsymbol{\nabla}\times\mathbf{E} = \frac{i\omega}{c}\mathbf{B}. \tag{47.46}$$

Now let us consider a circular guide of radius a. The boundary conditions are

$$\text{E:} \qquad \phi = 0 \text{ when } \rho = a, \tag{47.47a}$$

$$\text{H:} \qquad \frac{\partial\psi}{\partial\rho} = 0 \text{ when } \rho = a. \tag{47.47b}$$

The solutions to the mode equations are of the general form

$$\left\{\begin{matrix}\phi\\ \psi\end{matrix}\right\} = e^{im\phi} J_m(\gamma\rho), \tag{47.48}$$

where the boundary conditions are implemented by

$$\text{E:} \quad J_m(\gamma' a) = 0, \tag{47.49a}$$

$$\text{H:} \quad J_m'(\gamma'' a) = 0. \tag{47.49b}$$

The values of the lowest few zeroes for the E modes have been given already in Chapter 21, see (21.111b), (21.112b), (21.113b), for example:

$$\gamma_{01}' a = 2.405, \tag{47.50a}$$

$$\gamma_{02}' a = 5.520, \tag{47.50b}$$

$$\gamma_{11}' a = 3.832. \tag{47.50c}$$

For the H modes, $m = 0$ provides nothing new, because [see (16.71)]

$$J_0'(x) = -J_1(x), \tag{47.51}$$

so $\gamma_{0n}'' = \gamma_{1n}'$, implying

$$\gamma_{01}'' a = 3.832. \tag{47.52}$$

A new situation emerges, however, for $m = 1$. We could estimate the zeroes in this case by using the method described in Section 21.4; a better approximation is given in Problem 47.5. The result is

$$\gamma_{11}'' a = 1.8412. \tag{47.53}$$

Thus the lowest mode for a circular waveguide is the H mode with $m = n = 1$, H_{11}. The cutoff wavelength for such a waveguide is $\lambda_c = 3.4126a$.

47.4 Problems for Chapter 47

1. In regions where there are no currents, show that for E modes,

$$\mathbf{E} = \frac{1}{\epsilon(z)} \boldsymbol{\nabla}\times \left(\boldsymbol{\nabla}\times \frac{ic}{\omega} I' \mathbf{n} \right), \tag{47.54a}$$

$$\mathbf{B} = \boldsymbol{\nabla}\times (I'\mathbf{n}), \tag{47.54b}$$

whereas H-mode fields are given by

$$\mathbf{E} = -\boldsymbol{\nabla}\times (V''\mathbf{n}), \tag{47.55a}$$

$$\mathbf{B} = \boldsymbol{\nabla}\times \left(\boldsymbol{\nabla}\times \frac{ic}{\omega} V'' \mathbf{n} \right). \tag{47.55b}$$

2. Consider the rectangular waveguide sketched in Section 47.3. Suppose the mode functions are normalized by

$$\int_0^a dx \int_0^b dy [\boldsymbol{\nabla}_\perp \psi(x,y)]^2 = \frac{4\pi}{c}. \tag{47.56}$$

Compute the flux of energy from the Poynting vector

$$\mathbf{S} = \frac{c}{8\pi} \operatorname{Re} \mathbf{E}\times\mathbf{H}^*. \tag{47.57}$$

(Why if the factor of 1/2 present?) Determine the flux of energy

$$\int_S dx\, dy\, S_z \tag{47.58}$$

extended over the cross section of the waveguide, in terms of the function $I(z)$ and $V(z)$.

3. This problem refers to a coaxial waveguide. Let the inner radius be b, the outer radius a. Assume $a - b \ll a$, and *find* the cutoff wavenumbers γ', γ'' for the E and H modes. [Hint: You do not need Bessel functions. Look at the differential equations for the mode functions. Which mode has the lowest cutoff wavenumber (other than the T mode)?]

4. Again, consider a waveguide consisting of the region between two concentric perfectly conducting cylinders of radius a and b, $a > b$, but now with no restriction on $a - b$. The region between the cylinders is filled with material of permittivity ϵ and permeability μ, both of which are independent of frequency. Let the common axis of the cylinder be the z axis. Consider a T mode characterized by a function $\phi(x, y)$ satisfying the two-dimensional Laplace's equation

$$\nabla_\perp^2 \phi(x,y) = 0, \tag{47.59}$$

subject to the boundary condition that the potential ϕ be constant on each boundary (but not necessarily the same constant).

   (a) Solve the equation for ϕ, and determine the integration constants as much as possible from the normalization condition

$$\int (d\mathbf{r}_\perp)[\boldsymbol{\nabla}_\perp \phi(\mathbf{r}_\perp)]^2 = \frac{4\pi}{c}. \tag{47.60}$$

(b) Once ϕ is determined, the electric and magnetic fields are given by

$$\mathbf{E}_\perp(\mathbf{r}) = -\boldsymbol{\nabla}_\perp\phi(x,y)V(z), \tag{47.61a}$$
$$\mathbf{H}_\perp(\mathbf{r}) = -\hat{\mathbf{z}}\times\boldsymbol{\nabla}_\perp\phi(x,y)I(z), \tag{47.61b}$$
$$E_z(\mathbf{r}) = H_z(\mathbf{r}) = 0, \tag{47.61c}$$

where the "voltage" and "current" functions are determined by

$$\frac{dI(z)}{dz} = i\frac{\omega\epsilon}{c}V(z), \quad \frac{dV(z)}{dz} = i\frac{\omega\mu}{c}I(z). \tag{47.62}$$

If

$$I(z) = Ie^{i\kappa z}, \quad V(z) = Ve^{i\kappa z}, \tag{47.63}$$

use (47.62) determine the relation between κ and ω, and if $V = ZI$, determine the impedance Z in terms of ϵ and μ.

(c) Use the result (47.61) to compute the electric and magnetic fields in the waveguide in terms of the constant current I. Verify that the appropriate boundary conditions are satisfied:

$$\boldsymbol{\nu}\times\mathbf{E} = \mathbf{0}, \quad \boldsymbol{\nu}\cdot\mathbf{B} = 0, \tag{47.64}$$

where $\boldsymbol{\nu} = \hat{\boldsymbol{\rho}}$ is a normal to the cylindrical surface (ρ is the radial cylindrical coordinate).

(d) Compute the power transmitted by the waveguide,

$$P = \frac{c}{8\pi}\int(d\mathbf{r}_\perp)\hat{\mathbf{z}}\cdot\mathbf{E}\times\mathbf{H}^*, \tag{47.65}$$

where the integration is over the cross section of the waveguide, and verify that this equals $\frac{1}{2}VI$.

(e) Further compute the electric and magnetic energies per unit length,

$$W_E = \frac{1}{16\pi}\int(d\mathbf{r}_\perp)\epsilon|\mathbf{E}|^2, \quad W_H = \frac{1}{16\pi}\int(d\mathbf{r}_\perp)\mu|\mathbf{H}|^2, \tag{47.66}$$

in terms of I, and verify that they are equal.

(f) The surface current density on the cylindrical walls is given by

$$\frac{4\pi}{c}\mathbf{K}_{a,b} = \pm\hat{\boldsymbol{\rho}}\times\mathbf{H}\Big|_{\rho=a,b}. \tag{47.67}$$

Why does the sign reverse for the outer conductor? Compute these densities, and the corresponding currents flowing on the two cylindrical surfaces. What is the total conduction current? Show that this result follows from the Maxwell equation

$$\boldsymbol{\nabla}\times\mathbf{H} = \frac{1}{c}\frac{\partial}{\partial t}\mathbf{D} + \frac{4\pi}{c}\mathbf{j}, \tag{47.68}$$

when integrated over the cross section of the guide.

5. This problem illustrates an approximate method for finding the lowest cutoff wavelength for a circular waveguide, that is, for finding the first zero of J_1'.

(a) Derive the following formula for the eigenvalue from the differential equation satisfied by the mode function ψ from (27.24):

$$(\gamma a)^2 = \frac{\int_0^1 dt\, t \left[\left(\frac{d}{dt}\psi(t) \right)^2 + m^2 \left(\frac{\psi(t)}{t} \right)^2 \right]}{\int_0^1 dt\, t\, (\psi(t))^2}. \tag{47.69}$$

Notice that if this is varied with respect the ψ we recover the differential equation

$$\left[\frac{d^2}{dt^2} + \frac{1}{t}\frac{d}{dt} + (\gamma a)^2 - \frac{m^2}{t^2} \right] \psi = 0. \tag{47.70}$$

(b) We obtain an estimate for γ by choosing a trial function which satisfies the appropriate boundary conditions,

$$\psi(0) = 0, \quad \psi'(1) = 0. \tag{47.71}$$

(Where do these boundary conditions come from?) [Actually, the estimate so obtained is an upper bound to γ. See Chapter 27.] The function $\psi(t) = t(1 - t^2/3)$ satisfies these boundary conditions; show that if it is inserted into the formula for γ it gives

$$\gamma'' a \approx 1.842265. \tag{47.72}$$

(c) Show that if, in the spirit of Section 27.2, we now write the differential equation as

$$\left(\frac{d^2}{dt^2} + \frac{1}{t}\frac{d}{dt} - \frac{1}{t^2} \right) \psi = -(\gamma a)^2 \psi \approx -(\gamma a)^2 \left(t - \frac{t^3}{3} \right), \tag{47.73}$$

we can approximately solve for ψ, up to an overall factor, as

$$\psi \approx 22t - 9t^3 + t^5. \tag{47.74}$$

Put this in the formula for γ to find

$$\gamma'' a = 1.841198, \tag{47.75}$$

in remarkable agreement with the exact answer, 1.841184.

6. Show generally that in source-free regions the electric and magnetic fields can be represented by two Hertz vectors $\mathbf{\Pi}$ and $\mathbf{\Pi}^*$ (note that * does not signify complex conjugation),

$$\mathbf{E} = \boldsymbol{\nabla} \times (\boldsymbol{\nabla} \times \mathbf{\Pi}) + \frac{i\omega}{c} \boldsymbol{\nabla} \times \mathbf{\Pi}^*, \tag{47.76a}$$

$$\mathbf{B} = -\frac{i\omega\epsilon}{c} \boldsymbol{\nabla} \times \mathbf{\Pi} + \boldsymbol{\nabla} \times (\boldsymbol{\nabla} \times \mathbf{\Pi}^*). \tag{47.76b}$$

7. Show that the fields in a cylindrical waveguide are represented by Hertz vectors where

$$\mathbf{\Pi} = -\mathbf{n} \frac{E_z}{k_z^2 - \epsilon\omega^2/c^2}, \tag{47.77a}$$

$$\mathbf{\Pi}^* = -\mathbf{n} \frac{B_z}{k_z^2 - \epsilon\omega^2/c^2}. \tag{47.77b}$$

(47.77c)

What is the connection between $\mathbf{\Pi}$, $\mathbf{\Pi}^*$ and the functions I', I'', V', V''?

8. Use the transmission-line equations (47.32) and the corresponding mode functions that satisfy (47.31) to calculate the normal electromagnetic modes inside a rectangular cavity with edge having length a, b, and c, with the cavity walls being perfect conductors. What are the boundary conditions on the electric and magnetic fields? Show that each mode is characterized by a triplet of integers, $\{l, m, n\}$. Calculate the frequency of each mode, ω_{lmn}. Assuming $a < b < c$, which mode has the lowest frequency? Is the lowest frequency mode unique? In general, is there degeneracy between the modes, that is, do two different modes have the same frequency? In solving this problem this way, one makes a particular choice of the z axis, in terms of which the notion of E and H modes are defined. For example, the E modes have no H_z component. But if the x direction had been singled out, one would have an E′ mode with no H_x component. How is such a mode described in terms of the original E and H modes? Provide as much detail as possible.

9. A resonant cavity consists of a perfectly conducting circular cylinder with conducting endcaps. Calculate the resonant frequencies in terms of Bessel functions, and the explicit form of the fields for the lowest H and E modes.

10. A perfectly conducting cavity consists of a region of empty space V bounded by a conducting surface S. Consider the radiation gauge. Define mode functions $\mathbf{A}_\alpha$, which satisfy

$$\boldsymbol{\nabla}\times(\boldsymbol{\nabla}\times\mathbf{A}_\alpha(\mathbf{r})) = k_\alpha^2\mathbf{A}_\alpha(\mathbf{r}), \quad \boldsymbol{\nabla}\cdot\mathbf{A}_\alpha(\mathbf{r}) = 0, \tag{47.78}$$

with the boundary condition

$$\mathbf{n}\times\mathbf{A}_\alpha(\mathbf{r}) = \mathbf{0} \text{ on } S, \tag{47.79}$$

where $\mathbf{n}$ is the normal to S. If we use real mode functions, show that the following orthonormality condition holds,

$$\int_V (d\mathbf{r})\,\mathbf{A}_\alpha(\mathbf{r})\cdot\mathbf{A}_\beta(\mathbf{r}) = \delta_{\alpha\beta}. \tag{47.80}$$

Show that these mode functions are complete, for divergenceless functions,

$$\sum_\alpha \mathbf{A}_\alpha(\mathbf{r})\mathbf{A}_\alpha(\mathbf{r}') = [\mathbf{1}\delta(\mathbf{r}-\mathbf{r}')]_\perp, \tag{47.81}$$

where the latter notation means that

$$\boldsymbol{\nabla}\cdot[\mathbf{1}\delta(\mathbf{r}-\mathbf{r}')]_\perp = \mathbf{0}. \tag{47.82}$$

11. Now let the vector potential be given in terms of the current by a Green's dyadic $\mathbf{G}$,

$$\mathbf{A}(\mathbf{r},\omega) = \int_V (d\mathbf{r}')\mathbf{G}(\mathbf{r},\mathbf{r}';\omega)\cdot\frac{1}{c}\mathbf{J}(\mathbf{r}',\omega). \tag{47.83}$$

Show that, in vacuum,

$$\left[\boldsymbol{\nabla}\times(\boldsymbol{\nabla}\times\quad) - \left(\frac{\omega}{c}\right)^2\right]\mathbf{G}(\mathbf{r},\mathbf{r}';\omega) = 4\pi[\mathbf{1}\delta(\mathbf{r}-\mathbf{r}')]_\perp. \tag{47.84}$$

12. Prove that

$$\mathbf{G}(\mathbf{r},\mathbf{r}';\omega) = 4\pi \sum_\alpha \frac{\mathbf{A}_\alpha(\mathbf{r})\mathbf{A}_\alpha(\mathbf{r}')}{k_\alpha^2 - (\omega/c)^2}. \tag{47.85}$$

Show that, with $\omega \to \omega + 0i$, the resulting Green's dyadic is the retarded function,

$$\mathbf{G}(\mathbf{r},t;\mathbf{r}',t') = \begin{cases} \mathbf{0}, & t-t'<0, \\ 4\pi \sum_\alpha \mathbf{A}_\alpha(\mathbf{r})\mathbf{A}_\alpha(\mathbf{r}')\frac{c}{k_\alpha}\sin[k_\alpha c(t-t')], & t-t'>0. \end{cases} \tag{47.86}$$

13. A current $\mathbf{J}(\mathbf{r},t)$ is turned on and eventually off. Demonstrate that the total amount of electromagnetic energy created in the cavity is

$$\Delta E = \int (d\mathbf{r})\,dt\,(d\mathbf{r}')\,dt'\,\frac{1}{c}\mathbf{J}(\mathbf{r},t)\cdot\frac{\partial}{\partial t}\mathbf{G}(\mathbf{r},t;\mathbf{r}',t')\cdot\frac{1}{c}\mathbf{J}(\mathbf{r}',t'). \tag{47.87}$$

14. Show that ΔE can also be presented as

$$\Delta E = 2\pi \sum_\alpha \left| \int (d\mathbf{r})\,dt\,\mathbf{A}_\alpha(\mathbf{r})e^{ik_\alpha ct}\cdot\mathbf{J}(\mathbf{r},t)\right|^2. \tag{47.88}$$

15. A resonant cavity, composed of a homogeneous dielectric medium enclosed by perfectly conducting walls, allows an infinite number of electromagnetic oscillation modes. Prove that the wavenumber associated with a mode is given by

$$k^2 = \frac{\int (d\mathbf{r})\,(\boldsymbol{\nabla}\times\mathbf{E})^2}{\int (d\mathbf{r})\,\mathbf{E}^2}, \quad \text{or} \tag{47.89a}$$

$$k^2 = \frac{\int (d\mathbf{r})\,(\boldsymbol{\nabla}\times\mathbf{H})^2}{\int (d\mathbf{r})\,\mathbf{H}^2}. \tag{47.89b}$$

Verify that the first expression is stationary with respect to variations that do not violate the boundary condition, while the second form is stationary with respect to unrestricted variations.

16. A resonant cavity is constructed from a uniform waveguide by placing perfectly conducting walls at the planes $z = 0$ and $z = L$. Determine the characteristic wavenumbers of the cavity in terms of the cutoff wavenumbers of the various guide modes. Prove that the skin effect dissipation Q for a mode of this cavity is related to the Q of the guide by

$$\text{H mode}: \quad \frac{1}{Q_{\text{cavity}}} = \frac{1}{Q_{\text{guide}}} + 2\frac{\delta}{L}\left(\frac{\kappa}{k}\right)^2, \tag{47.90a}$$

$$\text{E mode}: \quad \frac{1}{Q_{\text{cavity}}} = \frac{1}{Q_{\text{guide}}} + 2\frac{\delta}{L}, \tag{47.90b}$$

where δ is the skin depth, and κ is the guide propagation constant for the mode in question. [For an oscillator, Q is defined as the ratio of the resonant frequency to the dissipation constant, $Q = \omega_0/\gamma$. Here, this is 2π times the average energy stored in the cavity divided by the power lost per cycle.]

17. (a) A waveguide is filled with a homogeneous dielectric of permittivity ϵ_1 in the half space $z < 0$, and a homogeneous dielectric of permittivity ϵ_2 in the half space $z > 0$. Prove that the decomposition into independent modes is still valid. Discuss the effect of the dielectric discontinuity on the modes.

(b) A dielectric medium of permittivity ϵ_2 and length L is inserted into a waveguide between two other media of permittivity ϵ_1 and ϵ_3, respectively. Determine under what condition waves will be transmitted without reflection from medium 1 into medium 3.

(c) Now, for the same geometry, suppose $\epsilon_1 = \epsilon_3 > \epsilon_2$, and the frequency is so chosen that $\kappa = \sqrt{\omega^2\epsilon - \gamma^2}$ is real in the middle slab, which occupies the region $0 < z < L$, but is imaginary outside this range. The longitudinal mode functions for this dielectric waveguide are of the form $\cos\kappa(z - L)$ for $0 < z < L$ and decrease exponentially, as given by the imaginary part of κ for $z \to -\infty$ and $z \to +\infty$. What are the implications of the boundary conditions at $z = 0$ and $z = a$? What are $I(z)$ and $V(z)$ in the three regions? Consider both polarizations.

18. For a given frequency ω, prove the complex Poynting vector theorem for ϵ, μ real:

$$\boldsymbol{\nabla}\cdot(\mathbf{E}(\omega)\times\mathbf{H}^*(\omega)) = i\frac{\omega}{c}(\mu|\mathbf{H}(\omega)|^2 - \epsilon|\mathbf{E}(\omega)|^2), \tag{47.91}$$

and the energy theorem (including dispersion)

$$\begin{aligned}&\boldsymbol{\nabla}\cdot\left(\frac{\partial\mathbf{E}(\omega)}{\partial\omega}\times\mathbf{H}^*(\omega) + \mathbf{E}^*(\omega)\times\frac{\partial\mathbf{H}(\omega)}{\partial\omega}\right)\\ &= \frac{i}{c}\left[\left(\frac{d}{d\omega}\omega\epsilon(\omega)\right)|\mathbf{E}(\omega)|^2 + \left(\frac{d}{d\omega}\omega\mu(\omega)\right)|\mathbf{H}(\omega)|^2\right] \equiv \frac{16\pi}{c}iU,\end{aligned} \tag{47.92}$$

The transmission line form of the Poynting theorem is obtained by integrating over the cross section of the waveguide, using a suitable normalization, so show, for a single mode,

$$\frac{d}{dz}\left[\frac{1}{2}V(z)I(z)^*\right] = \frac{d}{dz}P = 2i(W_H - W_E), \tag{47.93a}$$

where W_H and W_E are the magnetic and electric linear energy densities, and P is the complex power. If we disregard dispersion, we obtain from the energy theorem

$$\frac{d}{dz}\left\{\frac{1}{2}\left[\frac{\partial V(z)}{\partial\omega}I(z)^* + V(z)^*\frac{\partial I(z)}{\partial\omega}\right]\right\} = 2i(W_H + W_E) = 2iW. \tag{47.93b}$$

19. How are the Poynting theorem (47.91) and the energy theorem (47.92) modified if no relation is assumed between **D** and **E** and between **B** and **H**?

20. *Uniqueness theorem:* The electric and magnetic fields in a region are completely determined by specifying the electric (or magnetic) field tangential to the closed surface bounding that region. This generalizes the uniqueness theorem for electrostatics proved in Chapter 1. Prove this by defining $\mathcal{E} = \mathbf{E}_1 - \mathbf{E}_2$ and $\mathcal{H} = \mathbf{H}_1 - \mathbf{H}_2$, which satisfy, for harmonic time dependence

$$\boldsymbol{\nabla}\times\mathcal{E} = i\frac{\omega\mu}{c}\mathcal{H}, \quad \boldsymbol{\nabla}\times\mathcal{H} = -i\frac{\omega\epsilon}{c}\mathcal{E}, \tag{47.94}$$

and on the boundary surface, with normal $\mathbf{n}$, $\mathbf{n}\times\mathcal{E} = \mathbf{0}$ and $\mathbf{n}\times\mathcal{H} = \mathbf{0}$. Then show from the energy theorem (47.92) that

$$\mathcal{E} = \mathbf{0}, \quad \mathcal{H} = \mathbf{0}, \tag{47.95}$$

provided both $d(\omega\epsilon)/d\omega$ and $d(\omega\mu)/d\omega$ are positive, which is usually the case. However, this argument depends on continuity in ω, so it can and does fail for the normal modes in a cavity.

21. Use the results of Problem 18 to describe a propagating wave:

$$V(z) = Ve^{i\kappa z}, \quad I(z) = Ie^{i\kappa z}, \quad V = ZI. \tag{47.96}$$

Show that the complex power is then real and independent of z:

$$P = \frac{1}{2}VI^* = \frac{1}{2}Z|I|^2, \tag{47.97}$$

Use the energy theorem to prove

$$P = vW, \tag{47.98}$$

where the velocity of the wave is

$$v = \frac{d\omega}{dk} = c'\frac{dk}{d\kappa} = c'\frac{\kappa}{k}, \quad k = \frac{\omega}{c'}, \quad c' = \frac{c}{\sqrt{\epsilon\mu}}, \tag{47.99}$$

the expected group velocity that vanishes at the cutoff frequency.

22. To discuss further the meaning of the transmission line equations (47.32), let us include the permeability μ as well as the permittivity ϵ, and write the equations in practical SI units, following the conversions sketched in the Appendix (noting that it is then **H** and not **B** that is derived by a gradient of I:

$$\text{E}: \quad \frac{dI}{dz} = i\omega\epsilon V, \quad \frac{dV}{dz} = \left(i\omega\mu + \frac{\gamma^2}{i\omega\epsilon}\right) I, \tag{47.100a}$$

$$\text{H}: \quad \frac{dI}{dz} = \left(i\omega\epsilon + \frac{\gamma^2}{i\omega\mu}\right) V, \quad \frac{dV}{dz} = i\omega\mu I. \tag{47.100b}$$

(a) If we define a series impedance per unit length as

$$-Z_s = i\omega L_s + \frac{1}{i\omega C_s}, \tag{47.101a}$$

and a corresponding shunt admittance by

$$-Y_\perp = i\omega C_\perp + \frac{1}{i\omega L_\perp}, \tag{47.101b}$$

we see the following distributed properties:

$$C_\perp = \epsilon, \quad \begin{cases} \text{E}: \; L_\perp = \infty, \\ \text{H}: L_\perp = \mu/\gamma^2, \end{cases} \tag{47.102a}$$

$$L_s = \mu, \quad \begin{cases} \text{E}: C_s = \epsilon/\gamma^2, \\ \text{H}: \; C_s = \infty, \end{cases} \tag{47.102b}$$

Draw a circuit diagram illustrating these lumped quantities.

(b) Now assume the mode functions are chosen to satisfy the following orthonormality relations:

$$\text{E modes}: \quad \int dS\, \boldsymbol{\nabla}_\perp \phi_a^* \cdot \boldsymbol{\nabla}_\perp \phi_b = \delta_{ab}, \tag{47.103a}$$

$$\text{H modes}: \quad \int dS\, \boldsymbol{\nabla}_\perp \psi_a^* \cdot \boldsymbol{\nabla}_\perp \psi_b = \delta_{ab}, \tag{47.103b}$$

No statement of orthogonality between E and H modes is required.

(c) Now consider the linear energy density, the energy per unit length, derived from the energy density, without dispersion,:

$$U = \frac{\epsilon}{2}E^2 + \frac{\mu}{2}H^2. \tag{47.104}$$

Show that the linear energy density arising from E_z is

$$W_{E_z} = \frac{1}{4}\sum_a \frac{1}{\omega^2 (C_s)_a}|I'_a(z)|^2, \tag{47.105a}$$

where the sum is over E modes, and that arising from H_z is

$$W_{H_z} = \frac{1}{4}\sum_a \frac{1}{\omega^2 (L_\perp)_a}|V''_a(z)|^2, \tag{47.105b}$$

For the transverse fields, show that is no mixing between E and H modes, and that the linear energy densities are

$$W_{\mathbf{E}_\perp} = \frac{1}{4}\sum_a C_\perp \left[|V'_a(z)|^2 + |V''_a(z)|^2\right] \tag{47.106a}$$

$$W_{\mathbf{H}_\perp} = \frac{1}{4}\sum_a L_s \left[|I'_a(z)|^2 + |I''_a(z)|^2\right]. \tag{47.106b}$$

Show that the complex power, or the flux of energy along the waveguide is

$$P = \frac{1}{2}\int dS\, \hat{\mathbf{z}}\cdot(\mathbf{E}\times\mathbf{H}^*) = \frac{1}{2}\sum_a [V'_a(z)I'_a(z)^* + V''_a(z)I''_a(z)^*]. \tag{47.107}$$

48

Scattering by Small Obstacles

When a plane electromagnetic wave interacts with a macroscopic dielectric, we talk of reflection and refraction. For the interaction of the wave with a small object, the appropriate word is scattering. Here, small refers to the size of the object in comparison to the wavelength of the radiation. In this chapter, we will consider the elementary theory of scattering—in the following, a systematic approach will be given.

48.1 Thomson Scattering

As a first illustration, we consider a single charge e, of mass m, which accelerates due to an applied electric field such as is present in a light wave:

$$m\frac{d\mathbf{v}}{dt} = e\mathbf{E}, \tag{48.1}$$

(as long as $|\mathbf{v}/c| \ll 1$). An accelerated charge radiates, which gives rise to the "scattered" radiation; the power radiated is given by the Larmor formula, (35.22),

$$P_{\text{scatt}} = \frac{2}{3}\frac{e^2}{c^3}(\dot{\mathbf{v}})^2 = \frac{2}{3}\frac{e^2}{c^3}\left(\frac{e}{m}\mathbf{E}\right)^2 = \frac{2}{3}\frac{e^4}{m^2c^3}\mathbf{E}^2. \tag{48.2}$$

This scattered power is to be compared to the incoming power per unit area of the plane wave, which is, in vacuum, given by the magnitude of the Poynting vector,

$$|\mathbf{S}| = \frac{c}{4\pi}|\mathbf{E}\times\mathbf{B}| = \frac{c}{4\pi}|\mathbf{E}|^2. \tag{48.3}$$

The ratio of these two quantities defines the cross section,

$$\sigma = \frac{P_{\text{scatt}}}{|\mathbf{S}|}, \tag{48.4}$$

which is the effective area presented by the obstacle to the wave. For the situation considered here, the cross section is

$$\sigma = \frac{8\pi}{3}\left(\frac{e^2}{mc^2}\right)^2, \tag{48.5}$$

the so-called Thomson[1] cross section. For an electron, this cross section is conveniently expressed in terms of the length, r_0,

$$r_0 = \frac{e^2}{mc^2} = 2.8179\times10^{-13}\text{cm}, \tag{48.6}$$

[1] J. J. Thomson.

DOI: 10.1201/9781003057369-48

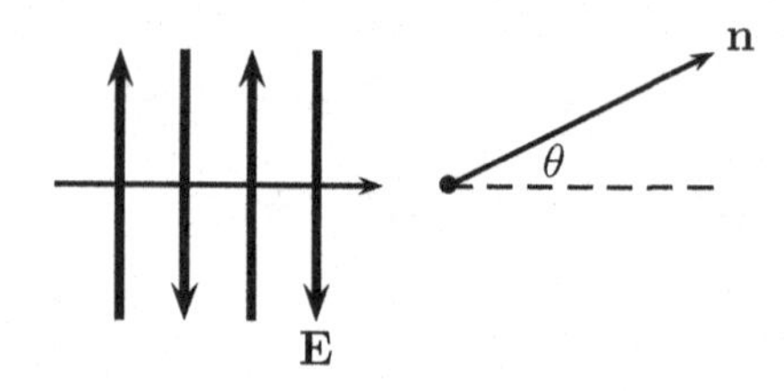

FIGURE 48.1
Scattering of a wave polarized in the scattering plane.

which, for historical reasons only, goes by the name "classical radius of the electron."[2] Numerically, the Thomson cross section for the electron is about

$$\sigma \approx \frac{2}{3}\times 10^{-24}\mathrm{cm}^2. \tag{48.7}$$

What is the angular distribution of the radiation scattered by the charge? We recall the nonrelativistic expression (35.20),

$$\frac{dP_{\text{scatt}}}{d\Omega} = \frac{1}{4\pi}\frac{e^2}{c^3}(\mathbf{n}\times\dot{\mathbf{v}})^2, \tag{48.8}$$

which gives the power radiated in the direction $\mathbf{n}$ per unit solid angle. Using (48.1), we rewrite this as

$$\frac{dP_{\text{scatt}}}{d\Omega} = \frac{1}{4\pi}\frac{e^4}{m^2c^3}(\mathbf{n}\times\mathbf{E})^2, \tag{48.9}$$

or, introducing $\mathbf{e}$ as a unit vector along the direction of $\mathbf{E}$,

$$\frac{dP_{\text{scatt}}}{d\Omega} = \frac{1}{4\pi}\frac{e^4}{m^2c^3}\mathbf{E}^2[1-(\mathbf{n}\cdot\mathbf{e})^2]. \tag{48.10}$$

There are two distinct possibilities here: If $\mathbf{e}$ is perpendicular to the scattering plane defined by $\mathbf{n}$ and the initial direction of propagation, then $\mathbf{n}\cdot\mathbf{e} = 0$. If $\mathbf{e}$ lies in that plane, then $|\mathbf{n}\cdot\mathbf{e}| = \sin\theta$, where, as Fig. 48.1 shows, θ is the scattering angle. Therefore, for the two choices of $\mathbf{e}$, the factor in the scattered power is

$$1-(\mathbf{n}\cdot\mathbf{e})^2 = \begin{cases} 1, & \mathbf{e}\perp \text{scattering plane,} \\ \cos^2\theta, & \mathbf{e}\parallel \text{scattering plane.} \end{cases} \tag{48.11}$$

Notice that there is no scattering at $\theta = \pi/2$ for the second choice, which just re-expresses the fact that there is no radiation emitted in the direction of the acceleration. If the incoming radiation is unpolarized, we have the average of the two possibilities given in (48.11) so that

$$1-(\mathbf{n}\cdot\mathbf{e})^2 = \frac{1+\cos^2\theta}{2}. \tag{48.12}$$

The differential cross section, the effective area for scattering into a given element of solid angle, is defined by

$$\frac{d\sigma}{d\Omega} = \frac{dP_{\text{scatt}}/d\Omega}{|\mathbf{S}|}. \tag{48.13}$$

[2]This is in no sense the size of an electron, which to an astonishing degree of accuracy, has no size at all—in the quantum mechanical sense, it is a point particle. See also (48.55).

For an unpolarized wave, the differential Thomson cross section is then

$$\frac{d\sigma}{d\Omega} = \left(\frac{e^2}{mc^2}\right)^2 \frac{1+\cos^2\theta}{2}. \tag{48.14}$$

The total Thomson cross section, (48.5), is recovered when (48.14) is integrated over all solid angles,

$$\sigma = \int \frac{d\sigma}{d\Omega} d\Omega. \tag{48.15}$$

48.2 Scattering by a Bound Charge

The preceding section described scattering by a free particle. We now suppose, instead, that the "electron" is bound by a damped harmonic oscillator force. As we discussed in Section 5.2, this interaction is characterized by a natural frequency, ω_0, and a damping constant, γ. The corresponding equation of motion of the electron is given by (5.15):

$$m\frac{d^2\mathbf{r}}{dt^2} + m\omega_0^2\mathbf{r} + m\gamma\frac{d\mathbf{r}}{dt} = e\mathbf{E}. \tag{48.16}$$

Suppose the electric field is due to an incoming light wave of a definite frequency ω,

$$\mathbf{E}(t) = \text{Re } \mathbf{E}e^{-i\omega t} = \frac{1}{2}\left(\mathbf{E}e^{-i\omega t} + \mathbf{E}^* e^{i\omega t}\right), \tag{48.17}$$

so the deviation of the particle from the center of the harmonic force is given by (5.17):

$$\mathbf{r}(t) = \frac{e}{m} \text{Re } \frac{\mathbf{E}e^{-i\omega t}}{\omega_0^2 - \omega^2 - i\gamma\omega}. \tag{48.18}$$

The corresponding radiated power,

$$P_{\text{scatt}} = \frac{2}{3}\frac{e^2}{c^3}(\dot{\mathbf{v}})^2 \tag{48.19}$$

involves

$$\ddot{\mathbf{r}} = -\frac{e}{m} \text{Re } \frac{\omega^2\mathbf{E}e^{-i\omega t}}{\omega_0^2 - \omega^2 - i\gamma\omega}. \tag{48.20}$$

We are interested, not in the instantaneous scattered power, but rather in its time average over one cycle. For a typical time varying quantity,

$$A(t) = \text{Re } Ae^{-i\omega t} = \frac{1}{2}(Ae^{-i\omega t} + A^* e^{i\omega t}), \tag{48.21}$$

the time average of its square is

$$\overline{(A(t))^2} = \frac{1}{2}|A|^2. \tag{48.22}$$

Thus, the time averaged power scattered is, from (48.19) and (48.20),

$$\overline{P_{\text{scatt}}} = \frac{2}{3}\frac{e^2}{c^3}\frac{1}{2}\frac{e^2}{m^2}\omega^4 \frac{|\mathbf{E}|^2}{(\omega_0^2 - \omega^2)^2 + (\gamma\omega)^2}, \tag{48.23}$$

while the average incident energy flux is

$$\overline{|\mathbf{S}|} = \frac{c}{4\pi}\overline{\mathbf{E}^2} = \frac{c}{4\pi}\frac{1}{2}|\mathbf{E}|^2, \tag{48.24}$$

both expressed in terms of the complex amplitude $\mathbf{E}$. The resulting cross section is

$$\sigma = \frac{8\pi}{3}\left(\frac{e^2}{mc^2}\right)^2 \frac{\omega^4}{(\omega_0^2 - \omega^2)^2 + (\gamma\omega)^2}. \tag{48.25}$$

This reduces to the Thomson cross section for a free particle when both ω_0 and γ are zero. More realistically, the above cross section approaches the Thomson limit,

$$\sigma \to \sigma_{\rm Th}, \tag{48.26}$$

when the frequency of light is large compared to the natural frequencies of the bound system,

$$\omega \gg \omega_0, \gamma. \tag{48.27}$$

Physically, in this limit, the time scale is set by the period of the incident light; if this is very short, the electron does not have time to be influenced by internal forces so it behaves as though it were free.

In the opposite limit of very low frequencies,

$$\omega \ll \omega_0, \tag{48.28}$$

the cross section (48.25) approaches

$$\sigma = \sigma_{\rm Th}\frac{\omega^4}{\omega_0^4}, \tag{48.29}$$

which is called the Rayleigh cross section. In this domain, the highest frequencies are scattered the most strongly. This frequency behavior is presumably one of the reasons the sky is blue.

We have the situation of *resonant* scattering when $\omega = \omega_0$, at which point the cross section becomes

$$\sigma = \sigma_{\rm Th}\frac{\omega_0^2}{\gamma^2}. \tag{48.30}$$

For small damping, $\omega_0 \gg \gamma$, (48.30) is much larger than the Thomson cross section. Of course, γ cannot be zero; energy is necessarily dissipated because, if for no other reason, the particle radiates due to its acceleration (recall Section 38.3). We will return to this issue shortly.

48.3 Scattering by a Dielectric Sphere

We do not require a microscopic description in order to calculate the scattering of electromagnetic waves by small objects. The scattering by such an object may be calculated from the dipole radiation formula (35.27),

$$\overline{P_{\rm rad}} = \frac{2}{3}\frac{1}{c^3}\overline{(\ddot{\mathbf{d}})^2} \to \frac{2}{3}\frac{\omega^4}{c^3}\overline{\mathbf{d}^2}. \tag{48.31}$$

As an example, recall from (25.81) that a dielectric sphere, of radius a, in a static electric field $\mathbf{E}$ acquires a dipole moment

$$\mathbf{d} = \alpha \mathbf{E}, \quad \alpha = \frac{\epsilon - 1}{\epsilon + 2} a^3, \tag{48.32}$$

in terms of the electric polarizability. This formula is applicable to a radiation field whenever $\mathbf{E}$ varies slowly on the scale set by the radius of the sphere, that is, if

$$\lambda\!\!\!^{-} \gg a, \tag{48.33}$$

where $\lambda\!\!\!^{-} = c/\omega$. Then the radiated power is

$$\overline{P_{\text{rad}}} = \frac{2}{3} \frac{\omega^4}{c^3} \alpha^2 \overline{\mathbf{E}^2} \tag{48.34}$$

so, by (48.24), the total cross section for scattering is

$$\sigma = \frac{8\pi}{3} \frac{1}{\lambda\!\!\!^{-4}} \alpha^2. \tag{48.35}$$

Again we see the ω^4 dependence of the cross section characteristic of the scattering of long wavelength radiation.

48.4 Radiation Damping

We now return to the oscillator and consider the damping due to radiation. From (36.8b), the radiation reaction force on the charged particle can be identified from $P = -\mathbf{F} \cdot \mathbf{v}$ to be

$$\mathbf{F} = \frac{2}{3} \frac{e^2}{c^3} \ddot{\mathbf{v}}. \tag{48.36}$$

For a definite frequency, this becomes

$$-\frac{2}{3} \frac{e^2}{c^3} \omega^2 \mathbf{v} = -m\gamma_r \mathbf{v}, \tag{48.37}$$

where here γ_r is the radiative part of the dissipation, which must be present in order that energy be conserved. It coincides with the previously determined damping constant for radiation by a harmonic oscillator, (38.47), as long as the characteristic radiated frequencies are near the resonant frequency ω_0. Recognizing that there may be other forms of dissipation (for example, collisions), we write the total dissipation constant as

$$\gamma = \frac{2}{3} \frac{e^2}{mc^3} \omega^2 + \gamma_d \equiv \gamma_r + \gamma_d. \tag{48.38}$$

The energy removed from the incident electromagnetic field feeds both γ_r and γ_d. The rate of energy transfer from this field is, according to (48.18),

$$\overline{(e\mathbf{E} \cdot \mathbf{v})} = \overline{e\text{Re}(\mathbf{E}e^{-i\omega t}) \cdot \text{Re}\left(\frac{e}{m} \frac{(-i\omega)\mathbf{E}e^{-i\omega t}}{\omega_0^2 - \omega^2 - i\gamma\omega} \right)}. \tag{48.39}$$

Generically, the time average of such a product is given by

$$\overline{\operatorname{Re} A(t) \operatorname{Re} B(t)} = \overline{\frac{1}{2}\left(Ae^{-i\omega t} + A^* e^{i\omega t}\right)\frac{1}{2}\left(Be^{-i\omega t} + B^* e^{i\omega t}\right)}$$
$$= \frac{1}{4}(AB^* + A^* B) = \frac{1}{2}\operatorname{Re} A^* B, \tag{48.40}$$

implying the power transferred to the oscillator,

$$\overline{(e\mathbf{E}\cdot\mathbf{v})} = \frac{e^2}{m}\frac{1}{2}|\mathbf{E}|^2 \operatorname{Re}\frac{-i\omega}{\omega_0^2 - \omega^2 - i\gamma\omega}. \tag{48.41}$$

Consequently, the total power removed from the incident field (not just scattered), is

$$\overline{P_{\text{tot}}} = \frac{e^2}{m}\frac{1}{2}|\mathbf{E}|^2 \frac{\gamma\omega^2}{(\omega_0^2 - \omega^2)^2 + (\gamma\omega)^2}, \tag{48.42}$$

which must exceed the scattered power, (48.23),

$$\overline{P_{\text{tot}}} \geq \overline{P_{\text{scatt}}}. \tag{48.43}$$

The corresponding total cross section is

$$\frac{\overline{P_{\text{tot}}}}{\overline{|\mathbf{S}|}} = \sigma_{\text{tot}} = 4\pi\frac{e^2}{mc}\frac{\gamma\omega^2}{(\omega_0^2 - \omega^2)^2 + (\gamma\omega)^2}. \tag{48.44}$$

If we use (48.38),

$$\frac{e^2}{mc}\omega^2 = \frac{3}{2}c^2\gamma_r, \tag{48.45}$$

we can rewrite (48.44) as

$$\sigma_{\text{tot}} = 4\pi\frac{3}{2}c^2\frac{\gamma_r\gamma}{(\omega_0^2 - \omega^2)^2 + (\gamma\omega)^2}. \tag{48.46}$$

At resonance, $\omega = \omega_0$, the total cross section becomes

$$\sigma_{\text{tot}}(\omega_0) = 4\pi\frac{3}{2}\left(\frac{c}{\omega_0}\right)^2\frac{\gamma_r}{\gamma} \tag{48.47}$$

and therefore if we write $c/\omega_0 = \lambda\!\!\!^-_0$, and note that

$$\frac{\gamma_r}{\gamma} \leq 1, \tag{48.48}$$

we have the inequality

$$\sigma_{\text{tot}}(\omega_0) \leq 6\pi\lambda\!\!\!^-_0{}^2. \tag{48.49}$$

We now break up σ_{tot} into two pieces corresponding to the two channels for energy loss, given in (48.38),

$$\sigma_{\text{tot}} = \sigma_{\text{scatt}} + \sigma_{\text{diss}}, \tag{48.50}$$

where the scattering cross section is identified as

$$\sigma_{\text{scatt}} = 4\pi\frac{3}{2}c^2\frac{\gamma_r^2}{(\omega_0^2 - \omega^2)^2 + (\gamma\omega)^2} = \frac{8\pi}{3}\left(\frac{e^2}{mc^2}\right)^2\frac{\omega^4}{(\omega_0^2 - \omega^2)^2 + (\gamma\omega)^2} \tag{48.51a}$$

[which is the same result found in (48.25)], while the dissipation cross section is

$$\sigma_{\rm diss} = 4\pi\frac{3}{2}c^2\frac{\gamma_r\gamma_d}{(\omega_0^2-\omega^2)^2+(\gamma\omega)^2}. \tag{48.51b}$$

If we ignore γ_d, and consider the high frequency limit $\omega \gg \omega_0$, the scattering cross section becomes

$$\sigma_{\rm scatt} = \sigma_{\rm Th}\frac{1}{1+(\gamma/\omega)^2} = \frac{\sigma_{\rm Th}}{1+\left[(2/3)(e^2/mc^2)(1/\lambda\!\!\!^{-})\right]^2}. \tag{48.52}$$

If $\lambda\!\!\!^{-} \gg e^2/mc^2$, the correction to the cross section is very small. But if we were to take literally the limit of very short wavelengths,

$$\lambda\!\!\!^{-} \ll \frac{e^2}{mc^2} = r_0, \tag{48.53}$$

the cross section would behave as

$$\sigma_{\rm scatt} \sim \frac{\frac{8\pi}{3}r_0^2}{\left(\frac{2}{3}r_0\frac{\omega}{c}\right)^2} = 6\pi\lambda\!\!\!^{-2}, \tag{48.54}$$

which has the same form as the upper bound (48.49). This limit is completely unbelievable, however, since quantum effects become significant already at the considerably larger distance, the Compton[3] wavelength,

$$\frac{\hbar}{mc} = 137\frac{e^2}{mc^2} = 2.426\times10^{-10}\,\text{cm}. \tag{48.55}$$

48.5 Problems for Chapter 48

1. Calculate the scattering cross section by a small, perfectly conducting sphere of radius $a \ll \lambda\!\!\!^{-}$. Notice that the result does not coincide with the $\epsilon\to\infty$ limit of (48.35), for the latter refers only to electric dipole radiation, while for a perfect conductor magnetic dipole radiation is important as well. See Problems 30.2 and 48.2.

2. (a) From the discontinuity conditions (28.19a) [for the surface current $\mathbf{K}$] and (13.58b) [for the surface charge density σ] show the conservation of charge,

$$\boldsymbol{\nabla}_\perp\cdot\mathbf{K} - i\omega\sigma = 0. \tag{48.56}$$

   (b) Calculate the leading contributions to $\mathbf{K}$ and σ in the long wavelength limit, $ka \ll 1$, and, from these, calculate the electric and magnetic dipole moments, $\mathbf{d}$ and $\boldsymbol{\mu}$, used in Problem 48.1. The results are

$$\mathbf{d} = a^3\mathbf{E}_{\rm inc}(\mathbf{r}=\mathbf{0}), \tag{48.57a}$$

$$\boldsymbol{\mu} = -\frac{1}{2}a^3\mathbf{B}_{\rm inc}(\mathbf{r}=\mathbf{0}), \tag{48.57b}$$

   which display the electric and magnetic polarizabilities of a perfectly conducting sphere.

[3] Arthur H. Compton, 1892–1962.

49

Partial-Wave Analysis of Scattering

A general approach to scattering is the so-called method of partial waves, in which there is a decomposition into modes of a definite angular momentum l. This technique is most commonly seen in a description of quantum-mechanical scattering, but we will see that the concept proves useful in the classical domain of electromagnetic scattering as well.

49.1 Mode Decomposition

We apply the formalism of Chapter 47 to the case of scattering. We begin by making a $2+1$ dimensional break up for a spherical geometry, where the transverse directions are represented by the spherical angles θ, ϕ, and the longitudinal one by the radial coordinate r. The Maxwell equations

$$\boldsymbol{\nabla}\times\mathbf{B} = -i\frac{\omega}{c}\epsilon\mathbf{E} + \frac{4\pi}{c}\mathbf{J}, \quad -\boldsymbol{\nabla}\times\mathbf{E} = -i\frac{\omega}{c}\mathbf{B}, \tag{49.1}$$

are projected into transverse and longitudinal parts by taking the cross product, and the dot product, with $\mathbf{r}$, respectively. Thus, for example, we easily see

$$\mathbf{r}\times(\boldsymbol{\nabla}\times\mathbf{B}) = \boldsymbol{\nabla}(\mathbf{r}\cdot\mathbf{B}) - \mathbf{B} - r\frac{\partial}{\partial r}\mathbf{B}, \tag{49.2a}$$

or, since this is purely transverse,

$$\mathbf{r}\times(\boldsymbol{\nabla}\times\mathbf{B}) = \boldsymbol{\nabla}_\perp(\mathbf{r}\cdot\mathbf{B}) - \mathbf{B}_\perp - r\frac{\partial}{\partial r}\mathbf{B}_\perp. \tag{49.2b}$$

If we then divide by r, the transverse parts of the two Maxwell equations above are

$$\boldsymbol{\nabla}_\perp B_r - \left(\frac{\partial}{\partial r} + \frac{1}{r}\right)\mathbf{B}_\perp = -\frac{i\omega}{c}\epsilon\frac{\mathbf{r}}{r}\times\mathbf{E} + \frac{4\pi}{c}\frac{\mathbf{r}}{r}\times\mathbf{J}_\perp, \tag{49.3a}$$

$$\boldsymbol{\nabla}_\perp E_r - \left(\frac{\partial}{\partial r} + \frac{1}{r}\right)\mathbf{E}_\perp = \frac{i\omega}{c}\frac{\mathbf{r}}{r}\times\mathbf{B}. \tag{49.3b}$$

The longitudinal components are simplified according to

$$\frac{1}{r}\mathbf{r}\cdot\boldsymbol{\nabla}\times\mathbf{B} = \frac{1}{r}\mathbf{r}\times\boldsymbol{\nabla}\cdot\mathbf{B} = \mathbf{r}\times\boldsymbol{\nabla}\cdot\frac{1}{r}\mathbf{B} = -\boldsymbol{\nabla}\times\frac{\mathbf{r}}{r}\cdot\mathbf{B} = -\boldsymbol{\nabla}\cdot\frac{\mathbf{r}}{r}\times\mathbf{B}, \tag{49.4}$$

to give

$$-\boldsymbol{\nabla}_\perp\cdot\frac{\mathbf{r}}{r}\times\mathbf{B} = -\frac{i\omega}{c}\epsilon E_r + \frac{4\pi}{c}J_r, \tag{49.5a}$$

$$-\boldsymbol{\nabla}_\perp\cdot\frac{\mathbf{r}}{r}\times\mathbf{E} = \frac{i\omega}{c}B_r. \tag{49.5b}$$

DOI: 10.1201/9781003057369-49

We can combine these equations to obtain

$$\left(\frac{\partial}{\partial r}+\frac{1}{r}\right)\mathbf{B}_\perp=\frac{i\omega\epsilon}{c}\left(\mathbf{1}_\perp+\frac{c^2}{\omega^2\epsilon}\boldsymbol{\nabla}_\perp\boldsymbol{\nabla}_\perp\right)\cdot\frac{\mathbf{r}}{r}\times\mathbf{E}_\perp-\frac{4\pi}{c}\frac{\mathbf{r}}{r}\times\mathbf{J}, \tag{49.6a}$$

$$\left(\frac{\partial}{\partial r}+\frac{1}{r}\right)\mathbf{E}_\perp=-\frac{i\omega}{c}\left(\mathbf{1}_\perp+\frac{c^2}{\omega^2\epsilon}\boldsymbol{\nabla}_\perp\boldsymbol{\nabla}_\perp\right)\cdot\frac{\mathbf{r}}{r}\times\mathbf{B}_\perp-\frac{4\pi i}{\omega\epsilon}\boldsymbol{\nabla}_\perp J_r. \tag{49.6b}$$

As we did in the cylindrical geometry of Chapter 47, we eliminate the transverse electric and magnetic fields in favor of the scalar functions V and I:

$$\mathbf{E}_\perp=-\boldsymbol{\nabla}_\perp V'-\boldsymbol{\nabla}_\perp\times\frac{\mathbf{r}}{r}V'', \tag{49.7a}$$

$$\mathbf{B}_\perp=\boldsymbol{\nabla}_\perp\times\frac{\mathbf{r}}{r}I'-\boldsymbol{\nabla}_\perp I'', \tag{49.7b}$$

and write

$$\frac{\mathbf{r}}{r}\times\mathbf{J}=\boldsymbol{\nabla}_\perp\times\frac{\mathbf{r}}{r}J'-\boldsymbol{\nabla}_\perp J''. \tag{49.8}$$

Now using

$$\frac{\partial}{\partial r}\boldsymbol{\nabla}_\perp=\boldsymbol{\nabla}_\perp\frac{\partial}{\partial r}-\frac{1}{r}\boldsymbol{\nabla}_\perp, \tag{49.9}$$

we obtain the following equations for the modes:

$$\text{E modes:}\quad \frac{\partial}{\partial r}V'=\frac{i\omega}{c}\left(1+\frac{c^2}{\omega^2\epsilon}\nabla_\perp^2\right)I'+\frac{i4\pi}{\omega\epsilon}J_r, \tag{49.10a}$$

$$\frac{\partial}{\partial r}I'=\frac{i\omega\epsilon}{c}V'-\frac{4\pi}{c}J', \tag{49.10b}$$

$$E_r=-\frac{ic}{\omega\epsilon}\nabla_\perp^2 I'-\frac{i4\pi}{\omega\epsilon}J_r, \tag{49.10c}$$

$$\text{H modes:}\quad \frac{\partial}{\partial r}I''=\frac{i\omega\epsilon}{c}\left(1+\frac{c^2}{\omega^2\epsilon}\nabla_\perp^2\right)V''-\frac{4\pi}{c}J'', \tag{49.10d}$$

$$\frac{\partial}{\partial r}V''=\frac{i\omega}{c}I'', \tag{49.10e}$$

$$B_r=-\frac{ic}{\omega}\nabla_\perp^2 V''. \tag{49.10f}$$

The characteristic functions of the transverse Laplacian operator,

$$\nabla_\perp^2=\frac{1}{r^2}\left[\frac{1}{\sin\theta}\frac{\partial}{\partial\theta}\left(\sin\theta\frac{\partial}{\partial\theta}\right)+\frac{1}{\sin^2\theta}\frac{\partial^2}{\partial\phi^2}\right], \tag{49.11}$$

are the spherical harmonics,

$$\nabla_\perp^2 Y_{lm}(\theta,\phi)=-\frac{l(l+1)}{r^2}Y_{lm}(\theta,\phi). \tag{49.12}$$

So we write, for example,

$$V'(r,\theta,\phi)=\sum_{lm}Y_{lm}(\theta,\phi)V'_{lm}(r), \tag{49.13}$$

and similarly for I', V'', I''. Then the mode equations become

$$\text{E modes:}\quad \frac{d}{dr}V'_{lm}(r) = \frac{i\omega}{c}\left(1 - \frac{c^2}{\omega^2\epsilon}\frac{l(l+1)}{r^2}\right)I'_{lm} + \frac{i4\pi}{\omega\epsilon}J_{rlm}, \tag{49.14a}$$

$$\frac{d}{dr}I'_{lm}(r) = \frac{i\omega\epsilon}{c}V'_{lm}(r) - \frac{4\pi}{c}J'_{lm}(r), \tag{49.14b}$$

$$\text{H modes:}\quad \frac{d}{dr}I''_{lm}(r) = \frac{i\omega\epsilon}{c}\left(1 - \frac{c^2}{\omega^2\epsilon}\frac{l(l+1)}{r^2}\right)V''_{lm} - \frac{4\pi}{c}J''_{lm}, \tag{49.14c}$$

$$\frac{d}{dr}V''_{lm}(r) = \frac{i\omega}{c}I''_{lm}(r). \tag{49.14d}$$

Let us see what the differential equations are when we are outside the sources. Let us set $\mathbf{J}(\mathbf{r}) = \mathbf{0}$ and take ϵ to be a constant. Then we can turn the above coupled system of first-order equations into the second-order one,

$$\left(\frac{d^2}{dr^2} + \frac{\omega^2}{c^2}\epsilon - \frac{l(l+1)}{r^2}\right)\left\{\begin{matrix} I'_{lm} \\ V''_{lm} \end{matrix}\right\} = 0. \tag{49.15}$$

49.2 Interior of Conducting Sphere

In the interior of a perfectly conducting sphere of radius a, filled with material of dielectric constant ϵ, the boundary condition that must be satisfied is

$$\mathbf{E}_\perp = 0 \text{ for } r = a. \tag{49.16}$$

This implies that

$$V'_{lm}(r = a) = 0, \quad V''_{lm}(r = a) = 0, \tag{49.17}$$

or

$$\text{H modes:}\quad V''_{lm}(r = a) = 0, \tag{49.18a}$$

$$\text{E modes:}\quad \left.\frac{d}{dr}I'_{lm}\right|_{r=a} = 0. \tag{49.18b}$$

The solutions of the differential equation (49.15) are $rj_l(kr)$, where j_l is the spherical Bessel function, and $k = \omega^2\epsilon/c^2$. For the two modes, then, the frequencies are determined by

$$\text{H mode:}\quad j_l(ka) = 0, \tag{49.19a}$$

$$\text{E mode:}\quad \left.\frac{d}{dr}(rj_l(kr))\right|_{r=a} = 0. \tag{49.19b}$$

Let us make quick estimates of the characteristic frequencies. First, in the spirit of Chapter 21, we recall the asymptotic behavior

$$j_l(t) \sim \frac{\cos(t - (l+1)\pi/2)}{t}, \quad t \to \infty, \tag{49.20}$$

so the H mode zeroes occur approximately where $t = \gamma''$,

$$\gamma'' \approx (l+1)\frac{\pi}{2} + (2n-1)\frac{\pi}{2}, \quad n = 1, 2, 3, \ldots, \tag{49.21}$$

where we are never concerned with $l = 0$ because then $\nabla_\perp^2 Y_{00} = 0$. Thus, the lowest H mode has

$$k''a = \gamma''_{l=1,n=1} \approx \frac{3\pi}{2} = 4.71. \tag{49.22}$$

And, for E modes, where it is the zero of $\frac{d}{dt} t j_l(t)$ that is sought, we have approximate zeroes at

$$\gamma' = (l+1)\frac{\pi}{2} + n\pi, \quad n = 0, 1, 2, \ldots, \tag{49.23}$$

so the lowest mode is

$$k'a = \gamma'_{l=1,n=0} = \pi = 3.14. \tag{49.24}$$

Let us improve on these estimates using the methods of Chapter 27. The eigenvalue functional corresponding to the differential equation

$$\left(\frac{d^2}{dt^2} + 1 - \frac{l(l+1)}{t^2}\right) u(t) = 0 \tag{49.25}$$

is

$$\gamma^2 = \frac{\int_0^1 dt \left[\left(\frac{du}{dt}\right)^2 + \frac{l(l+1)}{t^2} u^2\right]}{\int_0^1 dt\, u^2}. \tag{49.26}$$

(Regard $\gamma^2[\phi]$ as stationary under ϕ variations: Then (49.25) follows as the coefficient of $\delta\phi$ upon integration by parts, assuming $u(0) = u(1) = 0$, or for E modes, $u'(0) = u'(t) = 0$. The boundary conditions at the origin are necessary to ensure finiteness of the fields there.) The small t behavior of the solution is $u \sim t^{l+1}$. For the H modes, consider the test function, for $l = 1$,

$$u = t^2(1 - t^2). \tag{49.27}$$

When this is inserted into the functional (49.26), we obtain the upper bound $\gamma'' = 4.74$, which is worse than our previous estimate. (The exact answer is 4.49341.) For the E mode we can try

$$u = t^2 - \frac{1}{2}t^4, \tag{49.28}$$

for this has vanishing derivative at $t = 1$. This does give a considerably improved bound, $\gamma' = 2.7514$, compared to the exact root 2.74371. In Problem 49.1 these bounds are improved by iteration.

49.3 Spherical Hankel Functions

The fundamental solution to Helmholtz' equation,

$$(\nabla^2 + k^2)\phi = 0, \quad k^2 = \frac{\omega^2}{c^2}, \tag{49.29}$$

is the Green's function

$$G_0(\mathbf{r} - \mathbf{r}') = \frac{e^{ik|\mathbf{r}-\mathbf{r}'|}}{|\mathbf{r} - \mathbf{r}'|}, \tag{49.30}$$

as long as $\mathbf{r} \neq \mathbf{r}'$. For $r < r'$ we can write this as a Taylor expansion,

$$\frac{e^{ik|\mathbf{r}-\mathbf{r}'|}}{|\mathbf{r} - \mathbf{r}'|} = e^{-\mathbf{r}\cdot\boldsymbol{\nabla}'} \frac{e^{ikr'}}{r'}. \tag{49.31}$$

Now the operand here itself satisfies Helmholtz' equation,

$$(\nabla'^2 + k^2)\frac{e^{ikr'}}{r'} = 0, \quad r' > 0, \tag{49.32}$$

so, in the spirit of Chapter 20, we can regard $i\boldsymbol{\nabla}'/k$ as a symbolic unit vector when acting on $e^{ikr'}/r'$. Thus, we rewrite the Taylor expansion (49.31) as

$$\frac{e^{ik|\mathbf{r}-\mathbf{r}'|}}{|\mathbf{r}-\mathbf{r}'|} = \exp\left[ik\mathbf{r}\cdot(i\boldsymbol{\nabla}'/k)\right]\frac{e^{ikr'}}{r'}. \tag{49.33}$$

Now recall the expansion of the plane wave in terms of spherical Bessel functions, (23.57), or (23.68a),

$$e^{i\mathbf{k}\cdot\mathbf{r}} = 4\pi\sum_{l=0}^{\infty}\sum_{m=-l}^{l} Y_{lm}^*\left(\frac{\mathbf{r}}{r}\right)Y_{lm}\left(\frac{\mathbf{k}}{k}\right)i^l j_l(kr). \tag{49.34}$$

When this is used to express the operator $\exp[ik\mathbf{r}\cdot i\boldsymbol{\nabla}'/k]$ we have

$$\frac{e^{ik|\mathbf{r}-\mathbf{r}'|}}{|\mathbf{r}-\mathbf{r}'|} = \sum_{lm} 4\pi i^l j_l(kr)Y_{lm}^*\left(\frac{\mathbf{r}}{r}\right)Y_{lm}\left(\frac{i}{k}\boldsymbol{\nabla}'\right)\frac{e^{ikr'}}{r'}. \tag{49.35}$$

What is the latter symbolic operator? We recall the generating function for the spherical harmonics, (22.22),

$$\frac{(\mathbf{a}\cdot\frac{\mathbf{r}}{r})^l}{l!} = \sum_{lm}\psi_{lm}\sqrt{\frac{4\pi}{2l+1}}Y_{lm}\left(\frac{\mathbf{r}}{r}\right), \tag{49.36}$$

where $\mathbf{a}$ is a null vector, $\mathbf{a}\cdot\mathbf{a} = 0$. It is easy to see that

$$\frac{(\mathbf{a}\cdot\frac{i}{k}\boldsymbol{\nabla}')^l}{l!}\frac{e^{ikr'}}{r'} = \frac{(\mathbf{a}\cdot\frac{i}{k}\mathbf{r}')^l}{l!}\left(\frac{1}{r'}\frac{d}{dr'}\right)^l\frac{e^{ikr'}}{r'}. \tag{49.37}$$

Thus, we recognize, when acting on $e^{ikr'}/r'$, that

$$Y_{lm}\left(\frac{i}{k}\boldsymbol{\nabla}'\right) = \left(\frac{i}{k}\right)^l r'^l Y_{lm}\left(\frac{\mathbf{r}'}{r'}\right)\left(\frac{1}{r'}\frac{d}{dr'}\right)^l, \tag{49.38}$$

and so we have the expansion

$$\frac{e^{ik|\mathbf{r}-\mathbf{r}'|}}{|\mathbf{r}-\mathbf{r}'|} = \sum_{lm} 4\pi j_l(kr)Y_{lm}^*\left(\frac{\mathbf{r}}{r}\right)Y_{lm}\left(\frac{\mathbf{r}'}{r'}\right)(kr')^l\left(-\frac{1}{kr'}\frac{d}{d(kr')}\right)^l\frac{e^{ikr'}}{r'}. \tag{49.39}$$

Let us define a set of functions by

$$h_l(t) = t^l\left(-\frac{1}{t}\frac{d}{dt}\right)^l\frac{e^{it}}{it}; \tag{49.40}$$

these functions, which are defined analogously to the spherical Bessel functions j_l (see Problem 23.3), are called the spherical Hankel functions (of the first kind, often denoted by $h_l^{(1)}$). As with the cylinder functions $H_l^{(1)}$, we have the decomposition into real and imaginary parts,

$$h_l(t) = j_l(t) + in_l(t), \tag{49.41}$$

the imaginary part being the spherical Neumann function. Thus we have deduced the addition theorem

$$h_0(k|\mathbf{r}-\mathbf{r}'|) = \sum_l (2l+1) P_l(\cos\gamma) j_l(kr_<) h_l(kr_>), \tag{49.42}$$

(γ being the angle between $\mathbf{r}$ and $\mathbf{r}'$) the real part of which is

$$j_0(k|\mathbf{r}-\mathbf{r}'|) = \frac{\sin k|\mathbf{r}-\mathbf{r}'|}{k|\mathbf{r}-\mathbf{r}'|} = \sum_l (2l+1) P_l(\cos\gamma) j_l(kr_<) j_l(kr_>), \tag{49.43}$$

the spherical analog to the addition theorem for the cylinder functions, (18.65). We might also note the leading asymptotic behavior of the spherical Hankel function:

$$h_l(t) \sim (-i)^l \frac{e^{it}}{it} = \frac{e^{i(t-(l+1)\pi/2)}}{t}, \quad t \gg 1, \tag{49.44}$$

the real part of which coincides with (49.20).

49.4 Scattering

We set the stage for our discussion by recalling the result from the previous chapter for the scattering by a dielectric sphere. In particular, if we take the limit $\epsilon \to \infty$ of (48.35), we would expect to recover the situation of a perfectly conducting sphere, which limit gives

$$\sigma = \frac{8\pi}{3} \frac{a^6}{\lambda\!\!\!^{-4}}. \tag{49.45}$$

However, the well-known result for the scattering cross section by a perfectly conducting sphere is that of Gustav Mie[1] and Peter Debye (1908–1909),

$$\sigma = \frac{10\pi}{3} \frac{a^6}{\lambda\!\!\!^{-4}}. \tag{49.46}$$

What is the reason for this discrepancy? It is simply that in this case there is also magnetic dipole scattering, which gives the additional contribution $2\pi a^6/3\lambda\!\!\!^{-4}$ (see Problem 48.1). So this poses a nice problem for our general formulation.

Consider a circularly polarized electromagnetic wave propagating in the z direction, for which

$$E_x = e^{ikz}, \quad E_y = ie^{ikz}, \quad k = \frac{2\pi}{\lambda} = \frac{1}{\lambda\!\!\!^{-}}. \tag{49.47}$$

The time dependence, $e^{-i\omega t}$, is implicit. The corresponding magnetic field is $\mathbf{B} = \hat{\mathbf{z}} \times \mathbf{E}$, or

$$B_x = -E_y, \quad B_y = E_x. \tag{49.48}$$

Of course, $E_z = B_z = 0$. We wish to calculate the scattering from a spherical conducting shell, centered at the origin, so it is natural to use spherical coordinates,

$$E_r = \frac{\mathbf{r}}{r} \cdot \mathbf{E} = \frac{x+iy}{r} e^{ikz}, \tag{49.49a}$$

$$B_r = \frac{\mathbf{r}}{r} \cdot \mathbf{B} = -i\frac{x+iy}{r} e^{ikz}. \tag{49.49b}$$

[1] 1868–1957.

We express this in terms of spherical harmonics and spherical Bessel functions as follows: First we recall the expansion of the plane wave, (23.57),

$$e^{ikz} = \sum_{l=0}^{\infty}(2l+1)i^l j_l(kr)P_l(\cos\theta), \tag{49.50}$$

so, using the addition theorem for the spherical harmonics, we have

$$(x+iy)e^{ikz} = \frac{1}{i}\left(\frac{\partial}{\partial k_x} + i\frac{\partial}{\partial k_y}\right)\sum_{lm} 4\pi i^l j_l(kr)Y^*_{lm}(\hat{\mathbf{k}})Y_{lm}(\hat{\mathbf{r}})\bigg|_{\mathbf{k}_\perp=\mathbf{0}}. \tag{49.51}$$

We note that

$$\frac{\partial}{\partial k_x}f(k)\bigg|_{k_x=0} = \frac{k_x}{k}\frac{\partial f(k)}{\partial k}\bigg|_{k_x=0} = 0, \tag{49.52}$$

so in taking the derivative in (49.51), we can neglect its action on $j_l(kr)$. In working out

$$\left(\frac{\partial}{\partial k_x} + i\frac{\partial}{\partial k_y}\right)Y^*_{lm}(\hat{\mathbf{k}}) \tag{49.53}$$

we are looking for structures like $(k_x+ik_y)/k = \sin\theta e^{i\phi}$ in $Y_{lm}(\hat{\mathbf{k}})$ so that picks out $m=1$; from (22.26), we therefore have only

$$Y^*_{l1}(\theta,\phi) = -\sqrt{\frac{2l+1}{4\pi}}\frac{1}{\sqrt{l(l+1)}}\frac{k_x-ik_y}{k}\left(\frac{d}{d\mu}\right)^{l+1}\frac{(\mu-1)^l(\mu+1)^l}{2^l\, l!}\bigg|_{\mu=1}, \tag{49.54}$$

where the $\mu=1$ condition comes from setting $k_x=k_y=0$ after the differentiation. Thus the required derivative is

$$\left(\frac{\partial}{\partial k_x} + i\frac{\partial}{\partial k_y}\right)Y^*_{l1}(\hat{\mathbf{k}}) = -\frac{1}{k}\sqrt{\frac{2l+1}{4\pi}l(l+1)}. \tag{49.55}$$

Thus, the radial electric and magnetic fields are

$$E_r = iB_r = -\sum_{l=1}^{\infty}\sqrt{4\pi}i^{l-1}\frac{j_l(kr)}{kr}\sqrt{(2l+1)l(l+1)}Y_{l1}(\hat{\mathbf{r}}). \tag{49.56}$$

Now we return to the formalism of Section 49.1 and write, for $\epsilon=1$, using (49.10c),

$$E_r = -\frac{i}{k}\nabla^2_\perp I', \tag{49.57}$$

where the effect of $\nabla^2_\perp$ on a spherical harmonic is to divide by r^2 and multiply by $l(l+1)$. Therefore,

$$I' = \sum_{l=1}^{\infty}\sqrt{4\pi}i^l Y_{l1}(\hat{\mathbf{r}})\sqrt{\frac{2l+1}{l(l+1)}}rj_l(kr), \tag{49.58}$$

which represents an incoming wave. Also, since $B_r = -(i/k)\nabla^2_\perp V''$, (49.10f), we have also

$$V'' = -iI'. \tag{49.59}$$

The boundary conditions are that

$$\mathbf{E}_\perp = -\boldsymbol{\nabla}_\perp V' - \boldsymbol{\nabla}_\perp\times\frac{\mathbf{r}}{r}V'' = 0 \quad \text{at } r=a, \tag{49.60}$$

from which follows

$$V' = 0, \quad V'' = 0, \text{ at } r = a. \tag{49.61}$$

Because we further have (49.10b),

$$\frac{\partial}{\partial r} I' = ikV', \tag{49.62}$$

we have

$$\frac{\partial}{\partial r} I' = 0 \text{ at } r = a. \tag{49.63}$$

Let us consider the H modes, where in order to satisfy the boundary condition we must have not only the incident wave, but the scattered wave as well:

$$V'' = -i \sum_{l=1}^{\infty} \sqrt{4\pi} i^l Y_{l1}(\hat{\mathbf{r}}) \sqrt{\frac{2l+1}{l(l+1)}} \left[r j_l(kr) - r h_l(kr) \frac{j_l(ka)}{h_l(ka)} \right]; \tag{49.64}$$

h_l represents the outgoing spherical wave. Let us express the coefficient of the scattered wave,

$$\frac{j_l(ka)}{h_l(ka)} = \frac{j_l(ka)}{j_l(ka) + i n_l(ka)}, \tag{49.65}$$

in terms of an angle δ_l'', defined by

$$\tan \delta_l'' = \frac{j_l(ka)}{n_l(ka)}, \tag{49.66}$$

so that

$$\frac{j_l(ka)}{h_l(ka)} = -i \sin \delta_l'' e^{i\delta_l''}. \tag{49.67}$$

The scattered wave, the second term in (49.64), is then expressed asymptotically by the large t behavior given in (49.44):

$$V''_{\text{sc}} \sim -i \frac{e^{ikr}}{k} \sum_{l=1}^{\infty} \sqrt{4\pi} Y_{l1}(\hat{\mathbf{r}}) \sqrt{\frac{2l+1}{l(l+1)}} \sin \delta_l'' e^{i\delta_l''}. \tag{49.68}$$

Similarly, for the E modes, the mode function which satisfies the appropriate boundary condition, and therefore contains both the incident and scattered waves, is

$$I' = \sum_{l=1}^{\infty} \sqrt{4\pi} i^l Y_{l1}(\hat{\mathbf{r}}) \sqrt{\frac{2l+1}{l(l+1)}} \left[r j_l(kr) - r h_l(kr) \frac{(r j_l(kr))'|_{r=a}}{(r h_l(kr))'|_{r=a}} \right]. \tag{49.69}$$

As before, we define an angle in terms of the coefficient of the scattered wave,

$$\frac{(r j_l(kr))'|_{r=a}}{(r n_l(kr))'|_{r=a}} = \tan \delta_l', \tag{49.70}$$

and then, exactly as before, we find

$$I'_{\text{sc}} \sim V'_{\text{sc}} \sim \frac{e^{ikr}}{k} \sum_{l=1}^{\infty} \sqrt{4\pi} Y_{l1}(\hat{\mathbf{r}}) \sqrt{\frac{2l+1}{l(l+1)}} \sin \delta_l' e^{i\delta_l'}. \tag{49.71}$$

The transverse electric field is now given by

$$\mathbf{E}_\perp = -\boldsymbol{\nabla}_\perp V' - \boldsymbol{\nabla}_\perp \times \frac{\mathbf{r}}{r} V'', \tag{49.72}$$

which consists of an incident part and a scattered part. The radial component of the Poynting vector is proportional to

$$S_r = \frac{\mathbf{r}}{r}\cdot\mathbf{S} \propto \mathrm{Re}\,\frac{\mathbf{r}}{r}\cdot\mathbf{E}^*\times\mathbf{B} = |\mathbf{E}_\perp|^2, \tag{49.73}$$

while the incident flux, which is in the z direction, is proportional to

$$S_{z(\mathrm{inc})} \propto |E_{x\mathrm{inc}}|^2 + |E_{y\mathrm{inc}}|^2 = 2. \tag{49.74}$$

The total scattered power is proportional to

$$P_{\mathrm{sc}} \propto r^2 \int d\Omega\,|\mathbf{E}_\perp|^2 = r^2 \int d\Omega \left|\boldsymbol{\nabla}_\perp V' + \boldsymbol{\nabla}_\perp\times\frac{\mathbf{r}}{r}V''\right|^2, \tag{49.75}$$

where by partial integration we may show

$$\int d\Omega\,\boldsymbol{\nabla}_\perp V'^*\cdot\boldsymbol{\nabla}_\perp\times\frac{\mathbf{r}}{r}V'' = 0, \tag{49.76}$$

and further

$$\int d\Omega\,\frac{\mathbf{r}}{r}\times\boldsymbol{\nabla}_\perp V''^*\cdot\frac{\mathbf{r}}{r}\times\boldsymbol{\nabla}_\perp V'' = \int d\Omega\,\boldsymbol{\nabla}_\perp V''^*\cdot\boldsymbol{\nabla}_\perp V''. \tag{49.77}$$

By another partial integration we obtain

$$P_{\mathrm{sc}} \propto r^2 \int d\Omega\left[V'^*(-\nabla_\perp^2)V' + V''^*(-\nabla_\perp^2)V''\right], \tag{49.78}$$

in which, for each angular momentum mode, we may now replace the transverse Laplacian by its eigenvalue $-l(l+1)/r^2$. Using the above expressions for V'_{sc} and V''_{sc} in terms of the phases δ'_l and δ''_l, and exploiting the orthogonality of the spherical harmonics, we get the following expression for the scattered power,

$$P_{\mathrm{sc}} \propto \frac{1}{k^2}\sum_{l=1}^{\infty} 4\pi(2l+1)(\sin^2\delta'_l + \sin^2\delta''_l). \tag{49.79}$$

The cross section is obtained by dividing this by $|\mathbf{E}_{\perp\mathrm{inc}}|^2 = 2$:

$$\sigma = \frac{2\pi}{k^2}\sum_{l=1}^{\infty}(2l+1)(\sin^2\delta'_l + \sin^2\delta''_l). \tag{49.80}$$

Now we want to look at two special cases, where $ka = a/\lambda\!\!\!{}^{-}$ is either much smaller, or much larger, than one. In the former case, where $ka \ll 1$, the dominant term is $l = 1$, for which

$$j_1(t) = -\frac{d}{dt}\frac{\sin t}{t} \approx \frac{t}{3}, \quad t \ll 1, \tag{49.81a}$$

$$n_1(t) = \frac{d}{dt}\frac{\cos t}{t} \approx -\frac{1}{t^2}, \quad t \ll 1, \tag{49.81b}$$

so

$$\tan\delta''_1 \approx \delta''_1 \approx -\frac{1}{3}t^3 = -\frac{1}{3}(ka)^3. \tag{49.82}$$

Similarly,

$$\tan\delta'_1 = \frac{(t^2/3)'}{(-1/t)'} = \frac{2}{3}t^3, \quad \text{or} \quad \delta'_1 \approx \frac{2}{3}(ka)^3. \tag{49.83}$$

If we add together the contributions of the E modes and the H modes, we find for the cross section in the long wavelength limit

$$\sigma \sim \frac{2\pi}{k^2} 3 \left[\left(\frac{2}{3}(ka)^3 \right)^2 + \left(\frac{1}{3}(ka)^3 \right)^2 \right] = \frac{10}{3} \frac{a^6}{\lambda\!\!\!^{-4}}, \quad \lambda\!\!\!^{-} \gg a, \tag{49.84}$$

the result stated in (49.46). (Note that the E mode contribution agrees with (49.45).)

In the opposite limit, $ka \gg 1$, we use the asymptotic forms of the spherical Bessel functions,

$$j_l(t) \sim \frac{\sin(t - \pi l/2)}{t} = \frac{\cos(t - \pi(l+1)/2)}{t}, \quad t \gg 1, \tag{49.85a}$$

$$n_l(t) \sim \frac{\sin(t - \pi(l+1)/2)}{t} = -\frac{\cos(t - \pi l/2)}{t}, \quad t \gg 1, \tag{49.85b}$$

Thus, for the H modes, we have

$$\tan \delta_l'' = -\frac{\sin(t - \pi l/2)}{\cos(t - \pi l/2)}, \quad \Rightarrow \delta_l'' = -\left(ka - \frac{\pi l}{2} \right), \tag{49.86a}$$

and, for the E modes, we have

$$\tan \delta_l' = -\frac{\sin(t - \pi(l+1)/2)}{\cos(t - \pi(l+1)/2)}, \quad \Rightarrow \delta_l' = -\left(ka - \frac{\pi(l+1)}{2} \right) = \delta_l'' + \frac{\pi}{2}. \tag{49.86b}$$

So in the short wavelength limit, the cross section is formally divergent:

$$\sigma = \frac{2\pi}{k^2} \sum_{l=1}^{\infty} (2l+1)(\cos^2 \delta_l'' + \sin^2 \delta_l'') = \frac{2\pi}{k^2} \sum_{l=1}^{\infty} (2l+1). \tag{49.87}$$

In reality this is an erroneous conclusion, because the asymptotic expansion is only valid when $t \gg l$. We can estimate the effect by cutting off the sum at a value of $L \gg 1$ such that $L^2 \sim t^2$, as we can see from the differential equation,

$$\left(\frac{d^2}{dt^2} + 1 - \frac{l(l+1)}{t^2} \right) t j_l(t) = 0. \tag{49.88}$$

Therefore,

$$\sigma \sim \frac{2\pi}{k^2} \sum_{l=1}^{L} (2l+1) \sim \frac{2\pi}{k^2} \int_0^L dl\, 2l = \frac{2\pi}{k^2} L^2 \sim 2\pi a^2. \tag{49.89}$$

This result is not unexpected. In the short wavelength regime we should see just the cross sectional area presented to the beam by the sphere, πa^2. The extra factor of 2 comes from the fact that we have a shadow behind the sphere. The sphere must throw enough energy forward (with opposite sign) to cancel the incident light beam. Thus we get

$$(\pi a^2)_{\text{scattered}} + (\pi a^2)_{\text{shadow}} = 2\pi a^2. \tag{49.90}$$

This idea is explored in greater detail in Problem 53.10.

49.5 Problems for Chapter 49

1. Use the iteration method discussed in Sections 30.2 and 27.3 to improve on the estimates for the characteristic frequencies given in Section 49.2.

2. In Chapter 53, we will define the scattering amplitude f by

$$\mathbf{e}\cdot\mathbf{E}_{\text{sc}} \sim \frac{e^{ikr}}{r} f(\theta), \tag{49.91}$$

where $\mathbf{e}$ is the polarization vector for the scattered wave. Obtain an expression for the scattering amplitude in terms of Legendre's function, $P_l(\cos\theta)$, and the phase shifts δ_l' and δ_l''. Show that the differential cross section is given by

$$\frac{d\sigma}{d\theta} = |f(\theta)|^2. \tag{49.92}$$

Show that if we sum over final polarizations and integrate over all angles of scattering, we obtain the total cross section (49.80). Show that the optical theorem is satisfied,

$$\text{Im } f(0) = \frac{k}{4\pi}\sigma. \tag{49.93}$$

3. Derive an expression for the differential cross section for scattering by a perfectly conducting sphere, $d\sigma/d\Omega$. Show that in the long wavelength limit it becomes

$$\frac{d\sigma}{d\Omega} \to a^2(ka)^4\left[\frac{5}{8}(1+\cos^2\theta) - \cos\theta\right], \tag{49.94}$$

which becomes (49.46) when integrated over all scattering angles.

4. The analysis of Section 49.4 considers a particular circularly polarized incident wave, $\mathbf{E} = (1, i, 0)e^{ikz}$, $\mathbf{B} = (-i, 1, 0)e^{ikz}$. Carry out the same analysis for the other possibility, $\mathbf{E} = (1, -i, 0)e^{ikz}$, $\mathbf{B} = (i, 1, 0)e^{ikz}$.

50

Diffraction I

We now turn to the consideration of the scattering of electromagnetic radiation by objects large compared to the wavelength. We begin with a scatterer, consisting of a perfectly conducting screen, with a (macroscopic) hole in it. As we shall see, the predictions of geometrical optics are altered because of the finite wavelength of the radiation. This phenomenon is called diffraction. We first derive a general expression for the diffracted electric field in terms of its boundary values. Although it is an exact expression, we will apply it to three examples of near forward scattering.

50.1 Diffracted Electric Field

Away from the sources, in vacuum ($\epsilon = \mu = 1$), the pertinent Maxwell equations are

$$\begin{aligned} \boldsymbol{\nabla}\times\mathbf{B} &= \frac{1}{c}\dot{\mathbf{E}}, \quad \boldsymbol{\nabla}\cdot\mathbf{E} = 0, \\ -\boldsymbol{\nabla}\times\mathbf{E} &= \frac{1}{c}\dot{\mathbf{B}}. \end{aligned} \tag{50.1}$$

A consequence of these equations is that $\mathbf{E}$ satisfies the wave equation,

$$-\boldsymbol{\nabla}\times(\boldsymbol{\nabla}\times\mathbf{E}) \equiv -\boldsymbol{\nabla}(\boldsymbol{\nabla}\cdot\mathbf{E}) + \nabla^2\mathbf{E} = \frac{1}{c^2}\ddot{\mathbf{E}}, \tag{50.2}$$

or

$$\left(\nabla^2 - \frac{1}{c^2}\frac{\partial^2}{\partial t^2}\right)\mathbf{E} = \mathbf{0}. \tag{50.3}$$

Specifically, we consider a plane polarized wave normally incident on a perfectly conducting screen, with an aperture. We choose the coordinate system (see Fig. 50.1) such that x is in the direction of propagation and z is along the direction of polarization of the incident wave. From our choice of coordinate system, far from the source, the incident electric field has only a z-component. For radiation of a particular frequency ω (with $k = \omega/c$), the wave equation for E_z, for example, becomes

$$(\nabla^2 + k^2)E_z = 0. \tag{50.4}$$

We are interested in finding the electric field to the right of the screen ($x > 0$), subject to the boundary condition that the tangential electric field vanishes on the screen (since it is a perfect conductor).

The solution to the differential equation (50.4) can be expressed in terms of Green's function, for $x > 0$, which satisfies (34.19),

$$(\nabla^2 + k^2)G(\mathbf{r}, \mathbf{r}') = -4\pi\delta(\mathbf{r} - \mathbf{r}'), \tag{50.5}$$

DOI: 10.1201/9781003057369-50

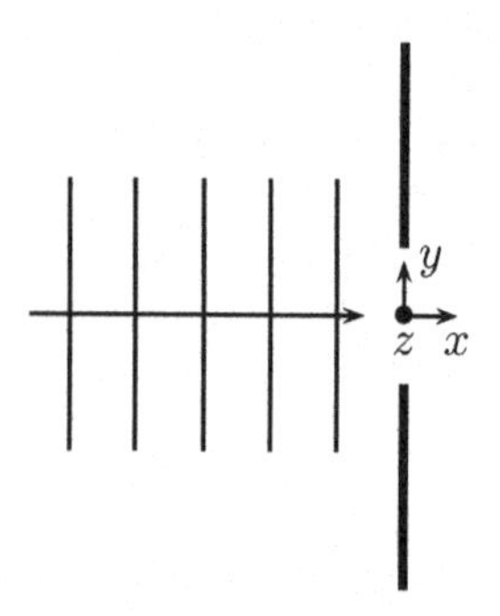

FIGURE 50.1
Plane wave incident on an aperture. Here, the z direction, the direction of the electric field, is out of the page.

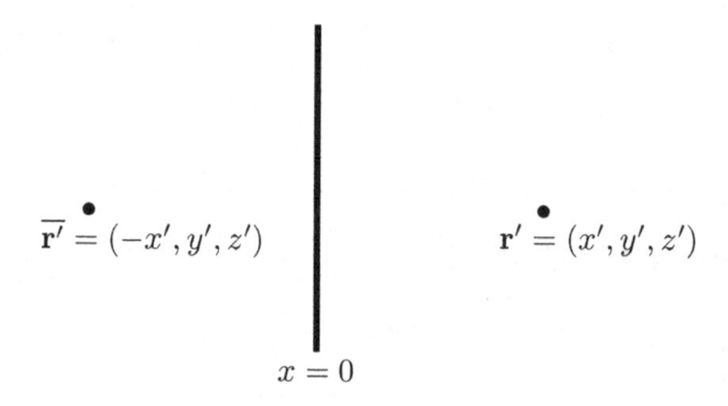

FIGURE 50.2
Image charge induced by a point charge adjacent to a perfectly conducting plane.

subject to the boundary condition that $G = 0$ on the surface $x = 0$. This Green's function refers to the field of a unit point charge in the presence of a perfect conductor lying in the entire $x = 0$ plane, as shown in Fig. 50.2. The solution in the region to the right of the conducting plane is the desired Green's function, which may be found by the method of images (*cf.* Chapter 17). If the coordinates of the source point $\mathbf{r}'$ are (x', y', z'), the image point is located at

$$\overline{\mathbf{r}'} = (-x', y', z'), \tag{50.6}$$

in terms of which the solution to (50.5) is

$$G(\mathbf{r}, \mathbf{r}') = \frac{e^{ik|\mathbf{r}-\mathbf{r}'|}}{|\mathbf{r}-\mathbf{r}'|} - \frac{e^{ik|\mathbf{r}-\overline{\mathbf{r}'}|}}{|\mathbf{r}-\overline{\mathbf{r}'}|}. \tag{50.7}$$

This is a generalization of the electrostatic Green's function (17.32).

To express the electric field in terms of this Green's function, we first multiply (50.4) by $G(\mathbf{r}, \mathbf{r}')$ and (50.5) by $E_z(\mathbf{r})$. Then subtracting the two resulting expressions, and interchanging $\mathbf{r}$ and $\mathbf{r}'$, we obtain, (dropping the z subscript on E_z for simplicity)

$$\begin{aligned} 4\pi E(\mathbf{r}')\delta(\mathbf{r}-\mathbf{r}') &= -E(\mathbf{r}')(\nabla'^2 + k^2)G(\mathbf{r}', \mathbf{r}) + G(\mathbf{r}', \mathbf{r})(\nabla'^2 + k^2)E(\mathbf{r}') \\ &= \boldsymbol{\nabla}' \cdot [G(\mathbf{r}', \mathbf{r})\boldsymbol{\nabla}' E(\mathbf{r}') - E(\mathbf{r}')\boldsymbol{\nabla}' G(\mathbf{r}', \mathbf{r})]. \end{aligned} \tag{50.8}$$

The electric field to the right of the screen can now be obtained in terms of its boundary values by integrating (50.8) over the volume to the right of the plane ($x' > 0$) and using the divergence theorem:

$$4\pi E(\mathbf{r}) = -\int_{\text{screen+aperture}} dS' \left[G(\mathbf{r}',\mathbf{r})\frac{\partial}{\partial x'}E(\mathbf{r}') - E(\mathbf{r}')\frac{\partial}{\partial x'}G(\mathbf{r}',\mathbf{r})\right]$$
$$+\int_{S_\infty} dS' \left[G(\mathbf{r}',\mathbf{r})\frac{\partial}{\partial r'}E(\mathbf{r}') - E(\mathbf{r}')\frac{\partial}{\partial r'}G(\mathbf{r}',\mathbf{r})\right]. \tag{50.9}$$

The first integral extends over the entire $x' = 0$ plane, the minus sign appearing because the outward normal to the volume is in the $-x'$ direction. The second integral is over a surface which can be taken to be a hemisphere with radius tending toward infinity. Far from the aperture, which acts as the source of the scattered wave, the leading behavior of the electric field, an outgoing spherical wave, is the same as that of the Green's function:

$$G(\mathbf{r}',\mathbf{r}) \sim \frac{e^{ikr'}}{r'}, \quad E(\mathbf{r}') \sim \frac{e^{ikr'}}{r'}, \quad \text{as} \quad r' \to \infty. \tag{50.10}$$

Then, the r' derivative can be effectively replaced by

$$\frac{\partial}{\partial r'} \to ik, \tag{50.11}$$

so that the integral in (50.9) over the hemisphere at infinity vanishes. Physically, this follows from the fact that the source of the electric field is not at infinitely remote points, but at the aperture.

Now we incorporate the boundary conditions that $G = 0$ on the entire $x' = 0$ surface, while $E = 0$ on the surface $x' = 0$ except for the aperture (since we are talking about E_z here). Thus the electric field to the right of the screen is

$$4\pi E(\mathbf{r}) = \int_{\text{aperture}} dS'\, E(\mathbf{r}')\frac{\partial}{\partial x'}G(\mathbf{r}',\mathbf{r}). \tag{50.12}$$

The derivative of Green's function can be evaluated from (50.7) by noting that, for the first term,

$$\frac{\partial}{\partial x'} = -\frac{\partial}{\partial x}, \tag{50.13a}$$

since $|\mathbf{r}' - \mathbf{r}|$ depends on $x' - x$, while in the second term,

$$\frac{\partial}{\partial x'} = \frac{\partial}{\partial x}, \tag{50.13b}$$

since $|\mathbf{r}' - \bar{\mathbf{r}}|$ depends on $x' + x$. Therefore, at $x' = 0$, the two terms contribute equally. Of particular interest is the electric field far away from the aperture ($kr \gg 1$, $r \gg r'$), for which we may use the expansion (35.1) to write[1]

$$\frac{1}{|\mathbf{r} - \mathbf{r}'|}e^{ik|\mathbf{r}-\mathbf{r}'|} \sim \frac{1}{r}e^{ikr}e^{-ik\mathbf{n}\cdot\mathbf{r}'}, \tag{50.14}$$

where $\mathbf{n}$ is the unit vector in the direction of observation, $\mathbf{r}$. If we note that

$$\frac{\partial r}{\partial x} = \frac{x}{r} = \cos\theta, \tag{50.15}$$

[1]This is called the Fraunhofer approximation, after Joseph Ritter von Fraunhofer, 1787–1826.

the normal derivative of the Green's function is seen to be

$$\frac{\partial}{\partial x'}G(\mathbf{r}',\mathbf{r})\bigg|_{x'=0} \sim \frac{-2ik}{r}\cos\theta e^{ikr}e^{-ik\mathbf{n}\cdot\mathbf{r}'}. \tag{50.16}$$

(Alternatively, we could use (50.14) directly.) Therefore, far from the aperture, the electric field is

$$E(\mathbf{r}) \sim -\frac{ik}{2\pi}\cos\theta\frac{e^{ikr}}{r}\int_{\text{aperture}} dS'\, e^{-ik\mathbf{n}\cdot\mathbf{r}'}E(\mathbf{r}'). \tag{50.17}$$

This expression is valid for any shape of the aperture. This form holds for E_y as well as E_z, but for E_x one must integrate over the whole plane. In the following sections, we will apply this result to discuss the diffraction by a circular hole, a slit, and a straight edge. We will do this in the approximation that the wavelength of the radiation is small compared to the dimensions of the aperture. Consequently, the wave travels mostly forward, with small angular deviation. We may further reasonably suppose that only E_z is present, and that it has very little z-dependence,

$$E_x \approx 0, \quad E_y \approx 0, \quad \Rightarrow \frac{\partial E_z}{\partial z} \approx 0. \tag{50.18}$$

In the same approximation, which might be dubbed the Kirchhoff approximation,[2] we may assume that the field in the hole is just that of the incident wave,

$$E_{\text{hole}} \approx E_{\text{inc}}, \tag{50.19}$$

which is valid except near the edges. Since only relative amplitudes enter into our discussion, we take the normally incident wave to have unit amplitude, $E_{\text{inc}} = 1$.

50.2 Diffraction by a Circular Aperture

We here specialize the aperture to be of circular shape, having a radius a, with $\lambda \ll a$. Using polar coordinates with the origin at the center of the hole, we can then express the electric field to the right of the hole as ($\cos\theta \approx 1$)

$$E \sim -ik\frac{e^{ikr}}{r}\int_0^a \rho\,d\rho\int_0^{2\pi}\frac{d\phi}{2\pi}e^{-ik\rho\sin\theta\cos\phi}, \tag{50.20}$$

where θ is the angle between $\mathbf{n}$ and the x-axis, and ϕ is the angle between $\mathbf{r}'$ and the projection of $\mathbf{n}$ on the plane defined by the hole (see Fig. 50.3). Because the integration over ϕ is identified as J_0, according to (18.4), the electric field is proportional to

$$\int_0^a \rho\,d\rho\, J_0(k\rho\sin\theta) = \frac{1}{(k\sin\theta)^2}\int_0^{ka\sin\theta} z\,dz\,J_0(z). \tag{50.21}$$

The remaining z-integration can be performed with the aid of Bessel's equation of zeroth order, (18.17b),

$$\frac{d}{dz}\left[z\frac{d}{dz}J_0(z)\right] + zJ_0(z) = 0, \tag{50.22}$$

[2]Gustav Robert Kirchhoff, 1824–1887.

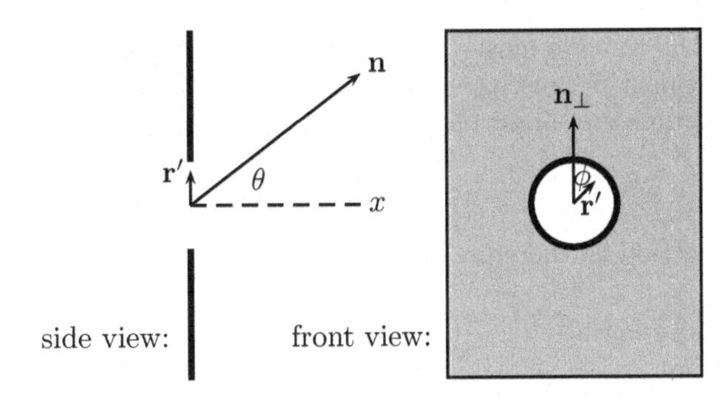

FIGURE 50.3
Geometry of diffraction by a circular aperture. $\mathbf{r}'$ denotes a point in the aperture, and $\mathbf{n}$ the direction of observation. The side view, on the left, shows the angle θ between the normal direction (x) and the direction of observation. The front view, on the right, shows the angle ϕ between a point in the aperture and the projection of $\mathbf{n}$, $\mathbf{n}_\perp$, on the plane $x = 0$.

that is,

$$\int_0^{\overline{z}} dz\, z\, J_0(z) = -\overline{z}\frac{d}{d\overline{z}}J_0(\overline{z}) = \overline{z}J_1(\overline{z}), \tag{50.23}$$

where we have used the recurrence relation

$$-\frac{d}{dz}J_0(z) = J_1(z), \tag{50.24}$$

which follows from (18.24) and (18.69a). Thus the relative diffracted electric field is

$$E \sim -ik\frac{e^{ikr}}{r}\frac{a}{k\sin\theta}J_1(ka\sin\theta), \tag{50.25}$$

or, consistently making the small angle approximation, $\sin\theta \approx \theta$,

$$E \sim -ik\frac{e^{ikr}}{r}a^2\frac{J_1(ka\theta)}{ka\theta}. \tag{50.26}$$

The ratio of diffracted to incident energy fluxes is

$$\frac{|E|^2}{|E_{\text{inc}}|^2} = \frac{1}{r^2}(ka^2)^2\left(\frac{J_1(ka\theta)}{ka\theta}\right)^2, \tag{50.27}$$

which, when multiplied by r^2, yields the differential cross section,

$$\frac{d\sigma}{d\Omega} = (ka^2)^2\left(\frac{J_1(ka\theta)}{ka\theta}\right)^2. \tag{50.28}$$

We anticipate that the total cross section, in the small wavelength limit,

$$ka \gg 1, \quad \text{or} \quad k^{-1} = \lambda\!\!\!^{-} \ll a, \tag{50.29}$$

will be

$$\sigma \approx \pi a^2, \tag{50.30}$$

since this is just the geometrical area of the circular hole, and we are in the domain of "geometrical" optics. This can be shown explicitly by integrating (50.28) over all solid angles. The element of solid angle is approximately

$$d\Omega = 2\pi\sin\theta\, d\theta \approx 2\pi\,\theta\, d\theta, \tag{50.31}$$

since physically the wave is mostly scattered near the forward direction. Mathematically, this last fact is evident because the important values of θ are those for which the argument of the Bessel function is of order one, that is,

$$\theta \sim \frac{1}{ka} = \frac{\lambda}{a} \ll 1. \tag{50.32}$$

The total cross section is, therefore, approximately,

$$\sigma \approx 2\pi \int_0^{\pi/2} d\theta\, \theta\, (ka^2)^2 \left(\frac{J_1(ka\theta)}{ka\theta}\right)^2 \approx 2\pi a^2 \int_0^{ka\pi/2 \to \infty} dz\, z \left(\frac{J_1(z)}{z}\right)^2. \tag{50.33}$$

The expectation, (50.30), will be borne out if

$$\int_0^\infty \frac{dz}{z} (J_1(z))^2 = \frac{1}{2} \tag{50.34}$$

is true.

To prove (50.34), we first need an integral representation for J_1^2. This can be achieved by starting from the addition theorem for Bessel's functions, (18.65),

$$J_0(\sqrt{z^2 + z'^2 - 2zz' \cos\phi}) = \sum_{m=-\infty}^{\infty} J_m(z) J_m(z') e^{im\phi}, \tag{50.35}$$

setting $z = z'$,

$$J_0\left(2z \sin\frac{\phi}{2}\right) = \sum_{m=-\infty}^{\infty} [J_m(z)]^2 e^{im\phi}, \tag{50.36}$$

and extracting the $m = 1$ harmonic:

$$\int_0^{2\pi} \frac{d\phi}{2\pi} e^{-i\phi} J_0\left(2z \sin\frac{\phi}{2}\right) = [J_1(z)]^2. \tag{50.37}$$

We next insert this representation into the integral in (50.34). In order to be able to interchange the order of integration, we introduce a vanishingly small lower limit for z, $z_0 \to 0$, and obtain

$$\int_{z_0}^\infty \frac{dz}{z} [J_1(z)]^2 = \int_0^{2\pi} \frac{d\phi}{2\pi} e^{-i\phi} \int_{x_0}^\infty \frac{dx}{x} J_0(x), \tag{50.38}$$

where z and x are related by

$$2z \left|\sin\frac{\phi}{2}\right| = x. \tag{50.39}$$

Because of the orthonormality properties of the functions $e^{im\phi}$, (18.28), we are only interested in the coefficient of $e^{i\phi}$ in the Fourier series expansion of the x integral. Since $J_0(x \ll 1) \approx 1$, the x integral is

$$\int_{2z_0|\sin\phi/2|}^\infty \frac{dx}{x} J_0(x) \approx \ln \frac{1}{2z_0 \left|\sin\frac{\phi}{2}\right|} + \text{constant}. \tag{50.40}$$

By means of the following expansion,

$$\begin{aligned} \ln \frac{1}{|\sin\frac{\phi}{2}|} &= \frac{1}{2} \ln \frac{4}{(1 - e^{-i\phi})(1 - e^{i\phi})} \\ &= \ln 2 + \frac{1}{2}\left[\left(e^{-i\phi} + \frac{1}{2} e^{-2i\phi} + \dots\right) + \left(e^{i\phi} + \frac{1}{2} e^{2i\phi} + \dots\right)\right], \end{aligned} \tag{50.41}$$

we obtain the coefficient of $e^{i\phi}$ to be $1/2$, verifying the identity (50.34).

TABLE 50.1
Table of the first three zeroes of J_1, compared with the estimate from the asymptotic form (50.46).

z_1 (asymptotic)	z_1 (exact)
$\frac{3\pi}{4}+\frac{\pi}{2}=3.927$	3.832
$\frac{3\pi}{4}+\frac{3\pi}{2}=7.069$	7.016
$\frac{3\pi}{4}+\frac{5\pi}{2}=10.210$	10.172

Let us turn to a discussion of the angular distribution given by (50.28). For such small angles that $ka\theta \ll 1$, we need to know the behavior of $J_1(z)$ for small z, which is easily inferred from the integral representation, (18.29),

$$J_1(z) = \frac{1}{2\pi i}\oint d\phi\, e^{iz\cos\phi}e^{-i\phi} \approx \frac{1}{2\pi i}\oint d\phi\, iz\frac{e^{i\phi}+e^{-i\phi}}{2}e^{-i\phi} = \frac{z}{2}, \quad \text{for} \quad z \ll 1; \tag{50.42}$$

this, of course, follows from the series representation (18.26). Therefore, very near to the forward direction, the differential cross section is

$$\frac{d\sigma}{d\Omega} \approx \frac{1}{4}(ka^2)^2 = \frac{a^2}{4}\left(\frac{a}{\lambda\!\!\!^{-}}\right)^2. \tag{50.43}$$

Comparing this with the total cross section, (50.30), we see that, since $a/\lambda\!\!\!^{-} \gg 1$, the diffraction is mostly forward, consistent with our assumption.

For scattering angles such that $z = ka\theta \gg 1$, we make use of the asymptotic behavior of $J_1(z)$, which is given in (18.37) and used as a starting point for the asymptotic analysis of the zeros of the Bessel functions in Sec. 21.4. A direct derivation can be obtained by using the integral representation, (18.29), with $\phi = \phi' - \pi/2$:

$$J_1(z) = \oint \frac{d\phi'}{2\pi}e^{i(z\sin\phi'-\phi')} = \int_0^\pi \frac{d\phi'}{\pi}\cos(z\sin\phi' - \phi'). \tag{50.44}$$

For $z \gg 1$, the main contribution to (50.44) comes from the region near the stationary phase point, $\phi' \approx \pi/2$; therefore, we reintroduce $\phi = \phi' - \pi/2$ to obtain

$$J_1(z) = \int_{-\pi/2}^{\pi/2}\frac{d\phi}{\pi}\cos(z\cos\phi - \phi - \pi/2) \approx \mathrm{Re}\int_{-\pi/2}^{\pi/2}\frac{d\phi}{\pi}e^{i(z-\pi/2)}e^{-iz\phi^2/2}, \tag{50.45}$$

where we have expanded $z\cos\phi - \phi$ about the stationary phase point, $\phi \approx -1/z$. Evaluating the Gaussian integral, we find (for $z \gg 1$)

$$J_1(z) \sim \mathrm{Re}\, e^{i(z-\pi/2)}\frac{1}{\pi}\sqrt{\frac{2\pi}{iz}} = \sqrt{\frac{2}{\pi z}}\cos\left(z - \frac{\pi}{2} - \frac{\pi}{4}\right). \tag{50.46}$$

This is just (18.37). This asymptotic formula is already a good approximation for relatively small z.

The behavior of $2J_1(z)/z$, as suggested by (50.42) and (50.46), is given in Fig. 50.4, as well as the corresponding differential cross section. Here we see quite clearly a diffraction pattern, with minima occurring at the zeroes of $J_1(z)$, denoted by z_1. These zeroes can be found approximately from the asymptotic form (50.46). A comparison of these with the exact values is shown in Table 50.1 for the first three zeroes.

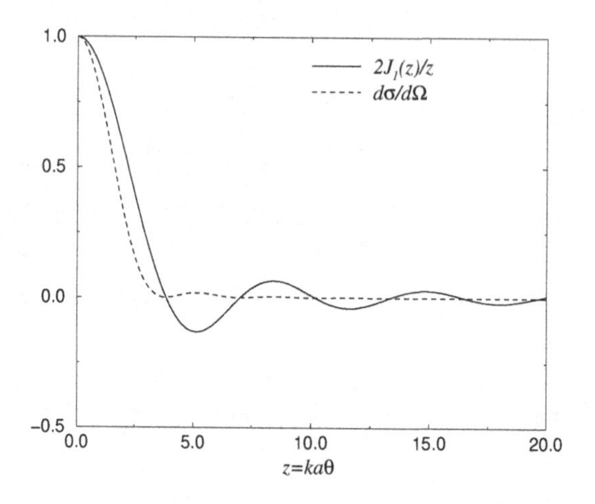

FIGURE 50.4
Behavior of $2J_1(z)/z$ and the differential cross section (normalized to unity at $\theta = 0$) given by (50.28) as a function of $z = ka\theta$.

50.3 Diffraction by a Slit

We now consider another example of diffraction, in which the aperture is a slit running along the entire z-axis. Because of this symmetry in z, when the electric field has no initial z-dependence, it acquires none subsequently. Therefore, we have an essentially two-dimensional problem, for which the preceding analysis applies, leading to (50.12): [Note that we cannot use (50.14) because the slit is infinitely long.]

$$E(x,y) = \frac{1}{4\pi}\int_{x'=0} dy'\,dz'\,E(y')\frac{\partial}{\partial x'}G(\mathbf{r}',\mathbf{r}). \tag{50.47}$$

(The justification for omitting the contribution of the surface at infinity is given in Problem 50.1.) Green's function is given by (50.7), or explicitly,

$$G(\mathbf{r}',\mathbf{r}) = \frac{e^{ik[(z'-z)^2+(\mathbf{r}'_\perp-\mathbf{r}_\perp)^2]^{1/2}}}{[(z'-z)^2+(\mathbf{r}'_\perp-\mathbf{r}_\perp)^2]^{1/2}} - \frac{e^{ik[(z'-z)^2+(\mathbf{r}'_\perp-\overline{\mathbf{r}}_\perp)^2]^{1/2}}}{[(z'-z)^2+(\mathbf{r}'_\perp-\overline{\mathbf{r}}_\perp)^2]^{1/2}}, \tag{50.48}$$

where

$$\mathbf{r}_\perp = (x,y), \quad \overline{\mathbf{r}}_\perp = (-x,y). \tag{50.49}$$

The z' integral here is identified as the Hankel function, (46.30),

$$\int_{-\infty}^{\infty} dz'\frac{e^{ik\sqrt{\rho^2+(z'-z)^2}}}{\sqrt{\rho^2+(z'-z)^2}} = i\pi H_0^{(1)}(k\rho). \tag{50.50}$$

(The Hankel function typically emerges in two-dimensional problems.) Thus the electric field to the right of the slit is

$$E(x,y) = \frac{i}{2}\int_{\text{slit}} dy'\,E(y')\left(-\frac{\partial}{\partial x}\right)H_0^{(1)}\left(k\sqrt{x^2+(y-y')^2}\right). \tag{50.51}$$

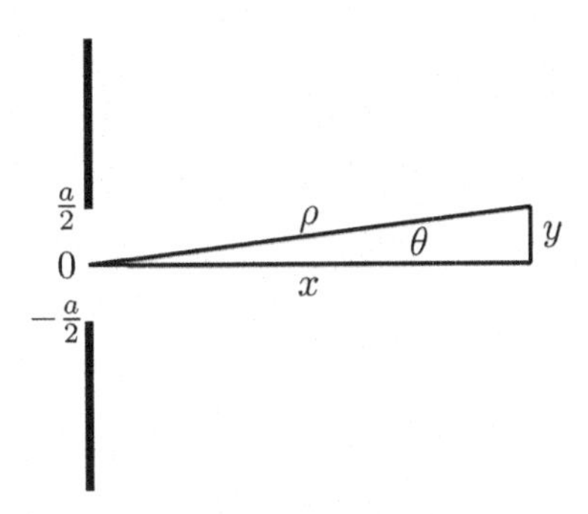

FIGURE 50.5
Geometry for diffraction by a slit.

We will be concerned with the diffracted field far away from the slit, for which we can use the asymptotic form (46.32),

$$H_0^{(1)}\left(k\sqrt{x^2+(y-y')^2}\right) \sim \sqrt{\frac{2}{i\pi k}}\frac{e^{ik[x^2+(y-y')^2]^{1/2}}}{[x^2+(y-y')^2]^{1/4}}, \quad (kx \gg 1). \tag{50.52}$$

To represent diffraction by a slit of width a, we adopt the coordinate system shown in Fig. 50.5. For a finite slit, since $a \ll x$, as well as

$$y' \sim a \ll \rho = \sqrt{x^2+y^2}, \tag{50.53}$$

we make the following approximation:

$$\sqrt{x^2+(y-y')^2} \approx \sqrt{\rho^2-2yy'} \sim \rho - \frac{y}{\rho}y' = \rho - \sin\theta\, y', \tag{50.54}$$

where θ is the angle of diffraction. Correspondingly, the asymptotic form of the Hankel function becomes

$$H_0^{(1)}\left(k\sqrt{x^2+(y-y')^2}\right) \sim \sqrt{\frac{2}{i\pi k\rho}}e^{ik\rho}e^{-ik\sin\theta y'}. \tag{50.55}$$

[We see here again a remnant of the factor

$$e^{-ik\mathbf{n}\cdot\mathbf{r}'}, \tag{50.56}$$

characteristic of radiation fields. See, for example, (50.16).] When the wavelength is small compared to the slit, $a \gg \lambda$, we may again approximate the field in the slit by the incident field,

$$E(y') \approx E_{\text{inc}}(y') = 1. \tag{50.57}$$

We also recognize that the radiation is predominately forward, implying

$$\begin{aligned} -\frac{\partial}{\partial x}e^{ik\rho} &= -ik\frac{x}{\rho}e^{ik\rho} \\ &\approx -ike^{ik\rho}. \end{aligned} \tag{50.58}$$

Thus, for $x \gg a \gg \lambda$, the field, (50.51), may then be approximated by

$$E(x,y) \sim \sqrt{\frac{i}{2\pi k\rho}}\int_{-a/2}^{a/2} dy'\,(-ik)e^{ik\rho}e^{-ik\sin\theta y'} = \sqrt{\frac{i}{2\pi k\rho}}(-ik)e^{ik\rho}\frac{2\sin\left(\frac{ka}{2}\sin\theta\right)}{k\sin\theta}, \tag{50.59}$$

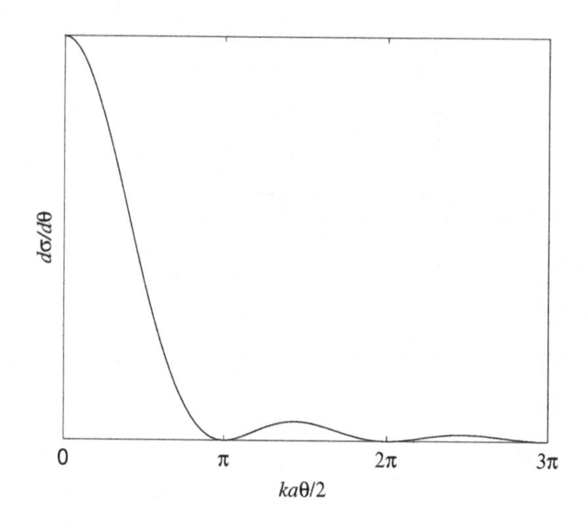

FIGURE 50.6
Differential cross section for diffraction by a wide slit, $a \gg \lambda$.

representing an outgoing cylindrical wave.

The differential cross section per unit length of the slit is determined by

$$d\sigma = \frac{|E|^2}{|E_{\text{inc}}|^2}\rho\, d\theta, \tag{50.60}$$

which, upon use of the asymptotic field, implies

$$\frac{d\sigma}{d\theta} = \frac{2}{\pi k}\left(\frac{\sin\left(\frac{ka}{2}\sin\theta\right)}{\sin\theta}\right)^2 \approx \frac{2}{\pi k}\left(\frac{\sin\left(\frac{ka\theta}{2}\right)}{\theta}\right)^2. \tag{50.61}$$

The latter holds for $\theta \ll 1$, which is the only region of validity for our result. The resulting diffraction pattern is shown in Fig. 50.6, where the zeroes occur at equally spaced points:

$$ka\theta = n2\pi, \quad n = \pm 1, \pm 2, \ldots. \tag{50.62}$$

At $\theta = 0$, the differential cross section is

$$\left(\frac{d\sigma}{d\theta}\right)_{\theta=0} = \frac{1}{2\pi k}(ka)^2. \tag{50.63}$$

In the short wavelength limit, we anticipate that the total cross section per unit length is given by the width of the slit. In fact, for $ka \gg 1$, most of the contribution arises from values of θ near zero so that we indeed have

$$\sigma \approx \int_{-\infty}^{\infty} d\theta\, \frac{2}{\pi k}\left(\frac{\sin\left(\frac{ka}{2}\theta\right)}{\theta}\right)^2 = \frac{2}{\pi k}\frac{ka}{2}\int_{-\infty}^{\infty} dz\left(\frac{\sin z}{z}\right)^2 = a, \tag{50.64}$$

where we have used the integral (37.17).

As a check of consistency, we now turn to the limit $a \to \infty$, for which the slit disappears and the incident plane wave should propagate undisturbed. The field far away from the "slit" is given by (50.51) and (50.52),

$$E(x,y) = \frac{i}{2}\int dy'\, E(y')\left(-\frac{\partial}{\partial x}\right)\left[\left(\frac{2}{i\pi k}\right)^{1/2}\frac{e^{ik\sqrt{x^2+(y-y')^2}}}{[x^2+(y-y')^2]^{1/4}}\right]. \tag{50.65}$$

If we put $E(y') = 1$ for all y', we must recover the incident plane wave, $E(x,y) = e^{ikx}$. We expect that $y' \sim y$ gives the major contribution to the integral, since the field should just advance with constant phase in y, thereby allowing us to use

$$\sqrt{x^2 + (y-y')^2} \approx x + \frac{(y-y')^2}{2x}. \tag{50.66}$$

Inserting this approximation into (50.65), we have the following asymptotic evaluation ($x \gg ƛ$):

$$\begin{aligned} E(x,y) &\sim \frac{i}{2}\int_{-\infty}^{\infty} dy' \left(-\frac{\partial}{\partial x}\right)\left[\left(\frac{2}{\pi i k x}\right)^{1/2} e^{ikx} e^{ik(y-y')^2/2x}\right] \\ &\sim -\frac{i}{2}\int_{-\infty}^{\infty} dy' \left(\frac{2}{\pi i k x}\right)^{1/2} ike^{ikx} e^{ik(y-y')^2/2x} \\ &= \sqrt{\frac{k}{2\pi i x}} e^{ikx} \int_{-\infty}^{\infty} dy'\, e^{ik(y-y')^2/2x} = e^{ikx}, \end{aligned} \tag{50.67}$$

as is expected. Here we note, as a check of the approximation $|y-y'| \ll x$, that the significant contributions to the Gaussian integral come from the values of y satisfying

$$\frac{|y-y'|}{x} \sim \sqrt{\frac{ƛ}{x}} \ll 1. \tag{50.68}$$

50.4 Diffraction by a Straight Edge

Finally, we consider the diffraction produced by a semi-infinite plane conductor, lying in the region defined by $x = 0$ and $y < 0$. The field produced by such a half plane conductor can be again described by (50.65), where, as a first approximation, we take

$$\begin{aligned} &E(y') \approx E_{\text{inc}} = 1, \quad y' > 0, \\ &E(y') \approx 0, \quad y' < 0. \end{aligned} \tag{50.69}$$

Using the approximation (50.66), we arrive at the expression [cf. (50.67)]

$$E(x,y) \sim \sqrt{\frac{k}{2\pi i x}} e^{ikx} \int_0^{\infty} dy'\, e^{-k(y-y')^2/2ix}, \tag{50.70}$$

which is valid for

$$x \gg ƛ \quad \text{and} \quad |y-y'| \ll x. \tag{50.71}$$

It is convenient to shift the origin of the y' integration:

$$\begin{aligned} E(x,y) &\sim \sqrt{\frac{k}{2\pi i x}} e^{ikx} \int_{-\infty}^{y} dy' e^{-ky'^2/2ix} = \sqrt{\frac{k}{2\pi i x}} e^{ikx} \left(\int_{-\infty}^{\infty} - \int_y^{\infty}\right) dy'\, e^{-ky'^2/2ix} \\ &= e^{ikx}\left\{1 - \sqrt{\frac{k}{2\pi i x}} \int_y^{\infty} dy'\, e^{-ky'^2/2ix}\right\}. \end{aligned} \tag{50.72}$$

For sufficiently large y,

$$y \gg \sqrt{ƛ x}, \tag{50.73}$$

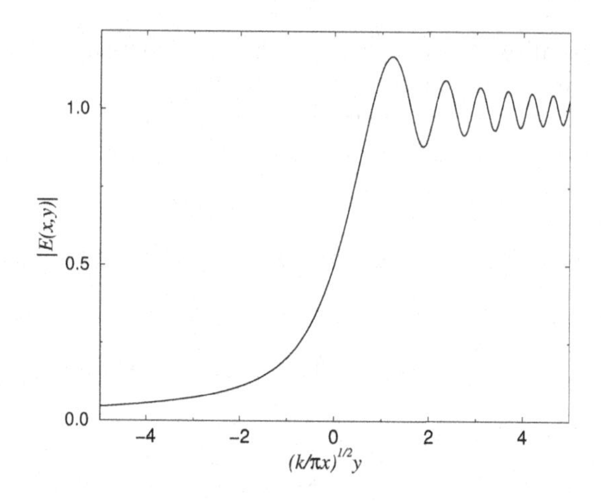

FIGURE 50.7
Magnitude of the electric field resulting from diffraction by a straight edge.

the second term in (50.72) may be neglected, and the wave travels undisturbed:

$$E(x, y) \sim e^{ikx}. \tag{50.74}$$

On the other hand, for $y = 0$ the integral in (50.72) is

$$\sqrt{\frac{k}{2\pi i x}} \int_0^\infty dy' \, e^{-ky'^2/2ix} = \frac{1}{2}, \tag{50.75}$$

and, in consequence, the amplitude of the wave there is reduced by half:

$$E(x, 0) \sim \frac{1}{2} e^{ikx}. \tag{50.76}$$

Far below the edge,

$$y < 0, \quad |y| \gg \sqrt{\lambda x}, \tag{50.77}$$

the diffracted field vanishes,

$$E(x, y) \sim 0. \tag{50.78}$$

To summarize the above results, we present a picture of the magnitude of the field strength as a function of y in Fig. 50.7. The region over which the intensity varies significantly has width

$$\Delta y \sim \sqrt{\lambda x}, \tag{50.79}$$

a distance small compared to x,

$$\frac{\Delta y}{x} \sim \sqrt{\frac{\lambda}{x}} \ll 1. \tag{50.80}$$

A quite different limit of diffraction by a straight edge occurs when both x and y are large, while the diffraction angle θ is fixed:

$$\frac{y}{x} = \tan\theta. \tag{50.81}$$

We anticipate that the dominant contribution to the scattered field comes from the region near the edge, where y' is small, and in consequence [see (50.54)]

$$\sqrt{x^2 + (y - y')^2} \approx \rho - y' \sin\theta, \tag{50.82a}$$

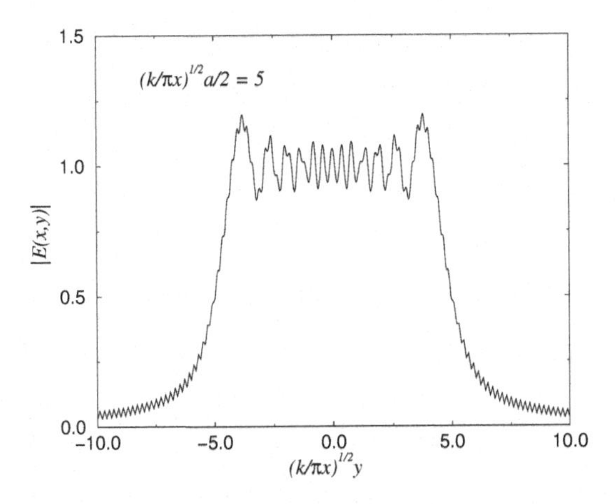

FIGURE 50.8
Diffraction by a slit as seen on a nearby screen.

where

$$\rho = \sqrt{x^2 + y^2}. \tag{50.82b}$$

Using this approximation in (50.65), we obtain, for $\theta \ll 1$,

$$E_{\text{scatt}} \sim \sqrt{\frac{i}{2\pi k\rho}}(-ik)e^{ik\rho} \int_0^\infty dy'\, e^{-iky'\sin\theta} \sim -\sqrt{\frac{i}{2\pi k\rho}} \frac{e^{ik\rho}}{\theta}. \tag{50.83}$$

We see the appearance of a cylindrical wave, originating from the edge, $x' = 0$, $y' = 0$. The corresponding differential scattering cross section per unit length is

$$\frac{d\sigma}{d\theta} = \frac{|E_{\text{scatt}}|^2}{|E_{\text{inc}}|^2}\rho \approx \frac{1}{2\pi k}\frac{1}{\theta^2}. \tag{50.84}$$

[In the next chapter, we will provide an exact treatment of this diffraction problem.]

The above discussion of diffraction by an edge provides a clarification of our earlier consideration of diffraction by a slit. If we make observations by use of a screen sufficiently close to the slit so that

$$a \gg \sqrt{\lambda x}, \tag{50.85}$$

we see a geometrical image of the slit modulated by edge diffraction (see Fig. 50.8). However, if we move the screen so far back that

$$\sqrt{\lambda x} \gg a, \tag{50.86}$$

the two patterns overlap to form the diffraction pattern derived in (50.61), see Fig. 50.9. Mathematically, the origin of the requirement (50.86) can be traced to the fact that the expansion (50.54) is only valid when terms of order a^2/ρ are negligible, that is

$$\frac{ka^2}{\rho} = \frac{1}{\lambda}\frac{a^2}{\rho} \ll 1, \tag{50.87a}$$

or

$$\lambda\rho \gg a^2. \tag{50.87b}$$

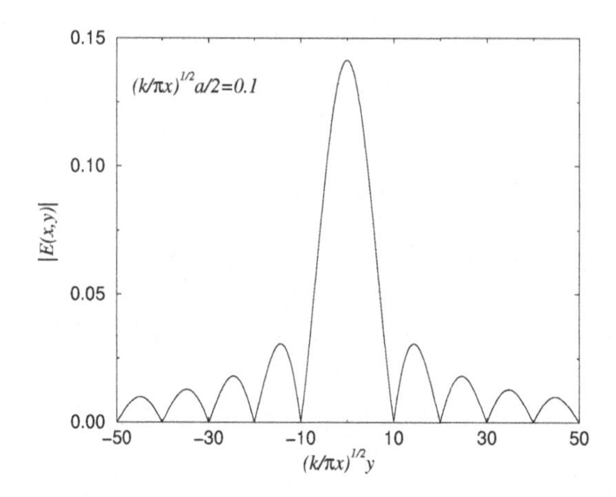

FIGURE 50.9
Diffraction pattern far away from a slit.

50.5 Problems for Chapter 50

1. Verify, in the situation of diffraction by a slit, that it is permissible to drop the contribution of the surface at infinity from the expression for the electric field given by (50.47).
2. Show that (50.72) can be represented in terms of Fresnel[3] integrals

$$C(x) = \int_0^x dt \cos\left(\frac{\pi}{2}t^2\right), \quad S(x) = \int_0^x dt \sin\left(\frac{\pi}{2}t^2\right), \tag{50.88}$$

as

$$E(x, y) = e^{ikx}\left\{\frac{1}{2} + \sqrt{\frac{1}{2i}}\left(C\left(\sqrt{\frac{k}{\pi x}}y\right) + iS\left(\sqrt{\frac{k}{\pi x}}y\right)\right)\right\}. \tag{50.89}$$

3. In the same approximation, give expressions for the electric field produced by a slit, and verify the results shown in Figs. 50.8 and 50.9.
4. Show that when the slit disappears ($a \to \infty$), (50.51) identically reduces to e^{ikx}.
5. A perfectly conducting sheet has two parallel infinitely long slits, of width a, that are separated by a distance $l \gg a$. An electromagnetic field of wavelength λ, polarized parallel to the slit edges, falls normally on the sheet. Using a reasonable physical assumption that does not involve the magnitude of a/λ, and the formalism applied in the text to one slit, compare the differential transmission cross section (per unit length) for the two slits to that of one slit. What are the new features of the diffraction pattern, particularly for $l/\lambda \gg 1$? How does this pattern alter if there are a very large number of equidistant slits?
6. Extend the argument in Section 50.2 to show that

$$\int_0^\infty \frac{dz}{z}[J_m(z)]^2 = \frac{1}{2m}, \quad m = 1, 2, \ldots,$$

which generalizes (50.34).

[3] Augustin-Jean Fresnel, 1788–1827.

51

Diffraction II

We now adopt a more physical approach to diffraction in which the currents that give rise to the scattered wave are made explicit. We will reconsider diffraction by a semi-infinite metal conductor, for which this method is capable of giving an exact solution. The geometry of the situation is as given in Fig. 51.1. We consider $\mathbf{E}$ to possess only a z-component (z subscript suppressed), and decompose the electric field into incident and scattered parts,

$$E = E_{\text{inc}} + E_{\text{scatt}}. \tag{51.1}$$

We assume the incident field is a normalized plane wave with frequency $\omega = kc$,

$$E_{\text{inc}} = e^{ikx}. \tag{51.2}$$

The scattered field arises from the induced current that flows on the metal plate, which by symmetry can have only a z-component, $J = J_z$, and has no dependence on z,

$$\frac{\partial}{\partial z} J_z(x, y) = 0, \tag{51.3}$$

in conformity with the properties of the incident electric field. Consequently, there is no charge density and the scattered electric field is expressed by

$$E_{\text{scatt}} = ik \int (d\mathbf{r}') \frac{e^{ik|\mathbf{r}-\mathbf{r}'|}}{|\mathbf{r}-\mathbf{r}'|} \frac{1}{c} J(\mathbf{r}'), \tag{51.4}$$

which follows from (38.4b) and (38.11a). We consider an infinitesimally thin conductor for which we may reduce the three-dimensional integral in (51.4) to a two-dimensional one by introducing the surface current,

$$\int dx'\, J_z = K. \tag{51.5}$$

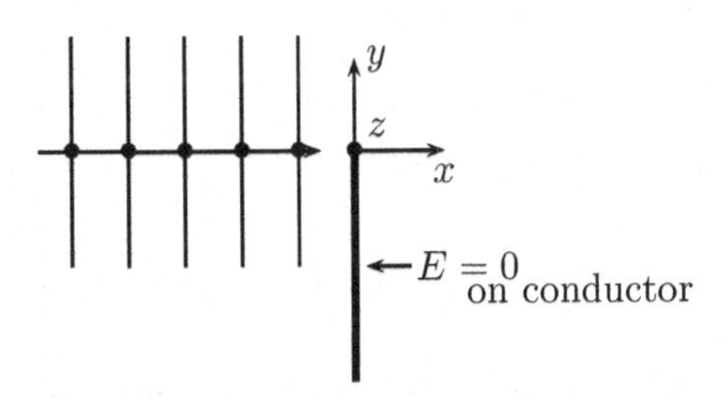

FIGURE 51.1
Geometry of diffraction by a perfectly conducting straight edge. The z axis is perpendicular to the page.

DOI: 10.1201/9781003057369-51

Further, by identifying the z integral with the representation of the Hankel function (46.30), we may express the result as an integration over y' alone:

$$E = E_{\text{inc}} + ik \int_{-\infty}^{0} dy' \, \pi i H_0^{(1)} \left(k\sqrt{x^2 + (y-y')^2}\right) \frac{1}{c} K(y'). \tag{51.6}$$

For the semi-infinite perfect conductor being considered here, (51.6) is subject to the boundary condition

$$E = 0 \quad \text{for} \quad x = 0,\, y < 0. \tag{51.7}$$

51.1 Approximate Solution

Before finding an exact solution to (51.6), we will first solve this equation by an approximate treatment, based on the fact that, on the conductor and far from the edge, the conducting sheet appears to be infinite. There, the incident wave is totally reflected,

$$E_z = e^{ikx} - e^{-ikx}, \quad B_y = -e^{ikx} - e^{-ikx}, \tag{51.8}$$

which we will assume hold for all $x < 0$, $y < 0$, and that the fields vanish for $x > 0$, $y < 0$. From these forms for the electric and magnetic fields, we can find the induced current through the use of Maxwell's equation

$$\boldsymbol{\nabla}\times\mathbf{B} = \frac{1}{c}\dot{\mathbf{E}} + \frac{4\pi}{c}\mathbf{J}, \tag{51.9}$$

which becomes here

$$\frac{4\pi}{c} J_z = \partial_x B_y - \frac{1}{c}\dot{E}_z. \tag{51.10}$$

By integrating (51.10) across the conducting surface from just to the left to just to the right,

$$B_y(x=+0) - B_y(x=-0) = \frac{4\pi}{c} K, \tag{51.11}$$

[this is (28.19a)], and noting that, in our approximation, (51.8),

$$y < 0: \quad B_y(x=+0) = 0, \quad B_y(x=-0) = -2, \tag{51.12}$$

we find the current appropriate to an infinite conducting sheet to be

$$K = \frac{c}{2\pi}. \tag{51.13}$$

Then from (51.6), the electric field everywhere is approximately given by

$$E = e^{ikx} - \frac{k}{2}\int_{-\infty}^{0} dy' \, H_0^{(1)}\left(k\sqrt{x^2 + (y-y')^2}\right). \tag{51.14}$$

If we further use the asymptotic form (50.52) for the Hankel function, together with the approximation that when $|x|$ is large, only small values of $|y - y'|$ are significant,

$$\sqrt{x^2 + (y-y')^2} \approx |x| + \frac{(y-y')^2}{2|x|}, \tag{51.15}$$

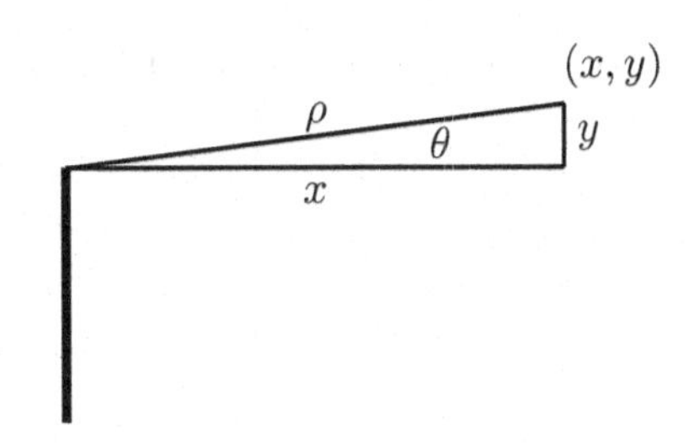

FIGURE 51.2
Diffraction by a straight edge at a fixed angle θ.

the expression for the electric field (51.14), becomes

$$E \sim e^{ikx} - \sqrt{\frac{k}{2\pi i|x|}} e^{ik|x|} \int_{-\infty}^{0} dy' \, e^{ik(y-y')^2/2|x|}. \tag{51.16}$$

With the substitution $y' - y \to y'$, the integral in (51.16) has the form

$$\int_{-\infty}^{-y} dy' \, e^{iky'^2/2|x|}. \tag{51.17}$$

When $y > 0$ and far away from the edge, $y \gg \sqrt{\lambda|x|}$, the integral (51.17) is negligible, and the wave propagates undisturbed,

$$E \sim e^{ikx}. \tag{51.18}$$

On the other hand, sufficiently below the edge, $y < 0$, $|y| \gg \sqrt{\lambda|x|}$, the integral is $\sqrt{2\pi i|x|/k}$, and

$$\begin{aligned} E &\sim e^{ikx} - e^{ik|x|} \\ &= \begin{cases} 0, & x > 0, \\ e^{ikx} - e^{-ikx}, & x < 0, \end{cases} \end{aligned} \tag{51.19}$$

reproducing the boundary condition (51.8). On line with the edge, $y = 0$, the electric field is

$$E \sim e^{ikx} - \frac{1}{2} e^{ik|x|}. \tag{51.20}$$

These results are identical with those found earlier in Section 50.4, but now a physical picture has been provided: The total electric field is the sum of the incoming field plus the field produced by the currents induced in the metal.

We next turn to the limit of large x and y where the angle of diffraction θ is fixed, as shown in Fig. 51.2. In this limit, the dominant contribution at (x, y) is the sum of the incident wave plus the wave scattered by the edge. In terms of the description employing currents, this is given approximately by (51.14), with

$$\sqrt{x^2 + (y - y')^2} \approx \rho - y' \sin\theta. \tag{51.21}$$

As in (50.83), the scattered field is

$$E_{\text{scatt}} \sim -\frac{k}{2} \int_{-\infty}^{0} dy' \sqrt{\frac{2}{i\pi k\rho}} e^{ik\rho} e^{-ik\sin\theta y'} \sim -\sqrt{\frac{i}{2\pi k}} \frac{1}{\sqrt{\rho}} e^{ik\rho} \frac{1}{\sin\theta}, \tag{51.22}$$

leading to a differential cross section per unit length (valid only for $\theta \ll 1$)

$$\frac{d\sigma}{d\theta} = \frac{1}{2\pi k}\frac{1}{\theta^2}, \tag{51.23}$$

in agreement with (50.84).

51.2 Exact Solution for Current

We now seek an exact solution to (51.6) subject to the boundary condition (51.7). This solution was first found by Sommerfeld. Equation (51.6) is an integral equation since $K(y')$ and $E(y)$ are interrelated functions. To find the surface current, we consider the electric field on the $x = 0$ plane,

$$E(y) = 1 - \pi k \int_{-\infty}^{\infty} dy'\, H_0^{(1)}(k|y - y'|)\, \frac{1}{c} K(y'). \tag{51.24}$$

In order to solve this equation, we must recognize that an incident plane wave is an over-idealization, one that can be removed by introducing an exponential cutoff in y:

$$E_{\text{inc}}(x = 0) = 1 \to e^{-\epsilon|y|}, \quad \epsilon \to +0. \tag{51.25}$$

Note in (51.24) that we have two conditions:

1. $K(y') = 0$ if $y' > 0$ (since there is no conductor there),
2. $E(y) = 0$ if $y < 0$ (since the conductor is perfect).

We have introduced an infinite range of integration in (51.24) in order to employ Fourier transforms:

$$E(\zeta) = \int_{-\infty}^{\infty} dy\, e^{-i\zeta y} E(y), \tag{51.26a}$$

$$K(\zeta) = \int_{-\infty}^{\infty} dy'\, e^{-i\zeta y'} K(y'). \tag{51.26b}$$

The Fourier transform of the incident field on the surface is

$$\int_{-\infty}^{\infty} dy\, e^{-i\zeta y} e^{-\epsilon|y|} = -\frac{i}{\zeta - i\epsilon} + \frac{i}{\zeta + i\epsilon}, \tag{51.27}$$

which, as expected, becomes $2\pi\delta(\zeta)$ as $\epsilon \to 0$. Furthermore, we require the Fourier transform of the Hankel function. Starting from its integral representation, (46.30), using the three-dimensional Fourier transform of Green's function, (46.17), and then integrating over z, we have

$$\begin{aligned}
\pi i H_0^{(1)}(k|y - y'|) &= \int_{-\infty}^{\infty} dz \frac{e^{ik\sqrt{(y-y')^2+z^2}}}{\sqrt{(y-y')^2 + z^2}} \\
&= 4\pi \int_{-\infty}^{\infty} dz \int \frac{dk_x dk_y dk_z}{(2\pi)^3} \frac{e^{ik_y(y-y')} e^{ik_z z}}{k_x^2 + k_y^2 + k_z^2 - \left(\frac{\omega + i\epsilon}{c}\right)^2} \\
&= 4\pi \int \frac{dk_x dk_y}{(2\pi)^2} \frac{e^{ik_y(y-y')}}{k_x^2 + k_y^2 - \left(\frac{\omega+i\epsilon}{c}\right)^2}.
\end{aligned} \tag{51.28}$$

(This is just the two-dimensional retarded Green's function.) Employing the simple contour integral

$$\int_{-\infty}^{\infty}\frac{dk_x}{2\pi}\frac{1}{k_x^2+k_y^2-\left(\frac{\omega+i\epsilon}{c}\right)^2}=\frac{i}{2\sqrt{(k+i\epsilon)^2-k_y^2}},\qquad k=\frac{\omega}{c},\tag{51.29}$$

(where the cuts are chosen not to cross the real k axis), we arrive at the following representation for the Hankel function,

$$\pi i H_0^{(1)}(k|y-y'|)=2\pi i\int_{-\infty}^{\infty}\frac{dk_y}{2\pi}\frac{e^{ik_y(y-y')}}{\sqrt{(k+i\epsilon)^2-k_y^2}},\tag{51.30}$$

which supplies the desired Fourier transform

$$\int_{-\infty}^{\infty}d(y-y')\,e^{-i\zeta(y-y')}H_0^{(1)}(k|y-y'|)=\frac{2}{\sqrt{(k+i\epsilon)^2-\zeta^2}}.\tag{51.31}$$

We now find for the Fourier transform of the integral equation (51.24),

$$E(\zeta)=-\frac{i}{\zeta-i\epsilon}+\frac{i}{\zeta+i\epsilon}-\pi k\frac{2}{\sqrt{(k+i\epsilon)^2-\zeta^2}}\frac{1}{c}K(\zeta),\tag{51.32}$$

where, if we make explicit the regions in which the integrands are nonzero,

$$E(\zeta)=\int_0^{\infty}dy\,e^{-i\zeta y}E(y),\tag{51.33a}$$

$$K(\zeta)=\int_{-\infty}^{0}dy'\,e^{-i\zeta y'}K(y').\tag{51.33b}$$

If these integrals exist for real ζ, they will also exist for complex values of ζ. Anticipating that $E(y)$, $K(y)$ fall off like $e^{-\epsilon|y|}$ as $|y|\to\infty$, we see that

$$E(\zeta)\quad\text{exists for}\quad \text{Im}\,\zeta<\epsilon,\tag{51.34a}$$

$$K(\zeta)\quad\text{exists for}\quad \text{Im}\,\zeta>-\epsilon.\tag{51.34b}$$

We will call the half planes $\text{Im}\,\zeta<\epsilon$, $\text{Im}\,\zeta>-\epsilon$ the lower half plane (LHP) and upper half plane (UHP), respectively. It is essential to observe that the UHP and the LHP overlap in a strip,

$$-\epsilon<\text{Im}\,\zeta<\epsilon.\tag{51.35}$$

Our physical requirements of boundedness ensure that

$$E(\zeta)\quad\text{is regular in the LHP,}\tag{51.36a}$$

$$K(\zeta)\quad\text{is regular in the UHP.}\tag{51.36b}$$

In order to examine clearly the analytic properties of (51.32), we multiply it by $\sqrt{k+i\epsilon-\zeta}$:

$$\sqrt{k+i\epsilon-\zeta}\,E(\zeta)=-i\frac{\sqrt{k+i\epsilon-\zeta}}{\zeta-i\epsilon}+i\frac{\sqrt{k+i\epsilon-\zeta}}{\zeta+i\epsilon}-2\pi k\frac{\frac{1}{c}K(\zeta)}{\sqrt{k+i\epsilon+\zeta}}.\tag{51.37}$$

The factors in (51.37) can be chosen to be regular in the following regions,

$$\sqrt{k+i\epsilon-\zeta}:\text{LHP},\tag{51.38a}$$

$$\sqrt{k+i\epsilon+\zeta}:\text{UHP},\tag{51.38b}$$

$$\frac{\sqrt{k+i\epsilon-\zeta}}{\zeta-i\epsilon}:\text{LHP},\tag{51.38c}$$

$$\frac{\sqrt{k+i\epsilon-\zeta}}{\zeta+i\epsilon}:-\epsilon<\text{Im}\,\zeta<\epsilon.\tag{51.38d}$$

The last combination can be written as the sum of terms regular in the LHP and the UHP, respectively,

$$\frac{\sqrt{k+i\epsilon-\zeta}}{\zeta+i\epsilon}=\frac{\sqrt{k+i\epsilon-\zeta}-\sqrt{k+2i\epsilon}}{\zeta+i\epsilon}+\frac{\sqrt{k+2i\epsilon}}{\zeta+i\epsilon}. \tag{51.39}$$

Thus we can reorganize (51.37) into parts that are regular in the LHP and in the UHP:

$$\begin{aligned}\sqrt{k+i\epsilon-\zeta}E(\zeta)+i\frac{\sqrt{k+i\epsilon-\zeta}}{\zeta-i\epsilon}-i\frac{\sqrt{k+i\epsilon-\zeta}-\sqrt{k+2i\epsilon}}{\zeta+i\epsilon}\\ =i\frac{\sqrt{k+2i\epsilon}}{\zeta+i\epsilon}-2\pi k\frac{\frac{1}{c}K(\zeta)}{\sqrt{k+i\epsilon+\zeta}}.\end{aligned} \tag{51.40}$$

The right-hand side of (51.40) is regular in the UHP, the left-hand side in the LHP. Since the two functions are regular in a common region [the strip (51.35)], they may be analytically continued into a function regular for all ζ.

We will now show that this function vanishes at infinity so that it vanishes everywhere. To this end, we examine (51.33b) in the limit $\zeta\to\infty$, for which only the behavior of $K(y')$ for $y'\to 0$ is significant. Because there is no intrinsic length scale in this limit the current must behave as a power of y',

$$K(y')\sim(-y')^{-\alpha},\quad y'\to -0, \tag{51.41}$$

where, in order that the integral (51.33b) exist,

$$\alpha<1. \tag{51.42}$$

The behavior of the corresponding Fourier transform of the current for large ζ is therefore

$$K(\zeta)\sim\int_{-\infty}^{0}e^{-i\zeta y'}(-y')^{-\alpha}dy'\sim\frac{1}{\zeta^{1-\alpha}}\to 0,\quad \zeta\to\infty. \tag{51.43}$$

Similarly, since $E(y)=0$ if $y<0$, the continuity of E requires the following power law for E near the edge,

$$E(y)\sim y^{\beta},\quad y\to +0,\quad \beta>0, \tag{51.44}$$

which is equivalent to the asymptotic statement

$$E(\zeta)=\int_{0}^{\infty}e^{-i\zeta y}E(y)\,dy\sim\frac{1}{\zeta^{1+\beta}}\to 0,\quad \zeta\to\infty. \tag{51.45}$$

Hence, the function represented by the analytic continuation of either side of (51.40) vanishes at infinity, and thus, by Cauchy's theorem[1] (mistakenly attributed to Liouville) is zero everywhere. Thus we have solutions for $K(\zeta)$ and $E(\zeta)$:

$$\frac{2\pi}{c}K(\zeta)=\frac{i}{\sqrt{k}}\frac{\sqrt{k+i\epsilon+\zeta}}{\zeta+i\epsilon} \tag{51.46a}$$

$$\sim\frac{1}{\sqrt{\zeta}},\quad |\zeta|\gg 1, \tag{51.46b}$$

$$E(\zeta)=i\left(\frac{1}{\zeta+i\epsilon}-\frac{1}{\zeta-i\epsilon}\right)-\frac{i\sqrt{k}}{\zeta+i\epsilon}\frac{1}{\sqrt{k+i\epsilon-\zeta}} \tag{51.46c}$$

$$\sim\frac{1}{\zeta^{3/2}},\quad |\zeta|\gg 1. \tag{51.46d}$$

[1] Proved in 1844.

Comparison with (51.43) and (51.45) determines the powers

$$\alpha = \frac{1}{2}, \quad \beta = \frac{1}{2}, \tag{51.47}$$

implying the following spatial dependences,

$$K(y') \sim \frac{1}{\sqrt{|y'|}}, \quad y' < 0, \tag{51.48a}$$

$$E(y) \sim \sqrt{y}, \quad y > 0, \tag{51.48b}$$

as $|y'|$, $y \to 0$. We will supply a physical discussion of the meaning of the behavior near the edge in Section 51.4.

51.3 Exact Diffraction Cross Section

The exact current and field in the plane $x = 0$ are given by (51.46a) and (51.46c). By rewriting the expression for the surface current in the form

$$\frac{1}{c}K(\zeta) = \frac{i}{2\pi}\left\{\frac{1}{\zeta + i\epsilon} + \frac{1}{\sqrt{k}}\frac{\sqrt{k+\zeta+i\epsilon} - \sqrt{k}}{\zeta + i\epsilon}\right\}, \tag{51.49}$$

we can identify the first term as the Fourier transform of

$$\frac{1}{2\pi}\begin{cases} e^{-\epsilon|y|}, y < 0, \\ 0, \quad y > 0, \end{cases} \tag{51.50}$$

which is the current, (51.13), we used in the first approximate solution to this problem, corresponding to the neglect of edge effects. [Here, (51.50) includes the exponential cutoff.] The second term in (51.49) thus gives the correction that must be added in order to obtain the exact current. By considering the behavior of (51.49) for ζ small (which, as we will see below, corresponds to small diffraction angles), we note that the first term is singular as $\zeta \to 0$, while the second is finite. Thus, the first approximation is valid for small angles, as we have previously asserted.

The asymptotic scattered field at fixed angle θ follows by use of (50.55) in (51.6):

$$E_{\text{scatt}} \sim -\sqrt{\frac{k}{2\pi i\rho}}e^{ik\rho}\int_{-\infty}^{0} dy'\, e^{-ik\sin\theta y'}\frac{2\pi}{c}K(y') = -\sqrt{\frac{k}{2\pi i\rho}}e^{ik\rho}\frac{2\pi}{c}K(\zeta), \tag{51.51}$$

where we have used the Fourier transform (51.33b) with

$$\zeta = k\sin\theta. \tag{51.52}$$

Then from the solution (51.46a) for the surface current, we find for the exact asymptotic scattered field

$$E_{\text{scatt}} \sim -\sqrt{\frac{i}{2\pi k\rho}}e^{ik\rho}\frac{\sqrt{1+\sin\theta}}{\sin\theta}. \tag{51.53}$$

As we anticipated, it agrees with the first approximation (51.22) when $\theta \ll 1$. The corresponding exact differential cross section per unit length is

$$\frac{d\sigma}{d\theta} = \frac{1}{2\pi k}\frac{1+\sin\theta}{\sin^2\theta}, \tag{51.54}$$

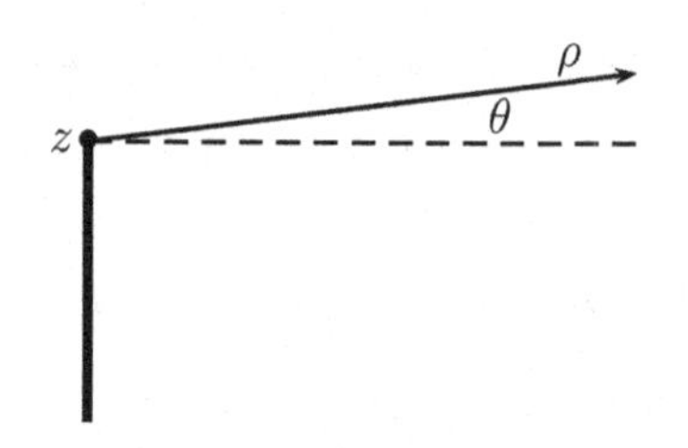

FIGURE 51.3
Cylindrical coordinate system for diffraction by a straight edge.

generalizing (51.23). We recognize that the new factor in (51.53), $\sqrt{1+\sin\theta}$, is present in order to enforce the boundary condition that the electric field vanish at $\theta = -\frac{\pi}{2}, \frac{3\pi}{2}$.

Finally, we wish to make contact with the other method, in which the field in the "aperture," not the current on the conducting plate is employed. That is, we use (50.51), relating the scattered field to the field in the aperture, and (50.55), the asymptotic form of the Hankel function, to give the scattered electric field for finite diffraction angle θ, $|\theta| < \frac{\pi}{2}$,

$$E_{\text{scatt}} \sim \sqrt{\frac{k}{2\pi i}} \frac{e^{ik\rho}}{\sqrt{\rho}} \cos\theta \int_0^\infty dy\, e^{-ik\sin\theta y} E(y), \tag{51.55}$$

where $E(y)$ is the exact electric field in the aperture. The integral in (51.55) is the Fourier transform (51.33a),

$$\int_0^\infty dy\, e^{-ik\sin\theta y} E(y) = E(\zeta = k\sin\theta) = -i\frac{1}{k\sin\theta}\frac{1}{\sqrt{1-\sin\theta}}, \tag{51.56}$$

according to (51.46c). The resulting scattered electric field coincides with (51.53).

51.4 Field Near Edge

Here we wish to examine the form of the field near the edge, (51.48b), from a different point of view. We may alternatively derive this result by solving the wave equation, (50.4),

$$\left(\nabla^2 + \frac{1}{\lambda^2}\right) E = 0, \tag{51.57}$$

near the edge. Since there the field is rapidly varying over a distance small compared to the wavelength, we can omit the $1/\lambda^2$ term. Thus our problem is the electrostatic one of finding the field near the edge of a plane conductor. Using the cylindrical coordinate system shown in Fig. 51.3, we write the Laplacian as

$$\nabla^2 = \left(\frac{1}{\rho}\frac{\partial}{\partial\rho}\rho\frac{\partial}{\partial\rho} + \frac{1}{\rho^2}\frac{\partial^2}{\partial\theta^2}\right), \tag{51.58}$$

since there is no z dependence. By separating variables ($\partial^2/\partial\theta^2 \to -m^2$), we find the characteristic solutions to Laplace's equation to be

$$E \sim \begin{Bmatrix} \sin m\theta \\ \cos m\theta \end{Bmatrix} \rho^m. \tag{51.59}$$

The boundary conditions that the field must vanish on the conducting plane,

$$E = 0 \text{ at } \theta = -\frac{\pi}{2} \text{ and } \frac{3\pi}{2}, \tag{51.60}$$

imply that the characteristic solutions are

$$E \sim \sin m(\theta + \pi/2)\rho^m, \tag{51.61}$$

with

$$\sin 2\pi m = 0. \tag{51.62}$$

The smallest value of m consistent with (51.62) is

$$m = \frac{1}{2}, \tag{51.63}$$

giving the electric field

$$E = C\sqrt{\rho}\sin\frac{1}{2}(\theta + \pi/2), \tag{51.64}$$

where C is a constant. [Note that the solution (51.53) exhibits this same behavior, since $\sqrt{1+\sin\theta} = \sqrt{2}\sin\frac{1}{2}(\theta + \pi/2)$.] For $x = 0$, $y > 0$ ($\theta = \pi/2$), the electric field is

$$E(\rho, \frac{\pi}{2}) = C\sqrt{\rho} = C\sqrt{y}, \tag{51.65}$$

which is (51.48b). Another way of writing (51.64) is

$$E = C\ \mathrm{Im}\left[\sqrt{\rho}e^{\frac{i}{2}(\theta+\pi/2)}\right] = C\ \mathrm{Im}\left[\sqrt{\rho e^{i\theta}}\sqrt{i}\right] = C\ \mathrm{Im}\sqrt{i(x+iy)}. \tag{51.66}$$

We note that a general solution to the two-dimensional Laplace's equation,

$$\left(\frac{\partial^2}{\partial x^2} + \frac{\partial^2}{\partial y^2}\right)E = 0, \tag{51.67}$$

has the form

$$E = \mathrm{Im}\ f(x+iy), \tag{51.68}$$

where f is a locally regular function. Equation (51.66) has this property away from the origin, and satisfies the appropriate boundary condition,

$$E = 0 \quad \text{for} \quad x = 0,\, y < 0, \tag{51.69}$$

since there

$$E = C\ \mathrm{Im}\sqrt{i(x+iy)} \to C\ \mathrm{Im}\sqrt{-y} = 0. \tag{51.70}$$

Next, we apply the above ideas to the situation of a slit, as shown in Fig. 51.4. When the slit is small compared to the wavelength,

$$a \ll \lambda\!\!\!^{-}, \tag{51.71}$$

the solution to Laplace's equation becomes relevant here. Since the appropriate boundary conditions are

$$E = 0 \quad \text{when} \quad x = 0, |y| > \frac{a}{2}, \tag{51.72}$$

the electric field is found by generalizing our previous result, (51.66),

$$E = C\ \mathrm{Im}\left[i\left(x + i(y + a/2)\right)i\left(x + i(y - a/2)\right)\right]^{1/2}. \tag{51.73}$$

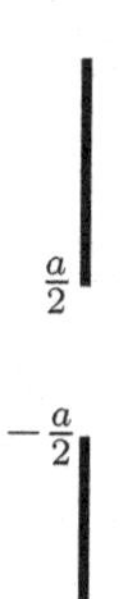

FIGURE 51.4
Geometry of a slit.

Explicitly, we note that on the surface $x = 0$, this reduces to

$$E = C \text{ Im } \sqrt{y^2 - (a/2)^2} = C \text{ Im } i\sqrt{a^2/4 - y^2}$$
$$= \begin{cases} 0, & |y| > \frac{a}{2}, \\ C\sqrt{a^2/4 - y^2}, & |y| < \frac{a}{2}. \end{cases} \tag{51.74}$$

To determine the constant C for the diffraction problem, we consider the fields far from the slit,

$$|x| \gg z, |y|, \tag{51.75}$$

where the static result, (51.73), becomes

$$E \sim C \text{ Im } i|x| = C|x|. \tag{51.76}$$

In order to reproduce the static limit in the absence of the slit ($a \to 0$), we superimpose on this field another solution to Laplace's equation, $-Cx$, which satisfies the same boundary conditions, (51.72), so that we have the following asymptotic solution,

$$E \sim -Cx + C|x| = \begin{cases} 0, & x > 0, \\ -2Cx, & x < 0. \end{cases} \tag{51.77}$$

Far from the surface, we can no longer neglect $k^2 = 1/\lambda^2$ in (51.57). Consequently, the approximate diffracted wave, the static limit of which is (51.77), is

$$E = -\frac{2C}{k} \sin kx, \quad x < 0. \tag{51.78}$$

This is exact for zero aperture, $a = 0$, since it represents a standing wave due to total reflection by the conducting plane. The associated magnetic field,

$$ikB_y = -\frac{\partial}{\partial x} E_z = 2C \cos kx, \tag{51.79}$$

does not vanish on the conducting plane,

$$ikB_y(x = 0) = 2C, \tag{51.80}$$

thereby determining the constant C in terms of the unperturbed magnetic field (that is, in terms of the magnetic field present when the slit is absent). Thus from (51.74), the electric field in the aperture, $|y| < a/2$, is determined by the magnetic field at $x = 0$,

$$E_z(x=0) \approx \frac{ik}{2} B_y(x=0)\sqrt{a^2/4 - y^2}, \tag{51.81}$$

for a narrow slit satisfying (51.71). The magnitude of the electric field in the aperture is small compared to the magnetic field there,

$$|E_z| \sim ka|B_y| \ll |B_y|, \tag{51.82}$$

and would be zero if (51.78) were exact. Equation (51.81) is the basis for treating the diffraction by a narrow slit.

51.5 Historical Note

As noted above, the exact solution to the problem of diffraction by a straight edge was worked out in 1896 by Sommerfeld [50]. He also obtained a solution for a wedge geometry, but not so explicit. Remarkably, the Green's dyadic for a wedge can be obtained in closed form [51].

51.6 Problems for Chapter 51

1. Use the development in the last Section to calculate the diffraction pattern produced by a narrow slit, $a/\lambda \ll 1$.
2. In this chapter, we obtained the exact electric field diffracted by a straight edge. Use this result to obtain the limiting result for the regime

$$\frac{\Delta y}{x} \sim \sqrt{\frac{\lambda}{x}} \ll 1 \tag{51.83}$$

   discussed in Section 50.3.
3. A slot of finite length l and very small width a is cut in a perfectly conducting sheet. An electromagnetic field of wavelength λ, polarized parallel to the long axis of the slot (z-axis), falls normally on the sheet. Find the angular (θ, ϕ) distribution of the transmitted radiation in an approximation valid for $l/\lambda \gg 1$. Compare the θ dependence with that of the uniformly excited antenna discussed in Section 37.1; explain the difference in the ϕ distribution.

52

Babinet's Principle

In the preceding chapter, we considered the diffraction of a plane wave with a particular polarization, in which the electric field has only a z component. We will show here how we can obtain results for the other polarization, in which the magnetic field has only a z-component, without additional effort.

We first recognize that because of the linearity of Maxwell's equations we can write the solution to our original problem, in which the incident wave approaches the screen from the left, as the superposition of the solutions of two problems, in each of which the incident wave approaches the screen from both the left and the right, as illustrated by Fig. 52.1. The arrows indicate the sense of propagation, while the signs denote the sense of E_z. The electric field corresponding to the first term here is odd under reflection in the $x = 0$ plane, so it satisfies the boundary condition

$$E = 0 \quad \text{at} \quad x = 0. \tag{52.1}$$

That is, the entire $x = 0$ surface can be considered to be a perfect electric conductor. The electric field in the second term is even under reflection,

$$E(x) = E(-x). \tag{52.2}$$

In this case we have two regimes:

$$y < 0: E_z = 0 \quad \text{at} \quad x = 0, \tag{52.3a}$$

$$y > 0: \frac{\partial E_z}{\partial x} = 0 \Rightarrow B_y = 0 \quad \text{at} \quad x = 0, \tag{52.3b}$$

where we have used the continuity of the derivative for $y > 0$. Thus the second term corresponds to a perfect magnetic conductor (for which the tangential magnetic field vanishes) for $y > 0$ and to a perfect electric conductor for $y < 0$. The situation for $x > 0$ is shown in Fig. 52.2, where a solid line represents a perfect electric conductor, and a dashed line a perfect magnetic conductor.

We now consider another situation (the "dual" of the above), in which the incoming wave has polarization B_z, and strikes a conducting plane $x = 0$, $y > 0$. The decomposition into two terms may be made as before, as shown in Fig. 52.3, where the +, − signs now

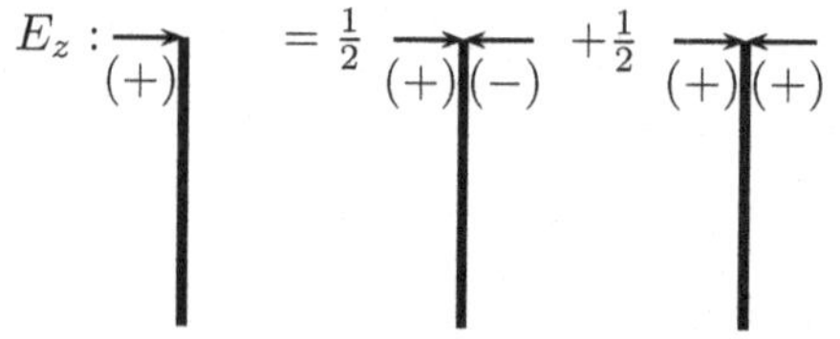

E_z : (+) $= \frac{1}{2}$ (+)(−) $+\frac{1}{2}$ (+)(+)

FIGURE 52.1
Superposition of solutions for E_z polarization.

DOI: 10.1201/9781003057369-52

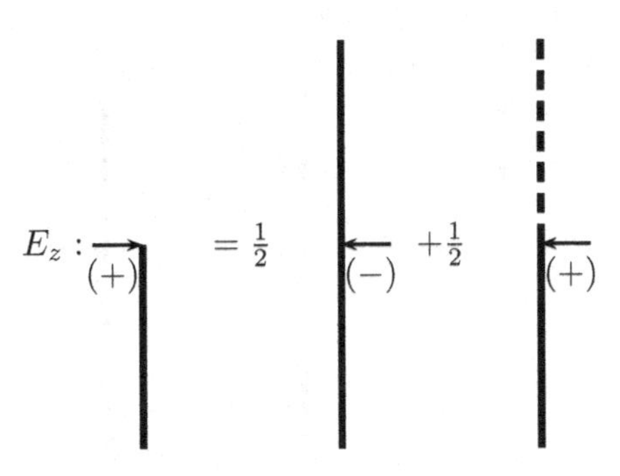

FIGURE 52.2
Electric fields for $x > 0$, for the E_z polarization.

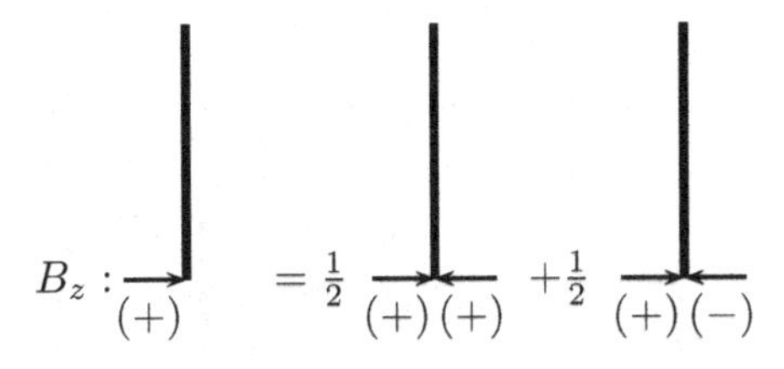

FIGURE 52.3
Superposition of solutions for the B_z polarization.

refer to the sense of B_z. In the first term here, B_z is even, so at $x = 0$, $y < 0$,

$$\frac{\partial}{\partial x} B_z = 0 \Rightarrow E_y = 0. \tag{52.4}$$

Correspondingly, the entire $x = 0$ plane acts as a perfect electric conductor. In the second term, B_z is odd under reflection in the $x = 0$ plane, so we may, as in (52.3), describe that situation in two regions,

$$y < 0 : \; B_z = 0 \quad \text{at} \quad x = 0, \tag{52.5a}$$
$$y > 0 : \; E_y = 0 \quad \text{at} \quad x = 0, \tag{52.5b}$$

since B_z is continuous for $y < 0$. Thus the $x = 0$ plane acts as a perfect electric conductor for $y > 0$, and as a perfect magnetic conductor for $y < 0$. So for this polarization the situation for $x > 0$ appears as shown in Fig. 52.4. For both the above circumstances, reflection from the infinite conducting planes corresponds to exactly forward scattering, and can therefore be ignored when we calculate the scattering cross section. The remaining contributions differ only in that electric and magnetic quantities are interchanged,

$$E \to B, \quad B \to -E. \tag{52.6}$$

Since Maxwell's equations are invariant under this transformation (recall Chapter 2), the cross sections for these two problems are identical.

An immediate generalization of the above arguments shows that, for normal incidence, the problems (E_z, aperture, screen) and (B_z, screen, aperture) correspond to the same cross sections, but with orthogonal polarizations. See Fig. 52.5. This is an example of what is called Babinet's[1] principle.

[1] Jacques Babinet, 1794–1872.

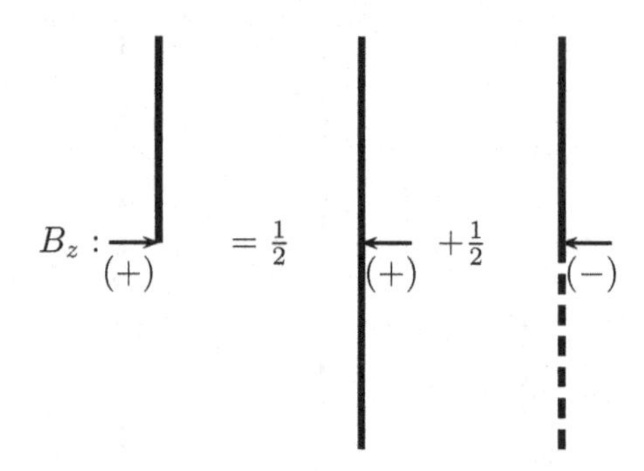

FIGURE 52.4
Magnetic fields for $x > 0$, for the B_z polarization.

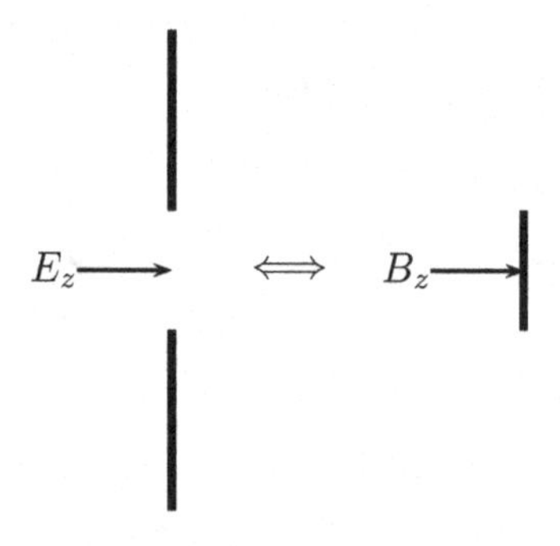

FIGURE 52.5
Equivalence between dual diffraction problems.

52.1 Problems for Chapter 52

1. Compute the cross section for the scattering of a normally incident plane polarized electromagnetic wave by a circular disk, in the spirit of Section 50.2, and thereby verify Babinet's principle.
2. Consider a perfectly conducting screen with a circular annulus cut in it, of inner radius a and outer radius b, as shown in Fig. 52.6. A plane wave, of frequency ω, polarized in the plane of the screen, is normally incident on the aperture. Using the approximation of Section 50.2, based on $\lambda \ll a$, b, or $b - a$, calculate
   (a) the electric field far to the right of the screen,
   (b) the differential cross section, $d\sigma/d\Omega$, and its value at zero scattering angle,
   (c) the total cross section. Is the latter what you expect geometrically? Is this consistent with Babinet's principle?
   (d) What is the electric field directly behind the central disk, *i.e.*, at $\theta = 0$, $r \to \infty$? Is it what you expect? Plot the electric field as a function of $kb\theta$ for $a/b = 1/2$.

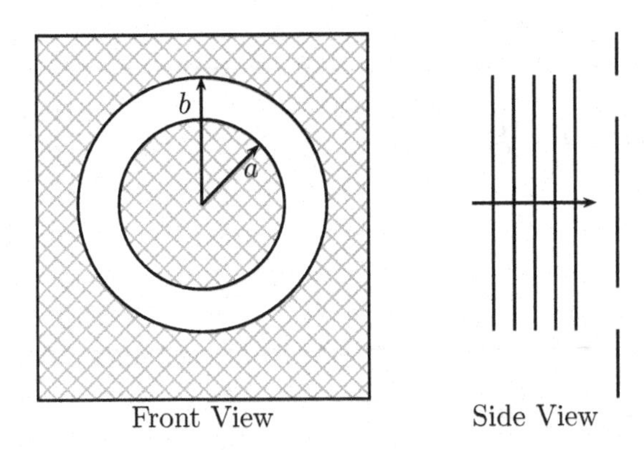

FIGURE 52.6
Perfectly conducting screen with a circular aperture

53

General Scattering

53.1 Integral Equation

Previously, for simplicity, we have considered special geometries and polarizations. (The formalism of Chapter 49 is completely general, but applied there in only simple situations.) Here we wish to give an outline of a method for treating more complicated problems. We start from Maxwell's equations in vacuum, which for radiation of a definite frequency ω, are in part ($k = \omega/c$)

$$\boldsymbol{\nabla}\times\mathbf{B} = -ik\mathbf{E} + \frac{4\pi}{c}\mathbf{J}, \tag{53.1a}$$

$$\boldsymbol{\nabla}\times\mathbf{E} = ik\mathbf{B}, \tag{53.1b}$$

which can be combined to yield

$$\boldsymbol{\nabla}\times(\boldsymbol{\nabla}\times\mathbf{E}) = k^2\mathbf{E} + ik\frac{4\pi}{c}\mathbf{J}. \tag{53.2}$$

The divergence of (53.1a) is

$$ik\boldsymbol{\nabla}\cdot\mathbf{E} = \frac{4\pi}{c}\boldsymbol{\nabla}\cdot\mathbf{J}, \tag{53.3}$$

which also follows from current conservation. Correspondingly, (53.2) can be written as

$$-(\nabla^2 + k^2)\mathbf{E} = \frac{4\pi}{c}ik\left(\mathbf{J} + \frac{1}{k^2}\boldsymbol{\nabla}(\boldsymbol{\nabla}\cdot\mathbf{J})\right), \tag{53.4}$$

which is the inhomogeneous wave equation satisfied by $\mathbf{E}$. (It is equivalent to the result given in (34.40a).)

For an incoming wave satisfying

$$(\nabla^2 + k^2)\mathbf{E}_{\text{inc}} = \mathbf{0}, \tag{53.5}$$

we can immediately write down the solution to (53.4) in terms of the retarded Green's function,

$$\mathbf{E}(\mathbf{r}) = \mathbf{E}_{\text{inc}}(\mathbf{r}) + ik\int(d\mathbf{r}')\frac{e^{ik|\mathbf{r}-\mathbf{r}'|}}{|\mathbf{r}-\mathbf{r}'|}\frac{1}{c}\left(\mathbf{J}(\mathbf{r}') + \frac{1}{k^2}\boldsymbol{\nabla}'(\boldsymbol{\nabla}'\cdot\mathbf{J}(\mathbf{r}'))\right), \tag{53.6}$$

or, if we integrate by parts,

$$\mathbf{E}(\mathbf{r}) = \mathbf{E}_{\text{inc}}(\mathbf{r}) + ik\left(\mathbf{1} + \frac{1}{k^2}\boldsymbol{\nabla}\boldsymbol{\nabla}\right)\cdot\int(d\mathbf{r}')\frac{e^{ik|\mathbf{r}-\mathbf{r}'|}}{|\mathbf{r}-\mathbf{r}'|}\frac{1}{c}\mathbf{J}(\mathbf{r}'), \tag{53.7}$$

assuming a confined current distribution. This is the three-dimensional generalization of (51.4). Alternatively it may be derived from the radiation gauge potential, (35.49).

DOI: 10.1201/9781003057369-53

We now consider the scattering due to a perfectly conducting surface S, which imposes the boundary condition

$$\mathbf{E}_{\text{tang}} = \mathbf{0} \quad \text{on} \quad S. \tag{53.8}$$

Because the conductor is perfect, only a surface current $\mathbf{K}$ is present. When we apply the boundary condition (53.8) to (53.7), we obtain an integral equation for the surface current, which uniquely determines $\mathbf{K}$. Once the surface current is known, the scattered field is determined from (53.7). This is a generalization of the procedure we followed in Chapter 51.

In terms of the surface current, we may easily write down the scattered field at a large distance from the surface S:

$$\mathbf{E}_{\text{scatt}} \sim ik\frac{e^{ikr}}{r}(\mathbf{1} - \mathbf{nn}) \cdot \int (d\mathbf{r}')e^{-ik\mathbf{n}\cdot\mathbf{r}'}\frac{1}{c}\mathbf{J}(\mathbf{r}'), \tag{53.9}$$

where we have used the asymptotic form of the Green's function, (50.14), with $\mathbf{n}$ being the direction of the scattered wave, and

$$\boldsymbol{\nabla} e^{ikr} = ik\mathbf{n}e^{ikr}. \tag{53.10}$$

We assume an incident plane wave of unit amplitude with polarization $\mathbf{e}_0$ and propagation vector $\mathbf{k}_0$,

$$\mathbf{E}_{\text{inc}} = \mathbf{e}_0 e^{i\mathbf{k}_0\cdot\mathbf{r}}. \tag{53.11}$$

We are interested in the scattered wave with polarization $\mathbf{e}$ and propagation vector $\mathbf{k} = k\mathbf{n}$. Since the polarization is transverse,

$$\mathbf{e}\cdot\mathbf{n} = 0, \tag{53.12}$$

we have

$$\mathbf{e}\cdot\mathbf{E}_{\text{scatt}} \sim \frac{e^{ikr}}{r}f, \tag{53.13}$$

where f is called the scattering amplitude, and depends on the emission direction and the polarization,

$$f = ik\mathbf{e}\cdot\int (d\mathbf{r}')e^{-i\mathbf{k}\cdot\mathbf{r}'}\frac{1}{c}\mathbf{J}(\mathbf{r}') = \frac{ik}{c}\mathbf{e}\cdot\mathbf{J}(\mathbf{k}). \tag{53.14}$$

Note that the integral here is the three-dimensional Fourier transform of $\frac{1}{c}\mathbf{J}(\mathbf{r}')$. The differential cross section for scattering is immediately given by the absolute square of f,

$$\frac{d\sigma}{d\Omega} = r^2|\mathbf{E}_{\text{scatt}}\cdot\mathbf{e}|^2 = |f|^2. \tag{53.15}$$

53.2 Optical Theorem

There is a very general theorem that relates the total cross section for scattering to the forward scattering amplitude, the optical theorem.[1] We will derive this theorem under restricted conditions. If we write the differential cross section for a particular polarization, (53.15), explicitly,

$$\frac{d\sigma}{d\Omega} = k^2\left|\mathbf{e}\cdot\int (d\mathbf{r}')e^{-i\mathbf{k}\cdot\mathbf{r}'}\frac{1}{c}\mathbf{J}(\mathbf{r}')\right|^2, \tag{53.16}$$

[1]Attributed to Lord Rayleigh, John William Strutt, 1871.

the total cross section is obtained by summing over polarizations and scattering angles,

$$\sigma = \int d\Omega \sum_{\mathbf{e}} k^2 \int (d\mathbf{r}) \frac{1}{c} \mathbf{J}^*(\mathbf{r}) e^{i\mathbf{k}\cdot\mathbf{r}} \cdot \mathbf{e}\mathbf{e} \cdot \int (d\mathbf{r}') e^{-i\mathbf{k}\cdot\mathbf{r}'} \frac{1}{c} \mathbf{J}(\mathbf{r}'). \tag{53.17}$$

Since $\mathbf{e}$ is perpendicular to $\mathbf{n}$, we have the completeness relation,

$$\sum_{\mathbf{e}} \mathbf{e}\mathbf{e} + \mathbf{n}\mathbf{n} = \mathbf{1}, \tag{53.18}$$

which can be used to write the sum over polarizations in (53.17) as

$$\sum_{\mathbf{e}} \mathbf{e}\mathbf{e} = \mathbf{1} - \mathbf{n}\mathbf{n} \to \mathbf{1} + \frac{1}{k^2} \boldsymbol{\nabla}\boldsymbol{\nabla}, \tag{53.19}$$

where the latter form is assumed to act on $e^{i\mathbf{k}\cdot(\mathbf{r}-\mathbf{r}')}$. The angular integral in (53.17) is easily found to be [cf. (41.32)]

$$\int d\Omega\, e^{i\mathbf{k}\cdot(\mathbf{r}-\mathbf{r}')} = 4\pi \frac{\sin k|\mathbf{r}-\mathbf{r}'|}{k|\mathbf{r}-\mathbf{r}'|} = 4\pi \,\mathrm{Im}\, \frac{e^{ik|\mathbf{r}-\mathbf{r}'|}}{k|\mathbf{r}-\mathbf{r}'|}, \tag{53.20}$$

and in consequence the total cross section is

$$\sigma = \frac{4\pi}{k} \,\mathrm{Im}\, k^2 \int (d\mathbf{r})(d\mathbf{r}') \frac{1}{c} \mathbf{J}^*(\mathbf{r}) \cdot \left(\mathbf{1} + \frac{1}{k^2} \boldsymbol{\nabla}\boldsymbol{\nabla} \right) \frac{e^{ik|\mathbf{r}-\mathbf{r}'|}}{|\mathbf{r}-\mathbf{r}'|} \cdot \frac{1}{c} \mathbf{J}(\mathbf{r}'). \tag{53.21}$$

We next use the boundary condition (53.8) together with the fact that $\mathbf{J}$ is confined to the surface and therefore has only a tangential component to infer that

$$\frac{1}{c} \mathbf{J}^* \cdot \mathbf{E} = 0. \tag{53.22}$$

Consequently, if we make use of the general expression for the field, (53.7), we deduce the identity

$$\begin{aligned} 0 = \int (d\mathbf{r}) \frac{1}{c} \mathbf{J}^* \cdot \mathbf{E} &= \int (d\mathbf{r}) \frac{1}{c} \mathbf{J}^* \cdot \mathbf{E}_{\mathrm{inc}} \\ &+ ik \int (d\mathbf{r})(d\mathbf{r}') \frac{1}{c} \mathbf{J}^*(\mathbf{r}) \cdot \left(\mathbf{1} + \frac{1}{k^2} \boldsymbol{\nabla}\boldsymbol{\nabla} \right) \frac{e^{ik|\mathbf{r}-\mathbf{r}'|}}{|\mathbf{r}-\mathbf{r}'|} \cdot \frac{1}{c} \mathbf{J}(\mathbf{r}'). \end{aligned} \tag{53.23}$$

Therefore, the total cross section, (53.21), is simply

$$\sigma = \frac{4\pi}{k} \,\mathrm{Im} \left[ik \int (d\mathbf{r}) \frac{1}{c} \mathbf{J}^* \cdot \mathbf{E}_{\mathrm{inc}} \right] = \frac{4\pi}{k} \,\mathrm{Im} \left[ik \int (d\mathbf{r}) \mathbf{E}^*_{\mathrm{inc}} \cdot \frac{1}{c} \mathbf{J} \right]. \tag{53.24}$$

Since $\mathbf{E}_{\mathrm{inc}}$ is simply a plane wave, (53.11), this is

$$\sigma = \frac{4\pi}{k} \,\mathrm{Im} \left[ik \int (d\mathbf{r}) e^{-i\mathbf{k}_0\cdot\mathbf{r}} \mathbf{e}_0 \cdot \frac{1}{c} \mathbf{J}(\mathbf{r}) \right], \tag{53.25}$$

or

$$\sigma = \frac{4\pi}{k} \,\mathrm{Im}\, f(\text{forward scattering}), \tag{53.26}$$

where we recall the expression (53.14) for the scattering amplitude. This result is the "optical theorem:" the total cross section is proportional to the imaginary part of the forward scattering amplitude. Note that this theorem is here a consequence of the conservation of energy; the result follows from

$$\int (d\mathbf{r})\mathbf{J}\cdot\mathbf{E} = 0, \tag{53.27}$$

which just says that the induced currents do no work in a perfect conductor.

As an indication that this theorem is actually much more general than the derivation that we have supplied, let us return to the scattering of radiation by an electron bound by a harmonic oscillator potential. The forward scattering amplitude, (53.14), is, in this circumstance

$$f = ik\mathbf{e}_0\cdot\frac{1}{c}e\mathbf{v}(\omega) = k^2 e\mathbf{e}_0\cdot\mathbf{r}(\omega), \tag{53.28}$$

since $\mathbf{k}_0\cdot\mathbf{r} = 0$, where [compare (48.18)]

$$\mathbf{r}(\omega) = \frac{e}{m}\frac{\mathbf{E}(\omega)}{\omega_0^2 - \omega^2 - i\omega\gamma}. \tag{53.29}$$

Since the incident field has unit amplitude, $\mathbf{e}_0\cdot\mathbf{E} = 1$, (53.28) becomes

$$f = k^2\frac{e^2}{m}\frac{1}{\omega_0^2 - \omega^2 - i\omega\gamma}, \tag{53.30}$$

implying

$$\frac{4\pi}{k}\,\mathrm{Im}\,f = 4\pi\frac{e^2}{mc}\frac{\omega^2\gamma}{(\omega_0^2 - \omega^2)^2 + (\omega\gamma)^2} = \sigma, \tag{53.31}$$

which is the total cross section given by (48.44).

53.3 Born Approximation for Scattering by Dielectric

As a final example we consider scattering by a dielectric. For radiation of a definite frequency, Maxwell's equation involving the displacement current is, when no free current is present,

$$\boldsymbol{\nabla}\times\mathbf{B} = -ik\epsilon\mathbf{E} = -ik\mathbf{E} - ik(\epsilon - 1)\mathbf{E}. \tag{53.32}$$

Comparison with (53.1a) now identifies the current effective in scattering to be

$$-ik(\epsilon - 1)\mathbf{E} = \frac{4\pi}{c}\mathbf{J}. \tag{53.33}$$

Inserting this in the formula (53.7) for the electric field, we obtain the following integral equation determining the electric field in the dielectric,

$$\mathbf{E}(\mathbf{r}) = \mathbf{E}_{\text{inc}}(\mathbf{r}) + ik\left(\mathbf{1} + \frac{1}{k^2}\boldsymbol{\nabla}\boldsymbol{\nabla}\right)\cdot\int (d\mathbf{r}')\frac{e^{ik|\mathbf{r}-\mathbf{r}'|}}{|\mathbf{r}-\mathbf{r}'|}(-ik)\frac{\epsilon(\mathbf{r}') - 1}{4\pi}\mathbf{E}(\mathbf{r}'). \tag{53.34}$$

Once $\mathbf{E}$ is known in the dielectric, the scattering amplitude, describing the field far from the dielectric, can then be found from (53.14),

$$f = ik\mathbf{e}\cdot\int (d\mathbf{r}')e^{-i\mathbf{k}\cdot\mathbf{r}'}(-ik)\frac{\epsilon(\mathbf{r}') - 1}{4\pi}\mathbf{E}(\mathbf{r}'). \tag{53.35}$$

When the dielectric constant is near unity,

$$|\epsilon - 1| \ll 1, \tag{53.36}$$

we can employ a simple approximation based on the fact that inside the dielectric, $\mathbf{E}$ is then roughly equal to the incident field,

$$\mathbf{E} \approx \mathbf{E}_{\text{inc}} = \mathbf{e}_0 e^{i\mathbf{k}_0 \cdot \mathbf{r}}. \tag{53.37}$$

Under this approximation ("the Born[2] approximation"), the scattering amplitude is

$$f = k^2 \int (d\mathbf{r}) \frac{\epsilon(\mathbf{r}) - 1}{4\pi} e^{i(\mathbf{k}_0 - \mathbf{k}) \cdot \mathbf{r}} \mathbf{e} \cdot \mathbf{e}_0, \tag{53.38}$$

which is expressed as the Fourier transform of the electric susceptibility,

$$f = k^2 \mathbf{e} \cdot \mathbf{e}_0 \chi(\mathbf{k}_0 - \mathbf{k}), \tag{53.39}$$

corresponding to the differential cross section

$$\frac{d\sigma}{d\Omega} = (\mathbf{e} \cdot \mathbf{e}_0)^2 k^4 \left|\chi(\mathbf{k}_0 - \mathbf{k})\right|^2. \tag{53.40}$$

As a simple example, consider a uniform uncharged dielectric sphere of radius a, small compared to the wavelength, $ka \ll 1$. The differential cross section is then

$$\frac{d\sigma}{d\Omega} = (\mathbf{e} \cdot \mathbf{e}_0)^2 k^4 \left(\frac{4\pi}{3} a^3\right)^2 \chi^2, \tag{53.41}$$

which, when averaged over incident polarizations, summed over final polarizations, and integrated over all angles, agrees with (48.35) for ϵ near unity.

For a plasma, the high frequency behavior of the susceptibility, from (5.20), is

$$\chi = -\frac{ne^2}{m\omega^2}. \tag{53.42}$$

The corresponding differential cross section, in this approximation, is expressed in terms of the Fourier transform of the number density,

$$\frac{d\sigma}{d\Omega} = \left(\frac{e^2}{mc^2}\right)^2 (\mathbf{e} \cdot \mathbf{e}_0)^2 \left| \int (d\mathbf{r}) n(\mathbf{r}) e^{i(\mathbf{k}_0 - \mathbf{k}) \cdot \mathbf{r}} \right|^2. \tag{53.43}$$

Note that for a single particle, for which $n(\mathbf{r}) = \delta(\mathbf{r} - \mathbf{R})$, we obtain the Thomson cross section, given by (48.14) [*cf.* (48.10)], when the sum over final polarizations is performed.

53.4 Problems for Chapter 53

1. Derive (53.7) from (35.49).

[2]Max Born, 1882–1970.

2. Consider an infinitely long, perfectly conducting cylinder with radius a and axis coinciding with the z axis. An incident plane wave, propagating in the $-x$ direction and polarized in the z direction,

$$\mathbf{E}_{\text{inc}} = \hat{\mathbf{z}}e^{-ikx}, \tag{53.44}$$

scatters from the cylinder. Because the cylinder is perfectly conducting, a surface current $\mathbf{K}$ is present with the physically obvious property

$$\mathbf{K} = \hat{\mathbf{z}}K(\phi) = \hat{\mathbf{z}} \sum_{m=-\infty}^{\infty} K_m e^{im\phi}, \tag{53.45}$$

where the latter form is the Fourier decomposition of $K(\phi)$.

(a) Using (53.7), calculate the electric field everywhere ($\rho \geq a$) in terms of the coefficients K_m.

(b) Apply the boundary conditions (53.8) and determine K_m.

(c) Determine expressions for $\mathbf{B}$ for $\rho \geq a$ and check the consistency of K_m by means of the boundary condition (28.19a).

3. (a) Derive the integral equation for $\mathbf{B}(\mathbf{r})$ equivalent to (53.7).

(b) Consider scattering from the cylinder of the previous problem by an incident plane wave polarized in the y direction so that the incident magnetic field is

$$\mathbf{B}_{\text{inc}} = -\hat{\mathbf{z}}e^{-ikx}. \tag{53.46}$$

Here, the surface current will have the form

$$\mathbf{K} = \hat{\boldsymbol{\phi}} \sum_{m=-\infty}^{\infty} K_m e^{im\phi}. \tag{53.47}$$

Calculate the magnetic field everywhere ($\rho \geq a$) in terms of the coefficients K_m and the ρ behavior of $H_m^{(1)}(k\rho)$.

(c) Determine K_m by means of (28.19a).

(d) Check this result by verifying (53.8).

4. (a) Show that in the region exterior to a bounded current distribution, the electric and magnetic fields can be expressed as

$$\mathbf{E} = \boldsymbol{\nabla}\times(\boldsymbol{\nabla}\times\mathbf{r}\Pi_1) + \frac{i\omega}{c}\boldsymbol{\nabla}\times\mathbf{r}\Pi_2, \tag{53.48a}$$

$$\mathbf{B} = -\frac{i\omega}{c}\boldsymbol{\nabla}\times\mathbf{r}\Pi_1 + \boldsymbol{\nabla}\times(\boldsymbol{\nabla}\times\mathbf{r}\Pi_2), \tag{53.48b}$$

where the scalar functions Π_1 and Π_2 are called Debye potentials. This is a specialization of Problem 47.6.

(b) Show that an alternative way of writing the result of part (a) in empty space is

$$\mathbf{E}(\mathbf{r}) = \sum_{lm} \left[f_l(r)\mathbf{X}_{lm}(\Omega) + \frac{ic}{\omega}\boldsymbol{\nabla}\times g_l(r)\mathbf{X}_{lm}(\Omega) \right], \tag{53.49a}$$

$$\mathbf{B}(\mathbf{r}) = \sum_{lm} \left[g_l(r)\mathbf{X}_{lm}(\Omega) - \frac{ic}{\omega}\boldsymbol{\nabla}\times f_l(r)\mathbf{X}_{lm}(\Omega) \right], \tag{53.49b}$$

where $\mathbf{X}_{lm}$ is a vector spherical harmonic,

$$\mathbf{X}_{lm}(\Omega) = \frac{1}{\sqrt{l(l+1)}}\mathbf{L}Y_{lm}(\Omega), \tag{53.50}$$

$\mathbf{L}$ being the angular momentum operator,

$$\mathbf{L} = \mathbf{r}\times\frac{1}{i}\boldsymbol{\nabla}. \tag{53.51}$$

(The l summation starts at $l = 1$.)

(c) Establish the following properties of the vector spherical harmonics

$$\int d\Omega\, \mathbf{X}^*_{l'm'}(\Omega)\cdot\mathbf{X}_{lm}(\Omega) = \delta_{ll'}\delta_{mm'}, \tag{53.52a}$$

$$\sum_{m=-l}^{l} |\mathbf{X}_{lm}(\Omega)|^2 = \frac{2l+1}{4\pi}, \tag{53.52b}$$

$$\int d\Omega\, [f(r')\mathbf{X}_{l'm'}(\Omega)]^*\cdot[g(r)\mathbf{X}_{lm}(\Omega)] = f(r')^* g(r)\delta_{ll'}\delta_{mm'}, \tag{53.52c}$$

$$\int d\Omega\, [f(r')\mathbf{X}_{l'm'}(\Omega)]^*\cdot[\boldsymbol{\nabla}\times g(r)\mathbf{X}_{lm}(\Omega)] = 0, \tag{53.52d}$$

$$\begin{aligned}&\int d\Omega\, [\boldsymbol{\nabla}'\times f(r')\mathbf{X}_{l'm'}(\Omega)]^*\cdot[\boldsymbol{\nabla}\times g(r)\mathbf{X}_{lm}(\Omega)]\\ &= \frac{1}{rr'}\left[\frac{d}{dr'}(r'f(r')^*)\frac{d}{dr}(rg(r)) + l(l+1)f(r')^* g(r)\right]\delta_{ll'}\delta_{mm'},\end{aligned} \tag{53.52e}$$

and for an arbitrary vector function V,

$$\begin{aligned}&\frac{1}{r^2}\left[\frac{d}{dr}r + l(l+1)\right]\int d\Omega\, \mathbf{X}^*_{lm}(\Omega)\cdot\mathbf{V}(\mathbf{r})\\ &= \int d\Omega\, [\boldsymbol{\nabla}\times\mathbf{X}^*_{lm}(\Omega)]\cdot\boldsymbol{\nabla}\times\mathbf{V}(\mathbf{r}),\end{aligned} \tag{53.52f}$$

$$\begin{aligned}D_l\int d\Omega\, \mathbf{X}^*_{lm}(\Omega)\cdot\mathbf{V}(\mathbf{r}) &= \int d\Omega\, \mathbf{X}^*_{lm}(\Omega)\cdot(\nabla^2\mathbf{V}(\mathbf{r}))\\ &- \int d\Omega\, \mathbf{X}^*_{lm}(\Omega)\cdot[\boldsymbol{\nabla}\times(\boldsymbol{\nabla}\times\mathbf{V}(\mathbf{r}))],\end{aligned} \tag{53.52g}$$

where

$$D_l = \frac{1}{r}\frac{d^2}{dr^2}r - \frac{l(l+1)}{r^2}. \tag{53.53}$$

5. Electromagnetic Green's dyadics are defined by the system of equations

$$\boldsymbol{\nabla}\times\boldsymbol{\Gamma}(\mathbf{r},\mathbf{r}') = i\frac{\omega}{c}\boldsymbol{\Phi}(\mathbf{r},\mathbf{r}'), \tag{53.54a}$$

$$-\boldsymbol{\nabla}\times\boldsymbol{\Phi}(\mathbf{r},\mathbf{r}') = i\frac{\omega}{c}\boldsymbol{\Gamma}(\mathbf{r},\mathbf{r}') + i\omega\mathbf{1}\delta(\mathbf{r}-\mathbf{r}'). \tag{53.54b}$$

In terms of

$$\boldsymbol{\Gamma}' = \boldsymbol{\Gamma} + \mathbf{1}\delta(\mathbf{r}-\mathbf{r}'), \tag{53.55}$$

show that

$$\mathbf{\Gamma}'(\mathbf{r},\mathbf{r}') = \sum_{lm} \left[f_l(r,\mathbf{r}')\mathbf{X}_{lm}(\Omega) + \frac{c}{\omega}\boldsymbol{\nabla}\times g_l(r,\mathbf{r}')\mathbf{X}_{lm}(\Omega) \right], \tag{53.56a}$$

$$\mathbf{\Phi}(\mathbf{r},\mathbf{r}') = \sum_{lm} \left[\tilde{g}_l(r,\mathbf{r}')\mathbf{X}_{lm}(\Omega) - \frac{ic}{\omega}\boldsymbol{\nabla}\times \tilde{f}_l(r,\mathbf{r}')\mathbf{X}_{lm}(\Omega) \right], \tag{53.56b}$$

where (use the results of the previous problem)

$$\tilde{g}_l = g_l, \tag{53.57a}$$

$$\tilde{f}_l = f_l - \frac{1}{r^2}\delta(r-r')\mathbf{X}^*(\Omega'), \tag{53.57b}$$

$$(D_l + k^2)\tilde{g}_l = i\frac{\omega}{c}\int d\Omega''\, \mathbf{X}^*_{lm}(\Omega'')\cdot[\boldsymbol{\nabla}''\times \mathbf{1}\delta(\mathbf{r}''-\mathbf{r})], \tag{53.57c}$$

$$(D_l + k^2)\tilde{f}_l = -\int d\Omega''\, \mathbf{X}^*_{lm}(\Omega'')\cdot[\boldsymbol{\nabla}''\times(\boldsymbol{\nabla}''\times \mathbf{1}\delta(\mathbf{r}''-\mathbf{r}))]. \tag{53.57d}$$

Show that the solution to these equations are

$$\tilde{g}_l(r,\mathbf{r}') = -i\frac{\omega}{c}\boldsymbol{\nabla}'\times G_l(r,r')\mathbf{X}^*_{lm}(\Omega), \tag{53.58a}$$

$$\tilde{f}_l(r,\mathbf{r}') = \frac{\omega^2}{c^2}F_l(r,r')\mathbf{X}^*_{lm}(\Omega), \tag{53.58b}$$

where F_l and G_l are scalar Green's function satisfying

$$(D_l + k^2)\Delta_l(r,r') = -\frac{1}{r^2}\delta(r-r'). \tag{53.59}$$

Then show

$$\begin{aligned}\mathbf{\Gamma}(\mathbf{r},\mathbf{r}') = \sum_{lm} \{ & k^2 F_l(r,r')\mathbf{X}_{lm}(\Omega)\mathbf{X}^*_{lm}(\Omega') \\ & - \boldsymbol{\nabla}\times[G_l(r,r')\mathbf{X}_{lm}(\Omega)\mathbf{X}^*_{lm}(\Omega')]\times\overleftarrow{\boldsymbol{\nabla}}'\} \\ & + \frac{1}{r^2}\delta(r-r')\sum_{lm}\mathbf{X}_{lm}(\Omega)\mathbf{X}^*_{lm}(\Omega') - \mathbf{1}\delta(\mathbf{r}-\mathbf{r}').\end{aligned} \tag{53.60}$$

6. What are the boundary conditions imposed on F_l and G_l by a perfectly conducting surface? Compute F_l and G_l in the region exterior to a perfectly conducting spherical shell.
7. How is the formalism of Problem 5 modified in the presence of a spatially varying dielectric constant, $\epsilon(\mathbf{r})$?
8. Show that Green's function for Helmholtz' equation in empty space,

$$(\nabla^2 + k^2)G_k(\mathbf{r},\mathbf{r}') = -4\pi\delta(\mathbf{r}-\mathbf{r}'), \tag{53.61}$$

may be written in the form

$$G_k(\mathbf{r},\mathbf{r}') = 4\pi ik\sum_{lm} j_l(kr_<)h_l(kr_>)Y_{lm}(\Omega)Y^*_{lm}(\Omega'). \tag{53.62}$$

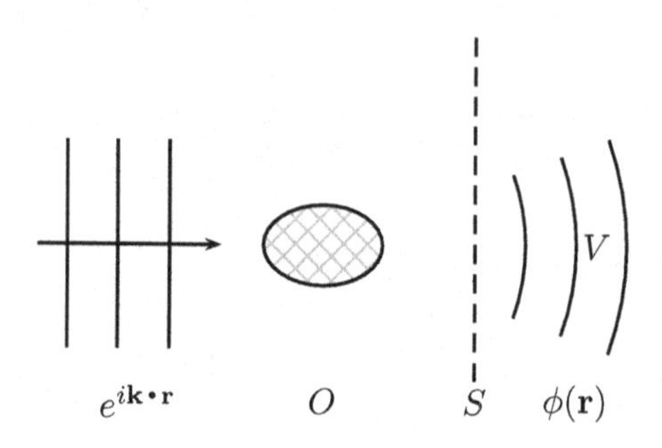

FIGURE 53.1
Scattering of a scalar field ϕ by an object O.

9. A scalar plane wave, $e^{i\mathbf{k}\cdot\mathbf{r}}$, is scattered by an object O, as shown in Fig. 53.1. Because of the scattering, the wave field is a superposition of the plane wave plus the scattered wave ϕ_{sc},
$$\phi(\mathbf{r}) = e^{i\mathbf{k}\cdot\mathbf{r}} + \phi_{\text{sc}}(\mathbf{r}). \tag{53.63}$$
Consider the volume V to the right of the imaginary plane S (O is to the left of S). In this region ϕ_{sc} satisfies
$$(\nabla^2 + k^2)\phi_{\text{sc}}(\mathbf{r}) = 0. \tag{53.64}$$
In terms of Green's function of Helmholtz' equation in unbounded space, derive, for $\mathbf{r}$ in V,
$$\phi_{\text{sc}}(\mathbf{r}) = \int_S \frac{d\mathbf{S}'}{4\pi}\cdot\left\{\phi_{\text{sc}}(\mathbf{r}')\boldsymbol{\nabla}'G(\mathbf{r},\mathbf{r}') - G(\mathbf{r},\mathbf{r}')\boldsymbol{\nabla}'\phi_{\text{sc}}(\mathbf{r}')\right\}. \tag{53.65}$$
This expresses the scattered wave in V in terms of its values, and normal gradients, on S. [It is necessary to show there is no contribution from the hemisphere "at infinity" also bounding V.] Now make the "Fraunhofer approximation" (see Section 50.1) that the point of observation $\mathbf{r}$ is far away from the portions of S that contribute significantly to this integral:
$$\mathbf{r} \gg \mathbf{r}', \text{ so } G(\mathbf{r},\mathbf{r}') \approx \frac{e^{ikr}}{r}e^{-i\mathbf{k}'\cdot\mathbf{r}'}, \quad \mathbf{k}' = k\frac{\mathbf{r}}{r}. \tag{53.66}$$
Then writing
$$\phi_{\text{sc}}(\mathbf{r}) = \frac{e^{ikr}}{r}f(\theta,\phi), \tag{53.67}$$
where f is the scattering amplitude, show that
$$f(\theta,\phi) = \frac{1}{4\pi}\int_S d\mathbf{S}'\cdot\left\{\phi_{\text{sc}}(\mathbf{r}')\boldsymbol{\nabla}'e^{-i\mathbf{k}'\cdot\mathbf{r}'} - e^{-i\mathbf{k}'\cdot\mathbf{r}'}\boldsymbol{\nabla}'\phi_{\text{sc}}(\mathbf{r}')\right\}. \tag{53.68}$$

10. Apply the result of the previous problem to calculate the scattering amplitude for a sphere in the "black" approximation, where
$$\phi = e^{i\mathbf{k}\cdot\mathbf{r}} + \phi_{\text{sc}} = 0 \tag{53.69a}$$
directly behind the sphere (the "shadow") and
$$\phi_{\text{sc}} = 0 \tag{53.69b}$$
everywhere else. Express the answer as a Bessel function. Compare the cross section with (50.28). Why is there a similarity?

54

Charged Particle Energy Loss

Hitherto we have concentrated on the radiation produced by an accelerated charged particle, although we have certainly included the radiation reaction back on the particle. Here we wish to examine the energy loss of such a particle passing through a medium. The mechanisms of such loss can include bremsstrahlung when the particle passes close to ions or nuclei, Čerenkov radiation, and excitation and ionization of atoms. Since the interactions are all electromagnetic, the effects should largely be be describable in terms of the electromagnetic susceptibility.

We will apply the general dispersion relations given in Chapter 6 for the dielectric constant to discuss the energy loss by a charged particle when it passes through a dielectric medium. The interaction between a charged particle and a medium is described by the macroscopic Maxwell equations, (4.56). As we remarked before, (4.56) can be transformed, formally, into the vacuum equations, (1.61), by the redefinitions given in Section 8.2. It is simplest, therefore, to start with a vacuum description and then transform to that of a medium.

54.1 General Expression

For a purely electrically charged particle, the rate at which the charge does work on the medium is

$$-\frac{dE}{dt} = -\int (d\mathbf{r}) \left[\rho_e \frac{\partial \phi}{\partial t} - \frac{1}{c}\mathbf{j}_e \cdot \frac{\partial \mathbf{A}}{\partial t}\right], \tag{54.1}$$

according to (10.29). In the Lorenz gauge, the potentials are related to the charge and current densities by (34.15a) and (34.15b), in terms of Green's function, $G(\mathbf{r}-\mathbf{r}', t-t')$. Therefore, the rate of energy loss by the particle can be rewritten as

$$\begin{aligned}-\frac{dE}{dt} = \int (d\mathbf{r})(d\mathbf{r}')dt' \bigg[&\frac{1}{c}\mathbf{j}_e(\mathbf{r},t) \cdot \frac{\partial}{\partial t} G(\mathbf{r}-\mathbf{r}', t-t') \cdot \frac{1}{c}\mathbf{j}_e(\mathbf{r}',t') \\ &-\rho_e(\mathbf{r},t)\frac{\partial}{\partial t}G(\mathbf{r}-\mathbf{r}',t-t')\rho_e(\mathbf{r}',t')\bigg].\end{aligned} \tag{54.2}$$

Occurring here is the retarded Green's function given by (34.18b) and (46.17):

$$G(\mathbf{r}-\mathbf{r}', t-t') = 4\pi \int \frac{(d\mathbf{k})}{(2\pi)^3} \frac{d\omega}{2\pi} \frac{e^{i[\mathbf{k}\cdot(\mathbf{r}-\mathbf{r}')-\omega(t-t')]}}{\mathbf{k}^2 - \left(\frac{\omega+i\epsilon}{c}\right)^2}. \tag{54.3}$$

The charge and current distributions for a single particle of charge Ze are given by

$$\rho_e(\mathbf{r},t) = Ze\delta(\mathbf{r}-\mathbf{v}t), \quad \mathbf{j}_e(\mathbf{r},t) = Ze\mathbf{v}\delta(\mathbf{r}-\mathbf{v}t), \tag{54.4}$$

DOI: 10.1201/9781003057369-54

where $\mathbf{v}$, the velocity of the particle, is approximated by a constant. Substituting (54.3) and (54.4) into (54.2), and carrying out the now trivial t' integration, we obtain

$$-\left(\frac{dE}{dt}\right)_{\text{vac}} = 4\pi(Ze)^2 \int \frac{(d\mathbf{k})d\omega}{(2\pi)^4} \frac{2\pi\delta(\mathbf{k}\cdot\mathbf{v}-\omega)}{\mathbf{k}^2 - \left(\frac{\omega+i\epsilon}{c}\right)^2} \left(\frac{v^2}{c^2}-1\right)(-i\omega). \tag{54.5}$$

Although this expression vanishes, as it must since it refers to the vacuum, from it we can easily obtain the result for a particle traversing a medium by making the substitutions,

$$\frac{1}{c^2} \to \frac{\epsilon}{c^2}, \quad Ze \to \frac{Ze}{\sqrt{\epsilon}}, \tag{54.6}$$

leading to the following formula for energy loss,

$$\left(-\frac{dE}{dt}\right)_{\text{medium}} = 4\pi(Ze)^2 \int \frac{(d\mathbf{k})d\omega}{(2\pi)^4} \frac{-i\omega}{\mathbf{k}^2 - \frac{\epsilon}{c^2}\omega^2} \left(\frac{v^2}{c^2} - \frac{1}{\epsilon}\right) 2\pi\delta(\mathbf{k}\cdot\mathbf{v}-\omega), \tag{54.7}$$

where now $\epsilon(\omega)$ supplies the necessary imaginary part in the denominator. The angular integration here is trivial:

$$\int_{-1}^{1} d(\cos\theta)\delta(kv\cos\theta - \omega) = \frac{1}{kv}\eta(kv - |\omega|), \tag{54.8}$$

in terms of the step function (33.11). We see here an intimation of Čerenkov radiation [recall Chapter 39]. This step function supplies the lower limit for the $|\mathbf{k}|$ integration. As for the upper limit, we note that momentum transfers must be limited in this description, since we have ignored recoil. Large momentum transfer events must be dealt with differently, as collisions with individual electrons (δ-rays). We therefore cut off the $|\mathbf{k}|$ integration at some characteristic momentum transfer K, reflecting the boundary between different ways of measuring energy loss. The energy loss rate is therefore

$$\begin{aligned} -\frac{dE}{dt} &= \frac{(Ze)^2}{2\pi v} \int_{-\infty}^{\infty} d\omega \int_{\omega^2/v^2}^{K^2} dk^2(-i\omega)\left(\frac{v^2}{c^2} - \frac{1}{\epsilon}\right)\frac{1}{k^2 - \frac{\epsilon\omega^2}{c^2}} \\ &= \frac{(Ze)^2}{2\pi v} \int_{-\infty}^{\infty} d\omega(-i\omega)\left(\frac{v^2}{c^2} - \frac{1}{\epsilon(\omega)}\right) \ln \frac{K^2}{\omega^2\left(\frac{1}{v^2} - \frac{\epsilon(\omega)}{c^2}\right)}, \end{aligned} \tag{54.9}$$

where we have assumed that the important values of ω which contribute to the integral satisfy $K^2 \gg \epsilon\omega^2/c^2$. The energy loss per unit distance traveled, which is more accessible experimentally, is

$$\begin{aligned} -\frac{dE}{dz} &= \frac{(Ze)^2}{2\pi c^2} \int_{-\infty}^{\infty} d\omega(-i\omega)\left(1 - \frac{c^2}{v^2\epsilon(\omega)}\right) \ln \frac{K^2 v^2}{\omega^2\left(1 - \frac{v^2}{c^2}\epsilon(\omega)\right)} \\ &= \frac{(Ze)^2}{\pi c^2} \operatorname{Im} \int_0^{\infty} d\omega\, \omega \left(1 - \frac{c^2}{v^2\epsilon(\omega)}\right) \ln \frac{K^2 v^2}{\omega^2\left(1 - \frac{v^2}{c^2}\epsilon(\omega)\right)}, \end{aligned} \tag{54.10}$$

where we have used the reality condition, (6.8), for $\epsilon(\omega)$. Note that if $\epsilon(\omega)$ is real, this implies the formula (39.16b) for Čerenkov radiation.

54.2 Evaluation in Terms of Spectral Functions

The remaining task is to express the above ω integral in terms of the dispersion relations for the dielectric constant given in Chapter 6. To do this, we first decompose (54.10) into two pieces:

$$-\frac{dE}{dz} = \frac{(Ze)^2}{\pi c^2}(\mathrm{I}+\mathrm{II}), \tag{54.11}$$

where

$$\mathrm{I} = \mathrm{Im} \int_0^\infty d\omega\, \omega \left(1 - \frac{c^2}{v^2 \epsilon}\right) \ln \frac{K^2 v^2}{\omega^2 \left(1 - \frac{v^2}{c^2}\right)}, \tag{54.12a}$$

$$\mathrm{II} = \mathrm{Im} \int_0^\infty d\omega\, \omega \left(1 - \frac{c^2}{v^2 \epsilon}\right) \ln \frac{1 - \frac{v^2}{c^2}}{1 - \frac{v^2}{c^2}\epsilon}. \tag{54.12b}$$

The imaginary part in (54.12a) arises entirely from the imaginary part of $-1/\epsilon(\omega)$, which, from (6.34), is described by the dispersion relation

$$\frac{1}{\epsilon(\omega)} = 1 - \omega_p^2 \int_0^\infty d\omega' \frac{q(\omega')}{\omega'^2 - (\omega + i\epsilon)^2}, \tag{54.13}$$

where the spectral function $q(\omega')$ satisfies the conditions

$$q(\omega') > 0, \tag{54.14a}$$

$$\int_0^\infty q(\omega')\, d\omega' = 1. \tag{54.14b}$$

Therefore, the imaginary part of the inverse of the permittivity is proportional to the spectral function $q(\omega)$,

$$\mathrm{Im}\left(-\frac{1}{\epsilon(\omega)}\right) = \frac{\pi}{2}\frac{\omega_p^2}{\omega} q(\omega). \tag{54.15}$$

If we define an average frequency ω_e by

$$\int_0^\infty d\omega\, q(\omega) \ln \frac{1}{\omega^2} = \ln \frac{1}{\omega_e^2}, \tag{54.16}$$

the integral (54.12a) becomes

$$\mathrm{I} = \frac{\pi}{2}\omega_p^2 \left(\frac{c^2}{v^2}\right) \ln \frac{K^2 v^2}{\omega_e^2 \left(1 - \frac{v^2}{c^2}\right)}. \tag{54.17}$$

Here, ω_e is an effective atomic excitation energy produced by an electrically charged particle, which, in the Fermi model, (6.33), is (see Problem 54.1)

$$\omega_e = \omega_0 \sqrt{\epsilon(0)}, \tag{54.18}$$

if $\gamma \ll \omega_0, \omega_p$.

The remaining integral (54.12b) is most conveniently evaluated by replacing the integration along the positive real axis by a contour integral in the complex-ω plane, as shown in Fig. 54.1. Since $1/\epsilon(\omega)$ is regular in the first quadrant, the integration along the contour C_1 is equivalent to that along $C_2 + C_3$. Furthermore, we notice that along the imaginary axis, $\omega = i\nu$, the dielectric constant is real, which follows from the dispersion relation (6.27) [or from (6.34)]:

$$\epsilon(i\nu) = 1 + \omega_p^2 \int_0^\infty d\omega' \frac{p(\omega')}{\omega'^2 + \nu^2} \leq \epsilon(0). \tag{54.19}$$

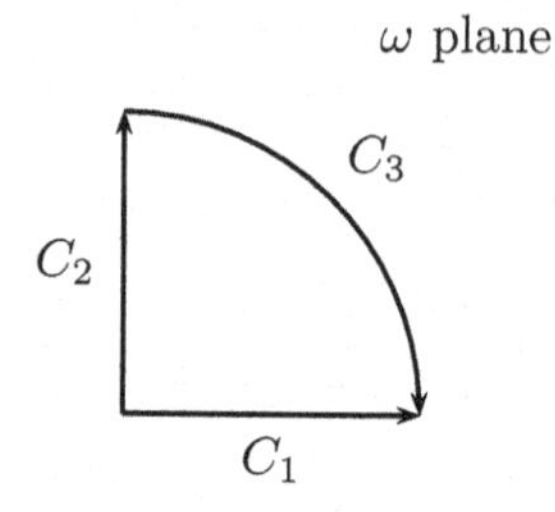

FIGURE 54.1
Contour used to re-express (54.12b).

This last inequality shows that whenever

$$\frac{v^2}{c^2}\epsilon(0) < 1, \quad \text{or} \quad \frac{v}{c} < \frac{1}{\sqrt{\epsilon(0)}}, \tag{54.20}$$

there is no contribution to II from the integration along the imaginary axis, C_2. This means that we may replace the integral II by the contour integral along the quarter circle, C_3, where $|\omega| \to \infty$. Since, in this circumstance, $|\omega|$ is large compared to atomic frequencies, we have there

$$\epsilon(\omega) \approx 1 - \frac{\omega_p^2}{\omega^2}. \tag{54.21}$$

Consequently, we may approximate the logarithm in the integrand by

$$\ln \frac{1-\frac{v^2}{c^2}}{1-\frac{v^2}{c^2}\epsilon(\omega)} \approx -\frac{\omega_p^2}{\omega^2}\frac{v^2}{c^2}\frac{1}{1-\frac{v^2}{c^2}}, \tag{54.22}$$

which results in

$$II = \text{Im}\int_{C_3} d\omega\,\omega\left(1-\frac{c^2}{v^2}\right)\left(-\frac{\omega_p^2}{\omega^2}\frac{v^2}{c^2}\frac{1}{1-\frac{v^2}{c^2}}\right) = \omega_p^2\left(\text{Im}\int_{C_3}\frac{d\omega}{\omega}\right) = \frac{\pi}{2}\omega_p^2. \tag{54.23}$$

Therefore, under the condition (54.20), we obtain the energy lost by the particle per unit distance traveled,

$$-\frac{dE}{dz} = \frac{1}{2}\frac{\omega_p^2(Ze)^2}{v^2}\left[\ln\frac{K^2v^2}{\omega_e^2\left(1-\frac{v^2}{c^2}\right)} - \frac{v^2}{c^2}\right], \quad \text{for} \quad \frac{v}{c} < \frac{1}{\sqrt{\epsilon(0)}}. \tag{54.24}$$

When the velocity of the particle is so large that (54.20) no longer holds,

$$\frac{v}{c} > \frac{1}{\sqrt{\epsilon(0)}}, \tag{54.25}$$

the integration along the imaginary axis C_2, is no longer zero, since the argument of the logarithm, $1 - \frac{v^2}{c^2}\epsilon(\omega)$ goes through zero. Now, there is an additional contribution arising from C_2:

$$\left(-\frac{dE}{dz}\right)_{\text{add}} = \frac{(Ze)^2}{\pi c^2}\,\text{Im}\int_0^{i\infty} d\omega\,\omega\left(1-\frac{c^2}{v^2\epsilon(\omega)}\right)\ln\frac{1-\frac{v^2}{c^2}}{1-\frac{v^2}{c^2}\epsilon(\omega)}, \tag{54.26}$$

where $\omega = i\nu$ $(\nu > 0)$. The value of ν, ν_v, for which $1 - \frac{v^2}{c^2}\epsilon(\omega)$ vanishes can be determined by the dispersion integral

$$\omega_p^2 \int_0^\infty d\omega' \frac{q(\omega')}{\omega'^2 + \nu_v^2} = 1 - \frac{v^2}{c^2}. \tag{54.27}$$

The two extreme situations are, from (54.13),

$$\frac{v^2}{c^2} = \frac{1}{\epsilon(0)}: \quad \nu_v = 0, \tag{54.28a}$$

$$\frac{v^2}{c^2} \to 1: \quad \nu_v^2 \approx \frac{\omega_p^2}{1 - \frac{v^2}{c^2}}, \tag{54.28b}$$

the latter of which follows from (54.14b). Equation (54.26) can be rewritten in terms of ν_v,

$$\left(-\frac{dE}{dz}\right)_{\text{add}} = -\frac{(Ze)^2}{\pi v^2} \operatorname{Im} \int_0^{\nu_v} d\nu\, \nu \left(\frac{v^2}{c^2} - \frac{1}{\epsilon(i\nu)}\right) \ln \frac{1 - \frac{v^2}{c^2}}{(-1)\left(\frac{v^2}{c^2}\epsilon(i\nu) - 1\right)}. \tag{54.29}$$

To determine the branch of the logarithm involved here, we return to the general spectral representation for $\epsilon(\omega)$, (6.27), and let ω approach the imaginary axis as follows:

$$\omega \to \nu e^{i(\pi/2 - \delta)}, \tag{54.30}$$

implying for the spectral denominator

$$\omega'^2 - \omega^2 \to \omega'^2 - \left(\nu e^{i(\pi/2-\delta)}\right)^2 = \omega'^2 + \nu^2 - i\delta, \quad \delta \to +0. \tag{54.31}$$

The imaginary part of $\epsilon(\omega)$, in this limit, is positive and tending toward zero, which implies that

$$\operatorname{Im}\left(\ln \frac{1}{1 - \frac{v^2}{c^2}\epsilon}\right) = \operatorname{Im}\left(\ln \frac{1}{(1 - \frac{v^2}{c^2}\operatorname{Re}\epsilon) - i\frac{v^2}{c^2}\operatorname{Im}\epsilon}\right) = \operatorname{Im}\ln \frac{1}{e^{-i\pi}} = \pi. \tag{54.32}$$

The additional term now may be expressed as follows:

$$\begin{aligned}
\left(-\frac{dE}{dz}\right)_{\text{add}} &= \frac{(Ze)^2}{2v^2} \int_0^{\nu_v^2} d\nu^2 \left[1 - \frac{v^2}{c^2} - \omega_p^2 \int_0^\infty d\omega' \frac{q(\omega')}{\omega'^2 + \nu^2}\right] \\
&= \frac{(Ze)^2}{2v^2} \left\{\left(1 - \frac{v^2}{c^2}\right)\nu_v^2 - \omega_p^2 \int_0^\infty d\omega'\, q(\omega') \ln \frac{\omega'^2 + \nu_v^2}{\omega'^2}\right\} \\
&= \frac{1}{2}\frac{\omega_p^2 (Ze)^2}{v^2} \left[\ln \frac{\omega_e^2}{\omega_{ve}^2} + \left(1 - \frac{v^2}{c^2}\right)\frac{\nu_v^2}{\omega_p^2}\right],
\end{aligned} \tag{54.33}$$

where ω_e^2 is defined by (54.16), while ω_{ve}^2 is given by the analogous equation,

$$\int_0^\infty d\omega' q(\omega') \ln \frac{1}{\omega'^2 + \nu_v^2} \equiv \ln \frac{1}{\omega_{ve}^2}. \tag{54.34}$$

Therefore, by adding (54.33) to (54.24), we obtain the energy lost by a fast charged particle to be

$$-\frac{dE}{dz} = \frac{1}{2}\frac{\omega_p^2 (Ze)^2}{v^2} \left[\ln \frac{K^2 v^2}{\omega_{ve}^2 \left(1 - \frac{v^2}{c^2}\right)} - \frac{v^2}{c^2} + \left(1 - \frac{v^2}{c^2}\right)\frac{\nu_v^2}{\omega_p^2}\right], \quad \text{for} \quad \frac{v}{c} > \frac{1}{\sqrt{\epsilon(0)}}. \tag{54.35}$$

A somewhat more convenient form of (54.35) can be obtained by expressing $\ln 1/\omega_{ve}^2$ in terms of $\ln 1/\omega_e^2$. We first observe that, from (54.27) and (54.34),

$$\frac{d}{d(v^2/c^2)}\ln\omega_{ve}^2 = \int_0^\infty d\omega'\, q(\omega')\frac{1}{\omega'^2+\nu_v^2}\frac{d}{d(v^2/c^2)}\nu_v^2 = \frac{1-\frac{v^2}{c^2}}{\omega_p^2}\frac{d\nu_v^2}{d(v^2/c^2)}, \tag{54.36}$$

which implies that

$$\frac{d}{d(v^2/c^2)}\left[\ln\frac{1}{\omega_{ve}^2}+\left(1-\frac{v^2}{c^2}\right)\frac{\nu_v^2}{\omega_p^2}\right] = -\frac{\nu_v^2}{\omega_p^2}. \tag{54.37}$$

Integrating this, we obtain the identity

$$-\int_{1/\epsilon(0)}^{v^2/c^2} d\left(\frac{v'^2}{c^2}\right)\frac{\nu_{v'}^2}{\omega_p^2} = \ln\frac{1}{\omega_{ve}^2} - \ln\frac{1}{\omega_e^2} + \frac{1-\frac{v^2}{c^2}}{\omega_p^2}\nu_v^2. \tag{54.38}$$

Therefore, an alternative form for the energy loss per unit distance for $v/c > 1/\sqrt{\epsilon(0)}$ is

$$-\frac{dE}{dz} = \frac{1}{2}\frac{\omega_p^2(Ze)^2}{v^2}\left[\ln\frac{K^2v^2}{\omega_e^2\left(1-\frac{v^2}{c^2}\right)} - \frac{v^2}{c^2} - \int_{1/\epsilon(0)}^{v^2/c^2} d\left(\frac{v'^2}{c^2}\right)\frac{\nu_{v'}^2}{\omega_p^2}\right], \tag{54.39}$$

where, in this form, we recognize that the last term is the added contribution when v/c becomes larger than $1/\sqrt{\epsilon(0)}$.

54.3 High Energy Limit

The energy lost by a ultrarelativistic charged particle passing through a dielectric medium is given by (54.35) where ν_v^2 is given by (54.28b). For a sufficiently large speed, ν_v is large compared to the characteristic frequencies that determine the dielectric constant, and in consequence

$$\ln\frac{1}{\omega_{ve}^2} \approx \int_0^\infty d\omega' q(\omega')\ln\frac{1}{\nu_v^2} = \ln\frac{1}{\nu_v^2}. \tag{54.40}$$

The energy lost in this ultrarelativistic limit is therefore

$$-\frac{dE}{dz} = \frac{\omega_p^2}{2}\frac{(Ze)^2}{c^2}\ln\frac{K^2c^2}{\omega_p^2}, \quad \text{as} \quad \frac{v}{c}\to 1. \tag{54.41}$$

Notice that this means that the energy loss saturates at a constant value not much above the minimum ionization loss rate. This phenomenon, often called the density effect, was first pointed out by Enrico Fermi in 1941.

Let us see, in the Fermi model, how this constant value (the Fermi plateau) is approached. From the definition of ν_v, the value of ν for which $1/\epsilon(i\nu) = v^2/c^2$, we obtain from (6.33),

$$1-\frac{v^2}{c^2} = \frac{\omega_p^2}{\omega_0^2\epsilon(0)+\gamma\nu_v+\nu_v^2}, \tag{54.42}$$

because $\epsilon(0) = 1+\omega_p^2/\omega_0^2$, or, neglecting γ in comparison to ω_0,

$$\left(\nu_v+\frac{1}{2}\gamma\right)^2 = -\omega_0^2\epsilon(0) + \frac{\omega_p^2}{1-\frac{v^2}{c^2}}. \tag{54.43}$$

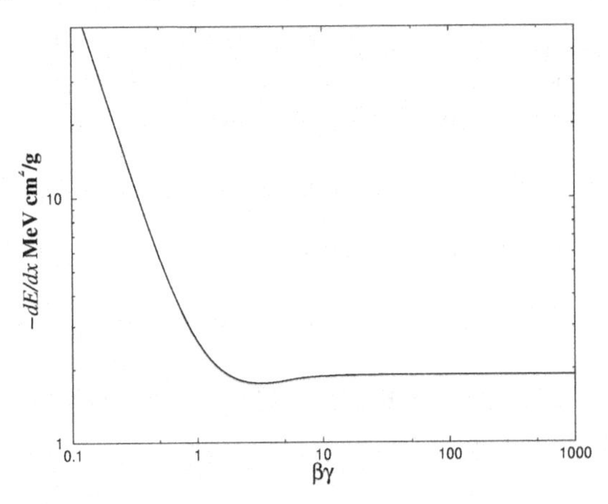

FIGURE 54.2
Energy loss by a charged particle, incorporating (54.24) and (54.45). Here $x = \rho z$, where ρ is the density of the medium, and the abscissa is $\beta\gamma = (v/c)(1-(v/c)^2)^{-1/2}$. We have taken representative values for the parameters, $Kc/\omega_e = 100$, $\omega_e^2/\omega_p^2 = 30$, and the damping constant $\gamma = 0$.

From this explicit expression for ν_v, we can then perform the v'^2/c^2 integration in (54.39):

$$\int_{1/\epsilon(0)}^{v^2/c^2} d\left(\frac{v'^2}{c^2}\right)\left[-\frac{\omega_0^2}{\omega_p^2}\epsilon(0)+\frac{1}{1-\frac{v'^2}{c^2}}-\frac{\gamma}{\omega_p^2}\left(\frac{\omega_p^2}{1-\frac{v'^2}{c^2}}-\omega_0^2\epsilon(0)\right)^{1/2}\right]$$
$$\approx \frac{\omega_e^2}{\omega_p^2}\left(1-\frac{v^2}{c^2}\right)+\ln\frac{\omega_p^2}{\omega_e^2\left(1-\frac{v^2}{c^2}\right)}-1-\frac{\pi}{2}\frac{\gamma}{\omega_e}+2\frac{\gamma}{\omega_p}\left(1-\frac{v^2}{c^2}\right)^{1/2}, \qquad (54.44)$$

where we have used the Fermi model expression for ω_e, (54.18), which implies $1-1/\epsilon(0) = \omega_p^2/\omega_e^2$, and retained terms through order $(1-v^2/c^2)$. Substituting (54.44) into (54.39), we obtain the energy loss per unit distance, in this model, for $1 \approx \frac{v^2}{c^2} > \frac{1}{\epsilon(0)}$,

$$-\frac{dE}{dz} = \frac{1}{2}\frac{\omega_p^2(Ze)^2}{v^2}\left[\ln\frac{K^2c^2}{\omega_p^2}-\left(1-\frac{v^2}{c^2}\right)\frac{\omega_e^2}{\omega_p^2}+\frac{\pi}{2}\frac{\gamma}{\omega_e}-2\frac{\gamma}{\omega_p}\left(1-\frac{v^2}{c^2}\right)^{1/2}\right]. \qquad (54.45)$$

This approach to the Fermi plateau [deviation proportional to $(1-v^2/c^2)^{1/2}$] is not unrealistic.

We conclude this Section by sketching the result for the energy loss by a charged particle passing through matter in Fig. 54.2. This figure should be taken as a qualitative indication only. For detailed discussion of the comparison with experiment, and full references to the literature, the reader is referred to Ref. [52].

54.4 Energy Loss by a Magnetic Monopole

We now discuss the energy loss by a purely magnetically charged particle (a magnetic monopole) of strength g, when it passes through a linear, isotropic, non-magnetic medium. (This calculation would be relevant for an experimental discovery of a cosmic magnetic monopole, such as that claimed by Buford Price in 1975 [10].) The formula can be immediately obtained from that given above by exploiting the symmetry of Maxwell's equations discussed in Chapter 2. Specifically, we have the replacements, in vacuum,

$$\mathbf{j}_e \to \mathbf{j}_m, \quad \rho_e \to \rho_m \quad (\text{or} \quad Ze \to g); \quad \mathbf{E} \to \mathbf{B}, \tag{54.46}$$

while, in going from vacuum to a non-magnetic medium, we have the further substitutions

$$\mathbf{j}_m \to \mathbf{j}_m, \quad g \to g, \quad \text{and} \quad c \to \frac{c}{\sqrt{\epsilon}}. \tag{54.47}$$

The net effect of these substitutions is obtained by multiplying (54.10) by $\epsilon(\omega)g^2/(Ze)^2$, yielding

$$-\frac{dE}{dz} = \frac{g^2}{\pi c^2} \operatorname{Im} \int_0^\infty d\omega\, \omega \left(\epsilon(\omega) - \frac{c^2}{v^2}\right) \ln \frac{K^2 v^2}{\omega^2 \left(1 - \frac{v^2}{c^2}\epsilon(\omega)\right)} \tag{54.48}$$

for the rate of energy loss by a magnetically charged particle.

Following the same decomposition given in Section 54.2, we find in place of (54.24)

$$-\frac{dE}{dz} = \frac{1}{2}\frac{\omega_p^2 g^2}{c^2}\left[\ln \frac{K^2 v^2}{\omega_m^2 \left(1 - \frac{v^2}{c^2}\right)} - 1\right], \quad \text{for} \quad \frac{v}{c} < \frac{1}{\sqrt{\epsilon(0)}}, \tag{54.49}$$

where, because of the appearance of $\epsilon(\omega)$, instead of $1/\epsilon(\omega)$, in (54.48), we encounter, from the dispersion relation (6.27),

$$\epsilon(\omega) - 1 = \omega_p^2 \int_0^\infty d\omega' \frac{p(\omega')}{\omega'^2 - (\omega + i\epsilon)^2}. \tag{54.50}$$

which then allows us to define

$$\int_0^\infty d\omega\, p(\omega) \ln \frac{1}{\omega^2} \equiv \ln \frac{1}{\omega_m^2}. \tag{54.51}$$

Comparing (54.49) with (54.24), we see that the energy loss by a magnetically charged particle is approximately obtained from that of an electrically charged particle by the substitution

$$\frac{Ze}{v} \to \frac{g}{c}, \tag{54.52}$$

provided that

$$\omega_e^2 \approx \omega_m^2. \tag{54.53}$$

In the Fermi model, we recognize from (6.32)

$$p(\omega') = \frac{2}{\pi} \frac{\omega'^2 \gamma}{(\omega'^2 - \omega_0^2)^2 + (\omega'\gamma)^2} \to \delta(\omega' - \omega_0), \tag{54.54}$$

where the latter is the limit as $\gamma \to 0$, that

$$\omega_e \approx \omega_0 \sqrt{\epsilon(0)} > \omega_m \approx \omega_0. \tag{54.55}$$

The general validity of this inequality can be established as follows. From (54.50) and (54.13), we easily infer that

$$\epsilon(i\nu) = 1 + \omega_p^2 \int_0^\infty d\omega' \frac{p(\omega')}{\omega'^2 + \nu^2} > 1, \tag{54.56a}$$

and

$$\frac{1}{\epsilon(i\nu)} = 1 - \omega_p^2 \int_0^\infty d\omega' \frac{q(\omega')}{\omega'^2 + \nu^2} < 1, \tag{54.56b}$$

from which follow the integral statements for $L \gg \omega_{\text{atomic}}$,

$$\int_0^L d\nu^2(\epsilon - 1) = \omega_p^2 \int_0^\infty d\omega' \, p(\omega') \ln \frac{L}{\omega'^2} = \omega_p^2 \ln \frac{L}{\omega_m^2}. \tag{54.57a}$$

$$\int_0^L d\nu^2(\epsilon - 1) > \int_0^L d\nu^2 \left(1 - \frac{1}{\epsilon}\right) = \omega_p^2 \int_0^\infty d\omega' \, q(\omega') \ln \frac{L}{\omega'^2} = \omega_p^2 \ln \frac{L}{\omega_e^2}, \tag{54.57b}$$

in view of (54.56a). Therefore, the desired inequality is proved,

$$\omega_e > \omega_m. \tag{54.58}$$

Just as in the electrically charged particle situation, there is here an additional contribution when $\frac{v}{c} > \frac{1}{\sqrt{\epsilon(0)}}$,

$$-\frac{dE}{dz} = \frac{1}{2}\frac{\omega_p^2 g^2}{c^2} \left[\ln \frac{K^2 v^2}{\omega_m^2 \left(1 - \frac{v^2}{c^2}\right)} - 1\right] + \left(-\frac{dE}{dz}\right)_{\text{add}}, \tag{54.59}$$

where, instead of (54.26), we have

$$\left(-\frac{dE}{dz}\right)_{\text{add}} = \frac{g^2}{\pi c^2} \operatorname{Im} \int_0^{i\infty} d\omega \, \omega \left[\epsilon(\omega) - \frac{c^2}{v^2}\right] \ln \frac{1 - \frac{v^2}{c^2}}{1 - \frac{v^2}{c^2}\epsilon(\omega)}. \tag{54.60}$$

The imaginary part arises from $\nu = \frac{1}{i}\omega < \nu_v$, where ν_v satisfies

$$\frac{c^2}{v^2} = \epsilon(i\nu_v) = 1 + \omega_p^2 \int_0^\infty d\omega' \frac{p(\omega')}{\omega'^2 + \nu_v^2}. \tag{54.61}$$

We can simplify (54.60) to obtain a form for the energy loss analogous to (54.39),

$$-\frac{dE}{dz} = \frac{1}{2}\frac{\omega_p^2 g^2}{c^2} \left[\ln \frac{K^2 v^2}{\omega_m^2 \left(1 - \frac{v^2}{c^2}\right)} - 1 - \int_{c^2/v^2}^{\epsilon(0)} d\left(\frac{c^2}{v'^2}\right) \frac{\nu_{v'}^2}{\omega_p^2}\right], \quad \text{for} \quad \frac{v}{c} > \frac{1}{\sqrt{\epsilon(0)}}. \tag{54.62}$$

[For the approach to the limiting high energy value, see Problem 54.4.]

54.5 Problems for Chapter 54

1. Derive the formula (54.18) for the effective excitation energy in the Fermi model. Similarly, derive (54.55), that is, $\omega_m \approx \omega_0$.
2. Fill in the steps leading to (54.44) and (54.45).
3. Supply the steps necessary to derive (54.62).
4. Find the approach to the high energy limiting value of the energy loss rate for a magnetically charged particle, in the Fermi model, in analogy to (54.45).

A

Units

This book has been written entirely in the Gaussian (G) system of units, which is most convenient for theoretical purposes. The Heaviside-Lorentz (HL) system of units, which is called "rationalized." is also very convenient particularly for problems involving radiation. However, for practical uses, and for engineering applications, it is essential to make contact with the SI (Système International) units, in which the meter, kilogram, second, and ampere are the fundamental units. Fortunately, it is very easy to transform equations written in the Gaussian system to the SI system. Here we will give that transformation, as well as explain the Gaussian and Heaviside-Lorentz units. Thus, evaluations may be performed in either system, or by using dimensionless quantities, such as the fine structure constant.

The SI system as of 2019, is based on *definitions* of observable constants of nature, corresponding to the seven base units, as shown in Table A.1. This makes it possible for laboratories anywhere to calibrate their experiments. For details, see Ref. [53].

The macroscopic Maxwell equations, (4.56), in in the three systems mentioned above, and be written in the form

$$\begin{aligned} \boldsymbol{\nabla}\cdot\mathbf{D} &= \kappa_1\rho, & \boldsymbol{\nabla}\cdot\mathbf{B} &= 0, \\ \boldsymbol{\nabla}\times\mathbf{H} &= \kappa_2\frac{\partial}{\partial t}\mathbf{D} + \kappa_2\kappa_3\mathbf{J}, & -\boldsymbol{\nabla}\times\mathbf{E} &= \kappa_2\frac{\partial}{\partial t}\mathbf{B}, \end{aligned} \tag{A.1a}$$

where

$$\begin{aligned} \mathbf{D} &= \kappa_3\mathbf{E} + \kappa_1\mathbf{P}, \\ \mathbf{H} &= \kappa_4\mathbf{B} - \kappa_1\mathbf{M}. \end{aligned} \tag{A.1b}$$

In addition, the Lorentz force law is

$$\mathbf{F} = e\left(\mathbf{E} + \kappa_5\times\mathbf{B}\right). \tag{A.2}$$

Here, the values of the κ's are given in Table A.2.

In SI, the constants ϵ_0 and μ_0 formerly had defined values,

$$\epsilon_0\mu_0 = c^{-2}, \tag{A.3a}$$

$$\mu_0 = 4\pi\times 10^{-7}, \tag{A.3b}$$

where the speed of light has been *defined* for many years to be exactly

$$c = 299\,792\,458 \text{ m/s}. \tag{A.4}$$

However, although (A.3a) continues to hold in the 2019 SI, (A.3b) does not, the permeability of free space being given implicitly by the measurement of the fine structure constant, which in SI has the form

$$\alpha = \frac{e^2}{4\pi\epsilon_0\hbar c}. \tag{A.5}$$

DOI: 10.1201/9781003057369-A

TABLE A.1
Definition of values of the seven fundamental constants used to define base SI units.

Name	SI unit	Defined constant of nature
second	s	^{133}Cs hyperfine ground-state transition frequency: $\nu_{\mathrm{HF}}^{^{133}\mathrm{Cs}} = 9\,192\,631\,770\ \mathrm{s}^{-1}$
meter	m	speed of light in vacuum: $c = 299\,792\,458\ \mathrm{m\ s}^{-1}$
kilogram	kg	Planck's constant: $h = 2\pi\hbar = 6.626\,070\,15 \times 10^{-34}\mathrm{kg\ m}^2\ \mathrm{s}^{-1}$
ampere	A	electronic charge: $e = 1.602\,176\,634 \times 10^{-19}\ \mathrm{A\ s}$
kelvin	K	Boltzmann constant: $k_B = 1.380\,649 \times 10^{-23}\ \mathrm{kg\ m}^2\ \mathrm{s}^{-2}\ \mathrm{K}^{-1}$
mole	mol	Avogadro's constant: $N_A = 6.022\,140\,76 \times 10^{23}\ \mathrm{mol}^{-1}$
candela	cd	luminous efficacy of radiation of frequency $540 \times 10^{12}\ \mathrm{s}^{-1}$: $K_{\mathrm{cd}} = 683\ \mathrm{cd\ sr\ kg}^{-1}\ \mathrm{m}^{-2}\ \mathrm{s}^3$

TABLE A.2
Constants occurring in Maxwell's equations in Gaussian (G), Heaviside-Lorentz (HL), and SI system of units. Note that the Gaussian and Heaviside-Lorentz systems have one nontrivial constant, $1/c$. while SI has two, ϵ_0 and $1/\mu_0$, and consequently, **E**, **D**, **B**, and **H** have different units, which complicates discussions of relativity. On the other hand, HL and SI are both "rationalized" systems, in that the 4π's occur not in the equations but in the solutions.

Factor	G	HL	SI
κ_1	4π	1	1
κ_2	$1/c$	$1/c$	1
κ_3	1	1	ϵ_0
κ_4	1	1	$1/\mu_0$
κ_5	1	1	$1/c \equiv \sqrt{\epsilon_0\mu_0}$

Thus, (A.3b) is true only to a very good approximation.[1]

To convert from the equations in Gaussian form to those in SI form, we multiply each quantity in the former system by a suitable constant:

$$\begin{aligned}
\mathbf{D}^{\mathrm{G}} &= k_D\mathbf{D}^{\mathrm{SI}}, & \mathbf{E}^{\mathrm{G}} &= k_E\mathbf{E}^{\mathrm{SI}},\\
\mathbf{H}^{\mathrm{G}} &= k_H\mathbf{H}^{\mathrm{SI}}, & \mathbf{B}^{\mathrm{G}} &= k_B\mathbf{B}^{\mathrm{SI}},\\
\mathbf{P}^{\mathrm{G}} &= k_P\mathbf{P}^{\mathrm{SI}}, & \mathbf{M}^{\mathrm{G}} &= k_M\mathbf{M}^{\mathrm{SI}},\\
\rho^{\mathrm{G}} &= k_\rho\rho^{\mathrm{SI}}, & \mathbf{J}^{\mathrm{G}} &= k_J\mathbf{J}^{\mathrm{SI}}.
\end{aligned} \tag{A.6}$$

Maxwell's equations, together with the relation between **D**, **E**, and **P**, and between **H**, **B**, and **M**, determine the ratios of all the k's; the Lorentz force law determines the overall

[1]The current best value of α is $7.2973525693(11) \times 10^{-3} = 1/137.035999084(21)$ [54].

scale. The results are

$$k_D = \sqrt{\frac{4\pi}{\epsilon_0}}, \qquad k_E = \sqrt{4\pi\epsilon_0},$$
$$k_H = \sqrt{4\pi\mu_0}, \quad k_B = \sqrt{\frac{4\pi}{\mu_0}},$$
$$k_\rho = k_J = k_P = \frac{1}{\sqrt{4\pi\epsilon_0}},$$
$$k_M = \sqrt{\frac{\mu_0}{4\pi}}. \tag{A.7}$$

So this means that all that is necessary to convert a formula in Gaussian units into the corresponding formula in SI units is to make the scaling of the electromagnetic quantities by the factors given in (A.7). To convert from Gaussian to Heaviside-Lorentz units, the conversion factors are as in (A.7), with ϵ_0 and μ_0 replaced by unity, because, like SI, the HL-system is rationalized, that is, no 4π's appear in Maxwell's equations, but they reappear in the solutions, such as Coulomb's potential.

Because the electric and magnetic susceptibilities in either system are defined by

$$\mathbf{P} = \chi_e \mathbf{E}, \quad \mathbf{M} = \chi_m \mathbf{H}, \tag{A.8}$$

the susceptibilities in the two systems are related by

$$4\pi\chi^{\mathrm{G}} = \chi^{\mathrm{SI}}, \tag{A.9}$$

reflecting

$$\epsilon^{\mathrm{G}} = 1 + 4\pi\chi_e^{\mathrm{G}} = \epsilon^{\mathrm{SI}}/\epsilon_0 = 1 + \chi_e^{\mathrm{SI}}, \tag{A.10a}$$
$$\mu^{\mathrm{G}} = 1 + 4\pi\chi_m^{\mathrm{G}} = \mu^{\mathrm{SI}}/\mu_0 = 1 + \chi_m^{\mathrm{SI}}. \tag{A.10b}$$

To find the actual values of electromagnetic quantities in the two systems, such as the charge on the electron, it is further necessary to convert the length and mass units (time is measured in seconds in both systems),

$$x \to \lambda_1 x, \quad m \to \lambda_2 m, \tag{A.11}$$

where here, of course, $\lambda_1 = 10^2$ and $\lambda_2 = 10^3$. Thus, for example, an energy scales by the factor $\lambda_1^2\lambda_2$, and hence the Coulomb potential of a electron at the origin is, in the two systems, related by

$$\frac{e_{\mathrm{G}}^2}{r_{\mathrm{G}}} = \lambda_1^2\lambda_2 \frac{e_{\mathrm{SI}}^2}{4\pi\epsilon_0 r_{\mathrm{SI}}}, \tag{A.12}$$

where the subscripts now indicate the system of units. Thus

$$e_{\mathrm{G}} = e_{\mathrm{SI}} \frac{1}{\sqrt{4\pi\epsilon_0}} \sqrt{\frac{r_{\mathrm{G}}}{r_{\mathrm{SI}}}} \sqrt{\lambda_1^2\lambda_2} = e_{\mathrm{SI}}\sqrt{10^{-7}}c\sqrt{10^2}\sqrt{10^7}$$
$$= 1.602 \times 10^{-19} \times 2.998 \times 10^9 = 4.803 \times 10^{-10}\mathrm{esu}, \tag{A.13}$$

where the name of the electrostatic unit is no longer of much significance. In this way we can easily work out the equivalence between SI units and their Gaussian counterparts, which are given in Table A.3.[2]

[2]It might be noted that the Gaussian system is an amalgamation of the earlier cgs electrostatic and electromagnetic systems of units. Here esu units are used for electrical quantities and emu units are used for magnetic. Inductance sits on the fence: We use the esu unit, but some authors use the emu unit, which differs by a factor of $(3 \times 10^{10})^{-2}$. A single factor of c gives the ratio of emu to esu units of charge. The other commonly used set of electrical units is the Heaviside-Lorentz system, which is "rationalized" like the SI system, the units differing, therefore, from the Gaussian units by various powers of 4π. See the reference at the end of this Appendix for more detail.

TABLE A.3
Relation between SI electromagnetic units and Gaussian units. Here 3 is an approximation for 2.99792458. Here the abbreviations for the electromagnetic units are standard: V for volt, C for coulomb, A for ampere, F for farad, T for tesla, and H for henry. The unit for conductance, the siemens (S), was formerly called mho, ohm spelled backward.

Quantity	SI unit	Corresponding number of Gaussian units
$\mathbf{E}$	V/m	$\frac{1}{3} \times 10^{-4}$
ϕ	V	$\frac{1}{300}$
e	C	3×10^{9}
ρ	$\mathrm{C/m^3}$	3×10^{3}
I	A	3×10^{9}
$\mathbf{J}$	$\mathrm{A/m^2}$	3×10^{5}
$\mathbf{P}$	$\mathrm{C/m^2}$	3×10^{5}
$\mathbf{D}$	$\mathrm{C/m^2}$	$12\pi \times 10^{5}$
σ	S/m	$9 \times 10^{9}\ \mathrm{s^{-1}}$
C	F	9×10^{11} cm
$\mathbf{B}$	T	10^{4} gauss
$\mathbf{H}$	A-turn/m	$4\pi \times 10^{-3}$ oersted
$\mathbf{M}$	A-turn/m	10^{-3}
L	H	$\frac{1}{9} \times 10^{-11}$

Finally, we turn to extraction of numbers from electromagnetic formulas, since this sometimes leads to confusion. We illustrate three different methods.

- The formula in Gaussian units may be evaluated directly.
- The formula may be transformed to SI units, using (A.7), and then evaluated using familiar SI quantities.
- The formula may be evaluated by eliminating electromagnetic quantities in favor of dimensionless numbers, and then the resulting formula is computed with the aid of dimensional analysis. This method, which may seem less direct, has the virtue that one need not keep track of factors of c, which is merely a conversion factor between time and space intervals and is naturally set equal to unity in relativistic calculations. ("Natural units" used in high-energy physics set $\hbar = c = 1$.)

We illustrate these three methods by evaluating the following formula, derived in Problem 35.1, for the mean lifetime of the orbital motion of an electron, having charge e and mass m, moving in a uniform magnetic field $\mathbf{B}$, whose motion decays due to the radiation emitted:

$$\tau = \frac{3m^3c^5}{4e^4B^2}. \tag{A.14a}$$

Suppose $B = 10^4$ gauss. Then, in Gaussian units, where $m = 9.11 \times 10^{-28}$ g, $c = 3.00 \times 10^{10}$ cm/s, and $e = 4.80 \times 10^{-10}$ esu, this evaluates to $\tau = 2.58$ s. On the other hand, we may convert the formula first to SI:

$$\tau = \frac{3\pi\epsilon_0 m^3c^3}{e^4B^2}. \tag{A.14b}$$

Now $m = 9.11 \times 10^{-31}$ kg, $c = 3.00 \times 10^8$ m/s, and $e = 1.60 \times 10^{-19}$ C, so for a magnetic field of 1 T, we obtain, again, $\tau = 2.58$ s.

The dimensionless method consists, first, in replacing the electric charge by the fine structure constant, in Gaussian units,

$$\alpha = \frac{e^2}{\hbar c} \approx \frac{1}{137}, \tag{A.15}$$

and the magnetic field is to be expressed in terms of its ratio to the characteristic (or critical) field strength,

$$B_c = \frac{m^2 c^3}{e\hbar} = 4.41 \times 10^{13} \text{gauss}. \tag{A.16}$$

Thus our formula becomes

$$\tau = \frac{3}{4}\left(\frac{B_c}{B}\right)^2 \frac{\hbar}{\alpha m_e c^2}, \tag{A.17}$$

so, because $\hbar = 6.58 \times 10^{-22}$ MeV s, and $m_e c^2 = 0.511$ MeV, we get the same result, $\tau = 2.58$ s. Although this latter method may seem unnatural because it introduces quantum quantities into a classical calculation, it has the virtue of allowing the powers of $\hbar$ and c to be determined at the end of the calculation by elementary considerations of dimensional analysis. Here because mc^2 has dimensions of energy, exactly one factor of $\hbar$ must be present in the numerator to obtain a quantity having the dimensions of time.

This advantage is lost if one uses a purely classical method of eliminating e and B, for example, by eliminating the former in favor of the electron mass in terms of the classical electron radius, (48.6),

$$r_0 = \frac{e^2}{mc^2} = 2.818 \times 10^{-13} \text{cm}, \tag{A.18}$$

and expressing B in terms of the cyclotron frequency for a field B_0 of 1 gauss,

$$\omega_0 = \frac{eB_0}{mc} = 1.759 \times 10^7 \text{rad/s}, \tag{A.19}$$

so that our formula becomes

$$\tau = \frac{3}{4}\frac{c}{r_0 \omega_0^2}\left(\frac{B_0}{B}\right)^2. \tag{A.20}$$

This, of course, gives the same answer for $B = 10^4$ gauss, namely 2.58 s, but now in terms of two-dimensional quantities, r_0 and ω_0, instead of the one, m, in the atomic system. The method using $\hbar$ is obviously preferable.

Another useful conversion factor is

$$\hbar c = 1.97327 \times 10^{-5} \text{eV cm}. \tag{A.21}$$

From this we can verify the Gaussian charge value in (A.13) again:

$$e^2 = \alpha \hbar c \Rightarrow e = \lambda_1 \frac{1.97327 \times 10^{-5}}{2.9979258 \times 10^5} \frac{1}{137.036} = 4.8032 \times 10^{-10}. \tag{A.22}$$

For a rather complete discussion of the vexing issue of electromagnetic units, the reader is referred to Ref. [55] Although this reference is 60 years old, and so does not reflect the amazing advances in metrology and the modern SI, its discussion of the framework for the description of electromagnetic quantities remains unsurpassed.

Bibliography

[1] K. A. Milton and J. Schwinger, *Electromagnetic Radiation: Variational Methods, Waveguides and Accelerators* (Springer, Berlin, 2006).

[2] J. A. Stratton, *Electromagnetic Theory* (McGraw-Hill, New York, 1941).

[3] A. Sommerfeld, *Electrodynamics* (Academic Press, New York, 1964).

[4] L. D. Landau and E. M. Lifshitz, *The Classical Theory of Fields*, 4th revised English edition, translated from the 6th revised Russian edition of *Teoriya Polia, Nauka, Moscow 1973* (Butterworth-Heinemann, Oxford, 1975).

[5] L. D. Landau and E. M. Lifshitz, *Electrodynamics of Continuous Media*, 1st English edition (Pergamon, Oxford, 1960).

[6] J. D. Jackson, *Classical Electrodynamics*, 1st, 2nd, and 3rd Eds. (Wiley, New York, 1962, 1975, 1999).

[7] J. Schwinger and D. Saxon, *Discontinuity in Waveguides* (Gordon and Breach, New York, 1968).

[8] J. Schwinger, "On the Classical Radiation of Accelerated Electrons," *Phys. Rev.* **75**, 1912 (1949).

[9] J. Schwinger, "On Radiation by Electrons by a Betatron," transcription by M. Furman, LBNL-39088/CBP Note-179, reprinted in *A Quantum Legacy: Selected Papers of Julian Schwinger*, ed. K. A. Milton (World Scientific, Singapore, 2000), p. 307.

[10] P. B. Price, E. K. Shirk, W. Z. Osborne, and L. S. Pinsky, "Evidence for Detection of a Moving Magnetic Monopole," *Phys. Rev. Lett.* **35**, 487 (1975).

[11] L. W. Alvarez, in *Proceedings of the 1975 International Symposium on Lepton and Photon Interactions at High Energies, Stanford, August 21–27, 1975*, ed. W. T. Kirk, p. 967.

[12] B. Cabrera, "First Results from a Superconductive Detector for Moving Magnetic Monopoles" *Phys. Rev. Lett.* **48**, 1378 (1982).

[13] M. Faraday, Laboratory journal entry, 1849, published in *Life and Letters of Faraday*, ed. Bence Jones (Longmans, Green & Co., 1870, Cambridge University Press, 2010).

[14] B. Acharya et al., MoEDAL Collaboration, "Search for Highly-Ionizing Particles in *pp* Collisions at the LHC's Run-1 Using the Prototype MoEDAL Detector," *Eur. Phys. J. C* **82**, 694 (2022).

[15] B. Acharya et al., MoEDAL Collaboration, "Search for Magnetic Monopoles Produced via the Schwinger Mechanism," *Nature* **602**, 62 (2022).

[16] A. F. Rañada and J. L. Trueba, "Ball Lightning: An Electromagnetic Knot?," *Nature*, **383**, 32 (1996).

[17] R. P. Cameron, "Monochromatic Knots and Other Unusual Electromagnetic Disturbances: Light Localised in 3D," *J. Phys. Commun.* **2**, 015024 (2018).

[18] J. S. Høye, I. Brevik, J. B. Aarseth, and K. A. Milton, "How Does the Transverse Electric Zero Mode Contribute to the Casimir Effect for a Metal?" *Phys. Rev.* E **67**, 056116 (2003).

[19] A. Lambrecht and S. Reynaud, "Casimir Force Between Metallic Mirrors," *Eur. Phys. J. D* **8**, 309–318 (2000).

[20] I. Brevik, M. Chaichian, and I. I. Cotăescu, "Remarks on the Abraham-Minkowski Problem, from the Formal and from the Experimental Side," *Int. J. Mod. Phys. A* **36**, 2150063 (2021).

[21] L. Brillouin, *Wave Propagation and Group Velocity* (Academic Press, New York, 1960).

[22] J. Schwinger,"Electromagnetic Mass Revisited," *Found. Phys.* **13**, 373 (1983).

[23] R. L. Workman *et al.*, (Particle Data Group), "Review of Particle Physics", *Prog. Theor. Exp. Phys.* **2022**, 083C01 (2022).

[24] S. Earnshaw, "On the Nature of the Molecular Forces which Regulate the Constitution of the Luminiferous Ether," *Trans. Camb. Phil. Soc.* **7**, 97 (1842).

[25] J. D. van der Waals, *Over de Continueiteit van den Gas- en Vloeistoftoestand*, Ph.D. thesis, Leiden, 1873.

[26] H. B. G. Casimir and D. Polder, "The Influence of Retardation on the London-van der Waals Forces," *Phys. Rev.* **73**, 360–372 (1948).

[27] K. A. Milton, "Theoretical and Experimental Status of Magnetic Monopoles," *Rep. Prog. Phys.* **69**, 1637–1711 (2006).

[28] G. R. Kalbfleisch, W. Luo, K. A. Milton, E. H. Smith, and M. G. Strauss, "Limits on Production of Magnetic Monopoles Utilizing Samples from the D0 and CDF Detectors at the Fermilab Tevatron," *Phys. Rev. D.* **69**, 052002-1-18 (2004) [arXiv:hep-ex/0306045].

[29] J. V. Jelley, *Čerenkov Radiation and its Applications* (Pergamon, New York, 1958).

[30] B. M. Bolotovskii, "Vavilov-Cherenkov Radiation: Its Discovery and Application," *Physics-Uspekhi*, **52**, 1099–1110 (2009).

[31] W. Pauli, *Theory of Relativity* (Pergamon, Oxford, 1958), p. 93.

[32] F. Rohrlich, "The Definition of Electromagnetic Radiation," *Il Nuovo Cimento* **21**, 811 (1961).

[33] F. Rohrlich, "The Principle of Equivalence," *Ann. Phys. (N.Y.)* **22**, 169 (1963).

[34] D. G. Boulware, "Radiation from a Uniformly Accelerated Charge," *Ann. Phys. (N.Y.)* **124**, 169 (1980).

[35] A. G. S. Landulfo, S. A. Fulling, and G. E. A. Matsas, "Classical and Quantum Aspects of the Radiation Emitted by a Uniformly Accelerated Charge: Larmor-Unruh Reconciliation and Zero-Frequency Rindler Modes," *Phys. Rev. D* **100**, 045020 (2019).

[36] G. A. Schott, "On the Motion of the Lorentz Electron," *Philos. Mag.* **29**, 49–69 (1915).

[37] D. R. Rowland, "Physical Interpretation of the Schott Energy of an Accelerating Point Charge and the Question of Whether a Uniformly Accelerating Charge Radiates," *Eur. J. Phys.* **31**, 1037 (2010).

[38] A. K. Singal, "Poynting Flux in the Neighbourhood of a Point Charge in Arbitrary Motion and Radiative Power Losses," *Eur. J. Phys.* **37**, 045210 (2016).

[39] H. Ren and E. J. Weinberg, "Radiation for a Moving Scalar Source," Phys. Rev. D **49**. 6526–6533 (1994).

[40] C. G. Callen, Jr., Sidney Coleman, and Roman Jackiw, "A New Improved Energy-Momentum Tensor," *Ann. Phys, (N.Y.)* **59**, 42–73 (1970).

[41] A. A. Sokolov and I. M. Ternov, "Polarizations and Spin Effects in the Synchrotron Radiation Theory," *Dokl. Akad. Nauk SSSR* **153**, 1052-4 (1963).

[42] J. Schwinger and W.-y. Tsai, "Radiative Polarization of Electrons," *Phys. Rev. D* **9**, 1843–5 (1974).

[43] J. S. Bell and J. M. Leinaas, "The Unruh Effect and Quantum Fluctuations in Storage Rings," *Nucl. Phys.* **B284**, 488–508 (1987).

[44] J. Schwinger, "Electron Radiation in High Energy Accelerators," *Phys. Rev.* **70**, 798 (1946).

[45] D. Iwanenko and I. Pomeranchuk, "On the Maximal Energy Attainable in a Betatron," *Phys. Rev.* **65**, 343 (1944).

[46] D. Ivanenko and A. A. Sokolov, "On the Theory of the 'Luminous' Electron," *Dokl. Akad. Nauk SSSR [Sov. Phys. Dokl.]* **59**, 1551 (1948).

[47] A. Sokolov and I. M. Ternov, *Synchrotron Radiation* (Akademie-Verlag, Berlin; Pergamon Press, Oxford, 1968).

[48] H. Wiedemann, *Synchrotron Radiation* (Springer, Berlin, 2003).

[49] A. Hofmann, *The Physics of Synchrotron Radiation* (Cambridge University Press, Cambridge, 2004).

[50] A. Sommerfeld, "Mathematische Theorie der Diffraction," *Math. Ann.* **47**, 317–374 (1896).

[51] W. Lukosz, "Electromagnetic Zero-point Energy Shift Induced by Conducting Surfaces," *Z. Phys.* **262**, 327-348 (1973).

[52] *The Review of Particle Physics*, Particle Data Group Prog. Theor. Exp. Phys. **2022**, 083CC01 (2022), Chapter 34, "Passage of Particles Through Matter," by D. E. Groom and S. R. Klein, https://pdg.lbl.gov/2022/reviews/rpp2022-rev-passage-particles-matter.pdf.

[53] E. Tiesinga, P. J. Mohr, D. B. Newell, and B. N. Taylor, "CODATA Recommended Values of the Fundamental Physical Constants: 2018," *Rev. Mod. Phys.* **93**, 025010 (2021).

[54] https://physics.nist.gov/cgi-bin/cuu/Value?alph

[55] F. B. Silsbee, *Systems of Electrical Units*, National Bureau of Standards Monograph 56 (U.S. Government Printing Office, Washington, 1962).

Index

Printed in the United States
by Baker & Taylor Publisher Services